S0-BLA-815

Concepts and Methods of Biostratigraphy

Edited by

Erle G. Kauffman

Smithsonian Institution

and

Joseph E. Hazel

U. S. Geological Survey

with the assistance of

Barbara Duffy Heffernan

Library of Congress Catalog Card Number: 76-42248
ISBN: 0-87933-246-8

79 78 77 1 2 3 4 5

Manufactured in the United States of America.

Library of Congress Cataloging in Publication Data

Main entry under title:
Concepts and methods of biostratigraphy.
Includes bibliographies and index.
1. Geology, Stratigraphic. 2. Paleontology, Stratigraphic. I. Kauffman, Erle Galen, 1933– II. Hazel, Joseph E., 1933– III. Heffernan, Barbara Duffy. IV. Title: Biostratigraphy.
QE651.C67 551.7'01 76-42248
ISBN 0-87933-246-8

Exclusive distributor: **Halsted Press,**
A Division of John Wiley & Sons, Inc.
ISBN: 0-470-99006-6

GEOLOGY

Preface

The term "biostratigraphy" has been applied to almost every conceivable scheme of organizing fossils found in rocks, including a variety of physical, evolutionary, and ecological classifications of biotic assemblages. However, in this book we are concerned with the principal purpose of biostratigraphy–zonal biostratigraphy, the zonation and correlation of strata, allowing the interpretation of earth history within a precise framework of geological time. Other bioassociational aspects of paleontology or stratigraphy in which the goal is *not* the delineation of geographically repetitive sequences of biologically consistent, spatially widespread, and temporally restricted units with near-isochronous boundaries is not our primary concern. It is the meaning and construction of these biozones, their biologic components, their correlation, and the basic factors that permit or retard such efforts that are emphasized.

Whereas zonal biostratigraphy has been widely practiced since the time of William Smith, and the results universally applied to problem solving and interpretation in the earth sciences and related fields, there is no comprehensive treatment of the subject in English. Existing discussions in stratigraphy, paleontology, and historical geology textbooks and research papers inadequately explain the broad spectrum of concepts and methods of biostratigraphy. In particular the biological concepts behind the selection of organisms for zonal biostratigraphy, and the logic behind the construction and use of zones, has been sparsely treated. This book has been conceived to at least partially fill this void, and to provide both an advanced text and a professional reference to contemporary biostratigraphy, as well as a source of new ideas.

The volume is designed to provide an assessment of biozone construction and correlation, to review biological, evolutionary, and ecological factors that affect the formulation and successful application of biostratigraphic systems, and to provide an up to date survey of major groups of organisms used in zonal biostratigraphy. Authors have been selected accordingly to reflect international experience and diversity of interests. We have attempted to be objective and achieve balance in our coverage–biologically, geographically, stratigraphically, and methodologically. Readers will find emphasis on both standard and newer concepts and methods in biostratigraphic analysis, from which they should be able to formulate an approach suitable to their special problems. For the student, the teacher, and the nonspecialist we have, in general, asked the authors to follow a central theme–to discuss the concepts and methods of their approach to biostratigraphy, and to illustrate this with a working example based on diverse organisms in varied environmental situations throughout the world. We have tried to organize these contributions in a logical manner which reflects the thought processes commonly involved in constructing a detailed biostratigraphic system. What factors should be considered in selecting organisms for biostratigraphy? What organisms are most valuable in zonation and correlation in diverse geological and biological situations? What methods and tools are used to construct and test a biostratigraphy? We have attempted to answer some of these questions by soliciting related papers on these critical subjects. To the extent that we have succeeded, we must credit the cooperation and patience of our contributing authors, and the strong and continuing support of our colleagues.

The principal objective of zonal biostratigraphy is to provide the mechanism by which the relative timing of geological and biological events can be accurately determined and correlated throughout the world. This objective has been achieved to a great extent because fossils are one of the most common and continuously occurring constituents of sedimentary rocks. Further, because organic evolution is irreversible, but variable in rate and pattern, even small segments of geologic time are characterized by unique biotas. Thus, biostratigraphy presents a more consistently applicable system of dating and correlation than either magnetic or radiometric stratigraphy. Despite the recent advances in these and related fields, biostratigraphy will remain for the foreseeable future the principal dating technique applicable to Phanerozoic rocks. As such, it is perhaps the single most important tool in sedimentary geology, and one which is critical to all interpretation.

Biostratigraphy is not without its problem areas, and the papers herein address most of these as well. Variations in the inferred relationship between absolute time and biostratigraphic zones comprise a major area of controversy. The frequent failure to consider the biological attributes of all potential organisms for construction of a biostratigraphic system is another problem. One problem not specifically addressed in the text that deserves special mention here (and such mention we hope will serve as a warning to the uninitiated) is that of variation in the use of nomenclature and, to a certain extent, the concepts underlying the diverse biostratigraphic terminology. These vary widely from worker to worker and region to region. It is impossible for us to resolve or even compare adequately the differences, for example, between the terms and concepts espoused by Schindewolf (1970), the International Committee on Stratigraphic Classification (Hedberg, 1971), the Soviet school (Khalfin, 1971), or the American Code (ACSN, 1961), to mention only a few. Nomenclatural discord has served as a vehicle for the use of much printer's ink and paper, some wisely used, some not. Nevertheless, despite these differences of opinion, the goals of all were and are the same– to unravel geologic history by using fossils to date and correlate rocks.

Because of their contrasts, these various sytems are a burden to the student and the practitioner. But this confusion is greatly reconciled when one bears in mind "... that rocks are measured in meters, time in days or years, and wine in liters.

As long as a stratigraphic sequence is not described in liters, time in meters, and wine in days, there is no confusion whatsoever" (de Cserna, 1972).

There is developing a widespread re-evaluation of historical concepts and methods of biostratigraphy among its practitioners and interested scientists in the related fields of biology, ecology, and in geology. Also, a new computer technology gives us the ability to analyze immense amounts of data. The geological revolution centered around widespread proof of plate tectonics has had a profound effect on biostratigraphic and biogeographic concepts; in many cases we can now expect to be able to construct detailed intercontinental systems of zonation and correlation utilizing the same suites of species; in other cases we can better explain biostratigraphic discordance of stratigraphic sequences which are now closely situated. Recent changes in paleontological research with greater emphasis on population systematics, evolutionary studies, and ecological analysis linked to a growing interest in living organisms have also forced a hard new look at some biostratigraphic concepts and methodology. The papers in this book broadly reflect the changes in thought and the conceptual problems that recently have been brought to light.

The production of this book has been greatly enhanced by the dedicated efforts of many individuals. In particular we would like to express our appreciation to Barbara Duffy Heffernan of the U.S. National Museum, who has provided superb editorial and technical assistance throughout the entire production of the volume, and to Ellen E. Compton and Barbara A. Bedette of the U.S. Geological Survey for their patient and careful proofing of the book. Lawrence B. Isham, artist for the Department of Paleobiology, U.S. National Museum, created the cover and part opening page designs. Donna Copeland of the National Museum deserves special thanks for flawless typing of large portions of the manuscript. Finally, our special thanks to our authors for the excellence of their contributions and for their patience with the editors, and to Charles Hutchinson and Shirley A. End of Dowden, Hutchinson and Ross, Inc. for their encouragement, guidance, and support in the production of this volume.

Erle G. Kauffman
Joseph E. Hazel

Contents

List of Contributors

R. L. Austin
Department of Geology, University of Southampton, Southampton, England

W. A. Berggren
Department of Geology and Geophysics, Woods Hole Oceanographic Institution, Woods Hole, Massachusetts 02543

William B. N. Berry
Department of Paleontology, University of California, Berkeley, California 94720

Tove Birkelund
Institut fur Historisk Geologi og Palaeontologi, Østervoldgade 10, DK-1350, København K, Denmark

W. A. Cobban
Paleontology and Stratigraphy Branch, U. S. Geological Survey, Denver, Colorado 80225

Raymond C. Douglass
Paleontology and Stratigraphy Branch, U. S. Geological Survey, E-114 U. S. National Museum, Washington, D. C. 20244

James A. Doyle
Department of Ecology and Evolutionary Biology, Museum of Paleontology, University of Michigan, Ann Arbor, Michigan 48104

Niles Eldredge
Department of Fossil Invertebrates, The American Museum of Natural History, New York, New York 10024

Gundolf Ernst
Institut für Geologie und Paläontologie, Technische Universität Braunschweig, 33 Braunschweig, Postfach 7050, Pockelstrasse 4 (Hochhaus), Braunschweig, Germany

Stephen Jay Gould
Museum of Comparative Zoology, Harvard University, Cambridge, Massachusetts 02138

J. M. Hancock
Department of Geology, University of London, Kings College, Strand, London WC2R 2LS, England

Joseph E. Hazel
Paleontology and Stratigraphy Branch, U. S. Geological Survey, E-502 U. S. National Museum, Washington, D. C. 20244

Jeremy B. C. Jackson
Department of Earth and Planetary Sciences, The Johns Hopkins University, Baltimore, Maryland 21218

Erle G. Kauffman
Department of Paleobiology, Smithsonian Institution, E-307 U. S. National Museum, Washington, D. C. 20560

W. J. Kennedy
Department of Geology and Mineralogy, University of Oxford, Parks Road, Oxford OX1 3PR, England

Bernard Mamet
Départment de Géologie, Université de Montréal, Case Postale 6128, Montréal, Canada

F. X. Miller
Amoco Production Company, Security Life Building, Denver, Colorado 80202

C. Wylie Poag
U. S. Geological Survey, Woods Hole Oceanographic Institution, Woods Hole, Massachusetts 02543

Frank H. T. Rhodes
Department of Geology, University of Michigan, Ann Arbor, Michigan 48104

William J. Sando
Paleontology and Stratigraphy Branch, U. S. Geological Survey, E-318 U. S. National Museum, Washington, D. C. 20244

Donald E. Savage
Museum of Paleontology, University of California, Berkeley, California 94720

Rudolf S. Scheltema
Woods Hole Oceanographic Institution, Woods Hole, Massachusetts 02543

Ekbert Seibertz
Institut für Geologie und Paläontologie, Technische Universität Braunschweig, 22 Braunschweig, Postfach 7050, Pockelstrasse 4 (Hochhaus), Braunschweig, Germany

Norman F. Sohl
Paleontology and Stratigraphy Branch, U. S. Geological Survey, E-309
U. S. National Museum, Washington, D. C. 20244

Fritz F. Steininger
Institute of Paleontology, University of Vienna, Vienna, Austria

Finn Surlyk
Mineralogisk Museum, Østervoldgade 7, København, Denmark

P. D. Sylvester-Bradley
Department of Geology, University of Leicester, Leicester, England

Michael E. Taylor
Paleontology and Stratigraphy Branch, U. S. Geological Survey, E-117
U. S. National Museum, Washington, D. C. 20244

James W. Valentine
Department of Geology, University of California, Davis, California 95616

John A. Van Couvering
Department of Geological Sciences, University of Colorado Museum, Boulder, Colorado 80302

J. B. Waterhouse
Department of Geology and Mineralogy, University of Queensland, Brisbane, Australia

Introduction

The Historic Development of Concepts of Biostratigraphic Correlation

J. M. Hancock **University of London**

In stratigraphic procedure, it is not what terms an author uses that matters, but whether he knows what he is talking about (Arkell, 1956a, p. 466).

THE DISCOVERY THAT FOSSILS CAN BE USED FOR CORRELATION

Only rarely nowadays can it be said that a new concept originates with one man alone, but the science of biostratigraphy was founded by William Smith, and he owed nothing to earlier writers. It is true that the law of superposition is contained within the seventeenth-century ramblings of Steno (1669), but to the end of his life Smith had probably never heard of Steno. By 1796, Smith had discovered from his wanderings around Bath in Somerset that " . . . everywhere throughout this district, and to considerable distances around, it was a general law that the 'same strata were found always in the same order of superposition and contained the same peculiar fossils' " (Eyles, 1969; Phillips, 1844). In true British style, he first made note of his ideas while in a pub, the Swan near Dunkerton. This momentous dual discovery became known through two Wiltshire friends of Smith, the Reverend Benjamin Richardson of Farleigh Hungerford and the Reverend Joseph Townsend of Pewsey. It was at Townsend's house in 1799 that Richardson drew up, at Smith's dictation, the succession of strata near Bath from the Chalk down to the Coal Measures, and listed "the most remarkable fossils which had been gathered in the several layers of rock" (Phillips, 1844). An outline of Smith's succession was first published in 1801 in Warner's *The History of Bath.* But for many years Smith's ideas spread by conversation. In March 1808, Smith showed his collection of fossils, arranged stratigraphically in his London house, to members of the Geological Society. It was not until 1813 that Smith's full table of 1799 was published by Townsend, who enthusiastically acknowledged that both the facts and ideas originated with his teacher Smith, who ". . . made known his discoveries to everyone who wished for information."

Smith, a civil engineer, found writing a great labor. It was not until 1815 that his results appeared under his own name as an accompaniment to his great geological map of England.

The accurate surveys and examinations of the strata . . . have enabled me to prove that there is a great degree of regularity in the position and thickness of all these strata; and although considerable dislocations are found in collieries and mines, and some vacancies in the superficial courses of them, yet that the general order is preserved; and that each stratum is also possessed of properties peculiar to itself, has the same exterior characters and chemical qualities, and the same extraneous or organized fossils throughout its course (Smith 1815, p. 2.).

Smith's best known works, from which he can now be judged, were published between 1815 and 1819. In June 1816 he started to publish *Strata Identified by Organised Fossils, containing Prints on Coloured Paper of the most characteristic specimens in each Stratum* (1816-1819), which was never completed. In many respects this contains the purest and most complete account of his own observations; but the nearest to a general account in his own words appeared in 1817 in his *Stratigraphical System of Organized Fossils*:

By the tables it will be seen which Fossils are peculiar to any Stratum, and which are repeated in others (p. iv).

The organized Fossils which may be found, will enable him to identify the Strata of his own estate with those of others . . . (p. v).

My observations on this and other branches of the subject are entirely original, and unencumbered with theories, for I have none to support . . . (p. vi).

Note that the "Strata" had first been delineated by mapping, and then their characteristic fossils collected. The fossils could be used to distinguish two strata of similar lithology that lay at different positions in the succession, e.g., the Oaktree Clay (Kimeridgian) from the Clunch Clay (Callovian-Oxfordian), but the idea of different facies being correlated by fossils in common did not as yet exist.

On the other hand, it is clear that Smith had further recognized two important principles that were long overlooked by many subsequent geologists. First, he realized that even a single lithic formation could be subdivided on fossil distribution alone:

The Cornbrash, though altogether but a thin rock, has not its organized fossils equally diffused, or promiscuously distributed. The upper beds of stone which compose the rock, contain fossils materially different from those in the under (1819, p. 25–26).

Second, he recognized that fossils from the same facies of different ages could resemble one another:

The upper part of this thick Stratum contains large incurved oysters or Gryphaea, so much resembling others I have collected from remote parts, of a clay which now appears to be Oak-tree clay, as to be distinguished with difficulty; but this is only one of the many instances of the general resemblances of organized Fossils, where the Strata are similar (1817, p. 22).

The importance of Smith's work is greater than that of any subsequent contributor to the theory of our science. His lack of a formal education and his very limited reading of earlier authors meant that he was remarkably unbiased by previous ideas. His approach was completely empirical, and he never tried to draw conclusions beyond the evidence he had found himself, let alone to speculate why each stratum had its own peculiar fossils. His work is completely free from grandiloquent attempts at a unitary theory of all geology, and equally free from any interaction with theology. He himself pointed out that a person did not need to be able to read and write to use his methods, and as Arkell (1933, p. 8) has aptly mentioned, "That he could turn fossils to such useful account although he had no names for most of them is an object-lesson which many modern workers might with advantage take to heart."

Smith's achievements were well advertised in 1818 by an article in the *Edinburgh Review,* which, although unsigned, was known to have been written by the widely respected W. H. Fitton (who, nevertheless, suffered socially for his courage). The textbook by Conybeare and Phillips (1822) shows how much the new principles had become common knowledge, and the respect that was felt for William Smith as the discoverer that strata could be correlated by fossils. One would never realize this from the books by Mantell (1822) or Young and Bird (1822). Mantell was a surgeon at Lewes in Sussex who felt too strongly about his station in life. Smith was socially beneath him, and so there is but one reference to him among the hundreds of observations and opinions of "gentlemen," and then he does not even accord him the title of "Mr"! Mantell makes no reference to using fossils for correlation, and his lack of understanding is shown in his sole discussion on correlation in the whole work. He is commenting on Buckland's table of Tertiary strata in England with their probable continental equivalents (based on marine or nonmarine faunas):

It is not implied that the above five subdivisional parts of the Tertiary formations, maintain the same relative order of succession in England, and on the Continent;

most of them probably alternate, but they are all more recent than the chalk of England, France and Italy (Mantell 1822, p. 249).

This is a far cry from Smith's words of 1815!

Young and Bird (1822) were clearly familiar with Smith's works and some of their phrases are very close to his, although they do not mention him by name and did not accept Smith's ideas:

The attempt to identify the several strata by their respective fossils must be confined within narrow limits. . . . the assistance thus furnished must be very limited; for if some of the fossils appear peculiar to certain beds, there are others very extensively diffused; and how can we be sure, that such as we deem peculiar to the beds which they occupy, may not be discovered, like their companions, in other beds of a very different description? (Young and Bird 1822, p. 300).

The same year saw a publication by Cuvier and Brongniart (1822) that shows as well as any contemporary work the possibilities of Smith's techniques. Some authors have accorded Cuvier an equal, or even superior, share in the honor of discovering the basic principles of biostratigraphy (e.g., Grabau, 1913), and the impression is repeated in the recent study by Wilson (1972). The error seems to originate by giving a modern meaning to Cuvier and Brongniart's undoubtedly original and outstandingly important discovery that the Tertiary deposits of the Paris basin contain alternations of marine and freshwater fossil assemblages, and hence that for the Paris basin one can draw up a stratal succession which shows that there have been "revolutions" of life on earth (Cuvier and Brongniart, 1808). It is clear from reading the original work that Cuvier was not concerned with correlating strata, but with the history of life on earth:

If there is any circumstance thoroughly established in geology, it is, that the crust of our globe has been subjected to a great and sudden revolution, the epoch of which cannot be dated much further back than five or six thousand years ago . . . (Cuvier, 1817, p. 171).

It would certainly be exceedingly satisfactory to have the fossil organic productions arranged in chronological order, in the same manner as we now have the principal mineral substances. [He is referring to the succession of strata according to Werner.] By this the science of organisation itself would be improved; the developments of animal life; the succession of its forms; the precise determination of those which have been first called into existence; the simultaneous production of certain species, and their gradual extinction—all these would perhaps instruct us fully as much in the essence of organisation, as all the experiments that we shall ever be able to make upon living animals. And man, to whom only a short space of time is allotted upon the earth, would have the glory of restoring the history of thousands of ages which preceded the existence of the race, and of thousands of animals that never were contemporaneous with his species (p. 181).

As for stratigraphy, even in the revised edition of 1822, Cuvier was thinking in terms of fitting the strata into a Wernerian scheme. He went to great trouble to prove that the three main *terrains – la craie, le calcaire marin grossier,* and *le gypse avec les sables* – were not parallel, and hence were truly separate formations according to the principles of Werner. There was no mention of fossils, even though Cuvier knew that different levels in *le gypse avec les sables* contained different fossils; these alone were not sufficient to make them separate formations.

This interpretation of Cuvier's place in the history of stratigraphy is supported by Young and Bird (1822), who discuss the theories of Cuvier and Smith as separate, unconnected subjects, and by Brongniart himself, who acknowledges the advances of Smith and other geologists in England. But Brongniart, as distinct from Cuvier, realized the tremendous possibilities of making long-distance correlation by fossils independently of facies (although the word "facies" did not yet exist in the literature). Brongniart compared the fauna of the white chalk of Gravesend and Brighton with that of the white chalk of Meudon and Dieppe; the fauna of the greensand of Folkestone was compared with that of the chloritic chalk of Rouen, Le Havre, and Honfleur. He goes on to compare certain beds in Poland with "... the system of the lower or chloritic chalk with which I believe it can be correlated. In one of these beds the mineralogical characters disappear entirely, the stratigraphic position is obscure, it [the correlation] depends only on the zoological characters" (Cuvier and Brongniart, 1822, pp. 326-327). In a footnote following this paragraph, Brongniart argues in favor of using fossils over all other characters, particularly over lithologies, to correlate rocks. He points out that at the present day you can see different types of rock forming at the same time in different places, and that equally in Calabria, over 38 years, it had been possible to see beds formed lying at an angle on earlier sediments, yet no one would treat them as belonging to different stratigraphic epochs.

The following year (1823) Brongniart showed that nummulitic facies of the Vicentine Alps in Italy were of the same age as the Tertiary of the Paris basin.

FACIES

The development of the concept of facies is more elusive than that of correlation by fossils. Young and Bird (1822) remarked,

> Some parts of the strata are so nearly allied, that they often pass into one another, and occupy the same beds. This observation applies particularly to sandstone and sandy shale, to coal and bituminous shale; and, we may add, to oolite and grey limestone; which, as we have observed (p. 270), pass into one another, at Kirkdale and other places (pp. 294, 295).

But there was no mention of fossils in this context.

The same concept of facies in a lithic sense was worked out independently by Eaton in 1828 for the Devonian of New York State, where he recognized that the red Catskill rocks are a lateral facies of the marine gray beds to the west (Wells, 1963).

Similarly, Constant Prévost in France recognized in 1839 that the Wealden of England must be the freshwater equivalent of the marine Neocomian of France and Switzerland, because they were both sandwiched between the Jurassic and the Chalk. In 1841 he pointed out that the Muschelkalk (of Germany) was absent over much of Europe, but must be represented in these other regions by clays and sands. Prévost's logic was sound, and historians such as Zittel (1899) and Hoelder (1960) have honored him as a major innovator of the idea of facies. This is unjustified: not only had Young and Bird and Eaton preceded him, but Prévost, like these earlier geologists, had little influence on his contemporaries (Wegmann, 1963), and other geologists already had much the same ideas, e.g., De La Beche (1839, p. 129):

Where we find limestones terminating abruptly against slates, and corals form a large portion of such limestones, we may be led to infer that the limestones were formed in the manner of modern coral reefs, mud and silt having accumulated against them laterally . . . .

We have seen how Brongniart correlated different facies with fossils, and similar ideas can be found in the work of Fitton (1827) on the Cretaceous of southeast England, and Phillips (1829) on the Jurassic of Yorkshire. However, it was not until 1838 that the word "facies" appeared in the literature and the concept was discussed in any detail. This was the work of Amand Gressly on the Middle and Upper Jurassic of the Soleurois Jura in Switzerland.

Gressly completed his monumental work in the summer of 1837, when he was only 23, and took it to Neuchâtel with the intention of reading it at the reunion of the Natural Science Society of Switzerland. He was so excited by everything that he saw and heard there that he forgot to read his own paper! Luckily, in spite of being timid, quiet, and withdrawn, he had the courage to show the manuscript to the famous Louis Agassiz. Agassiz was highly impressed; he got the manuscript published and invited Gressly to join him as his co-worker (Wegmann, 1963).

To begin with it is two main facts, which characterise the whole harmony of modification that I call *facies* or *aspects of the formation:* one means that *such or such a petrographic aspect of a formation necessarily implies, wherever it is encountered, the same palaeontological assemblage;* the other that *such or such palaeontological assemblage rigorously excludes genera and species of fossils frequent in other facies* (Gressly, 1838, p. 11).

These are somewhat sweeping statements, and reading the full memoir shows that Gressly had knowingly exaggerated his basic definition. In fact, he recognized almost all the basic problems of facies known today. He showed that each of his facies reflected a different environment of deposition. He emphasized that the vertical and horizontal relations of facies must be compatible in terms of the environments each represents, this 56 years before the publication of Walther's law of the correlation of facies. Unhappily, more have paid lip service to Gressly than have taken his ideas into their own work.

FAUNAL SUCCESSIONS NOT TIED TO A LITHIC SEQUENCE

Although Smith had noted that there were distinct fossils from the lower and upper Cornbrash, most of his faunal lists were tied to undivided formations. The idea of being able to subdivide a broadly homogeneous formation on fossils alone did not exist in Conybeare and Phillips's general study (1822); the Carboniferous Limestone, the Lias, the Oxford Clay, the Chalk were all left undivided.

The general realization that rock successions could be divided on their fossils alone really came from studies of Tertiary sediments. Cuvier had realized the desirability of this, but no one man has the honor here; within just three years, Deshayes in France (1830), Bronn in Germany on the basis of the Italian Tertiary (1831), and Lyell in England on the basis of Deshayes's tables (1833) had all proposed divisions of the Tertiary based on fossils alone. Indeed, as is well known, Lyell's names for the Tertiary systems were based on the proportions of extant living species the rocks contained.

In theory, at least, stratigraphy was now free from a lithological control, and the possibility of a pure biostratigraphy had been attempted. In practice, the next introductions, the concepts of the stage and the zone, were developed without reference to Tertiary rocks. This was the beginning of a split approach to stratigraphy by students of the Mesozoic and the Tertiary, which has lasted to the present day.

CONCEPT OF THE STAGE

Although it was seen during the 1830s that one could find a succession of fossil assemblages within a single lithic formation, the erection of a systematic pile of fossil assemblages, independent of facies, and for the whole geologic column, was introduced by the French geologist Alcide d'Orbigny. D'Orbigny saw that a consequence of Gressly's "facies" was that stratigraphy must be freed of the vagaries of formation names, each of which was based on a local facies body.

> Geologists in their classifications allow themselves to be influenced by the lithology of the beds, while I take for my starting point . . . the annihilation of an assemblage of life-forms and its replacement by another. I proceed solely according to the identity in the composition of the faunas, or the extinction of genera or families (d'Orbigny, 1842–1851, p. 9).

From this, d'Orbigny introduced the concept of the stage; but in discussing his contribution to biostratigraphy it is as well to separate his practice from the ideas behind it. The theoretical vision behind d'Orbigny's stages was conservative, even backward looking, in his own day; by present-day standards it is positively peculiar. Happily, even his contemporaries ignored his idea of repeated catastrophic destructions of life on earth, with repeated new creations, each of his stages representing a state of rest between one creation and the next destruction.

To understand what d'Orbigny meant in practice by a stage, let us take the Cenomanian in the Cretaceous, and read his mature account of it. This passage by d'Orbigny has been much abbreviated, but the terse style, the seminote form, and the odd use of capitals, or lack of them, are all true to the original.

20th Stage: Cenomanian, d'Orb. First appearance of the genera [list of 13 genera]. Reign of the order of cyclostegous Foraminifera, of the genera [8 names, chiefly bivalves]. Zone of *Nautilus triangularis,* of *Ammonites Rhotomagensis, Mantelli, varians;* of *Turrilites costatus,* of *Strombus inornatus,* of [10 species, chiefly bivalves]. Second zone of Rudists.

Derivation of the name. Here again it is the petrography which has served as a basis for the different names given to this stage, and which have prevented it being clearly distinguished as a geological horizon. One has called it *chloritic Chalk,* . . . when it is full of green grains, as at Havre, at Honfleur; but . . . the albian stage of . . . Escragnolles is of the same lithology, whilst the beds of this same geological level are elsewhere bluish, marly, or represented by white chalk and by quartzose sands, red, green or white . . . . The Sandstones . . . belong, by their stratigraphic position and by their palaeontologic characters, to all cretaceous stages . . . . We give to the . . . beds the stage name *cenomanian,* the town of Le Mans (*Cenomanum*) being built on the most characteristic type and the most complete of the stage which concerns us . . . .

Synonymy . . . it is . . . a part of the *chloritic Chalk,* of the upper green Sand, of the glauconitic Chalk, of several french geologists; . . . it is the *Tourtia* of Tournay of belgian miners; . . . nervian system . . . of Mr. Dumont; the *Chalk-marl* and *Fireston of Upper-Green-sand* of Mr. Mantell (Sussex); . . . *french type.* Le Mans, Saint-Calais (Sarthe); cap la Hève (Seine-Inférieure); [four more French type-localities]. *english type,* at Blackdowne . . . *portugese type,* banks of the Tage, near Lisbonne, etc. (1849–1852, pp. 630–652).

D'Orbigny follows this with a long list of localities where Cenomanian occurs, a discussion of its upper and lower limits, its relations to the overlying and underlying stages, its thickness and variety of lithologies, the paleogeography. He rounds off his essay with details of the fossil content: he lists numbers of genera of the underlying Albian and the overlying Turonian, which are absent in the Cenomanian; in more detail he gives all the genera of the Cenomanian unique to the stage; he notes that there are 841 species characteristic of the Cenomanian stage, which, he says, is quite enough to enable one to recognize the stage whatever facies it presents.

The inclusion, in the stage, such as we conceive it, of all the points indicated of its geographical extent is based on stratigraphical considerations of superposition, and on the assemblage of palaeontological characters of all these places. To prove that this reunion is nothing arbitrary, but that it is properly the result of the identity of contemporaneous species, we are going to give here the names of the most common, the most characteristic species which are found everywhere, not merely around the anglo-paris basin, in the Pyrenean basin, but in the Mediterranean basin [here follow localities from England to north Africa] (d'Orbigny, 1849–1852, p. 644).

This example has been given at some length because there is still argument today over what a stage is. When one examines d'Orbigny's own account, it is clear that his Cenomanian stage was conceived and based on its fossil content. The same is true of all his other stages: each is the major body of strata less than a system (anywhere in the world) that contains at least some of a long list of fossils which are peculiar to that piece of the total stratigraphic column.

To emphasize the certainty of his results, d'Orbigny gave multiple definitions of each stage. This has led to different usages because some authors have emphasized the use of type localities*; others have noted d'Orbigny's own emphasis on the use of fossils independently of lithologic variations; still others have read sedimentological cycles into his palaeogeography and stratigraphy. In recent years some have even, in effect, gone back to d'Orbigny's catastrophic theory: the stages are "the expression of the divisions which nature has delineated with bold strokes across the whole earth" (d'Orbigny, 1842-1851, p. 603). Hence the curious idea that the stages are terms of "time-scope" independent of their fossil content (e.g., Hedberg, 1959); no such meaning was given to d'Orbigny's stages by his contemporaries, nor for the next hundred years.

CONCEPT OF ZONES

The medieval words "zone" and "stage" were used by a number of geologists in the first half of the nineteenth century. The use of zone as a biostratigraphic term originates, like stage, with d'Orbigny. In this sense, he used it as an alternative word to stage: "I . . . became convinced that the Jurassic rocks were divisible into ten zones or stages . . . " (d'Orbigny, 1842-1851, p. 601). However, he also continued to use it, although less commonly, in the usual nontechnical sense of a band or belt, or natural division of the earth's geography, as when he wrote

In each stage, in fact, some fossils are characteristic of terrestrial deposits, others of marine deposits; in these latter deposits, floating species characterise the inshore sediments, at the level of high seas; others belong to zones deposited a little below tide-level; whilst some series belong entirely to the deep zones of the oceans (d'Orbigny, 1849–1852, p. 258).

*D'Orbigny invented the concept of type localities, yet he never used the phrase, but simply referred to "types." He was so confident in his correlations that he quoted many types for each stage. The more recent use of a single stratotype is not in d'Orbigny's own works, although he may refer to one of them as "the best type" or "the most beautiful type."

Various meanings for zone in stratigraphy might have continued for years, but in 1856 there started in Germany a work that gave the word a definite meaning, and which was to alter stratigraphic practice forever – Albert Oppel's *Die Juraformation Englands, Frankreichs und des Südwestlichen Deutschlands* (1856-1858). Oppel was only 27 when the whole of this work had been published, but his ability to observe and to draw conclusions from those observations, combined with tremendous energy to travel around Europe, had not been found in any previous stratigrapher since Sedgwick and Murchison. He died of typhoid fever when he was only 34.

Even today a brief perusal of Oppel's book impresses with its spread of detail. In eight separate districts of western Europe, the Jurassic rocks are subdivided into 33 zones correlated on the basis of their fossil content. Oppel's contemporaries outside Germany were completely bowled over; even the French admitted that it was pertinent to France and published a tabular summary (Laugel, 1858). Oppel explained his method within a mere seven-page forward:

> Comparison has often been made between whole groups of beds, but it has not been shown that each horizon, identifiable in any place by a number of peculiar and constant species, is to be recognised with the same degree of certainty in distant regions. This task is admittedly a hard one, but it is only by carrying it out that an accurate correlation of a whole system can be assured. It necessarily involves exploring the vertical range of each separate species in the most diverse localities, while ignoring the lithological development of the beds; by this means will be brought into prominence those zones which, through the constant and exclusive occurrence of certain species, mark themselves off from their neighbours as distinct horizons. In this way is obtained an ideal profile, of which the component parts of the same age in the various districts are characterised always by the same species (Oppel, 1856, p. 3; translated in Arkell, 1933, p. 16).

Today we can recognize that these words mark the birth of biostratigraphy as a separate discipline. Whatever d'Orbigny may or may not have meant by his stages, it was Oppel who introduced the "ideal profile" based on fossil successions independent of any local palaeontologic or lithic succession. And in that same hot summer of 1858, Alfred Russell Wallace and Charles Robert Darwin read their joint paper to the Linnean Society of London, "On the Tendency of Species to Form Varieties, and on the Perpetuation of Varieties and Species by Natural Means of Selection." Immediately, Oppel's ideal profile becomes synonymous with the record of irreversible evolution of life on earth. Oppel himself remarked that the more accurately the fossils are examined and species defined, the greater the number of zonal divisions that could be recognized.

Each zone was named after a fossil, generally an ammonite species, but Oppel said that they could equally well have been named after places. The index species was merely a name for the zone: it was only one species in the assemblage of fossils that actually defined the zone, although one must be careful to avoid the mistake of Woodward (1892, p. 298), who wrote, "Zones are assemblages of organic remains of which one abundant and characteristic form is chosen as an index," which, as Buckman tartly commented, is a definition that embraces fossils in museum drawers! In practice, Oppel quoted 10 to 30 species as characteristic of each of his zones.

D'Orbigny's stages were fitted into Oppel's scheme: groups of zones were gathered to form each stage. Thus the zones of *Ammonites macrocephalus, A. anceps,* and *A. athleta* together form the Callovian stage. It is important to understand that Oppel's approach was to build up each stage from zones, not to take each stage and then subdivide it into zones (Miller, 1965). At the same time, he recognized that each stage named at that time already had a particular meaning attached to it, and occasionally a zone did not fit into one of d'Orbigny's stages as then defined. Thus his zone of *Diceras arietina* fell between the Oxfordian and Kimeridgian stages.

CONSOLIDATION DURING THE LATE NINETEENTH CENTURY

D'Orbigny had first introduced stage names in 1843, when he used them in describing Gastropoda from the Cretaceous, but it was not until the publication of his textbook of stratigraphy from 1849 to 1852 that his method became well known. Since it was only four years after this that Oppel began to use zones in a modern sense, the gradual acceptance of stages and zones into stratigraphy forms a single subject. It is a tangled subject, because even before d'Orbigny there were geologists who used stages names as geologic jargon for lithic formations, or some combined lithic and fossil unit. Merely putting "-ien" endings on place names had been introduced by Brongniart in 1829 (e.g., Oxfordien and Portlandien), and the practice spread independently of d'Orbigny (e.g., Néocomien by Thurman in 1836, Séquanien by Marcou in 1848, and Maestrichtien by Dumont in 1850; but these authors did not give them the biostratigraphic meaning that d'Orbigny did). On the other hand, Trautschold in Russia was still using plain Kimmeridge and Portland as late as 1877, but giving them the meaning of Kimeridgian and Portlandian.

Equally, because authors quoted Oppel's zones when writing about the Jurassic is no proof that they understood them (e.g., Woodward, 1876). In France, as early as 1857, Hébert was referring to the "zone of *Am. primordialis*" in the Jurassic, and he clearly only gave a stratigraphical meaning to "zone." But Hébert used *zone, couche, assise,* and *horizon* interchangeably, more or less at random, throughout his life; sometimes he made them lithic divisions (e.g., *ces assises glauconieuses,* (Hébert, 1857b). Oppel himself in the subheadings of his chapters referred to "Strata of Ammonites exus" not Zone of *Ammonites exus.*

The history can be summarized by saying that stages were introduced more rapidly than zones, but were more often misunderstood, and met with more opposition than zones.

Outside France and Belgium, d'Orbigny's stages were either ignored, or received with hostility, for many years by most geologists. Quenstedt, in Germany, wrote,

Of what avail is it if a man has seen the whole world, and he does not understand aright the things which lie in front of his own doors? . . . To compare faithfully two beds, each a hand in height, one on top of the other in their true order, can effect a more fruitful development of science than the use of stratigraphic catalogues from the furthest regions of the earth. Right from the outset one has to admit that such records are not reliable (Quenstedt, 1856, pp. 23–24; translation based on Arkell, 1933, with corrections).

Let us not weary of searching our strata; let each one of us collect as much as he can in his own neighbourhood, labelling the specimens exactly with their localities, and compare then with material collected by others; then at least the first goal of all geological research should not remain far from our reach – *a true table of the succession of the strata* (Quenstedt, 1857, p. 823; translated in Arkell, 1933, p. 14).

Note that this critic equally believed that there was some natural "true table of the succession of the strata." His real complaint was that d'Orbigny's foresight had forestalled those with more patience who would produce an accurate record. In fact, Oppel's improvement on d'Orbigny's work was published in the same years as these remarks of Quenstedt.

In Britain and Ireland, the stage names of d'Orbigny and other French geologists came into use only very slowly. Tate (1865) used them for the Upper Cretaceous of Ireland but referred to them as "formations." In 1867 he used Renevier's stage name Hettangian but he still refrained from using Sinemurian for the next beds upward in the Jurassic (Tate, 1867). As late as 1896, Strahan could remark about the Cenomanian stage " . . . that English geologists were having reason to repent the introduction of Continental names into their Cretaceous nomenclature" (Jukes-Browne and Hill, 1896, p. 178).

In the United States, the concept of stages was completely ignored even as late as 1912 in the comprehensive *Index to the Stratigraphy of North America* (Willis, 1912).

These three countries – Germany, Britain, and the United States – were the slowest to introduce stages into their stratigraphy. By contrast, in Russia, Trautschold (1877) was only the first of a line of stratigraphers who used stages, and in their proper sense. Resistance to the use of stages was partly because outside France they were not always the self-evident divisions; partly because d'Orbigny was ahead of his time in recognizing the need for standard international divisions smaller than systems (d'Orbigny's Paleozoic stages were, in fact, synonymous with systems); and partly because in England and Germany local divisions were already strongly established. In the United States, correlation of Mesozoic faunas with their European counterparts was not developed until the 1920s, and without this the stages could not be identified in America.

Such opposition was not encountered by "zones," although like most novel concepts its adoption was relatively slow. Almost all stratigraphic theory was initially developed from work on Jurassic, Cretaceous, and Tertiary rocks. For these systems, and especially in France, Belgium, and Germany, the recognition and use of zones came in smoothly, limited only by the small number of stratigraphers to do the necessary detailed work. It is in this context that the work of Lapworth (1878, 1879–1880, 1879) is important. In this series of papers he showed how graptolite distributions could provide reliable zonal divisions of the Ordovician and Silurian; indeed, it was in the last of these papers that he introduced the Ordovician as a system. This was the first systematic erection of zones in the Paleozoic; it disproved Barrande's theory of "colonies," which could have meant that graptolites were without correlative value; and the British could no longer consider zones to be merely some sort of foreign idea (for many years the only detailed work on Cretaceous zones in Britain was by the French geologist Barrois, 1876). Lapworth recognized 19 zones grouped into seven stages, although these were not always given "-ian" endings. He explained how you could find the top and bottom of each of his zones. He also emphasized the chronological value of fossils: ". . . we have no reliable chronological scale in geology but such as is afforded by the relative magnitude of zoological change . . ." (1879, p. 3). It was still a slow business to get such an idea into the British; in the country of William Smith one could still find Woodward in 1887 being openly reluctant to use zones: he regarded lithology as safer for cross-country correlation!

EARLY ATTEMPTS AT INTERNATIONAL AGREEMENT

At the first International Geological Congress in Paris in 1878, Stephanesco of Roumania appealed for the establishment of a uniform nomenclature in stratigraphy. A 14-man commission, representing 16 countries, was set up to report on this. It met in Paris in April 1880, and recommended that the definitions of general terms be first attempted.

The congress at Bologna, Italy, in September–October 1880 decided on definitions of stratigraphic words like series and stage, and listed their synonyms in several languages (pp. 196–197). Rocks, considered from the point of view of their origin, were formations; the term was not part of stratigraphic nomenclature at all, but concerned how the rock had been formed (e.g. marine formations, chemical formations). Stratigraphic divisions were placed in an order of hierarchy, with examples, thus: group (Secondary Group), system (Jurassic System), series (Lower Oolitic Series), stage (Bajocian Stage), substage, assise (Assise à *A. Humphresianus*), stratum. A distinction was made between stratigraphic and chronologic divisions. The duration of time corresponding to a group was an era, to a system a period, to a series an epoch, and to a stage an age.

Some terms never got to be discussed at Bologna, including zone, horizon, and deposit, but subsequently the secretary of the commission, G. Dewalque of Belgium, attempted to draw up a consensus from reports of individual national committees, and this was published with the *comptes rendu* of the congress in 1882 (pp. 549–559). A revised version of this consensus report by Dewalque was published for the congress in Berlin (in 1886), and an English translation of this revision was published by the American committee in the same year (Frazer, 1886).

This appearance of unity was deceptive. The individual reports from the national committees show that even the agreed resolutions taken at Bologna were unacceptable to geologists in many countries. The British committee complained that those who had actually got to attending meetings of the commission were only a few of those who should have been there. This is the prime lesson of these early congresses: unless real agreement exists, international rules will be ignored.

TWENTIETH-CENTURY DEVELOPMENTS

Most of the concepts used in biostratigraphy today had already been developed by 1900. Three widely used textbooks of that time, Lapparent (4th edition, 1900), Geikie (4th edition, 1904) and Haug (1908-1911), show that stages and zones were being used as a matter of course in Europe for the Mesozoic and Tertiary, but for the Paleozoic their adoption had been haphazard; in America, lithostratigraphy still dominated. The gradual spread of biostratigraphic methods for other systems, and into other countries, would hardly be of general interest. Suffice to say that it would be a long story; in America, as recently as 1970, Berry and Boucot felt it necessary to publish a lucid explanation of the basic techniques of biostratigraphy. This was 114 years after Oppel had done the same for the Jurassic of western Europe!

Only the more important refinements of the twentieth century are discussed below.

Special Sorts of Zones

For many years no one worried that some zones were defined in one way, perhaps by the occurrence of a floral assemblage, whereas others were defined in some quite different way, such as by the stratigraphic range of an individual ammonite species. That such different sorts of zones existed was fully realized by Woodward in 1892, but it was not until 10 years later that a restricted type of zone was introduced by Buckman (1902, p. 557). This was the "faunizone," but by it Buckman meant no more than what many geologists of the time (and today) meant by the plain word "zone." Indeed, in defining faunizone, Buckman started by taking Marr's widely quoted definition of a zone, thus:

> Faunizones are, to paraphrase Mr. Marr, "belts of strata, each of which is characterized by an assemblage of organic remains," with this provision, that faunizones may vary horizontally or vertically, or the strata may not vary and yet may show several successive faunas. So faunizones are the successive faunal facies exhibited in strata, (Buckman, 1902, p. 557).

The first special type of zone that was quite distinct from Oppel's zones was the "epibole" introduced by Trueman in 1923 (p. 200) for the strata accumulated during the acme of a dominant species. The next year Frebold (1924) introduced "teilzone" for the strata at one locality deposited during the range of one species

at that locality. Both types of zones were aimed at the finest possible divisions of strata on the basis of fossils. Whereas epiboles have little use except in regions where fossils are abundant through considerable thicknesses of sediments, teilzones can be used anywhere; with what effect is a matter of dispute.

More zonal terms have been introduced since 1924, a plethora by the late 1960s. For those who enjoy "geologese" there is a glossary of them patiently compiled by Hedberg (1971). There are discussions of their use by Arkell (1933), Weller (1960), and Shaw (1964). In spite of Arkell's statement as long ago as 1933 that "The term 'zone' by itself has now become a kind of family term, which may be very ambiguous unless qualified," a quick perusal of almost any stratigraphic journal will show that unqualified zones are still the norm.

Stages Named After Fossils

In 1898, Buckman named the ages, corresponding to stages, after groups of ammonites (e.g., Parkinsonian, after *Parkinsonia,* for the age equivalent of the Bathonian stage). Later he elaborated this scheme to fit several ammonite ages into each stage. Since Buckman believed in the nearly instantaneous world migration of new species, and hence in the absolute synchronity of beds correlated by ammonites, his ages became a sort of substage without a type locality, and usually without a rigid definition.

This habit was followed by Spath (e.g., 1923-1943) and others, but in a rather half-hearted way. The technique is now dead, largely because older stage names were too well entrenched, but more importantly because it presupposes that ammonites are the most perfect, or even sole, group of fossils for correlation.

Confusion of Stratal and Time Terms in Biostratigraphy

The earliest biostratigraphers were little concerned with time in geology, but with the practical problem of subdividing the stratigraphic column and correlating its parts from place to place. They had little idea of the immensity of geologic time. When Huxley (1862) wrote that ". . . neither physical geology nor palaeontology possesses any method by which the absolute synchronism of two strata can be demonstrated. All that geology can prove is local order of succession," nobody of the time could disprove him. Hence the need for a distinction between "homotaxis" or "similarity of arrangement" and "synchrony" or "identity of date." But as Huxley himself observed, "It may be so; it may be otherwise," and the example he offered must have been as nearly as provocative to his audience as it is to us today:

> For anything that geology or palaeontology are able to show to the contrary, a Devonian fauna and flora in the British Islands may have been contemporaneous with Silurian life in North America, and with a Carboniferous fauna and flora in Africa. Geographical provinces and zones may have been as distinctly marked in the Palaeozoic epoch as at present, and those seemingly sudden appearances of

new genera and species, which we ascribe to new creation, may be simple results of migration (Huxley, 1862, p. xlvi).

Subsequent nineteenth-century geologists sometimes acknowledged that their correlations were homotaxial rather than temporal, but, in the absence of a reliable means of measuring absolute ages of rocks, there was little more they could do about it. Of course, they realized that each thickness of sediment must represent a certain quantity of time, and there gradually developed a set of terms to represent the time equivalents of stratal divisions. D'Orbigny had used "époque" for the time of one of his stages, but by 1880, at the Bologna congress, it was agreed that the geologic time equivalent of "stage" was "age"; and for "zone" the Swiss committee had suggested "moment." However, in 1901 the code of the Paris congress made "phase" the time equivalent of a zone (Renevier, 1901), a decision apparently unknown to Jukes-Browne (1903), who suggested "secule" for the time equivalent of a zone. As early as 1893, Buckman had introduced "hemera" to be "used in a chronological sense as a subdivision of an 'age' " (1893, p. 518), and in 1902 he introduced (but apparently never used) the word "biozone," "to signify the range of organisms in time as indicated by their entombment in the strata" (1902, pp. 556–557). In America, a similar concept was expressed by Williams (1901): ". . . the time-value of the species *Tropidoleptus carinatus* would be the *Tropidoleptus* biochron."

All these introductions of new words prove that stratigraphers around the turn of the century had a clear distinction between stratal terms (system, stage, zone) and time terms (period, age, phase). In fact, in 1898, when Buckman submitted a paper to the Geological Society of London in which he used stage names, such as Bajocian and Bathonian, and called them "ages," the council of the society objected that he was using stratigraphic terms in a chronologic sense. Equally, the lack of use of "phase," "moment," and "secule" shows that no need was felt for them. As for "hemera," very few geologists either then or later ever used it, and for Buckman himself it was synonymous with "subzone," because he believed his ammonites to be perfect chronologic correlators.

No real confusion between time and stratal terms seems to have developed until the 1950s. When Arkell published his classic essay in 1933 (on which we all depend so much), he believed that there was a subtle distinction between stage and zone, but they were both based on fossil distributions, and hence both were biostratigraphic terms. He expressed the distinction more bluntly in 1946:

A stage is an artificial concept transferable to all countries and continents; but a zone is an empirical unit. If the zonal index species and its associated fauna are absent we cannot record the zone as present. . . . In summary zones are of more restricted function than stages. Attempts to give them universal application are misdirected; such attempts merely make zones synonymous with subdivisions of stages and at the same time deprive them of their special qualities as the basis of correlation from one province to another (Arkell, 1946, p. 10).

Ten years later he seems to have realized that, in emphasizing that zones are not subdivisions of stages and neglecting to add to the passage that stages are built from zones, he left the impression that the two are quite different sorts of stratigraphic units.

> Just as it is convenient to group together formations into Series, so it is convenient to group like zones together and reduce the numbers for practical purposes, and above all to have a grouping which enables several zones to be correlated in a general way over long distances when the zones individually are too precise. Such groupings of zones are Stages (Arkell, 1956b, p. 7).

But it was too late! In 1952, at the international congress at Algiers, Hedberg proposed the setting up of an international commission "to establish principles and harmonize practice in stratigraphic nomenclature and terminology," and Hedberg became the first and, to date, only chairman.

There is now a series of papers by Hedberg, later to become a series edited (but still unmistakably dominated) by Hedberg (1948, 1951, 1954, 1959, 1961, 1964, 1965, 1968, 1971, 1972a, 1972b). These papers have provoked a lively discussion and for this reason alone they have been immensely valuable (e.g., Schindewolf, 1955, 1970; Arkell, 1956a; Hupé, 1960; Wang, 1964; Verwoerd, 1964, 1967; Callomon and Donovan, 1966; Hancock, 1966; Wiedmann, 1968, 1970a; and codes of stratigraphic nomenclature in 16 countries in all parts of the world, listed in Hedberg, 1972a, 1972b). Hedberg's chief innovation is the proposal that there should be set of "chronostratigraphic" units, which of itself is outside the scope of the history of biostratigraphy. Unhappily, among the formal units that he picked for his chronostratigraphic scale was "stage." Really, if Hedberg were not the gentleman I know him to be, I should be tempted to accuse him of a form of scientific theft. Whether "zone" and "stage" are different grades of the same concept, or whether they are different concepts, there is no question but that they are both biostratigraphic. As justly emphasized by Wiedmann (1968, 1970a), this was decided a long time ago, both by international agreement and common practice. Confusion now exists because a number of biostratigraphers have rushed to the defense of certain groups of fossils by claiming that their distribution is chronostratigraphic in Hedberg's sense (e.g., Sylvester-Bradley, 1967; Callomon and Donovan, 1966, for zones based on ammonities; Berry and Boucot, 1970, for graptolite zones). This is poor logic. Zones are not merely the most accurate known scale of stratigraphy, and by implication of stratigraphic time; they are the sole scale, other than radiometric dates in years, but they are still based on the distribution of fossils, and as such are biostratigraphic. The fact is that chronostratigraphic units are an imaginary entity of no value, and to claim an ammonite zone to be chronostratigraphic is to debase a practical stratigraphic unit, as well as to deny the biological characteristics of the origin, dispersal, and extinction of species populations – none of which is likely to be isochronous.

Making Biostratigraphic Units Objective

It will be remembered that one way by which d'Orbigny defined each of his stages was to quote type localities where a particular formation yielded a fauna that would enable one to recognize the same stage elsewhere. This objective manner of defining a stage has always appealed to many geologists, partly because most zones, in contrast, have been subjective; the recognition of any type of zone depends on the correct identification of certain fossils. French gelogists, in particular, have long placed emphasis on the use of type localities (stratotypes).

Attractive as this sort of stratotype approach might sound, it has met with practical difficulties. The top of a stage at one locality is not likely to coincide with the base of the overlying stage at *its* type locality; they may overlap or there may be a gap. To get over this difficulty, a group of British geologists suggested in the 1960s that one should define only the base of each stage and that this should be done by inserting a "golden spike" as a marker in an actual section (Ager, 1964). Hence one would have type localities to fix the base of each stage. Having defined the base of the stage, all strata are included in that stage as high as the base of the next stage upward. The idea was inserted in the first report of the stratigraphical subcommittee of the Geological Society of London (George et al., 1967), and the idea of "boundary stratotypes" as well as "unit stratotypes" is included in the report on stratotypes published by the Montreal Congress (Hedberg, 1970). The first examples of boundary stratotypes ("type horizons") were introduced by Sylvester-Bradley (1964), but at that time he did not use the principle of defining only the base of each stage. The Silurian-Devonian boundary is now fixed by a golden spike hammered in at Klonk in Czechoslovakia (Chlupáč et al., 1972).

Stages are groupings of zones. Therefore, the concept of defining only the base of each unit should start at zonal level. Thus one defines, with the help of a boundary stratotype, the base of each zone. The top of that zone is then defined by the base of the overlying zone, itself fixed at another boundary stratotype. The base of each stage is defined by the base of its bottom zone. Type localities for zones were introduced by Callomon (1964).

DISTINCTIONS BETWEEN CONCEPT AND PRACTICE

(Students of the cut and dried should jump to the summary).

When it came to writing this paper, I realized that some of the great innovators in biostratigraphy have put their conceptions into practice in such a way as to allow subsequent workers to make quite different interpretations of the supposed underlying concepts. The difficulty is a classic one for the historian: how do you distinguish what people say from what they do? Moreover, both may be ambiguous or inconsistent. Hence the student who seeks definite dates for the introduction of concepts is chasing a chimera.

The editors asked me to include some statements pointedly showing the historical derivations of various types of zonation, but there is no single point in history for the introduction of concepts like "concurrent-range zones" or "lineage zones." Certainly, the first time that these phrases were printed does not mark the first use of such zones, any more than to state that there was no biostratigraphy before 1910 when Dollo invented the word. Nor are such concepts introduced only once; most stratigraphic concepts are periodically rediscovered, so that the literature is littered with superfluous synonyms. Thus "biomere" of the 1960s and 1970s is only a new word for d'Orbigny's "stage." The current controversy over "chronostratigraphy" is largely an atavism: over the years 1850–1862 the same problems were discussed, albeit without the benefit of such twentieth-century words as "biostratigraphy," "epibole," or "negative association zone."

Oppel's ideas were so comprehensive that no simple definition of "zone" as used by Oppel is possible. He had already introduced, and therefore must have conceived without explicitly stating it, special sorts of zones such as assemblage zones and concurrent-range zones. He saw no need to distinguish such different types of zones, and did not feel it necessary even to define "zone." He visualized "an ideal profile," a concept sufficiently charismatic to attract, and sufficiently vague to allow us each to construct our own zonal succession, while thinking that we are getting closer to some fixed truth. European stratigraphers have been doing this for a century. Many of them, perhaps most, have never bothered themselves with just what sort of zones they have been erecting or handling. Some German stratigraphers (e.g., Schindewolf, 1950a, 1955) have even interpreted a zone as a time term to designate an interval during which was deposited the sediment that contains certain index fossils.

Such mystical inexactitude has dissatisfied twentieth-century American pragmatic stratigraphers. Hence the introduction of a variety of types of zones with precise meanings, among which the usual European practice is left as an almost undefinable "Oppelzone." To what degree this movement toward exact meanings for zones will improve our stratigraphic correlations remains to be seen; much of Shaw's apparently novel quantitative zonation (1964) is to be found in earlier European arguments. But it can hardly be doubted that exact definitions of the zones we use will improve our knowledge of their accuracy and limitations.

SUMMARY

1796	William Smith recognizes that formations are arranged in a regular order and each is characterized by its own peculiar fossils.
1833	Charles Lyell, with the help of Paul Deshayes, successfully subdivides the Tertiary of western Europe on fossil assemblages. He calls his Eocene, Miocene, and Pliocene divisions "systems," but because the divisions could be applied equally to separate basins of deposition, he and Deshayes had really anticipated d'Orbigny's invention of "stages."
1848	Amand Gressly introduced the word "facies," and shows that different facies of the same stratigraphic age yield different faunas.

1849–1852	Alcide d'Orbigny systematically subdivides all the Phanerozoic rocks of the world into stages based on the succession of their fossil assemblages.
1856–1857	Albert Oppel conceives stratigraphic divisions based on the general succession of fossils independently of the actual succession at any one place. He calls such divisions "zones" and applies them to the Jurassic.
1862	Thomas Huxley points out to geologists that their correlations with fossils show identity of arrangement (homotaxis); they do not prove synchronous deposition.
1879–1880	Charles Lapworth successfully applies Oppel's zonal methods to the Lower Paleozoic.
1880	Bologna: first international agreement on definitions of stratigraphic terms.
1893	Sydney Buckman introduces the "hemera," the first unit of geologic time, but it is unrelated to any stratigraphic division then in use.
1899	Arthur Rowe provides an early example of lineage zonation in his study of the evolution of the echinoid *Micraster* in the Upper Cretaceous of England.
1954	Hollis Hedberg appeals for a renewed effort to reach international agreement on procedure and terminology in stratigraphic classification, but purloins the "stage" from biostratigraphy for his new "chronostratigraphy."

Acknowledgments

Most of this paper was written while I was the guest of the George Washington University, Washington, D.C. I am most grateful to Dr. A. G. Coates, who organized this visit, and to Dr. Geza Teleki and his colleagues of the Department of Geology for their excellent hospitality. I should like to acknowledge with gratitude the help of the editors in the improvement of the manuscript.

Note on References Cited

The actual date of publication of many nineteenth-century works was not that given on the title page. Quite commonly, an author would republish in book form, with a new date, work that had first appeared in parts, but nothing in the book shows this. In general, the dates on the title pages in English and German works give the latest year in which any of the pages were published. In France, this rule did not apply; d'Orbigny, in particular, would often publish the title page before the contents, all of which are later than the date given; when the work was complete he would re-issue it with a new date on the cover and the original date on the title page.

Biological Concepts in Biostratigraphy

Evolutionary Models and Biostratigraphic Strategies

Niles Eldredge — American Museum of Natural History and Columbia University

Stephen Jay Gould — Museum of Comparative Zoology Harvard University

INTRODUCTION

Radioactive decay and biological evolution are the two processes that have led to the present level of refinement in the geological time scale. These two irreversible processes create products that can be analyzed in terms of developmental stages. Information about time emerges either directly (by radiochemistry) or in a relative way (through biostratigraphy).

For radiometry, theory necessarily preceded practice: once the schema of radioactivity had been established both in general and specifically for the decay of uranium, Boltwood was able to formulate the general method as early as 1907 (Knopf, 1949). But in biostratigraphy, practice preceded a grasp of the underlying process. To oversimplify, William Smith preceded Charles Darwin. It is therefore reasonable to ask whether the use of fossils for temporal correlation depends nearly as much upon evolutionary theory as isotope dating depends upon the theory of radioactivity. *All* the many kinds of biostratigraphic units ever proposed share the simple assumption that similar organisms in different outcrops imply some kind of equivalence of their enclosing matrices. Even if we take this equivalency to imply time, is it really necessary for us to realize that evolution underlies our attempts to subdivide geologic time and accounts for the vertical, and even the horizontal, changes in the fossil content of strata? A thorough grasp of evolutionary theory has not been essential to the working biostratigrapher; we have come too far in biostratigraphy, despite the near absence of explicit analysis of the relationship between evolutionary theory and biostratigraphic method, for us to claim otherwise (Hayami and Ozawa, 1975). Some of the most intriguing, durable, and successful work in biostratigraphy has been performed by paleontologists holding rather dubious views on the nature of the evolutionary process.

But we do believe that a more accurate and complete picture of evolutionary mechanisms would benefit biostratigraphy by sharpening our practices and helping us to weed out techniques based on idealized, if not downright spurious, notions of evolution. We might draw an analogy with the science of systematics, which, like biostratigraphy, ultimately depends upon the process of evolution. In terms of its basic methodology, its hierarchical format, and even the basic taxonomic arrangement of organisms, systematics has remained distinctly recognizable from its pre-Darwinian days. After 1859, scientists knew why the Linnean system worked, but they did not need to change it. Subsequent advances in evolutionary theory have fed back definite pressures for change in the methods of systematics; examples are the "new systematics" of the 1940s and the current emphasis on cladism (e.g., Schaeffer et al., 1972). It is precisely this type of refinement that might come to biostratigraphic methodology from a more critical look at the nature of the evolutionary process. Thus, we shall examine the basic evolutionary models currently available, characterize their properties, and see how they fit with past and present methods of research in biostratigraphy.

EVOLUTIONARY STUDIES IN PALEONTOLOGY

Evolutionary change is the modification through time of genes and gene frequencies. What we see, of course, is the phenotypic expressions of genes, since selection acts on the phenotype in any case; our fundamental source of information about any organism is its morphology. Evolutionary paleontologists and biostratigraphers are thus united in a common pursuit – to document and analyze the distribution of morphologies in space and time, the raw material for evolutionary studies and for the elaboration of biostratigraphic correlations. Paleoecology has taught us to consider physical and biotic parameters in trying to explain the presence or absence, or the particular morphologies, of individual taxa. Likewise, we have come to accept the new biogeographies, and to regard them as important adjuncts to our methodology to explain the distributions of taxa. What then of the process that subsumes all others – evolution? How many of us think in terms of character displacement when we confront two distinct "forms," usually never found together, and wonder why one of them is one and a half times larger than usual? Character displacement is but one process of many that have affected the morphologies of extant species. All such phenomena can, potentially, be documented in the fossil record (Eldredge, 1971); they should be actively considered by anyone attempting to explain the distribution of morphological data in the fossil record, and by those that use these distributions to zone and correlate strata.

But such matters have little meaning outside the context of the biology of populations, including its largest extension, the biology of species. For the simple purpose of resolution, biostratigraphers are rarely concerned with higher-level taxa. The "teilzones" of classes, orders, families, and even most genera are too long to apply to biostratigraphic zonation. Biostratigraphers are interested in refinement, and if their raw data be taxa, they must be taxa of the lower categorical ranks. This implies that we need to concern ourselves solely with models of "microevolutionary" phenomena, i.e., with processes which act directly on the phenotypes of populations and which determine the origin of aggregates of individuals and populations at or around the level of species.

There are two alternative ways of looking at microevolutionary studies in the fossil record. Both have weight, and their resolution leads dialectically to a basic strategy that we believe is both rational and functional. One view stresses the spotty and otherwise inadequate nature of both the fossil record and of geologic time itself as preserved in sediments. It is easy to be overwhelmed by this consideration, and to despair about ever accomplishing anything in evolutionary studies of fossils.

The alternative view is expressed by those of us who see huge populations of well-preserved fossils (e.g., the brachiopod *Platystrophia ponderosa*) at stratigraphically closely spaced intervals over a fairly widespread area. Here, confidence can soar, since correlations are fairly secure, collecting is easy, and we have, in charming contrast to the *Drosophila* geneticist, truly long intervals of time to work with. In such a context, it is easy to feel that we have encountered actual evolutionary experiments, already performed by nature, and frozen in the rocks awaiting our judicious analysis. The second of these two views is the more stimulating, simply because it is positive. The first was negative, if not depressingly weighty.

As paleontologists, it is unfortunately true that we cannot be sure that a collection of fossils is a truly representative sample of a biological population. Bedding plane clusters of sessile organisms are possible exceptions, but even these may be lumped over many years without clearly differentiated size classes as clues. Nonetheless, it is true that even a single ragged specimen constitutes a sample of a former local population. In general, we can collect samples carefully localized in time and space; these may fairly approximate samples of biological populations. Two examples of such collecting were recently given by Kauffman (1970) for Cretaceous bivalves.

What can we do with a series of such samples? The optimist might indulge in some elementary "population genetics" by comparing the distribution of variation in one or more characters through successive populations. The difficulty with such attempts, of course, lies in their interpretation. All experimental work in population genetics depends on distinguishing descendant generation F_1 from F_2. Even though the simulation of thousands of generations is now possible (and we could compare the resultant distributions with those seen in a fossil example), in point of fact, these simulations are computed with algorithms based on equilibrium formulas and controlled parameters predicated on short-term phenomena. There is a problem in order of magnitude here; the fossil record is too gross, or the models of population genetics are too fine, to permit meaningful analysis of one in terms of the other (see Eldredge and Gould, 1974, for a further discussion of this point). This is as much a criticism of population genetics as it is a denigration of the fossil record as a laboratory of evolution.

What then can we do with reasonable samples of local populations? We all know that the local biotas of Recent habitats are subdivided into component clusters of organisms, albeit with some fuzziness. Local populations always display variation, both continuous and discrete, in many phenotypic characters. We also know that many populations fluctuate in size and occasionally disappear within our own lifetimes; the current reduction in the size of frog populations in the northern United States is a case in point. Other populations seem more stable.

If an observer moves from the initial point where he observed a local population, he may encounter another suite of local populations containing one similar to that encountered at point 1. In other words, there are clusters of local populations so similar that they define a unit, called a "species." Here, boundaries may be even more fuzzy than for local populations. *Bison bison,* at least today, is clear-cut. *Cannabis sativa* is ambiguous; *Rana pipiens* is a mess. There are many definitions of "species." Some are genetical, others ecological, still others morphological. However defined and recognized, species are real and discrete biological entities (in contrast with higher and lower taxa), and we assume that species can generally be recognized. This reiteration is necessary in view of recent attacks on the reality of species in both the neontological (Sokal and Crovello, 1970) and paleontological (Shaw, 1969) literature. Species are important at this point of our discussion because most of the raw data of biostratigraphy (as well as much of the language of evolutionary theory) are expressed in terms of species. Any common ground between them must consider species.

Species, and carefully defined subspecies, are the best raw material for correlation. Species are real. As units, they are small enough to allow good biostratigraphic resolution. And they are large enough to surmount the problems of geographic variation, sexual dimorphism, other forms of polymorphism, etc., which control the features of such smaller units as "races" and local populations. The evolution of species is the key issue. How do species originate? Can we say anything about the general nature of their subsequent histories?

PHYLETIC GRADUALISM AND BIOSTRATIGRAPHY

Our entire biostratigraphic system is based on vertical ranges of "species." In the last 20 years, authors have insisted that biostratigraphers and many paleontologists are incapable of recognizing "biological" species (e.g., some papers in the symposium edited by Sylvester-Bradley, 1956). This may be so; but Hall and Clark

(1888) postulated (as it turns out) a 10 million year stratigraphic range for the trilobite *Phacops rana,* and more recent work (Stumm, 1953; Eldredge, 1972), far from destroying the earlier concept, has confirmed it in explicitly biological terms. In fact, one of the greatest sources of evolutionary data, nearly totally ignored to date, is the biostratigraphic literature. Implicit in the stratigraphic range of any "species" in the literature is the existence of one or more morphological features serving simultaneously to (1) cluster a series of population samples over a certain segment of time and space, and (2) form a basis for distinguishing such a cluster from other similar ones. This leads to an important conclusion: by the mere recognition of *any* nontrivial stratigraphic range of *any* morphologically defined taxon at or near specific rank, we are necessarily implying a stability or stasis in species-specific *differentia.* All characters that are modified in the process of speciation and allow us to recognize a "new" species must remain in a more or less recognizable state throughout the biochron of the species, or we would not be able to recognize it through time.

These statements, expressed in rather extreme terms, mirror reality sufficiently well to prompt us to wonder why evolutionary paleontologists have continued to seek, for over a century and almost always in vain, the "insensibly graded series" that Darwin told us to find. Biostratigraphers have known for years that morphological stability, particularly in characters that allow us to recognize species-level taxa, is the rule, not the exception. It is time for evolutionary theory to catch up with empirical paleontology, to confront the phenomenon of evolutionary nonchange, and to incorporate it into our theory, rather than simply explain it away. For in the face of a formidable bulk of contrary evidence, paleontologists have doggedly persisted in viewing the origin of new species in terms of gradual, progressive modification of an entire ancestral species. We (Eldredge, 1971; Eldredge and Gould, 1972) have characterized this model and labeled it "phyletic gradualism." This model is familiar to us all. It is the relatively simple notion of phyletic transformation within a population of an entire species of one or more phenotypic characters. The transformation is gradual and usually is considered to proceed at fairly constant and rather slow rates, although some paleontologists recognize periods of slow and accelerated change within lineages. New subspecies, species, and even genera are simply subsets (arbitrarily delineated) of an evolving continuum. Under this view, the difficulty of defining taxa is a consequence of the evolutionary process itself: ideally, the boundaries of taxa are wholly arbitrary; in the real world, we rely on convenient breaks in the fossil record.

We have already claimed that the fossil record speaks strongly against such a view as a general pattern. Even when we allow for gaps in time and preservation, there should still be far more evidence for directional change within species when good population samples are compared at intervals throughout their stratigraphic range. Some examples *are* recorded (e.g., Kauffman, 1970, on two species groups of inoceramid bivalves), but most careful analyses, while they do provide evidence of change, show patterns more oscillatory than unidirectional (e.g., Spencer's 1970 study of *Neochonetes granulifer*). There are also theoretical reasons for denying phyletic gradualism as a general mechanism in the evolution of species. We have discussed these at length elsewhere (Eldredge and Gould, 1972, 1974); the major problems involve the origin and maintenance of appropriate linear selection forces, especially those affecting an entire, far-flung species over thousands of generations and perhaps millions of years. Such selection regimes ("orthoselection") are difficult to envisage. A constant unidirectional change in physical or biotic variables is implausible. Most adaptive changes in populations are relatively rapid adjustments to local, newly encountered environmental conditions. Faced with truly long term changes in the environment, a population is more likely to move away or become extinct. Although *in situ* adaptation to changing conditions will always be possible to some extent, such adaptive modification will rarely persist to the length of time required, since migration or extinction will eventually interrupt any process of linear change.

Stage of Evolution and Shaw's "Analytic Paleontology" as Biostratigraphic Strategies

If phyletic gradualism is the dominant evolutionary model in paleontology, we should examine its implications for biostratigraphic methodology. We have already claimed that the methods of biostratigraphy have developed in near independence from evolutionary theory, and, in light of our feelings about phyletic gradualism, this is all to the good. Strict adherence to paleontological dogma about evolution would have effectively prevented a biostratigraphy based on the comparative distribution of vertical ranges of species. Instead, we would be wrestling with a sort of character correlation technique proposed from time to time over the years, most recently by Shaw (1969) and more moderately (without abandoning the species concept) by Kauffman (1970) and Hayami and Ozawa (1975). We do not argue that careful analysis of evolutionary change in character states could not make great contributions to studies of correlation. We do claim that such techniques, when they also abandon species as raw data, represent an illusion based on a view of the fossil record (and of the evolutionary process) that has been proved false by a voluminous literature compiled by paleontologists over the past century and a half.

Many applications of this "modernist" strategy (as opposed to "old-fashioned" recording of simple species ranges) are now thoroughly discredited. Trueman's brave proposal for lower Liassic *Gryphaea* (1922) is problably the most widely cited example of stratigraphic correlation by simultaneous change in a set of biologically correlated characters. Since the paper became so famous as an evolutionary study, many people forgot that Trueman presented his sequence as an exemplar for a biostratigraphic strategy. His celebrated paper bears the title "The Use of *Gryphaea* in the Correlation of the Lower Lias." Trueman really believed that his correlated trends in shell coiling, size, and thickness could be used in precise stratigraphy. He presented his famous histograms of shell coiling at given horizons not as biological arguments, but as a method for the determination of stratigraphic position in the field; one had merely to collect 50 *Gryphaeas* at any one spot and construct a histogram for coiling. Since mean coiling increased unilinearly with time, a statistical determination of this parameter established exact stratigraphic position. So inexorable and universal was his phyletic gradualism that he did not even consider the possibility of extensive geographic variation at a time plane. Trueman's story has now been thoroughly discredited (Hallam, 1959; Burnaby, 1965; Gould, 1972); *Gryphaea* displays no temporal change in coiling at all throughout lower Liassic rocks of Great Britain. It does show extensive geographic variation at each time plane.

Character states and their transformations cannot be analyzed apart from the concepts of population biology. Populations evolve, not individuals or, still less, anatomical parts of individuals. Microevolution can only be understood in terms of phylogenetic relationships among populations. We must understand interrelationships among species in order to pinpoint what segment of any species is most closely related, and perhaps ancestral, to another species. Species are the units that must be carefully delineated and understood in phylogenetic terms, although evolutionary change itself occurs in local populations. Phylogenies must be constructed by careful analysis of characters; monophyletic taxa can only be recognized, if not entirely defined, by the identification of shared, derived character states. Character analysis cannot be understood and applied to biostratigraphy unless it is first used to frame the most likely, or least objectionable, hypothesis of relationship among all population samples. Only after we determine such a plausible configuration for an evolving series of populations can we return to the character states themselves and determine if there is any temporal significance to them.

Infraspecific trends in vertical outcrop of one local area may not be repeated in an adjacent region; even when we find a similar trend elsewhere, we cannot be sure that we deal with time equivalents. One of several subtle trends detected statistically among population samples of *Phacops rana* serves as an excellent example. Most changes occur allopatrically, but Eldredge (1972) detected some pervasive trends within the subspecies *P. rana rana,* when 30 cephalic measurements were subjected to a factor analysis (see Eldredge, 1972, p. 71 ff., and fig. 9 for a more complete discussion of this example). One trend, reproduced here as Text-figure 1, involves changes in eye size and, in particular, in the total number of lenses in the eye in relation to overall cephalic size. High scores imply relatively smaller eyes and fewer lenses relative to size of the cephalon. Although data points are a bit scarce, there seems to be a perceptible trend from the Upper Cazenovian through the Upper Tioughniogan (a time interval of roughly 6 to 8 million years), in which lens number in relation to head size increased progressively.

Can we use this trend for a finer subdivision of geologic time? We can assume, for the sake of argument, that the measurements are repeatable and that any new sample can be incorporated into the scheme. For each time horizon, there is a roughly constant variation in mean values of scores on this factor. Although there is great overlap between horizons, we can potentially surmount this problem by an analysis of variance on the scores themselves and, statistically at least, place any new sample roughly in its proper stratigraphic setting. So far, so good. But, as

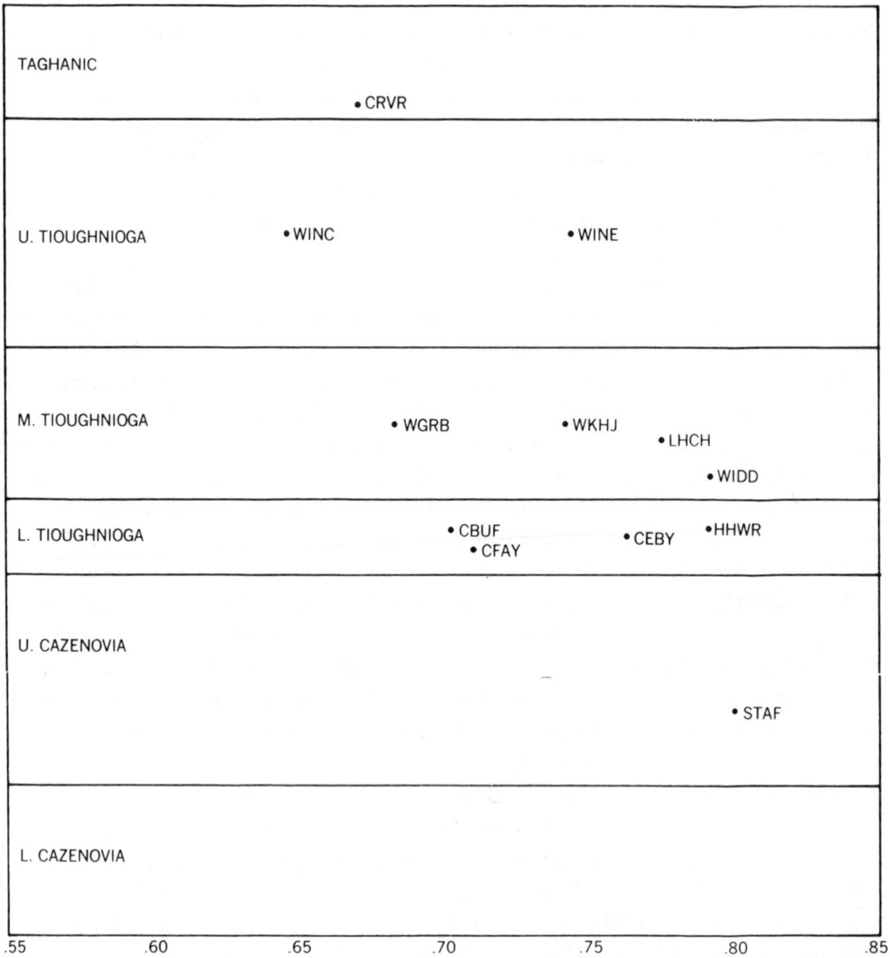

Text-Figure 1

Factor analysis of *Phacops rana* cephala. Data normalized by samples. Mean score of each sample for the first rotated factor plotted against approximate stratigraphic position. High scores imply relatively few lenses in the eye in relation to overall cephalic size. Thus there is an apparent trend toward increasing the number of lenses in relation to head size during the Upper Cazenovian–Lower Taghanic portion of the stratigraphic range of this species. Sample abbreviations and sizes are given in Eldredge (1972, p. 74, table 5). The samples do *not* come from a single, local rock column, but were collected along an approximate east–west line of outcrop from central New York westward through Ontario, Michigan, and Iowa.

discussed in Eldredge (1972, p. 72 ff.), the variation among quasi-contemporaneous population samples within the Tioughniogan seems to be correlated with gross sedimentological features of the various formations. At any one time, samples from the

Ludlowville (LHCH) and Moscow (WINE) formations in central New York have higher mean scores than samples from equivalent formations further west. In central New York, both formations are variable shale-siltstone, "near-shore" sequences; their equivalents to the west are fine-grained argillaceous sediments with a high calcareous content. Specimens from purer, more clastic limestone units also tend to score high. We are dealing with geographic variation in number of lenses in relation to cephalon size, which apparently involves the physiological requirements of phacopid vision in water over different substrates. But, since we cannot be sure of such an interpretation, we are left with an apparent lag effect where, at any one point in time, the trend is further advanced in some environments than in others. A stage of the trend "catches up" in the clastic, near-shore, and in the limestone facies at some point in time later than its attainment in other environments. The biostratigraphic utility of such a trend is not at all straightforward; if we are to use it at all, we must first have a very clear idea about the relationships of population samples and their relative distributions in space and time. To do this, we must already have a correlation scheme based on other criteria!

Before leaving this example, we should also note that the values for HHWR and WIDD are probably best understood when we examine the interactions of *P. rana* with *P. iowensis.* We have an apparent case of character displacement that distorts the position of these samples within the trend. This serves as further illustration that the direct analysis of isolated characters for purposes of correlation can be meaningless. Only in the context of microevolutionary phenomena and population biology can such data be useful to biostratigraphy. If such problems as these routinely occur when we deal with closely related population samples within a single depositional basin, then the wholesale application of such a research strategy to problems of international correlation (as Shaw, 1969, and Barnett, 1972, have recently done with conodont data) presents many problems and carries little prospect that self-correcting results can emerge. The assumptions underlying such a procedure are simply too vast and too ill-founded. It can't work.

A more informal use of character analysis in biostratigraphy, one which has played an important role in mammalian biostratigraphy (Tedford, 1970, p. 687), is embodied in the notion of "stage of evolution." This approach is a complex mixture of empiricism and a priori assumption. On the one hand, we may readily expect to find morphologically differentiated segments of monophyletic lineages (species within genera; even genera within families) that are restricted to certain segments of the biochron of the lineage. In his study of Cambrian trilobites from the boulders at Levis, Rasetti (1948) noted that Lower Cambrian pygidia of *Kootenia* possess 14 marginal spines, while Middle Cambrian pygidia have 12 or fewer. Shaw (1955) cited this observation and treated it as a general indicator of age. Examination of the literature supports the generalization, and Kay and Eldredge (1968) found it useful in determining the age of a small sample of fossils from the volcanic belt of central Newfoundland. In this instance, empirical observation underlies the general idea of "stage of evolution"; such an approach may be useful.

On the other hand, we cannot possibly develop a biostratigraphic strategy based largely on the "stage-of-evolution" style of argument, for we would inevitably be drawing more on a false theory than on empirical data when following this approach. In relying on phyletic gradualism as the underlying mechanism, we can extend the "stage-of-evolution" concept until "intermediacy" in one or more features will constitute *prima facie* evidence that a taxon comes from sediments intermediate in age in the stratigraphic sequence containing all taxa. Correlations effected in this manner are more a priori assumptions than the products of careful analysis. In general, the "stage-of-evolution" approach illustrates the most pervasive and nefarious influence that phyletic gradualism has had on the development of biostratigraphic methodology.

SPECIATION AS AN ALTERNATIVE EVOLUTIONARY MODEL

We shall support an alternative view of microevolutionary phenomena in paleontology. It incorporates the empirical data of biostratigraphy on the stability of species-level taxa with modern neontological notions of the process of speciation. We have elaborated this model elsewhere (Eldredge, 1971; Eldredge and Gould, 1972), and shall discuss its properties only briefly before examining its implications for biostratigraphy.

The basic idea is simple: as a result of the intricate and closely interdependent processes of morphogenesis and of opposing forces that create equilibria in the genetics of populations, populations and, by extension, species, are actually highly conservative, i.e., highly resistant to change. Although single mutations may produce striking changes in the phenotypes of individuals, these do not create species. Most significant morphological change involves a more drastic rearrangement of the genotype, which must be incorporated into the morphogenetic process. Evolutionary change is a difficult thing to accomplish and an event of relatively low probability. Such probabilities only become high with the passage of long periods of time. It is simply very difficult to overhaul the collective genotype of a population to a point equal to the average difference in genetic structure and composition between any two closely related (including parent–daughter) species. The larger the number of individuals involved, the more difficult such a "revolution" of the genotype (Mayr, 1963) will be.

The theory of speciation, as established during the past 40 years (summarized in Mayr, 1963), has emphasized the key role played by geography in the establishment of full genetic isolation between populations. Geographic isolation has been considered a prerequisite to the initiation of speciation, primarily because gene flow only ceases when populations are separated. Endler (1973) and Ehrlich and Raven (1969) have recently challenged the importance of gene flow as a homogenizing, stabilizing influence within species, and it may well be that homeostasis, not gene flow, plays the major role in maintaining the integrity of species. But this much is clear: new species originate from portions of ancestral species, i.e., from one segment of a species in one portion of the geographic range of the ancestral species. We have already argued that the required genetic changes are more easily developed in small populations. We further believe, following Mayr (1963), that such populations will normally occupy a position at the outer fringes of the geographic range of the ancestral species. We expect that the greatest amount of genetic and phenotypic divergence will have already occurred at these fringes. All species display sòme geographic variation representing the close adjustment of local populations to particular requirements of their local habitat. Almost by definition, the most severe and extreme variants of a species' habitat are at the limits of its geographic range (because we presume that one or more ecologic factors at the margins are responsible for defining limits to the range). Thus there is good reason to identify small populations at the margin of a species' range – the "peripheral isolates" – as those most likely to produce new species. Most isolates, of course, suffer extinction.

These considerations have led us (1972, p. 94 ff.) to the following generalization: most morphological differences between parent and daughter species (1) already exist (through geographic variation) prior to full speciation, when the populations are still allopatric, (2) are further developed and perfected during a relatively short phase after genetic isolation is completed, and (3) are accentuated rapidly in initial stages of sympatry between parent and daughter species, through the phenomenon of character displacement. There is good reason to believe that most morphological differences between parent and daughter species arise prior to, during, and right after a speciation event. Relative to the average duration of species, this period of time is very short. The history of any species at any locality, even if it is not interrupted by local extinction and replacement by other populations from adjacent areas, should be more oscillatory than directional. We shall always be able to document differences among successive population samples in a vertical sequence, but these will, at most, represent adaptive shifts to minor environmental changes rather than a dominant evolutionary force inexorably leading to the creation of new species.

SPECIATION AND BIOSTRATIGRAPHY

This view of speciation, so drastically different from the notion of phyletic gradualism, has rather different implications for biostratigraphic methodology. First, the importance of geographic variation immediately raises problems in correlation with local populations. It is the biological equivalent of the facies problem in lithostratigraphy, and its dangers are readily imagined. It leads again to the conclusion that species-level clusters of populations are, in general, the most appropriate taxa for biostratigraphic data.

This model also implies that first occurrences of new species are, by definition, diachronous to some extent. Although the origin and subsequent spread of a new species within a depositional basin may well be "instantaneous" in geologic time, the matter is usually not so simple. The parent species, and other closely related taxa, are usually still present in some areas of the basin, and they may well impede the quick distribution of a new species. Returning for a moment to *Phacops rana*, Eldredge (1972) argued that *P. rana rana* arose in the Lower Cazenovian somewhere in the Appalachian exogeosyncline and rapidly spread through it. Throughout the Upper Cazenovian, *P. rana rana* was confined to this area, while *P. rana crassituberculata* and *P. rana milleri* occupied equivalent niches over the cratonal interior. Only after *P. rana milleri* and *P. rana crassituberculata* were eradicated by a widespread regression near the end of the Upper Cazenovian times was *P. rana rana* able to invade the cratonal interior during the Centerfield transgression. Correlation of Upper Cazenovian rocks cannot be accomplished between the cratonal interior and the Appalachians (roughly a 2 million year span) by using *P. rana*. To complicate matters further, *P. iowensis*, a coeval species found predominantly within the Michigan Basin, also hindered the spread of *P. rana rana* (the two species, with a single exception, are never found in the same stratigraphic unit). Finally, on the specific level, *P. rana* is also known from the Spanish Sahara in rocks thought to be Eifelian in age on other criteria (Burton and Eldredge, 1974). In North America, *P. rana* is known only from Givetian and Frasnian formations. Although no one seriously contends that a single species can completely determine a biostratigraphic correlation, this example further questions the use of conjoint first appearances to define the bases of biostratigraphic zones.

Beyond this point, evolutionary theory has little to offer in resolving the old dilemma of whether first or last occurrences are more reliable for defining tops and bottoms of zones. When vertical ranges are compared, we agree with Shaw (1964): a priori, speciation and extinction should be random. However, Moore (1954, 1955), Kauffman (1972), and Spiller (1973) have related higher rates of both speciation and extinction within large depositional basins to different phases in an overall series of transgressions and regressions. Eldredge (1971) related distributions and extinctions, but not originations, to similar geographic and ecologic controls. We suspect that such correlations exist, but the patterns will differ in each situation and no generalizations are possible, at least at the present time. We agree that major ecological changes can cause profound alterations in biogeographic distributions; it is not hard to envision near-simultaneous extinctions affecting many species as a result of such changes. It is far more difficult to imagine how such changes affect rates of speciation, since directions and intensities of selection regimes have never been directly correlated with speciation rates (Eldredge, 1974).

Our model of speciation insists that species are real and recognizable entities that may well persist for several millions of years. As we have said, most field geologists and biostratigraphers have long known this, and their observation underlies much biostratigraphic practice. At the risk of sounding conservative, we feel that the basic procedure of using fossil species to define biostratigraphic units, i.e., the principle of overlapping range zones generally credited to Albert Oppel (Berry, 1968b), is the approach most consistent with evolutionary theory. Shaw's (1964) technique of regression analysis, which amounts to a statistical analysis of teilzones, is based on this model (and is, paradoxically, diametrically opposed to his later dubious

[1969] proposal). Shaw's (1964) technique simultaneously frees us from the necessity of defining small biostratigraphic units (so heavily dependent upon the vagaries of preservation and ecological control of distributions), and permits an impressively minute subdivision of time without recourse to detailed analysis of purported evolutionary character changes within lineages. The only data needed are first and last occurrences (assuming that accurate systematics have already been established) and the stratigraphic positions of these occurrences. And Hay (1972) has recently proposed an even simpler technique based on relative, rather than absolute, stratigraphic positions of species. We leave the evaluation of such innovations to others in this symposium, but the last word has clearly not been written on the elaboration of powerful techniques, affording great biostratigraphic resolution and using nothing more than the stratigraphic ranges of a series of fossil species.

PHYLOGENIES AND BIOSTRATIGRAPHY

We have already discussed the problems of using population phylogenies within species – or, really, character transformation gradients. A related idea advocates the establishment of phylogenies among species as an aid to increasing biostratigraphic resolution. But what can phylogenies do for us? For if we agree that rocks should be correlated by analyzing the common occurrences of the same taxa in two or more areas, then phylogenies are useful only in correlating rocks sufficiently far apart that only closely related *different* species are present in the different areas.

Phylogenies can be expressed in either of two ways. As paleontologists, we are more familiar and comfortable with relationships expressed in terms of ancestors and descendants. Some paleontologists (e.g., Schaeffer et al., 1972) have recently argued that phylogenies are better expressed in terms of sister taxa, related by recency of common ancestry. These two ways of expressing relationship are rather different. Neither has much to recommend it as a springboard for biostratigraphic analysis. Ancestor-descendant relationships are never truly verifiable: they convey only one definite piece of information related to time; the ancestor must have been living before the descendant appeared. But cases abound in which supposed ancestors were coeval with, or even outlived, descendants. It is difficult to develop any coherent biostratigraphic strategy using phylogenetic relationships of this sort.

Furthermore, ancestor-descendant phylogenies are frequently based on stratigraphic position and a tacit acceptance of phyletic gradualism (Schaeffer et al., 1972; Eldredge and Tattersall, 1975). In such cases, either an accurate scheme of correlation must exist already, or we run the danger of falling into a vicious circle, interpreting stratigraphic relationships by phylogenetic hypotheses based themselves on stratigraphic interpretations.

As an example, suppose we argue that *Phacops logani* (Gedinnian-Siegenian), *P. cristata* (Emsian-Eifelian), and *P. iowensis* (Givetian) form an ancestral-descendant sequence. In all sequences containing two or more of these species, they always occur in the same relative position. Suppose that we had no scheme of correlation for the Lower and Lower Middle Devonian of eastern North America. We could then use these trilobites in a first attempt to frame such a hypothesis. But all we could do would be to correlate all *P. logani* bearing rocks with each other, all *P. cristata* bearing units, etc. The phylogenetic relationships of these taxa are sublimely irrelevant to the problem. In short, there seems to be no way to use ancestor–descendant relationships in biostratigraphic research.

Phylogenies expressed in terms of sister taxa have problems of their own. As Schaeffer et al. (1972) argue, the few attempts at this kind of analysis in the paleontological literature have ignored stratigraphic position and rely solely on comparative character analysis of the "basic taxa." If there were some temporal significance to such relationships, the problem of circular feedback would be avoided in subsequent biostratigraphic applications. But these relationships merely state that taxa A and B shared a more recent common ancestry than either did with taxon C. We can never be sure that any two "sister species" are the actual products of a single split in an ancestral species. Two, three, or a dozen species may have existed that were more closely related to taxon A than A was to B. But if we do not know these taxa, we cannot incorporate them into our scheme.

If two species of a sister group were daughter species (i.e., "true" sister species in an evolutionary sense), we could say that they were coeval when they split from the hypothetical common ancestor. But in the real world, species A may become extinct immediately, become so changed that it constitutes a third species, or persist relatively unchanged as species A, alongside its daughter, species B. Furthermore, species A may have existed for a million years or more before it gave rise to species B. Since we have already conceded that "true" sister species are difficult to find and impossible to ascertain, the use of such information in biostratigraphy seems hopeless.

The concept of sister taxa does suggest a way to correlate widely separated rock units sharing related but not identical species. First, we must search for pairs of sister species from two or more different regions. The first occurrences of each species in a sister group give a rough approximation to the time of splitting of a lineage. In any one case, the error may be enormous, perhaps a million years or more. But a plot of all known sister species between any two sections should yield a sloppy version of Shaw's (1964) straight line for teilzones of the same species. The probability that two remote sections would contain a significant number of sister groups is small, so this approach is most suitable for large-scale attempts at simultaneous correlation of many sections (e.g., of all Upper Devonian rocks within a continent). Each section would be represented on a separate Cartesian coordinate, and the straight line of Shaw's regression technique becomes a hyperplane. The hyperplane of a correlation web would emerge in the first few principal components (or coordinates) of a multivariate analysis. Successive axes would be useful in detecting directed biases in the data, e.g., possible consistently earlier or later appearances of sister species in specific areas, suggesting patterns of dispersal. This procedure is analogous to Stehli's (1965) use of residuals from spherical harmonics to map paleocurrents, and Shaw's (1964) use of regular discrepancies in his plotted data to detect variable sedimentation rates and faults.

In plainer language, the closer the relationship of two species, the closer they will approximate each other in time. This is why "stage of evolution" has worked in the past as a broad gauge of relative stratigraphic position. Now that we are accustomed to radically different configurations of crustal plates in remote times, the difficulty of accepting sister species on different continents has disappeared. Shaw's (1964) techniques can be extended, using first occurrences of species judged to be *most* closely related to each other after careful phylogenetic analysis of all available taxa. Some of our more intractable problems of intercontinental correlation might be attacked in this manner.

CONCLUSIONS

The theme of this paper might seem paradoxical, if not reactionary. We really are defending the oldest technique of empirical biostratigraphy against all the subtleties supposedly introduced under the aegis of "modern" evolutionary theory; i.e., we believe that the vertical ranges of species treated as static entities are the proper basis of biostratigraphy.

We do not offer this argument because we mistrust biological theory and yearn for a completely "objective" science of stratigraphy based only upon pure observation and free from the fetters of a priori dogma. Such a thing is not only undesirable; it is impossible. Theory inheres everywhere. The naive empiricist who fancies that he works by observation alone is even more fettered to his theories because he does not recognize them explicitly and cannot, therefore, assess their hold upon him. We do not mistrust theory; we merely claim that the biological theory underlying most "modernist" approaches to biostratigraphy is simply wrong, and that the theory supporting "old-fashioned" biostratigraphy is correct.

The paradox arises because the incorrect theory (phyletic gradualism) has such a hold upon the minds of paleontologists that it has been treated as an ineluctable consequence of the fact of evolution itself. Thus empirical biostratigraphers, who know by experience that most species do not change progressively during their existence, are forced into the uncomfortable position of regarding evolutionary theory as irrelevant, if not subversive. And biologically minded paleontologists, when they indulge in problems of correlation, spend their lives in search of the chimera of "insensibly graded sequences," while knowing in their heart of hearts that most species look pretty much the same at their inception and extinction. We merely wish to alleviate the anxiety of both groups by arguing that a more correct theory of biological speciation predicts what is actually observed; i.e., "old-fashioned" biostratigraphy is the approach most consistent with proper evolutionary theory. We also reiterate that a defense of "old-fashioned" concepts does not tie us to the old-fashioned method of qualitative species lists and purely subjective assessment. Shaw's method of correlation (1964) is but one way to dress our favored old concepts in the quantitative cloak of modern technology.

We may, therefore, summarize our specific claims. We believe that species are real biological entities (although they may be difficult to recognize in practice).

Among models for their origin, phyletic gradualism proposes the progressive modification of an entire lineage (with occasional slow splitting to produce the recognized diversity). Our alternative model emphasizes speciation (splitting) and claims that most morphological differences between two species appear in conjunction with the speciation process itself, whereas most of a species' history involves little further change, at least of a progressive nature. Although little attention has been paid to evolutionary theory in the development of biostratigraphy, nevertheless, biostratigraphers are in the best position to make a choice between the two models; both make claims about the nature of microevolutionary phenomena, and their products as preserved in the fossil record. We believe that, unconsciously, biostratigraphic methodology *has* been evolutionarily based all along, since biostratigraphers have always treated their data as if species do not change much during their teilzones, are tolerably distinguishable from their nearest relatives, and do not grade insensibly into their close relatives in adjacent stratigraphic horizons. These time-honored, if largely implicit, observations can easily be incorporated into a speciation model. It is far less easy, perhaps impossible, to reconcile phyletic gradualism with these data. Biostratigraphers, thankfully, have ignored theories of speciation, since the only one traditionally available to them has not made much sense. To date, evolutionary theory owes more to biostratigraphy than vice versa. Perhaps in the future evolutionary theory can begin to repay its debt.

Biostratigraphical Tests of Evolutionary Theory

P. C. Sylvester-Bradley **University of Leicester**

PALEONTOLOGY AND THE SCIENTIFIC REVOLUTION

Kuhn (1962) claims that revolutions occur in science when generally accepted beliefs are abandoned, and a new set of paradigms is substituted instead. Darwin (1859) claimed that, if his theory of the origin of species were true, "numberless intermediate varieties, linking closely together all the species of the same group, must assuredly have existed. . . ." Evidence of their former existence could, he goes on to suggest, "be found only among fossil remains, which are preserved . . . in an extremely imperfect and intermittent record." Darwin's theory became the first paradigm of evolutionary theory; it has formed the basis of most phylogenetic theories derived from paleontological data. The paradigm has recently been christened "phyletic gradualism," and in a series of lively and well-argued papers (Eldredge, 1971; Schaeffer et al., 1972; Eldredge and Gould, 1972) Eldredge and his colleagues have attacked the concept, claiming that it is time we had a paleontological revolution. They substitute a new paradigm termed "punctuated equilibria," in which new species arise abruptly. They suggest that their new picture "is more in accord with the process of speciation as understood by modern evolutionists," but also admit "that the data of paleontology cannot decide which picture is more adequate" (Eldredge and Gould, 1972, p. 99). As this controversy strongly affects the basic concepts of species, their determination and their use in biostratigraphical zonation and correlation, it is of primary interest to this volume.

I believe that this attack on the old ideas is fully justified. But I do not agree that the new paradigm is much better than the old. Nor am I prepared to accept a defeatist attitude, which suggests that paleontology can never decide which theory is wrong. On the contrary, I propose here a third paradigm, combining features of the other two, which I believe provides a better model of many situations than either of the others. This third paradigm is not new; it has been implicit in the writing of many evolutionists. I have already, some 15 years ago, attempted to give it expression (Sylvester-Bradley, 1960, 1962). It may be termed "reticulate speciation," and suggests that new species arise in the course of an expanding migration, with hybridization between geographical races.

In this chapter, the three paradigms are tested against each other. The conclusion reached is not that one is right and the others wrong. The three models turn out to be no more than particular examples. We are not yet in sight of any general law. New species arise in a number of ways, of which these are three.

THREE PHYLOGENETIC MODELS

To compare the three models, we shall see how each postulates the origin of new species not only by the splitting of the phyletic line ("cladogenesis"), but also by modification arising during the course of geological time in a single phyletic line ("phylogenesis"). We shall consider how a single species can become ancestral to two or more descendant species.

The model of "phyletic gradualism" is illustrated in Text-figure 1. The ancestral species A_1 is at first confined to region A, but on reaching time plane 1, it expands in numbers and geographical extent, and comes to occupy an area comprising both regions A and B, as well as the area between. Its morphological variation increases during the expansion; this variation can be characterized by referring it to the two morphotypes A_1 and B_1. At time plane 2, more rigorous conditions set in, and a peripheral population gets cut off in region B, and thus remains isolated from the main population in region A. During the course of time, both populations continue to evolve gradually, that in region A passing gradually through a series of subspecies A_1, A_2, A_3, and A_4, that in region B diverging and forming the series B_1 to B_7. The rate of evolution in the two regions may differ; the peripheral population may change more quickly than the main population. If, at some subsequent period the populations again expand and occupy the same geographical area, they may either have become genetically incompatible, so that both continue to exist as sympatric species in competition, or they may still be able to hybridize, in which case they would fuse and form a single interbreeding population exhibiting a high degree of morphological variation (cf. Sylvester-Bradley, 1951).

In the model of punctuated equilibria (Text-fig. 2), the species inhabiting each region do not evolve; they are held in "homeostatic equilibrium" (Eldredge and

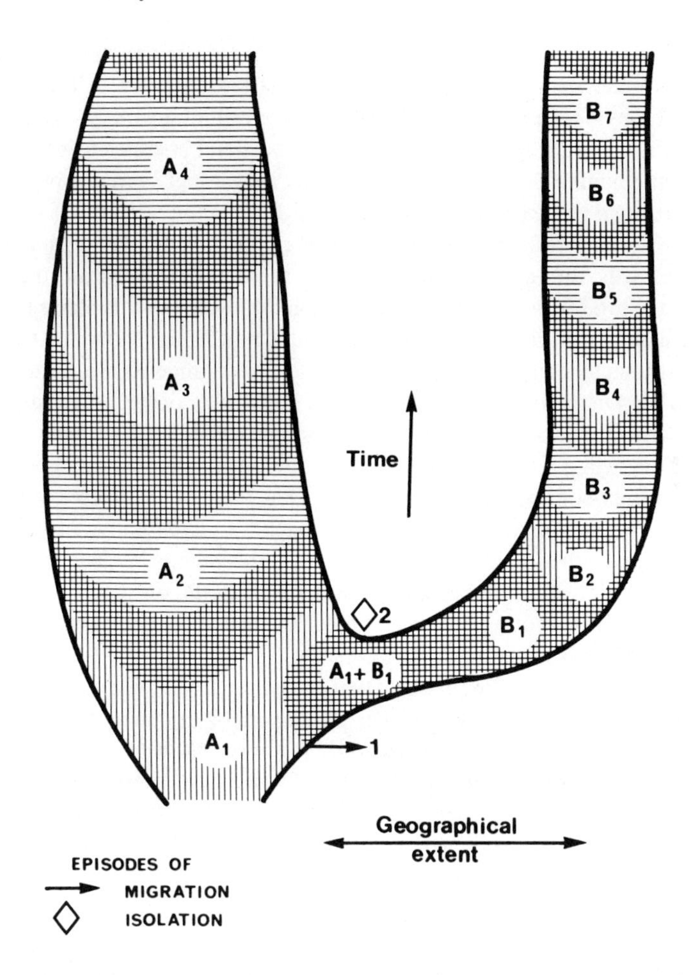

Text-Figure 1
Model A: speciation by phyletic gradualism. Both phylogenesis and cladogenesis occur. The two branches show unequal rates of evolution. Migration is initiated at time level 1. Geographical isolation is established at level 2. Ultimately, the two branches may become genetically incompatible. Until they do, they must be regarded as geographical subspecies. Each time-separated grade is a chronological subspecies.

Gould, 1972). When migration is followed by isolation, cladogenesis occurs, and new species arise. The old species are replaced as they die out. New species can only succeed each other as a result of migration; they can only be modified by changing selection pressure if they become isolated from their parental stock.

The third model, demonstrating reticulate speciation, is shown in Text-figure 3. In this model, phylogenesis and cladogenesis proceed simultaneously. The process starts when the ancestral species increases in numbers and migrates to new areas, meanwhile maintaining gene flow throughout the increasingly variable population. This is the *eruptive* phase (Table 1); it leads straight into the *polytypic* or *reticulate* phase, with the establishment of races separated by varying degrees of isolation. As evolution proceeds, fluctuating numbers in each geographical race lead to an alternation between interracial hybridization and isolation. During this phase there is a marked increase of morphological modification. New characters may be developed. Preadaptive characters may be selected by the new environments encountered after migration. *Anagenesis* (in the sense of Huxley, 1959) may occur. A third phase is established when the races remain isolated for long enough to attain genetic incompatibility. They will continue to diverge along the trends selected by previous adaptation to their contrasting environments. This, then, is the *divergent* phase, characterized by directional selection and phylogenesis. Eventually, a fourth phase will be established, when stabilizing selection pressures take over. The species remain, from the point of view of evolution, in a steady state, until a new expansion of population leads to another eruption, one of the species becomes polytypic, and the whole procedure is repeated.

It will be seen that the second model (that of punctuated equilibria) is the antithesis of the first (phyletic gradualism). The third model (reticulate speciation) is a compromise, a synthesis of the other two. If, as I have suggested, all three models are valid, if all three have been active through geologic time, what difference do their various effects have on biostratigraphic procedures? Before this question can be answered, one must assess the comparative time scales involved. Reticulate speciation can only be recognized in the geological record if the species is preserved over a whole geographic region, maybe across a whole continent. Present-day polytypic species in the northern hemisphere show that the speciation process has been active during postglacial times. New species must have arisen within the last

Table 1
Evolutionary Phases According to the Reticulate Speciation Model (explanation in text)

	Phase	*Selection Pressure*	*Type of Evolution*
1.	Eruptive	Reduced	Cladogenesis
2.	Polytypic or reticulate	Disruptive	Anagenesis
3.	Divergent	Directional	Phylogenesis
4.	Stabilized	Stabilizing	Stasigenesis

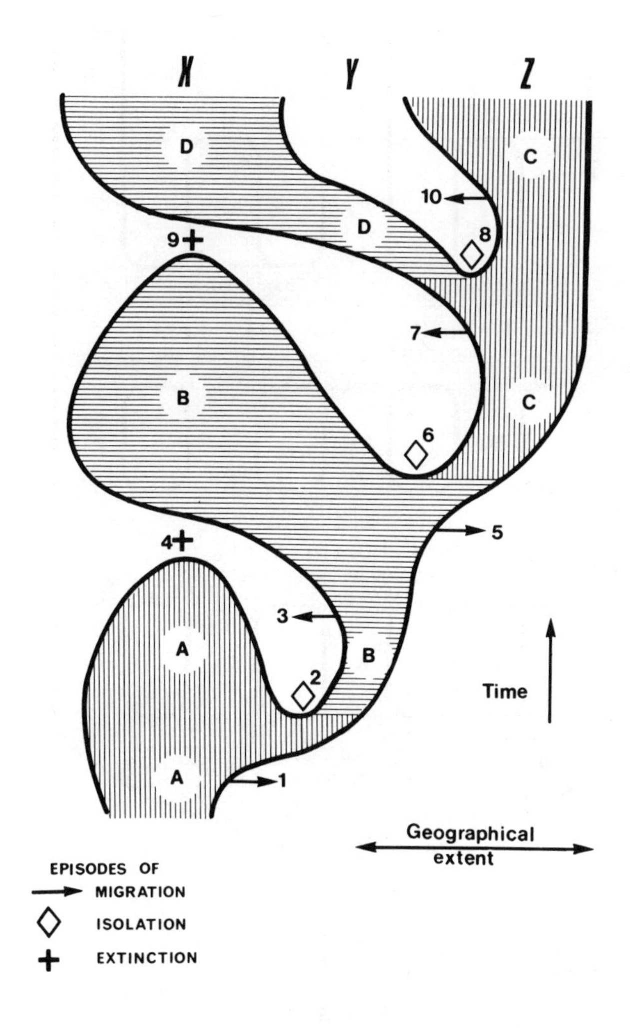

Text-Figure 2
Model B: speciation by punctuated equilibria. Species A in region X gives rise at time level 1 to a branch, which migrates to region Y; when it becomes isolated at level 2, it changes morphologically into species B. This species migrates back to region X at level 3, replacing species A, which becomes extinct at level 4. Species B also migrates to region Z at level 5, and changes to species C after isolation at level 6. Similarly, a branch of species C migrates back to region Y at level 7 and changes to species D, with isolation at level 8, and afterward invades region X, and replaces species B. Region X is thus characterized by three successive species, A, B, and D, which abruptly replace each other.

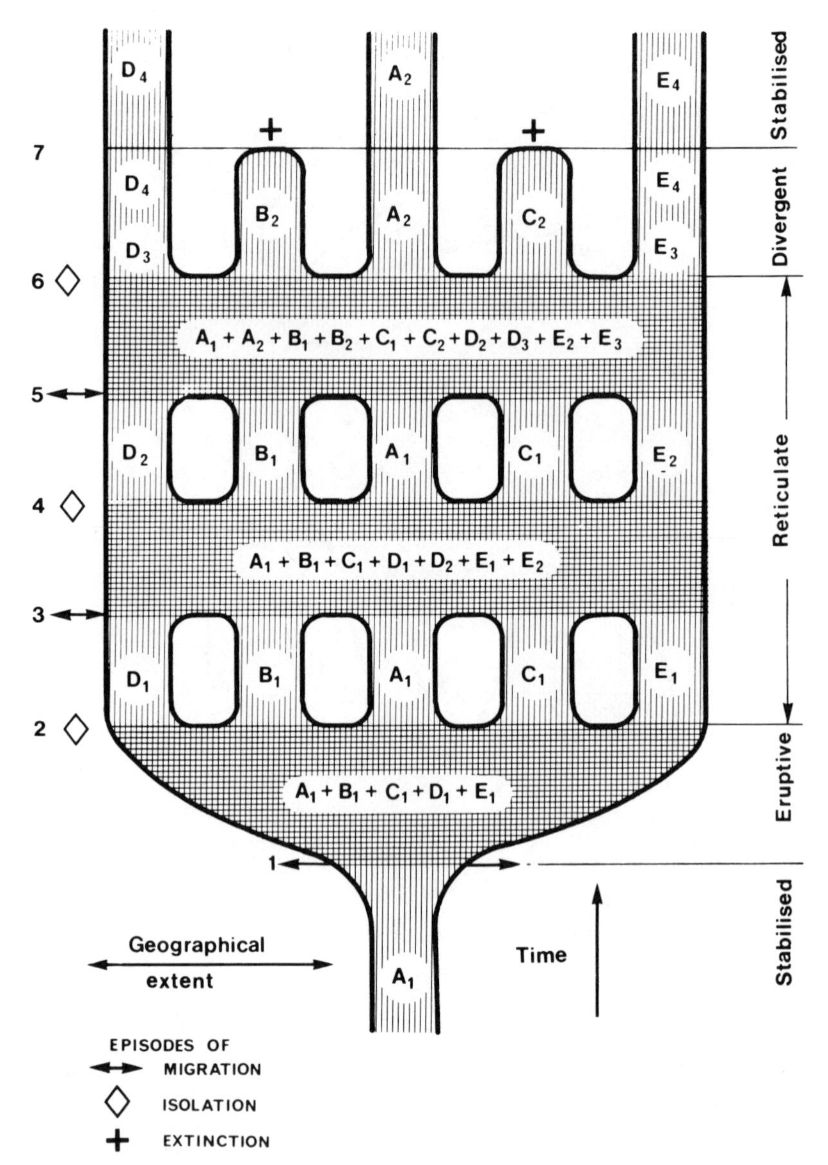

Text-Figure 3

Model C: reticulate speciation. Four phases succeed each other. In the eruptive phase the ancestral species increases in numbers and in variety (becoming polymorphic), and extends its area of occupation (levels 1 to 2). During the reticulate phase, isolation and hybridization alternate, and the species becomes polytypic (levels 3 to 6). Of the five phyletic lines that survive this phase, two become extinct during the divergent phase (levels 6 to 7). The three survivors persist into a stabilized phase without further modification.

500,000 years. That is a very short space of time in the geological record. So we might expect that reticulate speciation would only be recognizable in the geological record on one time plane, and that its distinction from the other two models is only the result of a scale effect; that, in fact, cladogenesis as it occurs in either the model of phyletic gradualism or that of punctuated equilibria will always appear reticulate if it is examined in minute detail. This expectation is not, however, borne out when actual examples are analyzed. It will be seen that reticulation is characteristic not only in rapidly evolving Holocene vertebrates, such as the red-backed voles described below, but also in the much more slowly evolving Jurassic oysters of northwestern Europe, where the story takes a whole stage of, say, 5 million years duration to unfold.

If scale effects are not significant in comparing the three models, it is possible to claim that speciation processes are important to biostratigraphy in two contrasting ways: both phyletic gradualism and reticulate speciation provide a tool that can be coupled with the statistical evaluation of any evolving population to give a precise point on a curve correlating evolutionary modification against stratigraphical horizon. The model of punctuated equilibria, on the other hand, provides a series of index fossils each of which characterizes one of a series of consecutive zones.

To test the models against actual situations, we shall consider four examples: the first is a polytypic complex of rodents embracing both Recent and fossil material; the second and third are both Jurassic ammonites, fossils well known for their high rate of evolutionary modification; the fourth is a complex of Jurassic oysters, equally well known for their low rate of evolutionary change.

TESTING THE MODELS

Polytypic Speciation in the Red-Backed Voles

The red-backed voles (Text-fig. 4) are abundant and widely distributed throughout the northern hemisphere. They form the highly polytypic genus *Clethrionomys* (family Microtidae); more than 100 local races have been named, although there is no general agreement on the taxonomic validity of most of them. It is an abundant fossil in Quaternary deposits of the same area. It provides a splendid example of the way in which statistical, morphological, ecological, behavioral, genetic, distributional, and paleontological data can be used to test phylogenetic theories.

Clethrionomys is known from the beginning of Pleistocene time, and the history of the Ice Age must have played a major part in the control of its distribution. At present its range extends from the Arctic to as far south, in the Far East, as 30°N, or, in America, to 35°N in North Carolina. Although so polytypic, its multitudinous races can be grouped into as few as three species (Ellerman, 1951). It was supposed that the three ancestral groups were initially confined to North America (*C. rutilus*), Central Asia (*C. rufocanus*), and Europe (*C. glareolus*). These were then supposed to spread, so that *C. rutilus* invaded northern Eurasia, *C. rufocanus* spread westward

Text-Figure 4
Red-backed vole, *Clethrionomys glareolus,* X0.85. (Photo by John Markham, copyright.)

into Europe, and *C. glareolus* invaded both Asia and North America (Chaworth-Musters in Ellerman, 1951). At present, the three species have ranges that overlap in northern Eurasia; but in general *C. rutilus* is Arctic, *C. glareolus* characterizes Europe, and *C. rufocanus,* Asia; *C. rutilus* also occurs in North America, with a number of other species.

Voles are perhaps more valuable than any other fossils (except perhaps elephants) as stratigraphical indexes in the Pleistocene (Kowalski, 1966), but this does not apply to *Clethrionomys,* which despite its variability is a conservative genus. *Clethrionomys* is distinguished from other voles by its cheek teeth, which, in the adult, develop roots instead of remaining open. This is a primitive feature. The subspecies are mainly differentiated by coat color, body size, and details of the teeth. The details of the third upper molar are particuarly useful (Text-fig. 5); they may be simple, intermediate, or complex. The main competitor of *Clethrionomys* is its close relative the meadow vole, *Microtus.* On small islands, *Clethrionomys* and *Microtus* seem to be incompatible; very seldom do both genera occur together. It seems that in the history of the postglacial colonization of these islands, the first-comer vole successfully expands its habitat to include both grassland and woodland, and is thereafter able to exclude its competitor. Thus, of the two neighboring Scottish islands of Skye and Raasay in the Inner Hebrides, Skye is occupied by *Microtus*,

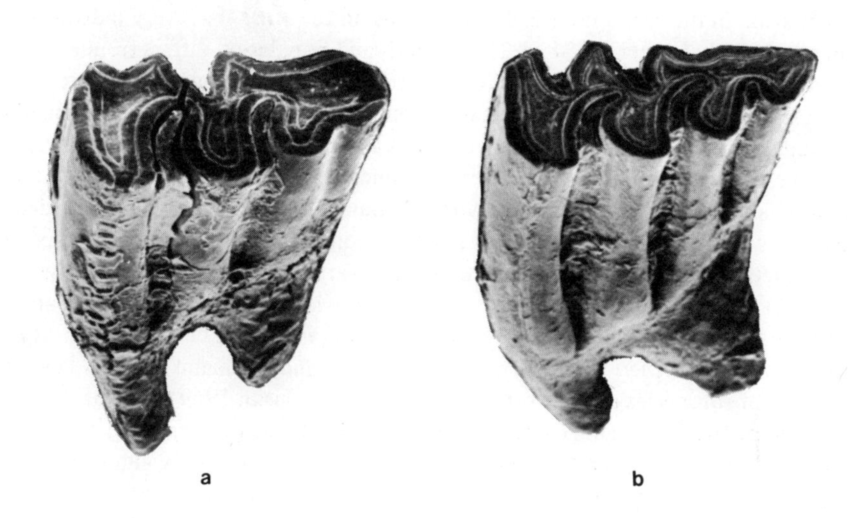

Text-Figure 5
Scanning electron micrographs of third upper molars of *Clethrionomys glareolus:* (a) simple; (b) complex. X25.

Raasay by *Clethrionomys.* Yet the islands are only 1 mile apart and provide very similar habitats (Cameron, 1965). In America, *Clethrionomys* has not successfully invaded any island; they are all occupied by *Microtus*. In Europe there are five island races of *Clethrionomys*, distinguished from the mainland forms mainly by size; they are all larger. As many characters are allometric, size is correlated with proportional differences in various parts of the skeleton (Corbet, 1964). On one island (Skomer, off south Wales), the coat color is quite distinctive. Morphological characters are not the only diagnostic features. There are behavioral differences. Steven (1953) notes that all the island forms of *Clethrionomys* are, when brought into captivity, relatively docile, whereas the mainland form remains extremely wild. This difference is genetically controlled. According to Corbet (1961), the geological evidence leads one to suppose that the islands were colonized by the voles long after their separation from the mainland. Corbet believed that their migration was the result of accidental human introduction, and that the morphological divergence is the result of the contrast in environment provided by the islands, notably by the absence of competitors and predators. Although there is little evidence to suggest that man is responsible (Cameron, 1965), Corbet's other contentions seem perfectly valid. Genetic experiments show that the island races are all interfertile with both each other and with the mainland form (Steven, 1953, 1955). They must all, therefore, be regarded as different races of one species.

Variation in the island races is discontinuous, in the sense that every individual ("nearly 100%"; Corbet, 1964, p. 214) of an island population differs from every individual from the mainland. This is in contrast to the mainland races, where variation is continuous. For example, it has been proposed that the race inhabiting the mainland of Britain should be distinguished as a separate subspecies (*britannicus*) from the race inhabiting the neighboring continent, but Corbet has shown that the variation among various populations within Britain is almost as great as that which exists between those from Britain and from the continent, and abandons this specific distinction. Within Eurasia, variation shows every degree of continuity. The mountain forms, like the island forms, are normally larger than those of the plains, and separate subspecies are recognized, e.g., *C. glareolus nageri* in the western Alps and *C. glareolus ruttneri* in the eastern Alps (Gruber and Kahmann, 1968). Tracing the chain of subspecies eastward through Poland (Serafinski, 1969), Jugoslavia (Rossolimo, 1964), Bulgaria (Mitev, 1968), and into Russia (Bolshakov, 1968), every kind of division is present between races, from a discontinuity almost as abrupt as that which distinguishes island forms, to clinal gradients controlled by the environment. The analyses presented by Serafinski (1969) and Bolshakov (1968) strongly suggest that the variation is, at least in part, *ecotypic*; the same environmental forces if exerted on the same gene complex in different areas elicit the same morphological response; the result is an ecological race (Heywood, 1959; Mayr, 1963, pp. 351-357). Thus Serafinski finds that body length is strongly controlled by altitude; there is a regular "altitude cline." Consequently, there is no reason to suppose that the various mountain races, distinguished as they are by larger size, are necessarily more closely related to each other than they are to the neighboring lowland forms. Likewise, the island races share their large size with other island races as a result of having undergone a similar environmental response rather than because they share a common ancestor.

Fossil teeth of *Clethrionomys* are abundant in suitable Quaternary deposits; these are occasionally supplemented by other skeletal parts. The first detailed study of fossils was that by Hinton (1910, 1926); these studies are now being revised by Kowalski and Sutcliffe and, although their paper is not yet published, Sutcliffe has been kind enough to let me know their results as far as they concern *Clethrionomys;* I am indebted to him for the following information. Four successive horizons can be recognized. The lower three, which range in age according to present views from the upper part of the Early through to the Middle Pleistocene, all carry races similar to the smaller, mainland races of *C. glareolus* that today occupy the lowland regions of Europe. In the succeeding horizon, the Ightham Fissure deposits of Late Pleistocene age, Hinton recognized two races, one larger and one smaller; but it now transpires from the work of Kowalski and Sutcliffe that only the larger one was truly Pleistocene. The smaller one was based on Recent bones of the present-day race of *C. glareolus* that had become mixed with the Pleistocene fossils. This larger race compares with the forms now found on the islands, and with the subspecies *C. glareolus nageri* characteristic of the Alps. Kowalski and Sutcliffe suggest that the development of greater body size was likely to be an environmental response to the difficult climatic conditions of glacial periods. Body size decreases in forms found in interglacial or postglacial deposits. Sickenberg (1939) records a similar association in Belgium, small forms being associated with a forest fauna, large forms with an Arctic fauna. The fossil history, as it is now being unraveled (e.g., Kowalski, 1966, 1970; shows that the voles first reached eastern Europe in a late phase of Early Pleistocene time (Tiglian [or Tegelian] interglacial of Poland). By the beginning of the Middle Pleistocene (Cromerian), it had reached England, and it continues as a common fossil of interglacial deposits throughout Europe (Kurtén, 1968). In contrast to other voles and to lemmings, it seems always to have been a woodland genus. In its Pleistocene history, "it vanished from the tundra regions only in the proximity of glaciers, but reappeared as soon as the climatic conditions had allowed the development of thickets and forests" (Kowalski, 1966). Today, the Arctic species (*C. rutilus*) has a circumpolar distribution in the tundra (Manning, 1956; MacPherson, 1965), where it occupies scrubland on the banks of lakes; it has not, however, succeeded in colonizing the vast areas covered with dwarf birch (Pjastolova, 1972).

The two variable characters that are monitored most easily from fossil material are body size and the complexity of the molar teeth. It is interesting that in high mountains, in peripheral island groups, and in periglacial regions, *Clethrionomys* develops ecotypes of larger body size than those inhabiting lowland regions. This variation in body size is clinal on the mainland, and discontinuous on the islands. In the case of molar teeth (Text-fig. 6), the variation is apparently a case of balanced genetic polymorphism (Kowalski, 1970).

The proportion between simple and complex types varies much from population to population, both in fossils and in Recent races (Zejda, 1960; Corbet, 1963; Kowalski, 1970). The way that genetic polymorphism is controlled is not fully understood; it forms a topic of much current research (Smith, 1970). It has been suggested that disruptive selection pressures may be involved, and may sometimes lead to sympatric speciation (Smith, 1962).

The size of the teeth and the length of the tooth row varies clinally (Voronov, 1961). Kowalski ascribes this to adaptation to different foodstuffs; the more northern populations eat a higher proportion of the green parts of plants, and hence develop larger teeth.

Text-Figure 6
Crown-view diagrams of third upper molars of *Clethrionomys:* (a) simple; (b) intermediate; (c) complex. (After Corbet, 1963.)

Like all voles and lemmings (Elton, 1942), *Clethrionomys* is characterized by periodic fluctuations in population size, and consequently by frequent alternations between migration and retraction (Bergstedt, 1965; Grant, 1969, 1970; Gaines and Krebs, 1971). This contributes to the pattern of phylogenetic reticulation.

The fossil history and present distribution of the bank voles provide us with a clear example of the evolution of a polytypic species. Although discontinuous variation is found with the establishment of island races, and with some of the peripheral races on the Eurasian continent (e.g., in southern Italy and in several of the mountain ranges of Central Asia; Corbet, pers. comm.), this is atypical of the complex as a whole, which is characterized by clinal gradation. There is every reason to believe that the genus was as polytypic during Pleistocene time as it is today. A succession of fossil forms, if complete, would show as much gradation between successional forms as exists at any one time between geographical races and ecotypes. It is true that variation is gradual in the middle of the area of distribution and punctuated at the edge. But the middle is involved to a much greater extent in evolutionary change of the whole complex than are the miniscule peripheral island populations.

The patterns that emerges is reticulate. It is a meshwork in three dimensions. A polytypic species weaves through space and time, the gaps between its races opening and closing with isolation and migration. Morphological variation is controlled by environmental response and marked by ecological convergence. The contrast between the Arctic voles and races found in more southern latitudes today is the same as that found between the voles of glacial and interglacial periods during Pleistocene times. The variation exhibited is discontinuous in some places, clinal in others. It is very probable that it will also have been discontinuous at some times, gradual at others.

Successional Speciation and Polymorphism in Ammonites

Some paleontologists have claimed that, among macrofossils, ammonites provide incomparably the best zonal indexes. It has been suggested that they evolve faster, have a wider geographical range, and are more facies tolerant than any of their rivals (Schindewolf, 1950b; Arkell, 1956b). Nevertheless, within the Jurassic System, the zonal sequence established on ammonites seldom depends on the recognition of stages of evolution in an ancestor–descendant relationship. On the contrary, each zone more often represents the replacement of one species by an unrelated immigrant invader, and it is this pattern that gave rise to the theory of iterative evolution in Jurassic ammonites (Arkell, 1949). It is not a case of "punctuated equilibria," because the immigrant species are not related to the species they replace.

To test speciation theory, one must therefore look for evidence for evolution within an ammonite zone. Within the Jurassic System, ammonites are found most abundantly in condensed deposits that are often known as "cephalopod beds." These are the very worst horizons to examine for evidence of speciation, for the zones are crowded together, and nonsequence is the rule. This would not matter in more slowly evolving organisms, but with ammonites only the most expanded sequence will provide the degree of resolution required. Two such examples are presented.

***Middle Jurassic genus* Kosmoceras.** This example affords what is perhaps the most famous example of ammonite evolution known in the geological record. It formed the subject of the classic researches of Brinkmann (1929a, 1929b), and has been incorporated in textbooks of evolution, stratigraphy, and paleontology ever since (e.g., Simpson, 1944, 1953; Woodford, 1965; Raup and Stanley, 1971). Brinkmann was one of the first to apply statistical methods to stratigraphical paleontology. He collected his material from the Oxford Clay (Middle Jurassic) of Peterborough, England, and systematically recorded all the ammonites found at 1-cm intervals through 13 m (43 ft).

The most striking feature among the many fossils to be found in the Lower Oxford Clay is the profusion of ammonites which, although crushed, are otherwise usually complete up to the final aperture, with test often still preserved in iridescent white aragonite. By far the overwhelming proportion belong to the genus *Kosmoceras,* and the zonal classification is based on the evolutionary change observed in this genus (Callomon, 1968).

Brinkmann found some 3,000 specimens of the genus *Kosmoceras* and distinguished five phyletic lines. In each, he quantified a number of morphological characters (e.g., maximum diameter, diameter of position at which ventral tubercles disappear, secondary rib ratio, "bundling" index) and plotted these against stratigraphical position. For each character, each specimen was represented by a point on a graph plotting character against stratigraphical level on the 1,300-cm scale. The direction of evolution was then represented by a regression line (Text-fig. 7). Brinkmann's work was far ahead of its time. In many ways it remains the most valuable work ever performed on the recording of changing morphology through time. But it is important to be able to recognize what, in his data, is objective, and what is the result of subjective interpretation. Callomon (1968) has revised the stratigraphy and paleontology of the Oxford Clay and has verified Brinkmann's data in almost every particular.

The at first glance monotonous uniformity of the Oxford Clay is only apparent, and closer inspection reveals a rapid succession of distinct beds separated by sharp lithological boundaries. These boundaries often coincide with thin shell-beds or breccias, crowded with crushed ammonites ("ammonite-plasters") or bivalves, chiefly *Nucula,* and are sometimes pyritic. Two types of such shell-beds can be distinguished; basal shell-beds, marking the onset of a cycle of sedimentation; and terminal shell-beds, marking its end.... Similar discontinuities were found in the succession of *Kosmoceras* coinciding where they occurred, with lithological breaks. . . . In the meantime, the breaks in development of *Kosmoceras* are very convenient stratigraphically in defining the precise limits of ammonite zones and subzones (Callomon, 1968, p. 269).

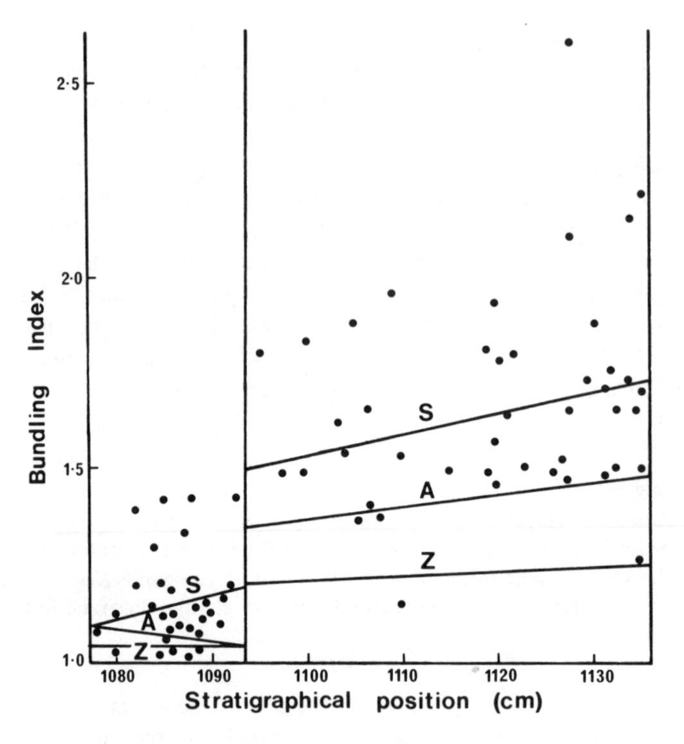

Text-Figure 7
"Bundling index" (ratio of secondary ribs to outer tubercles) in three morphotypes of *Kosmoceras* plotted against stratigraphical position above a certain datum line in centimeters. Upper regression line (S) and scatter, a microconch; middle line (A), another "species" of microconch, here treated as a polymorphic form; lower line (Z), the macroconch. Vertical lines, lithological discontinuities. Brinkmann placed each of these forms in a separate subgenus (S, *"Spinikosmoceras";* A, *"Anakosmoceras";* Z, "*Zugokosmoceras*"). Note that the scatter represents not the whole fauna, but only those selected by Brinkmann as "*Spinikosmoceras.*" (From data of Brinkmann, 1929a.)

But it is in the paleontology that the subjective element was introduced. For Brinkmann claimed to be able to recognize five phyletic lines. The first thing that he did on recording a specimen was to assign it to one of these five groups. This preliminary subjective sorting was thus impressed on all the subsequent statistics. As Callomon (1963) points out, having thus imposed a classification on his material, it is hardly surprising that the statistics confirmed it! And Callomon suggests an alternative hypothesis – that all the specimens from one bed are representatives of a single polymorphic species: "The most telling single piece of evidence in favour of the genetic unity of the whole group lies in the onset of bundling in secondary ribs in *all* forms simultaneously at the base of the Athleta Zone (level 1094 cm at Peterborough)" (Text-figs. 7 and 8). Tintant (1963), in his extensive monograph of the genus, also notes the prevalence of parallel evolution and considers the possibility that at each horizon we have only one highly polymorphic species. If, in this way, the variety of discontinuous polymorphic forms were interfertile, we would have a ready explanation of how so often they are subject in their evolution to parallel modifications.

There is a further complication; in addition to the general polymorphy, there is imposed a clear dimorphy. Two *kinds* of *Kosmoceras* exist at every horizon, "macroconchs," which are large forms with simple apertures, and "microconchs," which are small, with lappets (Text- fig. 8). It has now been demonstrated beyond reasonable doubt that this dimorphy is sexual (Makowski, 1963; Callomon, 1963); probably the macroconchs are female, the microconchs male.

Taking Brinkmann's data as they stand, however, we find that quite often he has been able to demonstrate a statistically significant correlation between changing morphology and stratigraphical position. The resulting regression line (Text-fig. 7) never continues at the same slope for very long, but insofar at it measures gradually

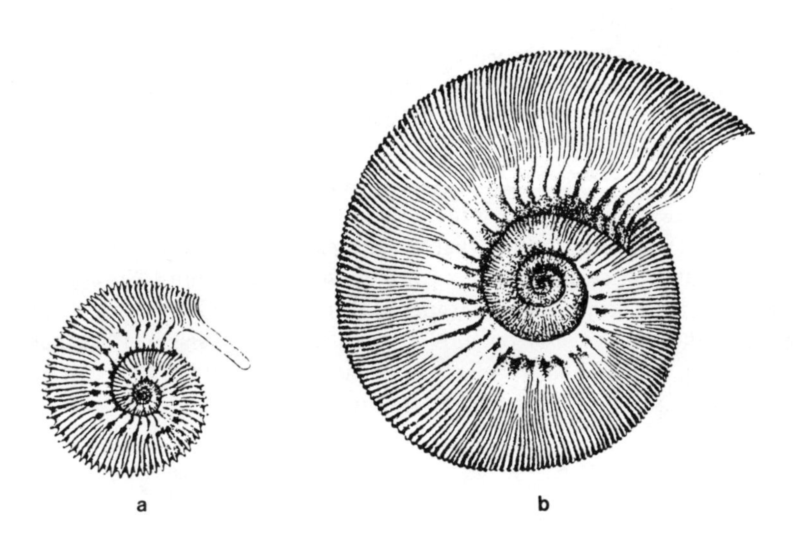

Text-Figure 8
Sexual dimorphs of *Kosmoceras:* (a) microconch (probable male); (b) macroconch (probable female). (After Brinkmann, 1929b, and Callomon, 1963.)

changing morphology, it is demonstrating gradualism. If it were to continue as a straight line through a considerable thickness of strata, it would illustrate the establishment of an evolutionary trend. Instead, we find frequent reversals of trend, or periods when evolution of that character ceases.

***Lower Jurassic genus* Dactylioceras.** A second example of ammonite evolution in an expanded sequence comes from the Lower Jurassic, an analysis of the genus *Dactylioceras* from the Upper Liassic Grey Shales of the Yorkshire coast in Britain (Howarth, 1973b). The ammonites in these beds are abundantly preserved in calcareous nodules. They contrast with those from the Oxford Clay in having many more characters preserved for analysis (Plate 1). These nodule beds occur in seven successive horizons distributed through 10 m (32 ft) of shale. Crushed ammonites also occur in the shales between and above the nodule beds. Howarth (1973, p. 247) claims that the ammonites

> have been collected with a degree of stratigraphical control upon which it would hardly be possible to improve. . . . Collections from these beds do not represent single populations in a biological sense, but they do represent single populations in the sense that it is almost certain that no evolution has occurred within the collection from each bed.

The ammonites are highly variable at each horizon, particularly in whorl proportions, in rib density, in the "bundling" or "fibulation" of the ribs, and in the development of lateral tubercles. Some of these variable features therefore parallel those which later developed in *Kosmoceras,* particularly the pattern usually called "fibulation" in *Dactylioceras,* but referred to by Brinkmann as "bundling" in *Kosmoceras.* In *Dactylioceras,* this is caused by primary ribs coming together at one of the ventrolateral tubercles (Text-fig. 9); in *Kosmoceras,* it is the secondary ribs that unite in the same way (Text-fig. 8). At each horizon, variation seems to be continuous. It is interesting that there is more variation at adolescence than in juvenile or adult stages. Thus Howarth finds that the greatest variation occurs at diameters of between 50 and 75 percent of that reached by the average adult. The degree of variation also changes during the succession, so that it is less in horizons 3, 4, and 5 than it is in those above or below; this change affects whorl shape rather than rib density, which remains highly variable throughout. Yet there is some correlation between rib density and whorl shape, for at horizons 1 and 2, and again in 6 and 7,

---

Plate 1
Polymorphic species of *Dactylioceras (Orthodactylites*) at four successive horizons in the Lower Jurassic Grey Shales of Yorkshire, England. Figs. 1, 2: *D. semicelatum* from top horizon; figs. 3, 4: *D. tenuicostatum* from third horizon; figs. 5, 6: *D. clevelandicum* from second horizon; figs. 7–9: *D. crosbeyi* from bottom horizon. (After Howarth, 1973.)

1a
1b
2a
2b
3a
3b
4a
4b
5a
5b
6a
6b
7a
7b
8a
8b
9a
9b

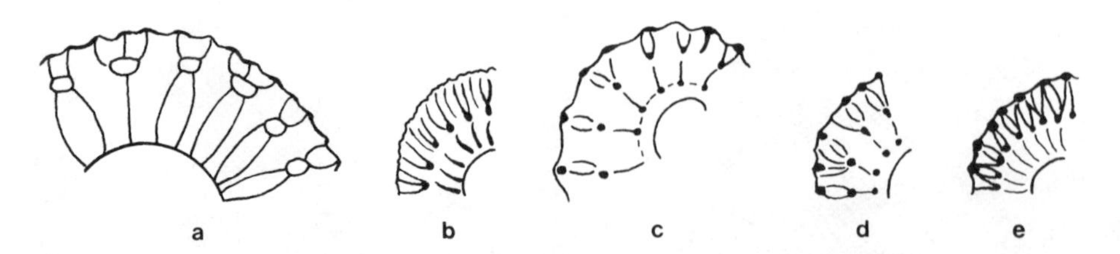

Text-Figure 9
"Fibulation" in *Dactylioceras* and "bundling" in *Kosmoceras:* (a) *Dactylioceras;* (b–e) *Kosmoceras.* (After Tintant, 1963.)

specimens with depressed whorls have coarser ribs than those with compressed whorls. There are no specimens with depressed whorls in horizons 3, 4, or 5.

Sexual dimorphism has been claimed to occur in *Dactylioceras* (Lehmann, 1968; Guex, 1971); but Howarth believes that this is in error, for adult specimens of *Dactylioceras* from the Grey Shales show a size range with no bimodality. Yet contemporary ammonites of the family Hildoceratidea are clearly dimorphic.

An orthodox and traditional taxonomic treatment of these Grey Shale ammonites would probably, on the base of the variaiton present, divide the complex into at least two genera, one for those with depressed whorls (which would be referred to *Kedonoceras*), one for those with compressed whorls (*Orthodactylites*). Moreover, one would expect that each horizon would be characterized by three or four species distinguished by rib density. But Howarth is convinced that the continuous variation present shows that the forms at each horizon are more closely related to each other than they are to those that precede or succeed them. Consequently, he puts them all in one genus, and recognizes four successive species: *D. crosbeyi* at horizon 1, *D. clevelandicum* at horizon 2, *D. tenuicostatum* ranging from horizon 3 to 5, and *D. semicelatum* from 6 to 7. All four species show a wide range of rib density, but *D. tenuicostatum* differs from the others in having no forms with depressed whorls, no forms with tubercles, and no forms with fibulate ribs.

Of the four species described, *D. crosbeyi* and *D. clevelandicum* are each confined to a single horizon; *D. tenuicostatum,* on the other hand, occurs abundantly, both crushed and solid in nodules, through 2.4 m (8 ft) of strata, and *D. semicelatum,* although somewhat less abundant, through nearly 5 m (17 ft). Although all four species are so variable, there is no overlap between them; they are quite distinct. Yet they are clearly related, and all four are referred by Howarth to the same subgenus, *Orthodactylites.* As we have seen, traditional taxonomy would recognize 3 or 4 species at each of the levels. When grouped together, they form polymorphic species. Yet the change over from one such polymorphic species to another is abrupt. There is no sign of gradation. *Datylioceras* provides us with an excellent example of punctuated equilibria.

Howarth believes that the succession is not a local phenomenon; he was able to use the succession in Yorkshire to establish three new ammonite subzones, one based on *D. crosbeyi* and *D. clevelandicum,* one based on *D. tenuicostatum,* and one based on *D. semicelatum,* and to trace them through Europe. The *D. semicelatum* subzone seems to be widely distributed throughout Europe and even perhaps beyond; it often contains another species of the same subgenus, *D. directum.* The lower two subzones (those of *D. tenuicostatum* and *D. clevelandicum*) certainly occur in Europe, but are either condensed or are not well characterized. The *D. clevelandicum* zone seems to be well developed in Siberia. All these records have been summarized by Howarth (1973b).

There is no sign at all in *Dactylioceras* of the establishment of geographical races even though its species are highly polymorphic. Its phylogenetic pattern contracts strongly with that of *Clethrionomys.* Yet we are dealing with a comparable length of geological time in the two instances, probably something under 1 million years: from halfway through the Pleistocene to the present in the one case; for the duration of three ammonite subzones in the other.

Successional and Polytypic Speciation in Oysters

Oysters contrast well with ammonites. They include slowly evolving groups and are extremely variable. They are probably the most abundant macrofossils of Mesozoic time.

At the beginning of the Bathonian Stage in Middle Jurassic time, the oyster *Catinula* was represented in western Europe by a widely distributed species known from Britain, Normandy, Alsace, the Swiss Jura, the Franconian Jura in Germany, and northward through Germany and Poland. It is a small, ribbed oyster, highly variable but showing no geographical differentiation. This Early Bathonian oyster can be referred to the species *C. knorri,* but it occurs at a lower horizon than the type specimen of that species, and differs from it in having slightly finer ribs, and, on the average, rather a smaller attachment area. It is therefore distinguished as a subspecies, *C. knorri lotharingica* (Sylvester-Bradley, 1954, 1958, 1959). Throughout its range, it occurs at the same horizon in the Lower Bathonian.

In eastern France, Switzerland, and Germany, this subspecies is replaced at higher horizons (Middle Bathonian) by the typical subspecies *C. knorri knorri.* At still higher horizons, a third subspecies is characterized by a somewhat different shape and a slightly larger size. This subspecies is still more restricted in distribution, and is found only in eastern France. It seems to be the last of its line, and no descendants are known.

In western France (Normandy) and England, the widely distributed *C. knorri lotharingica* is followed in the Middle Bathonian by a subspecies with a rather larger attachment area, and with coarser and fainter ribs; the populations also include specimens in which the ribbing disappears during some stages of growth. This subspecies is referred to *C. knorri mendipensis,* and differs from both the ancestral race *C. knorri lotharingica,* and from the contemporary but geographically distinct race *C. knorri knorri.* In Normandy and England, higher horizons (Upper Bathonian and Lower Callovian) yield a succession of forms in which four trends are developed:

(1) they increase in size; (2) the ribbing gets coarser and less distinct, and in later forms is restricted to early growth stages, or is absent altogether; (3) the left valve deepens so that there is an increase in the angle between the first growth line and the last (θ in Text-fig. 10); (4) a minor but increasing proportion of specimens develop a posterior radial sulcus. A whole series of subspecies can be established throughout the late Bathonian and early Callovian zones, but the later subspecies differ so much from the Lower Bathonian ancestors that they must not only be referred to different species, but even a different genus (*Gryphaea*). Subsequently, this genus expands its provenance and covers an even wider geographical area than the ancestral *C. lotharingica.* Indeed, it seems itself to be the ancestor of all the Upper Jurassic species of *Gryphaea* that spread through the epicontinental seas of Europe and Asia (Text-fig. 11).

This succession suggests a split in the phyletic line; the widely distributed Lower Bathonian ancestor divided in Middle Bathonian times into two geographically distinct groups. Each diverged by phyletic gradualism. The eastern branch prospered at first, but subsequently declined and became extinct at the end of Bathonian time. The western branch did better. In Upper Bathonian time, the rate of evolution increased, and during Callovian time immigrant branches spread out over all the northern seas of the eastern hemisphere.

The pattern is one of a polytypic species influenced, in the later stages of evolution, by evolutionary trends. Selection must have been directional and anagenesis occurred (see Table 1 and Text-fig. 3). This pattern most nearly fits the model of reticulate speciation.

PHYLOGENETIC PATTERNS IN BIOSTRATIGRAPHY

I have attempted in these pages to use paleontological data to test phylogenetic theory. We have examined the origin of species as set out in three hypothetical

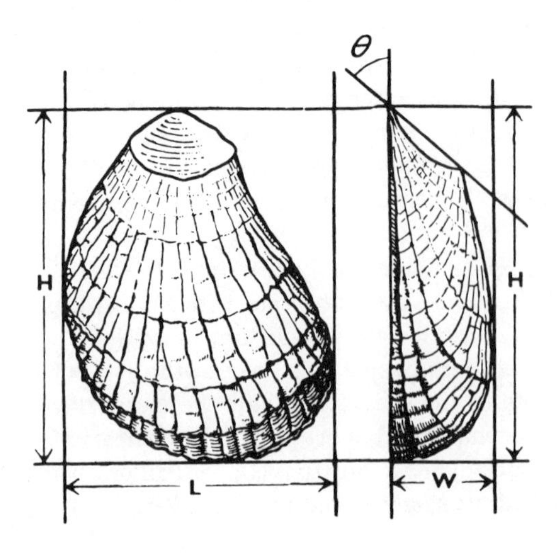

Text-Figure 10
Catinula knorri: diagram showing dimensions used in discrimination of species. (After Sylvester-Bradley, 1958.)

models. To test them we have used evolutionary patterns established in four genera. *Clethrionomys* is polytypic, and its phylogenetic pattern approximates most nearly to the model of reticulate speciation. Both *Kosmoceras* and *Dactylioceras* are polymorphic species; *Kosmoceras* provides a good example of phyletic gradualism, *Dactylioceras* of punctuated equilibria. *Catinula* is a polytypic fossil; its distribution supports the model of reticulate speciation.

It seems, then, that the nature of intraspecific variation has an important bearing on the kind of phylogenetic pattern displayed. Polytypic species must, of necessity, have arisen as a result of a reticulate phylogeny. Polymorphic species may differentiate continuously, as a result of phyletic gradualism, or discontinuously with the establishment of punctuated equilibria.

There can be no doubt that the three phylogenetic patterns discussed have quite different functions in biostratigraphy. A polytypic complex, with coexisting geographical subspecies and a morphology that is modified in different directions and at different rates in different geographical areas, can only be used for correlation with grave danger of confusion. If the species is characterized also by ecotypic variation, so that the same modification is controlled by the same environment at different times, the danger is greatly increased. Only a specialist with a knowledge of the variation present over the whole area colonized can hope to use such a species for stratigraphical zonation.

But polymorphic species are quite another case. Discontinuous speciation characterized by punctuated equilibria yields results that are particularly amenable to the establishment of zonal bundaries. The effect is one of replacement; one zone is replaced suddenly and simultaneously by another. The zonal succession of ammonites first selected by Oppel for the subdivision of the Jurassic System more than a century ago is of this type. Nature seems almost to have followed the pattern that d'Orbigny detected when he propounded the concept of stages: "I was therefore bound to adopt them for the double reason that there is nothing arbitrary about them and that they are, on the contrary, the expression of the divisions which Nature has delineated with bold strokes across the whole Earth" (Arkell, 1933, p. 10). The changes evident at each "punctuation" may be discerned even by a nonspecialist. A stratigrapher can be his own paleontologist.

Not so, however, with phyletic gradualism. The detection of the precise stage reached in an evolving lineage may not only require specialist knowledge of the taxonomic group in question, but perhaps also a large collection and statistical analysis. Most commonly, this can only be provided by micropaleontology, and most stratigraphers in these circumstances need to call in the help of specialist experts. Despite this disadvantage, the statistical analysis of a lineage displaying phyletic gradualism can yield a precision in stratigraphy far surpassing that obtainable by any other single technique. It would not be surprising, therefore, if this method proved to be the one preferred before all others by commercial undertakings concerned with the search for oil and gas fields. That this is not the case is an indication of how rarely phylogeny proves to be gradual. In micropaleontology, as in macropaleontology, the succession of species is more often discontinuous than gradual.

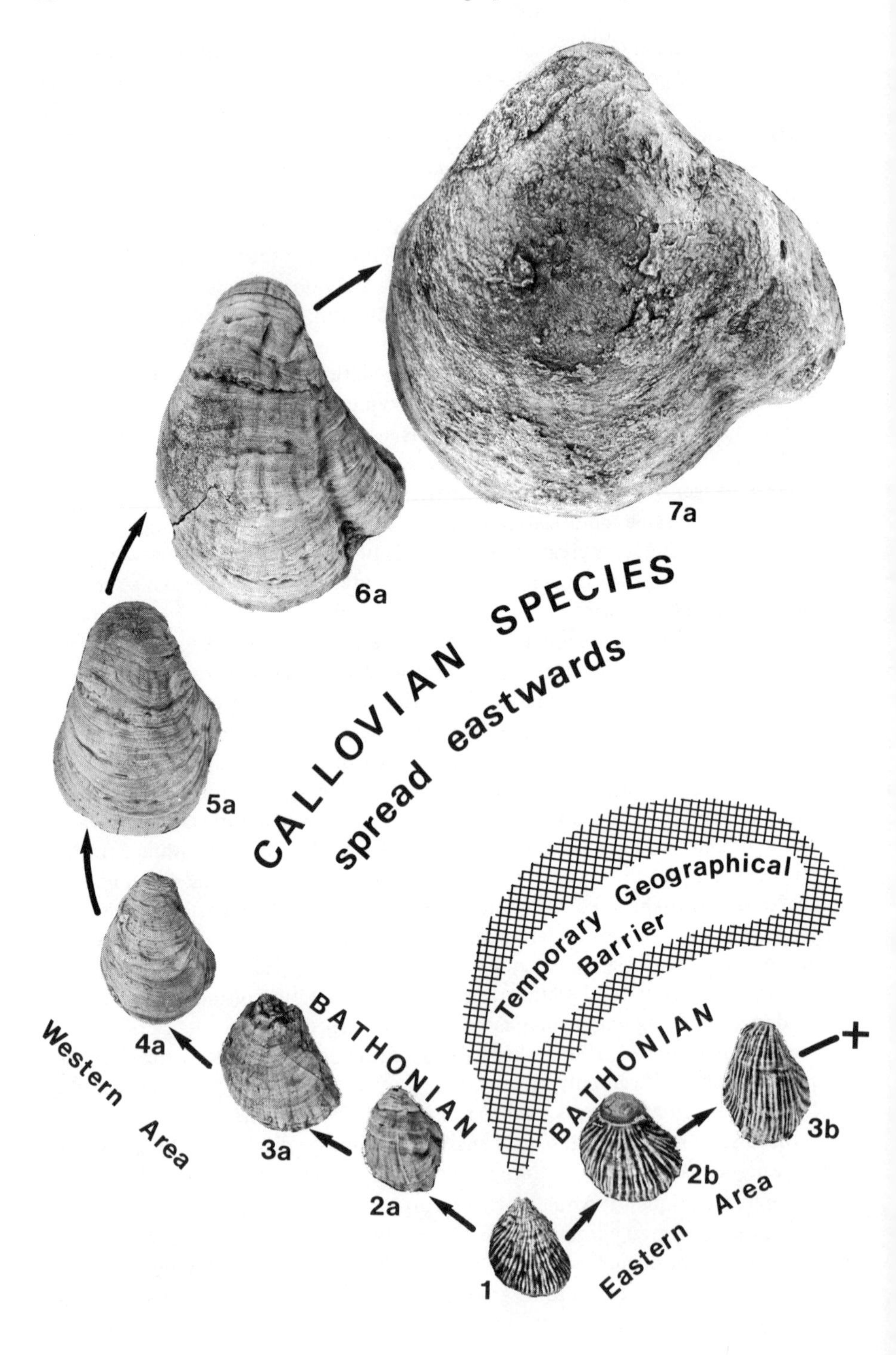
7a
6a
5a
4a
3a
2a
1
2b
3b
CALLOVIAN SPECIES
spread eastwards
Temporary Geographical
Barrier
BATHONIAN
BATHONIAN
Western Area
Eastern Area

So Eldredge and his colleagues were right when they attacked the preconceived paradigm of phyletic gradualism. Yet it would be a mistake to suppose that new species arise in only one way. Nevertheless, punctuation in biostratigraphy is real. Replacement zones are based on natural discontinuities, and are not just the result of an arbitrary response to the human obsession to classify. And if the punctuation is real, so is the grammar and syntax. For it is the punctuation of biostratigraphy that sets a limit to the zones. These are the sentences of the story that is told. It is the story of constant evolutionary change, sometimes episodic, sometimes gradual, always to be continued.

Acknowledgments

The first draft of this paper was read through by a number of reviewers, colleagues, and friends, and I have benefited much from their advice. They are R. S. Boardman of the Smithsonian Institution; G. B. Corbet, M. K. Howarth, and A. J. Sutcliffe of the British Museum (Natural History); R. Semeonoff of the University of Leicester; and the editors.

Text-Figure 11
Evolution of *Catinula knorri* and its descendants, showing cladogenesis and division into two allopatric branches.

Some Relationships Between Habitat and Biostratigraphic Potential of Marine Benthos

Jeremy B. C. Jackson
The Johns Hopkins University

INTRODUCTION

The ideal fossil species for biostratigraphy would be easily identifiable, consistently well preserved and abundant, widely distributed in a variety of habitats (facies), and of short geological time range. Except for preservation, all these characters reflect basic attributes of the ecological strategies of organisms and are strongly covariant (Valentine, 1973b). Species living in different kinds of environments may be controlled in their distribution and abundance by fundamentally different processes. As a result, they often show significant differences in geographic range and intensity of biological interactions with other species; in turn, these differences influence the probability of further speciation or extinction.

An an example of the importance of ecological strategies in determining the biogeographic distribution of organisms and their evolutionary history, I will briefly review what is known of the kinds of factors that control the distributions of modern tropical marine infaunal bivalves, characterize the nature of these distributions, and try to point out some of their more important implications to biostratigraphy.

LOCAL DISTRIBUTIONS OF TROPICAL INFAUNAL BIVALVES

The term "environmental stress" refers to all those physical and chemical aspects of the environment (temperature, salinity, etc.) that may limit the distributions and abundance of organisms. It is a collective term for three aspects of physical and chemical environmental variability. For any single variable, these are (1) the mean value, (2) the variance, and (3) the pattern or "predictability" of the variance. In all these aspects, environmental stress decreases significantly in the oceans with increasing water depth (Sanders, 1968; Sanders and Hessler, 1969; Jackson, 1972). Moreover, the environmental-stress gradient is very steep, and intertidal and shallow subtidal environments of about 1-m depth or less are much higher stress environments than are areas with depths of only a few meters.

These trends are clearly evident in shallow tropical environments overgrown by the marine angiosperm *Thalassia testudinum* (J. B. C. Jackson, 1972, 1973). Bottom areas supporting *Thalassia* growth comprise a highly distinctive set of environments. *Thalassia* often forms an extensive root–rhizome mat, which may be up to 25 cm or more in thickness (Ginsburg and Lowenstam, 1958). This mat is of profound importance in stabilizing bottom sediments against erosion, and contributes to the maintenance of extreme physical and chemical gradients across the sediment–water interface. Daytime temperature differences across the gradient may exceed 3°C (Jackson, 1972). Overlying waters may have daytime oxygen concentrations of 200 percent of saturation, whereas sediments only 1 cm beneath the interface generally lack oxygen and show strong sulfide ion potentials (Jackson, 1972).

Reef-flat *Thalassia* environments (⩽1 m) are characterized by unpredictable environmental variability of very high variance (Jackson, 1972). Diurnal variations in temperature, salinity, dissolved oxygen, etc., may occur very rapidly, and are almost as great as their entire yearly range. The chemical gradient across the interface is extreme. These environments support very large numbers of comparatively few species of infaunal bivalves, with the complete absence of entire groups commonly found in only slightly deeper, shallow-water *Thalassia* environments. Lucinidae generally comprise in excess of 90 percent of the number of individuals and biomass of infaunal bivalves. In the Caribbean, one species (*Codakia orbicularis*) may contribute more than 80 percent of the total abundance; in the Indo-Pacific, two related *Codakia* spp. may be similarly dominant (Taylor, 1968, 1971).

In contrast, *Thalassia* environments in shallow bays (2 to 4 m) are characterized by predictable (seasonal) variations in temperature and salinity of very low variance (Jackson, 1972, and unpublished data). Diurnal variations of most environmental variables are far lower than annual variations, and the chemical gradient across the interface is comparatively moderate. For example, sediments are often oxygenated in the upper few centimeters, and dissolved oxygen concentrations in the water column are only slightly above saturation values. No one species of infaunal bivalves dominates these environments. Lucinidae are often still the most important family; but below depths of 2 to 3 m, Cardiidae and Veneridae are usually more abundant.

Similarly contrasting environmental and faunal conditions exist between other kinds of shallow tropical sedimentary environments (Taylor, 1968; Vohra, 1971). In all these cases, tropical infaunal bivalves living in depths of about 1 m or less must tolerate a far wider range of weather conditions (temperature, salinity, water movements, etc.) than do the same and related species found in waters only slightly deeper than 1 m. They also tolerate a wider range of substrate conditions, and consequently are found in a greater variety of environments (J. B. C. Jackson, 1973).

Interspecific depth zonation patterns of infaunal bivalve species are generally poorly defined curves characterized by broad overlap in depth ranges and habitat distributions of symaptric congeneric or otherwise closely related species (Taylor, 1968; Eltringham, 1971; Vohra, 1971; J. B. C. Jackson, 1972, 1973). This is in striking contrast with the sharply zoned distributions of various rocky substrate organisms, such as barnacles, mussels, and algae (Ricketts and Calvin, 1968; Stephenson and Stephenson, 1972), which have been experimentally demonstrated to be strongly influenced by spatial competition (Connell, 1961; Dayton, 1971). The absence of sharp zonation patterns for infaunal bivalves and other distributional data provide strong circumstantial evidence that competition is not a major factor in limiting the distributions of closely related suspension-feeding bivalve species, and most bivalves are at least facultative suspension feeders (Stanley, 1973; Jackson, 1974). However, at least in temperate seas, deposit-feeding bivalves may exclude infaunal suspension feeders in soft-mud environments by decreasing the stability of the sediment surface and increasing turbidity (Rhoads and Young, 1970).

Except in the extreme upper limits of their depth range, tropical infaunal bivalves from depths less than about 1 m exhibit considerably higher population densities than do species limited to deeper-water environments. For example, on Jamaican reef flats, the lucinids *Codakia orbicularis* and *Ctena orbiculata* commonly reach densities of 50 to 60 individuals/m^2 (1-mm sieve) (J. B. C. Jackson, 1972, 1973). Similar densities of other species occur in *Thalassia* beds at depths $\leqslant$1 m in Kingston Harbour, Jamacia (J. B. C. Jackson, 1973). In contrast, *Thalassia* bottoms in shallow bay areas (4 to 7 m) show maximum infaunal bivalve species densities of about 1 to 1.5/m^2, and most of the clams are juveniles (Jackson, 1972, and unpublished data). Furthermore, the most abundant species at these depths are often those abundant in depths $\leqslant$1 m. Species limited to depths >1 m (various Tellinacea, Veneridae, and Cardiidae) are commonly present, primarily as juveniles, in densities of only about 0.1/m^2 (Jackson, unpublished data). Unvegetated sands to depths of about 50 m may show similarly low densities (Jackson, unpublished data). On the north coast of Jamaica, even the commonest species in these depths (various tellinaceans and *Lucina pensylvanica*) occur in densities of only 0.1 to 0.2/m^2. Again, most individuals are juveniles.

Thus tropical soft-bottom environments $\leqslant$1 m commonly exhibit infaunal bivalve densities 50 to 500 times greater than those commonly found in shallow bays and at shelf depths. For any length of coastline, these differences in densities mean that the more abundant clam species living in a narrow (100 m wide) bottom area of depths ⩽1 m probably have total population sizes equal to those of the more abundant species limited to a 10-km-wide area of bottom in deeper waters. Since species found in depths ⩽1 m are also often among the more common bivalves in greater depths (e.g., the Caribbean *Parvilucina costata* and *Diplodonta punctata*) (J. B. C. Jackson, 1973), it is evident that such species may possess larger total population sizes than do species limited to deeper waters. This is in spite of the fact that the area of shelf bottoms deeper than about 1 m is far greater than that of shallower areas.

Increasing predation rates upon tropical infaunal bivalves with increasing depth seem to be a major reason for these trends (Jackson, 1972) and apparently are responsible for keeping population densities below levels at which resource limitations and competition might occur. Similar predation effects have been reported for Danish (Muus, 1973) and Mediterranean (Massé, 1971) infaunal bivalves and North Sea polychaetes (Kirkegaard, 1969) and are probably of general occurrence.

GEOGRAPHIC DISTRIBUTIONS OF TROPICAL INFAUNAL BIVALVES

Large differences exist between the geographic ranges of tropical infaunal bivalves from high- and low-stress environments. Thus species of Lucinidae, Veneridae, and Tellinidae able to live in depths of 1 m or less have significantly wider longitudinal and latitudinal distributions than do infaunal bivalves restricted to deeper waters (not including deep-sea species) (Jackson, 1974). Such differences may be very great. For example, species of Indian Ocean Lucinidae able to live in depths ⩽1 m commonly range from East Africa and the Red Sea to halfway around the world in the Tuamotu Archipelago, a distance of some 20,000 km. One intertidal species, *Ctena divergens*, extends from South Africa and the Gulf of Suez all the way to Hawaii and Easter Island. In contrast, Indian Ocean Lucinidae limited to waters deeper than 1 m rarely range more than about 5,000 to 7,000 km. These differences result because the shallower species can tolerate a wider range of environmental conditions and may have longer-lived larvae than do species limited to depths greater than 1 m.

The above arguments do not pertain to the deep sea. This is because comparative physiological tolerances are apparently not important factors in the geographic distirbution of deep-sea species owing to the near constancy of environmental conditions over widespread areas of the deep ocean floor (Sanders and Hessler, 1969). Thus, in spite of probably narrow temperature (or pressure?) tolerances indicated by narrow vertical distributions (Sanders and Hessler, 1969), many deep-sea infaunal bivalve species are widely distributed (Knudsen, 1967). Geographic ranges of tropical infaunal bivalve species within any group having representatives from the intertidal zone into the deep sea (e.g., Lucinacea) are therefore greatest with the upper 1 m of water and in the deep sea (perhaps increasing below the level of the thermocline) and minimal for middle- to deep-shelf species. All this suggests that infaunal bivalves from shallow water and intertidal environments might have greater potential in both regional and interfacies correlation than those of deeper-shelf environments.

It is not yet known whether similar environmental differences in geographic range exist between representatives of other kinds of benthic marine organisms. As far as is known, the vast majority of tropical-shelf-depth infaunal bivalves display planktotrophic larval development (Mileikovsky, 1971; Ockelmann, 1965; Thorson, 1950). This uniformity of development may simplify the depth pattern of the geographic distributions of these bivalves compared with other invertebrate groups (e.g., echinoderms, gastropods, polychaetes), which display a variety of developmental strategies (Mileikovsky, 1971). For example, many high intertidal gastropods exhibit direct development. The result is a significant opposite trend to that for tropical infaunal bivalves; high intertidal gastropod species display narrower geographic ranges than do lower intertidal snails (Vermeij, 1972). Whether lower intertidal gastropods have wider or narrower distributions than their subtidal relatives is not known. Such differences point out the need for data on the comparative distributions of all major groups from different types of marine and marginal marine environments.

EFFECTS OF GEOGRAPHIC RANGE AND COMPETITION ON SPECIES LONGEVITY

Geographic Range

I have argued elsewhere (Jackson, 1974) that, because of their differences in geographic distribution, infaunal bivalves able to live as shallow as about 1-m depth or less should be less likely to speciate or to become extinct than species restricted to deeper waters. Preliminary data compiled for first occurrences of Caribbean Veneridae appear to support this conclusion (Jackson, 1974). Species were assigned a rank number value ranging from 1 to 8 corresponding to their geological age (Recent to Early Miocene) (Table 1). The median geolgical age of species able to live in 1-m depth or less is Middle Pliocene, whereas the geological age of species limited to waters deeper than 1 m is only Pleistocene. These differences are highly significant ($P < 0.001$; Mann-Whitney U Test). In addition, the Kendall coefficient of rank correlation is $\tau = 0.44$, which is also significant at $P < 0.001$. Bretsky (1973) has shown that a similar positive correlation exists between geographic range and evolutionary longevity for genera of Paleozoic Bivalvia, but there are possible difficulties in the biological interpretation of such trends for groups above the species level (Jackson, 1974).

Similar trends may exist among species of Brachiopoda. For example, of the 43 species of brachiopods found in the Danish Chalk (Upper Cretaceous, Maastricktian), *Terrebratulina chrysalis* was the only species found in nearshore deposits (Surlyk, 1972). It is also the most abundant brachiopod species in the Chalk, occurring in

Table 1
Earliest Reported Geological Age of 40 Species of Caribbean Veneridae[a]

Species	*Earliest Geological Appearance*
†*Anomalocardia brasiliana*	Late Miocene
†*A. cuneimeris*	Pleistocene
A. membranula	Recent
Antigona callimorpha	Recent
†*A. listeri*	Early Pliocene
A. rigida	Middle Pliocene
A. rugatina	Middle Pliocene
Callista eucymata	Recent
†*Chione calcellata*	Late Miocene
†*C. granulata*	Middle Miocene
†*C. grus*	Late Miocene
†*C. intapurpurea*	Pliocene
C. latilirata	Late Miocene
C. mazycki	Recent
C. minor	Recent
†*C. paphia*	Pleistocene
C. pinchoti	Recent
†*C. pygmaea*	Early Pliocene
†*Cyclinella tenuis*	Late Miocene
Dosinia concentrica	Early Pliocene
†*D. elegans*	Late Miocene
Gemma purpurea	Pleistocene
†*Gouldia cerina*	Pleistocene
G. insularis	Middle Miocene
Macrocallista maculata	Early Miocene
†*Parastarte triquetra*	Early Miocene
†*Pitar albida*	Late Pliocene
P. aresta	Recent
P. circinata	Early Miocene
P. dione	Early Pliocene
†*P. fulminata*	Late Pliocene
P. simpsoni	Pliocene
P. subaresta	Pleistocene
†*Tivela abaconis*	Recent
†*T. mactroides*	Pleistocene
T. trigonella	Recent
†*Transennella conradina*	Middle Pliocene
T. cubaniana	Pleistocene
T. culebrana	Recent
T. gerrardi	Recent

[a]Species able to live in depths of 1 m or less designated by †. References for data in Jackson (1974).

all 47 localities studied by Surlyk, including offshore as well as onshore deposits. It is thus significant that *T. chrysalis* "is a member of a morphologically very stable group which started in Cenomanian and continued through the Tertiary to recent times" (Surlyk, 1972). This is in striking contrast to the very rapid evolution of the other Danish Chalk brachiopods. Further investigations of the evolutionary longevity of different fossil groups from high-stress (marginal marine) and low-stress (offshore) environments provide an area in which paleontologists could make major contributions to ecology and evolutionary biology, and will be necessary for stratigraphers to realize the full stratigraphic potential of different kinds of animals.

Competition

The effects of environmental stress and associated geographic range on evolutionary longevity should be apparent within any major taxonomic group. Intensity of competition may also strongly influence evolutionary turnover. Other factors being equal, a high degree of interspecific competition, as among mammals, may favor much more rapid evolutionary rates than for organisms such as infaunal bivalves not subject to intense competition (Stanley, 1973). Among the Scleractinia and Bivalvia, a positive correlation between competition for space and high evolutionary rate (at least on the generic level) is suggested for the hermatypic corals and Cretaceous reef-forming Hippuritacea (rudists) (Stanley, 1973). Lang (1971, 1973) has studied interactions among hermatypic corals and has demonstrated interspecific competition to be a major factor in determining the distributions of corals in modern reef environments; by analogy, competitive interactions (different mechanisms) were probably of importance among the reef-forming rudists (Kauffman and Sohl, 1974).

Competition is apparently often an important process in limiting the distributions of many other benthic groups, especially those inhabiting hard substrata where living space becomes a limiting resource (Jackson, in press). This is especially true in cryptic (Jackson et al., 1971) or otherwise protected subtidal environments comparatively free from predation and other environmental disturbances, which tend to prevent resource saturation in many exposed and/or intertidal environments (Dayton, 1971; Paine, 1969, 1971). Such groups include almost all encrusting organisms of solid substrates (Kato *et al.*, 1961; Rützler, 1970; Sará, 1970; Stebbing, 1973) and various free-living, territorial organisms, such as stomatopod crustaceans (Kinzie, 1968). The possible effects of competition on the evolutionary rates of these groups has not been investigated.

IMPLICATIONS TO BIOSTRATIGRAPHY

If common among other benthic groups, the positive correlation between geographic range and evolutionary longevity of tropical infaunal bivalves imposes an important limitation on biostratigraphic correlation of widely separated regions. This is because the majority of environmentally and geographically widespread and abundant species are also organisms likely to persist for periods of longer than average duration.

Competition is apparently most important among Recent benthic groups inhabiting specialized, largely discontinuous environments (coral reefs and other rocky surfaces) or those characterized by poor preservation (sponges, crustaceans, polychaetes), or both. Stanley (1973) has suggested that competition was an important process among trilobites (by analogy with crustaceans) and ammonites (by analogy with teleost fishes); the biostratigraphic importance of these groups is of course well known. Further study of the relationships of geographic range and competition to the evolutionary rates of other major groups of Recent benthos, such as echinoids, bryozoans, and prosobranch gastropods, are needed to establish their fundamental biological limitations as biostratigraphic indicators.

Benthic species or higher taxa characteristic of shallow-water, high-stress environments (eurytopic generalist strategy) should be of particular value in regional correlation (because of their wide geographic ranges) and in interfacies correlation (because of their occurrence in many kinds of environments). Such species have the disadvantage for stratigraphy of great evolutionary longevity. Within more limited regions, benthic species characteristic of deeper-water, lower-stress environments (stenotopic specialist strategy) should provide the finest time-stratigraphic resolution. Such species have the disadvantages of more restricted geographic and environmental (facies) range. Biostratigraphic programs can be designed to take advantage of these consequences of ecological strategies.

Dispersal of Marine Invertebrate Organisms: Paleobiogeographic and Biostratigraphic Implications

Rudolf S. Scheltema **Woods Hole Oceanographic Institution**

INTRODUCTION

Sedimentary strata can be related to one another both spatially and in time by the geographical distribution and vertical occurrence of certain key taxa called *index fossils* or *guide fossils.* To be useful in identifying and relating strata to one another, two requirements in particular should characterize such fossil species: (1) they must have a wide geographic range, and (2) they must be restricted temporally over a short geologic time span.

A broad geographic range in both contemporary and fossil species can result only if a form has had an effective means of dispersal. Among sedentary marine benthic invertebrate species, there are two principal ways to disperse both rapidly and widely: (1) *passive transport* of planktonic larvae in ocean currents, and (2) *rafting* of adults upon floating objects adrift on the sea surface.

The temporal, or vertical, restriction of species in the fossil record may result in two ways: (1) early *extinction,* i.e., a speedy termination of a lineage, or (2) rapid *evolution* leading to morphological transformation (i.e., phyletic change) and subsequent speciation.

We consider here evidence for larval transport and for rafting and their significance to the distribution of marine benthic invertebrates both in space and time.

DISPERSAL AND THE GEOGRAPHIC RANGE OF BENTHIC AND SEDENTARY MARINE INVERTEBRATES

Larval Dispersal

That marine invertebrate species are dispersed by the drift of their planktonic larval stages was already proposed in the last century by Wallace (1876, p. 30) and Mortensen (1898, p. 112), and in 1904 by Gardiner, but not until recently has much evidence been available to either substantiate or refute this hypothesis. Ekman (1953) minimized the role of larval dispersal over long distances, such as across ocean basins, because he claimed that the duration of pelagic development of most shoal-water invertebrates was much too short. Thorson (1961) assembled information on the duration of pelagic larval life of 195 species of Temperate and Boreal bottom invertebrates and concluded that their period of larval development was usually too short to account for larval transport across major zoogeographic barriers. More recently, however, it has been shown that within Tropical regions there are many shoal-water or continental-shelf benthic invertebrates that possess long-distance larvae having a pelagic stage from 6 months to over 1 year. These I have termed *teleplanic* larvae from the Greek *teleplanos,* meaning "far wandering" (Scheltema, 1971a).

Dispersal in coastal regions and epicontinental seas. *Larvae along continental shelves.* Dispersal of larvae along the continental shelf and in nearshore waters has been observed mostly in Temperate and Boreal seas. The duration of pelagic development for Temperate species, when known, is usually 2 to 6 weeks (Thorson, 1961); but even in such a relatively short time a larva, if carried by a surface current with a velocity of only 0.5 k/h, can drift as far as 150 to 500 k. Such a moderate current speed is often exceeded in coastal areas, e.g., along the east coast of North America (see Bumpus, 1973).

Details of surface circulation in coastal regions are ordinarily complex, and seldom will larvae be carried over a direct route between two points. However, drift-bottle data can give not only a general notion of the direction and magnitude of currents, but also an estimate of probability that a larva will be retained in the surface water over the continental shelf. Just as surface currents may be determined from passively floating drift bottles, so net circulation near the bottom may be estimated by passively drifting, neutrally buoyant bottom drifters. Ten years of data for the Middle Atlantic coast of America between 36 and 40° north latitude (Bumpus, 1973), for example, show that (1) there is a marked shoreward drift along the sea floor of the continental shelf (Table 1), and (2) the return of both surface bottles and bottom drifters decreases with increasing distance of release from shore (Bumpus, 1965). Particularly marked landward components of water movements occur along the shelf bottom near the mouths of estuaries, e.g., Chesapeake Bay (Harrison et al., 1967). The consequence of the circulation in this region is that larvae near the sea floor will have a better chance for landward transport than those occurring near the surface (Scheltema, 1975).

Table 1
Drift Bottles and Seabed Drifters Released over the Atlantic Continental Shelf of the United States Between 36 and 40° North Latitude and Subsequently Recovered Along the Adjacent Coast[a]

	Number Released	*% Recovered on Coastline*	*% Recovered in Estuaries*
Surface drift bottles	76,326	12.5	None
Bottom seabed drifters	31,166	18.1	2.0

[a]Data from Bumpus (1973).

The behavior of larvae as it affects their vertical distribution is therefore related to the extent and direction of their dispersal. The early developmental stages of most coastal benthic species are usually found near the surface (e.g., 82 percent of those species summarized from the literature by Thorson, 1964); on the other hand, late-stage larvae are most often photonegative and positively geotropic and tend to be found just above the bottom. With such a behavior pattern, young larvae will be dispersed near the surface out over the shelf and along the coast by currents that parallel the shore; older larvae near the time of settlement will be returned shoreward by the net circulation along the sea bottom. The result will be an along-shore displacement of larvae before settlement. Not all larval species show differences in vertical distribution related to age, but the relationship between larval behavior and the hydrography between Cape Henlopen and Cape Hatteras explains how the pelagic stages of certain species can be retained within that region (Scheltema, 1975). The principle probably can be applied more generally to many other coastal regions, such as the coast of Kamchatka, where Makarov (1969) has shown "larval belts" parallel to the continental shelf.

Drift-bottle data also show that currents along the coast change direction seasonally. Consequently, along the eastern American coast south of Cape Hatteras, the larvae of species breeding in early spring can only be transported northeastward, whereas those breeding in late summer must be dispersed only in a southwesterly direction.

Examples of larval dispersal along coastlines exist for a number of species. Along the Atlantic coast of North America, successful larval dispersal is well documented in at least two cases. A southward spread of *Littorina littorea* from Canada, where it was first known from pre-Columbian midden heaps, to the New Jersey shore within the last 130 years is one dramatic example (Kraeuter, 1974). This species of prosobranch gastropod, now ubiquitous on Cape Cod, was unrecorded from this region only slightly more than 100 years ago. Its range extension has seemingly resulted from surface circulation, which is to the southwestward during the late winter and early spring when *Littorina littorea* is known to reproduce and its planktonic egg capsules and larvae are found adrift in the plankton.

A second example of coastal dispersal is the periodic larval settlement of *Mytilus edulis* at Cape Hatteras, North Carolina, during the winter season, about 200 miles south of its normal range; however, the young juveniles never survive the high temperatures of the following summer season (Wells and Gray, 1960).

Other direct evidence for broad dispersal of larvae over the continental shelf is their demonstrated existence in the overlying surface waters. Three examples from different major taxa can be given.

The larval development of contemporary articulate brachiopods is very short, with a planktonic stage of not more than 1 or 2 days, but inarticulate brachiopods belonging to the genera *Lingula* and *Glottidia* have a pelagic stage of several weeks. There are numerous accounts of inarticulate larvae (Text-fig. 1a) in continental-shelf waters throughout the world – off Brazil, West Africa, India, and Japan, the North Sea, and the shores of Singapore. Along the coast of the eastern United States the larvae of *Glottidia pyramidata* are regularly carried by the Gulf Stream along the coast and some even very far out to sea (Text-fig. 2).

Most Bryozoa have a very short nonfeeding (i.e., lecithotrophic or "yolk-feeding") development, with a pelagic stage ranging from a few hours to a few days. Transport for only a short distance is therefore possible. Only the contemporary cheilostomatous bryozoans *Membranipora, Conopeum, Electra,* and *Tendra* (about 20 species; Lagaaij and Cook, 1973) are known with certainty to have cyphonautes larvae that remain pelagic for several weeks, and that consequently are widely dispersed throughout coastal waters (Ryland, 1965; Nielsen, 1971). Cyphonautes (Text-fig. 1b) are found also throughout the Sargasso Sea, but these mostly belong to *Membranipora tuberculata,* a species exclusively found on sargassum weed.

Dispersal of bivalve larvae along the coastline of Norway (Text-fig. 3) has been well demonstrated (Mileikovsky, 1968). Even in such cold waters, where veligers are planktonic for less than 6 weeks, there is the possibility for extensive transport, notwithstanding Mileikovsky's opinion that "most larvae remain in water masses above zones inhabited by their parents." From hydrographic studies of the Norwegian Sea, we may assume an average surface current velocity of between 10 and 20 cm/s (8.6 to 17.2 k/day). From bathymetric surveys (see Lee, 1963), we also know that there are numerous shoal areas which may act as "stepping stones" for "island hopping." The pattern and velocity of surface circulation in the Norwegian Sea is such that any species with a pelagic larva may be expected eventually to occupy any habitat compatible to its postlarval survival.

Larvae in epicontinental seas. The North Sea is a contemporary example of an epicontinental sea; nowhere does its depth exceed 200 m. During the summer months, larvae of mollusks, echinoderms, bryozoans, and cirripeds occur throughout the surface waters (Rees, 1952, 1954). Among the 35 different kinds of bivalve veliger larvae encountered (Text-fig. 4), 15 species occur frequently. The 11 that are specifically identifiable all belong to widely distributed species, and five (*Mytilus edulis, Zirfaea crispata, Modiolus modiolus, Mya truncata,* and *Hiatella arctica*) are known from both the eastern and western Boreal Atlantic. Likewise, from a total of 13 recognizable echinoderm larvae, two, the ophiuroid *Ophiura robusta* and the echinoid *Strongylocentrotus dröbachiensis,* are also amphi-Atlantic species. Clearly, species with pelagic larvae of 6 weeks' duration or less can disperse

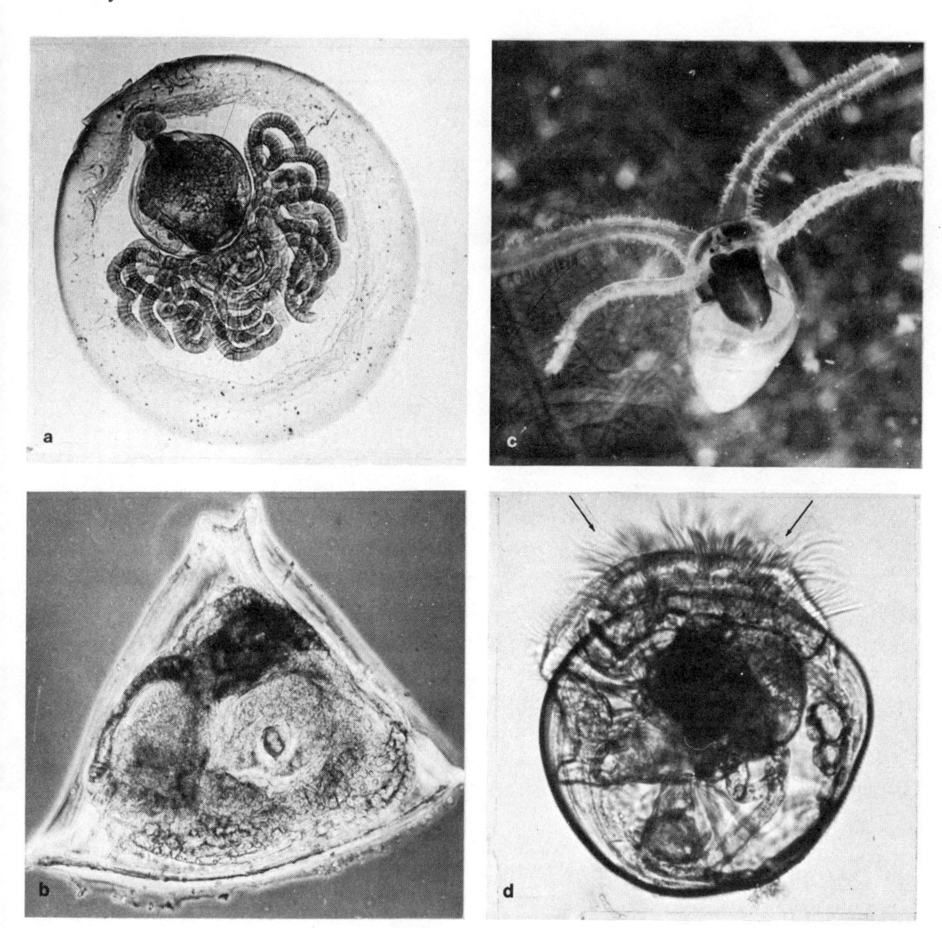

Text-Figure 1
Larvae of taxonomic groups commonly found as fossils: (a) inarticulate brachiopod *Glottidia pyramidata* from the Gulf Stream off eastern North America. These larvae were reported by Brooks in 1879, but complete description and duration of development is still unknown. Size approximately 0.9 by 0.8 mm; (b) minute cyphonautes (0.3 mm on edge) of a cheilostomatous bryozoa *Membranipora tuberculata* are found throughout the Gulf Stream and Sargasso Sea. Adults normally live on drifting *Sargassum* (see Text-fig. 9); (c) veliger of *Cymatium parthenopeum* from the Gulf Stream off North America. The large ciliated velar lobes are used in swimming and feeding. Shell is approximately 5 mm long and uncalcified until after settlement; (d) swimming veliger of the scallop *Placopecten magellanicus*. Most teleplanic bivalve veligers from oceanic waters have more extensive velar lobes (arrow) than this species (see Allen and Scheltema, 1972). Specimen approximately 275 μm in longest dimension. (Photo by J. Culliney.)

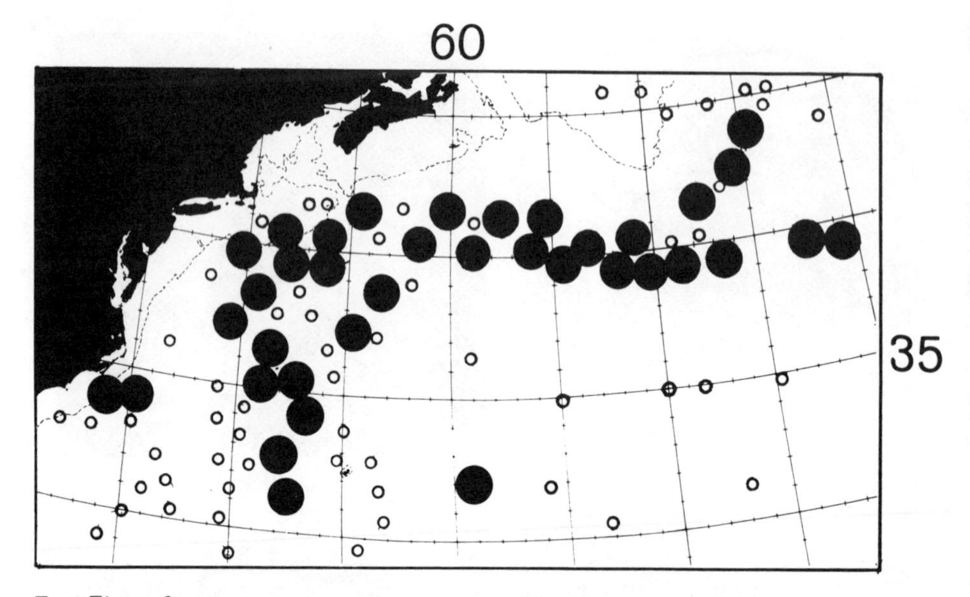

Text-Figure 2
Dispersal of inarticulate brachiopod larvae of *Glottidia pyramidata* off the Atlantic coast of the United States. Larvae of this Caribbean species are carried northward on coastal currents and the Gulf Stream; those found off the Grand Banks (upper right) will not survive to settlement because larval development is too short for transport to a region where the postlarvae can survive. Large circles are locations where brachiopod larvae were taken; open small circles are locations without brachiopod larvae.

anywhere throughout the North Sea. The only constraint is that the larvae settle in a place where the adult organisms can subsequently live and survive. Settlement in an environment where survival is likely is enhanced by the responses of the larvae to a suitable bottom (Thorson, 1946; Wilson, 1952; Scheltema, 1974a).

The Baltic presents a particularly interesting shallow inland sea. Dense saline water enters it along the sea floor from the Atlantic through the Kattegatt and the Danish sounds, periodically replenishing the water of 105-m-deep Bornholm Basin. However, at irregular intervals, owing to hydrographic conditions, the deep water of this basin becomes anaerobic, and the production of H_2S renders the entire bottom abiotic. When high-salinity, oxygenated water again enters along the basin floor, recolonization by benthic invertebrates begins during the next breeding season (Leppäkoski, 1971). Most of the early colonists are infaunal polychaetes introduced (with a few exceptions) by the upstream transport of their pelagic larvae in bottom water; the three bivalve immigrants, *Macoma calcaria, Macoma baltica,* and *Astarte borealis,* have pelagic larval forms. Repeated natural catastrophes of anaerobic conditions in this basin show that larval dispersal is important to recolonization after a return to normal conditions.

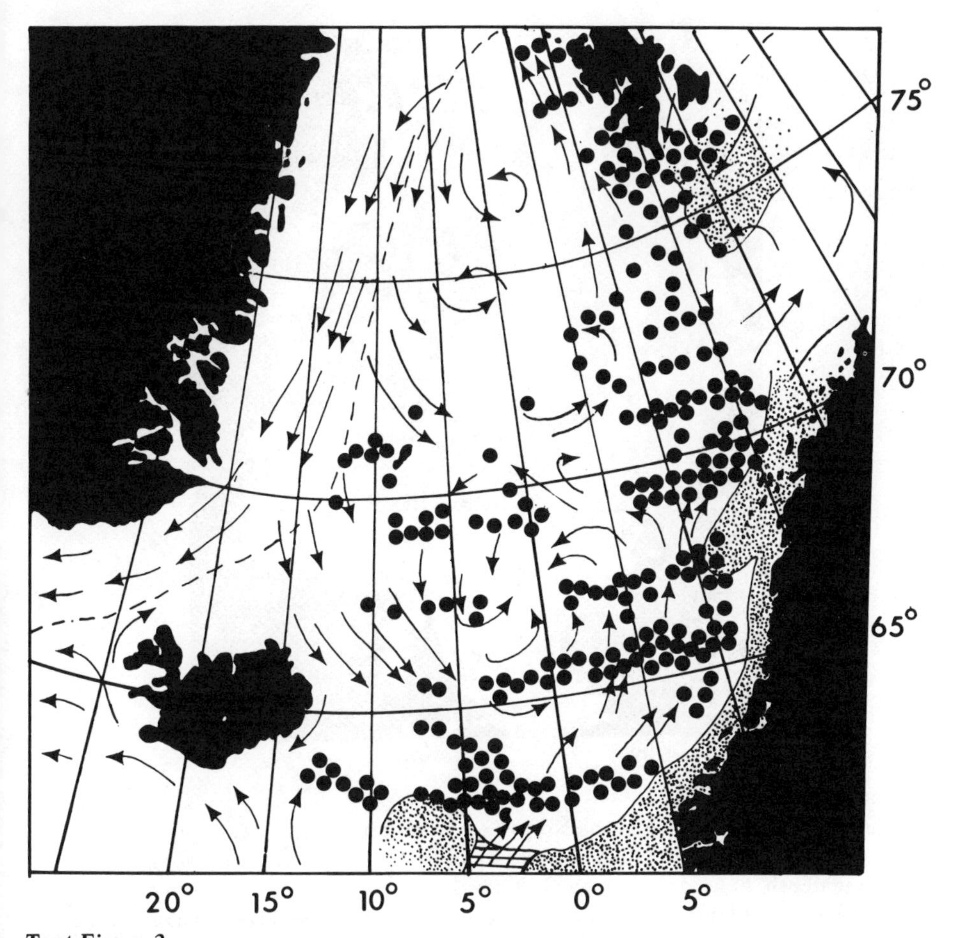

Text-Figure 3
Distribution of bivalve larvae in the Norwegian Sea; data based on 1,459 stations collected from 1958 through 1961 during all seasons of the year. Species were not identified. Stippled region is "zone of maximum abundance of larvae in the waters above the shallow shelf"; horizontally hatched area is "zone of common abundance of larvae drifted from the shallow shelf above the Faroe–Shetland Channel"; small black circles indicate locations where 1–10 larvae occurred in sample. Dashed line delineates cold arctic waters of the East Greenland current. Arrows indicate surface circulation. (Modified after Mileikovsky, 1968, Fig. 4; currents modified after Lee, 1963.)

It is concluded that constraints upon dispersal of pelagic larvae along coasts and in epicontinental seas are few. Only the direction and velocity of currents and the time of reproduction and length of pelagic life will place limits upon where, when, and how far larvae can be transported. Larvae effectively disperse species along

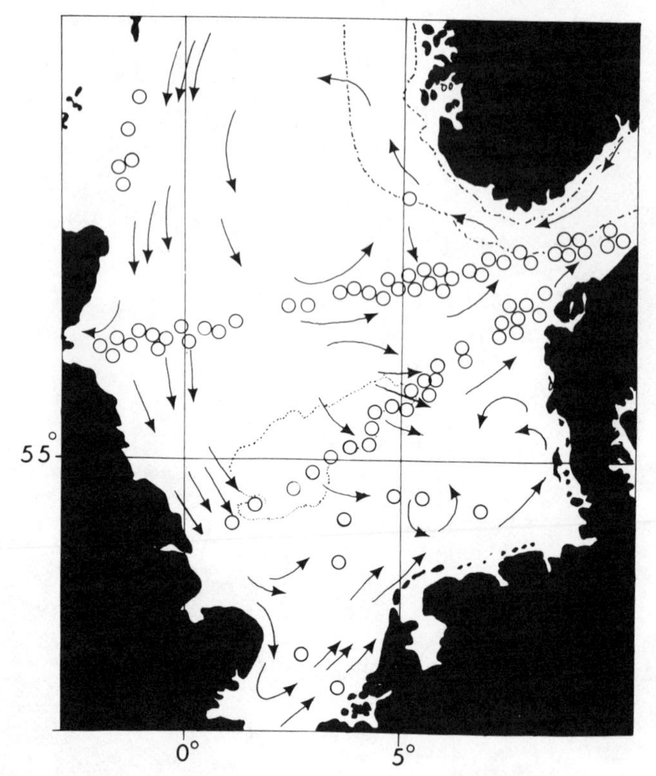

Text-Figure 4
Dispersal of bivalve larvae in the North Sea, a modern epicontinental sea with depths of no more than 200 m; Dogger Bank (bounded by stippled line) is less than 50 m in depth, but a deep trough (> 200 m) extends along the southern end of Norway (dashed line). Open circles indicate locations where bivalve larvae were taken; all common North Sea bivalve species with pelagic larval development are represented (Rees, 1952, 1954), but not individual species. Arrows show direction of surface currents. (Figure modified after Rees, 1954; currents after pilot charts, U.S. Hydrographic Office.)

spatially heterogeneous shorelines because they can drift across ecological barriers; transport from one estuary to another is also possible. Moreover, in most coastal regions and epicontinental seas, favorable environments are common and barriers relatively small. Species having planktonic larvae constantly tend to colonize new regions; the extent of their range is dynamic and ever changing. Marginal populations can arise from larval colonization, only to become extinct from inability to maintain themselves. Areas within the normal range of the species, where it has become extinct through natural catastrophes, can also be recolonized by immigrant larvae. To those species with pelagic larvae, dispersal within epeiric seas will be instantaneous in geological time; by utilizing such species, intrabasinal correlations should be readily possible whenever the fossil record is well preserved.

Dispersal over long distances and across ocean basins. Of greatest interest to the zoogeographer and biostratigrapher is the long-distance dispersal of pelagic larvae. Although the possibility of transport across ocean basins has been considered periodically by paleozoogeographers and marine biologists, it generally has been assumed that larvae of most shallow-water benthic organisms cannot be dispersed over very long distances (Ekman, 1953). Two kinds of evidence demonstrate long-distance larval dispersal: (1) the relationship of oceanic circulation to the geographical distribution of larvae, and (2) the duration of pelagic larval life and its relationship to the velocity of transoceanic currents.

Ocean circulation and geographical distribution of larvae in Warm-temperate and Tropical waters. The direction and route of larval dispersal are determined by the principal ocean currents. In the North Atlantic Ocean, the circulation forms an enormous anticyclonic (clockwise) gyre, bounded to the north by the North Atlantic Drift, to the east by the Canary Current, to the south by the North Equatorial Current, and to the west by the Gulf Stream. In this system the North Atlantic Drift provides a means of larval dispersal eastward toward the European continent and Northwest Africa coast, whereas the North Equatorial Current allows transport of larvae westward toward the Caribbean Sea and North American shores.

In the equatorial region between the westwardly moving North and South Equatorial currents is the diffuse, weakly developed, seasonally occurring Equatorial Countercurrent, which extends only partway across the Atlantic. This ephemeral surface current is inadequate to account for regular eastward transport of larvae across the Tropical Atlantic Ocean. However, directly beneath the superficial South Equatorial Current in the upper 50 m flows the strong eastwardly moving Equatorial Undercurrent or Lomonosov Current; its maximum velocity occurs at a depth of between 50 and 100 m and it flows from the coast of Brazil to the Island of São Tomé off the coast of West Africa (Metcalf et al., 1962). This undercurrent offers possible means of eastward larval dispersal.

Although two currents flow eastward and two westward across the North and Tropical Atlantic, are larvae actually carried over long distances by these currents? Direct evidence from plankton samples taken within the major surface currents of the Atlantic shows that teleplanic (long-distance) larvae are indeed found in the open sea within the upper 100 m, and that they are represented by all the major higher invertebrate phyla, including the Mollusca, Arthropoda, Annelida, and Echinodermata (Table 2). Only taxa that commonly leave a good fossil record need be considered here, i.e., gastropods, bivalves and echinoderms.

Teleplanic gastropod larvae (Text-fig. 1c) are present throughout the North Atlantic gyre and Tropical current system (Text-fig. 5), but only recently have the identity and distribution of the 10 most common species been established (Scheltema, 1971b). In the equatorial region, teleplanic gastropod larvae occur not only in the surface water of the South Equatorial Current but also in the Equatorial Undercurrent (Scheltema, 1968). For example, veliger larvae of the tropical genus *Bursa* occur both at the surface and at depths up to 100 m and, consequently, can be dispersed in either direction between South America and West Africa (Scheltema, 1972a, fig. 6, p. 872).

Table 2
Frequency of Teleplanic (Long-Distance) Larvae in Open Waters of the North Temperate and Tropical Atlantic Ocean and Adjacent Seas[a]

Taxon	*Number of Samples with Larvae*	*% Samples with Larvae*
Gastropoda	518	70.1
Bivalvia	463	62.7
Decapod Crustacea	396	53.6
Polychaeta	319	43.2
Echinodermata	303	40.1

[a]Data from 739 samples.

Fully half of those contemporary families within the prosobranch order Mesogastropoda that are known to have a pelagic development are represented by at least some species with larvae capable of long-distance dispersal. Even within the Neogastropoda, in which most species have a direct development (Radwin and Chamberlin, 1973), about one-fifth of all families are represented by one or more species having a teleplanic larva. Among the third order of prosobranchs, the Archaeogastropoda, only a single species of Neritidae is known from the teleplanos. Families of prosobranch gastropods thus far known to contribute larvae to the open ocean plankton include the Neritidae, Cymatiidae, Bursidae, Tonnidae, Naticidae, Cassidae, Cypraeidae, Triphoridae, Cerithiidae, Architectonicidae, Ovulidae, Thaidae, Muricidae, and Lamellariidae.

At present 40 species of teleplanic gastropod veligers are common and recognizable, although not all identified; additional species and examples from other families are likely to be found that have thus far been missed owing to the low sensitivity of the plankton sampling technique (Scheltema, 1971b, p. 313).

Bivalve veliger larvae (Text-fig. 1d) were found almost everywhere plankton samples were taken in the Atlantic (Text-fig. 6). Moreover, for those species studied in detail, while their size increases the number of larvae decreases with increasing distance from the continental shelf (Allen and Scheltema, 1972, Thiede, 1974), demonstrating thereby their coastal origin.

Larvae of two bivalve families, the Teredinidae and Pinnidae, have distinctive characters that make them readily identifiable. The widely distributed amphi-Atlantic species *Teredora malleolus* is dispersed throughout the North Atlantic (Text-fig. 6) by means of a pelagic veliger stage (Scheltema, 1971c). Howevere, one cannot generalize from a single species about the dispersal mechanism of an entire family. Pelagic development of some Teredinidae is only 1 day or less and the wide geographical range of such species can only be explained by dispersal of adults in wood (Turner, 1966).

The bivalve family Pinnidae is represented in the Atlantic by two genera and eight species. According to Rosewater (1961), both genera *Pinna* and *Atrina* form analog-

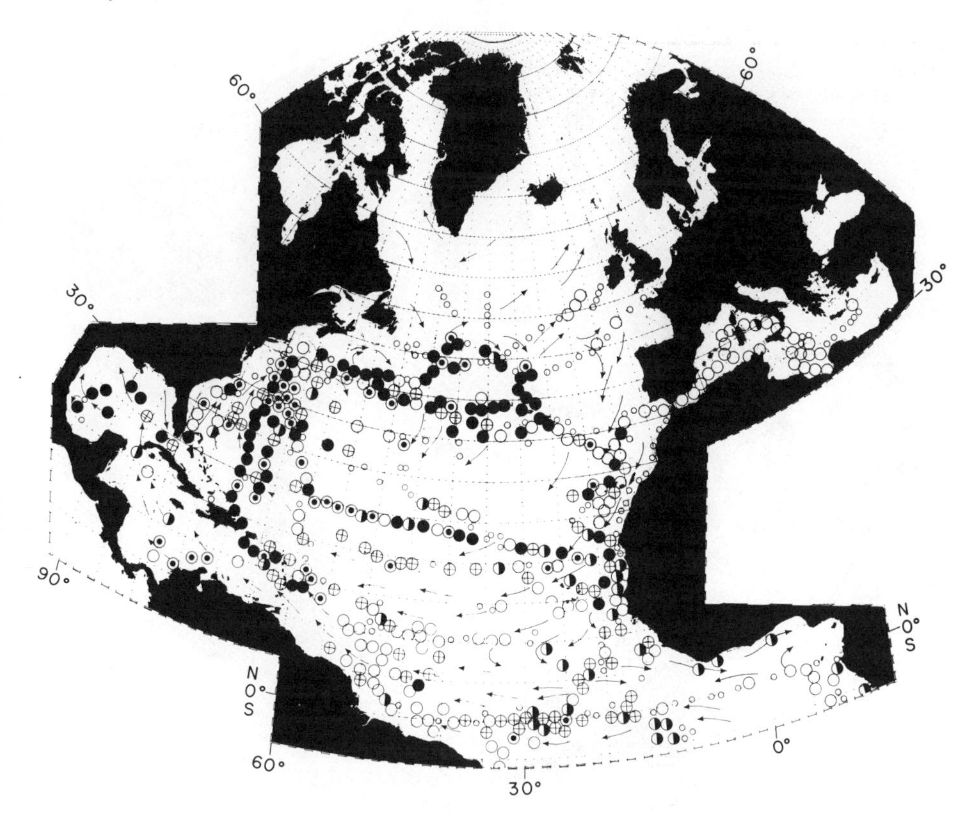

Text-Figure 5
Distribution of gastropod veliger larvae in the North and Tropical Atlantic Ocean. Large circles indicate collections containing larvae: large black circle, *Cymatium parthenopeum*; large half-blackened circle, *Charonia variegata*; large circle containing small black circle, both *Cymatium parthenopeum* and *Charonia variegata*; large circle containing cross, other species belonging to the family Cymatiidae; large open circle, other gastropod families. Small open circles are locations where gastropod larvae were absent. Samples were taken obliquely from approximately 100 m to the surface. Arrows indicate direction of surface currents.

ous species pairs in the North Atlantic, but the single species *Pinna rudis* is considered amphi-Atlantic in its geographic distribution. Larvae of the Pinnidae are recognized by their triangular-shaped shell valves (Ota, 1961); in life they have an unusual, large bilobed velum. The larvae of the family Pinnidae are found in open waters throughout the North Atlantic (Text-fig. 6).

Further work will be required rearing larvae captured at sea before most teleplanic bivalve veligers can be identified as to family or species. The importance of larval dispersal for establishing and maintaining the geographical limits of bivalve species is, however, already established.

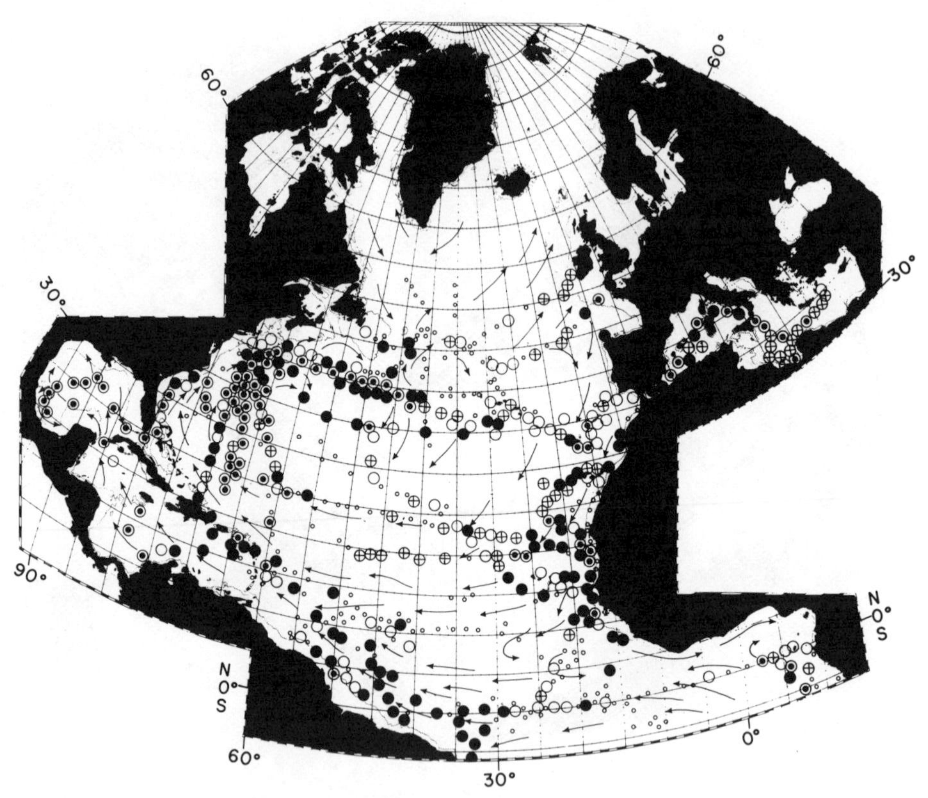

Text-Figure 6
Distribution of bivalve veliger larvae in the North and Tropical Atlantic Ocean. Large circles indicate collections containing larvae: large black circle, family Pinnidae; large circle containing cross, family Teredinidae; large circle containing small black circle, both families Pinnidae and Teredinidae; large open circle, other bivalve families. Small open circles are locations where bivalve larvae were absent. Samples were taken obliquely from approximately 100 m to the surface. Arrows show direction of surface currents.

The dispersal of certain echinoderm larvae over long distances has also been demonstrated (Text-fig. 7). Bipinnaria and brachiolaria larvae of asteroids were present in 23 percent of the open-sea plankton samples; echinoplutei and ophioplutei occurred in 22 and 12 percent, respectively. The plutei are endowed with long arms supported by a larval skeletal structure; the latter has much potential utility in identification of larvae (see Mortensen, 1921; Fell, 1967b).

Among the asteroids, a most remarkable teleplanic bipinnaria larva is that of the genus *Luidia,* which may be up to 24 mm in length. Of circum-Tropical and Temperate geographical distribution, three species of *Luidia* are known from both the eastern and western Atlantic (Downey, 1973).

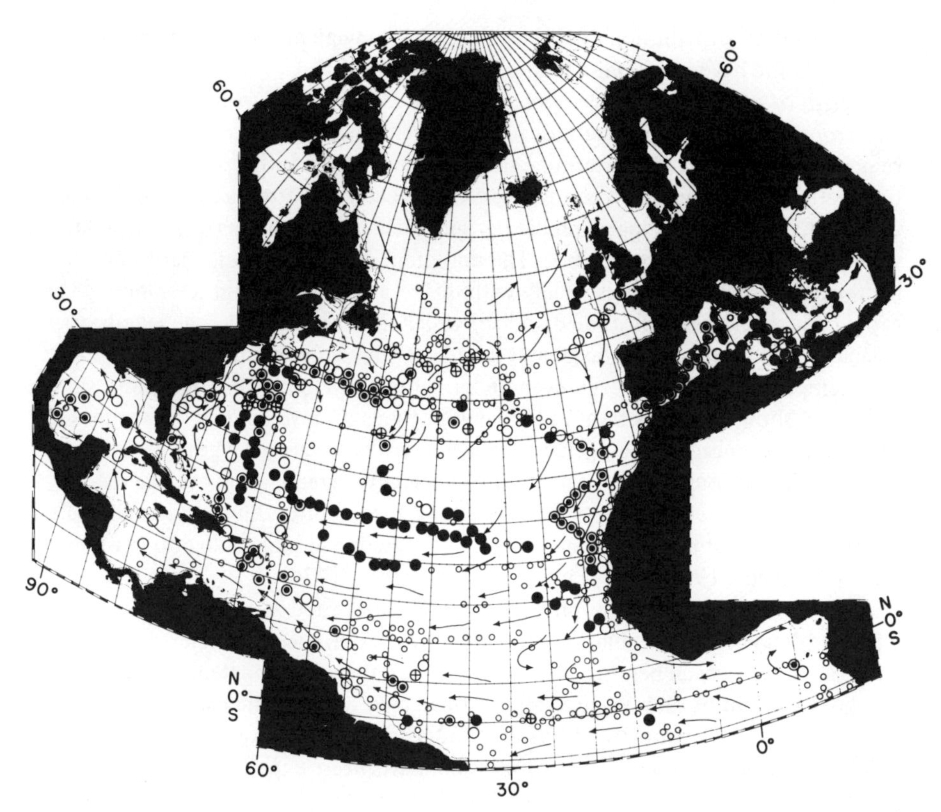

Text-Figure 7
Distribution of echinoderm larvae in the North and Tropical Atlantic Ocean. Large circles indicate collections containing larvae: large black circle, bipinnaria and brachiolaria larvae; large open circle, plutei larvae; large circle containing small black circle, plutei, bipinnaria, and brachiolaria larvae; large circle containing cross, auricularia larvae. Small open circles are locations where echinoderm larvae were absent. Samples were taken obliquely from approximately 100 m to the surface. Arrows show direction of surface currents.

The characteristic very long-armed echinopluteus larva of the ubiquitous tropical sea urchin *Diadema* is not infrequently found in the open sea, and the very large "*Auricularia nudibrachiata,*" more than 15 mm long, is a typical teleplanic larva of an unknown holothuroidian species (Chun, 1896).

In summary, the geographical distribution of the larval forms of biostratigraphical important taxa (i.e., gastropods, bivalves, and echinoderms) in the open Atlantic Ocean (Text-fig. 5-7) gives the most conclusive evidence yet compiled to suggest that pelagic larvae are dispersed over long distances. That this dispersal capability is important to the geographical range of the adults is demonstrated by the known wide geographic distribution of species that have a teleplanic larval stage; almost all in the Atlantic have an amphi-Atlantic distribution.

Length of pelagic larval development and current velocities in Warm-temperate and Tropical regions. The length of pelagic development and its relationship to the velocity of ocean currents determine the maximum distance that a larva can be transported. If pelagic life is too short or the velocity of currents too slow, larvae will be unable to cross zoogeographic barriers such as ocean basins or reach shoal water where settlement is possible. The minimum duration in which larval development can be completed varies widely with different species; but, at least among the gastropods that have been studied and for which data are available, the length of pelagic existence appears to be greater among certain Tropical species than in most Cold-temperate ones. However, even under controlled laboratory conditions, sibling larvae show large variations in the length of pelagic development, sometimes up to threefold or more. Siblings that have a greater-than-average length of development will have a longer opportunity for dispersal over greater-than-average distances. Physical factors (e.g., temperature) and the quantity, kind, and quality of food are important in determining the rate of growth and morphological development (Scheltema 1965, 1967; Pilkington and Fretter, 1970).

Between completion of morphological development and settlement there may be a further delay in the termination of pelagic life. Experimental studies show that a wide variety of coastal benthic invertebrates, even though competent to settle, can postpone the end of their pelagic stage until an environment amenable for post-larval existence is encountered (Wilson, 1952; Crisp, 1974; Gray, 1974; Scheltema 1974a). Included among those forms that have been experimentally shown to have settlement responses are gastropods (e.g., Scheltema, 1961), bivalves (e.g., Bayne, 1965), and echinoderms (e.g., Cameron and Hinegardner, 1974). However, knowledge on the delay of settlement has become too extensive to be considered in detail here and is adequately summarized in the articles cited.

The duration of larval development of 10 tropical teleplanic gastropod larvae has been estimated and related to the velocity of ocean currents in the North and Tropical Atlantic. Data on current velocities and the computed number of days required for trans-Atlantic dispersal by various routes are summarized in Table 3. If the times required for transoceanic dispersal are compared with the estimated periods of pelagic larval development (Table 4), it is shown that many contemporary larval species can be quite easily transported across the North Atlantic Basin. The estimates are naturally quite rough, and considerable variation may be expected in the values for both the current velocities and the length of larval life.

Gastropods are by no means the only forms with such a long duration of larval development. For example, settlement of larvae of the sea-star *Mediaster aequalis* has been delayed in laboratory culture for up to 14 months (Birkeland et al., 1971), and that of the polychaete *Spiochaetopterus* sp. for about an equal time (Scheltema, 1974b).

Experimental evidence on the long duration of pelagic development in teleplanic larval forms is particularly difficult to collect because of the long periods during

Table 3
Velocity of the Eastwardly and Westwardly Flowing Currents of the North and Tropical Atlantic Ocean[a]

Region	*Estimated Current Velocity (km/h)*	*Estimated Number of Days Required for Trans-Atlantic Drift*
North Atlantic Drift		
Bahamas to Azores	0.5 - 1.3	128 - 300
Bahamas to NW Africa	0.9	400
North Equatorial Current		
NW Africa to West Indies	0.9 - 1.2	128 - 171
South Equatorial Current		
Gulf of Guinea to Brazil	1.0 - 3.2	60 - 154
Equatorial Undercurrent		
Brazil to Gulf of Guinea	2.0	96

[a]Data from (1) Pilot Charts, U.S. Hydrographic Office; (2) direct measurements: Dickson and Evans, 1956, North Equatorial Current; Stalcup and Metcalf, 1966, Equatorial Undercurrent; and (3) drift-bottle data: Guppy, 1917; Bumpus and Lauzier, 1965; and W.H.O.I. files, courtesy of Mr. Dean F. Bumpus. (From Scheltema, 1971b.)

which larvae must necessarily be maintained in the laboratory. However, direct evidence from the geographic distribution of larvae in the ocean is compelling enough to make evidence from the laboratory unnecessary for many species.

Duration of pelagic life and the dispersal of larvae in Cold-temperate waters. Up to this point the discussion of long-distance larval dispersal has been centered wholly upon Warm-temperate and Tropical seas. To understand the role that the transport of larvae may play in Boreal regions, one must consider not only the bathymetry and surface circulation of the North Atlantic and the Norwegian and Labrador seas (see Lee, 1963; Collin and Dunbar, 1964), but also the relative frequency of pelagic and nonpelagic development in the fauna and the duration of the free-floating existence in species that have a planktonic stage.

Among the prosobranch gastropods, bivalves, and echinoderms, the number of species having phytoplanktotrophic pelagic larval development decreases sharply at high latitudes (Thorson, 1962; Ockelmann, 1958; Einarsson, 1948), apparently in response to the short period of phytoplankton production (Thorson, 1946). Sometimes, however, the type of development bears no obvious relationship to latitude; e.g., Radwin and Chamberlin (1973) concluded that "the mode of early development in the stenoglossa (= Neogastropoda) tends to follow phyletic lines, regardless of latitude or climatic conditions."

With a length of pelagic development of 4 to 6 weeks common to most Cold-temperate species (Thorson, 1961), and a current velocity of 10 to 20 cm/s

Table 4
Estimated Duration of the Pelagic Stage in 10 Gastropod Species Having Teleplanic Larvae

Species	*Estimated Minimum Days Required for Settlement*[a]	*Additional Days Larvae Delayed Settlement in Culture*	*Total Days Planktonic*
Cymatium nicobaricum	207	113[b]	320
Cymatium parthenopeum	155	138	293
Charonia variegata	219	57	276
Tonna galea	148	94	242
Tonna maculosa	198	–	198[c]
Phalium granulatum	107	–	107[c]
Thais haemastoma	62	28[b]	90
Philippia krebsii	67	7[d]	74
Smaragdia viridis	25	30	55
Pedicularia sicula	42	–	42[c]

[a]For method used to determine this value, see Scheltema, 1971b, p. 304.

[b]Species successfully held to settlement.

[c]No data on delay of settlement; this value probably an underestimate.

[d]Delay time appears to be an underestimate, since larvae seem to settle in response to surface of culture dish.

representative of surface waters in the Norwegian and Labrador seas (Lee, 1963; Collin and Dunbar, 1964), larvae may be transported 300 to 500 km (12 km/day). That this may well be an underestimate is shown by the data of Mileikovsky (1968, Table 1, p. 162) who records bivalve veligers 1,200 km (Text-fig. 3) and ophioplutei 925 km from the nearest coast.

The dispersal of larvae north of 50° north latitude in the Cold-temperate Atlantic Ocean has been recorded by work done at the Edinburgh Oceanographic Laboratory (1973) using a continuous plankton recorder. These data show that echinoderm larvae occur in large numbers and with great frequency in coastal areas, and decrease in number and frequency with increased distance from shore (Text-fig. 8). However, transport of echinoderm larvae extends into the central Boreal Atlantic Ocean, and other data show that bivalve veligers are also to be found thus widely dispersed.

Ekman (1953) ventures that "the distance between Europe and North America is divided up into smaller stages by the islands and the underwater ridges between Scotland, Iceland, Greenland and America . . .". Indeed, bounding both the Norwegian and Labrador seas, long continuous regions of continental shelf make the widespread dispersal of Temperate species with shorter pelagic development possible.

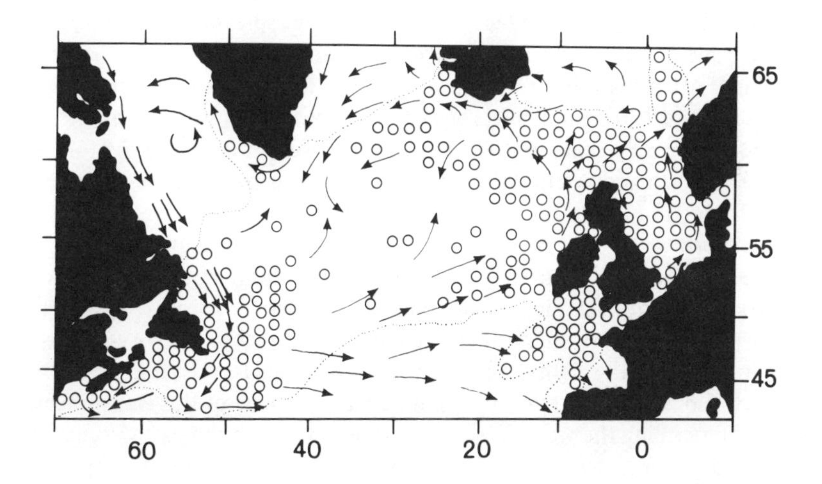

Text-Figure 8
Distribution of echinoderm larvae in the Cold-temperate North Atlantic Ocean. Each point records the occurrence of larvae in a square of 1° latitude and 2° longitude. Data are from monthly continuous plankton-recorder samples made from a depth of 10 m between 1958 and 1968. Stippled line denotes limits of area sampled; arrows show direction of surface circulation. (Modified from Figure 271, Edinburgh Oceanographic Laboratory, 1973.)

Significance of larval dispersal in the geologic past. Evidence from contemporary species of marine benthic invertebrates allows us to conjecture about the effectiveness of larval dispersal in past geological time.

The Late Cretaceous was a period of extensive epeiric seas. The most noteworthy of these from the paleobiogeographic standpoint was the Tethys Sea, which spanned the vast region between the Indo-Pacific and the eastern Atlantic Ocean. That this passage served as a dispersal route for Indo-Pacific fauna into the eastern Atlantic is abundantly documented in the paleobiogeographic literature (see Ekman, 1953; Fell, 1967b). It is easy to understand from the evidence of contemporary forms how larvae could be passively transported in the westward-flowing currents along the southern coast of the Tethys Sea (Fell, 1967b; Berggren and Hollister, 1974), and hence into the eastern Atlantic Ocean. Even species with larvae of relatively short pelagic development could readily have dispersed stepwise from the Indo-Pacific to the eastern Atlantic by the establishment of numerous successive populations.

After reaching the Atlantic, some of the fauna of Indo-Pacific origin, the "Tethyan" fauna, was further dispersed into the Caribbean off the western Atlantic Ocean. The ease with which Tethyan larvae could cross from the eastern to western Cretaceous and Neogene Atlantic depended upon the direction and velocity of the surface circulation and distance to be traversed between the continents at that time. According to the theory of plate tectonics and sea-floor spreading, it was between Late Triassic and Middle Jurassic (200 to 150 million years ago) that a narrow and shallow North Atlantic Basin first formed. Not until the Late Jurassic and Early Cretaceous (150 to 130 million years ago) did the southern continents of Africa and South America first separate. The sequence of events that led to the formation of the present North and South Atlantic basins is summarized and illustrated by Phillips and Forsyth (1972); important to note here is that by the end of the Cretaceous the North Atlantic was only 75 percent of its present width.

The paleocirculation in Late Cretaceous had similarities to contemporary Atlantic circulation: an anticyclonic gyre dominated the North Atlantic, with a branch continuing eastward across the northern epicontinental Tethys Sea. Westward-moving currents from the Tethys Sea formed a current analogous to the contemporary North Equatorial Current; but a considerable volume of this trans-Equatorial current, upon reaching the western Atlantic, flowed through the passage separating the North and South American continents (Berggren and Hollister, 1974, p. 170, fig. 16). Late Miocene and Early Pliocene closure of this passage (about 5 million years ago) served to intensify the northwesterly flowing Gulf Stream. Thus, during the Late Cretaceous and Early Cenozoic, currents were already favorable for transporting the larvae of Tethyan forms across the Atlantic to the West Indies, and even into the eastern Pacific.

Recent experimental tank studies of paleocirculation of Atlantic surface waters (Luyendyk et al., 1972) suggest that westwardly flowing Equatorial currents from the western Tethys to the Caribbean during Late Cretaceous had a velocity of 2 to

4 knots. Taking into account the smaller size of the Late Cretaceous Atlantic, it can be computed that such a current speed would transport a passively drifting larva across the equatorial Atlantic in 28 to 56 days, a time which is less than half that required to cross the present equatorial Atlantic (Table 2). Fell (1967a) suggests that many islands or shoals existed on the Mid-Atlantic Ridge during the Tertiary along that segment of the ridge which "follows the transatlantic flow or the South Equatorial Current." However, if estimates of Cretaceous current velocities are correct, it would not be necessary for most species to utilize such islands as stepping stones.

Kauffman (1975) believes that "circumferential migration" around marginal shelf areas by larval transport, if coupled with successive establishment of new populations, is an effective means of long-distance or even intercontinental dispersal. The extension northward of a Warm-temperate climate during the Cretaceous and Early Cenozoic (see Berggren and Hollister, 1974, for review) must also have made possible trans-Atlantic dispersal of warm-water species at high latitudes.

It is concluded that pelagic larvae during the Cretaceous or Early Cenozoic must have been fully as effective for long-distance dispersal as larvae of contemporary species. Indeed, in the geological past there may have been times that were more favorable than the present for widespread larval dispersal.

Rafting

Rafting has serious limitations as a means of dispersal for marine invertebrates; it can be successful only among those species able to remain attached to floating objects and to survive for long periods in the open sea. Most members of the level-bottom community are ill-adapted for such a "pseudo-pelagic" existence (see Thorson, 1957). It is fanciful to believe that benthic infaunal deposit feeders will have the slightest chance for dispersal by rafting. That some invertebrates are dispersed by attachment to drifting objects can scarcely be denied, however. Indeed, transport of marine species has long been known to occur upon objects afloat on the sea surface.

Rafting in contemporary species. Herbaceous plants and trees (Guppy, 1917), marine algae (Vallentin, 1895; Yeatman, 1962), and pumice from volcanic islands serve as floats for the dispersal of a variety of invertebrate organisms. For example, adult bryozoans and serpulid polychaetes are regularly carried out to sea upon detached and floating thallus algae (Text-fig. 9). In cold waters and at high latitudes, enormous kelps may carry moderately large species of gastropods and byssate bivalves in their holdfasts. Fell (1967a) has stressed the importance of "epiplanktonic" drift on algae to trans-Atlantic dispersal of certain echinoderms; he cites the common Indo-Pacific asteroid *Patiriella exigus,* which lacks a pelagic larval development, as an example of a species introduced into Saint Helena by rafting on the kelp *Ecklonia* upon which it lives.

Text-Figure 9
Organisms attached to the alga *Sargassum,* which floats in large mats in the Sargasso Sea. Many invertebrate species on this alga are habitual inhabitants and have taken on rafting as an exclusive way of life. Shown are the bryozoan *Membranipora tuberculata* Busk and the calcareous tubes of the polychaete worm *Spirorbis corrugatus* Mont.; both also have pelagic larvae.

Bivalves that are cemented to a surface (e.g., oysters, Stenzel, 1971) or that are attached by a byssus (e.g., mussels) can be rafted on almost any floating object. The shipworms (Teredinidae) are bivalves that burrow into wood and consequently can be dispersed in floating timber (Turner, 1966). *Lyrodus pedicellatus,* which has a pelagic larva of about 1 day, has nonetheless a circum-Tropical and Temperate distribution. Gastropod species too large to be rafted can attach egg masses to drifting debris, and the wide geographic distribution of the large volutid gastropods *Cymba* and *Adelonilon,* which lack a planktonic dispersal stage, may be explained by the dispersal of their egg capsules (Marche-Marchad, 1968).

Species most characteristically "epiplanktonic" are those known in nautical parlance as "fouling organisms" and are found attached to boat bottoms and dock pilings; examples are barnacles, hydrozoan coelenterates, serpulids, and bryozoans. The Lepadomorpha, or gooseneck barnacles, are often found on flotsam along open stretches of beaches after storms. Balanomorpha, or acorn barnacles, have also been widely dispersed by rafting in historical times on the hulls of ships. *Eliminius modestes,* now a common barnacle in Great Britain, was recently introduced there on ships from Australia (Crisp, 1958; Barnes et al., 1972). Some barnacle species have become so highly specialized that they are found exclusviely upon cetaceans and the carapaces of sea turtles, thereby living their entire life on self-propelled "rafts."

Knowledge of contemporary rafting of marine organisms is largely anecdotal, however, and based on scattered and coincidental observations made at sea. Seldom has any systematic attempt been made to study the importance of rafting on the geographic distribution of marine animals (see Yeatman, 1962). Whereas the observation has frequently been made that the amount of drifting algae originating from the coast generally decreases rapidly with increasing distance from shore, few quantitative data have been published. The occurrence of drifting debris and logs also is said to be greater near the shore and in shipping lanes offshore. Such a distribution of potential rafts suggests that much of this floating material is either never transported very far from shore or originates from jetsam cast overboard.

Evidence for rafting in fossil species. Many paleogeographers believe that rafting can adequately explain the distribution of fossil marine invertebrates. An example of probable rafting in past geological time is the dispersal of cheilostomatous bryozoans across the North Atlantic Ocean during the Paleocene, Eocene, and Oligocene epochs (Cheetham, 1960). Genera that showed the strongest proclivity for attachment to seaweed were also most commonly amphi-Atlantic in distribution. Of the 67 fossil bryozoan genera commonly attached or encrusted upon algae, 60 percent were inferred to be migrants; however, of the 51 genera with arborescent or erect zoaria that seldom are found attached to algae, only 33 percent appeared to be migrants. From these data, Cheetham (p. 251) concluded that there "is no basis for doubting that early Tertiary cheilostome faunas of the North Atlantic were dispersed by trans-oceanic currents," and that the reciprocal exchange of species between the eastern and western Atlantic occurred largely through transport on drifting algae.

Hallam (1967b) supposes that the apparent free migration of Cretaceous Inoceramid bivalves is entirely accounted for by rafting or a "pseudoplanktonic mode of life," and Westermann (1973) attributes worldwide distribution of the thin-shelled, plicate, Middle and Late Triassic bivalves *Daonella, Halobia,* and *Monotis* to "a nectonic [pseudoplanktonic] habitat with byssal attachment to floating objects." However, Kauffman (1975) concludes that it is not necessary automatically to assume that byssate bivlaves will always be widely dispersed by rafting. Although sometimes found attached to algae and floating wood, almost all contemporary forms with a byssus also have a pelagic larval stage, and veligers of some genera have been shown to be widely dispersed (see discussion on *Mytilus* and *Pinna* elsewhere).

One must conclude from the available evidence, I believe, that rafting, although not uncommon among certain kinds of marine invertebrates, is restricted to only a relatively small proportion of all species. On the other hand, the significance of planktonic larvae among the shelf faunae is apparent from the large proportion of benthic and sedentary species that have this kind of development (70 percent of all bottom invertebrates according to Thorson's estimate). Even among those species for which both larval transport and rafting are possible, the former will be the more common mode of dispersal, because (1) the total number of pelagic larvae produced by most bottom invertebrates is very high (from more than 100 in most prosobranch gastropods to 10^8 eggs/female/year in some bivalves), and because (2) the incidence of floating objects in the open sea or even in coastal regions is very modest in comparison to the number of larvae found there.

It appears that forms most likely to be transported by rafting include (1) fouling organisms and epiphytic species, and (2) wood-boring organisms such as teredinid bivalves and certain isopods. Most benthic infaunal species, and especially deposit-feeding organisms, are unlikely to be rafted.

The arguments presented in the previous section on the significance of larval dispersal in the geologic past are equally germane to rafting and need not be repeated here.

DISPERSAL AND THE TEMPORAL OR STRATIGRAPHIC DISTRIBUTION OF BENTHIC AND SEDENTARY MARINE INVERTEBRATES

In addition to biogeographical significance, another aspect of dispersal is its relationship to the longevity of species in geological time. It is proposed that there exists an inverse relationship between dispersal capability and the rate of evolutionary change and extinction.

Differences in Rates of Evolution Among Animal Species

The rate of phyletic change or modification of a species with time varies markedly among major taxa of organisms. Simpson (1944) showed in his now classic volume, *Tempo and Mode of Evolution,* that a striking difference exists in the rate of evolution between carnivorous terrestrial mammals and marine bivalve mollusks. He asserted (1953b, p. 37) that Late Tertiary beds, with ages on the order of 1 million years and upward, contain very few and usually only rather dubiously identified Recent mammal species, and that mammals living today are on the order of 0.5 to 1 million years old. Jackson (1974, p. 556, Table 2), on the other hand, found the mean age of 40 species of Caribbean Venerid bivalves to be approximately 6.5 million years, and Stanley (1973, p. 488, fig. 1) showed that Cenozoic Venezuelan marine molluscan species include 20 percent that extend back 15 million years. Among the marine prosobranch gastropod mollusks, it is known that a fair number of species extend into the Middle Miocene or even the Oligocene, between 10 and 30 million years ago. In the Tropical prosobranch family Architectonicidae there are at least three extant species, *Architectonica nobilis, Philippia krebsii,* and *Pseudomalaxis centrifuga,* that are found in the Middle and Early Miocene of Central America and Jamaica (Woodring, 1959). Zeuner (1958, p. 390) found that 63 percent of all marine molluscan species of the Lower Pliocene of Italy have survived up to the present time.

The problem of determining and comparing evolutionary rates is not a simple one, and its complexity has been summarized and critically discussed by Simpson (1944; see p. 10, Table I), and more recently considered by a large number of geologists interested in evolutionary problems. Although there is much controversy concerning evolutionary rates, that large differences do exist between different species cannot be questioned. Various explanations have been offered for these differences. Recently, Stanley (1973) proposed as a solution that "competition has . . . contributed to higher rates of taxonomic turnover among mammals than among bivalves by accelerating rates of extinction" in the former. Kauffman (1972, 1973) has suggested that evolutionary rates, both phyletic change and extinction, are related in Cretaceous marine bivalve mollusks to periods of transgression and regression, but that eurytopic and stenotopic species (i.e., species tolerating wide and narrow environmental changes, respectively) respond differently to such events. These arguments hinge largely on the kinds and intensity of natural selection. The differences in evolutionary rates cannot be wholly explained in this way.

Factors Determining Evolutionary Rate in Marine Species

Why do so many marine species evolve so slowly? What are the conditions which promote or prevent extinction, i.e., allow the continuation or bring about the end of a lineage? Part of the answer to these two queries can best be sought in the study of contemporary species.

A three-dimensional model in space and time may help clarify our questions. In Text-figure 10, each disc represents a population that is separated in space or time from other populations of the same species. The horizontal plane *ABC* represents two-dimensional space (e.g., the sea floor); *A, B,* and *C,* respectively, are three

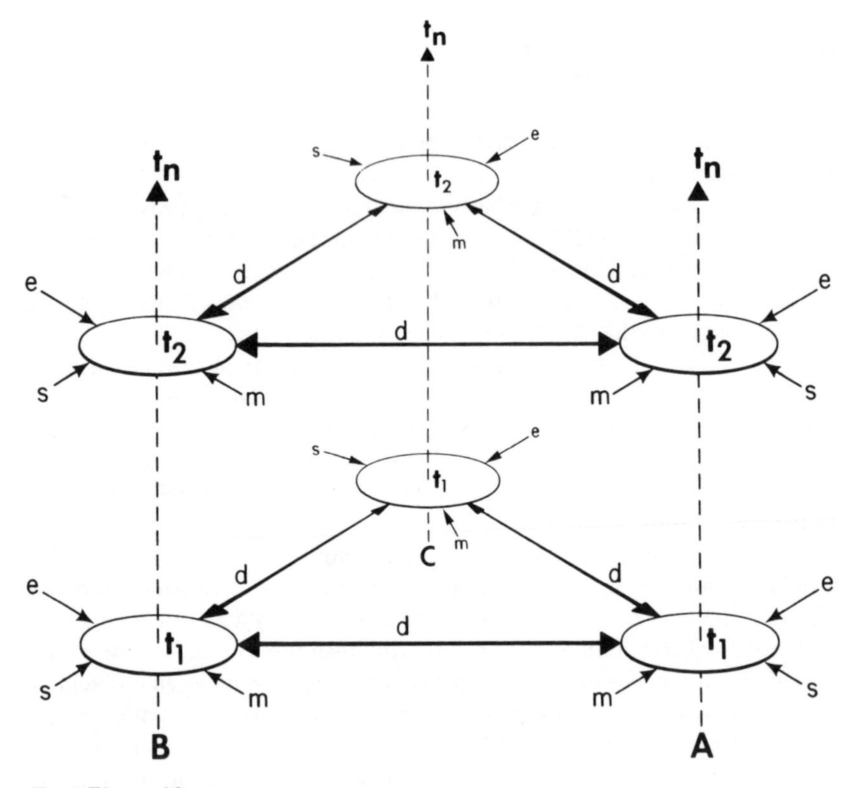

Text-Figure 10
Diagram showing continuity of gene flow through space and time among three geographically separated populations, *A*, *B*, and *C*. Lines *d* show genetic exchange in either direction by dispersal of pelagic larvae (or rafting) between populations; with an increase in number of populations, the maximum routes for gene flow will increase by $[(X - 1)X]/2$, where *X* equals the number of populations, although the realized number of paths for gene flow will almost always be less. Each population is affected by natural selection (*e*, factors of the physical environment; *s*, biological interactions) and spontaneous natural mutations (*m*). The vertical interval from t_1 to t_2 indicates a single generation, and t_n the nth generation.

populations separated by an ecological or biogeographic barrier. The vertical dashed axes represent the progression of time and lineage of each population; t_1 designates the population at some arbitrary time and t_2 the next succeeding generation, and so forth to t_n.

Each population will be acted upon by change in the physical environment (*e*) and by intra- and interspecific interactions (*s*), such as competition of different sorts or predator-prey relationships. Together these constitute natural selection; i.e., they bring about differential survival to the members of each population. Clearly, continuous interplay occurs between the physical and biological environments, and the two (*e* and *s*) are closely coupled to one another.

The rate, magnitude, and kind of changes that affect contemporary marine benthic populations may vary widely from place to place and time to time. If the fluctuations in the physical factors are large and aperiodic, the environment is regarded as unpredictable and is said to favor eurytopic species with high genetic diversity (see Sanders, 1968; Bretsky and Lorenz, 1969; Grassle, 1972). More predictable physical environments are said to favor stenotopic species that are better adapted to biological interactions such as competition or predation, and that also have a more restricted geographical range.

Selection, in a stable and homogeneous environment, is believed to reduce genetic variability but there are also mechanisms that tend to counteract this tendency (e.g., pleiotrophy, the capacity of a gene to affect several characters, and epistasis, the interaction between nonallelic genes). New genetic variation in a population can arise from two different sources: (1) it can originate from spontaneous natural gene mutations (*m*, Text-fig. 10); (2) it may be introduced by gene flow or the exchange of genetic information between populations by means of pelagic larval dispersal (*d*). To result in successful genetic exchange, however, an immigrant must survive to reproduce. If natural selection is too strong, high mortality will result among the immigrant postlarvae, and genetic exchange and the introduction of new genetic material into the recipient population will not occur. Levins (1964), on theoretical grounds, concluded that "the adaptive significance of gene flow is that it permits populations to respond under natural selection to long-term widespread fluctuations in the environment while damping the response to local ephemeral oscillations."

The genetic diversity within a species depends largely upon the extent of its geographic range, the amount of environmental heterogeneity in which populations are found (Powell, 1971), and probably the number and also to a degree the size of its populations and the amount of gene flow between them.

The model in Text-figure 10 shows how a dynamic equilibrium between natural selection and gene flow can determine the "flux" of genes through space and time (i.e., along *d* and *t*) and, consequently, the rate of phyletic change and extinction of marine species, although it oversimplifies and does not take into account random fluctuations in gene frequency within populations (e.g., genetic drift or the "founder effect").

To amplify my point, consider three hypothetical species, each with a different dispersal capability, and the consequences that these differences will hold for each.

1. Species with a very restricted geographic range and lacking a pelagic dispersal stage: Such a species is likely to have a restricted genetic diversity as a result of natural selection in the absence of gene flow. However, preliminary evidence suggests that some marine species may maintain genetic diversity even in a very stable environment such as the deep sea (Schopf and Gooch, 1972; Gooch and Schopf, 1973); these conclusions are still quite controversial (Doyle, 1972; Grassle, 1972). Spontaneous natural gene mutation will be the sole source for obtaining new genetic diversity. Under relatively benign and stable environments there will be little tendency for such species to change. If, however, the pressure of natural selection is very low, and the kind or intensity of selection changes but very slowly, it is

possible that gradual morphological transformation or phyletic speciation may occur (see Hecht et al., 1974, for discussion and critique). But it is also to be expected that the species will be particularly susceptible to extinction because rapid change or catastrophic events in its physical (*e*) or biological (*s*) environment may lead to the destruction of all members of the species at once. This case can be represented in the model (Text-fig. 10) by a single population when *d* is equal to zero.

2. Species with a wide geographic range but a restricted capacity for larval dispersal: Here the dispersal of larvae is sufficient for maintaining or even increasing the geographical range of the species, but too low or infrequent to result in significant gene flow or to prevent divergence between populations. The value for *d* will be very low. In the case where natural selection is intense and discriminates against immigrant larvae originating from other populations, gene flow will be even further reduced, and allopatric (or geographic) speciation may eventually occur. In Text-figure 10, three allopatric species, *A, B,* and *C,* would result.

3. Species with a wide geographic range and marked capacity for long-distance larval dispersal: Here larval transport (*d*) between geographically separated populations is high and the potential for gene flow is correspondingly great. Continuous gene flow between various populations of the species is possible; consequently, geographic speciation is unlikely to occur. At the same time, the total amount of genetic variation will be increased owing to a characteristic or perhaps unique natural selection that occurs in different regions of the species range (Grassle, 1972; Bretsky and Lorenz, 1969; Powell, 1971).

The first hypothetical species, although it may have a restricted temporal range resulting from early extinction, is of little value for correlating fossil assemblages (i.e., as guide or index fossils) because of its restricted geographic distribution. This example will not be further considered here.

The second case is an example of allopatric or geographic speciation, which results in a multiplication of species (or lineages) and must be distinguished in the fossil record from plyletic speciation or the temporal change and morphological transformation within a single lineage.

The third example consists of a single lineage represented by many populations and meets two of the three conditions required of a good guide fossil, abundance and wide geographic range. However, whether such a species will also satisfy a third requirement, vertical or temporal restriction in the geological record, needs yet be considered.

Dispersal Capability and the Evolutionary Rate of Marine Species

Species existing today that approximate the latter two proposed hypothetical cases can be examined to see how they fit the theoretical consequences of different dispersal ability. The tropical gastropod *Cymatium parthenopeum* is an example of a species with a marked capacity for long-distance larval dispersal. Although the concentration of the larvae in the Atlantic is very low, the volume of water in which they are found is very large. This volume may be computed for each major trans-Atlantic current by delimiting their surface areas and considering only the upper 50 m. Using this method, the volume of the North Atlantic Drift is estimated at 2.45×10^5 km^3. The average concentration of *Cymatium parthenopeum* larvae calculated from 81 plankton samples on a transect along the North Atlantic Drift was found to be 0.0036 larvae/m^3. Consequently, a total of 8.6×10^{11} larvae are estimated to exist in the North Atlantic Drift waters. Moreover, this number of larvae approaches a steady state since there appears to be no seasonal differences; as veligers are removed from the eastern Altantic, new ones are added from the western Tropical Atlantic. To compute the dispersal frequency (*d*) of *Cymatium parthenopeum,* given the fact that the larvae can delay settlement sufficiently long (Table 4), it is now only necessary to know the number of days required for a larva to be carried from its source in the western Tropical Atlantic to its landfall in the eastern Atlantic (e.g., the Azores). Drift-bottle data show that it requires between 125 and 350 days to cross this distance. A value midway between these two is 238 days, or approximately two-thirds of a year. The total number of *Cymatium parthenopeum* larvae transported across the North Atlantic can then be estimated at 13.2×10^{11} larvae/year.

The frequency at which some other larval species occur in the same plankton station series across the Atlantic Basin can be compared to that of *Cymatium parthenopeum* to give an index of their relative rate of transoceanic dispersal (*d*). Four additional species of gastropods, *Tonna galea, Phalium granulatum, Thais haemastoma,* and *Cymatium nicobaricum* are, like *Cymatium parthenopeum,* represented by adult populations in both the eastern and western Tropical and Warm-temperate Atlantic. The relative numbers (height of bars) and frequency of occurrence (number of bars) of larvae of four species are illustrated in Text-figure 11 for each of the three major east–west surface currents of the equatorial and North Atlantic Ocean. *Cymatium parthenopeum* is found most often and at the highest concentrations; *Thais haemastoma* and *Cymatium nicobaricum,* the latter not shown in Text-figure 11, are found least often and then only in very small numbers. *Cymatium nicobaricum,* although not uncommon in the Gulf Steam, is known from only a single record in the mid-North Atlantic Drift and not at all in the North and South Equatorial Currents.

Among the five gastropod species, there is a remarkable direct correspondence between the estimated frequency of larval dispersal and the degree of morphological similarity of eastern and western Atlantic populations (Table 5). Adults of species estimated to have a high frequency of larval dispersal show little or no morphological differences between the eastern and western Atlantic. Conversely, species having a restricted larval dispersal are usually represented by different subspecies (Scheltema, 1972b). The data suggest that veliger larvae are successfully bringing about gene flow between populations.

The five species from our examples now may be related to the hypothetical situations previously proposed. Thus *Thais haemastoma* and *Phalium granulatum* most

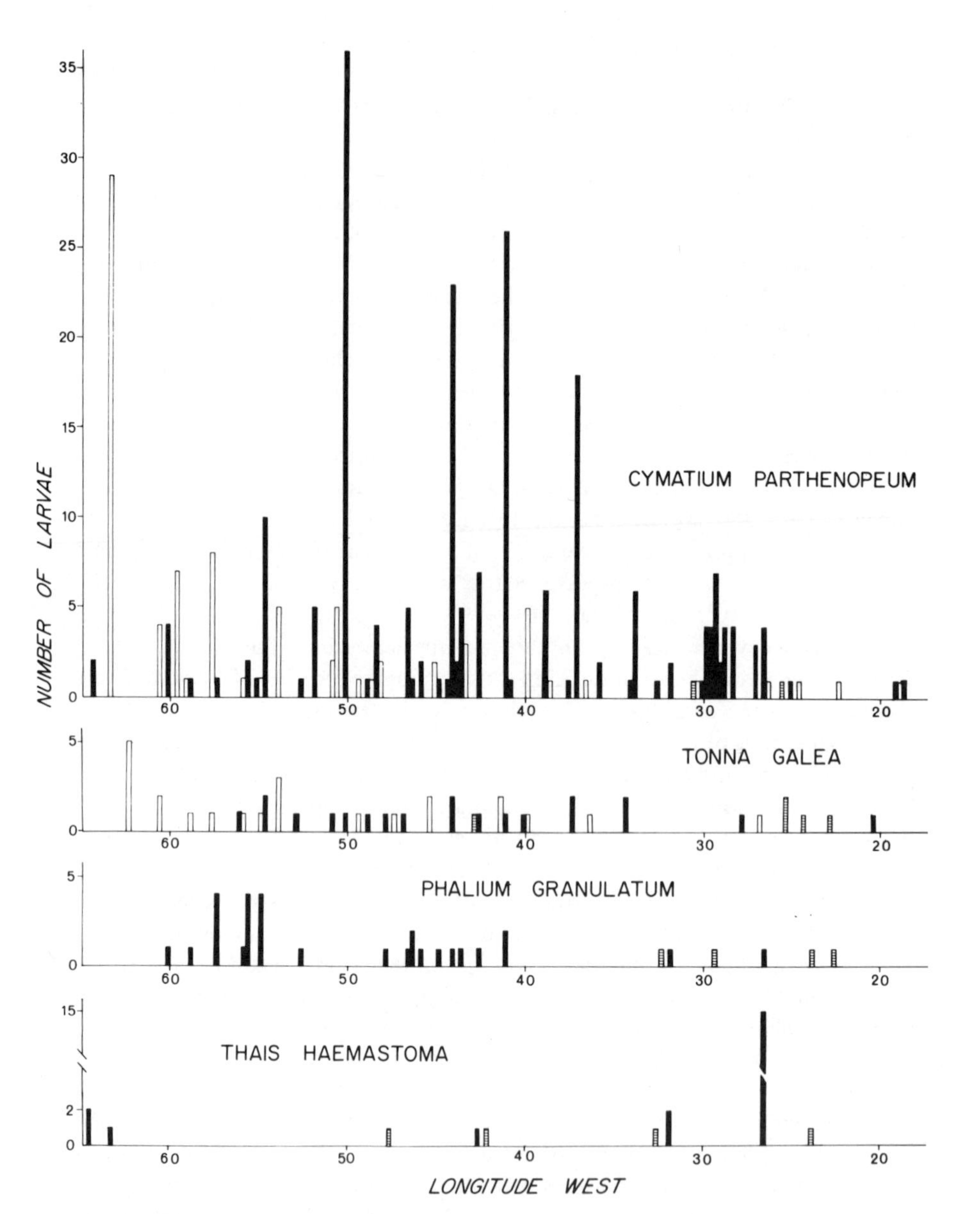

Text-Figure 11
Number of larvae from four species of gastropods found in plankton samples taken along three major east–west trans-Atlantic currents. Black bars, samples from the North Atlantic Drift; white bars, samples from the North Equatorial Current; horizontally lined bars, samples from the South Equatorial Current. Degrees of west longitude shown along abscissa (after Scheltema, 1972b).

Table 5
Estimated Number and Percent Occurrence of Veligers from Five Gastropod Species Found in All Samples Taken from the Three Major East–West Transoceanic Surface Currents of the North and Equatorial Atlantic Ocean[a]

Species	*% Occurrence (no. of samples)*	*Estimated No. of Larvae*[b]	*Remarks*
Cymatium parthenopeum	26.3	18.16×10^{11}	One Atlantic subspecies; probably much gene flow
Tonna galea	19.0	3.63×10^{11}	No subspecies; considerable gene flow likely
Phalium granulatum	12.7	1.14×10^{11}	Two Atlantic subspecies; some restriction in gene flow probable
Thais haemastoma	4.9	0.51×10^{11}	Four "subspecies" described in North Atlantic; considerable restriction in gene flow between populations likely
Cymatium nicobaricum	0.5	0.04×10^{11}	Probably only an infrequent colonizer in the eastern Atlantic

[a] After Scheltema, 1972b.

[b] The data are the combined number of larvae found in the North Atlantic Drift, North Equatorial Current, and South Equatorial Current and computed as explained in Scheltema (1971b).

closely resemble the second hypothetical case of species with a wide geographic range but a restricted capacity for dispersal; *Cymatium parthenopeum* and *Tonna galea,* on the other hand, present instances of species with wide geographic range and marked capacity for long-distance dispersal (case 3). A high capacity for larval dispersal indeed seems to inhibit allopatric speciation and the formation of new lineages.

But what about phyletic speciation, i.e., changes along a single lineage in time and its relationship to larval dispersal? What rate of evolutionary change do species with a high dispersal capability show? Are slow evolutionary rates and a tendency to resist extinction linked with dispersal capability and wide geographic range?

Evidence bearing on these questions is difficult to obtain; two kinds of information are required. First, it must be known if a species has a long pelagic larval development; second, the history of the species must be recorded in the fossil record. Some examples of contemporary forms with pelagic larvae and a record into the past can be drawn from the previous discussions on long-distance dispersal.

For example, the dispersal capability of the family Pinnidae has already been suggested from the geographical distribution of their larvae (Text-fig. 6). More than half the samples containing bivalve veligers include larvae belonging to this family. Contemporary Pinnidae – almost world-wide in their distribution – occur not only in the Temperate and Tropical Atlantic and Mediterranean, but also in the Indo- and eastern Pacific (Rosewater, 1961) and are represented worldwide by approximately 20 modern species. In the geologic record the family first appeared in the Late Paleozoic, and the genus *Atrina* in the Carboniferous and *Pinna* during the Jurrasic (Turner and Rosewater, 1958). Most species are large and fragile and seldom represented in museums by adequate series of modern forms, nor from the fossil record by more than fragments. Considerable ecotypic variation also occurs. Although series of fossil shells well enough preserved to describe phyletic change through a significant span of time have not been found, it is obvious that the family Pinnidae has a wide geographic range, an efficient means of disperal, and a long paleontological record.

Another example is the prosobranch gastropod family Cymatiidae. *Cymatium parthenopeum* is a species with a circumtropical range. Some subspecific differentiation has occurred in the Pacific, e.g., from Japan and the eastern Pacific, but the Indo-Pacific and the entire Atlantic are represented by a single form (Beu, 1970a). The dispersal capability of *Cymatium parthenopeum* and other Cymatiidae is illustrated by the wide geographic distribution of their veliger larvae (Text-fig. 5). The species *Cymatium parthenopeum* is known to extend to the Middle Miocene of southern Italy, and its dispersal capability is apparently linked to species longevity. Some other Cymatiidae with teleplanic larvae have changed but little since the Middle Miocene or earlier (see Woodring, 1959).

Data for prosobranch gastropods are summarized in Table 6, which shows only those species known to have teleplanic larvae and whose dispersal over long distances has been previously described (Scheltema, 1971b). These early (if perhaps numerically inadequate) results show that 53 percent of the contemporary species listed in Table 6 already existed in the Miocene in their present form, not evolving enough since then to be differentiated beyond the subspecific level. About 23 percent of the species belong to lineages that have changed sufficiently so that they are now regarded as different from their Miocene counterparts. Finally, only 15 percent of the species seem to have evolved at the generic level in the same time interval. Only one recent species with a teleplanic larva was not found represented in the fossil record, and this form is very fragile and probably would not preserve.

Table 6
Geographic and Stratigraphic Distribution of Contemporary Tropical and Warm-Temperate Gastropod Species Known to Be Dispersed over Long Distances by Teleplanic Veliger Larvae

	Geographic Distribution	*Geologic Range*
Cymatium parthenopeum (von Salis)	Subsp. *parthenopeum* *Western Atlantic*: N.C. south to Bahia, Brazil. *Eastern Atlantic*: Azores; Mediteranean south discontinuously to S. Africa. *Indo-Pacific.*	S. Italy (Calabria) Miocene to Pleistocene (Watson, 1886); Florida, Caloosahatchee formation, Pliocene (Weisbord, 1962; Beu, 1970a).
Cymatium nicobaricum Rõding	*Western Atlantic*: Florida; West Indies; Mexico; Honduras; Bahia, Brazil. *Eastern Atlantic*: Madeira, Canary Islands. *Indo-Pacific.*	Caimito formation, Panama; Oligocene (as *C. ?ogygium*) (Woodring, 1959). Probably same lineage.
Charonia tritonia (L.)	Subsp. *variegata* *Western Atlantic*: Bermuda south to Santos, Brazil. *Eastern Atlantic*: Cape Verde, Canary Islands; Portugal; Mediterranean.	Cabo Blanco, Venezuela, Pliocene?, genus known from Oligocene of Europe (Beu, 1970b), polytypic sp. closely related to *C. lampus.*
Architectonica nobilis Rõding	*Western Atlantic*: Cape Hatteras to Bahia, Brazil. *Eastern Atlantic*: Saõ Vincent, Cape Verde; Ivory Coast; Senegal; Saõ Tomé; Gabon; to Congo.	Early Miocene of Venezuela, Brazil, Haiti, Florida, Ecuador, Peru. Also in fossil record through Pleistocene throughout tropical western Atlantic and eastern Pacific (Woodring, 1959).
Pseudomalaxis centrifuga Monterosato	*Western Atlantic*: Florida to Bahia, Brazil. *Eastern Atlantic*: Mediterranean, Madeira, Canary Islands; Senegal; Saõ Tomé.	Bowden formation, Jamaica, Miocene; Calabria, Pliocene? (Woodring, 1959).
Philippia krebsii (Mõrch)	*Western Atlantic*: Cape Hatteras, Mexico, Antilles; tropical S. America. *Eastern Atlantic:* Canary, Cape Verde, Ascension and St. Helena islands.	Bowden formation, Jamaica, Middle Miocene; Dominican Republic, Miocene (Woodring, 1928; Dall, 1890; see Robertson, 1973).

Table 6 (Continued)

	Geographic Distribution	*Geologic Range*
Phalium granulatum (Born)	*Western Atlantic*: N.C. to Bahia, Brazil; *Eastern Atlantic*: Mediterranean, Azores, Madeira, NW Africa.	Culebra formation, Panama, Early Miocene (as *Semicassis aldrichi*) (Woodring, 1959). Also found in Jamaica and Venezuela.
Thais haemastoma (Say)	*Western Atlantic*: N.C., Gulf of Mexico, Trinidad, to Uruguay. *Eastern Atlantic*: NW Africa to Senegal and Congo.	Gatun formation, Panama, Middle Miocene (Woodring, 1959).
Tonna galea (L.)	*Western Atlantic*: N.C., Antilles, Brazil. *Eastern Atlantic*: Senegal to Angola, Mediterranean.	Genus extends probably into Miocene (as *Malea*); exceedingly variable in color, size, sculpture, and shape as determined from large series; shells fragile and seldom preserved (see Turner, 1946; Woodring, 1959, p. 208).
Tonna maculosa (Dillwyn)	*Western Atlantic*: N.C., Trinidad to Brazil. *Eastern Atlantic:* West Africa (?). *Indo-Pacific.*	
Bursa granularis (Röding) (=*B. cubaniana* Orbigny)	*Western Atlantic*: Florida, West Indies. *Eastern Atlantic*: West Africa (?). *Indo-Pacific.*	Miocene, Gatun formation, Panama, contains forms closely resembling this species (Woodring, 1959), and probably belonging to the same lineage or species complex.
Pedicularia siculus Swainson	Subsp. *decussata* (Gould) found in *Western Atlantic*: West Indies and N.C. Subsp. *suculus* (Swainson) found in *Eastern Atlantic*: Azores, Mediterranean.	No fossil record; shell fragile and probably therefore not preserved.
Smaragdia viridis (L.)	*Western Atlantic*: Columbia, Yucatan, West Indies, Florida, Bermuda. *Eastern Atlantic*: Madeira, NW Africa, Mediterranean.	Bowden formation, Jamaica, Mid-Miocene (Woodring, 1928).

These data suggest not only that species having teleplanic larvae have a wide geographic range but also that they belong to lineages that tend to evolve slowly.

If there is indeed a relationship betweeen larval dispersal, geographic range, and longevity of species, the question arises why this should be so. First, species with a large range and with pelagic larvae will tend to have greater genetic diversity. Extinction occurs when populations become too narrowly specialized, a condition that may result from natural selection in a relatively unvarying or stable environment (when *e* and *s* are approximately constant). Eurytopic species should tend to have greater genetic diversity because they live in temporally and spatially heterogeneous environments. Stenotopic forms, on the other hand, should tend toward less genetic diversity since they exist in a spatially homogeneous environment. Indeed, eurytopic forms usually have large populations, wide geographic ranges, and a means of widespread dispersal (see Grassle and Grassle, 1974). Stenotopic forms, however, tend to have more restricted ranges and a restricted capability for dispersal.

Second, species with a wide geographic range and pelagic larval development are not likely to become the victims of localized catastrophic events. Even if a single population becomes extinct, the species as a whole is little affected, and the region of catastrophe may soon be repopulated by immigrant larvae. Species with restricted range and lacking pelagic larvae are less likely to survive local catastrophic events or to repopulate barren regions.

A relationship between wide geographic range, a tendency to resist allopatric speciation, and a resilience against extinction have been demonstrated for marine bivalve mollusks (Jackson, 1974). It was predicted that shallow-water, low-diversity, eurytopic-species associations should be evolutionarily more stable than deeper-water ($>$ 1 m), high-diversity, stenotopic-species associations. This hypothesis was supported by the bivalve fossil record.

Clearly, much additional data will be needed to support the hypothesis that rates of evolution and extinction are closely linked with dispersal capability and hence to the possession or lack of a pelagic larval stage. In particular, information on the type of development of fossil species is required. Not only must the role of pelagic larvae be related to evolutionary rate and extinction, but also, conversely, it must be shown that species with nonpelagic development and restricted range ordinarily have a greater extinction rate.

A method whereby the type of development, pelagic or nonpelagic, can be determined from well-preserved fossil gastropod shells is accomplished by an examination of the protoconch or larval shell at the apex of the spire (Thorson, 1950, p. 33; Text-fig. 12). The protoconch characters that may be used to distinguish pelagic from direct development are summarized by Robertson (1974). The method has been recently applied to determine mode of larval development in Miocene, Pliocene, and Pleistocene fossil gastropods from the western and Indo-Pacific Ocean by Shuto (1974). A similar technique for contemporary bivalve mollusca has also been described by Ockelmann (1965) and attempted by Labarbera (1974) with five Miocene bivalve species with the use of a scanning electron microscope.

Biological evidence has recently begun to become available that bears on the relationship of pelagic and nonpelagic development to the genetics of marine gastropods and bivalves (gastropods: Berger, 1973; Snyder and Gooch, 1973; bivalves: Koehn and Mitten, 1972; Milkman et al., 1972; Levinton, 1973; Ayala et al., 1973; Boyer, 1974). This work does not yet allow a definitive conclusion about the relative amount of genetic diversity between species having and lacking pelagic larvae, but it does already suggest that both dispersal and natural selection must play a role (Struhsaker, 1968). The relative importance of dispersal versus selection will probably prove to differ in different species and genera, and will depend on magnitude and rates of change in selective processes and the changes in effectiveness of larval dispersal, i.e., in current patterns and velocities and distances of land masses from one another in past geological time (Text-fig. 10). To place larval dispersal into the context of evolution over geological time is, however, the problem left to the paleoecologist and biostratigrapher.

CONCLUSIONS

The evidence from marine invertebrate organisms suggests that dispersal capability is related directly to geographical range and inversely to the rate of phyletic change and species extinction. Hence species with a high dispersal capability show a tendency for a large geographic range, but a concomitant decrease in (1) allopatric or geographic speciation, (2) the rate of phyletic change, and (3) the likelihood for extinction. Those attributes that define species best suited for use in regional correlations are wide geographic range, rapid evolution and consequent temporal restriction, wide tolerance to the physical environment resulting in preservation over a large variety of facies, and great abundance so as to be readily found in the fossil record (Kauffman, 1975). Ultimately, it will be the relationship between the dispersal rate d and the magnitude of natural selection e, s that will determine the distribution of species in space and time (Text-fig. 10). Unfortunately, however, abundant, physically tolerant species with large geographic ranges (i.e., eurytopic species) are also forms unlikely to evolve very rapidly (Carlson, 1960, p. 89). Herein lies a paradox for the biostratigrapher!

Text-Figure 12
Protoconchs of two contemporary species of prosobranch gastropods having a pelagic larval development: (a) *Drupa nodulosa*, a species known from both the eastern and western tropical Atlantic coasts; teleplanic larvae from the South Equatorial Current have been identified by comparison of their shells with this protoconch. Specimen from Cabo Rojo, Puerto Rico; (b) *Litiopa melanostoma*, a species found exclusively inhabiting the drifting alga *Sargassum*. Veliger larvae with shells identical to this protoconch are common in plankton of the Sargasso Sea. Specimen from sargassum weed taken west of Bermuda.

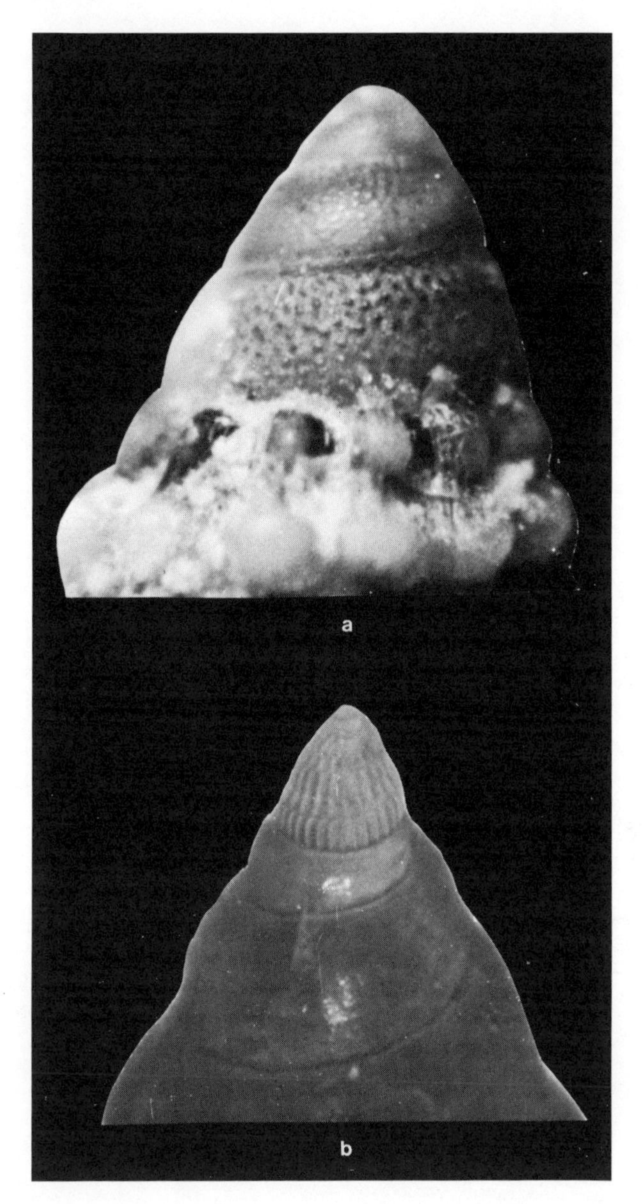

Acknowledgments

I should like to thank my wife, Amelie, for her encouragement and for critically reading the manuscript, and the editors, E.G. Kauffman and J.E. Hazel, for their suggestions and careful reading of the text. My research assistant, Isabelle Williams, and students, especially Alison Ament, have also read portions of the manuscript and made constructive suggestions. This article was written while I was a guest investigator at the Duke University Marine Laboratory, Beaufort, North Carolina, and during the tenure of grant 40144 from the National Science Foundation (Biological Oceanography). I am grateful to the director, Dr. John Costlow, of the Duke Marine Laboratory and to the National Science Foundation for their support.

Evolutionary Rates and Biostratigraphy

Erle G. Kauffman **United States National Museum**

INTRODUCTION

Detailed correlation of geological and biological events preserved in sedimentary rocks, and thus refined analysis of earth and life history, are highly dependent upon a system of precisely defined biostratigraphic zones with short time duration and broad, rapidly acquired, paleogeographic distribution. We ideally seek ways to construct the finest possible zonation, while maintaining regional application, by the use of two evolutionary factors: (1) differential rates of evolution within and between lineages, and (2) differential points of origin and extinction of taxa among lineages, reflecting varying evolutionary response by diverse organisms to natural selective factors and environmental perturbations. Thus the most refined biostratigraphic systems are those based on the most rapidly evolving groups of organisms (e.g., ammonites, mammals), and/or those constructed from massive range-zone data for diverse organisms arranged into complex assemblage zones, which emphasize numerous concurrent origin and extinction points instead of short temporal ranges in zonal construction. The most refined biostratigraphic systems combine these two factors in selecting rapidly evolving and widely dispersed organisms for the construction of complex assemblage zones; e.g., Kauffman (1970) documents regional composite assemblage zones in the American Western Interior Cretaceous ranging from 0.08 to 0.5 million years in duration (measured radiometrically), using this method.

Inasmuch as biostratigraphy is still the primary tool of dating, zoning, and correlating sedimentary rocks throughout the world, continued reevaluation of existing concepts and methods used in the construction of biostratigraphic systems is critical. In particular, we need to develop more efficient ways of constructing and applying highly refined zonal schemes in new areas of exploration (e.g., in deep-sea sediments), and to overcome some of the inherent problems in existing biostratigraphic systems.

Biostratigraphic systems that are largely based on one or a few lineages of rapidly evolving organisms, as are many ammonite zonations extant today (e.g., Cobban, 1951, 1958; Gill and Cobban, 1966; Imlay, 1971; Callomon, 1964; Mouterde et al., 1971), have certain inherent problems. First, the system is self-limiting; it can be no more refined than the maximum evolutionary rates of the one or few lineages emphasized. Ideally, this would be determined by an extremely detailed population systematic study of collections which were so closely spaced stratigraphically and temporally that within-species variation was mainly observed in comparing stratigraphically adjacent populations, and all species or subspecies level events were documented (e.g., Kauffman, 1970, Text-fig. 4). Comparison of species and subspecies events with a radiometric scale would allow evolutionary rates (the number of new taxa added in a lineage per unit time) to be determined. Where this type of study has been approached, it has clearly shown that evolutionary rates vary *within* lineages, commonly in response to the degree to which regional environmental stress is developing (Kauffman, 1970, 1972). This method is extremely time consuming; theoretically, all members of any lineage should be so analyzed, even though only certain portions of the evolutionary sequence are comprised of taxa whose evolutionary turnover is rapid enough to allow them to be used for refined zonation and correlation. In practice, more subjective and probably less refined taxa are commonly erected within lineages and used in biostratigraphic zonation, and the detail possible by recognizing maximum evolutionary rates within a lineage is never really achieved. A third drawback of this type of biostratigraphic system is that by concentrating on a single lineage, or a few lineages of organisms, the entire biostratigraphic system may lead to erroneous correlations *if* these lineages are strongly affected in their local occurrence by unrecognized paleoenvironmental factors. A biostratigraphic system based on only one or a few related lineages lacks a system of cross-checks for the degree of environmental influence on occurrence; such a system of checks exists only where varied lineages, representing diverse adaptive strategies, are used within the context of assemblage zonation.

Complex assemblage zone systems such as that proposed by Kauffman (1970) thus solve some of the potential problems of correlation discrepancies resulting from strong environmental control on the occurrence of taxa in single lineages. The best assemblage-zone systems are based on diverse taxa initially selected for their biostratigraphic potential, i.e., groups showing minimum environmental control on distribution, rapid wide dispersal potential, rapid evolution, and abundant representation in the fossil record. Assemblage zonation allows more slowly evolving taxa to be used in refined biostratigraphy because points of origin and extinction are the basis for defining the zonal scheme. The main drawback of this system is the fact that it takes massive amounts of reliable species range data from numerous localities to produce and test the zonation, or to identify many of the zones themselves, and again this may take years to develop.

The problem that faces biostratigraphers is obvious: new methods must be developed to allow further refinement of biostratigraphic systems, with greater confidence that the organisms being utilized are not strongly controlled in their distribution by local facies, and with increased ability to select lineages or parts of lineages for assemblage zonation that will be characterized by rapid evolutionary turnover and/or wide rapid dispersal. In addition, we are faced with the challenge of developing highly refined, durable, and regionally applicable biostratigraphic systems with greater efficiency, so that they can be applied to zonation and correlation of new areas of investigation in a relatively short period of time.

The solution to these problems lies largely in predictive ability: From our multitudinous observations of ecological and evolutionary patterns in living and fossil organisms, can we successfully develop criteria for the consistent prediction of taxa or adaptive strategies that will respond through rapid evolutionary turnover to certain environmental (natural selective) situations, and/or have biological and ecological characteristics that will allow them to be rapidly and widely dispersed into a variety of environments? With such predictive ability, varied organisms with high biostratigraphic potential might be preferentially selected from the very beginning in the construction of a biostratigraphic system, which would quickly and efficiently allow the definition of widely applicable, historically stable assemblage zones with short time durations and regional correlation potential. We might be able to accomplish in a short time what it now takes decades to produce, and speed up the whole process of geological interpretation in sedimentary rocks.

The purpose of this paper is to investigate the predictability in evolutionary rates of various organisms and their adaptive traits, centering around a Cretaceous example from the Western Interior United States, and the possible application of predictable evolutionary patterns to the construction of biostratigraphic systems. Several questions arise that have important connotations in evolutionary theory:

1. Are evolutionary rates constant or nonconstant within and between lineages subjected to the same general environmental (natural selective) phenomena? How can this be demonstrated?

2. If evolutionary rates are nonconstant, are they randomly variable or are they patterned in response to certain biological or environmental phenomena that are themselves patterned? Is there a different response between closely related as opposed to unrelated lineages? Are there certain taxa or adaptive traits that characteristically undergo accelerated rates of evolution during each phase of development in a marine cyclothem, especially those representing (1) major changes in global sea

level; (2) a long- or short-term climatic trend; (3) the development of a widespread competitive boundary; or (4) large-scale genetic isolation caused by plate tectonic movements, or the like?

3. Do diverse lineages with similar living habits, adaptive strategies, trophic grouping, behavior patterns, or even parallel morphological adaptations show similar evolutionary responses to a given set of environmental variables?

4. If relationships such as those in questions 1 to 3 can be shown to exist, to what degree can variations in rates of evolution for various kinds or organisms, and within certain adaptive zones, be predicted in response to a series of broadly acting environmental shifts? What would such predictability mean in terms of biostratigraphic strategies?

MEASURING EVOLUTIONARY RATES

Evolutionary rates are defined and measured in this paper as the number of species added within a lineage per unit absolute time (usually per million year interval). The measure is not to be confused with diversity (any definition). An evolutionary rate of five species per million years during a certain interval of time does not imply coexistence of any particular number of closely related taxa, and thus high or low lineage diversity. If this hypothetical evolutionary event involves simply replacement of one taxon by a derived taxon in a single general environment, then only one or, less commonly, two closely related taxa will be expected to coexist at any one time, and diversity within the lineage will be low. If evolution proceeds at the same rate by rapid radiation into various new habitats, lineage diversity may be high (three to five taxa in our example), but with individual taxa occupying distinct habitats at any one time. In this theoretical example rate is held constant, whereas in at least some actual cases the latter modes of evolution also produce more rapid rates of species turnover, suggesting a testable relationship between evolutionary rate and diversity within lineages.

The *species* is used here as the unit of measure instead of the more stable genus concept because it is most commonly the basic working unit of biostratigraphy and biological analysis. Dangers in measuring and comparing rates of speciation in paleontology, stemming from varied and inconsistent approaches to the species concept among different paleontologists (even for the same taxon) are fully recognized here, and the precision of the analyses and conclusions must be treated by the reader with this in mind. I have attempted to minimize this effect by utilizing species data in the succeeding Cretaceous example which are conceptually consistent to the extent that definition of species and lineages have been the work of single paleontologists for ammonites (W. A. Cobban), gastropods (N. F. Sohl), and bivalves (E. G. Kauffman). Lineages are narrowly defined here as phylogenetically closely related taxa directly derived from one another, or nearly so, in the evolutionary process. In most Cretaceous examples analyzed here, this relationship can be demonstrated by successive analysis of stratigraphically closely spaced species populations.

The time scale against which the production of new species within lineages is measured is dependent upon accurate and relatively consistent radiometric data throughout the interval studied, i.e., radioisotopic dates derived from good material, uncontaminated, analyzed with care, and cross-checked by multiple analyses of the same source rock, and of distinct minerals from each sample. The Western Interior Cretaceous time scale used in this paper is one of the most refined on record, because bentonites are common throughout the sequence; but even here many areas and source rocks do not lend themselves to such scrutinizing analysis. Furthermore, a certain degree of variation in accuracy may exist because of different techniques used by various dating laboratories in analysis, in the sampling process, and in the material itself. As an example, for a series of dates from a single Cretaceous bentonite used in the construction of the radiometric scale by which evolutionary rates are measured in this paper, the mean values and error factors for two minerals dated from the same sample are as follows: biotite, 93.8 ± 4.6 MYBP, and sanidine, 94.6 ± 6.1 MYBP (sample KP-63-17b-10; upper Lincoln Member, Greenhorn Limestone, Colorado; Text-fig. 1; and Kauffman, 1976, in press). A date by Obradovich and Cobban (1973) from the same biostratigraphic level has a mean value of 93 MYBP (corrected to 91.3 MYBP; Obradovich and Cobban, 1975). Most sets of radiometric dates are equally disparate. The reader is urged to consider both the variance in means and error factors that lie behind any averaged radiometric date (93.2 above). Time scales based on these data are *generalized*, and the rate measurements taken from them are equally generalized. They are, however, sufficient to show the principal trends in evolutionary rates for lineages and their relationship to environmental changes, and that is the concept being tested in this paper.

At present, radiometry, for all its potential sources of variance in any given date, is the best means for calculating geological and biological rates. Other, mainly older attempts to use sedimentary thickness versus known rates of sedimentation in modern sediments, or even biostratigraphic zones assuming a constant duration (and in many cases constant evolutionary rates, such as 1 million years per species), are much less satisfactory, and commonly involve circular reasoning.

The methodology involved in the construction of the radiometric time scale applied to the subsequent Cretaceous example is standard practice and is reviewed in detail by Kauffman (1977, in press). Using new and previously published dates for this area, and averaging all seemingly reasonable dates available for each ash (bentonite) bed, or regionally for each biostratigraphic zone or zone boundary where single bentonites were not continuous, a revised and detailed Cretaceous time scale (Text-figs. 1 and 2) was constructed that allowed a general measurement of evolutionary rates in diversely adapted lineages of Cretaceous mollusks. Stratigraphic intervals between known dates were divided into equal 0.25, 0.5, or 1.0 million year time intervals depending upon the number of major biostratigraphic zones, reflecting species-level evolutionary events, that occurred in each undated interval (Text-figs. 1 and 2). Ages so calculated between actually dated levels are conjecture, however, and simply treat these intervening biostratigraphic units and their

STAGES: WESTERN INTERIOR U.S.A. USAGE			STRATIGRAPHY, CENTRAL WESTERN INTERIOR U.S.A.		BIOSTRATIGRAPHY: MAIN ZONAL INDICES	GENERALIZED TIME SCALE; THIS PAPER	ESTIMATED, GILL AND COBBAN, 1966	OTHER DATA (SEE KEY BELOW)	NEW DATA: AVERAGED, THIS PAPER
UPPER CRETACEOUS	TURONIAN	UPPER	CARLILE SHALE	UPPER SHALE MEMBER ? = SAGE BREAKS MBR.	INOCERAMUS WALTERSDORFENSIS; I. KLEINI; I. FRECHI; MYTILOIDES FIEGI; M.? LUSATIAE	↑			
					INOCERAMUS PERPLEXUS, N. SUBSP. PRIONOCYCLUS N. SP.	87.25			
					SCAPHITES WHITFIELDI INOCERAMUS PERPLEXUS				87.4
				JUANA LOPEZ MBR.	PRIONOCYCLUS WYOMINGENSIS INOCERAMUS DIMIDIUS (LATE FORM)	87.5			
					PRIONOCYCLUS MACOMBI LOPHA LUGUBRIS (EARLY FORM)	↑			
		MIDDLE		CODELL SS MBR.	PRIONOCYCLUS HYATTI; LOPHA BELLAPLICATA BELLAPLICATA				
				BLUE HILL SHALE MBR.	PRIONOCYCLUS HYATTI; LOPHA BELLAPLICATA NOVAMEXICANA	88			
					INOCERAMUS FLACCIDUS PRIONOCYCLUS HYATTI	↓			
				FAIRPORT CHALKY SHALE MBR.	MYTILOIDES LATUS COLLIGNONICERAS WOOLLGARI ?	↑			
					MYTILOIDES HERCYNICUS COLLIGNONICERAS WOOLLGARI	88.25		89 (3)	
					MYTILOIDES SUBHERCYNICUS COLLIGNONICERAS WOOLLGARI	↓ 88.5		↑	
		LOWER	GREENHORN FM.	BRIDGE CREEK LS. MBR.	MYTILOIDES LABIATUS, S.S.	88.75		88.9 (1)	
					MYTILOIDES MYTILOIDES MAMMITES NODOSOIDES	89		↓ 90 (3)	
					MYTILOIDES OPALENSIS WATINOCERAS COLORADOENSE	89.5 90			
	CENOMANIAN	UPPER		HARTLAND SHALE MBR.	SCIPONOCERAS GRACILE	91			91.4
					DUNVEGANOCERAS ALBERTENSE	↑ -91.5			
					DUNVEGANOCERAS CONDITUM	↓			
				LINCOLN MBR.	DUNVEGANOCERAS PONDI	↑		91.3 (1)	94.1, 91.2 } AV.92
		?			PLESIACANTHOCERAS WYOMINGENSE	92 ↓			94.1, AV. 88.9 } 92.2
		MIDDLE	GRANEROS SH.	UPPER SHALE MBR.	ACANTHOCERAS AMPHIBOLUM	92.25		92.1 (1)	
					ACANTHOCERAS MULDOONENSE	92.5			
					ACANTHOCERAS GRANEROSENSE	↓			
				THATCHER LS. MBR.	CALYCOCERAS (CONLINOCERAS) GILBERTI				
		LOWER		LOWER SHALE MBR.	"INOCERAMUS" BELVUENSIS (LATE FORM: ? = "I" CRIPPSI, S.L.)	93 93.5			91.2 92.6
			MOWRY SHALE		"INOCERAMUS" DUNVEGANENSIS "I." ATHABASKENSIS; "I." BELVUENSIS, N. SUBSP.	93.75	ESTIMATED FOLLINSBEE ET AL 1963; CASEY, 1964		AV.92.8
LOWER CRETACEOUS	ALBIAN	UPPER			NEOGASTROPLITES MACLEARNI	94 95	94 (2)	94 (1)	
			UPPER DAKOTA GROUP		NEOGASTROPLITES AMERICANUS	96	96 (2)	↓	
					NEOGASTROPLITES MUELLERI	96.5			
				THERMOPOLIS SHALE	NEOGASTROPLITES CORNUTUS	97	98 (2)	98 (1) 95.3 (1)	
					NEOGASTROPLITES HAASI	98			
					UNZONED	98.5		?	
		MID.			UNZONED	99		98, 98, 108 (2)	
		LO.			UNZONED	100 101		102, 98 (2)	
	APTIAN	UP.	LOWER DAKOTA GROUP		UNZONED	102			
		MID.			UNZONED	104 105			
		LO.			UNZONED	106 107	←AVERAGE	115 (2) / 89, 109 (2)	
BARREMIAN			MISSING						

Text-Figure 1
Stratigraphy, biostratigraphy, and geochronology of the Lower and Middle Cretaceous sequence in the Western Interior United States, showing typical kinds of data, their spacing, and the methodology of constructing a generalized time scale for the measurement of evolutionary rates. This time scale is used for the case history given at the end of the paper. (From Kauffman, 1977, in press.)

STAGES: WESTERN INTERIOR U.S.A. USAGE			STRATIGRAPHY, CENTRAL WESTERN INTERIOR U.S.A.			BIOSTRATIGRAPHY: MAIN ZONAL INDICES	GENERALIZED TIME SCALE; THIS PAPER	ESTIMATED, GILL AND COBBAN, 1966	OTHER DATA (SEE KEY BELOW)	NEW DATA: AVERAGED, THIS PAPER
UPPER CRETACEOUS	MAASTRICHTIAN	M.U.	HELL CREEK FM.			TRICERATOPS SP.	63 - 66.5	63 - 67	64-66 (1) 64.0 (4) 65.5 (4)	
		LOWER	FOX HILLS SANDSTONE			DISCOSCAPHITES NEBRASCENSIS	67	68		
						HOPLOSCAPHITES NICOLLETTI	67.5		67.4 (1)	
						SPHENODISCUS (COAHUILITES)	↑	69		
			PIERRE SHALE	SOUTH: "TRANSI--TION" MBR.	NORTH: UPPER UNNAMED SHALE MBR.	BACULITES CLINOLOBATUS	68 ↓			
						BACULITES GRANDIS	↑	70	68.5 68.6 (4)	
						BACULITES BACULUS	68.5 ↓		68.5 (4)	
	CAMPANIAN	UPPER		"TEPEE BUTTE" ZONE	KARA BEN--TONITE M.	BACULITES ELIASI	↑	71		
					LOWER UNNAMED SHALE UNIT	BACULITES JENSENI	69			
						BACULITES REESIDEI	↓	72	68.7 (4)	
						BACULITES CUNEATUS	70		69.9 (4)	
						BACULITES COMPRESSUS	71	73	71.5 (1,4)	
						DIDYMOCERAS CHEYENNENSE	71.5			
						EXITELOCERAS JENNEYI	72	74	72.0 (1)	
						DIDYMOCERAS STEVENSONI	72.25			
						DIDYMOCERAS NEBRASCENSE	72.5	75	72.2 (1)	
				"RUSTY ZONE"	REDBIRD SILTY MBR.	BACULITES SCOTTI	73			
						BACULITES GREGORYENSIS	74	76		
						BACULITES PERPLEXUS (LATE FORM)	75			
						BACULITES GILBERTI	76	77		
					MITTEN BLACK SHALE MBR.	BACULITES PERPLEXUS (EARLY FORM)	77			
				SHARON SPRINGS MBR.		BACULITES SP. (SMOOTH)	77.5	78		
						BACULITES ASPERIFORMIS	78			
		LOWER			SHARON SPRINGS MBR.	BACULITES MCLEARNI	78.5	79	80 (3)	
						BACULITES OBTUSUS	↑		77.9 (1) ↕ 77.5 (1) 81 (3)	
				APACHE SS MBR.	GAMMON FERRUGIN--OUS MBR.	BACULITES SP. (WEAK FLANK RIBS)	79 ↓	80	78.2 (1)	
				"TRANSI--TION MBR."		BACULITES SP. (SMOOTH)	80			
						HARESICERAS NATRONENSE	80.5	81		
						HARESICERAS PLACENTIFORME	81			
			NIOBRARA FORMATION	SMOKY HILL MEMBER		HARESICERAS MONTANAENSE	82	82		
	SANTONIAN	UPPER				DESMOCAPHITES BASSLERI	82.5		82.5 (1)	
						DESMOCAPHITES ERDMANNI	83	83		
		MIDDLE				CLIOSCAPHITES CHOTEAUENSIS	83.5			
						CLIOSCAPHITES VERMIFORMIS	84	84		
						CLIOSCAPHITES SAXITONIANUS	84.5			
		L.				SCAPHITES DEPRESSUS	85	85		
	CONIACIAN	UP. MID.				SCAPHITES VENTRICOSUS	86			
		LOWER		FORT HAYS LIMESTONE MEMBER		SCAPHITES PREVENTRICOSUS INOCERAMUS DEFORMIS	↑	86	86.8 (1) 88 - 89 (3)	
						INOCERAMUS ERECTUS (LATE FORM)	87			
						INOCERAMUS ERECTUS, S.S. BARROISICERAS, PERONICERAS	↓			

xt-Figure 2

ratigraphy, biostratigraphy, and geochronology of the post-Turonian Upper Cretaceous sequence the Western Interior United States, showing typical kinds of data, their spacing, and the meth-lology of constructing a generalized time scale for the measurement of evolutionary rates. This ne scale is used for the case history given at the end of the paper. (From Kauffman, 1977, in ess.)

component speciation events as nearly equal in duration over short spans of time. The reader is therefore cautioned to view this as a truly *generalized* time scale. Nevertheless, the scale is based on more real data than any other pre-Pleistocene geochronologic system, and provides for the best possible approximation of the age and duration of geologic and biologic units, and the measurement of evolutionary rates.

DO EVOLUTIONARY RATES VARY?

Obvious evidence for varied evolutionary rates comes from biostratigraphic data. Cumulative plots of the range zones of diverse organisms found together in the same sedimentary sequence, and presumably subjected to the same types and magnitude of natural selective forces through time, commonly show that the ranges of species in one lineage will be longer or shorter than those of another, and that the points of origin and extinction of individual taxa within two lineages will not always coincide (e.g., Kauffman, 1970; Text-fig. 6). The implications are obvious: some taxa evolve faster than others in the same environmental circumstances, and the factors that trigger speciation or extinction do not affect all kinds or organisms equally. Recognition of this fact, and consequently detailed studies of the evolutionary patterns of diverse organisms in the same environmental sequences, might then provide the first level of predictability concerning evolutionary rates: given a certain set of environmental situations (a single stratigraphic sequence), we observe that groups *X, Y,* and *Z* will consistently have shorter range zones, i.e., time durations, than will co-occurring groups *A, B,* and *C.* It logically follows that, given an environmentally similar but unstudied situation involving the same types of organisms, groups *X, Y,* and *Z* will be researched *first* with the expectation, or better prediction, that they are among the most rapidly evolving groups; thus, a detailed biostratigraphic zonation can be constructed quickly and efficiently from them even *before* study of groups *A, B,* and *C.*

Demonstration that evolutionary rates vary within a single lineage through time is also possible. In a sedimentary sequence, the stratigraphic extent of range zones commonly varies among species within one lineage, in some cases in definite patterns, leading to the *inference* that rates of speciation within the lineage are variable in the same manner. However, variations in sedimentary rates, in diagenesis and compaction of different types of sediments, in the number and time extent of diastems and more extensive unconformities, and thus in continuity of sedimentation, are all factors which make it impossible to prove that the real-time duration of one range zone is different from another just simply because it is stratigraphically longer or shorter in a plot of a lithological succession. Integration of a real-time scale based on radiometric analysis with species range-zone data from any lineage is critical to the demonstration of variation in evolutionary rates within a lineage, or between any two taxa from different stratigraphic levels.

This has recently been done for a variety of lineage studies in the Western Interior Cretaceous sequence in North America. For example, Kauffman (1967b) not only showed differences in evolutionary rates between the generalized infaunal bivalve *Thyasira* (average species duration 1.58 million years) and the co-occurring ammonite *Baculites* (average species duration 0.6 MY; compared in Text-fig. 3) during the Campanian, but also distinct variation in length and duration of range zones for species of *Thyasira* in each of two studied lineages (Text-fig. 3). During times of rapid radiation of the *Thyasira rostrata* lineage, two to five species per million years were produced, with an average species duration of 0.50 million years. Following this radiation, species durations averaged 3.05 million years, with an average of 0.67 species produced per million year interval. In the *T. advena–T. becca* lineage, species durations before rapid radiation were approximately 1 or more million years (data incomplete), and 0.25 species were produced each million years. During radiation of this lineage, two species were produced per million years, with an average duration of 1.06 million years per species. Kauffman (1970, 1972; Text-figs. 3, 5, and, 6 this paper) further showed considerable fluctuations in speciation rates for diverse Cretaceous mollusks related to marine cyclothems and environmental changes characteristic of them, as well as to the life habit and adaptive breadth of various taxa.

Recognition that evolutionary rates vary within lineages allows an additional suite of hypotheses to be made and tested. These hypotheses suggest that various kinds of organisms will have certain biological traits – trophic group, degree of tolerance for stress factors, reproductive and dispersal potential, environmental range, habitat, etc. – which will cause *consistent, patterned variations in their evolutionary rate* in response to changing natural selective forces and variations in stress intensity. Recognition of these factors allows prediction of groups or adaptive strategies that would be expected to undergo relatively rapid speciation in response to a certain set of paleoenvironmental phenomena, other groups that will evolve rapidly in response to a second set, etc. This in turn will permit biostratigraphers to selctively use certain kinds of organisms for rapid and efficient construction of a refined biostratigraphic zonation, for each major phase of a long-term environmental change, e.g., in a major marine cyclothem such as those of the Western Interior Cretaceous (Kauffman, 1967a, 1970). It now remains to identify biological and environmental factors that trigger species-level evolution and determine its rate in various types of organisms.

CONTROLS ON EVOLUTIONARY RATES

A voluminous literature exists which deals specifically with the factors that determine evolutionary rates. Only the principal ideas are treated here to demonstrate the concept that various controls on evolutionary rates do exist, that they act differently on biologically different groups, and that the evolutionary response to these controlling factors is largely predictable. Major hypotheses regarding controls on evolutionary rates are tested against the Cretaceous example subsequently presented. These tests indicate that determination of differences in evolutionary response among varied organisms is an extremely complex process involving diverse, interacting controls.

Genetic Makeup and Evolutionary Rates

Many evolutionists have concentrated on genetic structure and its plasticity in nature as being paramount in determining rates of evolution. This has been extensively debated, and is difficult to evaluate in paleontology unless the genetics of modern counterpart taxa are well known (not usually the case). A simple prediction of expected evolutionary rates among fossil groups based on modern genetic data is not yet possible.

Haldane (1957) claimed that rates of evolution are set by the number of loci in a genome (a measure of organism complexity) and the number of stages through which they can mutate (e.g., reproductive rate and periodicity of generations). But many exceptions exist in nature; some slow-breeding organisms (e.g., elephants) have evolved faster than fast breeders (many invertebrates). Although many workers (e.g., Simpson, 1953b; Stanley, 1973) point to morphologically complex mammals as having faster evolutionary rates than structurally simple bivalves, the evolutionary rates of certain important Cretaceous ammonite and bivalve groups in the Cretaceous (e.g., Inoceramidae, Ostreidae) are equal to or faster than that reported on the average for Cenozoic mammal species (Kauffman, 1970, 1972). We cannot, therefore confidently predict that structurally complex and/or rapidly

---

Text-Figure 3
Evolutionary history of the Cretaceous bivalve *Thyasira* from the Western Interior United States and Canada (Kauffman, 1967b, text-fig. 13; radiometry from Text-fig. 2 of this report), showing: (a) comparative duration of baculitid ammonite species (average 0.64 MY; names in left column) and more slowly evolving species of *Thyasira* (average duration 1.58 MY; each pattern in each vertical bar equals stratigraphic duration of one species), both from the same sedimentary sequence (Pierre Shale, Bearpaw Shale, and equivalents) in the nothern United States and southern Alberta; (b) variations in evolutionary rates within single thyasirid lineages (species complexes); compare duration of species range bars against radiometric scale on left, with faster rates occurring during initial times of radiation; and (c) relationships between evolutionary history and one of the four major marine cyclothems in the Western Interior seaway. Radiation of the *T. rostrata* lineage takes place during the final phase of a transgressive pulse; peak transgression was at about *B. perplexus* (early form)–*B. gilberti* time. Abrupt replacement of this lineage by rapid speciation within the *T. advena–T. becca* lineage takes place during early low stress phases of regression. This moderately stenotopic group becomes abruptly extinct during mid-regression, just after its second major radiation, as stress factors increased and its principal living environment (deep basinal dark muds) was greatly diminished.

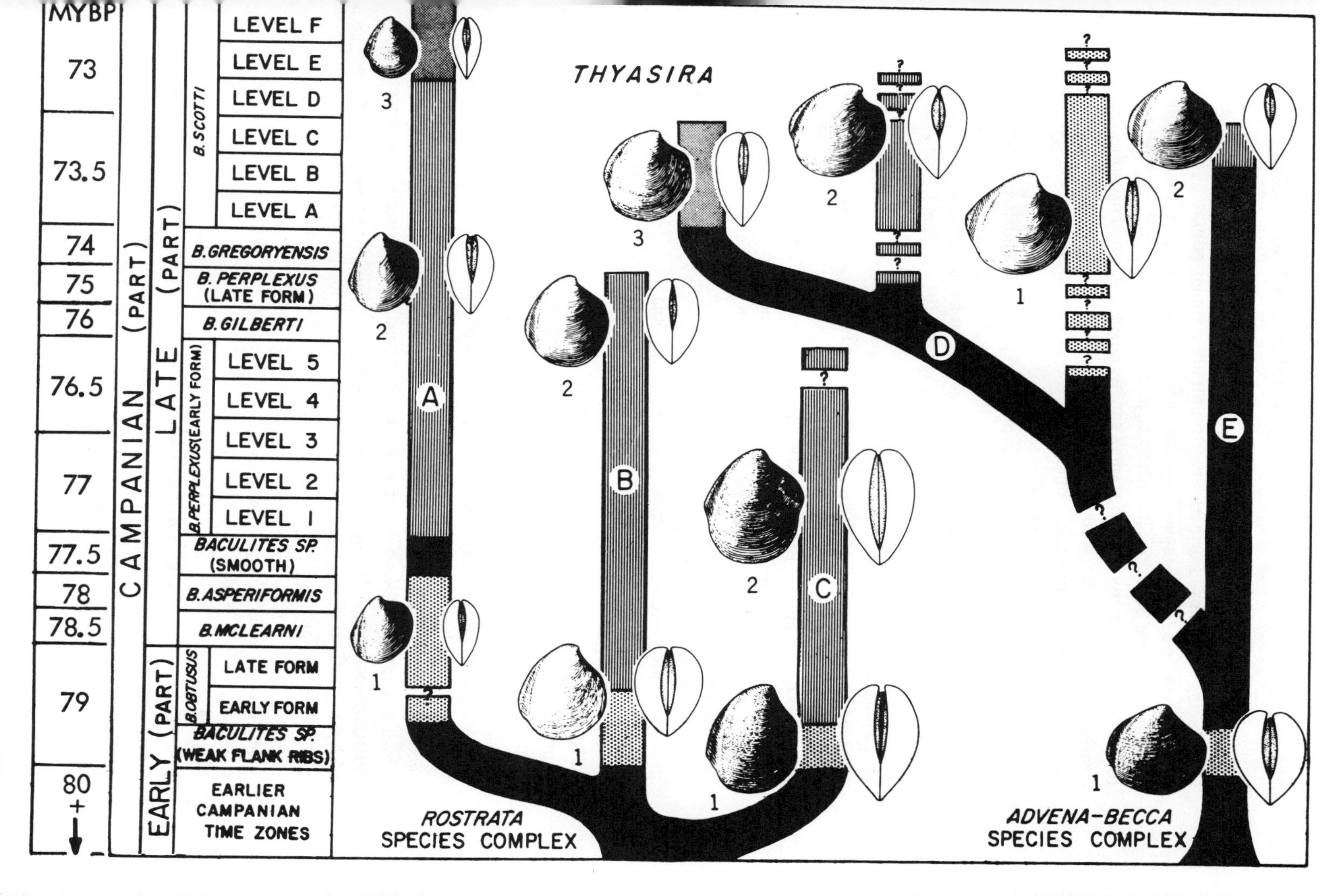

MYBP
73
73.5
74
75
76
76.5
77
77.5
78
78.5
79
80+
CAMPANIAN (PART)
LATE (PART)
EARLY (PART)
B.SCOTTI
LEVEL F
LEVEL E
LEVEL D
LEVEL C
LEVEL B
LEVEL A
B.GREGORYENSIS
B. PERPLEXUS (LATE FORM)
B.GILBERTI
B.PERPLEXUS (EARLY FORM)
LEVEL 5
LEVEL 4
LEVEL 3
LEVEL 2
LEVEL 1
BACULITES SP. (SMOOTH)
B.ASPERIFORMIS
B.MCLEARNI
B.OBTUSUS
LATE FORM
EARLY FORM
BACULITES SP. (WEAK FLANK RIBS)
EARLIER CAMPANIAN TIME ZONES
THYASIRA
ROSTRATA SPECIES COMPLEX
ADVENA-BECCA SPECIES COMPLEX
A
B
C
D
E
1
2
3

reproducing organisms will have faster evolutionary rates than simple and/or slowly reproducing organisms. Furthermore, structurally complex groups, by nature of small changes in a few of their many characters, may appear to "evolve" faster than structurally simple organisms like bivalves, merely because species concepts applied to the two are distinctly different; small genetic changes of equal magnitude may be clearly manifest in fossils of complex organisms and not at all in simple organisms. "Evolutionary rate," therefore, could merely be a measure of structural complexity rather than genetic change (Schopf et al., 1975).

Haldane's (1957) concept that the evolutionary process is inherently slow because of the large number of "genetic deaths" necessary to allow replacement of one gene by another, the fact that very few genes can be substituted simultaneously (Kimura, 1960), and the probability that two species differ at hundreds to thousands of gene loci led Haldane (1957) and Mayr (1963, p. 259) to the conclusion that a minimal estimate of 300,000 generations would be necessary for speciation to occur if the two species differed at 1,000 loci. This suggests some constancy of rate within any one lineage; however, that again is dependent upon the genetic complexity of the organism. But Mayr (1963) points out that evolutionary rates may be speeded up if (1) combinations of co-occurring genetic deaths are synergistic and collectively have a greater effect on loss of fitness than just the sum of the individual deaths; (2) the replacement of certain genes exercises a selective effect on others, causing a chain reaction; (3) populations are small and the spread of new genes is thus accelerated through the entire available gene pool; and (4) genetic change is mainly by immigration of new genes in gene substitution and recombination, rather than by mutation.

It cannot be assumed, either, that organisms prone to rapid mutation rates will consistently show rapid evolutionary rates, and thus allow prediction in the fossil record. Mayr states (1963, p. 567) that *rate of evolution depends only to the smallest extent on rate of mutation, but more on population size and gene flow characteristics.* Evolution is most rapid in small isolated populations. Evolutionary rate depends primarily on the frequency and effectiveness of barriers to gene flow, the rate at which they are emplaced to produce isolation of small gene pools, rates at which geographical isolates become genetically transformed to acquire their own isolating mechanisms, the degree of ecological diversity and availability of new niches to newly arising species, and the nature of selection pressures within the niche of the isolates. Mayr (1963) concludes that *ecological factors are far more important in determining rates of speciation than are genetic factors.*

We worry tremendously about not being able to see into the genetics of fossil organisms so as to study the evolutionary process. This is not necessary; Mayr's message is clear. The major controls on rates of evolution are *externally* manifest – population size, degree of isolation, rate of isolation, diversity of niches, variations in selective pressures, size and mobility of organisms and ecological controls on their distribution, trophic relationships, and breadth of ecological tolerance. Much of this can be clearly studied in the fossil record.

The study of intraspecies and interspecies genetic variability through electrophoretic analysis (e.g., Levinton, 1973, for bivalves) shows some promise in predicting evolutionary rates in that it demonstrates that genetic variation within and between species populations is related to environmental variability and stress gradients. Within some species populations, genetic variability increases toward the stressed margins of the species range. The reverse is true in other taxa. Between species of bivalves studied by Levinton (1973), variability increases with increasing exposure to environmental variability in progressively shallower water habitats, and among shallow as compared to deep burrowing bivalves. There may be a further link between genetic variability and heterozygosity, producing polymorphism; the latter condition greatly enhances the chance and rate of evolutionary change, and is commonly manifest in fossils.

If these relationships can be shown to exist widely in living species populations, and if the niche range and environmental tolerance limits of these organisms are well documented, times and sites of expected acceleration in the evolution of fossil taxa might be predicted. The recognition of certain paleoenvironmental conditions developing in space and time that are known to stress certain taxa, and/or recognition of increasing polymorphism expressed as increasing phenotypic variation in fossil populations, would possibly predict the onset of accelerated evolution in certain groups and focus the attention of the paleobiologist-biostratigrapher on these as potential zone fossils. This approach is worthy of more study.

On a larger scale, if increasing genetic variation, heterozygosity, and polymorphism can be consistently related to increasing environmental stress, then it can be predicted that large-scale, biologically unpredictable environmental fluctuations, like global marine regressions or cooling trends, which stress whole ecosystems, should be accompanied by widespread increases in evolutionary rates among diverse organisms. This appears to have been the situation in the Cretaceous of the Western Interior (Kauffman, 1970, 1972).

Trophic Group and Evolutionary Rates

In a recent paper, Levinton (1974) has proposed that a direct relationship exists between evolutionary rates and feeding habits among Bivalvia. He demonstrated appreciably faster evolutionary rates among tested genera of suspension feeders (Pectinacea, Pteriacea, Veneracea) than found among deposit feeders (Nuculoida) (mortality rates 1.2 to 1.5 percent per million years and 0.8 percent per million years, respectively). He related this to resource predictability; phytoplankton supply used by suspension feeders was much less predictable than intrasediment bacteria and other materials used by detritus feeders as a food source, and thus suspension-feeding organisms are more commonly stressed through time. This suggests that *environmental stress, defined and measured by the degree to which unpredictable environmental perturbations affect the species populations, plays a major role in determining evolutionary rate.* It would follow also from Levinton's data that the more generalized the scope of food or other resources used by the organism, the lower the probability that it will be stressed over long periods of geological time, and the slower will be its evolution. In theory, a trophically graded series of organisms from the most generalized to the most specialized feeding habits and food sources should be precisely reflected in a gradient of slow to fast evolutionary rates, all other things being equal. If this were true, it would allow certain organisms to be selected from the outset for construction of a biostratigraphic system because of their specialized feeding habits and thus potentially rapid evolutionary rates. This hypothesis is successfully tested in the subsequent Cretaceous example.

Isolation and Evolutionary Rates

High rates of gene flow among members of a species population exist when isolating mechanisms are not effective, producing a cohesive genetic framework throughout the population, keeping it evolutionarily conservative, and selecting against most new characters derived from mutation or recombination. Mayr (1963), Eldredge and Gould (1972, and this volume), and others point out the importance of isolating portions of the gene pool, preferably in small subpopulations, to achieve rapid evolutionary turnover. The number, nature, and rate of emplacement of barriers to gene flow are important factors in determining evolutionary rate, as are population size, selective pressures, and environmental possibilities in the area occupied by each isolate subpopulation. Most workers accept that the majority of evolution, and some of the fastest rates, occur in such marginal, isolated species populations and have throughout geological time. Some, like Eldredge and Gould (1972), suggest that virtually all evolution takes place in such populations, and that large populations with free gene flow remain totally conservative, damping even gradualistic change through long periods of time. This view may be somewhat severe in light of various demonstrations of gradualistic evolutionary response in closely spaced fossil populations (e.g., Kauffman, 1970; Text-fig. 4, this report) from single facies.

Theoretically, the more rapidly that subpopulations are genetically isolated and selection pressures changed, the more numerous the environmental opportunities for new forms evolving in isolated populations; and the smaller and more numerous the isolate populations, the more rapid will be evolution and speciation. The tendency for isolate populations to develop at the range margins of the parent population, in exactly the environmental situations where Levinton (1973) found increased genetic variability (possibly producing increased heterozygosity and polymorphism) in certain bivalve populations, may further accelerate the process of genetic change and speciation.

Well-defined spatial and temporal situations (facies) can be identified from physical stratigraphic and paleoecological observations that reflect environments with the potential mechanisms for genetic isolation, stressing of small subpopulations, and thus predictable periods of rapid evolution in certain elements of fossil biotas. The formation of islands by a transgressing epicontinental sea, or of isolated marine basins during regression, are good examples. The biostratigrapher might selectively employ taxa for zonation that occupied such areas in parts of the geological column characterized by the onset of widespread isolating mechanisms.

Environmental Stress and Evolutionary Rates

A recurrent theme in discussions of evolutionary change and rates is the degree of stress (i.e., intensity of natural selection) placed on species populations by changing biological, physical, and chemical environmental parameters. Increasing selective pressures accelerate the process of evolution by increasing death rates among poorly adapted individuals and by more rapidly selecting for new adaptations. This is testable in both living and fossil organisms, and is dealt with in the Cretaceous example.

Stress is the degree of nonpredictability of environmental perturbations for organisms living in any particular area. Thus, as Levinton (1973, 1974) pointed out, marine bivalves buffered from environmental perturbations by deep burial or deep-water habitats live in more predictable, low-stress environments and show lower genetic variability and lower potential for evolutionary change. Similarly, bivalves utilizing a predictable, slowly varying food source are relatively unstressed and evolve slowly. Conversely, bivalves exposed to unpredictable food shortages, turbid periods, salinity and temperature changes because of their shallow-water epifaunal or shallow infaunal habitat, and/or their adaptations for specialized suspension feeding, show higher genetic variability and faster rates of evolution.

The environmental adaptability of organisms also determines the degree to which they are stressed. In difficult environments like the deep sea, the intertidal zone, and brackish to freshwater habitats, life conditions are poor and/or highly variable from an environmental point of view. Nevertheless, for organisms native to these habitats, stress is *low* because they are *adapted to* low oxygen, poor sediment chemistry, severe competition, and wholly detrital food sources in the deep sea, or to large-scale fluctuations in temperature, salinity, exposure, predation, etc., in the intertidal zone. Most freshwater organisms are similarly adapted to high levels of environmental fluctuations.

Jackson (1972, 1973) documented much slower evolutionary rates for bivalve species of the families Lucinidae, Veneridae, and Tellinidae living in Tropical marine environments in waters less than 1 m deep than for similar infaunal bivalves found in deeper, but still shelf depth, waters. He suggests that the shallower-water species evolve more slowly because they can tolerate a wider range of environmental fluctuations (and thus are less likely to be stressed by unpredictable perturbations), have longer-lived larvae, and have broader biogeographic distribution (both increasing survival potential).

Stanley (1973) has documented evolutionary bursts among mammals and high rates of evolution in rudistid bivalves due to the effects of intense competition for space and resources. Kauffman (1970, 1972) has related increasing rates in the evolution in Cretaceous mollusks from North America to widespread stress within the marine ecosystem associated with major regressive pulses, decrease in ecospace,

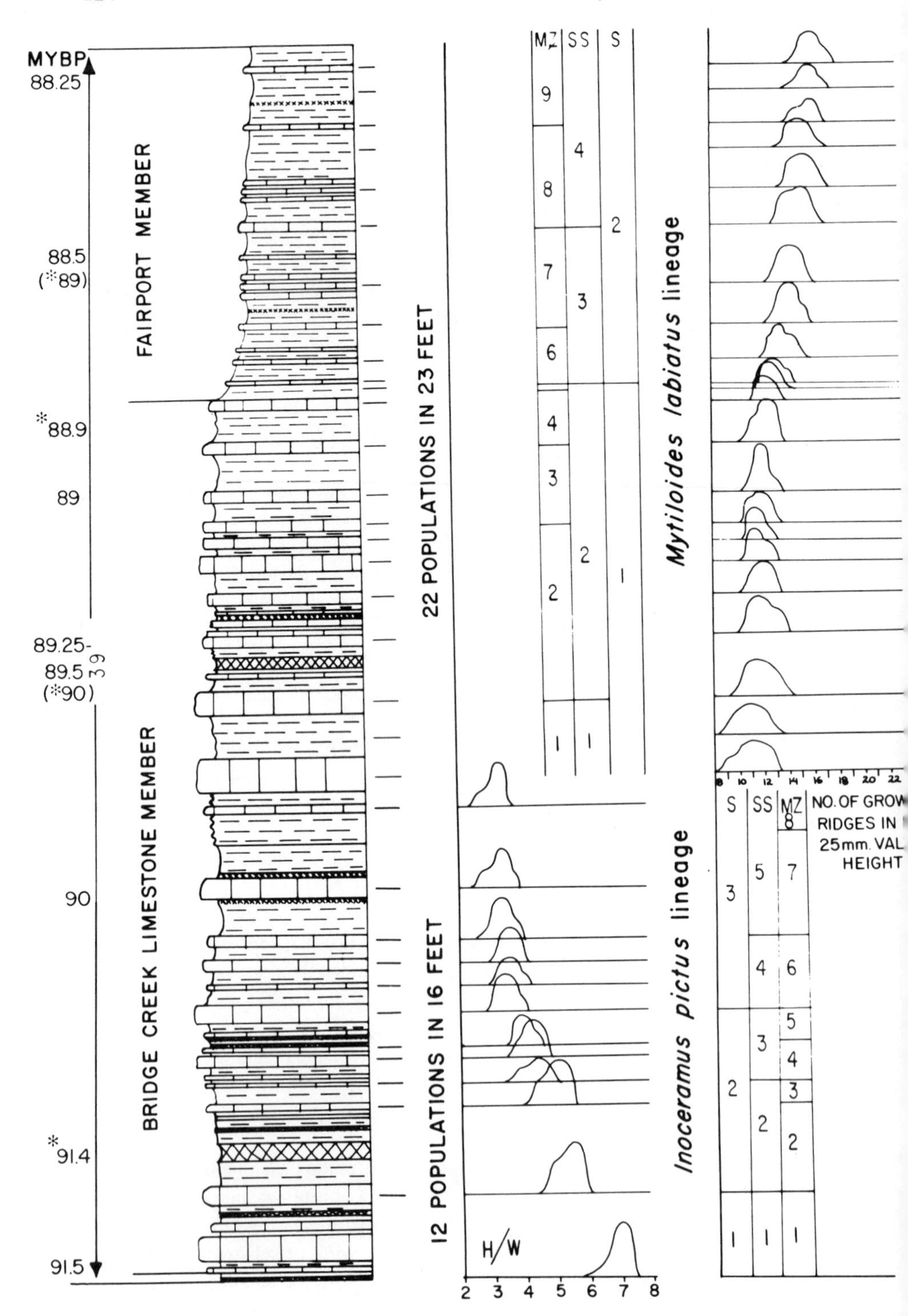
MYBP
88.25
88.5
(*89)
*88.9
89
89.25-
89.5
(*90)
90
*91.4
91.5
39
FAIRPORT MEMBER
BRIDGE CREEK LIMESTONE MEMBER
22 POPULATIONS IN 23 FEET
12 POPULATIONS IN 16 FEET
Mytiloides labiatus lineage
Inoceramus pictus lineage
MZ SS S
H/W
NO. OF GROW
RIDGES IN
25mm. VAL
HEIGHT

restriction of habitats, and increased competition. He further demonstrated that similar stress levels acted differentially on stenotopic and eurytopic organisms, triggering rapid evolution and increased extinction rates earlier in the regression (at lower stress levels) for stenotopic than for moderately eurytopic or highly eurytopic taxa (see subsequent discussion). Jackson (1974, and this volume) noted that widely ranging and environmentally tolerant eurytopic marine taxa had broader biogeographic distribution and less potential for rapid evolution than did environmentally sensitive stenotopic taxa with more restricted biogeographic ranges and smaller more readily stressed populations.

The connotations of these observations to biostratigraphy are profound. In selecting organisms for construction of a biostratigraphic system where rapid evolutionary rates leading to biostratigraphic refinement are of greatest importance, one can predict which groups would initially be excluded from consideration because of their slow evolutionary rates: (1) normal inhabitants of freshwater, brackish water, intertidal, and very shallow subtidal environments, all eurytopic organisms adapted to a broad range of environmental fluctuations; (2) deep-water and deep infaunal taxa

---

Text-Figure 4

Typical example of (a) occurrence and thickness of bentonites (lines of X's in the stratigraphic column) useful in construction of a radiometric time scale for measuring evolutionary rates of Cretaceous mollusks in the Western Interior United States, and ages obtained from them to date (asterisk indicates dated level; other values calculated); (b) spacing of fossil collections made in conjunction with sampling bentonites for dating, so as to tie each dating level to a biostratigraphic zone, and to provide material for the study of evolutionary rates and patterns through population systematic analysis in evolving lineages (in this case bivalves of the family Inoceramidae); (c) patterns of evolution within the opportunistic *Mytiloides labiatus* (Schlotheim) and more specialized *Inoceramus pictus* Sowerby lineages, as demonstrated diagrammatically by generalized histogram curves for one character in each lineage, at each level studied in the population analysis. Systematic decisions proposed for each lineage (and thus the biostratigraphic system based on them) are shown beside each row of population curves with numbers representing "taxa" within a species (S) column, a subspecies (SS) column, and a morphological zone (MZ) column; these divisions are based on the evolutionary analysis. Example from the Bridge Creek Member, Greenhorn Formation (latest Cenomanian, Early Turonian) and the lower part of the Fairport Member, Carlile Shale (Middle Turonian) along the central and southern Colorado Front Range (modified from Kauffman, 1970, fig. 4). Note both gradualistic (*M. labiatus*) and punctuated evolutionary patterns (*I. pictus*), as well as rate changes in the *I. pictus* lineage approaching the transgressive peak of the Greenhorn cyclothem (Middle Bridge Creek Limestone Member), marked (at maximum transgression) by flooding of Subtropical waters into the Interior, rapidly increasing temperature and salinity, and abrupt extinction of many more specialized Cenomanian organisms, as *I. pictus*.

protected from environmental perturbations by the buffering effects of sediments or a thick water column; and (3) environmental generalists, such as detritus feeders utilizing a broad food base and having high tolerance for chemically poor environments. All these taxa might later be incorporated into complex assemblage zones once their origin and extinction points were known.

Rapid evolutionary rates could be predicted, however, in shallow burrowing, epibenthic, or pelagic organisms of the marine shelf zone or in the upper water layers of open ocean or epicontinental sea environments, as well as for taxa with specialized feeding or other behavioral traits, all of which would be strongly affected by unpredictable environmental perturbations through time. These easily identified taxa could be selectively chosen for the initial construction of a refined biostratigraphic system, short-cutting the trail and error methodology.

Finally, it can be theorized, for the adaptive gradient between stenotopic and eurytopic organisms living in time-stressed areas, that certain unique suites of organisms will be expected to respond to each major phase or level of intensity in a stress gradient. Given a long-term environmental decline (lowering regional temperature, global marine regression, changing water chemistry, etc.), with progressively increasing intensity of stress, certain kinds of organisms (stenotypes) will undergo rapid evolutionary bursts during initial stages of environmental decline, others at a second stage, and so on, until the final evolutionary burst among the most adaptive eurytopes as stress factors become severe. If this theory is valid, it should be possible to predict and preferentially select fossil taxa for building a detailed biostratigraphic system, based on rapid evolutionary turnover, for every stage of a major environmental fluctuation, and thus to construct a system of zonation and correlation in the most efficient manner. It is this hypothesis that is primarily tested in the succeeding Cretaceous example.

TEST FOR THE PREDICTABILITY OF EVOLUTIONARY RATES IN MARINE ORGANISMS

During the last 40 million years of the Cretaceous (Late Albian–Early Maastrichtian stages), the Western Interior of the North American continent was occupied continuously by a large epicontinental seaway, 3,000 by 1,000 miles in extent during maximum flooding, connecting the Gulf of Mexico with the northern circum-Boreal sea through Arctic Canada. The seaway was predominently of shelf depth, but deepened in a series of north–south trending basins near its western margin. Diverse and widespread benthonic environments characterized the basin during widespread flooding associated with peak transgression (middle Late Albian, Late Cenomanian–Early Turonian, Coniacian–Early Santonian, and early Late Campanian times), and effective gene flow was possible throughout the basin. A gradual temperature gradient from Mid-temperate (north) to Warm-temperate–Subtropical (south) marine climatic zones existed during much of the history of the basin. During major regression of the seaway in the Late Albian, late Middle Turonian, Early Campanian, and Maastrichtian, the marine basin was reduced to less than half the size attained during transgression, and many islands, isolated deeper basins, and shallow discontinuous platform areas probably developed. Isolating mechanisms to gene flow were established. Major environmental fluctuations included (1) an overall cooling trend toward the end of the Cretaceous; (2) fluctuations between increasingly more normal salinity and warmer waters during maximum transgression, and slightly less than normal salinity with cooler waters during regression; and (3) the massive effects on the marine environment of four major transgressive-regressive cycles of sedimentation, which probably reflected global eustatic changes of sea level connected with Cretaceous plate tectonic movements (Kauffman, 1967a, 1969, 1973).

Specifically, during transgressions, marine climatic conditions ameliorated, with increasing ecospace, increasing size and diversity of habitats, warming temperatures culminating in a flood of Subtropical waters during at least the two major transgressive peaks (Late Cenomanian–Early Turonian; Coniacian–Early Santonian), the onset of more normal marine conditions, and possibly increasing productivity and food resources (as measured by increase in planktonic organisms). Conditions were optimal for the immigration, colonization, and radiation of diverse marine organisms. During regressive pulses, the basin was characterized by diminishing ecospace, restriction, isolation, and elimination of many prime habitats, shallowing of the water accompanied by increasing turbidity, increased competition, decreasing salinity, and somewhat lower water temperatures. These were increasingly high stress situations for the diverse organisms that occupied this seaway.

The biota of the Western Interior Cretaceous is numerically rich but, in general, not as diverse as in more normal marine situations owing to the poor representation of algae, sponges, corals, bryozoa, articulate brachiopods, many arthropod groups, planktonic foraminifera and coccoliths (except during transgressive maxima), echinoids, and certain mollusk groups. Low salinity during much of the basin's history is probably responsible for this taxonomic depletion. Consequently, paleocommunities of the basin are simple and largely dominated by Mollusca, especially bivalves and ammonites; benthonic foraminifera were common as well in shallower facies. Despite this low diversity, the fossil record of the Western Interior Cretaceous is excellent, and populations of dominant mollusks can be collected at closely spaced intervals for evolutionary studies (see Kauffman, 1967a, 1970; Text-fig. 4, this paper). Many lineages have now been studied, representing diverse taxonomic groups, ecological strategies, and varied habitats within the basin during its 40 million year marine history. These form the basis for the subsequent observations on evolutionary rates and patterns, and those results already published (Kauffman, 1967b, 1970, 1972). Biogeographically, these are mixtures of northern Mid-temperate zone elements and southern Warm-temperate to Subtropical elements; endemism is unusually high in the Western Interior basin as compared to other epicontinental seaways, possibly reflecting the presence of numerous isolating mechanisms during the history of the basin, especially with regression. Evolutionary rates for these lineages can be measured against one of the most detailed geochronologies yet developed, based on radiometric analysis of biotite and sanidine in numerous closely spaced bentonites throughout the column (Text-figs. 1 and 2, and previous discussion).

Against this extensive Cretaceous data base, it is possible to test the principal theorum of this paper: that consistent, repetitive evolutionary patterns exist, and can be predicted, among diverse organisms in response to stress variations in the surrounding paleoenvironment, and as a reflection of the biological characteristics of each group of organisms (trophic group, ecological tolerance, habitat, etc.). Knowledge of these relationships in turn will allow selection of specific groups of organisms evolving under specific sets of environmental conditions for their high biostratigraphic potential (rapid evolution, rapid widespread dispersal, etc.), thus greatly enhancing the efficient construction of detailed biostratigraphic systems throughout the geologic column.

Taxa Selection

Biostratigraphically important organisms share many characteristics, including rapid evolutionary rates, wide biogeographic distribution, abundance, ease of recognition, at least moderate tolerance for environmental variation, and mechanisms for rapid dispersal. Not all of these characteristics are commonly found in the same organism; as Jackson points out elsewhere in this volume, rapidly dispersed and geographically widespread species are unfortunately often slowly evolving, because their great spread also reflects high levels of gene flow. For a single basin, however, where broad regional spread of key taxa is somewhat less important in biostratigraphy, the characteristic of rapid evolution is most sought after in taxa applied to zonation, allowing the finest possible division of the stratigraphic column based on the fewest taxa, and thus the most efficient development of a biostratigraphic system.

Initially, therefore, we are concerned with the prediction of taxa that will be expected to demonstrate rapid evolution (one species per million year duration or less) as the base data for construction of a biostratigraphic system. Several hypotheses have been presented that should allow us to do this.

First, it has been proposed that structural complexity is the key to identification of organisms with the potential for rapid evolution. This should be true whether or not we are analyzing change in complex versus simple organisms due to genetic change consistently of species magnitude (e.g., the hypothesis of Haldane, 1957), or merely looking at lesser-scale genetic changes in certain groups, which are visible because of the structural complexity of the animal and not of equivalent grade to morphologic change manifest in a lineage of conservatively structured organisms (e.g., Schopf et al., 1975).

To test this hypothesis, evolutionary rates in complex marine organisms such as mosasaurs (Reptilia) and various ammonites were compared with those of moderate complexity (gastropods and bivalves) and relatively low complexity (foraminifera) in the Cretaceous of the Western Interior. Rates were determined by radiometric measurement (Text-figs. 1 and 2) of species duration for lineage-related taxa that replaced one another in time, within the same general environmental framework (the central part of the basin in Colorado, Wyoming, and Montana) and the same time span. The slowest evolutionary rates were obtained from mosasaurs (one species per 3 to 5 million years; durations of 0.75 to 9.5 MY, averaging 4.1 MY) and benthonic foraminifera (species durations of 0.75 to 26.75 million years, averaging 7.6 MY). Planktonic foraminifera (species durations of 0.5 to 18.5 MY, averaging 3.27 MY) and more generalized infaunal bivalves like *Nucula, Nuculana, Thyasira,* and *Lucina* had species with 0.25 to 8.0 million year durations (averages: *Lucina*, 2.8 MY; *Thyasira,* 1.95 and 1.07 MY for two lineages; *Nucula,* 3.1 MY). Specialized suspension-feeding infaunal gastropods (*Turritella*) and bivalves (*Ethmocardium*) had an average species range of 1.2 and 1.7 million years, respectively. On the other hand, most ornate ammonites, and the simply ornamented marine oysters and Inoceramidae, had average species durations of 0.06 to 2.0 million years in the interval sampled (averages: Inoceramidae, 0.43 and 0.19 for two lineages; ammonites, 0.42, 0.5, 0.54, and 0.9 MY for four lineages). These data conclusively show that structural grade, by itself, is not a good measure of potential evolutionary rate among American Cretaceous taxa, and possibly all organisms.

Turning to Levinton's (1974) hypothesis that trophic strategy, and thus exposure to varying degrees of stress (fluctuation of food source in an unpredictable manner), was reflected in evolutionary rates, Mollusca representing various trophic groups were compared in the Cretaceous sequence for their evolutionary rate.

Ammonites (specialized pelagic to saltating benthonic predators, ?filter feeders)

- *Baculites* (see Text-fig. 2 for species): species durations of 0.25 to 1 MY, averaging 0.54 MY per species.
- *Scaphites* s. s. (species from Cobban, 1951): species durations of 0.12 to 1.5 MY, averaging 0.42 MY per species.
- *Clioscaphites* and *Desmoscaphites* (Cobban, 1951): all durations and average, 0.5 MY per species.
- *Neogastroplites* (see Text-fig. 1 for species): species durations of 0.5 to 1.25 MY, averaging 0.9 MY per species.

Specialized suspension-feeding Bivalvia (epifaunal)

- *Inoceramus* s. s. (*rutherfordi, pictus, perplexus* lineages) (see Text-fig. 1 for species): species durations of 0.06 to 2.0 MY, averaging 0.43 MY per species; maximum rates of radiation, 11 species per million years.
- *Mytiloides* s. s. (Inoceramidae; see Text-figs. 1 and 4 for species): species durations of 0.08 to 0.5 MY, averaging 0.19 MY per species.
- *Lopha* (Ostreidae) (see Text-fig. 1 for species): species durations of 0.06 to 0.5 MY, averaging 0.17 MY per species. Average for all North American species, 0.61 MY.

Specialized suspension-feeding Bivalvia (infaunal)

- *Ethmocardium, E. pauperculum* group, shallow infaunal; species durations of 0.75 to 2.75 MY, averaging 1.7 MY per species.

Generalized suspension-feeding and partially detritus feeding infaunal Bivalvia

"Lucina" occidentalis group: species durations of 1 to 8 MY, averaging 4 MY per species (average 2.8 excluding one long-ranging species).

Thyasira rostrata lineage (Text-fig. 3: 0.25 to 4.5 MY species duration range, averaging 1.95 MY per species (see previous discussion of differences within evolutionary history).

Thyasira advena–T. becca lineage (Text-fig. 3): species durations of 0.25 to 2.75 MY, averaging 1.07 MY per species (see previous discussion of variation).

Detritus-feeding infaunal bivalves

Nucula (Western Interior and Gulf Coast species): species durations of 1 to 8 MY, averaging 3.1 MY per species.

Clearly, a gradient exists in evolutionary rates of Cretaceous mollusks that is correlative with trophic specialization; open-water carnivores, predators, and suspension-feeding mollusks and specialized epifaunal suspension feeders have the fastest evolutionary rates in our example, specialized shallow infaunal suspension feeders the next fastest rates, more generalized suspension–detritus feeders next, and wholly infaunal detritus feeders are characterized by the slowest evolution, as Levinton predicted (1974). This clearly defines several groups of benthic and pelagic marine organisms among mollusks that have higher biostratigraphic potential than the rest, and which should be selectively studied first in the construction of a zonal scheme with short-term individual zones, i.e., pelagic, epibenthic, and specialized infaunal to shallow infaunal molluscan taxa.

These data also support a second proposition by Levinton (1973): in normal marine situations, bivalves of deeper water, and of deeper infaunal habitats that are buffered from environmental perturbations, will have lower genetic variability (and therefore potentially slower evolutionary rates) than those of normal shallow-water habitats and shallow infaunal burial. The data presented above strongly suggest that the deeper infaunal taxa and those characteristic of cool and/or deep-water paleocommunities (*Lucinidae, Thyasira, Nucula,* and other protobranchs; see Kauffman, 1967b, Tables 1 and 2) have slower evolutionary rates than taxa that are characteristic of open-water (upper pelagic zone), epibenthic (shelf depth), or shallow infaunal (shelf depth) life habits. Furthermore, the data suggest a gradient among these last three, with pelagic and fully epibenthic mollusks having higher evolutionary rates than those of semi-infaunal or shallow infaunal habitats. This again allows definition of specific habitat groups of organisms (pelagic, epibenthic, shallow infaunal) that would be expected to have high rates of evolution and therefore be more useful in construction of a biostratigraphic system.

Finally, Cretaceous data are available that allow testing of Jackson's (1972, 1973) hypothesis that intertidal to shallow subtidal (less than 1-m depth) organisms, which are highly adapted to broad fluctuations in the environment and thus not easily stressed, should be more slowly evolving than organisms that occur in deeper shelf-depth waters. The taxa for which data have been previously given are all common inhabitants of waters deeper than 1 m, primarily in the shelf zone. By comparison, evolutionary rates among known intertidal Donacidae of the Western Interior Cretaceous show species durations ranging between 1.25 and 8 million years, and averaging 2.8 million years per species. Brackish-water and freshwater organisms, which are subjected to broad variations in environment similar to those of the intertidal zone, should have the same slow evolutionary rates for the same reasons. Cretaceous data bear this out. Among brackish-water bivalves, members of the *Crassostrea soleniscus–C. glabra* lineage have species ranges between 1 and 8 million years, and averaging 3.7 million years per taxon. Members of the *Brachidontes filisculptus* lineage range from 1 to 3 million years in species duration, and average 1.8 million years per species. Freshwater unionid bivalve species of the Cretaceous appear to be even longer ranging, from 1.8 to 22 million years per species, and averaging about 4 million years, possibly more, per taxa (data are still incomplete). These rates suggest exclusion of intertidal, shallowest subtidal, brackish water, and freshwater invertebrates from initial construction of a biostratigraphic system that is highly dependent on rapid evolutionary rates among component taxa.

Thus the marine organisms that can be predicted to evolve rapidly (species ranges averaging 1 million years or less) from their ecological and biological characteristics, and from their fossil record, are primarily pelagic organisms of photic, open-water habitats, and epibenthic, shallow infaunal, or semi-infaunal organisms of shallow epicontinental and continental shelf seas. They are also taxa with relatively specialized feeding habits (predatory or specialized suspension feeding). These are the groups most likely to be stressed by unpredictable environmental perturbations over long periods of geological time. These taxa have been selectively chosen to construct the basic biostratigraphic system of the Western Interior Cretaceous (Kauffman, 1970), and to be subsequently tested for repetitive patterns in their evolutionary history.

Fresh, brackish, intertidal, shallow subtidal, and deep ocean organisms, as well as those living deep in the substrate, are less likely to be stressed and will have predictably slower evolutionary rates. Their use in constructing biostratigraphic systems is secondary, and comes when the systems evolve to complex assemblage zonation (e.g., Kauffman, 1970), utilizing numerous coincident origin and extinction points rather than short species ranges as the primary tool of zonation. These are not further considered here.

Evolutionary Patterns Among Selected Taxa

Since different organisms can be shown to evolve at different rates under the same general set of environmental conditions, the identification of organisms with evolutionary rates that are predictably rapid in geological time is the first step in efficiently constructing a biostratigraphic system based on short zonal durations. The next major step is to demonstrate that different, rapidly evolving groups of organisms consistently respond to certain types of natural selective pressures, and from this to be able to predict certain suites of organisms that will probably undergo rapid evolution for each phase of a major environmental fluctuation through geologic time (e.g., for early, middle, and peak transgression, and for early, middle, and peak regression in a major sedimentary cycle). That this is possible is suggested by stratigraphic plotting and the observations that different suites of organisms and/or adaptive strategies react differently, undergoing rapid evolution at different times, under the same set of changing environmental parameters (e.g., a marine cyclothem). It takes far less stress to trigger rapid evolution and/or widespread extinction in most stenotopic taxa, with low tolerance for environmental change, than for the environmentally tolerant eurytopic taxa (Kauffman, 1972).

Kauffman (1970, 1972, 1973) has clearly demonstrated these relationships for diverse, rapidly evolving Cretaceous mollusk lineages (n = 45), which have been studiedin regard to details of their evolutionary history by Kauffman (bivalves), and by W. A. Cobban (ammonites) and N. F. Sohl (gastropods) of the U.S. Geological Survey. Measures of speciation rates (number of new species added in each lineage per 1 million years) relative to various environmental fluctuations in the Western Interior Cretaceous and in the Gulf Coastal Plain (Kauffman, 1970, 1972) show strong correlation between rate fluctuations and major environmental fluctuations connected with transgressive-regressive pulses in marine cyclothems. Typical plots are shown in Text-figures 5 and 6. A summary of this work (Text-fig. 7) shows the following relationships among taxa initially selected for their rapid evolutionary potential.

1. Early transgression is marked by low-diversity assemblages with numerous opportunistic species (e.g., simple ostreid, inoceramid, and pterioid bivalves, ubiquitous ammonites), and the rootstocks of the more stenotopic taxa that will dominate later phases of marine history. Eurytopic taxa are relatively abundant. Rapid evolution is limited to opportunistic forms (at rates that are less than those of selected taxa later in the cyclothem). Although environmental conditions are ameliorating during this time, the opportunists are stressed early in the history of the cyclothem by competition from more specialized taxa, which mounts during

---

Text-Figure 5
Speciation rates (number of new species added to the lineage per million year interval) among diverse Cretaceous mollusks from the Cretaceous of the Western Interior United States, plotted against a radiometric scale (bottom) and the transgressive (T)-regressive (R) marine history of the basin. Eurytopic (many inoceramid bivalves, scaphitid and baculitid ammonites) and moderately eurytopic ostreids are plotted. Note high rates of speciation associated with regression, especially during last part of the first (fastest; 2 MY) regression among all groups, and slower but rising evolutionary rates during transgression throughout. Baculitid ammonites show the relationship between evolutionary rates and low stress (transgressive phase, low evolutionary rate) and high stress (regressive phase, high evolutionary rate) most clearly. Ostreid data in Coniacian to Maastrichtian stages are too sparse to be meaningful. (After Kauffman, 1970.)

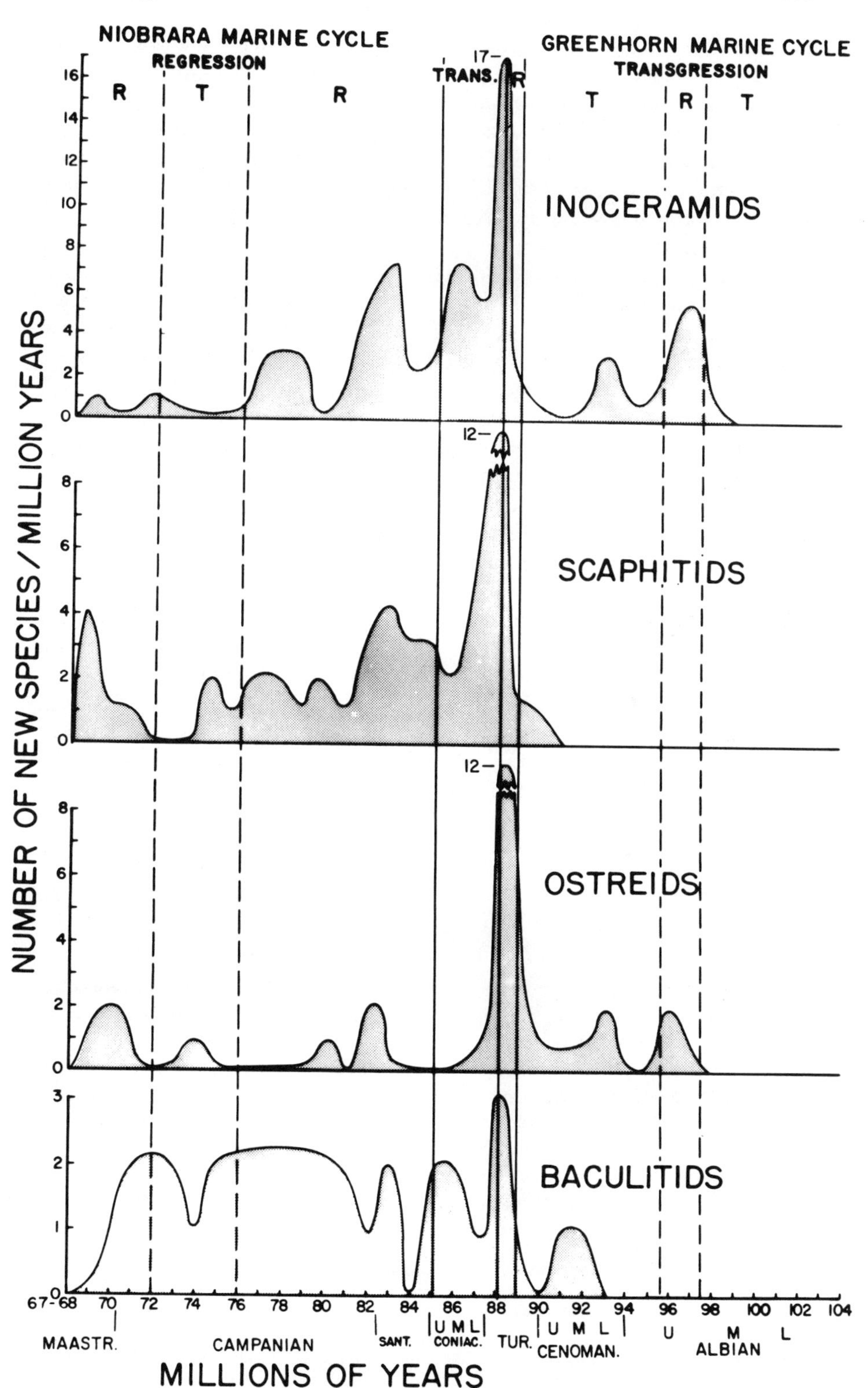
NIOBRARA MARINE CYCLE
REGRESSION
GREENHORN MARINE CYCLE
TRANSGRESSION
R
T
R
TRANS.
R
T
R
T
INOCERAMIDS
SCAPHITIDS
OSTREIDS
BACULITIDS
NUMBER OF NEW SPECIES / MILLION YEARS
MILLIONS OF YEARS
MAASTR.
CAMPANIAN
SANT.
U M L
CONIAC.
TUR.
U M L
CENOMAN.
U M L
ALBIAN

transgression and eliminates or greatly restricts most opportunist lineages within a relatively short time.

2. Middle transgression is characterized by steadily increasing but never very rapid rates of speciation and evolution among eurytopic and moderately stenotopic organisms (much less than later in the cycle) and introduction of specialized stenotopes, all in response to expansion of ecospace, diversification of niches, low levels of competition, and increasingly favorable environments, promoting increasing levels of radiation.

3. Maximum transgression is marked by rising speciation and evolutionary rates, especially among moderately to highly stenotopic organisms that are reaching the peak of their radiation (e.g., see Text-figs. 3 and 4). Near the peak of transgression in the Western Interior, floods of Tropical waters moved far into the Cretaceous seaway. This temperature change, possibly abrupt, constitutes the first major environmental stress crisis, and brings about widespread extinction of many more restricted stenotopes (e.g., specialized ammonites, offshore bivalves, gastropods), creating an almost catastrophic faunal boundary. In addition, it triggers rapid evolution among other stenotopic and more open water eurytopic taxa (Text-figs. 3 and 4). This accounts for the sharp rises in speciation rates in Text-figs. 5 and 6 among certain ammonites and bivalves.

4. Early regression triggers very rapid evolutionary rates among highly stenotopic taxa in all Cretaceous environments except the nearshore clastic zone, followed by widespread extinction of stenotopes toward middle regression. This occurs even though only minor change in environment is obvious from the sedimentary sequence; but it appears to coincide, in part, with an end to the brief flood of Subtropical waters into the Interior, and with marked lowering of water temperatures to Warm- and Mid-temperate levels. Planktonic microbiota, aberrant heteromorph ammonites, most Tropical and Subtropical bivalve stocks, and many specialized gastropods show their highest rates of evolution at this time, and soon become extinct. Kauffman (1972) illustrates this for many lineages of Cretaceous mollusks.

Text-Figure 6

Speciation rates (number of new species produced in lineages per million year intervals) among lineages (genera, subgenera; represented by different patterns) of the ammonite family Scaphitidae during the Cretaceous history of the Western Interior of North America, plotted against a radiometric scale (base, and Text-figs. 1 and 2) and the transgressive (T)–regressive (R) history of the basin. Note high speciation rates in this eurytopic group associated with regression (especially late regression), with maximum evolutionary rates during the first (Middle Turonian) and fastest (2 MY) regression. Note also proliferation of lineages during the Santonian–Early Campanian regression, suggesting that rapid evolutionary response to increasing stress conditions during regression extends in some cases above the species level. (From Kauffman, 1970.)

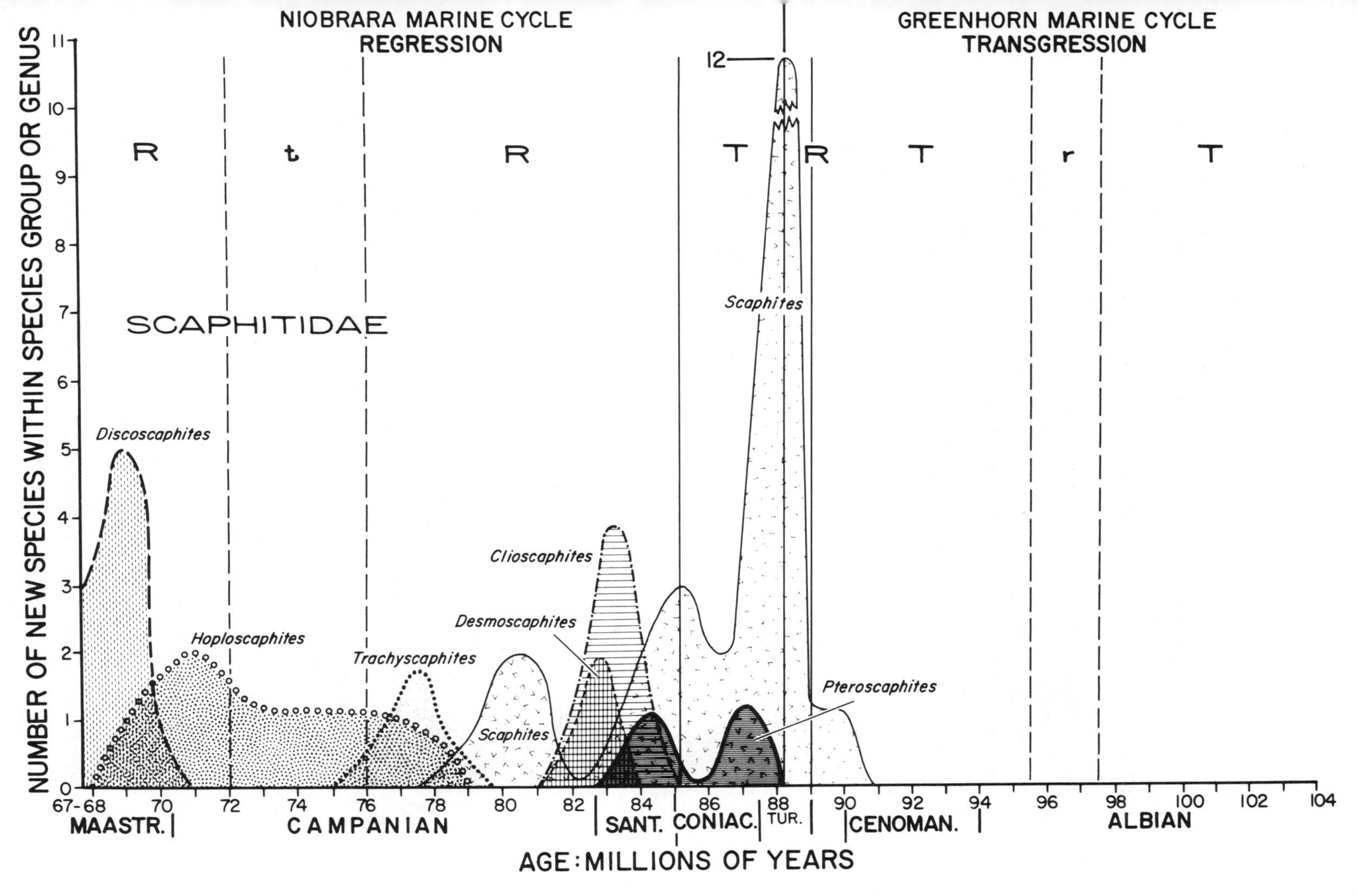

NIOBRARA MARINE CYCLE
REGRESSION
GREENHORN MARINE CYCLE
TRANSGRESSION
12
R
t
R
T
R
T
r
T
SCAPHITIDAE
Scaphites
Discoscaphites
Clioscaphites
Desmoscaphites
Hoploscaphites
Trachyscaphites
Pteroscaphites
Scaphites
NUMBER OF NEW SPECIES WITHIN SPECIES GROUP OR GENUS
0
1
2
3
4
5
6
7
8
9
10
11
67-68
70
72
74
76
78
80
82
84
86
88
90
92
94
96
98
100
102
104
MAASTR.
CAMPANIAN
SANT.
CONIAC.
TUR.
CENOMAN.
ALBIAN
AGE: MILLIONS OF YEARS

5. Middle regression shows the most rapid evolutionary response among a great diversity of benthonic invertebrates that are moderately specialized in habitat ("moderately eurytopic"). This is a time when the first obvious and severe environmental stress effects of the regression, involving decreasing ecospace, restriction and ultimate elimination of many habitats, increased competition, decreasing depth, increasing current action and turbidity, etc., take effect. Kauffman illustrates this for many gastropods and bivalves from the Gulf Coast Cretaceous (1972).

6. Finally, late regression brings the most severe stress situations yet encountered by organisms within the basin, and these reach even the tolerant eurytopic organisms, such as baculitid and scaphitid ammonites and many inoceramid and ostreid bivalves. These respond with extreme increases in rates of speciation and decreases in species duration (Text-figs. 5 and 6). So do the stenotopes and moderately eurytopic taxa of the nearshore clastic zone. The most rapid speciation rates known from the Cretaceous occur here, and, significantly, stress is highest. This is also a period where there may have been widespread isolation of gene pools throughout the shrinking basin.

These relationships are summarized in Text-figure 7 and modeled in Text-figure 8. Acceleration of the transgressive or regressive events in time merely accelerates the response of various organisms at different stages of cyclothem history. Very rapid transgression or regression may create massive radiations or extinctions as these distinct speciation events become crowded together in time, forming marked biostratigraphic boundaries.

CONCLUSIONS

Knowledge of stress levels inherent in each step of any large-scale environmental fluctuation, the reaction to stress gradients of various types of rapidly evolving organisms in relation to their ecological and biological characteristics, and the rate at which stress or nonstress situations are imposed on an area allow some important predictions to be made and applied to the quick, efficient construction of a detailed biostratigraphic system, as outlined in Text-figure 8.

First, among diverse fossil organisms, specific groups can be selected from the outset because they have biological and ecological characteristics which *predict* that they will evolve faster than other organisms. In the methodology modeled here, the initial biostratigraphic system utilizes these, and the remaining taxa are set aside for later study and possible integration into complex assemblage-zone systems, which are more dependent on origin and extinction points than on short-range zones in their construction. Among factors that control evolutionary rates in marine organisms, the nature of the habitat and the adaptive characteristics of the organism to that habitat are of primary importance, as is the nature of change in the surrounding environment(s). Organisms with specialized feeding mechanisms and food resources, and those which live in the upper water column or as benthos in shelf-depth habitats, where they are not clearly protected by deep burial or by a deep water column from unpredictable environmental perturbations approaching or

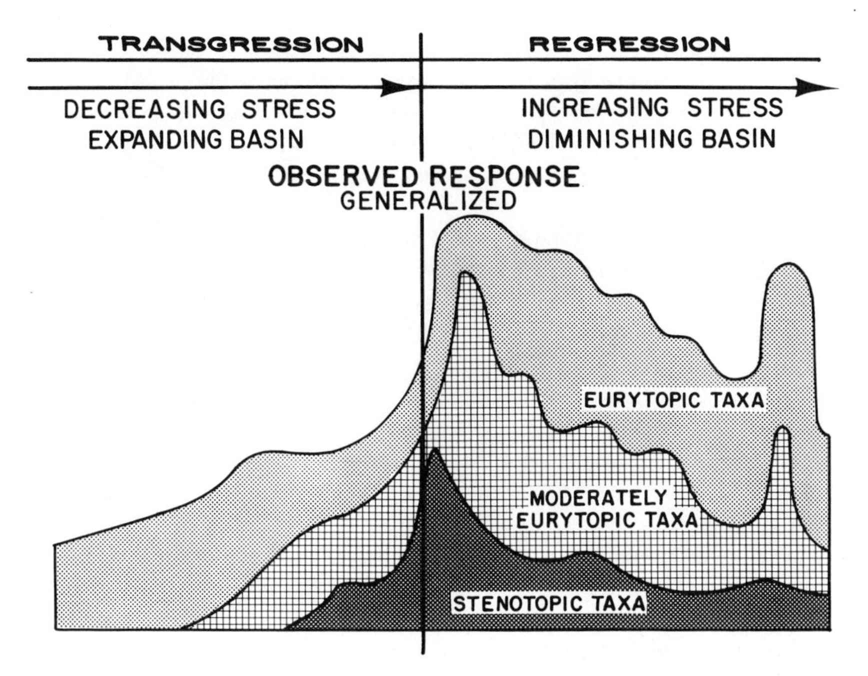

Text-Figure 7
Model of evolutionary response of diverse Cretaceous molluscan lineages with distinct ecological characteristics (eurytopic to stenotopic) and stress-tolerance limits to environmental (stress) changes associated with global marine cyclothems in the Western Interior Cretaceous basin (after Kauffman, 1972). Height of curve diagrammatically represents speciation rates (number of species per million years within lineages) and also reflects species duration. Slow rise in speciation rate during transgression is a combination of radiation among varied organisms into expanding environmental opportunities within basin, and stress reaction (competition), producing rapid evolution within early colonizing, opportunistic lineages. Note graded peaks of evolution from the most stenotopic to the most eurytopic organisms as regression proceeds and stress factors become more severe. Peak of evolution in moderately eurytopic taxa near peak regression is in nearshore clastic facies faunas.

exceeding their environmental tolerance limits, are specifically those organisms which can be predicted to evolve more rapidly than others. Ecologically more specialized organisms and environmentally less tolerant forms (stenotopes) will evolve faster at lower overall levels of stress in a marine basin (and thus more frequently) than will ecological generalists or eurytopic organisms. Eurytopes, however, are capable of rapid evolution under stress of high intensity. Organisms, and subpopulations within species, that have higher genetic variability are more likely to evolve at rapid rates than genetically conservative organisms; this relates to stress insofar as genetic variability, heterozygosity, and polymorphism may all be characteristics of subpopulations that are living at the ecological limits of the species, and thus in the zone of rapidly increasing stress gradients. Small population size and genetic isolation of these subpopulations enhance rapid evolution. Structural grade and morphological complexity alone are *not* good indicators of the potential for rapid evolution among organisms.

In the biostratigraphic method proposed here, selection of a suite of organisms with the potential for rapid evolution is followed by subdivision of that suite into groups whose characteristics predict rapid evolutionary response to varying degrees of stress and various kinds of stress factors (e.g., temperature change, competition, fluctuations in resources, etc.). Thus some groups (e.g., opportunists) might undergo their most rapid evolution at times of seemingly low stress in the physicochemical environment, but when they are subjected to severe biological competition by more specialized organisms migrating into an area of ameliorating marine climates and expanding environmental opportunities. Other groups (e.g., most eurytopic forms) might undergo rapid evolution only under the most severe stress conditions in a basin. *Each level of development of physicochemical and biologically controlled stress gradients during the long-term geological and environmental history of an area should be characterized by a certain suite of organisms that, for perhaps different reasons, will predictably undergo rapid evolution.*

Thus, in our Cretaceous example, a widespread transgression of the Western Interior epicontinental sea would seem to represent the most equitable possible environmental situation for marine organisms and a low stress period. Yet there are major bursts of rapid evolution among diverse organisms at and just after the peak of maximum transgression. Why? This suite of organisms (diverse mollusks, microbiota, etc.) seemingly have little in common ecologically, yet they are the backbone of this part of the biostratigraphic system because of their rapid evolutionary rates.

Text-Figure 8
Diagrammatic scheme of biostratigraphic system (zones numbered, right column) developed from concepts presented in this paper. With the basic data on lithostratigraphic succession, and thus changing paleoenvironments and natural selective forces (the cyclothem model, left), and with a radiometry (left column) for the measurement of evolutionary rates, distinct groups of diverse organisms (A–D) can be predicted from their ecological and biological characteristics to evolve rapidly in response to certain sets of environmental conditions and stress levels, and tested against the time–environment matrix. These are selectively applied to the construction of a biostratigraphic system at each phase of environmental change (difference in range-zone lengths indicate differences in time duration of species; model assumes no diastems and equal sedimentation rates throughout). Thus group A is primarily used to zone early transgression, group B peak transgression and early regression, group C middle regression, group D late regression, etc., because environmental conditions during each of these phases of the cyclothem are known to specifically trigger rapid evolution in members of each diverse group.

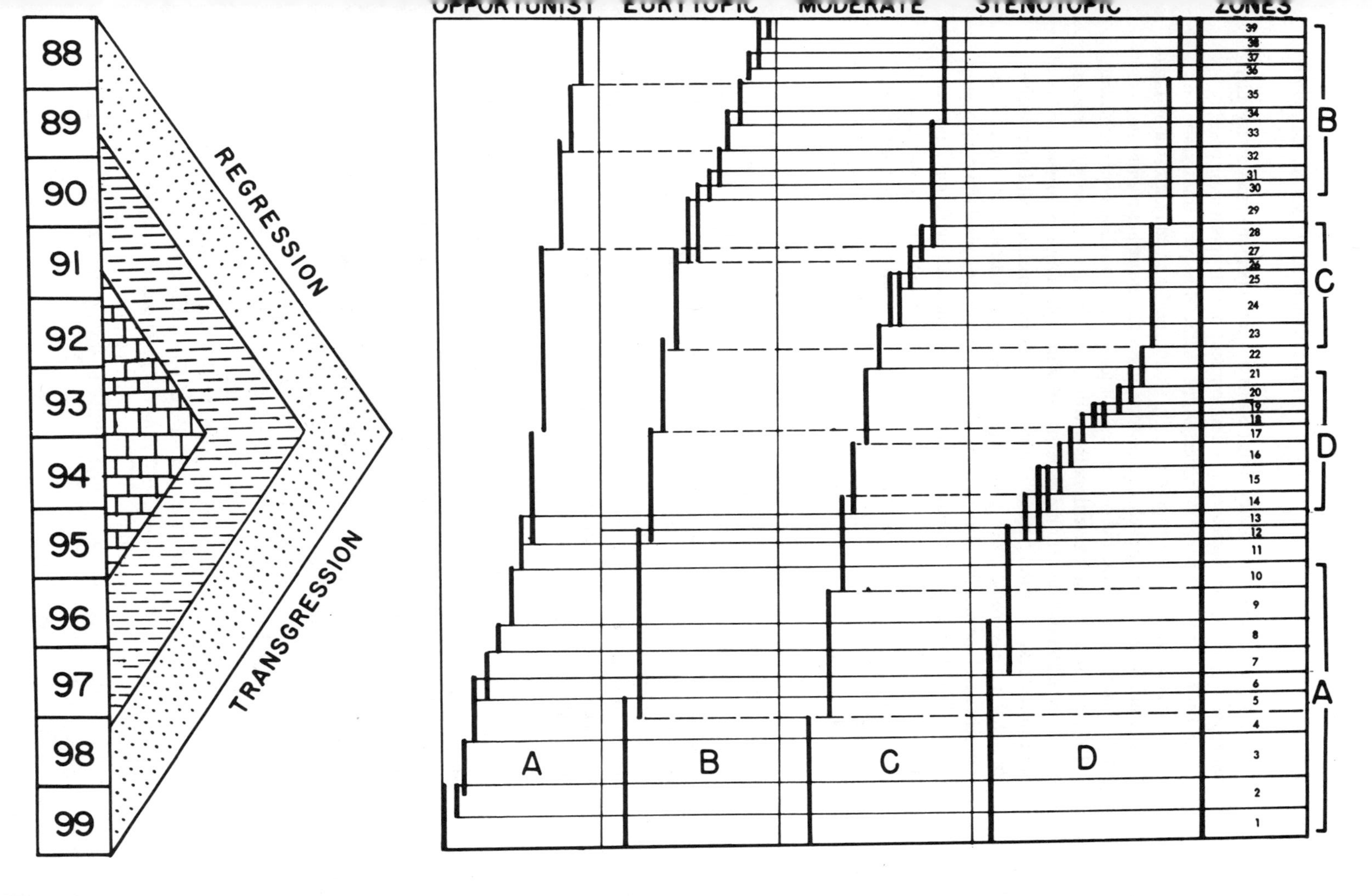
88
89
90
91
92
93
94
95
96
97
98
99
REGRESSION
TRANSGRESSION
A
B
C
D
1
2
3
4
5
6
7
8
9
10
11
12
13
14
15
16
17
18
19
20
21
22
23
24
25
26
27
28
29
30
31
32
33
34
35
36
37
38
39
B
C
D
A

The answer is complex but predictable. Some of these taxa are highly stressed by widespread temperature increases and development of more normal or possibly somewhat higher than normal salinities, which characterize peak transgression of Cretaceous seaways. Other taxa (especially stenotopes) are highly stressed by rapid though subtle environmental changes, which affect the entire basin with the first stage of marine regression and return to "normal" basinal environments immediately following peak transgressive flooding. Still other taxa are stressed by increasing competition and pressure for niche partitioning brought about by massive influx of Subtropical taxa on warm normal marine waters characterizing peak transgressive flooding. Finally, some taxa radiate rapidly at this time because their prime environment develops rapidly, or their prime food resource (e.g., planktonic microbiota) quickly flourishes during periods of peak transgression; their evolution is not related to stress but to rate of emplacement of prime environments. Possessed with a knowledge of those specific organisms among diverse fossil biotas that will be expected to evolve rapidly during their evolutionary history, and the conditions favoring their periods of rapid evolution, all of the above is broadly predictable.

This predictive ability allows the biostratigrapher to selectively apply certain types of organisms to zonation of nearly any set of changing geological environments, reflected in facies progressions, with the confidence that for the most part he or she will be sampling from the outset the best data (rapidly evolving species) for construction of a refined biostratigraphic system with zones of short duration in any given basin. Text-figure 8, based on the Western Interior Cretaceous data, idealistically represents this principle in a marine cyclothem and suggests that, for organisms which evolve rapidly in response to stress, four major divisions of the cyclothem may be made, with different suites of taxa characterizing each by their period of greatest evolutionary turnover: (1) an early transgressive phase of rapid evolution mainly among opportunistic taxa; (2) a peak transgressive phase of rapid evolution mainly among highly stenotopic taxa sensitive to a variety of environmental parameters (temperature, competition, etc., as explained above); (3) a middle regressive phase characterized by rapid evolution among a great variety of moderately eurytopic to moderately stenotopic taxa; and (4) a final peak regressive phase characterized by eurytopic taxa that react by rapid evolutionary turnover only to the most severe stress situations. Each division would then be selectively zoned by the biostratigrapher (Text-fig. 8, right column) initially on those select organisms with rapid evolutionary response to that set of environmental conditions. This system also has the advantage that, for the most part, each zone so constructed would probably have one or more "index" taxa, drawn from rapidly evolving lineages, whose stratigraphic-time range was restricted to the zone.

Following construction of this initial, but highly refined, biostratigraphic system, all other known taxa could then be studied as time became available and integrated into the biostratigraphic system, increasing the potential for complex assemblage zonation (as urged by Kauffman, 1970) and even greater stratigraphic refinement. But while this process was going on, a refined biostratigraphic zonation would be in operation.

Acknowledgments

I am greatly indebted to my coeditor, Joseph E. Hazel of the U.S. Geological Survey, for his monumental contribution to this entire volume. The paper was reviewed by Dr. J. B. C. Jackson of Johns Hopkins University and Dr. Jeff Levinton of the State University of New York, Stony Brook. I am indebted to these colleagues for their thoughtful criticism, which has done much to improve the quality of the paper. My thanks also to Ms. Barbara D. Heffernan, who assisted in the collection and plotting of data for this paper, and to Mr. Larry Isham, who did most of the art work; both are in the Department of Paleobiology, U.S. National Museum. This work was supported by grants from the Smithsonian Research Foundation (SRF 430057, SRF 430020, SRF 437216), and was largely written while I was Visiting Professor at the University of Tübingen under a German government grant to the Institut für Geologie und Paläontologie. This is Publication No. 40 of the research program "Fossil – Vergesellschaftungen" within the Sonderforschung bereich 53 (Palökologie) of the University of Tübingen. The financial support of these organizations is gratefully acknowledged.

Biogeography and Biostratigraphy

James W. Valentine **University of California, Davis**

INTRODUCTION

The fossil taxa and fossil assemblages of which biostratigraphic units are composed are clearly not found everywhere that rocks of appropriate ages occur. Instead, the vast majority are relatively restricted in geographic range, although at certain times some fossil groups range broadly throughout whole ecological realms. The chief purpose of this paper is to examine the factors that permit or require either limitations or extensions of the geographic ranges of taxa, communities, and provinces, and to explain the principal effect of biogeographic processes on the biostratigraphic record. As the fossil record is best for shallow marine invertebrates, the discussion is largely restricted to these organisms, although the principles should apply to all ecological realms.

TAXA AS BIOGEOGRAPHIC UNITS

Species as a Fundamental Biogeographic Unit

Species are usually considered to be the most fundamental biological unit in nature. Their internal dynamics, based on their gene pools, are at any time independent of all other species, and their evolutionary futures are distinct. They are the basic unit that undergoes evolution; the species niche is the basic functional unit involved in ecological interactions; and species are the basic units of biogeography. They are also the fundamental units of biostratigraphic zonation and correlation.

At any time, the unique gene pool of each species is realized in the genotypes of all living conspecific individuals. Each realized genotype has interacted with the environment to produce a phenotype, an individual organism with a distinctive set of environmental requirements and tolerances that we shall call its functional range. As genotypes can be recombined with each generation, there are nearly always many more genotypes possible than are actually realized at any time. The pooled functional ranges of all possible phenotypes that can develop from the gene pool of a species constitutes the prospective functional range of that species, its *prospective niche* (see Hutchinson, 1957; Valentine, 1969, 1973b). However, the realized functional ranges, *realized niches*, of species are far more restricted than their prospective niches. Partly this is because all potential genotypes will not be present at any time; partly it is because many phenotypes express only part of the functional range inherent in their genotypes; and, partly, environments in which some functions may be expressed do not coincide with the appropriate phenotypes (Text-fig. 1). For most purposes, the concepts of prospective and realized niches are adequate in considering the biogeographic and biostratigraphic patterns of species.

As every species has certain environmental requirements, it obviously cannot occur regularly in regions where these requirements are not met. No species (man just possibly excepted) appears capable of existence in all earthly environments; rather, each is restricted to regions where its appropriate environment prevails. The factors that prevent a species from occurring in any place are termed *limiting factors.* These may include such physical parameters as temperature and salinity, such biological parameters as food and predators, and such population parameters as a minimal habitat area capable of supporting a viable population structure. The total region where all these requirements are met for any species constitutes the *prospective range* of that species; the region that is actually occupied at any time or over any time period is the *realized range.*

Marine species vary greatly in their powers of dispersal; this affects their ability to spread into all suitable regions, and thus to fill out their prospective ranges and to maintain gene flow between populations that are separated geographically by unsuitable regions. Since the geographic range of a species greatly affects its usefulness in biostratigraphy, dispersal is of great interest, for it bears on problems of detailed regional and interregional correlations. There have been relatively few direct studies of marine invertebrate dispersal, although biogeographic evidence has suggested plausible generalities (Thorson, 1961), which have been supported by recent work. For example, Berger (1973) has shown that among western Atlantic species of the intertidal gastropod genus *Littorina,* those that lack a pelagic larval stage are differentiated geographically with respect to allele (gene) frequencies, whereas a species with a pelagic larval stage gave no evidence at all of such differentiation.

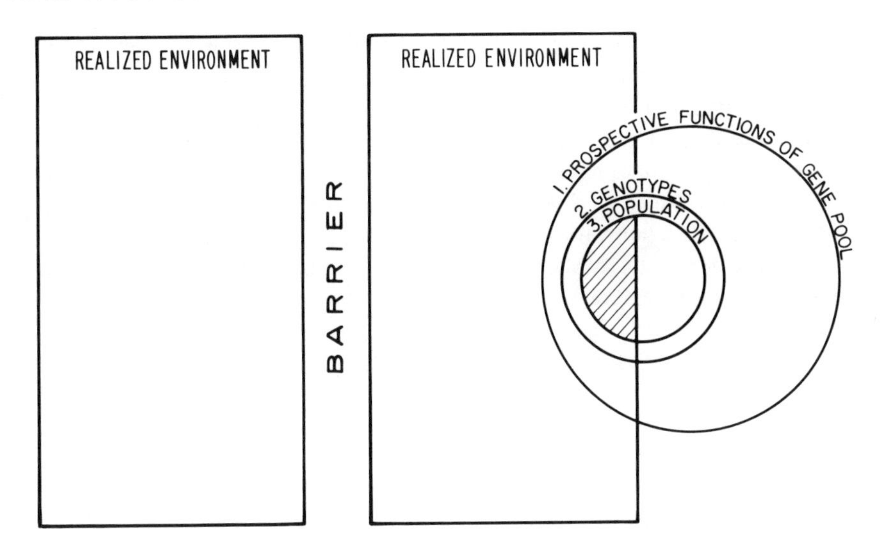

Text-Figure 1
Two-dimensional depiction of a multidimensional model of the functional ranges of various aspects of a population, including the prospective functions of the gene pool (circle 1), genotypes (circle 2), and actual phenotypes of a population (circle 3). Only functions that happen to overlap a realized environment are realized functions (crosshatched portion of circle 3). Realized environments may exist that could overlap with a population's functions, as at the left above, but from which the population is barred by intervening, unsuitable conditions.

This is interpreted to mean that pelagic larval dispersal favors interpopulation migration and, therefore, gene flow. Furthermore, Scheltema (1971b) has studied pelagic larvae from plankton tows in the Tropical and Warm-temperate North Atlantic, and has found that numbers of Tropical gastropods have particularly long-lived pelagic larvae, many of which have a developmental period exceeding 3 months and some exceeding 10 months (see also Thorson, 1961). Most of these long-lived larval forms can easily cross the Atlantic via the main surface currents; Scheltema has calculated that those species common enough to be detected in sampling must be common enough to complete the journey frequently. Indeed, there is a correlation between the commonness of the larvae collected and of the similarity of amphi-Atlantic populations of adults of these species. In general, species with the more common larvae have the more similar populations on either side of the Atlantic; those with the rarer larvae tend to be represented by different subspecies. Evidently, species with the more common oceanic larvae have significantly higher trans-Atlantic gene flow.

These observations tend to bear out plausible inferences and generalizations concerning the dispersal powers of marine invertebrates. Species with the more long-lived and hardy pelagic larvae have the greater chance to be widely dispersed after reproduction, and frequent reproduction of large broods also enhances migratory potential. Therefore, species with such attributes would commonly be able to colonize habitats that lie at some distance from their parental ranges, and would usually be able to maintain gene flow to such outlying populations. Species with shorter planktonic developmental periods, smaller broods, or more restricted larval requirements would tend to colonize only localities that are fairly close to their parental regions. If a population became established at any considerable distance from others, gene exchange might be sporadic or lacking altogether, leading to divergence between the colonists and the parental population, and a reduction in their usefulness in correlation.

For a species in a given locality, then, a geographic range exists for which colonization is essentially obligatory, as the region lies within the normal migratory range of the population; thus, by some standard time, occupation is virtually assured. This can be called the *local range.* It may not be filled at any particular time because of local extinctions due to temporal environmental fluctuations, to some seasonal rhythm in the biology of the population, or to some temporary reproductive failure or similar event, but usually most of the local range will be occupied. Of course, no species occurs absolutely everywhere within its local range either, for environments are heterogeneous; the habitat requirements of sessile species are more or less patchily distributed, while the population density of vagile forms must be regulated at some level well below saturation.

For at least some species, there are clearly many regions in the world where they could persist if they could reach them in appropriate numbers, but from which they are barred by the breadth of intervening areas with unsuitable habitat conditions. Such inhospitable areas are called *barriers* to species dispersal. Broaching of narrow barriers may occur commonly; broaching of wide barriers may be extremely unlikely. But given a long time period, even a few very broad barriers may be crossed by some species. The spread of species is thus a stochastic process, as Simpson (1953a) especially has emphasized. Once a barrier is crossed, the colonizers will expand their range at the new location until it is circumscribed by barriers, filling out their new local range. Subsequently, they may broach other barriers and create additional local ranges. The pattern of migration may be one of hopping from one habitable region to the next across barriers of varying strength, episodically expanding the total species range.

Just what percentage of species in the ocean are capable of living in areas that they have not reached is not certain; but judging from the occasional appearance of exotic species when opportunities are provided by man, it may be a significantly high number. Classic examples of the survival of exotic populations in new regions include the house sparrow in North America, the rabbit in Australia, and the giant African snail in Hawaii. There are many less spectacular cases that involve marine species (see Hanna, 1966). In these cases the barriers have been broached as a result of man's activities. Probably, most species have had enough chances to broach their present barriers so that few can be expected to do so naturally in the future, barring widespread environmental change. On the other hand, when environmental changes do occur, they are expected to permit some range expansions and often to require range contractions. In sum, the biogeographic pattern of a species results from an interplay between the species tolerances, requirements, and vagility on one hand, and the structure of the environment on the other – particularly upon the geography of the limiting factors for that species.

Species may have several different combinations of the factors that ordinarily lead to different geographical distribution patterns. For given environmental states, several patterns may be envisioned as outlined in Table 1. These patterns may nearly all be found in the sea today. For an example, the environment is relatively homogeneous in the present Indo-Pacific province, especially in climate and in the aspect of the coral reef community there, so that many species range from Africa east to the Marquesas and Tuamotu island groups, a distance of over 13,000 miles (about 21,000 km). Even relatively specialized forms, such as the "killer clam" *Tridacna maxima,* may range across nearly the entire reef belt; this species is widespread, not so much because it is extremely tolerant, but because the environment to which it is adapted happens to be widespread (Table 1, category 7). Of course, there are even more narrowly adapted forms restricted only to the central portions of the reef belt, such as *Tridacna gigas,* the giant relative of *T. maxima.* On the

Table 1
Simplified Relations of Species That Have Different Larval and Adult Adaptations to Biogeographically Homogeneous and Heterogeneous Conditions[a]

	Environment	
Larval Vagility	*Spatially Homogeneous*	*Spatially Heterogeneous*
High		
adult flexible	1. cosmopolitan ubiquitous	2. cosmpolitan ubiquitous
adult specialized	3. cosmopolitan nonubiquitous	4. widespread discontinuous nonubiquitous
Low		
adult flexible	5. cosmopolitan ubiquitous	6. restricted locally ubiquitous
adult specialized	7. cosmopolitan nonubiquitous	8. restricted patchy

[a]Conditions refer here to range-limiting factors and not to the conditions within the habitat mosaic that are so important within communities.

other hand, relatively flexible, broadly adapted species may be less widespread than either of these forms, because the particular range of environments that they tolerate, even though quite broad, happens to be restricted geographically or to be surrounded by nearly impenetrable barriers (Table 1, category 8). This situation is broadly true for shallow-water forms from Temperate provinces along north-south trending coastlines. In general, they tolerate broad environmental fluctuations, but have relatively narrow ranges circumscribed by climatic changes latitudinally and by habitat failure (to terrestrial and to deep-water conditions) longitudinally.

While many species have relatively continuous ranges, others are clearly disjunct, occurring in two or more regions with intervening gaps. On the basis of the patterns of marine shelf species along California and Baja California, disjunct ranges are not uncommon even along continuous coastlines. For example, some species of marine mollusks end their southern ranges along the California mainland in central California, but occur well to the south on the Channel Islands. Other forms that occur regularly in southern California disappear north of about Point Conception or Morro Bay, only to reappear well to the north in Monterey Bay. These disjunctions probably result from irregularities in the water temperature regime along the coast. Disjunct occurrences due to habitat failure are certainly widespread among Indo-Pacific species, which commonly occur on scattered islands separated by deep water. A few of the gaps are relatively spectacular, as in the case of populations of Indo-Pacific species that live also on the tropical west American shelf (for a list, see Hertlein, 1937). Disjunctions also occur on even grander scales, such as the phenomenon of antitropicality (Hubbs, 1952; also termed "bipolarity"). This pattern is characterized by the presence of separate conspecific populations inhabiting subtropical or temperate (or even more poleward) waters in both the northern and southern hemispheres while being absent in the intervening lower latitudes. Hubbs (1952) documents numbers of fish that display this pattern; it is also well-known among pelagic invertebrates, such as the chaetognaths *Sagitta minima* and *S. californica* that are widespread in northern and southern intermediate latitudes but tend to be missing near the equator (Text-fig. 2; Bieri, 1959). Amphi-Atlantic disjunctions in mollusks are well known, as mentioned above in discussing Scheltema's work, and occur in the Pacific also. Indeed, some pelagic species are widely disjunct longitudinally across the Pacific, occurring along the eastern and western sides but avoiding the vast central Pacific regions (Text-fig. 3; Bieri, 1959). Many additional examples of disjunct distributions may be found in Ekman (1953). The presence of a few common species in distant regions that are otherwise faunally dissimilar would be of great help in the correlation of fossil assemblages.

Range Borders of Species

Clearly, it is the local geography of limiting factors that determines the species local range border, and the combination of the species vagility with the nature of barriers that determines the success of a species in spreading into other areas, wherein the local geography of limiting factors will again determine the local range.

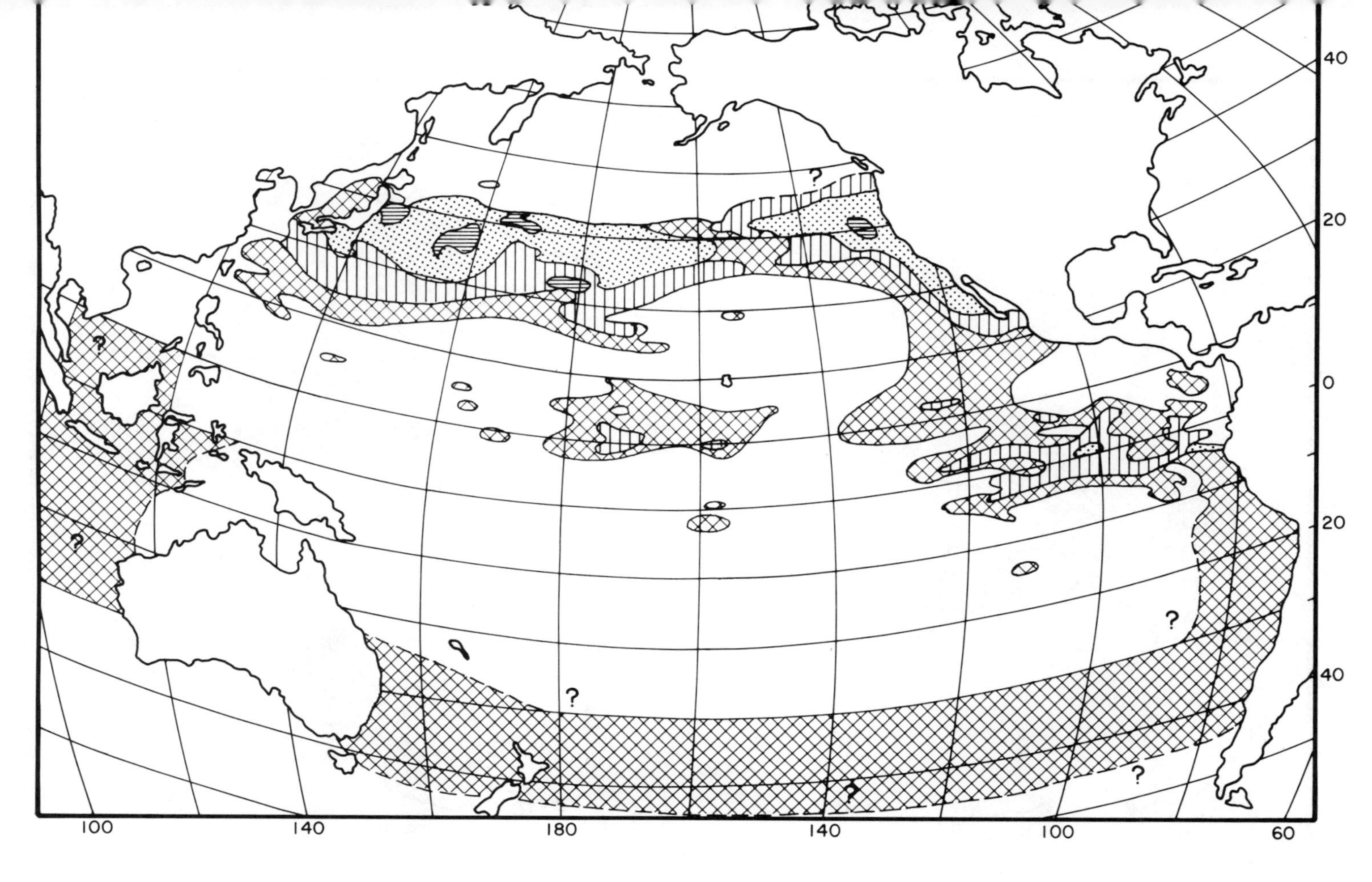

Text-Figure 2
Living distribution of the chaetognath *Sagitta minima* in the Pacific Ocean; it is distinctly antitropical. (After Bieri, 1959.)

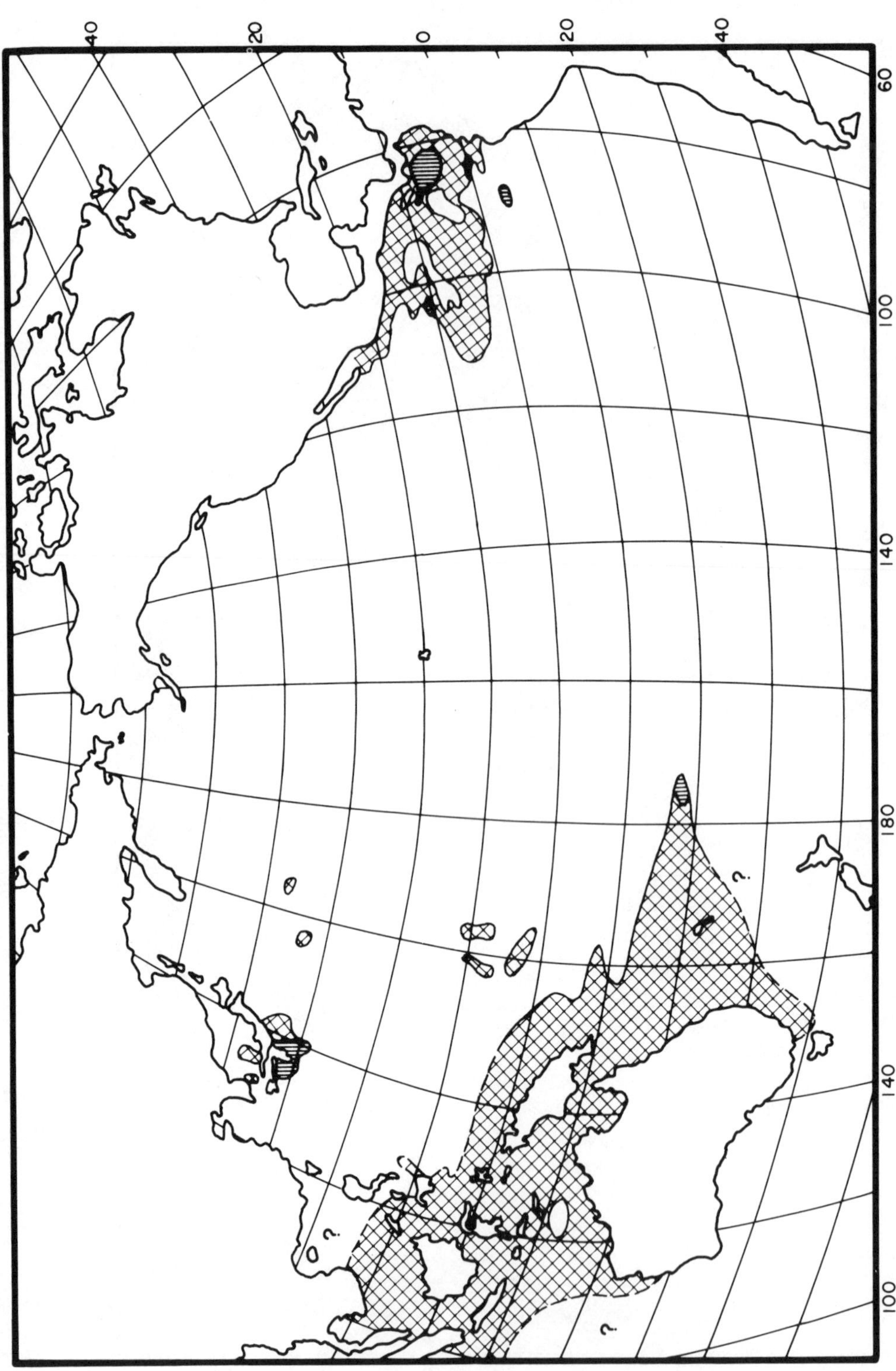

Text-Figure 3

However, species are evolutionary units as well as biogeographic ones, and they are capable of evolving. Certainly, the planetary environment has changed much during the Phanerozoic, and the evolution of new adaptations must have been proceeding much of the time within numerous lineages; the evolution and sudden expansion of new forms that have biostratigraphic utility have been documented (e.g., by Eldredge, 1971). Why then do most species not overcome the barriers at their borders simply by evolving new tolerances?

There appear to be at least two major reasons, which can only be reviewed briefly here. One, proposed by Mayr (see Mayr, 1963), is based partly on a presumed genetic coadaptation of gene pools. The members of a species that are most likely to be subjected to selective pressures for adaptation to conditions beyond the species border are those that actually inhabit the species range margins; they are thus subjected most often and most strongly to conditions that are common beyond the border. Here, novel genes, adaptive to the extralimital conditions, would be most likely to be incorporated into the gene pool. These unusual genotypes will be rare when compared with the vast reservoir of "normal" genotypes present in the central range regions, and they will therefore be broken up by recombination and swamped by "normal" genes. Although a few novel genes may be maintained in marginal populations, truly novel gene combinations that permit range extensions will thus be difficult to assemble and maintain. Furthermore, truly novel genes will usually not be well integrated within the normal genome, so that epistatic interactions would tend to be unfavorable and reduce the fitness of the new gene.

Another difficulty in the way of evolving new tolerances for range expansion is the presence of partial to relatively complete ecological analogues of species in neighboring regions. When two ecologically similar species begin to overlap at a range border, interspecific competition ensues there. The effects of this are added to those of the normal intraspecific competition with which individuals are regularly faced; thus individuals in the overlap region tend to be at a disadvantage. The results of such a confrontation evidently vary according to conditions, but the ranges of the two forms may become partitioned at a common border as a result of the competition. Such processes can act to place at a disadvantage those genotypes that might permit species range expansions, and species borders become stabilized within a particular environmental regime or range of regimes.

Large numbers of the ecological parameters in the sea may serve as barriers to species. Two are of primary importance, however, and seem to account for the majority of range restrictions: (1) habitat failure, as when shelf habitats give way to land or to deep-sea conditions, and (2) temperature, which forms barriers because of the temperature sensitivity of biological reaction rates, frequently affecting larvae more than adults. Because most of the world's shelves and coastlines have a strong north-south component today, habitat failure is now most important in longitudinal directions on a planetary scale, whereas temperature barriers are most important latitudinally. The precise determinants of range end points are truly legion, with just about every factor that can be limiting serving for one species or another, including biological factors such as competition, predation, or amensalism, as well as physical factors. However, it appears that many range limits imposed by these many factors correspond closely today with the changes in temperature and habitat that create the major features of biotic distribution patterns, blocking out the ecological framework of the environment.

Range Borders of Higher Taxa

The geographic ranges of genera are simply the combined ranges of the species of which they are composed, and commonly are larger in extent than any one of their species ranges. Because genetic discontinuities exist between all the congeneric species, the generic range is not controlled by the structure of a gene pool, except by those of the individual species; it is more or less a collection of independent entities. On the other hand, all congeneric species have descended from a common "trunk" species that founded the genus (assuming that taxonomists have correctly identified a monophyletic genus), and they are expected to share many common attributes. Some genera stretch from equator to poles in rather similar habitats, but are represented by chains of species, all of which have similar ecological habits but each of which occupies a relatively narrow temperature regime. However, many genera are restricted climatically, being entirely or principally tropical, or being found in high latitudes. A number of high-latitude genera extend their ranges equatorward by possessing a species that submerges into deeper, cooler water in lower latitudes. These are usually scavengers and detritus feeders. Durham (1950) has noted that their paleobiogeographic interpretations are complicated by this habit. Other temperature-related depth patterns have been documented by Hazel (1970a). Equatorward emergence occurs in some species that require low winter temperatures for reproduction when winter bottom temperatures nearshore are lower than those offshore, such as occurs off the southeastern United States. In this situation also, poleward submergence of more thermophilic species can occur.

The major longitudinal barriers of ocean deeps and continents separate many genera; many clusters or chains of conspecific species inhabit a given coastline or province but are not found across the continent or on opposite oceanic shores.

Families tend to be more widespread biogeographically and broader ecologically than genera, since they are composed of clusters of different though allied genera. Their range limits are therefore composed of numerous range limits of independent species. Families are less frequently restricted than genera to a given climatic zone or coastline, although some are endemic to rather small biogeographic regions. It is certainly true that the characteristic features of a family that permit it to occupy its distinctive adaptive zone (Simpson, 1944, 1953b) or ecospace (Valentine, 1969) sometimes restrict it to certain regions. Families of hermatypic corals with their symbiotic zooxanthella live only in tropical and subtropical waters. Orders and classes are on the average progressively more widespread than families, being represented by more species on the average with more diverse adaptations, and phyla are usually quite widespread. Nevertheless, many phyla (about one third of them) are nearly or quite restricted to the sea, and some small phyla, such as the Phoronida and Pogonophora, have ecological and biogeographical ranges no larger than many genera.

COMMUNITIES AND PROVINCES AS BIOGEOGRAPHIC UNITS

Since some biostratigraphic units are based upon single species or upon single taxa that consist of closely allied species, their usefulness is clearly a function of the extent and pattern of the geographic distribution of taxa involved. However, other biostratigraphic units are based upon assemblages of species that are not closely related phylogenetically, but that share ecological features that have brought them into consistent association. Usually, the assemblages are formed by species that lived in the same community. In this case, the factors that created community distribution patterns, and that formed provincial boundaries, also limited the geographic extent of biostratigraphic units, and are therefore of practical importance in biostratigraphy. We shall consider some biogeographic properties of formal biostratigraphic units later; it is appropriate to first examine the environmental controls on the distribution patterns of communities and provinces.

Communities are second-order biogeographic units. We shall regard communities as characteristic associations of species populations that inhabit some certain range of habitats, termed a *biotope.* The boundaries of communities are thus set by the factors that limit them to a certain number of component species, enough species so that the community composition beyond the boundaries is altered sufficiently that the characteristic species association is not realized. These limiting factors define the margin of the biotope; commonly, different factors combine to create the biotope margin in different places.

Community patterns are somewhat analogous to species patterns in that communities do not occur everywhere within their total ranges but only in local stands, sometimes markedly disjunct, where the biotope is realized. The total community range is limited by the occurrence of barriers that are too severe or too broad to be broached by enough of the species to permit recurrence of the community. In theory, a biotope, beyond the usual dispersal range of most species, could be inhabited by a characterizing set of species of the community if they happened to reach it, so a stand of the community could occur there if the barrier were broached by the appropriate species. In practice, this is very unlikely to happen. There are biotopes on opposite sides of the Tropical Atlantic Ocean wherein many species now restricted to one or the other sides of the ocean could evidently exist if given the opportunity. As we have seen, only those species with truly long-lived larvae manage the trip regularly enough to be established and indistinguishable on each side of the Atlantic. Of course, it may be that such biotic factors as competition would prevent the invasion of many new forms from one side to the other, even if many of the larvae were suddenly transported by some unusual event. At any rate, as matters stand, there are distinctively different communities in similar biotopes across the Atlantic, even though they share some species.

As with species, the chief determinants of the extreme range end points or range borders of communities are temperature, having chiefly a latitudinal effect, and habitat (biotope) failure, which may operate in any direction but has a largely longitudinal effect today. Latitudinal boundaries of shelf communities are chiefly localized in regions where the latitudinal temperature gradient is most steep and temperature changes are relatively abrupt. Usually, numbers of different communities will end their ranges in the same region, in response to such abrupt changes, so that the species composition of the entire fauna changes significantly. These changes are termed *provincial boundaries.*

Provinces are third-order biogeographic units composed of characteristic associations of communities and therefore of species. The differences in species composition between contiguous provinces at present clearly vary greatly from one pair of provinces to another. Living provincial biotas are not well enough known that we can calculate their differences precisely. From studying the best-known taxonomic groups, however, we can form useful estimates. Mollusks are the most diverse skeletonized phylum in the marine benthos; along the Pacific coast of North America the molluscan species compositions of adjoining provinces differ by well over 50 percent and by over 80 percent at the boundary of the tropical Panamanian province (Valentine, 1966). On the Atlantic coast, one region usually recognized as a province is the Virginian, lying between Cape Hatteras and Cape Cod (see Hazel, 1970a). The number of endemic molluscan species in this region is so low (10.5

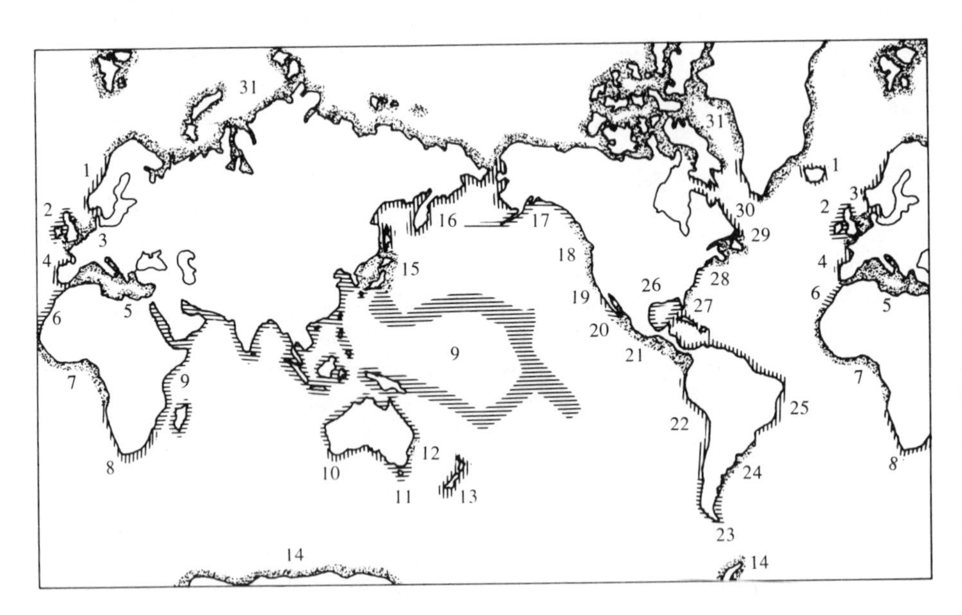

Text-Figure 4
Shelf and shallow-water provinces at present, based chiefly on marine mollusks. (After Valentine, 1973b.)

percent) that Coomans (1962) did not wish to recognize it as a separate province. Nevertheless, both its boundaries are limiting to large numbers of species, and in fact it appears to be well over 50 percent distinct in species composition from the next region to the south (the Carolinan Province) and over 40 percent distinct from the next region to the north (the Nova Scotian Province). Thus, although levels of endemism are certainly of great interest in biogeography, their use as definitive criteria in establishing biogeographic regions can mask important distributional patterns (see also Kauffman, 1973). Widely separated provinces may be essentially 100 percent different in species composition today.

At present, there are approximately 31 shelf provinces (Text-fig. 4), which represent a very high level of provinciality when compared with the average levels suggested by the fossil record. Usually, there have been only a few major provinces, and at times perhaps only a single province containing all the world's marine shelf fauna (Valentine, 1973b). Since the distributional patterns of provinces, communities, and species of the present are governed largely by the patterns of limiting factors, it is obvious that these patterns have changed enormously through geological time. This history of biogeographic patterns, then, should be primarily a history of changing barriers, although changes in the breadth of adaptation of species must also have played a role (Valentine, 1973a).

MAJOR FEATURES OF MARINE BIOGEOGRAPHIC UNITS DURING THE PHANEROZOIC

Major changes in the framework of the global environment must occur as the positions, sizes, and configurations of continents and ocean basins are altered by global tectonic processes. For example, consider only three major controls of distributional patterns: topographic barriers, temperature barriers, and ocean current patterns. The strength of deep-sea barriers between continental shelves will vary greatly as oceanic widths are altered by plate-tectonic processes. The latitudinal temperature gradient, very high at present, can be greatly reduced by the polar warming that can accompany changes in continental geographies. Ocean currents will run in different patterns as oceanic configurations are altered. Clearly, the distributional patterns of shelf species will change in response to such environmental changes, and this will in turn affect any biostratigraphic units to which they contribute their traces and remains.

To discuss all possible configurations of continents and ocean basins and their concomitant climates and currents, or even the major classes of them, is clearly impracticable. Instead we shall examine three cases that seem to be of particular importance in geological history, and consider the results of evolution from one to another in the order that they are believed to have occurred.

The first case (Text-fig. 5A) is a world composed of a large southern polar continent, Gondwana, with perhaps a few smaller continents chiefly in low latitudes and extending to intermediate latitudes in the opposite hemisphere. This seems to be a general description of conditions during much of the Paleozoic era (Valentine

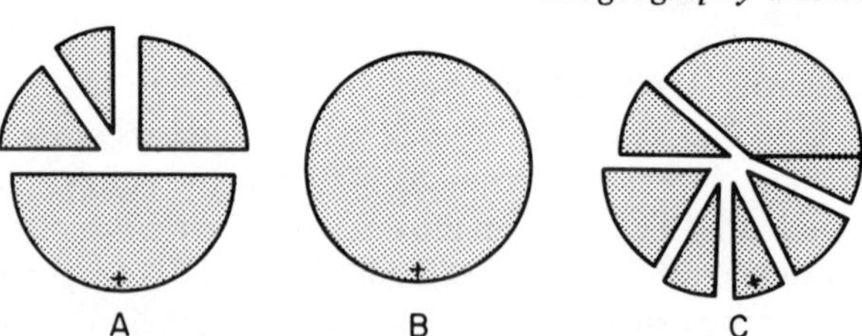

Text-Figure 5
Idealized diagrams of three continental configurations that have radically different biogeographic effects. Crosses represent the south pole: (A) the Paleozoic configuration (it happens to be closest to the early Paleozoic), with a large southern continent and a few others; the north pole was usually oceanic and widely open to the major oceans; (B) Pangaea, with a single supercontinent; (C) present configuration, with several small dispersed continents, including a small south polar continent; the north pole lies in a restricted ocean. (Modified after Valentine and Moores, 1972.)

and Moores, 1970, 1972; Smith et al., 1973; Valentine, 1973a, 1973b). In this situation, the latitudinal temperature gradient would probably be much lower than today; in particular, the southern shores of the polar continent would be washed by a relatively warm current, running south from low latitudes, that rounded the cape separating eastern from western Gondwana shelves and then returned northward, somewhat cooler but nevertheless far warmer than the frigid waters of present Antarctic and subantarctic regions. Little latitudinal provinciality would be expected, and indeed little is found in the fossil record (e.g., see Boucot et al., 1969; Cocks and McKerrow, 1973; and many papers in Hallam, 1973). At most times during the Paleozoic there are but two latitudinal provinces around Gondwana, as opposed to between four and seven along similar latitudinal distances today (Text-fig. 4). Indeed, Whittington and Hughes (1973) have found that for all continents, based on genera and families of trilobites, only four provinces are present in the Early Ordovician, and that they gradually merge so that by the Late Ordovician only a single worldwide province is present. Williams (1973) has erected a somewhat different scheme based on brachiopod genera, but also records a progressive diminution in provinciality from the Early Ordovician (five provinces) toward the Late Ordovician. The same trend is even found among the planktonic graptolites (Skevington, 1973). Distributional analyses on the species level are badly needed and represent a rewarding task for the future. It is at any rate clear that Paleozoic provinciality was only moderate, as most continents had little marine climatic variation, and there were few continents, some of which were closely grouped at times. Thus the average range of species was much broader than today latitudinally, and the percentage of species common to separate continents should have been much higher than today also. Communities would therefore have been many fewer than today, and analogous communities in separate provinces would have tended to share more species. Under these conditions, intercontinental correlations either by individual taxa or by faunal assemblages would be relatively easy. Patterns of disjunction would depend heavily upon current patterns and probably shifted markedly as the continental geographies, and therefore the shapes of ocean basins, were altered. Too little attention has been given to systematizing disjunction patterns, although they are recorded regularly from fossil evidence (e.g., see Coates, 1973; Waterhouse, 1967; Westermann, 1973; Wiedman, 1973).

Provinciality may certainly be affected by the land-sea geographies of continental platforms, as well as by continental numbers and dispersion patterns. Topographic relief on platforms during the Paleozoic was sufficient to break shallow epicontinental seas into distinctive arms or basins with much endemism in each arm. Cases of Paleozoic epicontinental provinciality are discussed by Johnson (1971) for the Devonian of North America and by Runnegar and Newell (1971) for the Permian of the Paraná Basin in South America. Such trends may make correlations difficult, even between adjacent basins.

The second case is that of the classic Pangaea configuration (Text-fig. 5B), when the northern continents had joined the southern Gondwana continent to form a supercontinent that reached from north of the equator to the south pole, which it usually covered or at least stood very near. In this configuration there could be no topographic barriers around the continental shoreline, and the only provinciality around Pangaea was associated with barriers due to climate. However, other marine provinces could have been present on island arcs or intraplate volcanic chains in open ocean waters at some distance from Pangaea. There is as yet no direct evidence to this effect, but Batten (1973) has shown that a number of Guadelupian gastropod genera disappear from the fossil record in uppermost Permian and lowermost Triassic, to reappear in the Ladinian. Perhaps the undiscovered intermediate populations inhabited oceanic island environments, as Batten has suggested (pers. comm., 1972). At any rate, it seems certain that provinciality around Pangaea was even lower than during most of the Paleozoic, and that the average range of species covered a larger proportion of the available continental shelf than previously, for barriers are fewer and the realized species ranges would frequently approximate their prospective ranges. Correlation between what are separate continents today should be relatively easy. This situation seems to have lasted through much of the Lower and Middle Mesozoic, during which time shelf species appear to have been exceedingly cosmopolitan.

From the middle Jurassic onward, a trend toward increasing apparent restriction of species ranges appears. At first, the chief range restrictions are latitudinal as the "Boreal" fauna of Middle and Late Mesozoic time develops (see Arkell, 1956b; Hallam, 1973; and appropriate papers in Hallam, 1973). Later, longitudinally separated communities on separate continents or along separate sides or seaways of continents become increasingly dissimilar as a result of increasing longitudinal species endemism. Both latitudinal and longitudinal provincial trends seem attributable to continental movements resulting from plate-tectonic processes; "Boreal" seas are restricted as continents spread northward away from the Polar Antarctic–Australia continent, leading to an increased latitudinal climatic gradient in the northern hemisphere. At the same time, longitudinal continental drift of the Americas relative to the Eurasian and African continents produced longitudinal endemism. These trends have long been apparent and in recent years have been carefully documented for several benthic taxa (Sohl, 1961, 1971; Coates, 1973; Kauffman, 1973). Thus, when the Pangaean supercontinent broke up, the biogeographic pattern changed and biostratigraphic correlations became more difficult, for isolation of biotas led to shrinking species ranges and increasing provinciality.

This led to the development of the third case, the continental and biogeographic pattern of the present (Text-fig. 5C). This is a configuration with numerous relatively small continents widely scattered but with important north–south topographic barriers, the alternation of terrestrial and deep-sea environments with continental shelves that inhibits the longitudinal dispersal of shallow-water organisms. Thus longitudinal endemism is high. Additionally, the latitudinal climatic gradient began to increase at least from the Late Jurassic and is very high at present, evidently owing to different factors in the north than in the south. In the north the Arctic Ocean is cut off from warm-water sources by the North American and Eurasian land masses and by islands and shoals; the result is a cool polar ocean. In the south, the circum-Antarctic west wind drift completely encircles the polar continent and effectivley cuts off any supply of warm water from low latitudes. This did not happen when Gondwana lay upon the pole and extended into intermediate or low latitudes (Valentine and Moores, 1973; Valentine, 1973b). Therefore, a high latitudinal temperature gradient has appeared in each hemisphere, and the average latitudinal range of marine invertebrates has been greatly restricted. Chains of provinces are present along each north–south trending coast (Text-fig. 4), and even analogous communities tend to be very distinctive in species composition from one province to the next. Correlations between faunas in different climatic zones are more difficult through this period of time.

In summary, there appear to have been two great marine biogeographic systems during the Phanerozoic, separated by a transition period (Valentine, 1973a). The first system is based on a large south-polar continent that extended equatorward so far that the boundary currents of the oceanic gyre to the east of it (the Tethyan gyre) supplied water for a warm coastal current that must have swept around its southernmost shores, creating relatively mild climatic conditions (Valentine and Moores, 1973). This effect is not dissimilar from that of the Gulf Stream off northwestern Europe today. With warm poles and few continents, provinciality was only moderate. This system came to an end with the assembly of all continents into Pangaea, a reduction in provinciality, and a concomitant increase in species ranges and in the similiarity of community compositions throughout the shelf realm. When Pangaea fragmented, a relatively small polar continental mass was isolated behind the northward-drifting major continents, and over the site of the circum-Antarctic ridge a circum-Antarctic current system appeared. This, coupled with the growing restriction of the small north polar ocean, led to increasingly low polar temperatures and, therefore, to rising latitudinal temperature gradients, and the framework for the second biogeographic system appeared – a system of very high provinciality.

It should be emphasized that the evolution of biogeographic systems is a slow process, involving geographic changes that arise from plate-tectonic processes. Although more local and rapid changes in provinciality occur, especially from topographic complications within epicontinental seas and from climatic fluctuations, they tend to produce variations on the major systems rather than causing fundamental biogeographic repatterning.

Clearly, the numbers, strengths, and patterns of barriers have changed, and this has permitted or required species and communities to become more or less widespread at different times in appropriate habitats; i.e., species have at times been able to spread so widely that their realized ranges have come to approximate their prospective ranges. However, these changes are even more complicated, for the patterns of the prospective ranges themselves have been vastly altered. At times during the presence of Pangaea, for example, the prospective ranges of many species probably ran right around the marginal shelf, so true cosmopolitanism was not difficult to attain. Today the prospective ranges of nearly all species are more widespread, but even so are restricted to small segments of the present shelves. Even if prospective ranges were realized, true cosmopolitanism would not be common. Even though the average prospective range may be much broader today when measured in terms of area encompassed by the prospective range border, it forms a smaller fraction of the inhabitable biospace available to shelf species, which has expanded greatly. Table 2 summarizes in a qualitative way the differences between realized and potential ranges within and between the three major biogeographic cases discussed above: the Paleozoic case with Gondwana, the early Mesozoic case with Pangaea, and the Cenozoic case with a small Antarctic continent, a small Arctic Ocean, and dispersing continents. Oceanic island provinces are omitted.

It is worth noting that each biogeographic system outlined here implies very different numbers of species in the marine shelf realm. The Paleozoic levels of provinciality imply moderate numbers, but Pangaea would have supported only relatively few species, owing to reduced provinciality alone. The development of the present high levels of provinciality would promote a rise in species diversity to new high levels because gene flow between many populations would be severed. Probably, the biogeographic factor alone stands as the underlying cause of a high percentage of the extinctions that accompanied the formation of Pangaea, and of the great diversifications, chiefly at low taxonomic levels, that accompanied the Late Mesozoic and Cenozoic continental dispersion (Valentine, 1969; Valentine and Moores, 1970, 1972; Valentine, 1973a, 1973b, 1973c).

BIOSTRATIGRAPHIC IMPLICATIONS OF THE BIOGEOGRAPHIC PATTERNS

The biostratigraphic nomenclature normally employed (e.g., American Commission on Stratigraphic Nomenclature, 1961–1970) uses three main concepts: the range zone, assemblage zone, and concurrent-range zone. The nature, extent, and

Table 2
General Patterns Expected of Prospective and Realized Ranges in Three Distinctive Paleogeographic Settings (see Text-fig. 5)

Geographic Case	*Prospective Range Pattern*	*Realized Range Pattern*
Gondwana and a few other continents	Fairly widespread on shelves and shelves moderately well dispersed themselves	Generally broad and forming high percentage of prospective range but restricted locally at barriers on continental platforms
Pangaea	Cosmopolitan on shelves but not widespread on globe owing to continental assembly	High proportion of prospective ranges realized; species widespread but only around Pangaea
Numerous continents, small S. polar continent, small N. polar sea	Restricted to portions of shelves but widespread owing to continental dispersion	Small proportion of prospective ranges realized, species restricted owing to topographic barriers, chiefly longitudinal

utility of each of these sorts of units is affected by biogeographic changes (Valentine, 1963).

Range zones are bodies of strata that comprise the total horizontal as well as vertical range of specified taxon, by definition; therefore, their geographic extent at any one time would appear at first glance to be a straightforward biogeographic effect. Temporal changes in the horizontal boundaries of range zones would result chiefly from environmental fluctuations that altered the distributional patterns of the taxa involved. However, it is now clear that the original geographic extent of a taxon may also be vastly altered by plate-tectonic processes.

Consider a species that is moderately widespread around Pangaea. Today continental masses bearing its fossil remains could easily be spread over twice the latitude and longitude of its original range. Range zones may also shrink; Paleogene species that lived on both the Indian and Asian sides of the Tethyan seaway across the present site of the Himalayas, for example, have experienced the collapse of the horizontal component of their original range zones as Tethys closed. A static concept of the range zone is clearly inadequate for a dynamic fossil record that is itself on the move. Perhaps the time for which a given range zone is employed should be specified but constant from original to present situation. This can be awkward; e.g., "Permian range zone of *Olenellus*" sounds incongruous. Perhaps the most frequent references to range zones will be to the original *living range zone,* and to the *present range zone.*

The average horizontal extents of living range zones of taxa has varied much during the Phanerozoic, as we have seen. It happens that they were probably broadest for species and genera while Pangaea existed and during its early fragmentation. Therefore, the present range zones of these taxa are exceedingly broad, and the taxa should be among the best index fossils available insofar as their horizontal distributions are concerned. Many ammonites, the classic Mesozoic index fossils, are examples of this sort, and early Mesozoic mollusks in general appear to be widespread. Whether or not such taxa would evolve rapidly and therefore have narrower vertical ranges than average is not certainly predictable at present from evolutionary theory. There is evidence that they would have relatively small amounts of genetic variability (e.g., see Ayala et al., 1974), which would be expected to retard their rates of evolution. It is possible but not certain that the more specialized and restricted taxa at any time evolve at a faster rate. Thus, within their more restricted ranges, such taxa should make the best index fossils. Before Pangaea, widespread index fossils would be expected to be somewhat less common, although at some times (Late Ordovician) they may be present in numbers; but after the fragmentation and separation of the elements of Pangaea, they would be in very short supply indeed. Even planktonic forms became more and more restricted during the Cenozoic.

Another zonal concept is that of the concurrent-range zone, a vertical zone of strata defined by the temporal overlap of the ranges of included taxa that do not necessarily coincide otherwise. It is clear that the geographic ranges of many of the taxa that characterize the concurrent-range zone must be overlapping as well, to bring them into association. The spatial overlap of species at certain times, and their partial segregation at earlier and later times, must be due to one of two types of changes: (1) the species actually evolved anew and/or evolved into new nominal species or become extinct; or (2) the species migrated into and/or out of the critical overlap area within or at the temporal boundaries of the zone. In the second circumstance, the zonal boundaries are controlled locally by biogeographically important parameters. These would usually be limiting factors such as temperature, salinity, or the presence of effective competitors or predators, changes in any of which may either exclude certain species from or permit the immigration of certain species into a region. It seems certain that the majority of local appearances or disappearances of species in concurrent-range zones are due to local geographic range changes. The boundaries of most concurrent-range zones thus vary in temporal significance from place to place as a function of changes in biodistributional patterns. The times when vertically overlapping species' ranges are most likely to be expanded over the broadest areas are simply those when the average living range zone of taxa is widest also. The boundaries of many concurrent-range zones are determined by erosion, or at least by nonsequences in many regions, but this does not really affect the general properties of the zone.

Assemblage zones are characterized by associations of fossil taxa, which may also range earlier or later elsewhere. Assemblages of species that are so consistent as to characterize belts of strata may include species drawn from a single community, or from two or more communities that lived in such circumstances (as above each other in the water column or sea bottom), such that their remains are regularly intermingled. In this case, assemblage zones are usually restricted to single provinces because community ranges commonly end at provincial boundaries. However, assemblage zones frequently include cosmopolitan and ubiquitous taxa that ranged through more than one province. Nevertheless, the geographic extent of assemblage zones has undergone about the same temporal changes as of range zones, although on a broader scale, and the same problems arise with the relations between their original biodistributional patterns and their subsequent patterns as modified by plate-tectonic processes. Although markedly disjunct assemblages are theoretically possible, they are unlikely, because assemblages are composed of so many species, most of which would have to penetrate the barriers intervening between the discrete occurrences. Therefore, the occurrence of compositionally similar assemblage zones in widely distant regions is very good evidence that the regions were essentially connected at the time the assemblages were living.

In conclusion, biostratigraphic units may be treated as biogeographic units and, in fact, are very useful in understanding paleobiodistributional patterns. This is especially true since the geographic framework upon which the distributional patterns are formed is now known to have been moving itself. These movements have commonly altered biogeographic patterns after the extinction of the taxa involved. They have also created environmental changes that have caused new living patterns, leading to changes in the extent and pattern of living range zones and assemblage zones. Judging from generalized geographic reconstructions, we would expect the early and middle Mesozoic to be the times that interregional correlations are easiest on a global average, the Paleozoic being somewhat less easy, and the late Mesozoic and Cenozoic being progressively more difficult. Clearly, the disciplines of biostratigraphy and biogeography have many common interests and goals.

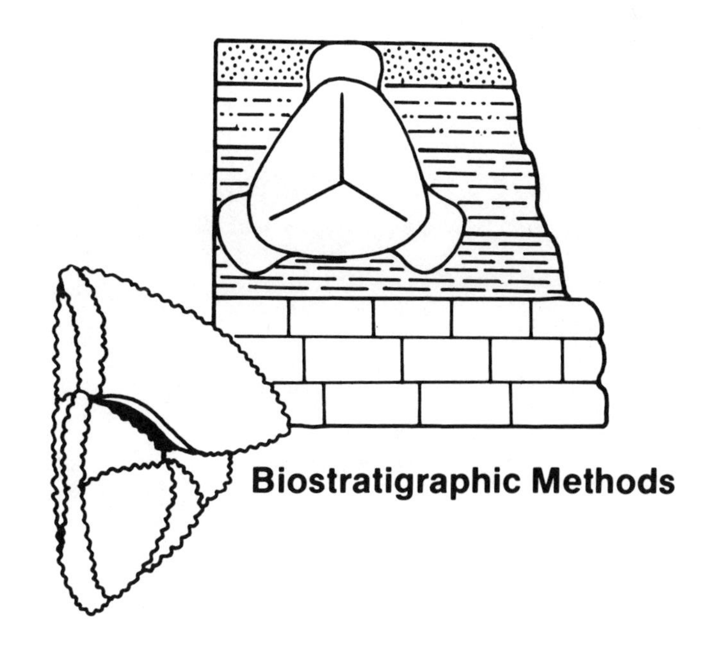

Biostratigraphic Methods

The Graphic Correlation Method in Biostratigraphy

F. X. Miller

Amoco Production Company

INTRODUCTION

In 1964, Shaw proposed a method of paleontologic correlation that utilizes the total range of fossils to develop time-stratigraphic control in sedimentary rock sequences. The method involves a graphic plot that visibly displays the best time correlation between two sections of rock of similar age, based on the sequence of occurrence and disappearance (i.e., stratigraphic range) of fossils present in both sections and the rate of rock accumulation at each section. The basic principles and theory resulting in the development of this method are explained in Shaw (1964).

The purpose of this paper is to review the graphic correlation method. The selection of a standard reference section (SRS), the graphing technique, line of correlation (LOC), changes in rate, the interpretation of graphs, the development of composite ranges and a chronologic scale, and the mathematical aspects of Shaw's method will be discussed and illustrated from a user's standpoint. Finally, the results of an actual palynologic study using graphic correlation will be presented in the form of stratigraphic cross sections taken from a comprehensive study of the Upper Cretaceous rocks of southwestern Wyoming.

GENERAL CONSIDERATIONS

The basic premise of Shaw's method is that fossils are excellent indicators of geologic time and events, and that fossils have determinable first and last occurrences in the geologic record. *Local stratigraphic range* is defined as the determinable range of a fossil in any one section. *Total stratigraphic range* is the total determinable stratigraphic interval through which a fossil is preserved in the geologic record. The total range is the sum of local ranges in correlatable sections. The distinction between local and total range must be kept in mind throughout this entire paper.

Many traditional methods of paleontologic correlation are capable of determining both local and total range of fossils. Most, however, are incapable of making a *precise* statement of the overall succession of the fossils because they lack a suitable chronologic scale. The graphic method proposes a practical and workable chronologic scale for the paleontologist. This is not an absolute scale using real time as a measure, but an alternative to it.

In a stratigraphic section, we observe only the sediment that was preserved and lithified, not the total amount of sediment that was originally deposited. The immeasurable effects of processes such as the original rate of sedimentation, subaerial and subaqueous erosion prior to burial, and compaction after burial are compensated for in the thickness of rock that we observe and measure. Shaw defines the product of what we actually observe in a stratigraphic section as *rate of rock accumulation,* as opposed to rate of sedimentation, and as such implies no relationship to absolute time.

Likewise, the fossils observed in a stratigraphic section are not a true measure of absolute time. Not only are they subject to the same shortcomings as the stratigraphic section in which they are entombed, but individuals may have lived prior to and after their determinable stratigraphic range, but were not preserved.

The graphic method is concerned only with that which the paleontologist can observe, determine, and measure: the rate of rock accumulation in a stratigraphic section and the local and total stratigraphic ranges of fossils. The chronologic scale and the resulting chronostratigraphic correlations are based on these factors.

The chronostratigraphic correlations accomplished by using the graphic method will not be acceptable to paleontologists who seek to make absolute time correlations based on fossils on the grounds that the chronologic scale is not exact. The proposed scale is superior to a relative scale in that it can be quantified and used as an accurate measure for time-stratigraphic correlations. It differs from an absolute scale in that it does not use units of real time as a measure, but a unit based on time as a substitute. For the paleontologist interested in biostratigraphy, the graphic method offers a technique to assimilate paleontologic data into meaningful total stratigraphic ranges, and to make chronostratigraphic correlations that show the diachronous nature of formations and emphasize facies relationships.

STANDARD REFERENCE SECTION

The paleontologist selects a single stratigraphic section as the standard reference section (SRS) for all his work. He determines the ranges of the fossils, and establishes the chronologic scale for the SRS. All other sections of similar age studied by the paleontologist are compared to that SRS. The uniqueness and workability of Shaw's method is that the chronologic scale and the total stratigraphic range of all the fossils are keyed to this one section.

The importance of the selection of the original stratigraphic section to be used as the SRS cannot be overemphasized. The ideal SRS would have the following characteristics: (1) a large and varied fossil content; (2) the thickest section possible, with the oldest and youngest rocks to be studied at its base and top, respectively; and (3) a complete and unfaulted stratigraphic section. In practice, one uses the section most closely approaching this ideal.

The SRS should be sampled as completely as possible. The sample density directly controls the degree of accuracy obtainable. The initial paleontologic study of these samples should be thorough and include the sequence of occurrence and disappearance of all fossils so that the total assemblage is represented. The ranges of the fossils determined from the SRS are by definition accurate, but not necessarily complete, because some ranges may be the result of facies control or poor preservation. Initially, the ranges of the fossils in the SRS are considered to be total stratigraphic ranges. Using the graphic method of correlation with information from other sections, the paleontologist can quickly determine whether the ranges in the SRS are total stratigraphic ranges or are incomplete and need to be extended.

GRAPHING TECHNIQUE

Once fossil ranges within the SRS have been determined, a second outcrop section or well is chosen to be compared to the SRS. The paleontologic methods are the same, i.e., to determine the stratigraphic range of as many fossils that were described from the original section as possible. Other fossils not observed in the SRS should also be recorded.

At this point, traditional paleontologic methods diverge to a variety of techniques to establish correlation. These methods are critically reviewed in Shaw (1964, Chaps. 13-17). Many use various statistical analyses to establish faunal similarity, faunal succession, or maximum abundance. Some rely on the local range of fossils by basing correlations on the hope that the range observed in one section is the exact time equivalent of the range observed in the next section. The graphic method provides a technique by which local range data can be assimilated into meaningful total stratigraphic ranges, and used to establish chronostratigraphic correlations over wide geographic areas.

The idea of using a graph to compare two sections of rock is unconventional but highly effective. Shaw's method is merely a departure from traditional motion problems; i.e., distance = rate × time. Motion problems can be solved mathematically or graphically. To use a graphical solution (Text-fig. 1), we plot distance on the vertical axis (Y) and time on the horizontal axis (X). Given any two of the variables in the equation $D = R \times T$, we could solve for the third. For example, given distance

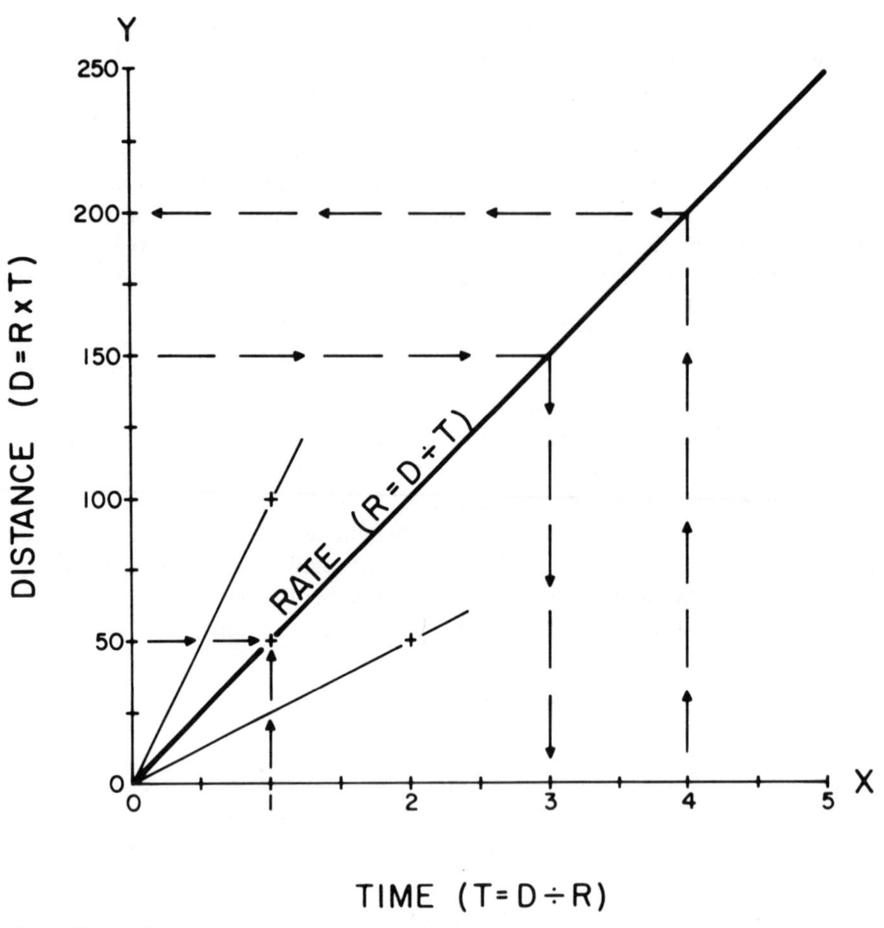

Text-Figure 1
Graphic solution to motion problems.

and time, we can find the rate by plotting D and T as a single point on the graph. A line drawn from the origin and extended through that point gives us the rate. The rate is expressed as distance per unit of time. All we have done is to solve graphically the equation $R = D \div T$. The rate line is very important to the graphic solution of motion problems because it allows us to *compare* the units on the X axis to the units on the Y axis, and vice versa. A change in either of the variables (D or T) will result in a change of the rate. This change will be reflected in the slope of the rate line. Three sloping lines are shown in Text-figure 1, each representing a different rate.

Furthermore, given the rate and time, we can determine the distance on the graph by projecting vertically from the time unit on the X axis to the rate line, then horizontally to the Y axis ($D = R \times T$). Conversely, given the distance and rate, we can determine the time ($T = D \div R$).

The graphic correlation technique uses the same reasoning and basic principles as the graphic solution to motion problems in order to compare two sections of rock: the SRS plotted on the horizontal axis (X) and a new section plotted on the vertical axis (Y). The paleontologist establishes the time units himself in the SRS. The distance is the thickness of a new section Y of rock to be compared to the SRS. The rate line must be determined from the two sections of rock. Similar to the graphic solution of motion problems, the paleontologist must establish two of the variables in the equation $D = R \times T$ as constant and then solve for the third. The graphic correlation method establishes time and rate as constants and solves for distance.

LINE OF CORRELATION

In motion problems, the rate is either known or can be determined using D and T, and expressed as a line on a graph. In the graphic correlation method, the rate must always be determined. It too can be expressed as a line on a graph and used to *compare* the units plotted on the X axis to the units plotted on the Y axis. In Shaw's method, the rate line is called the *line of correlation* (LOC).

Paleontologists know that the total stratigraphic range of an extinct fossil defines a specific interval of time, even though we do not know the length of that interval in numbers of years. We can also measure the thickness of rock contained between the first and last occurrence of an extinct fossil. The stratigraphic interval enclosed by the total range of a fossil may vary from place to place, depending on the amount of rock deposited, preserved, and lithified at each locality, but the interval of time remains the same. By using the total stratigraphic range of the same fossil (or fossils) in each locality studied, we have a means of comparing the thickness of rock preserved at each locality during a specific interval of time. This is the rate that the paleontologist must establish: the rate of rock accumulation (deposited, preserved, and lithified) per interval of time. In terms of motion problems, we are solving the equation $R = D \div T$. When this rate is expressed as a line on a graph, it is the LOC. The mechanics involved to establish the LOC are discussed below.

Using the total range of a single fossil to compare the SRS and the new section Y, the LOC would be easy to determine, but its accuracy would be questionable. The probability of a single fossil occurring at the maximum of its total range in every locality is very low because of local environmental conditions or poor preservation. Therefore, the graphic method of correlation uses the total range of as many fossils as possible to determine the LOC as accurately as possible. The paleontologist must decide which fossils in the new section are at or near their total stratigraphic range. The graphic method helps him to make that decision.

The graphing technique used to find the LOC is done on a conventional two-axis graph. The SRS is the X axis (horizontal) and the new section Y is the Y axis (vertical). The oldest rocks in both sections are plotted nearest to the origin of the graph. For convenience, the oldest occurrence of a fossil will be called the base, and the youngest occurrence, the top. The coordinates for the top of a fossil common to both sections are determined by measuring the footage above the base of the section in the SRS for the X coordinate and the footage above the base in the new section

Y for the Y coordinate. These two footages can then be plotted as a single point on the graph. Tops are plotted as plus signs (+) on all the accompanying graphs. Coordinates for bases are determined in the same manner and are shown as circles on the graphs.

Text-figure 2 is a graphic plot of the tops and bases of fossils from the SRS and the new section Y referred to above. For illustration purposes only, the tops (+) and the bases (o) of the fossils common to both sections are numbered and plotted along the X and Y axes to show their respective position in each section. This is not common practice. Normally, the data are derived from checklists, range charts, or data on individual samples. The graphic plot for the top of fossil number 7, for example, is the coincidence of the points $X = 350$, $Y = 355$. This procedure is

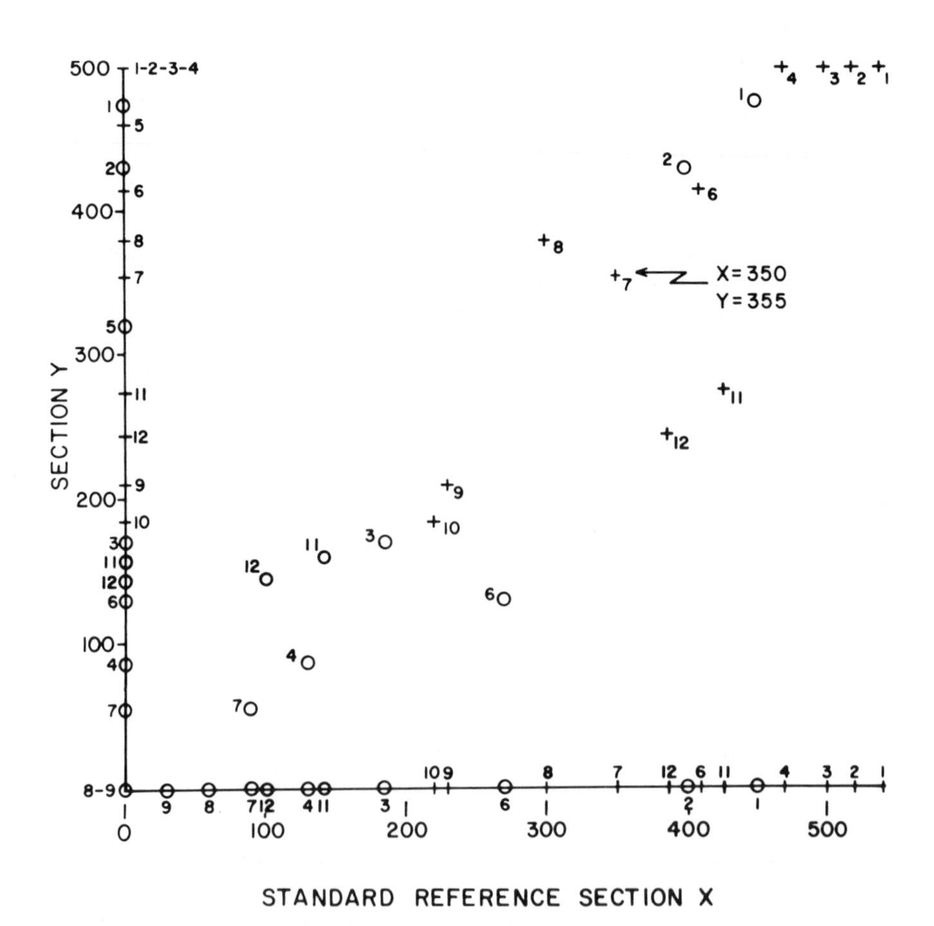

Text-Figure 2
Graphic plot of tops (+) and bases (o) of fossils in the standard reference section (SRS) and new section Y.

followed until the tops and bases of all fossils are plotted. The reader should remember that the ranges of the fossils in the SRS are always considered to be total stratigraphic ranges, even for the initial graph. Ranges that are not complete in the SRS will be pointed out by the graphic method and can be extended. This will be discussed later.

Our general knowledge of the two sections assures us that some points in section *Y* are the time equivalents of some points in the SRS. The graphic array helps to locate these equivalents. Now we must attempt to find the LOC ($R = D \div T$) by using the array of points that will give us the best possible correlation between the two sections. Except for two obvious points, which will be discussed later, an examination of the plot will show that the bases (o) are confined to the left side of the graph while the tops (+) are confined to the right. If the tops and bases we established for the fossils in the SRS are correct, i.e., at the maximum of their total stratigraphic range, then tops (+) will always plot on the right side of the graph and bases (o) on the left, because of the method used to plot them. A straight line drawn through the bases of fossils 8 and 4, the tops of fossils 9 and 7, the base of fossil 1, and the top of fossil 4 will exactly confine the bases to the left and the tops to the right side of the graph. This is the rate line that the paleontologist must determine. Similar to the rate line in motion problems, the LOC allows us to compare the SRS on the *X* axis to the stratigraphic section plotted on the *Y* axis. The LOC is used to determine the total stratigraphic range of fossils and to determine time equivalent points between the SRS and the new section *Y*. This will be discussed subsequently.

The LOC shown in Text-figure 3 is based on the total range of the majority of fossils in both sections. The LOC was drawn through the tops and bases of those fossils whose local range in section *Y* most closely approached the known total stratigraphic range in the SRS. The tops of fossils 8, 11, and 12 and the bases of fossils 6, 11, and 12 plot far off the LOC and were not used to determine the position of the LOC. These occurrences will be discussed in a subsequent section.

In practice, it is the paleontologist who must determine the best position of the LOC through the array of points. His knowledge of the fossils and their environmental restrictions, the sequence of fossil occurrence, fossil relationships, and the stratigraphic section will all play a role in his interpretation. The graphic display forces the paleontologist to consider each top and base before locating the LOC.

CHANGES IN RATE OF ROCK ACCUMULATION

After the paleontologist determines the best position for the LOC through the array of points on a graph, a straight line is used to show the correlation between the two sections. In theory, this line may show "doglegging"; i.e., the LOC may be a continuous series of short, straight lines each with a different slope reflecting minute, oscillatory, vertical changes in the rate of accumulation between the two sections. These minute differences in rate are more apparent where the thickness

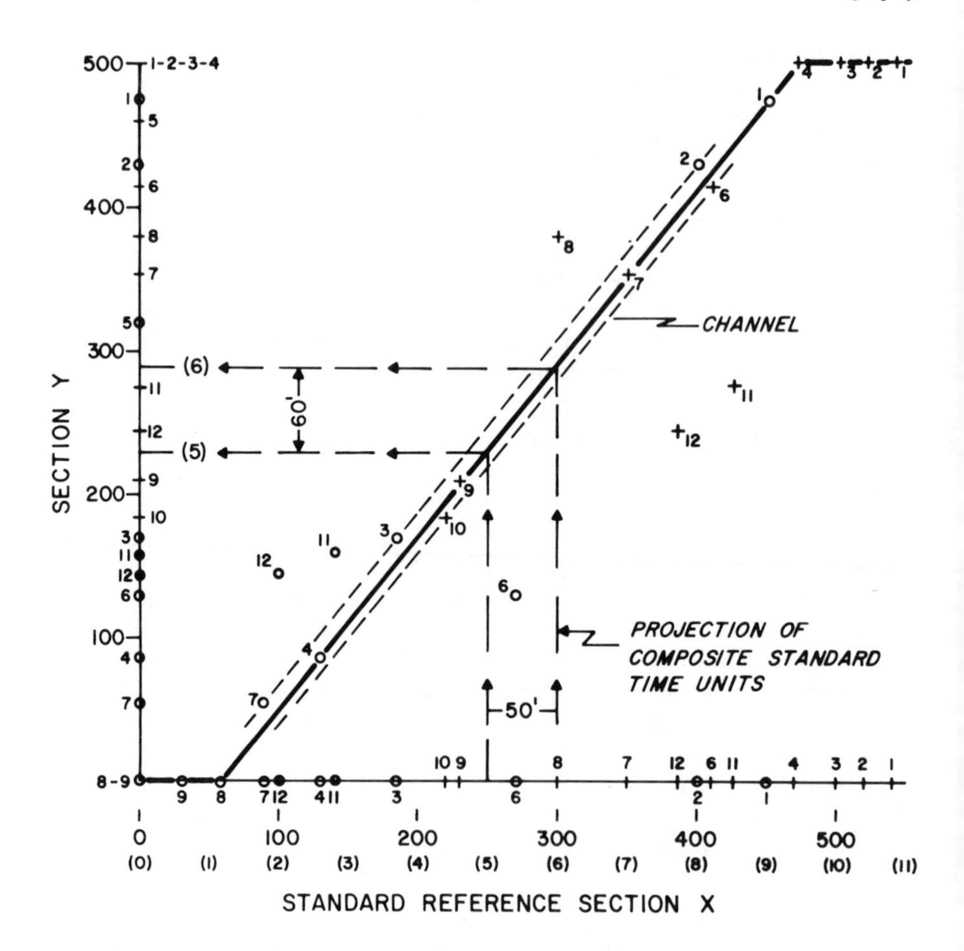

Text-Figure 3
Graphic plot illustrating the line of correlation (LOC), a channel, and points of time equivalence between the standard reference section (SRS) and the new section *Y*.

of rock in both sections is measured in inches or less. The amount of error introduced into this system by fitting a straight line to a doglegged array is negligible when the vertical, oscillatory changes in rate between the two sections are 10 percent or less (see Shaw, 1964, p. 138 and Chapter 20).

Working at a scale and the general level of accuracy that most stratigraphers use to measure sections and collect samples, the difference between a straight line and a line composed of a large number of small doglegs will have little or no effect on the overall accuracy of the correlations in practical applications.

The use of a straight line as the LOC rather than a line composed of a large number of small doglegs has several advantages. First, the graphic method is designed to be a practical and rapid technique to establish chronostratigraphic correlations. A straight line allows the paleontologist to make direct readings from his graph. Second, a straight line is easier to manipulate mathematically for computer applications.

A straight line should not be interpreted to mean that the rate of rock accumulation is constant through time at a particular locality. Small oscillatory changes can be expected, but these tend to average out over a given thickness of stratigraphic section. The straight line shows the average rate of rock accumulation through the entire section. The expediency of using a straight line far outweighs the disadvantages, with little effect on the overall accuracy of the correlations.

A gross change in the rate of rock accumulation between a new section and the SRS at a particular point in time will be clearly reflected by the graphic method, i.e., a change in the slope of the LOC. Changes of this sort can be expected in deltaic areas where a shifting river system would cause abrupt, local changes in the sedimentary rate. The LOC in an area of deltaic sedimentation may show one or more doglegs, with the straight-line segments between the "bends" having different slopes. This type of graphic pattern is common in the Tertiary rocks of the Gulf Coast of Texas and Louisiana. On the other hand, each of more than 100 graphs of Upper Cretaceous and Teritary rocks of the Rocky Mountains, western United States and Alaska, has produced a LOC without doglegs.

In choosing a section to be used as the SRS, a section that does not show a gross change in the rate of rock accumulation when plotted against another section is best. This is a matter of convenience, since each section of a doglegged pattern must be treated individually if computer applications are to be used. It will not affect the correlations.

If the first graph between the SRS and the new section *Y* shows an obvious change in slope in the LOC (i.e., a dogleg in the line), both sections should be graphed against a third section to determine which of the two sections does not have a change in rate. By process of elimination, we may be able to select a section that does not show a dogleg in the LOC to use as the SRS.

DETERMINATION OF COMPOSITE STANDARD RANGES

The term *composite standard range* is synonomous with total stratigraphic range as defined earlier; it is used by Shaw to indicate that the total range of a fossil in the SRS is the sum of its local range in correlatable sections.

In any section chosen by the paleontologist to be used as the SRS, some fossil ranges will probably be at their maximum total stratigraphic range. Others, because of environmental conditions or poor preservation at the SRS, will give incomplete ranges. Using the LOC and local range data from new sections, the graphic method shows the paleontologist which ranges in the SRS are total and which need to be extended.

In general, if the range of a fossil in the SRS is total, the top will always plot on the right side of the LOC and the base on the left when plotted against the local range of the same fossil in any other section. As the local range approaches the total range, the top and base will plot closer to the LOC. When the local range equals the total range, the top and base will fall on the LOC. On the other hand, if for any reason the range of a fossil in the SRS is incomplete, the top will plot to the left of the LOC and the base to the right. These are the ranges that need to be adjusted.

In Text-figure 3, the LOC was drawn in the best position through the array of points. Those fossils whose top and/or base falls on the LOC are considered to be at their maximum total range in the new section *Y* as previously determined in the SRS. Fossil 4 is an example. The fact that each top and/or base does not fall exactly on the LOC could be the result of the sample interval or poor preservation in section *Y*. Since the tops plot immediately to the right of the LOC and the bases to the left, the total ranges in the SRS are correct and do not need to be adjusted.

The top and base of fossils 11 and 12 plot far off the LOC. This indicates that the range of these fossils in section *Y* does not approach their known total stratigraphic range, and they cannot be used to determine the position of the LOC. Fossils whose occurrences are controlled by local environmental conditions normally plot in this manner. Since the tops plotted to the right side of the LOC and the bases to the left, their total range in the SRS is correct.

The top of fossil 8 and the base of fossil 6 plot on the opposite side of the LOC established by the rest of the array. This indicates that fossil 8 occurs in younger rocks in section *Y* than it does in the SRS. Likewise, the base of fossil 6 is older in section *Y* than it is in the SRS. Therefore, the total ranges established for fossils 6 and 8 in the SRS are incomplete. Text-figure 4 shows the method used to handle these occurrences. To find the more nearly correct top for fossil 8, project the top of fossil 8 in section *Y* to the LOC and then down to the SRS. This is the new top for fossil 8, an adjustment from 300 to 370 ft. The same procedure is followed to determine the corrected base for fossil 6.

Fossil 5, which ranges from 320 to 460 ft in section *Y*, did not occur in the SRS. To determine where fossil 5 should have occurred in the SRS, simply project the top and the base of fossil 5 in section *Y* to the LOC and down to the SRS. The top of fossil 5 in the SRS should have occurred at 437 ft and the base at 320 ft.

All we have done in the cases of fossils 5, 6, and 8 is to take information from section *Y* and transfer it (graphically; $T = D \div R$) to the SRS using the LOC as the best possible correlation between the two sections. In other words, we now have information in the SRS that is the composite of information from both section *Y* and the SRS. Herein lies the derivation of the name "composite standard." Each subsequent section graphed against the SRS may produce additional data on the range of fossils that can be compounded into the SRS. In this manner, we can eventually determine the total stratigraphic range of all fossils involved.

In many cases, fossils whose stratigraphic range is contained in the SRS were never found in the original reference section (e.g., fossil 5, Text-fig. 4). This could be the result of environmental conditions, poor sampling, poor preservation, or simply poor observation on the part of the paleontologist. The compounding of range data is an important feature of the graphic method, because it will eventually

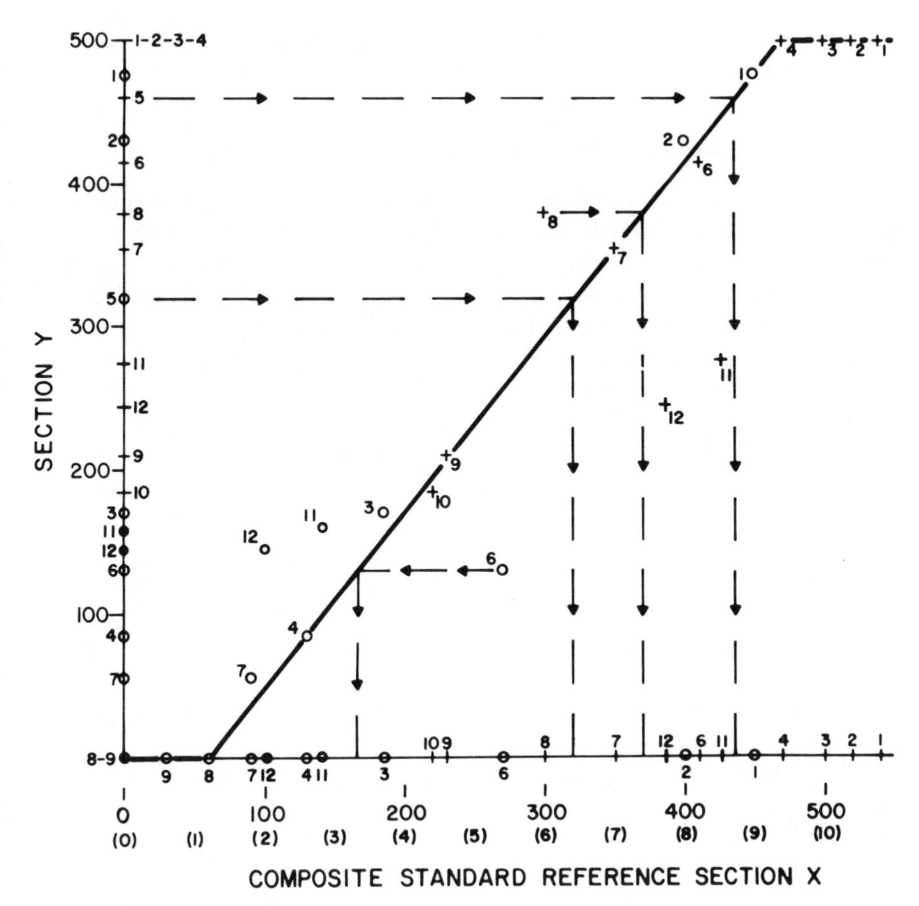

Text-Figure 4
Graphic plot illustrating method used to determine composite ranges.

allow us to determine the total stratigraphic range of fossils that are totally facies controlled. Miscorrelations based on facies-controlled fossils are not possible with the graphic method.

The range data gathered by compounding the information from other correlatable sections into the SRS are referred to as *composite standard ranges.* Once the compounding procedure begins, Shaw refers to the original SRS as the *composite standard reference section* (CSRS).

In practice, the total stratigraphic range of all the fossils will not be determined from the initial graph. Experience has shown that between six and twelve graphs are needed, with frequent adjustments, before the composite standard ranges can be considered reliable. When a point is reached where most of the tops and bases no longer require adjusting, the reliability of the composite standard ranges should be considered quite high. Range data on additional fossils can be added to the system at any time and are treated in the same manner as described above.

DEVELOPMENT OF A CHRONOLOGIC SCALE

For reasons previously stated, an absolute scale using real time is unattainable through paleontologic methods. Therefore, most traditional methods of paleontologic correlation use a relative chronologic scale with unequal units, such as geologic periods and epochs, as subdivisions on the scale. Using this method, one is able to determine the age of a rock and to place a series of rocks in their proper stratigraphic sequence. The time zones established by these methods are frequently broad, and the boundaries between the zones hazy.

Using the graphic method of correlation, the paleontologist can develop a chronologic scale that is capable of establishing fine time zones with definite boundaries. This is not an absolute geochronologic scale that uses units of real time as a measure, but an alternative to it. The method presented here is equally as usable as an absolute scale, but simply cannot be defined in terms of sidereal years. The difficulty in understanding and accepting the method arises from the fact that geologists and paleontologists are accustomed to thinking of time in terms of years. The reader is therefore cautioned against implying real-time units to this chronologic scale.

In Text-figure 3, for any point in the SRS, we can determine the time equivalent point in section Y by projecting vertically from the X axis to the LOC, then horizontally to the Y axis. For example, the point at 300 ft in the SRS is equivalent to the point at 290 ft in section Y. This method alone is sufficient to make chronostratigraphic correlations of points of specific interest to the stratigrapher in a local area. Outside the local area, these points may no longer be of interest.

To establish a scale in the CSRS, we must designate a series of equidistant points which can be projected into all sections that are compared to the CSRS. Shaw calls these points *composite standard time units* (CSTU); they serve as the chronologic scale for all future work. We can subdivide the CSRS into increments of equal thickness by establishing the entire thickness of the CSRS as our measure of time (as opposed to years). We know that it took a specific interval of time for the observable rock in the CSRS to accumulate; therefore, each increment is an equal part of the whole. This is the second variable, time, in the equation $D = R \times T$ that the paleontologist must establish as a constant before the complete interpretation of the graphical method can be accomplished.

From previous discussion, the reader is aware that the graphic method uses a straight line as the LOC rather than a line composed of a large number of small doglegs. By projecting the CSTU from the CSRS to a straight line, the equivalent CSTU in each new section will also be equally spaced, but the spacing will depend on the slope of the LOC. Since the LOC represents an average of the smaller doglegged lines, the spacing between the projected CSTU also represents an average. In theory, had we used a LOC composed of a large number of small doglegs, the interval between the projected CSTU would vary, some larger and some smaller than the CSTU in the CSRS.

Taken as a whole, the time-stratigraphic correlation of the entire thickness of the CSRS and the new section is accurate, but the correlation of equivalent CSTU within the sections is only approximate since they represent an average.

This does not mean that the system is unusable. We are only speaking of an error of 10 percent or less (10 ft per 100 ft of thickness at a maximum) which will average itself out over the entire thickness. When we consider the level of accuracy of most measured sections, this error is negligible, and the method is usable for practical biostratigraphic applications. Compared to a relative chronologic scale, which can only age date a stratigraphic sequence in broad intervals, the time zones established using CSTU are considerably more accurate and usable.

In Text-figure 3, the CSTU in the SRS are established at 50-ft intervals and are identified by the numbers in parentheses. These units will never change unless the CSRS is changed. The equivalent CSTU can be determined in each new section by projecting vertically from the CSRS to the LOC, then horizontally to the *Y* axis. The interval between the CSTU in a new section may vary depending on the slope of the LOC, which is established from the total range of fossils as previously discussed. For example, the interval between CSTU 5 and 6 in the SRS is 50 ft; the time-stratigraphic equivalent in section *Y* is 60 ft. *This is the only chronostratigraphic information that we can determine from the graphic method: the time equivalent points or intervals between the CSRS and any other section. We cannot determine an absolute time value for the CSTU, nor can we determine how long the intervals between CSTU's lasted.*

At this point, some geologists and paleontologists will argue that the intervals between each CSTU in the SRS are totally unequal. This conclusion is reached because of their instinctive compulsion to put an absolute time value on each CSTU. Granted, in terms of real time, their observations may be correct. The years between each CSTU may be unequal because the rate of sedimentation and the time for the processes prior to and after burial may be totally different for each interval. However, as pointed out previously, we cannot determine from the observable stratigraphic section the immeasurable effects of sedimentation rate and other sedimentary processes and therefore cannot apply real-time units to this scale.

A chronologic scale using CSTU is particularly useful for illustrating paleontologic data in the form of stratigraphic sections. A CSTU can be used as a datum for the cross section, rather than using sea level or a lithologic datum. In this manner, the diachronous nature of rock units can be easily displayed and facies relationships emphasized. The stratigraphic sections shown in Text-figures 9 and 10 are illustrated in this manner.

INTERPRETATION

The graphic pattern shown in Text-figure 3 is the classic pattern developed between two unbroken sections of similar age: a sloping line (LOC) with a flattening (horizontal terrace) at the top and bottom. The use of the LOC to determine the total stratigraphic range of fossils and to make chronostratigraphic correlations between the CSRS and a new section has already been discussed.

The horizontal terrace at the top of the graph indicates that the top of the SRS is younger than the top of section *Y*. The tops of fossils 1, 2, 3, and 4, which range through a 70-ft interval in the SRS, all occur in one sample in section *Y*. This indicates that the uppermost 80 ft of the SRS is not present in section *Y*. The horizontal terrace at the base of the graph indicates that 55 ft of older section is present at the base of the SRS which is not in section *Y*.

Another type of graphic pattern of importance is shown in Text-figure 5. Here the terracing occurs in the central part of the graph separating the LOC into two parallel segments. The flattening and offset of the LOC is caused by the omission of

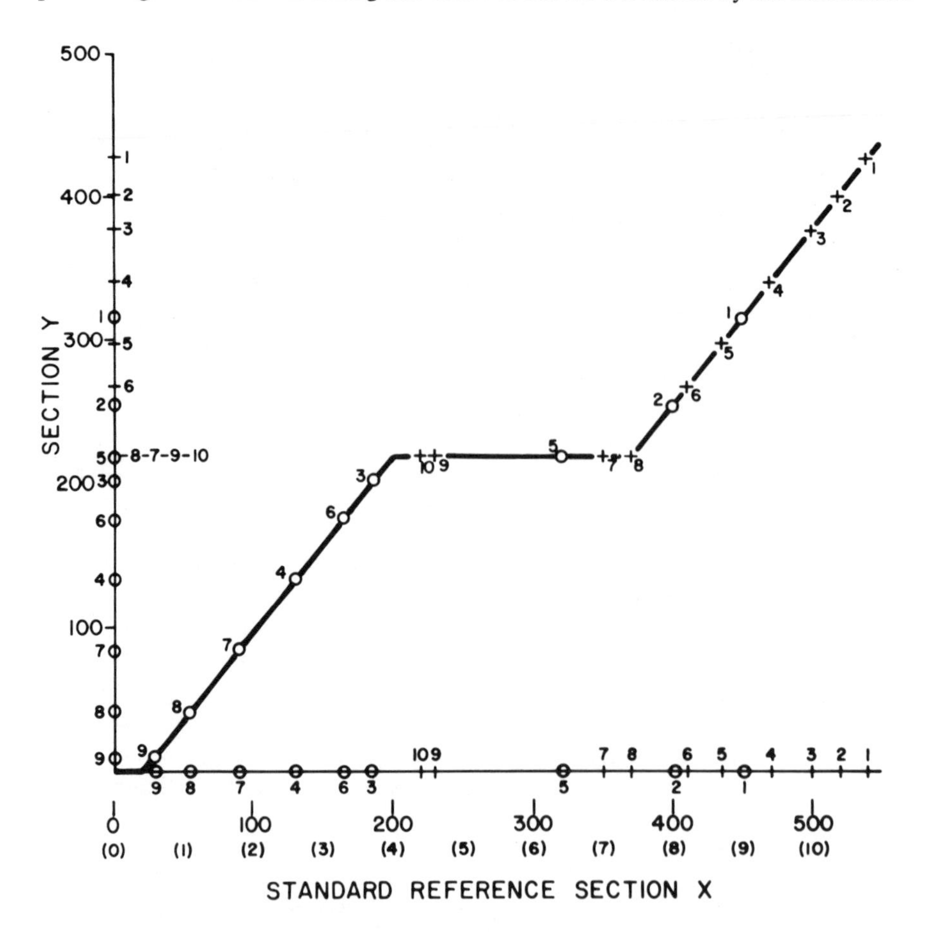

Text-Figure 5
Graphic plot showing line of correlation (LOC) with horizontal terrace caused by omission of strata.

strata in section *Y*. The tops of fossils 7 and 9, which are separated by 120 ft of strata in the SRS, occur in the same sample in section *Y*. Reading directly from the graph, the paleontologist can interpret that approximately 170 ft of section is missing from section *Y*.

The pattern shown in Text-figure 5 can be caused by faulting or unconformity. The pattern itself does not indicate the cause but merely alerts the paleontologist that there is an omission of section. The cause must be interpreted from other geologic evidence.

Another important graphic pattern with which the paleontologist should be familiar is called "channeling." This feature develops with the use of a mature CSRS, i.e., when the ranges of the fossils in the CSRS are complete total stratigraphic ranges or nearly so. Channeling is the tendency for the points in a graphic array to be aligned in a narrow band with the bases (o) confined to the left side and the tops (+) to the right. This feature is caused by plotting a new section with too few fossils against a mature CSRS or by using a new section that was inadequately sampled. In either case, the fossil ranges in the new section might be less than their known total ranges in the CSRS. Therefore, the tops will plot too low and the bases too high, forming a channel. Normally, if you can find one side of the channel, the other side will be parallel, or nearly so, to it. In Text-figure 3, the dashed lines parallel to the LOC outline a channel.

The channel is useful because we know that the LOC must fall within that channel. If a few tops or bases occur within the channel, as in Text-figure 3, they can be used to find the position of the line. At times there will be no tops or bases within the channel. Other geologic evidence, such as correlative bentonites, as well as the stratigrapher's knowledge of the section, will have to be used to find the correct position of the line within the channel. This may be a trial-and-error effort until the best position is found.

MATHEMATICAL ASPECTS

The mathematical aspects of the graphic method of correlation could be misinterpreted to mean that all correlations are strictly mathematical and could be done by a computer. This is not the case, however. The graphic method relies totally on the ability of the paleontologist to identify the fossils and to find the correct position of the LOC, and on the experience of the biostratigrapher to interpret and illustrate the meaning of the chronostratigraphic correlations. Throughout the preceding discussion, I have alluded to computer applications, but I have attempted to show that the graphic method can be used satisfactorily without the aid of mathematical equations and computations.

With continued use of the graphic method, data on the occurrences and total stratigraphic ranges of fossils will accumulate rapidly. The paleontologist can handle these data manually for a while by keeping a graph on each section or well, and by recording any apparent changes in the total stratigraphic range of the fossils. Eventually, such files will become overwhelming, and the only efficient way to handle the data for orderly filing and retrieval is through automatic data processing. I would strongly recommend that the paleontologist solicit the help of a qualified computer programmer with knowledge of geologic data processing to build the program to handle the data. It is at this point that the mathematical aspects of Shaw's method are invaluable.

SUMMARY

The graphic method of paleontologic correlation uses the basic principles of the graphic solution to motion problems—*distance* = *rate* × *time*—to establish time-stratigraphic correlation between two sections of rock of similar age. The paleontologist selects and studies a single stratigraphic section to be used as the *composite standard reference section* (CSRS) to which all future work will be compared. He determines the total stratigraphic range of all the fossils and establishes the *composite standard time units* (CSTU) for the CSRS.

New stratigraphic sections are then compared to the CSRS on a conventional two-axis graph with the CSRS plotted on the horizontal (X) axis and the new section plotted on the vertical (Y) axis. The local range of the fossils in the new section (Y axis) is plotted against the total range of the same fossils in the CSRS (X axis) as points on the graph. The *line of correlation* (LOC) is drawn through the tops and bases of those fossils whose local range in the new section most closely approaches their known total stratigraphic range in the CSRS. Fossils whose local range does not approach their known total range are not used to determine the best position of the LOC.

A straight line is used for the LOC rather than a line composed of a large number of small doglegs. The amount of error produced by fitting a straight line to a dog-legged array is negligible considering the scale at which most stratigraphers work.

The LOC is used to (1) determine points of time equivalence between the two sections of rock by projecting CSTU from the CSRS on the X axis into the new section on the Y axis, (2) determine the average relative rate of rock accumulation at the new section, (3) determine if the total ranges of the fossils established in the CSRS are complete or need to be adjusted, and (4) detect the omission of strata from the new section.

The chronologic scale developed by the graphic method is superior to a relative scale because it can be quantified. Despite a negligible amount of error, fine chronostratigraphic zones with definite boundaries can be established for local and regional correlations. This scale differs from an absolute scale only in that it does not use real time as a unit of measure.

The graphic method of correlation is most applicable in areas where thick, continuous outcrop sections are available, because the ranges of many fossils can usually be established within these sections. In areas where outcrops are thin or represent only a short interval of time, the method can only be used if sufficient first and last occurrences can be established within the interval. The lack of outcrops does not preclude the use of the method. There are several alternatives. First, samples from wells or coreholes can be used to determine the total fossil ranges in the CSRS. This alternative restricts the paleontologist almost exclusively to the use of microfossils. Second, a CSRS from outside the area of study can be used. For example, the CSRS for the Late Cretaceous of the Rocky Mountains is also applicable to rocks of similar age in Canada and Alaska.

EXAMPLE FROM THE UPPER CRETACEOUS OF SOUTHWESTERN WYOMING

Four wells located in the Green River Basin of southwestern Wyoming have been selected to illustrate the biostratigraphic application of the graphic method of correlation. The wells penetrated Upper Cretaceous rocks of the Mesaverde Group. The formations involved are, in ascending order, the upper part of the Blair, the Rock Springs, the Ericson, and the Almond. Tertiary rocks of Paleocene age unconformably overlie the Cretaceous in all the wells. These wells were selected because of wide facies variations within the formations. The names and locations of the wells are shown on the map of Text-figure 6.

The Upper Cretaceous CSRS used in developing Text-figures 7 to 10 is based on an outcrop on the eastern flank of the Washakie Basin, Carbon County, Wyoming. The CSRS is 10,000 ft thick. The total stratigraphic ranges of over 150 fossil pollen grains, spores, and dinoflagellate cysts have been compounded into the CSRS from approximately 42 wells and outcrops in Wyoming, Colorado, South Dakota, and Alaska. The CSRS has been divided into CSTU that are 100 ft apart and labeled from 0 to 100. Every tenth unit is identified on the graphs.

The graphs of the Mountain Fuel No. 1 Firehole (Text-fig. 7) and the Amoco No. 1 Spider Creek (Text-fig. 8) wells plotted against the Upper Cretaceous CSRS are included in this paper. The tops (+) and the bases (o) of the fossils identified from each well are plotted following the method described in the preceding discussion. The number beside the top or base is merely an identifying number given to each taxon.

The graphs shown in Text-figures 7 and 8 are essentially the same. Both have a channel containing a sloping line, which is broken into two segments by a horizontal terrace. Using the channel alone on each graph, we can determine, for example, that some point between 5,050 and 5,400 ft in the Mountain Fuel well (Text-fig. 7) is the time equivalent of some point between 10,550 and 10,950 ft in the Amoco well (Text-fig. 8). The slope of the LOC within the channel on each graph was determined after a careful evaluation of the occurrences of the taxa and an evaluation of the stratigraphic section involved. Using the LOC within each channel, we can more precisely determine that the point at 5,250 ft in the Mountain Fuel well is the time equivalent of the point at 10,625 ft in the Amoco well (20 CSTU).

The horizontal terrace on each graph indicates that a section is missing from each well. This could be a fault or an unconformity. Based on the entire palynologic study of the Green River Basin, this omission has been interpreted as an unconformity. Twenty-six wells have a section missing at approximately the same interval.

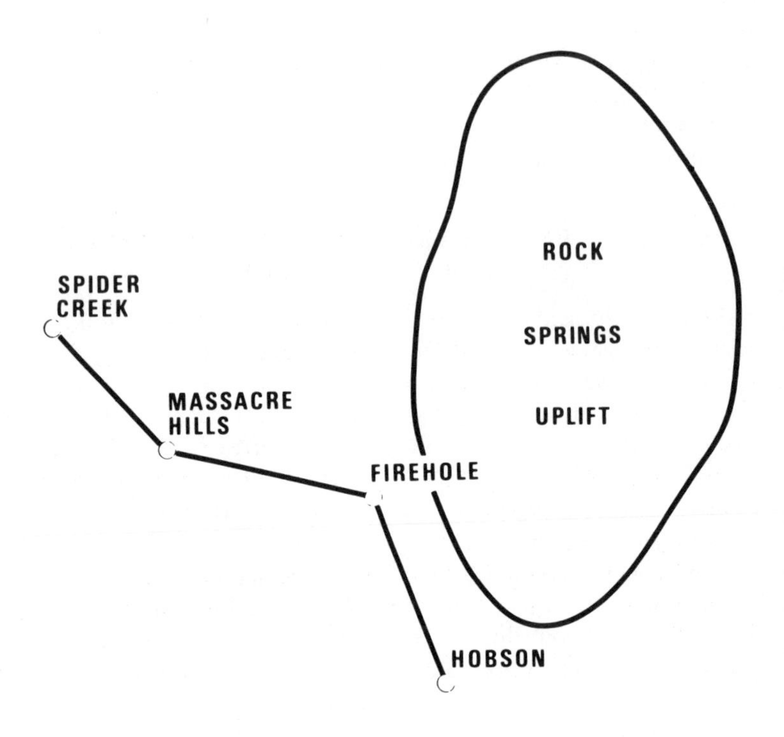

Text-Figure 6
Index map showing location of wells used in stratigraphic cross sections.

In the Amoco No. 1 Spider Creek (Text-fig. 8), the unconformity could occur anywhere between 9,200 and 9,400 ft. This interval is shown as the hachured area on the graph. Using stratigraphic information, the unconformity was interpreted to be at 9,400 ft at the base of the Ericson Sandstone.

It is evident from both graphs that all the fossils did not occur at the maxima of their known stratigraphic ranges or the tops and bases would have formed two straight-line segments above and below the terrace. This is a characteristic of the

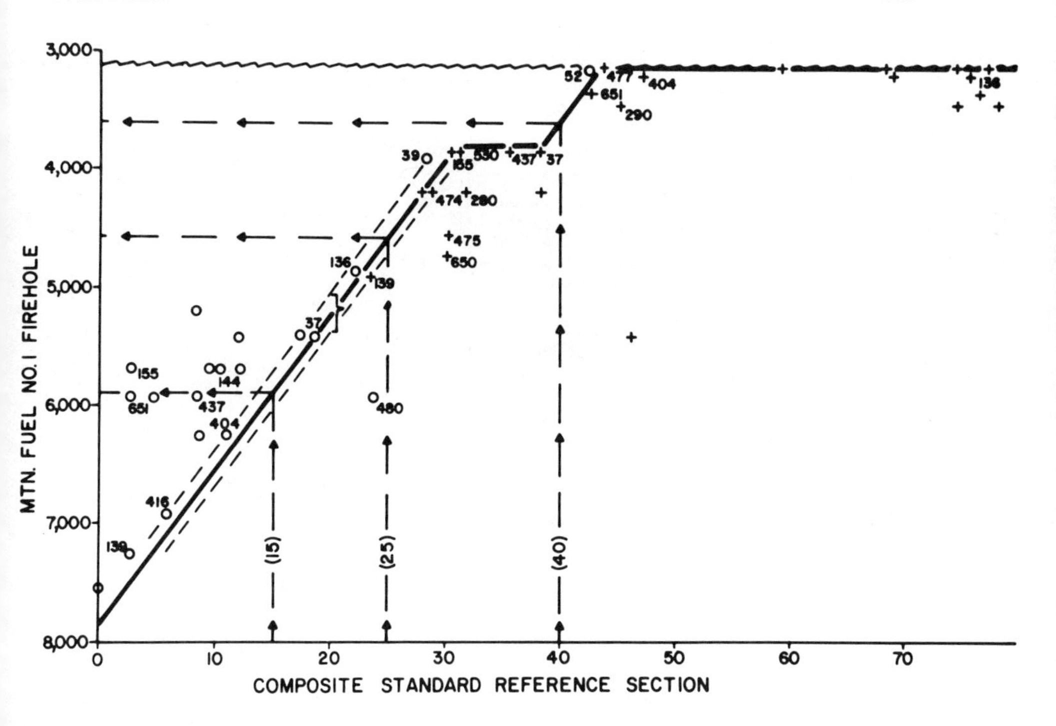

Text-Figure 7
Graphic plot of Mountain Fuel Supply Co. No. 1 Firehole well.

graphic method that many traditional paleontologic correlation methods cannot match. The fact that the graphic method relies on the total ranges of an entire assemblage rather than a few index fossils enabled these correlations to be made.

The tops plotted in the upper-right corner of both graphs form a terrace in the graphic pattern. We can interpret from this terracing that a part of the uppermost Cretaceous section present in the CSRS is missing from these wells. The terrace is formed because these fossils range into younger Cretaceous rocks.

To determine the CSTU in each well, project each unit from the CSRS vertically to the LOC and horizontally to the well. Projections of the 15, 25, and 40 CSTU are shown on both graphs.

Text-figure 9 is an interpretation of the stratigraphy of the Rock Springs Formation in the four wells. The datum for this cross section is the 20 CSTU. The Rock Springs Formation changes from a swampy sequence in the Spider Creek and Massacre Hills wells to a sandy shoreline facies in the Firehole well to a nearshore marine sequence in the Hobson well. The upper part of the Rock Springs Formation has been truncated and is not present in this part of the basin. The diachronous nature of the top of the Blair Formation is also evident.

Text-figure 10 is an interpretation of the Almond and Ericson formations. The datum for this cross section is the 40 CSTU. The facies relationship between the

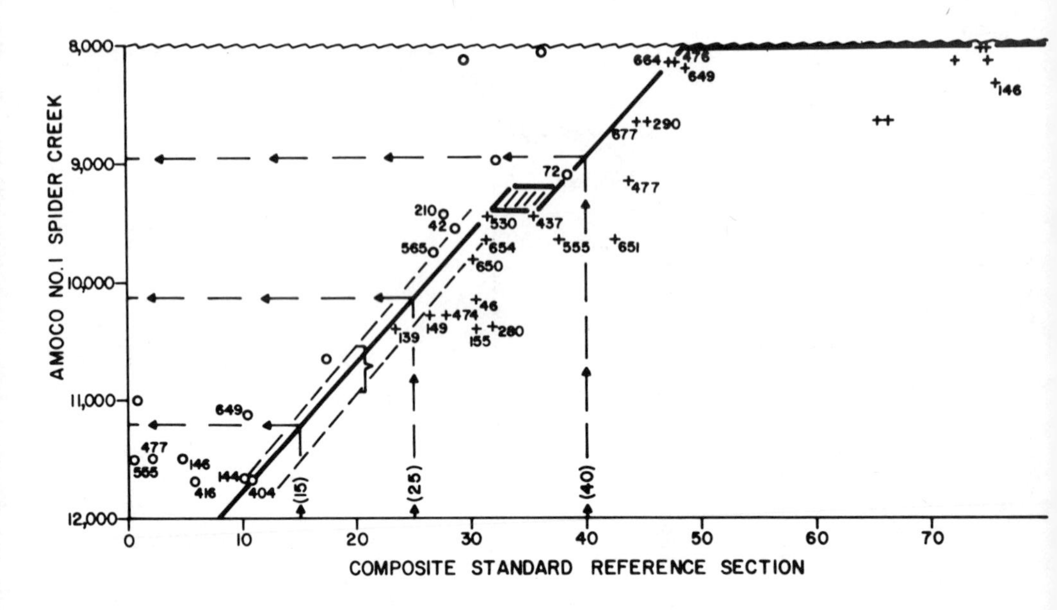

Text-Figure 8
Graphic plot of Amoco No. 1 Spider Creek well.

Ericson Sandstone and the shales of the Almond is evident. The Ericson Sandstone was deposited over the truncated top of the Rock Springs Formation. The Ericson and the Almond are interpreted to have been deposited in an inland, fluvial environment. The diachroneity of the two formations is evident.

For additional illustrations of the biostratigraphic application of the graphic method of correlation, the reader is referred to papers by Newman (1974), Upshaw et al. (1974), and Baron (1976).

CONCLUSIONS

The graphic correlation technique proposed by Shaw (1964) has widespread application in the field of geology. First, the method is important to the stratigrapher. A graph can be used to (1) determine the time relationships between two sections, (2) establish fine time zones with definite boundaries, (3) determine the average relative rate of rock accumulation at a given locality, (4) determine the diachronous nature of rock formations, and (5) determine the facies relationships between stratigraphic sections. Traditional correlative methods can give some of these results, but they are not able to give them all. Most methods cannot be quantified and are not capable of the same order of accuracy.

Second, the method is important to the paleontologist. It is the only method that offers the opportunity to accurately and quickly determine the total stratigraphic range of all fossils. Composite standard range data of all fossil groups can be keyed

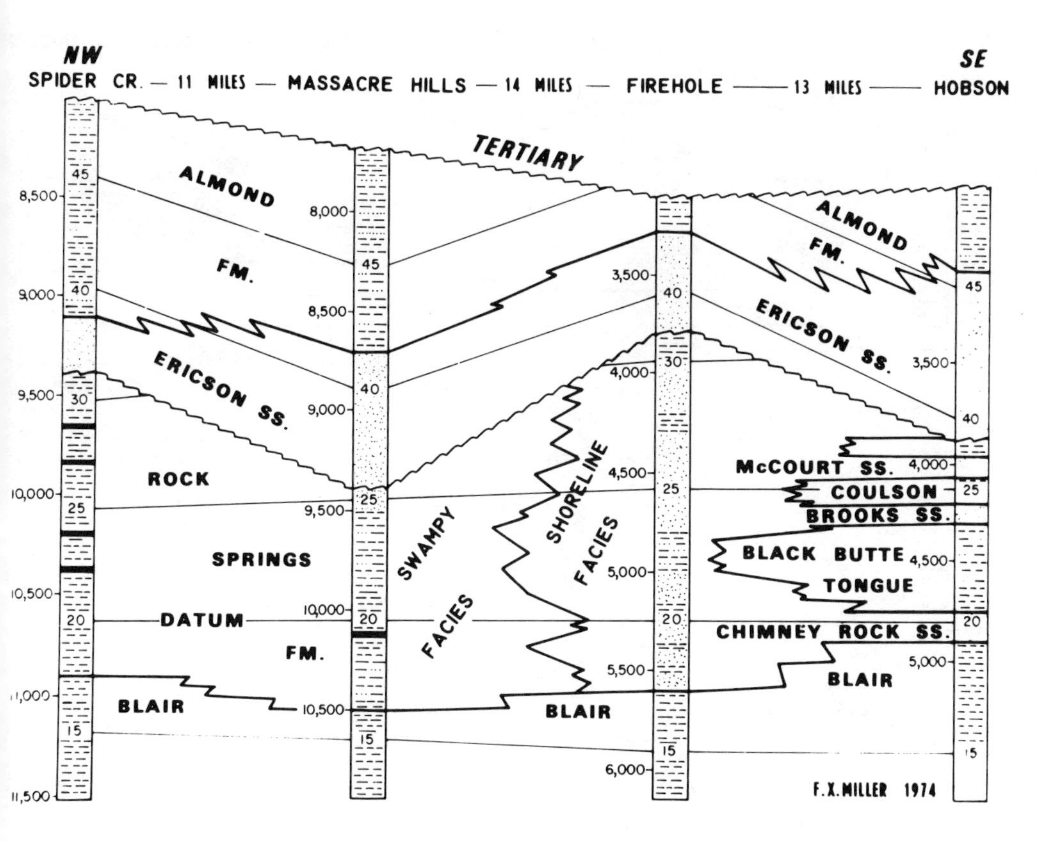

Text-Figure 9
Cross section illustrating the time-stratigraphic relationships of the Rock Springs Formation.

to the same CSRS and are usable immediately for time-stratigraphic correlations without years of data gathering. The chronostratigraphic correlations based on any group of fossils in the CSRS will be identical, thus avoiding disagreements between the various disciplines.

Third, the paleontologic, stratigraphic, and environmental information that can be accumulated is useful to the petroleum geologist. Transgressive and regressive sequences can be recognized, shoreline trends delineated, facies relationships determined, stratigraphic traps isolated, and productive intervals traced into new areas.

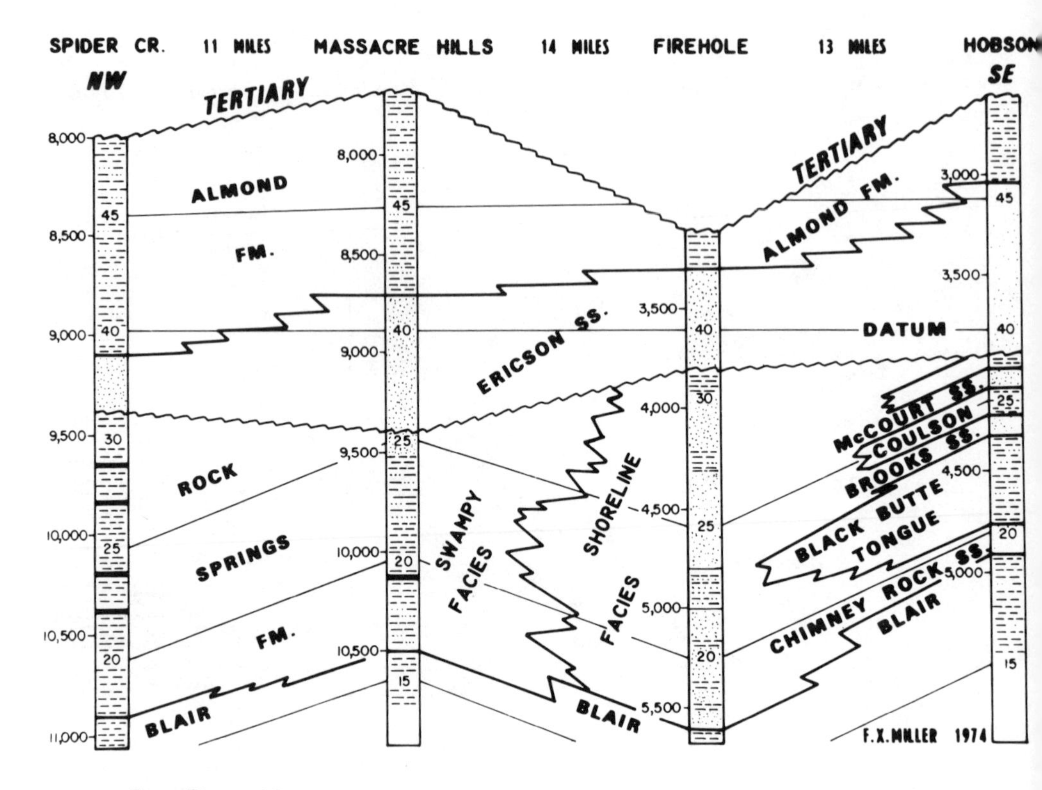

Text-Figure 10
Cross section illustrating the time-stratigraphic relationships of the Ericson and Almond formations.

Use of Certain Multivariate and Other Techniques in Assemblage Zonal Biostratigraphy: Examples Utilizing Cambrian, Cretaceous, and Tertiary Benthic Invertebrates

Joseph E. Hazel

U.S. Geological Survey, Washington, D.C.

INTRODUCTION

Biostratigraphy is concerned with the study and interpretation of fossils in sedimentary rocks, including the identification of taxa, the tracing of their lateral and vertical extent, and explanation of their distribution patterns. Although biogeography, biofacies analysis, and paleoecology can be construed to be different bioassociational aspects of biostratigraphy, to most workers the term connotes the correlation of sedimentary rocks by the use of fossils. The biostratigrapher whose study is concerned with correlation searches for temporally diagnostic patterns in data arrayed in matrixes consisting of N rows (samples) and k columns (taxa).

This paper addresses itself to the problems encountered when these samples $\times$ taxa ($N \times k$) matrixes become large enough to cause the worker to base his conclusions on only part of the available data base. It is pointed out herein, primarily by discussion of examples, how certain multivariate mathematical techniques offer consistent ways of searching for biostratigraphic patterns in large data matrixes.

Although these multivariate techniques are applicable to other types of biostratigraphic problems, we will be concerned here only with their use in the delineation of temporally significant assemblage zones utilizing benthic marine invertebrates. The philosophy behind and need for their usage in biostratigraphy are discussed. Other techniques that are useful in conjunction with multivariate analysis are also discussed and illustrated.

NEED FOR NEW APPROACHES

The fact that there are reasonable and useful biostratigraphic grids within which to operate in many if not most areas of the world is a tribute to biostratigraphers from William Smith to the present. In general, one cannot quarrel with the results of these workers. Even so, criticism of some methods is justified.

As mentioned above, because data matrixes in biostratigraphy are commonly large, conclusions have been based on only part of the data. For example, one may have the knowledge to identify accurately all the brachiopods from all the samples collected in a problem, but if the numbers of samples and taxa are so great that the mind cannot assimilate the data, the tendency is to reduce the number of taxa whose distributions will be studied, or to composite the samples. This approach is obviously less desirable than using all available data; it can lead to meaningless groupings if, for example, the favored taxa, chosen a priori, are biostratigraphically inconsistent.

One procedure open to criticism can be demonstrated by the use of Text-figure 1, in which the stratigraphic column represents a composite section for a hypothetical geologic area or region, sometimes called the standard section. Commonly, this "standard" has been constructed as follows: (1) fossils are identified from local sections; (2) the local sections are then correlated, using fossils and physical stratigraphic criteria to build a composite section; (3) the fossils used at least in part to make the composite section are plotted against the same section to determine their "range"; and (4) at points where one or more taxa apparently began or ended their existence, horizontal lines are drawn, and these are assigned biostratigraphic boundary status.

Such charts as these can be very informative; the writer and most readers of this paper have seen or constructed similar-looking figures. However, if constructed and interpreted as outlined above, they are logically unsound (the fallacy of *petitio principii*) because they beg the question. If the correlations of the local sections are based at least in part on the fauna, and then the distribution of the same taxa is plotted against the composite section to determine their ranges, the data have been justified by the model and the model by the data. Second, attention is focused on the apparent beginnings and ends of things – actually on the nothing between the beginning of one taxon or group and the ending of another taxon or group of taxa. It is often implied that more than one taxon originated or became extinct at the same time (the reason Shaw, 1964, referred to such charts as examples of catastrophism). The biostratigraphy may look rather clearcut on the chart, but is it in the *rocks*? Can one consistently identify the "zones" in the field or laboratory?

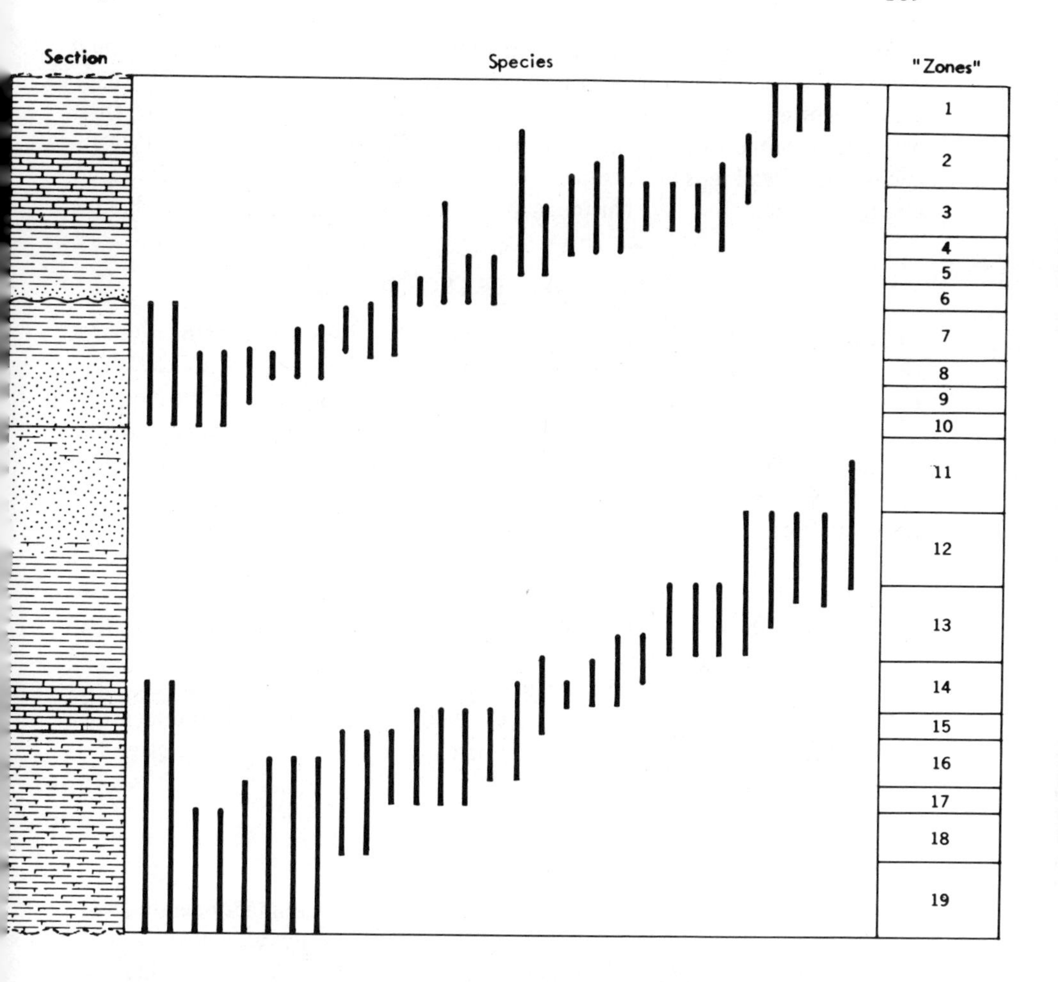

Text-Figure 1
Hypothetical section assumed to have been composited from local sections on the basis of biologic and physical criteria. Vertical black lines represent the composite ranges of the species found at the various local sections. The column at the right indicates one interpretation that could be defined from such data.

Also, repeatability is low. In the hypothetical example set up in Text-figure 1, no two people are likely to "zone" the section in exactly the same way. Also, for that matter, will the same worker "zone" different sections consistently? So much depends on one's innate conservatism or liberalism, taste, and sense of proportion. One is tempted facetiously to call such chart analysis "biochartigraphy," rather than biostratigraphy, and to define it as the study of the arrangement of black lines, representing taxa, on sheets of paper.

More consistently applicable alternatives to the grouping of samples or to the reduction in the number of taxa used to reduce the dimensionality in a biostratigraphic study are offered by various multivariate mathematical analytic techniques. These have been referred to as classification systems by Harbaugh and Merriam (1968) and others. Two kinds of classification systems are particularly appropriate for biostratigraphic data; these are cluster analysis (CA) and certain ordination techniques.

MULTIVARIATE TECHNIQUES

There are several varieties of cluster analysis (CA) (Sokal and Sneath, 1963; Gower, 1967; Jardine and Sibson, 1971; Davis, 1973; Sneath and Sokal, 1973), but the type that is becoming rather standard in bioassociational work is the hierarchical average-linkage or pair-group methods devised by Sokal and Michener (1958) and described by Sokal and Sneath (1963) and Sneath and Sokal (1973). The application of CA to biostratigraphy has been previously discussed and demonstrated by Hazel (1970b; 1971). Ordination includes those multivariate analytic techniques in which the coordinates of each object being compared are obtained for the smallest number of axes (i.e., smallest dimensional space) that preserves sufficient information about the distances between the objects (Kruskal, 1964). Three of these techniques that do not require any a priori assumption as to groups within the data are discussed and compared in this paper. These are principal components analysis (PCA), principal coordinates analysis (PCOORD), and nonmetric multidimensional scaling (MDSCALE); examples using the first two in biostratigraphy are given herein.

COEFFICIENTS

Principal components analysis (PCA) and ordination techniques operate on matrixes of coefficients of similarity or distance. Empirically, with bioassociational data, coefficients of similarity seem to consistently yield more readily interpreted results than do distance coefficients, despite the fact that some distance coefficients are simply the scaled complements of similarity coefficients. The presence of a species is less sensitive to environmental control than is its abundance, and in zonal biostratigraphy, it is temporally significant patterns in the data that are being searched for; because of this, binary (presence-absence) coeficients rather than multistate measures are used. Cheetham and Hazel (1969) have outlined the properties of 22 binary coefficients, and Hazel (1970b) has discussed the application of some of these in biostratigraphy.

Almost any coefficient can be used with CA, PCOORD, and MDSCALE. Principal components analysis is based on the extraction of eigenvectors and eigenvalues from a variance-covariance matrix or a product-moment correlation matrix. A variance-covariance matrix of standardized variables is a correlation matrix. The binary case of the product-moment correlation coefficient (r) is the phi coefficient (see Cheetham and Hazel, 1969); this measure is used in the PCA examples given.

In the given examples of CA and PCOORD, the Otsuka coefficient is used. This coefficient is somewhat intermediate in its properties and does not emphasize similarity or difference. The Otsuka coefficient is the binary equivalent of the multistate coefficient of proportional similarity (cos θ) of Imbrie and Purdy (1962).

CLUSTER ANALYSIS

The calculation of coefficients between all pairs of objects (t) being compared results in a matrix of coefficients with $t(t - 1)/2$ values. In effect, this places each object in space at distances from each other equivalent to the values of the calculated coefficients. Thus there are $t(t - 1)/2$ dimensions to the problem. Average-linkage or pair-group cluster analysis (CA) reduces this dimensionality by an averaging process that graphically results in two-dimensional nested hierarchial plots, most commonly called dendrograms. The two forms of pair-group CA are the unweighted (UPGM) and weighted (WPGM) methods (Sokal and Michener, 1958; Sokal and Sneath, 1963; Sneath and Sokal, 1973). The differences between these two are discussed in these references and in Hazel (1970b). Ordinarily, in bioassociational work, the UPGM is more desirable because it can be demonstrated that the resulting dendrograms are more faithful to the original matrix of coefficients. However, the WPGM may be preferred in cases where sampling is biased toward one of two or more units being compared. This is because in the WPGM, each new joiner to a cluster is given a weight equal to the stem of the existing cluster, whereas in the UPGM the numerical level at which a new member will join a cluster is based on the average similarities of all the existing members calculated from the original matrix of coefficients.

Cluster analysis is a most useful multivariate technique when compact groupings exist in the data. The within-group relationships in such cases are very accurately shown, and the between-compact-group relationships are equally clearly manifested graphically in the dendrogram. If, however, the data matrix and, therefore, the coefficient matrix do not contain well-defined groups, and the structure is a more subtle one of gradation or of loosely compacted groups, CA may not be the most appropriate technique for revealing the relationships. Because of the averaging process, more and more distortion is added as each cluster joins others. For convenience, this between-group distortion can be referred to as "down-the-dendrogram" distortion. The distortion can be measured by comparing the dendrogram with the original coefficient matrix by calculating a correlation coefficient between them (the "cophenetic" correlation coefficient) (Sokal and Rohlf, 1962; Rohlf, 1970; Hazel, 1970b). Clusters will be formed because they must be, but the relationships between major clusters may not be very accurately expressed. In cases of gradation or at least subtle faunal change, a change in coefficient or clustering method can result in quite different looking dendrograms.

Notwithstanding these disadvantages, CA is an attractive tool because of its graphic and mathematical simplicity; it is particularly useful in the analysis of data from measured sections. The application of CA to biostratigraphy has been discussed and examples given in Hazel (1970b; 1971) and Carbonnel (1973). An example utilizing Cambrian trilobites follows.

In 1962, Grant published the results of a study of the distribution of trilobites in sections measured in the upper part of the Franconia Sandstone (Upper Cambrian) at Lake Pepin in southeastern Minnesota. From 65 samples he identified 24 species of trilobites. Grant recognized four biostratigraphic units in the upper Franconia in this area on the basis of the first (lowest) occurrences of distinctive assemblages. He pointed out that the previously named zonal units were actually based on the acme of particular species and not their true ranges. The nomenclature that Grant used for his zonal units is as follows:

Ptychaspis - Prosaukia zone
- *Prosaukia* subzone
  - *P. striata* zonule
  - *P. granulosa* zonule
  - *Psalaspis* zonule

This compact data set is amenable to analysis by both cluster and ordination techniques, and so provides a base for comparing these techniques with each other and with the more classical method used by Grant.

Text-figure 2 shows the results of a *Q*-mode cluster analysis of the trilobite binary data using the range method (see p. 208), the Otsuka coefficient, and the UPGM method of clustering. The position of the samples, related to the level of Lake Pepin, and the lithostratigraphic units are also shown. There are four principal clusters of samples in the dendrogram. The hierarchical arrangement of samples indicates that two major groupings are related at about the 10 level and that one of these contains three subordinate groups, two of which are clearly more closely related to each other than to the third. The two major groupings correspond to Grant's *Prosaukia* and *Ptychaspis* subzones; the three subordinate groups of the large cluster correspond to the two *Ptychaspis* and the *Psalaspis* zonules. The only significant difference between the biostratigraphic interpretation based on the cluster results and that of Grant (1962) lies in the Rileys Coulee section, in which Grant drew the *Prosaukia-Ptychaspis* boundary between samples 20 and 21, whereas the cluster analysis results suggest that the entire section is *Prosaukia* subzone. Grant's interpretation is apparently based on the first appearance of *Chariocephalus whitfieldi*, and to a lesser extent *Saratogia*? *hamula*, which first appear in the beds assigned to the *Prosaukia* subzone at the Lake City section. However, on the basis of overall faunal similarity as calculated from presence-absence data, all the samples at Rileys Coulee are more similar to those assigned to the *Prosaukia* subzone at Lake City. It is not clear which interpretation is more biostratigraphically satisfactory; information from more than just these two sections is needed.

Nonetheless, overall, the two solutions are very similar and indicate that the conclusions drawn after a considerable length of time devoted to plotting data on charts and studying the plots (R. E. Grant, pers. comm., 1972) can be closely reproduced in a few hours of data preparation and a few minutes of computer manipulation, and without making a priori judgments as to which taxa are the best ones to rely upon.

If it is assumed that the solution suggested in Text-figure 2 is correct, it can be determined which of the variables (species) were primarily responsible for the clusters of samples seen in *Q*-mode. Text-figure 3 contains an *R*-mode dendrogram comparing the 24 trilobite species, and the calculated values for constancy and biostratigraphic fidelity of the species for the biostratigraphic units. Species with constancy values of 4 or more are also indicated graphically by symbols. (Grant's trilobites in his table 2 were numbered 1 to 24 from top to bottom.) It is quite clear which species are causing the samples to cluster. With these data, one could place other samples from this area into the biostratigraphic grid without further multivariate analyses.

PRINCIPAL COMPONENTS ANALYSIS

Principal components analysis (PCA) is an ordination technique in which the new axes (components) are orthogonal linear combinations of the original variables; PCA operates in *R*-mode but yields information on both the objects and the variables. A covariance matrix of standardized variables (a correlation matrix) is computed. The elements of this matrix can be visualized as defining points on a t-dimensional ellipsoid. By certain procedures, eigenvectors (components) and eigenvalues are extracted from the covariance matrix. The eigenvectors of the matrix yield the slope of the major axes of the ellipsoid relative to the original variables. As many eigenvectors are fitted as there are variables. The first eigenvector is fit to the greatest dimension of the ellipsoid and therefore represents the maximum variance in the variables. The next eigenvector is fit to the next greatest direction of variance at right angles to the first, and so on. The eigenvectors are at right angles (orthogonal). The eigenvalues represent the magnitude or lengths of the axes. More detailed discussions of the theory and mathematics of PCA can be found in Harman (1967), Morrison (1967), Davis (1973), and Jöreskog et al. (1976).

The output from a PCA consists of a correlation or covariance matrix and an eigenvector matrix whose dimensions are the same as the number of variables. The elements of the eigenvector matrix (the factor loadings) represent the projections of the variables onto each eigenvector. Thus it can be seen at a glance to which variables the variance of each eigenvector is related.

Eigenvalues represent the variance of the eigenvectors. If an eigenvalue is divided by the total variance (sum of the eigenvalues), the resultant is the percentage of variance explained by that eigenvector. For standardized variables, the total variance is equal to the number of variables. For a given eigenvector, the sum of the factor loadings squared is equal to either the eigenvalue or unity, depending on the computational method used to set the length of the eigenvector.

A most important item of output in a PCA from the standpoint of biostratigraphy is the coordinates on the fitted eigenvectors of each object whose variables were used to obtain the correlation matrix. If the eigenvectors are set to unity, the coordinates of an object for an eigenvector can be obtained by the multiplication of the values in the original $N \times k$ data matrix by the elements (factor loadings) of the eigenvector matrix for that eigenvector. The summed products for each object (sample) for each desired eigenvector give the coordinates. This procedure amounts to matrix multiplication of the data matrix by the eigenvector matrix.

In PCA analysis, the major structural patterns (general spacing between clusters of faunally related samples) are usually accurately depicted in a reduced space. The relationships between individual samples, however, are usually distorted, because these smaller distances may be related to variation on axes other than the first one to three eigenvectors. Thus, in contrast to cluster analysis, PCA distorts somewhat within compact groups but not between such groups.

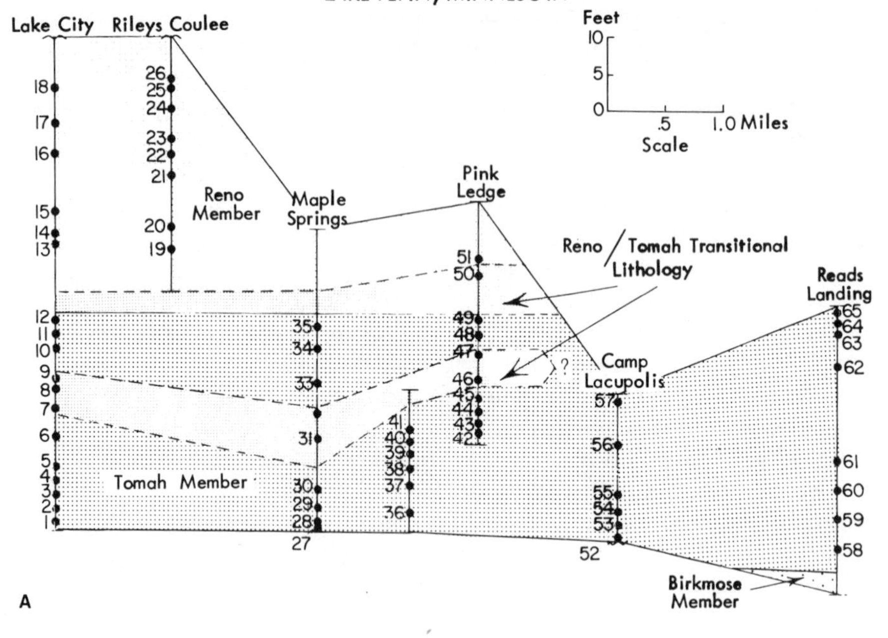

Text-Figure 2 *(above and following two pages)*
Cluster analysis of trilobite samples from the upper Franconia Sandstone (Upper Cambrian) of southeastern Minnesota: (A) lithostratigraphy and the position of the trilobite samples; (B) a Q-mode dendrogram resulting from a comparison of the 65 samples containing 24 trilobite species; (C) summary of the larger dendrogram; (D) biostratigraphic interpretation of the results. (Data from Grant, 1962.)

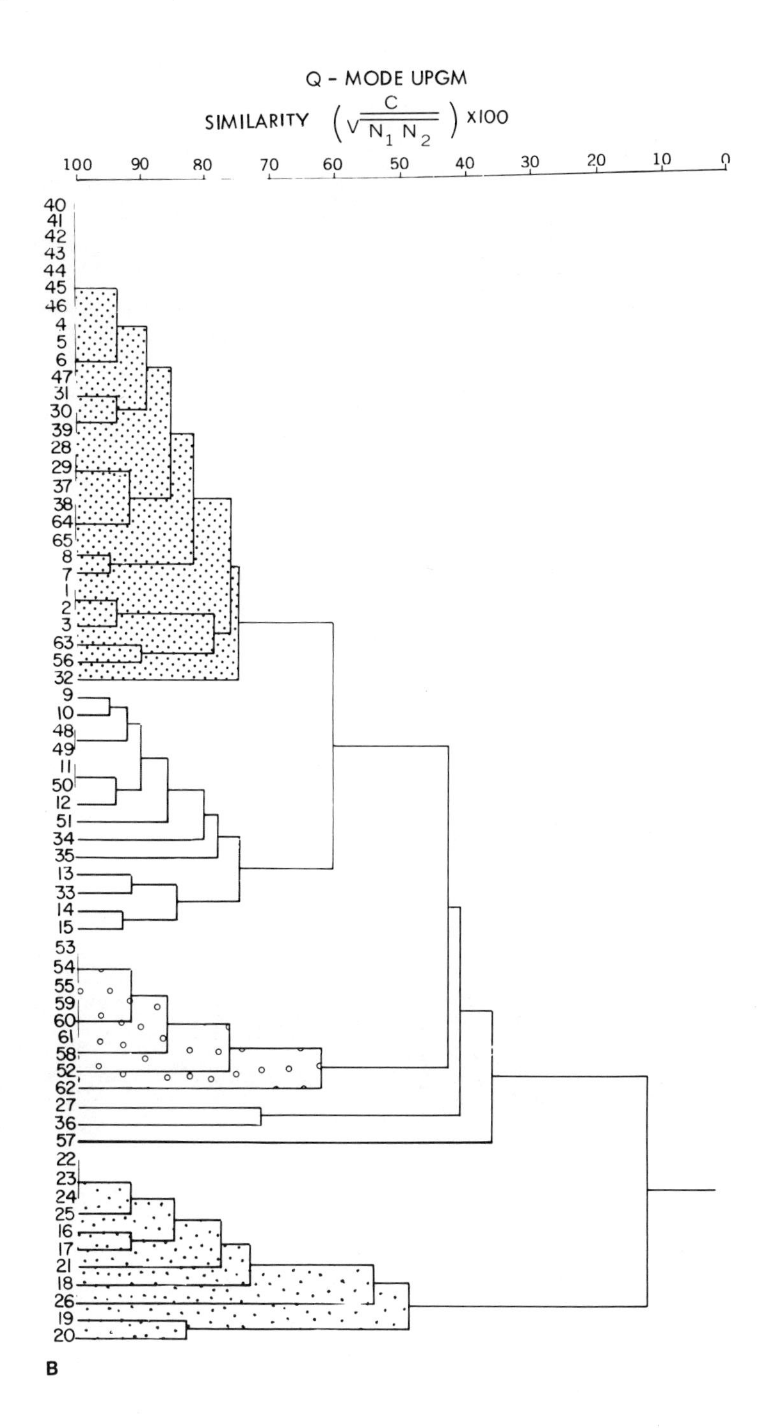
Q - MODE UPGM
SIMILARITY $\left(\sqrt{\frac{C}{N_1 N_2}}\right) \times 100$
100 90 80 70 60 50 40 30 20 10 0
40
41
42
43
44
45
46
4
5
6
47
31
30
39
28
29
37
38
64
65
8
7
1
2
3
63
56
32
9
10
48
49
11
50
12
51
34
35
13
33
14
15
53
54
55
59
60
61
58
52
62
27
36
57
22
23
24
25
16
17
21
18
26
19
20
B

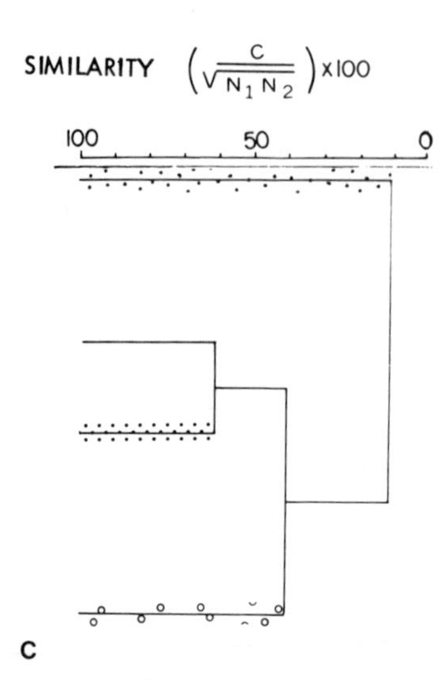

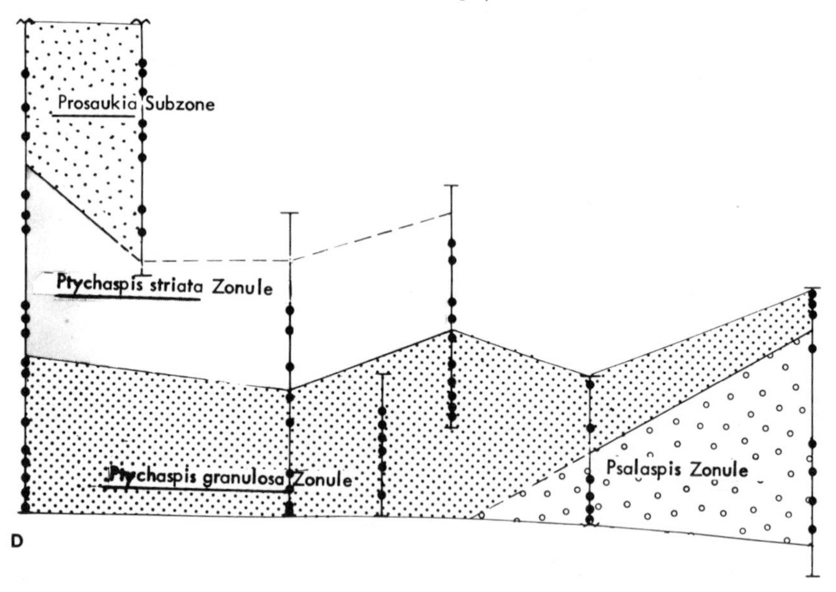

Principal components analysis has the advantage of being virtually unlimited with respect to the number of samples that can be manipulated, because it operates on variable x variable coefficient matrixes. However, this means it is a variable-limited technique, and biostratigraphic data sets with large numbers of species (greater than about 70) cannot be efficiently handled because of the computer core storage and

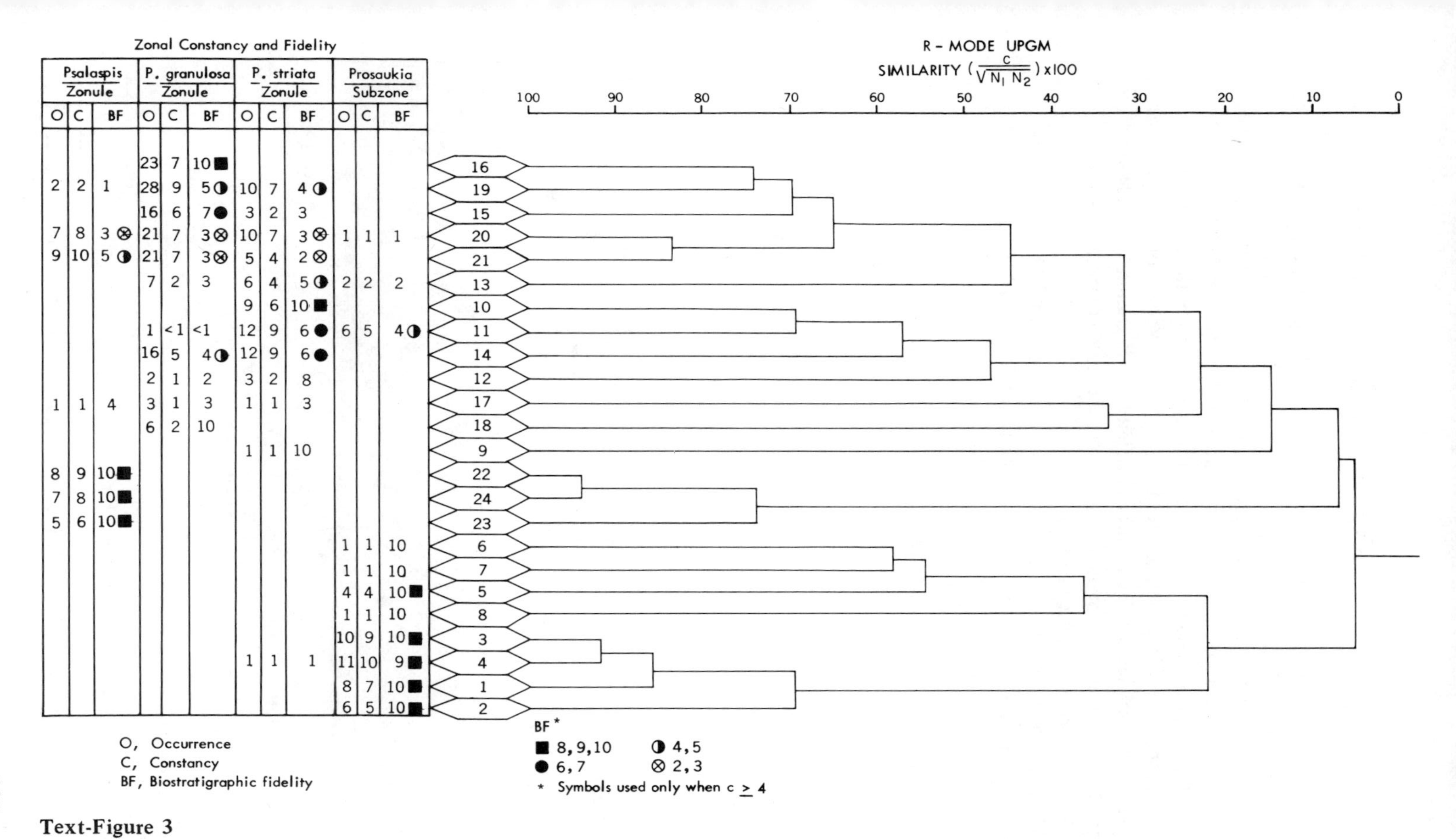

Species	Psalaspis Zonule O	C	BF	P. granulosa Zonule O	C	BF	P. striata Zonule O	C	BF	Prosaukia Subzone O	C	BF
16				23	7	10 ■						
19	2	2	1	28	9	5 ◑	10	7	4 ◑			
15				16	6	7 ●	3	2	3			
20	7	8	3 ⊗	21	7	3 ⊗	10	7	3 ⊗	1	1	1
21	9	10	5 ◑	21	7	3 ⊗	5	4	2 ⊗			
13				7	2	3	6	4	5 ◑	2	2	2
10							9	6	10 ■			
11				1	<1	<1	12	9	6 ●	6	5	4 ◑
14				16	5	4 ◑	12	9	6 ●			
12				2	1	2	3	2	8			
17	1	1	4	3	1	3	1	1	3			
18				6	2	10						
9							1	1	10			
22	8	9	10 ■									
24	7	8	10 ■									
23	5	6	10 ■									
6										1	1	10
7										1	1	10
5										4	4	10 ■
8										1	1	10
3										10	9	10 ■
4							1	1	1	11	10	9 ■
1										8	7	10 ■
2										6	5	10 ■

Text-Figure 3

R-mode cluster analysis of 24 trilobite species occurring in the 65 trilobite samples shown in Text-figure 2. Opposite the end points of the dendrogram, values for constancy, biostratigraphic fidelity, and occurrence of species in the biostratigraphic units of Text-figure 2D are given. See text for discussion.

time required. The same trilobite data used in the CA example are used to demonstrate PCA in biostratigraphy.

Text-figures 4 and 5 and Table 1 give the results of a PCA of the Grant trilobite data. Note that the first axis fitted delineates the *Prosaukia* and *Ptychaspis* subzones. The *Psalaspis* zonule samples are differentiated from the two *Ptychaspis* zonules along the second component (Text-fig. 4). However, not all the *Ptychaspis striata* zonule samples are clearly differentiated from *P. granulosa* zonule samples on the first and second components; but when the third component is added (Text-fig. 5), these two groups are distinct.

The *R*-mode information arrived at by the CA is present in the eigenvector matrix of the PCA (Table 1). Here it can be seen that the first component axis is

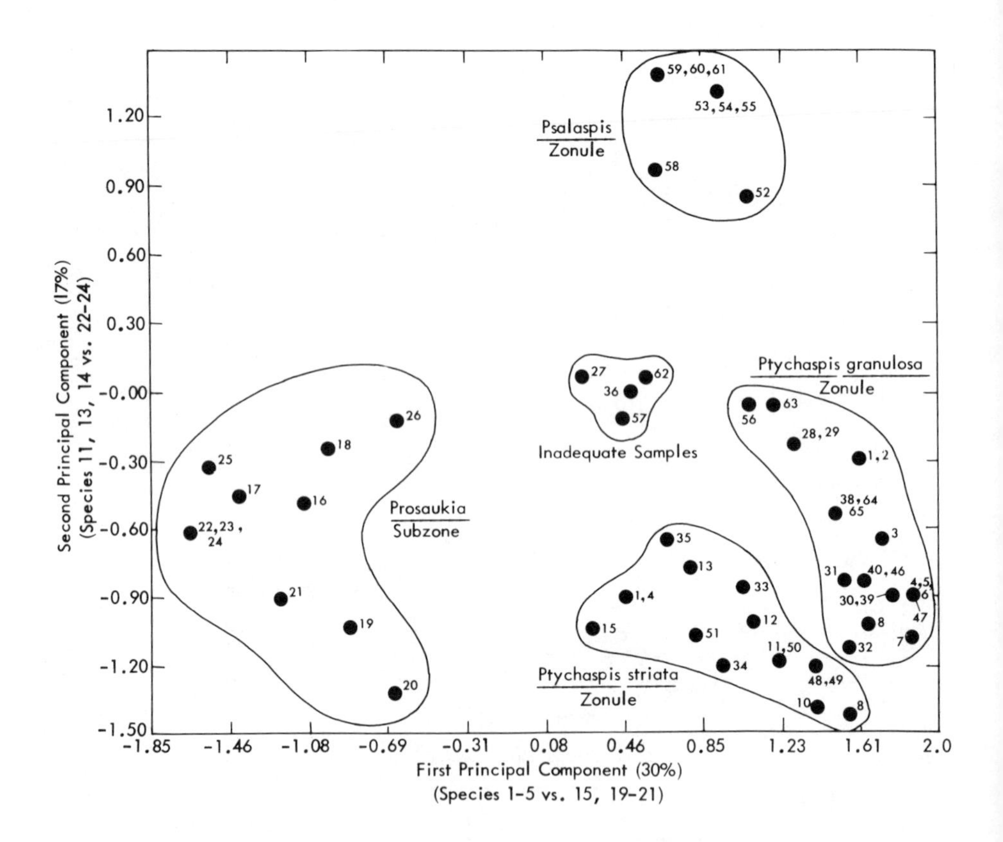

Text-Figure 4
Principal component analysis of Grant's (1962) trilobite data. Sixty-five samples are projected onto the first two eigenvectors. Note that the first eigenvector fitted differentiates clearly the *Prosaukia* subzone from the zonules of the *Ptychaspis* subzone, and note that the *Psalaspis* zonule is clearly separated on the second eigenvector.

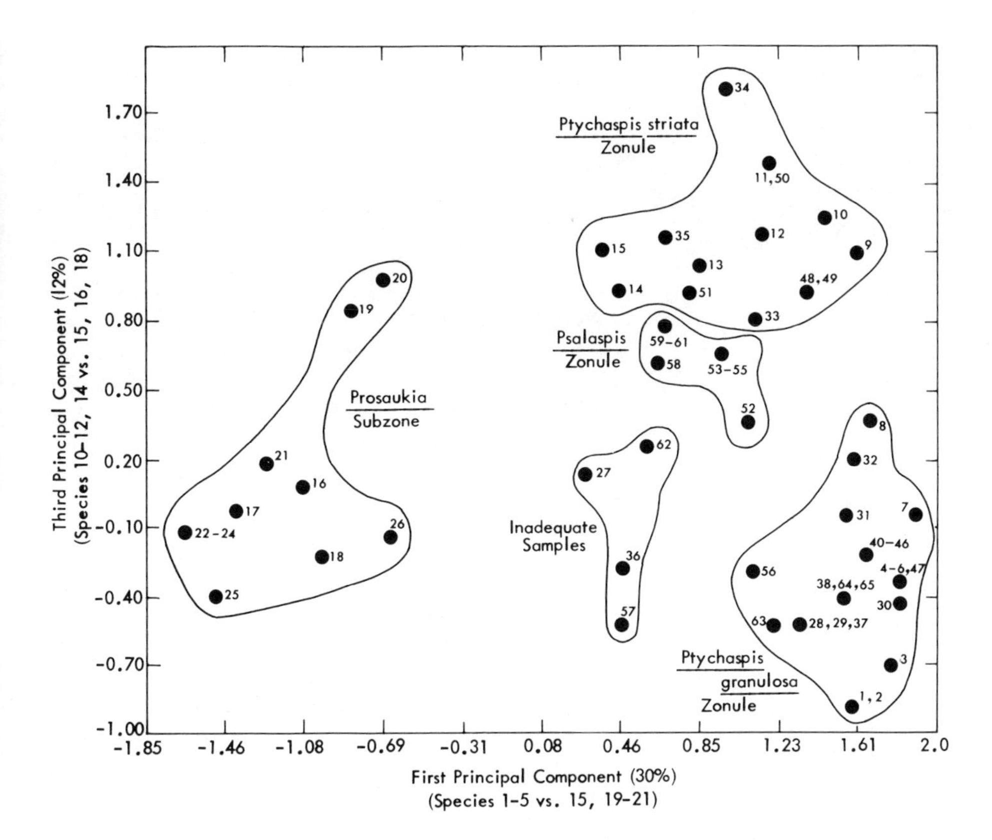

Text-Figure 5
Principal component analysis of Grant's (1962) trilobite data (Text-fig. 2A), projections on the first and third eigenvectors. The two *Ptychaspis* zonules are differentiated on the third eigenvector. Compare with Text-figure 4.

fitted primarily to the contrast between species 1 to 5 and species 15 and 19 to 21 (compare with Text-fig. 3). The second axis is fit to the contrast between species 11, 13, 14, and species 22-24. The two *Ptychaspis* zonules are differentiated on the third axis by the contrast between species 10 to 12, and 14, and species 15, 16, and 18.

The PCA yielded two-dimensional plots and other output information that result in biostratigraphic hypotheses to be tested against the stratigraphy. In this case, the conclusions are the same as those based on the CA. Empirically, it is generally true that, when the same biostratigraphic data set can be used, CA and PCA (or PCOORD analysis; PCOORD only in *Q*-mode) give very similar results. However, many data sets contain too many species for PCA.

Table 1
Eigenvalues and Eigenvector Matrix from a Principal Components Analysis on a Matrix of Phi Coefficients[a,b]

Eigenvector:		1	2	3	4	5
Eigenvalue:		7.123	4.031	2.865	2.190	1.134
Variance	% total:	30	17	12	9	5
	cumulative:	30	47	59	68	73
Variable						
1		-0.310	-0.041	-0.150	-0.223	-0.128
2		-0.284	-0.026	-0.149	-0.229	-0.106
3		-0.339	-0.091	-0.066	0.057	-0.150
4		-0.336	-0.099	-0.028	0.081	-0.050
5		-0.288	-0.102	-0.019	0.177	-0.168
6		-0.087	-0.064	0.095	0.380	-0.130
7		-0.111	-0.118	0.178	0.538	-0.001
8		-0.080	-0.090	0.114	0.383	0.032
9		0.011	-0.070	0.229	-0.169	-0.001
10		0.061	-0.181	0.386	-0.280	0.009
11		-0.158	-0.248	0.341	-0.095	-0.069
12		0.078	-0.164	0.285	-0.232	-0.320
13		0.171	-0.312	0.134	0.015	0.078
14		0.161	-0.323	0.204	0.106	-0.076
15		0.246	-0.173	-0.252	0.040	-0.134
16		0.197	-0.063	-0.398	0.109	0.174
17		0.137	-0.067	-0.138	0.022	-0.706
18		0.101	-0.037	-0.221	0.076	-0.384
19		0.289	-0.070	-0.133	-0.033	-0.110
20		0.311	0.051	0.104	-0.065	-0.043
21		0.296	0.039	0.148	0.255	0.007
22		0.011	0.447	0.183	0.049	-0.169
23		0.006	0.430	0.185	0.047	-0.090
24		0.013	0.426	0.173	0.047	-0.194

[a]Calculated from the Cambrian trilobite data (24 species in 65 samples) from Minnesota given by Grant (1962).

[b]Only the first five eigenvectors and corresponding eigenvalues and factor loadings are shown. Note that on the first eigenvector, species 1 to 5 are contrasted with species 15 and 19 to 21; on the second eigenvector, species 11, 13, and 14 are contrasted with species 22 to 24; and on the third eigenvector, species 10 to 12 and 14 are contrasted with species 15, 16, and 18. Compare this with the R-mode cluster analysis (Text-fig. 3). The results of ordination are given in Text-figures 4 and 5.

PRINCIPAL COORDINATES ANALYSIS

Principal coordinates analysis (PCOORD) was devised by Gower (1966). It is a method of relating the objects in an analysis to major axes (eigenvectors) to reduce the dimensionality of the problem and to obtain the same scatter plot results as in PCA. The eigenvectors and eigenvalues are extracted, however, from an object X object (*Q*-mode) matrix of coefficients which has been transformed so that the elements of the extracted eigenvectors represent the coordinates of the objects (Gower, 1966; Rohlf, 1972; Blackith and Reyment, 1971; Rowell et al., 1973).

The output from a PCOORD includes the values for a specified number of eigenvalues and eigenvectors and the coordinates for the objects on the eigenvectors. Because it works in *Q*-mode, PCOORD is virtually unlimited with respect to variables, but the number of objects is limited by available computer core storage. The usability of PCOORD analysis in biostratigraphy is demonstrated below with examples from the Cretaceous and Tertiary.

There have been arguments concerning correlations of the formations of the Austin and Taylor Groups (Coniacian to Campanian) in central Texas for many years, and a variety of interpretations have been published (Text-fig. 6). The time-stratigraphic position of the Austin-Taylor boundary in the region from Austin to Waco, particularly, has drawn a variety of opinions. The difficulties are in part related to the fact that the Austin-Taylor outcrop belt is affected by faulting in the Balcones fault system. This, coupled with subtle facies changes, has created many lithostratigraphic problems. Thus it has been very difficult to determine what is or is not equivalent to the upper formations of the Austin Group and the Sprinkle Formation (Young's, 1965, proposed term for the lower Taylor Group) north of the Austin area.

Seventy-four species of ostracodes have been identified in 94 samples from the upper Austin and Sprinkle in the region from Austin to Dallas (Paulson, 1960; Hazel, 1963; Hazel and Paulson, 1964). The samples are variably fossiliferous; thus there is considerable "noise" in the data related to sample size. The size of the data set precludes the use of principal components analysis. It is amenable to cluster analysis; however, the lack of compactness in the data causes the dendrogram to have many single sample "clusters," and it is not easily interpreted. A PCOORD of these data (Text-fig. 7), however, provides a good basis for a biostratigraphic interpretation. It might be added here that the number of sections, samples, and species makes an analysis of the data by plotting on charts very difficult.

Text-figure 7 indicates the position of the 94 samples relative to the first two coordinate axes fitted; these account for only 24 percent of the variance. Despite the considerable scatter in the points, there are three more or less distinct groups of samples. Group 1 consists of samples from rocks that were identified in the field in the Austin area as Sprinkle Formation (i.e., lower Taylor) or, north of Austin, the Sprinkle-Burditt transition. Group 2 is composed of samples assigned in the field to the Burditt Marl, Sprinkle-Burditt transition, Gober Formation (this term is used here for the "Upper Chalk" in the Dallas area), upper part of the Dessau Formation, Stephenson's "lower Taylor Chalky Marl" at Waco, and in three instances the Sprinkle Formation. Group 3 contains samples from the Dessau Formation, the Bruceville Formation at Dallas, and one sample from the basal Gober Formation at Dallas. On the third principal coordinate axis (not shown), there is a gradational

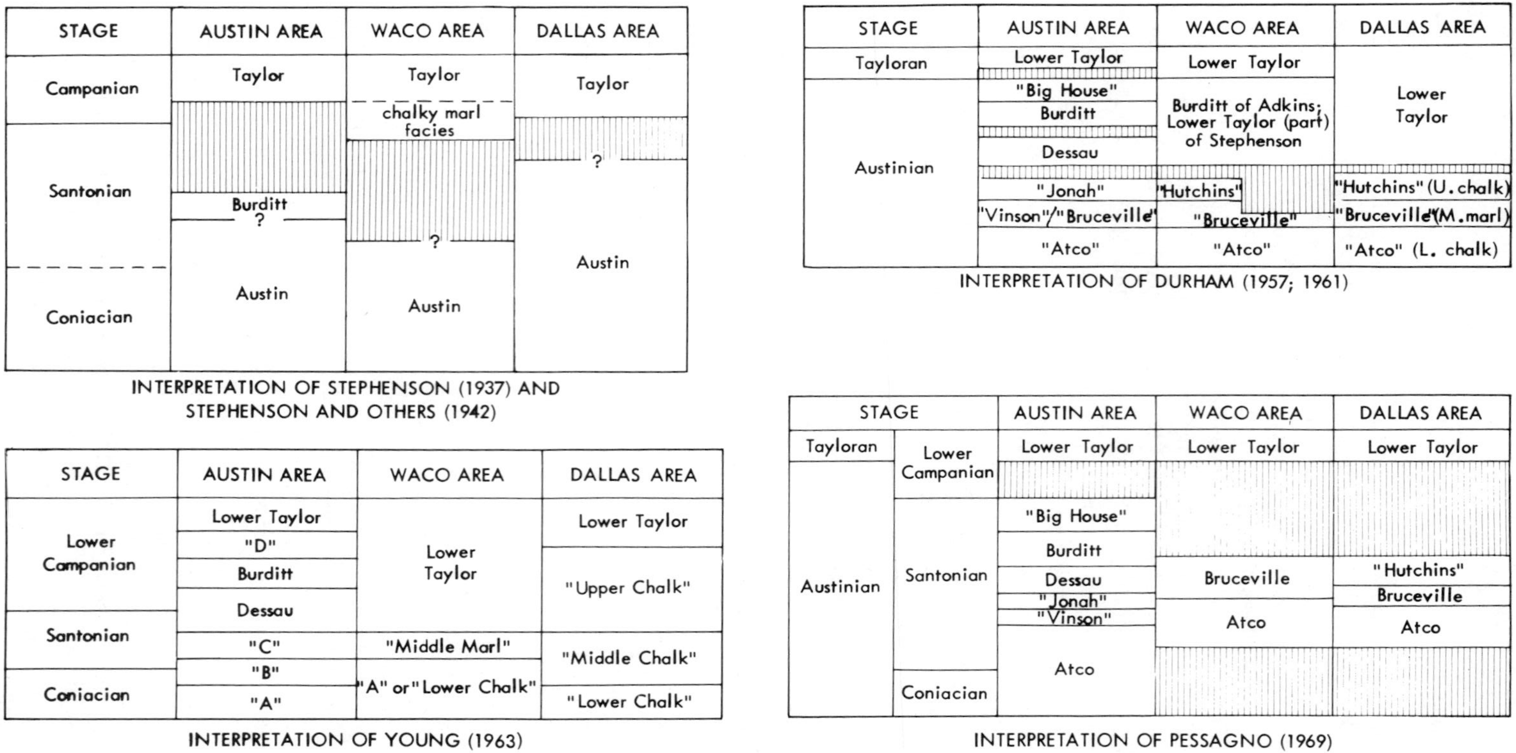

STAGE | AUSTIN AREA | WACO AREA | DALLAS AREA
Campanian
Santonian
Coniacian
Taylor
Burditt
?
Austin
Taylor
chalky marl facies
?
Austin
Taylor
?
Austin

INTERPRETATION OF STEPHENSON (1937) AND STEPHENSON AND OTHERS (1942)

STAGE | AUSTIN AREA | WACO AREA | DALLAS AREA
Tayloran
Austinian
Lower Taylor
"Big House"
Burditt
Dessau
"Jonah"
"Vinson"/"Bruceville"
"Atco"
Lower Taylor
Burditt of Adkins; Lower Taylor (part) of Stephenson
"Hutchins"
"Bruceville"
"Atco"
Lower Taylor
"Hutchins" (U. chalk)
"Bruceville" (M. marl)
"Atco" (L. chalk)

INTERPRETATION OF DURHAM (1957; 1961)

STAGE | AUSTIN AREA | WACO AREA | DALLAS AREA
Lower Campanian
Santonian
Coniacian
Lower Taylor
"D"
Burditt
Dessau
"C"
"B"
"A"
Lower Taylor
"Middle Marl"
"A" or "Lower Chalk"
Lower Taylor
"Upper Chalk"
"Middle Chalk"
"Lower Chalk"

INTERPRETATION OF YOUNG (1963)

STAGE | AUSTIN AREA | WACO AREA | DALLAS AREA
Tayloran
Austinian
Lower Campanian
Santonian
Coniacian
Lower Taylor
"Big House"
Burditt
Dessau
"Jonah"
"Vinson"
Atco
Lower Taylor
Bruceville
Atco
Lower Taylor
"Hutchins"
Bruceville
Atco

INTERPRETATION OF PESSAGNO (1969)

Text-Figure 6
Previous interpretations of the correlation of the Austin Group and lower part of the Taylor Group in the region from Austin to Dallas, Texas.

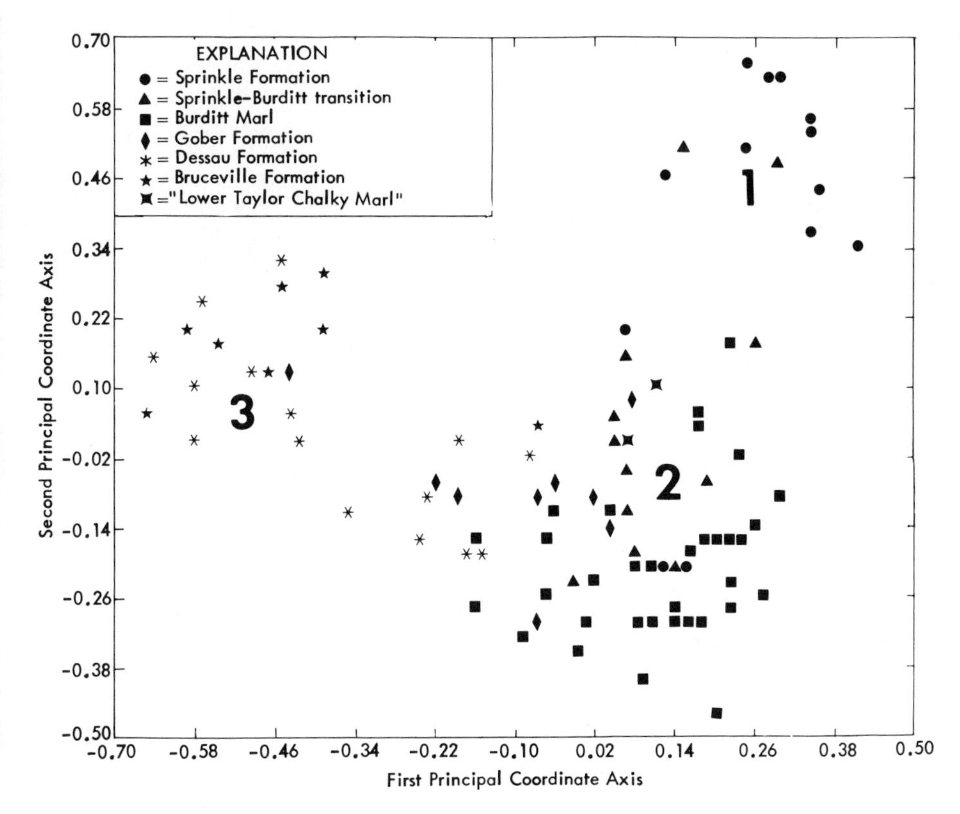

Text-Figure 7
Principal coordinate analysis of 94 samples containing 74 ostracode species from the upper Austin Group and the overlying Sprinkle Formation (of Young, 1965) of the Taylor Group. Symbols indicate the lithostratigraphic unit to which the sample was assigned in the field. Samples are projected onto the first and second coordinate axes. See text for discussion.

separation of the Dessau and the single Gober sample from the Bruceville samples. These data have been tested against the various lithostratigraphic and biostratigraphic interpretations, and the correlations for the upper Austin and Sprinkle given in Text-figure 8 are suggested. The writer's tentative conclusions as to the correlation of the lower Austin, where the data are much poorer, are also given for completeness, although they are not discussed.

Assemblage zones delineated by cluster analysis or ordination techniques using benthic invertebrates are necessarily useful only within faunal provinces. Correlations between provinces must be by means other than the recognition of this type of assemblage zone. Such correlations can be accomplished by use of range zones or concurrent-range zones of widely distributed species.

STAGE		OSTRACODE ASSEMBLAGE ZONES	AUSTIN AREA	WACO AREA	DALLAS AREA
Lower Campanian	Tayloran	"Cythereis" plummeri	Sprinkle Fm.	Sprinkle Fm.	Sprinkle Fm.
	Austinian	Alatacythere cheethami	Burditt Marl	Burditt Marl	Burditt Marl
		4	Dessau Fm.		Gober Fm.
Santonian		3	"Jonah" Fm.		
		2	"Vinson" Fm.	Bruceville Fm.	Bruceville Fm.
Coniacian		1	Atco Fm.	Atco Fm.	Atco Fm.

Text-Figure 8
Correlation of the Austin and lower part of the Taylor groups in the Austin to Dallas, Texas, region. Upper Austinian and lower Tayloran correlations are based on the results of the principal coordinates analysis (Text-fig. 7). For completeness, the writer's interpretation of the relationships of lower Austin formations is also shown; however, the data are less reliable and the conclusions only tentative. Use of the Burditt Marl follows Adkins (1933); Gober is applied by the writer to the "Upper Chalk" at Dallas, as well as to the tongue into northeast Texas; "Jonah" and "Vinson" are nude names for rock units used by Durham (1957; 1961).

For example, a shifting faunal province boundary existed in the North Carolina area during the Pliocene. North of the Neuse River, fossiliferous sediments assigned to the Yorktown Formation crop out. South of the Neuse, the terms Duplin, Waccamaw, James City (of Dubar and Solliday, 1963), and others are used for what seem to be Yorktown equivalents; but, although many species lived in both areas, the faunal compositions are different enough to make precise correlations difficult.

Using a CA of ostracode data, Hazel (1971) recognized three assemblage zones in the Yorktown Formation. One way of determining where units such as the Duplin and James City might fit biostratigraphically relative to these zones is to analyze samples from the Duplin and James City in conjunction with those from the Yorktown. Text-figure 9 represents a PCOORD of 43 Yorktown samples, one diverse sample from the Duplin type locality, and two samples from the James City type area.

From the two-dimensional plot (there was little or no biostratigraphic information on other axes), it is quite clear that the James City is faunally more similar to the youngest Yorktown zone, the *Puriana mesacostalis* zone, whereas the Duplin falls between the *Puriana mesacostalis* zone and the *Orionina vaughani* zone, the Duplin fauna being closest to some samples of the *Puriana mesacostalis* zone. This suggests

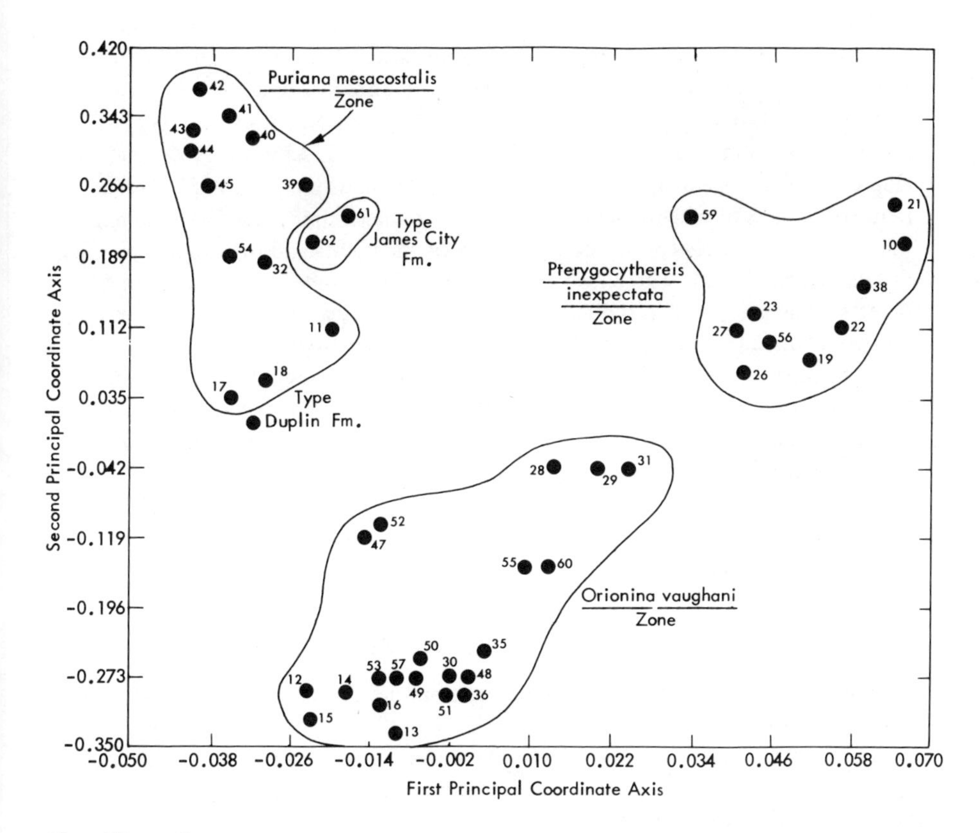

Text-Figure 9
Principal coordinate analysis of 43 ostracode samples from the Yorktown Formation of Virginia and northern North Carolina, one diverse sample from the type locality of the Duplin Formation, and two from the James City Formation in southern North Carolina. All three formations are Pliocene in age. Data and zone terminology for the Yorktown samples are from Hazel (1971). The *Pterygocythereis inexpectata* zone is the oldest and the *Puriana mesacostalis* zone the youngest of the ostracode assemblage zones. Samples are projected onto the first and second coordinate axes. The object of the analysis was to determine to which of the Yorktown zones the Duplin and James City formations, which are from a different faunal province, showed the greatest affinity.

a biostratigraphic position for the type Duplin equivalent to about the boundary between these two zones.

OTHER ORDINATION TECHNIQUES

Two other techniques that have been used or have potential for application in biostratigraphy are factor analysis (FA) and nonmetric multidimensional scaling

(MDSCALE). Factor analysis is related to PCA, but has a different set of underlying assumptions. Recent summaries of its applications in geology can be found in Harbaugh and Merriam (1968), Davis (1973), and Jöreskog et al. (1976).

Multidimensional scaling was developed by Kruskal (1964) and has been discussed and used in a variety of fields (Rohlf, 1970, 1972; Kruskal, 1972), including a paleobiogeographic study by Whittington and Hughes (1972). It is an iterative process of placing points in a space of reduced dimensions so as to minimize the "stress" between the configuration of the points and the original coefficient matrix (Rohlf, 1972). The output from MDSCALE includes coordinates for the objects being compared in reduced space, from which scatter plots for the objects on two or three (or more if necessary) axes can be constructed.

Multidimensional analysis has the advantage of distributing distortion evenly between the smaller and larger clusters. However, because of the iterative nature of the process, considerable core storage and time are required, and the number of samples that can be analyzed in one run is limited.

Space in this volume as well as the writer's inexperience with FA and MDSCALE precludes the inclusion of practical biostratigraphic examples of the application of these two techniques.

COMPARISON OF MULTIVARIATE TECHNIQUES

The major features of the multivariate techniques discussed above are summarized in Text-figure 10 (also see Rohlf, 1972; Rowell et al., 1973). Similar answers are obtained by using the same data set; however, because of the varying sizes of biostratigraphic data matrixes, all the techniques can seldom be used for comparative purposes on the same data set.

If graphic simplicity is desired or required, and if very accurate between-group relationships are not important, then pair-group CA would be a very appropriate technique in both *Q*- and *R*-mode. The number of samples or taxa that can be analyzed in CA is governed by the available core storage.

Principal components analysis offers the advantage of the ability to analyze a data set having almost any number of samples, but because the $t(t - 1)/2$ coefficient matrix and the eigenvector matrix must both be stored, the number of variables that can be handled is limited by core storage. If the number of variables is not a problem, PCA has the added advantage of yielding *R*-mode information in the eigenvector matrix along with *Q*-mode information in the coordinates of the objects on the fitted eigenvectors. Principal components analysis gives an accurate representation of the between-group distances, but the within-group relationships may be distorted. In PCA, a graphic representation of the variables in two or three dimensions can be obtained by plotting them as vectors by using the factor loadings as direction cosines. Loading plot diagrams, such as those illustrated in Davis (1973, p. 513, fig. 7-32) and Harbaugh and Merriam (1968, p. 187, fig. 7-20), can be obtained.

Principal coordinates analysis is in effect *Q*-mode PCA in that it yields the same

Technique	Input	Output	Information	Distortion	Size Limitations
Cluster analysis (CA)	Sample x sample or variable x variable similarity or distance matrix	Two-dimensional hierarchical plots (dendrograms)	Q- and R-mode (but not in same run)	Between group relationships distorted within compact group relationships accurate	Limited only by available core storage
Principal components analysis (PCA)	Variable x variable correlation or variance - covariance matrix	Scatter plots of samples in reduced space; eigenvector matrix; eigenvalues	Q-mode (scatter plots) and R-mode (eigenvector matrix) in same run	Between group relationships accurate; within group relationships distorted in reduced space	Variables limited by available core storage; samples practically unlimited
Principal coordinates analysis (PCOORD)	Sample x sample similarity or distance matrix	Scatter plots of samples in reduced space; eigenvalues	Q-mode (scatter plots)	Between group relationships accurate; within group relationships distorted in reduced space	Samples limited by available core storage; variables practically unlimited
Multidimensional scaling (MDSCALE)	Sample x sample similarity or distance matrix	Scatter plots of samples in reduced space	Q-mode (scatter plots)	Evenly distributed between large and small distances	Samples more severely limited by available core storage than above

COMPARISON OF SOME CLASSIFICATION TECHNIQUES

Text-Figure 10
Comparison of cluster analysis (CA), principal components analysis (PCA), principal coordinates analysis (PCOORD), and multidimensional scaling (MDSCALE), relative to type of input required, output, information yielded, distortion characteristics, and size limitations.

coordinate values for scatter plots of the objects but works with a coefficient matrix computed between the objects. Therefore, the number of variables that can be used is virtually unlimited; however, no information about the relationships between variables is given. If the number of variables is not too large, this information can be gained through an *R*-mode CA.

There is no a priori way of judging the "best" technique to use with biostratigraphic data sets that are amenable to all the types of analyses discussed above. As Davis (1973) and others have pointed out, the techniques are not statistical in the strict sense of that word. In simplest terms, they provide a way of reducing the dimensionality of problems so that more data can be used than otherwise could be, and they further assist the researcher in making inferences that can only be tested against the stratigraphy. Thus the use of such techniques does not take the decision-making process out of the hands of the biostratigrapher and give it to a machine, as some have naively said, but provides him with another sophisticated tool with which to do his work.

OTHER HELPFUL TECHNIQUES and PROCEDURES

Range-Through Method

One danger inherent in the process of biostratigraphically correlating rocks is that the similarity of assemblages may be the result not of coeval deposition but of similar environments. A natural factor mitigating against such an error is that animals continue to evolve irrespective of whether the environment changes or not. Second, major environmental changes that would be reflected in regional changes in successive lithologic types seem to happen at a rate slower than the average evolutionary rate of species in most groups. Nevertheless, precautionary measures can be taken to guard against making such a mistake, which should help to allay the fears of those who worry about facies control of assemblages.

The first measure, already discussed above, is the use of binary data. Species presences must be less susceptible to environmental control than are abundances. Second, the range-through method of calculation (Cheetham and Deboo, 1963) can be used in the analysis of samples from measured sections (see also discussion by Hazel, 1970b, p. 3240). In the comparison of different samples, it is the regional time of existence of species and assemblages as represented by the occurrences in samples that is important. In the range-through method, in measured sections the number of species counted as present in a sample for the purpose of calculation of coefficients of similarity includes not only the species actually present, but also those that are both stratigraphically above and below the sample in question in that section. The purpose of this, of course, is to counteract the bias toward whatever factor, environment, preservation, etc., is causing the species to be absent.

An example of the utility of the range-through method is offered in Text-figure 11. Part of an actual measured section in the lower Oligocene at St. Stephens Quarry, Alabama, is shown (locality 4 of Deboo, 1965). Here the Forest Hill is a sparsely fossiliferous sandy clay overlain and underlain by more fossiliferous units. Deboo (1965) found only 20 species of microfossils in the Forest Hill (sample 3) at this locality, whereas samples from the Marianna and Red Bluff are more diverse. This difference in sample size is reflected in the lower dendrogram of Text-figure 11; sample 3 does not cluster with either the Marianna or Red Bluff samples and therefore does not reflect faunally its intermediate stratigraphic position. This tells something about the Forest Hill: on the basis of what is actually in it, it is neither closely allied with the Marianna or with the Red Bluff. However, many species are not present in the Forest Hill, presumably for environmental reasons, that actually existed during Forest Hill time, because 16 species occur in both the Red Bluff and Marianna that are absent in the Forest Hill. When these are added to the 20 species actually found in the Forest Hill, the sample sizes become more equal. Thus, when the coefficients are recalculated and the matrix subjected again to cluster analysis, a better idea of the relationship of Forest Hill time is obtained, as demonstrated by the upper dendrogram. The microfaunal assemblage that lived during Forest Hill time is more similar to that which existed during Red Bluff time. We can infer then that the

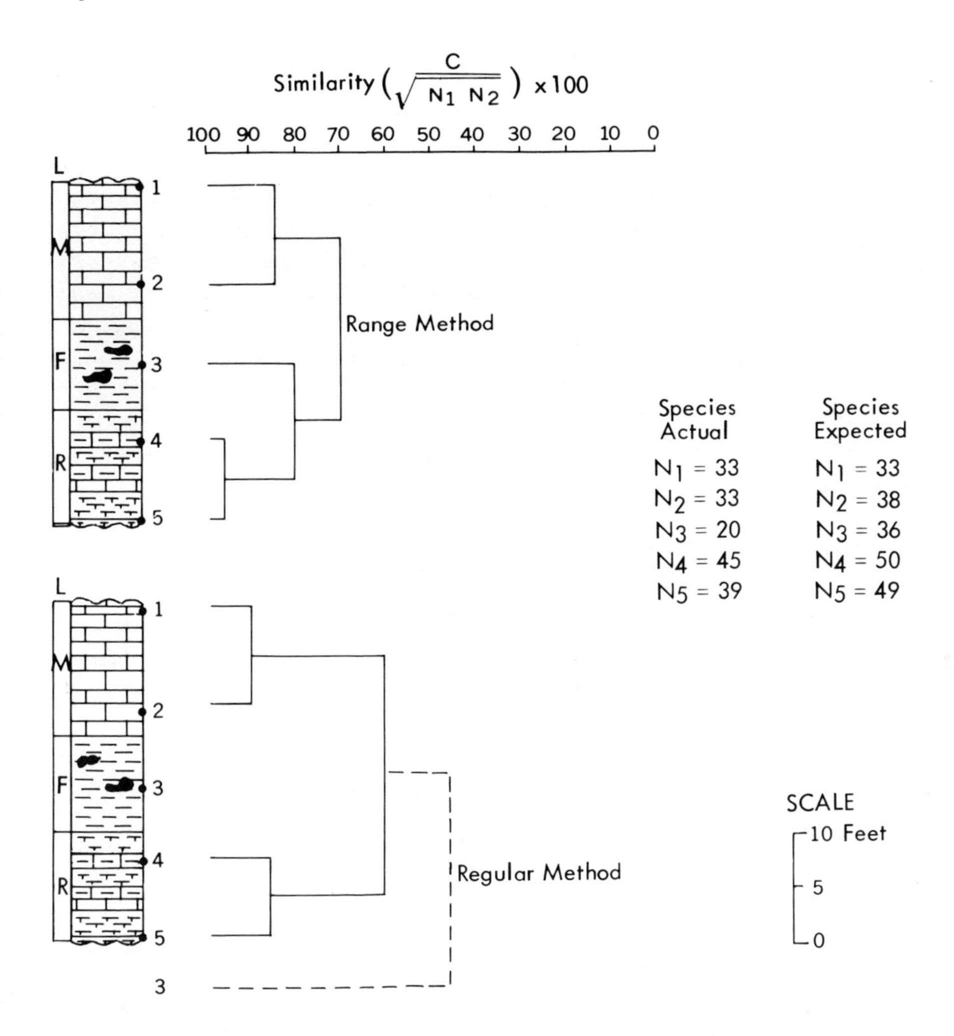

Text-Figure 11
Comparison of the range-through and regular calculation methods with the same real data. Section is from the lower Oligocene at the quarry at St. Stephens, Alabama. Lithologic units (L): R, Red Bluff Clay; F, Forest Hill Sand; M, Marianna Limestone. The species expected for sample 5 is 49 because of known occurrences in samples stratigraphically below those shown. Data from Deboo (1965); also see Hazel (1970b). See text for discussion.

Forest Hill is biostratigraphically more closely related to the Red Bluff, and suggest that the Forest Hill was deposited during a period of time closer to that represented by the Red Bluff than the Marianna. This is supported by other studies.

This example of the range-through method of calculation shows that it can be a useful device when environmental "noise" or poor preservation is a problem in measured sections.

Trellis Diagrams

Shaded trellis diagrams have been used in bioassociational studies by, for example, Valentine (1966), Valentine and Peddicord (1967), and Hazel (1971, 1972). They are simply graphic illustrations of the coefficient matrix in which the numbers have been replaced by symbols representing point classes. The difficulty in determining the best position of samples along the diagonal can be overcome by accepting the arrangement indicated in a dendrogram. The trellis diagram is particularly informative when used in conjunction with (set opposite) a dendrogram. The relationships between the clusters of the dendrogram can be seen in the body of the trellis diagram, thus overcoming to a certain extent the between-group distortion present in the dendrogram.

Biostratigraphic Fidelity and Constancy

The relative contribution of a species to a zonal unit can be determined simply by calculating the percentage of samples in which it occurs in each unit. This can be referred to as constancy. The percentages can be used to calculate the fidelity of a species for a zonal unit by the following formula (Hazel, 1970b; 1971):

$$BF_{j,i} = \frac{P_i}{\sum_{i=1}^{p} P_i} \times 10$$

where j is the species ($j = a, b, c, \ldots, n$), and P_i is the percentage of occurrences of a species in a biostratigraphic unit i ($i = 1, 2, 3, \ldots, p$). The BF of a species for a particular biostratigraphic unit is therefore calculated by dividing the percentage of occurrences of a species in a unit by the sum of the percentages of occurrences of that species in all other biostratigraphic units within the limits of the problem. Biostratigraphic fidelity and constancy are expressed in tens and rounded to the nearest whole number to indicate both the level of precision assigned to the method and to remove the decimal point. An example of the use of biostratigraphic fidelity in biostratigraphy follows. The same Pliocene ostracode data used in the second PCOORD example are utilized.

Twenty-eight ostracode species occur in the type Duplin that also occur in the Yorktown. Table 2 gives the constancy (C) and biostratigraphic fidelity (BF) of these taxa for the two younger ostracode zones of the Yorktown. The species are ranked by increasing fidelity for the *Orionina vaughani* zone. A perusal of the BF values indicates that the type Duplin probably correlates near the boundary between the *Puriana mesacostalis* zone (slightly favored) and *Orionina vaughani* zones (see Text-fig. 9 for comparison).

Table 2
Fidelity and Constancy of Ostracode Species from the Duplin Formation for the *Orionina vaughani* and *Puriana mesacostalis* Zones of the Yorktown Formation

	O. vaughani *Zone*		P. mesacostalis *Zone*	
Species	*BF*	*C*	*BF*	*C*
1	0	0	10	9
2	0	0	10	4
3	1	1	9	7
4	2	2	8	6
5	2	3	8	10
6	2	1	8	4
7	3	3	7	5
8	4	8	6	10
9	4	7	6	9
10	4	7	6	10
11	4	5	6	8
12	5	7	5	8
13	5	9	5	9
14	5	8	5	10
15	5	7	5	7
16	5	6	5	5
17	5	9	5	9
18	5	10	5	10
19	5	8	5	9
20	5	7	5	8
21	5	7	5	8
22	6	5	4	3
23	7	9	3	3
24	7	7	3	3
25	8	4	2	1
26	8	5	2	2
27	8	4	2	1
28	10	4	0	0

COMPUTER PROGRAMS USED

The cluster analyses for this paper were done using the program written by Bonham-Carter (1967) but modified so that it will calculate various binary and multistate coefficients and produce a trellis diagram as well as a dendrogram. The PCA and PCOORD programs used are those of Blackith and Reyment (1971), except that they have been modified for use on IBM equipment, and the PCOORD program has been modified to accept the correlation coefficient or the coefficient of proportional similarity (cos θ), or their binary equivalents, the phi and Otsuka coefficients. The analyses were all performed on the U.S. Geological Survey IBM System 360/65.

SUMMARY

Using different types of examples from the Paleozoic, Mesozoic, and Cenozoic, the writer has attempted to demonstrate the utility of multivariate techniques in assemblage zonal biostratigraphy with benthic invertebrates. When data matrixes are large, there is a tendency to introduce bias into the analysis by favoring certain taxa or by compositing samples and sections; multivariate techniques provide an alternative to this procedure without taking the decision making out of the hands of the researcher. Three techniques were emphasized; (1) pair-group cluster analysis (CA); (2) principal components analysis (PCA); and (3) principal coordinates analysis (PCOORD). Cluster analysis offers mathematical and graphic simplicity. Sample $\times$ sample (*Q*-mode) and species $\times$ species (*R*-mode) matrixes can be analyzed, but not in the same computer run. The number of samples and taxa is limited by the available core storage. Because of the reduction to two dimensions and the averaging process, between-group relationships are distorted most. However, within-group relationships are accurate, and if compact groupings are present in the data, between-group relationships are also accurately shown.

The advantages of PCA are that a virtually unlimited number of samples can be analyzed; the relationships between species (an eigenvector matrix) and samples (scatter plots) can be obtained in the same analysis; in addition, between-group distortion is minimal. However, the worker is constrained by the number of species in the data set, and there is some distortion of the within-group relationships in reduced space.

Principal coordinates analysis is a form of *Q*-mode PCA. Thus it operates on sample $\times$ sample coefficient matrixes and therefore is practically unlimited as to the number of species. It is particularly advantageous to use PCA and PCOORD when there is some structure to the data but when the groups are not compact.

Other useful techniques include the calculation of biostratigraphic fidelity, the use of the range-through method of coefficient calculation, and trellis diagrams.

Biostratigraphy in Gulf Coast Petroleum Exploration

C. Wylie Poag

U.S. Geological Survey,
Woods Hole, Massachusetts

INTRODUCTION

The ironies of scientific progress are many, but one of the most interesting in the field of biostratigraphy arose when the eminent paleontologist T. Wayland Vaughan (1923, p.517) asserted that "the smaller foraminifera possess the least value of any of the groups of organisms considered" for recognizing stratigraphic zones. In spite of Vaughan's negative evaluation, the Gulf Coast petroleum industry during the last 50 years has relied heavily upon the remains of these microscopic organisms to establish biostratigraphic zones for correlating subsurface sedimentary beds.

Gulf Coast paleontologists began earnestly to utilize foraminifers to solve subsurface stratigraphic problems during the early 1900s (Croneis, 1941). This early work is exemplified in the classic report of Applin et al., (1925), who described the stratigraphic sequence of benthic foraminiferids from wells drilled through Upper Eocene, Oligocene, and Miocene strata beneath the northern Gulf Coastal Plain. They documented a series of biostratigraphic assemblage zones and partial-range zones that still serve as the basic correlation framework in the middle Tertiary beds of this province. During the succeeding 25 years, the enlargement of industrial micropaleontological laboratories in the Gulf Coast brought about a consequent surge in the knowledge of biostratigraphic sequences and correlation of the thick subsurface beds. The primary goal was to obtain accurate correlations between well bores, and the principal means of attaining this goal was taxonomic analysis of fossil assemblages of smaller foraminiferids (Ellisor, 1933; Garrett and Ellis, 1937; Garrett, 1938; Smith, 1948). Benthic species were used almost exclusively, because the taxonomy and evolutionary lineages of planktonic foraminifers were not yet well understood; also, planktonic specimens were rare in many of the beds that represented inner and middle shelf accumulation.

After nearly 25 years of chiefly taxonomic study, the exploration potential of paleoecological studies became widely realized. Lowman (1949) was among the first to demonstrate that the living communities of Gulf Coast benthic foraminiferids occupy limited environments; he suggested that these communities and their ancient analogs could be recognized in subsurface sediment samples. At the same time, Israelsky (1949) helped to develop detailed paleoecological analyses by introducing his oscillation charts. He recorded vertical paleoenvironmental oscillations within a well bore on the basis of benthic foraminiferids, and used them biostratigraphically to identify equivalent fluctuations in surrounding wells. Paleoecological studies indicated that the stratigraphic distribution of many benthic guide fossils was controlled not by inherent genetic change but by ancient ecological factors. This brought about a keener awareness of the need to map ecologic facies carefully, especially where these facies fluctuated rapidly through time. As a result of a wealth of information on modern Gulf faunas (Phleger and Parker, 1951), paleoecologic analyses have become an integral part of the biostratigraphic effort and today are critical elements in the search for Gulf Coast petroleum reserves (Hoppin, 1953; Crouch, 1955; Tipsword et al., 1966; Stude, 1970; Poag and Valentine, 1976).

As drilling progressed steadily basinward into offshore Louisiana and Texas, many shallow marine facies became increasingly more pelagic. In consequence, the standard benthic zones, many of which were distinct only because of the nature of their enclosing lithofacies, became steadily less reliable in the downdip direction, and this complicated the solution of correlation problems. Eventually, though, detailed analyses of the planktonic foraminiferal assemblages led to reliable zonation of the deeper-water facies. Akers and Drooger (1957) used planktonic foraminiferids for regional correlation, and documented the relationships between the ranges of several important middle Tertiary planktonic species and the standard Gulf Coast subsurface benthic zones. In the outer shelf and deeper facies, benthic guide species occur together with abundant planktonic guides; their stratigraphic relationships can be established by direct observation. In the shallower-water facies, however, many planktonic species are rare or absent, thus requiring greater paleoecological detail in order to integrate the stratigraphic ranges of the two groups of foraminifers. Even so, some experienced Gulf Coast micropaleontologists have doubted the value of planktonic analyses (Tipsword, 1962).

In spite of the doubts of a few and the dearth of published accounts of subsurface investigations, the utility of planktonic foraminiferids for correlation in mid-shelf and deeper facies has been thoroughly established in most of the larger industrial laboratories since the mid-1960s (see Poag and Akers, 1967; Poag, 1971). The last few years have also witnessed the increased use of the calcareous nannofossils.

The following discussion will delve into the major stratigraphic problems encountered in Gulf Coast petroleum exploration and describe how biostratigraphers have helped to solve them. The foraminiferids are the most widely used fossils and are emphasized for this reason, but the principles set forth can be applied to other microfossil groups as well.

The Gulf Coast is a complex geologic province, encompassing more than 700,000 cubic miles of sedimentary strata of Mesozoic and Cenozoic age (Murray, 1961). The bulk of recent exploration has been concentrated in middle Tertiary to Quaternary sequences offshore, however, and the narrative herein applies to this group of strata.

The term "biostratigraphy" is broadly defined in the context of this discourse and includes paleoecological and paleotemperature analyses based on sequences of superposed fossil assemblages.

SAMPLING METHODS AND TECHNIQUES

Sample Collection

Perhaps the most critical stage in a subsurface biostratigraphic investigation, as in any other micropaleontological study, is collecting proper samples, because each subsequent analysis and interpretation depends upon the quality and accuracy of the sampling. The routine sampling method used in Gulf Coast wells is dependent upon the continuous circulation of a dense viscous drilling fluid (mud) down the drill pipe; this flushes the rotary cuttings back to the surface through the well casing, where they are separated from the mud by a series of sieves on the shale shaker (Text-fig. 1). A cuttings sample thus obtained does not represent a specific point within the well bore, but constitutes a trench-type sample derived from an interval at least several meters thick. Because of the potential mixing of clastic particles from throughout the entire length of the well bore, this type of sample has been criticized as being of little value for detailed biostratigraphic studies. The problem of contaminating the older samples by younger wall cavings is not usually serious in practice, however. It is initially minimized by the thick mudcake (composed of the solutes from the drilling mud) that forms an impervious sheath on the walls of the well. In addition, a series of casing pipes is usually cemented into the well after each penetration of 1,000 to 3,000 m, and this completely seals off the upper beds.

At the surface, contamination by particles suspended in the drilling mud is reduced by passing the sieved fluid through a series of settling tanks (Text-fig. 1), which allows unsieved particles to settle out before the fluid is recirculated down the drill pipe. The depths from which samples originate are constantly monitored by recording the pumping rate and flow velocity of the mud and by inserting tagging elements into the mud stream at regular intervals and measuring their return time. As a result of these procedures, serious contamination is infrequent and is easily detected by an experienced micropaleontologist.

The normal sampling interval for rotary cuttings is approximately 10 m, the length of one joint of drill pipe. Because of the rapid sedimentation rate in the Gulf

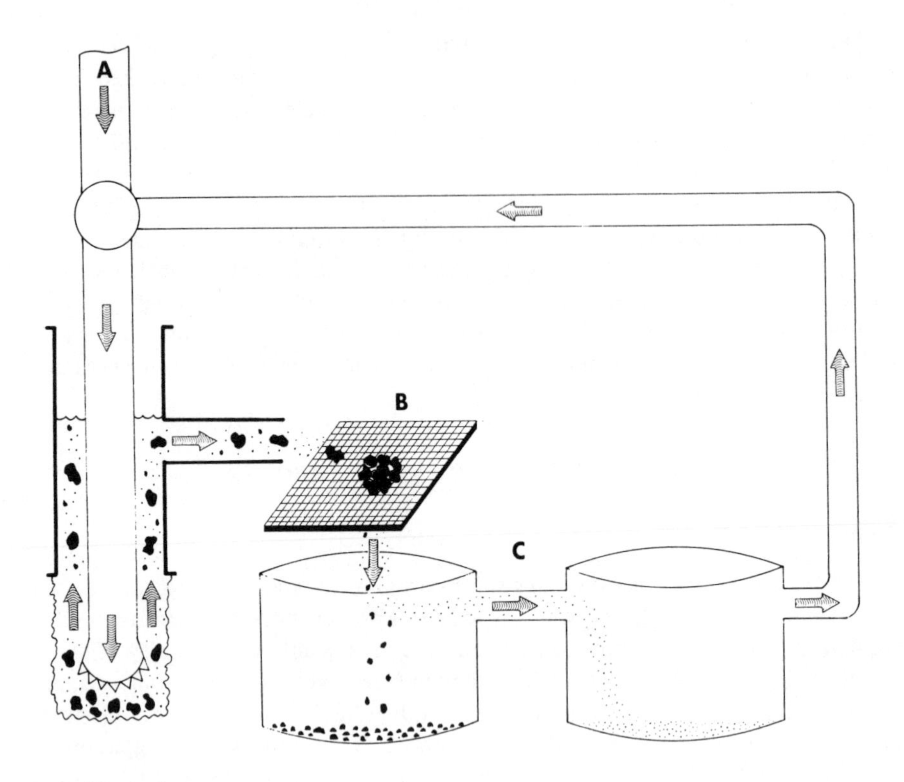

Text-Figure 1
Schematic illustration of the flow (arrows) of drilling fluid through a rotary drilling system. Viscous mud is pumped down the hollow drill pipe (A) and out through ports in the drill bit. It lifts the sediment cuttings from the bottom of the bore hole and flushes them back up around the outside of the drill pipe and onto the drill rig through the surface casing pipe. Sediment-laden mud is then passed through a series of sieves on the shale shaker (B), where the larger sediment chunks are retained and collected for biostratigraphic sampling. Partially cleansed mud then passes through several settling tanks (C) in order to eliminate recycling of the fine sediment particles back down the well, thus avoiding contamination of the older fossil faunas with geologically younger specimens. Sediment-free mud is then pumped back down the drill pipe to begin its cycle again.

Coastal province, the span of geologic time represented between samples of Pliocene and Pleistocene age is generally about 5,000 to 10,000 years (as measured by planktonic time scale); this is comparable to sampling a deep-sea core at 20-cm intervals. For the more compacted Miocene and older beds, the sample interval may represent as much as 30,000 to 40,000 years.

Reliable and consistent interpretation of data derived from rotary cuttings normally requires several years experience, but can be easily accomplished by anyone who has a firm background of basic micropaleontological and stratigraphic principles. There is a widely held misconception that the initial (oldest) stratigraphic occurrences of species cannot be accurately determined from rotary cuttings. This is indeed a problem in the underconsolidated sands and clays of the upper Pleistocene, but in consolidated older strata, these samples are entirely satisfactory for the degree of stratigraphic resolution generally required for exploratory drilling. When specimens are rare, even the upper (youngest) limits may be indeterminate, but when specimens are common, both upper and lower limits may be easily approximated. A clear example of this ability is illustrated in Text-figure 2, which shows the dramatic change in faunas that occurs as facies boundaries are crossed. In this example, the upper and lower range of mid-shelf species is terminated by a paleoecological fluctuation, but the same principle would apply in detecting the earliest and last appearance of chronospecies of an evolving lineage within a single shale unit.

Another factor that significantly reduces the chance of misinterpretation is the large number of data points (well bores) now available among which stratigraphic ranges can be checked. Presently, more than 15,500 wells have been drilled in the offshore region of Louisiana and Texas alone (E. B. Picou, pers. comm., Feb., 1974).

For special studies, sidewall cores and conventional cores are sometimes taken, but each of these methods requires special equipment, and the results seldom justify the tremendous expense involved.

To separate the microfossils, the sediment particles must be further comminuted. A series of steps involving thorough drying, emersing in solvents, agitation, and sieving is generally used. Zingula (1968) has described a method of using the solvent Quaternary 0, which is exceptional for cleaning the calcareous matrix from microfossil specimens.

Sample Examination

Foraminiferal samples may be rapidly examined by using a 30-power stereomicroscope and a sample tray (approximately 10 by 8 cm) lined with a rectangular grid, which enables one to examine the entire sample semiquantitatively while avoiding repetitious specimen counts. Several larger laboratories now use the scanning electron microscope to solve special taxonomic problems, but this instrument is not used routinely in exploration programs. The sample depth, sediment type, marker species present, estimate of relative abundance of various taxa, and estimate of paleobathymetry and relative paleotemperature are typically recorded by hand on paper well logs. An alternative method requires a tape recorder with a foot control; the recording microphone is attached to the base of the microscope, leaving the hands free for focusing the lenses and moving the sample tray. This method requires less time than manual recording and thereby permits assessment of more parameters within a given time span.

DATA DISPLAY AND SYNTHESIS

Each organization has its own special needs that dictate the precise way in which biostratigraphic data are reduced, displayed, and interpreted. Basic for all,

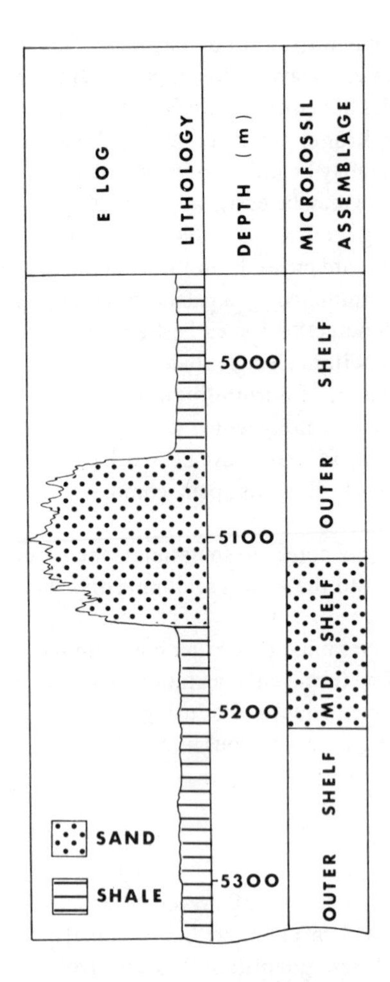

Text-Figure 2
Illustration of the distinct change in fauna that occurs as facies boundaries are crossed. Microfossil assemblages contained in the rotary cuttings are expressed in terms of their paleoenvironments (at right). These are compared to the spontaneous potential curve of an electrical log (at left), which is expressed in terms of lithology of the strata composing the walls of the well. As a result of wall cavings, the faunal changes appear to occur about 60 m deeper than the lithologic boundaries. As the mudcake seals off the walls of the well, however, the caving stops, and the upper and lower stratigraphic ranges of the midshelf fossils are clearly marked. The true depth of the faunal boundaries can then be derived from the electrical log.

however, are three time-honored items: logs, maps, and cross sections. Traditionally, these displays have been assembled by visually reviewing large quantities of data and reducing and plotting the results by hand.

The modern approach, using magnetic tape for recording, facilitates computer processing, which is especially suitable for comparing the biostratigraphic data with the myriad other parameters gathered by other groups within an exploration team (e.g., lithology; electrical, sonic, and radioactivity measurements; seismic records; formation pressure; gas–oil ratio; hydrocarbon chemistry; and many others, see Glaser and Jurasin, 1971). The computer output may take many forms; part of a simple but very effective printout in log form, scaled to depth, is illustrated in Text-figure 3. It is easy to compare such logs visually for interpretation, but the computer can help additionally when teamed with a mechanical plotter to produce maps, charts, and cross sections of great variety.

DEPTH	SAMPLE STATUS	SPECIES		FIRST SPECIMEN	MINERALS	ABUNDANCE	PALEOENVIRONMENT	PALEOTEMPERATURE
		GENUS	TRIVIUM	ABUNDANCE				
5000	3	GN	MT	[6]	PYRT	2	5	W

Text-Figure 3

Simple computer paleolog. The series of symbols along the bottom line are identified in the superjacent columns. The parameters printed are (1) depth of sample in well (1 in./100 ft is the usual vertical scale, because it may easily be compared with electrical logs of this scale); (2) sample status (e.g., an indication that the sample is too small to be representative; or the sample may be full of lost circulation materials introduced by the drillers); (3) species, or other taxa, present (in the example, only one species, *Globigerinoides mitra*, is listed, but most samples would include 10 to 50 taxa); (4) abundance of each taxon (in this example, the numeral 6 indicates that there are more than 100 specimens of *Globigerinoides mitra* in a single tray of sample); (5) indication of the uppermost or lowermost occurrence of a taxon in the well (in this case, the uppermost occurrence of *Globigerinoides mitra* is indicated by brackets around the abundance code); (6) significant mineral grains (in this case, pyrite); (7) relative abundance of each significant mineral (numeral 2 indicates that pyrite is common in the sample); (8) paleoenvironment (numeral 5 indicates an upper Continental Slope environment); and (9) paleotemperature (letter W indicates warm paleotemperature).

PROBLEMS FACING GULF COAST BIOSTRATIGRAPHERS

Three characteristic geologic features of the Gulf Coast province contribute most to the inadequacy of electrical logging and seismic surveying for producing reliable correlations. First are the repetitious vertical sequences of sandstone, siltstone, and shale produced by rapidly fluctuating shorelines (Text-fig. 4; see Bornhauser, 1947; Curtis, 1970).

The second perplexing characteristic is the widespread occurrence of growth faults (Ocamb, 1961; Meyerhoff, 1968), which were active contemporaneously with rapid sedimentation, causing the sediments on the downthrown blocks to accumulate in thicknesses many times greater than those on the upthrown blocks (Text-fig. 5).

Third, and perhaps most confusing, is the distortion of original bedding by the omnipresent salt diapirs (Text-fig. 6), whose locations and rates of movement may change through time (Atwater and Foreman, 1959; Murray, 1961; Halbouty, 1967).

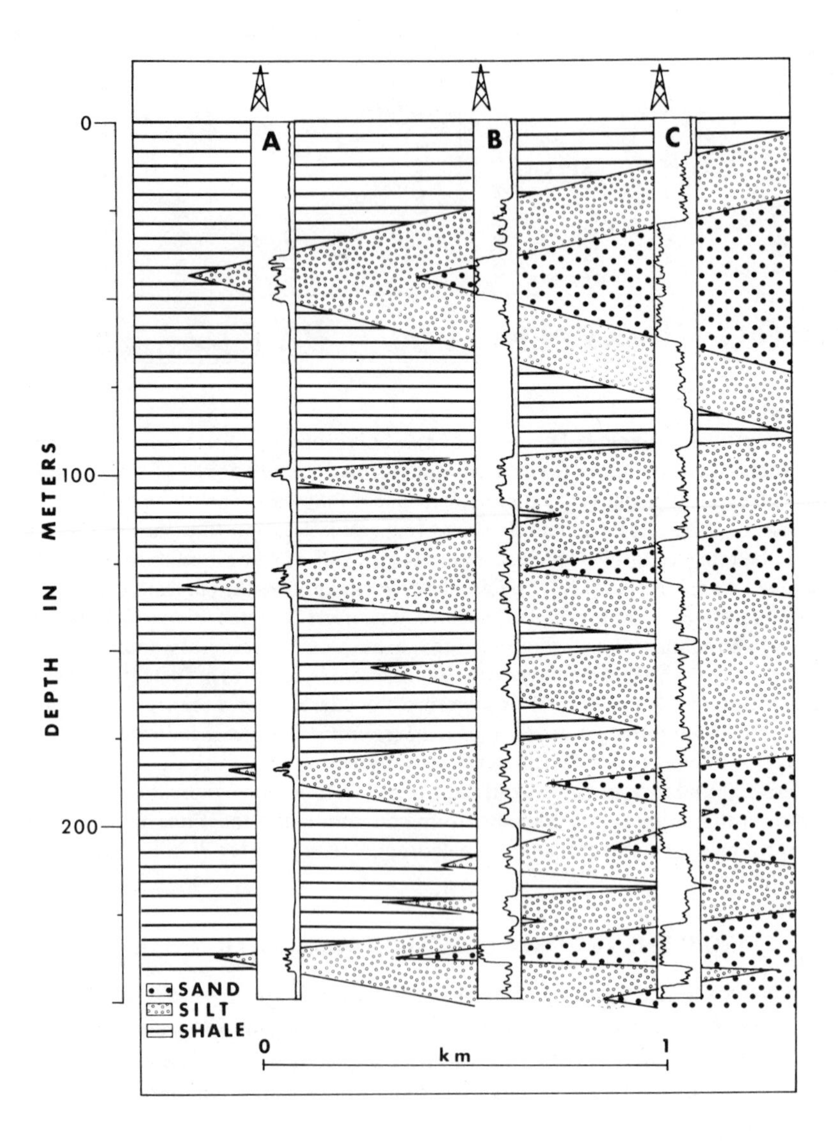

Text-Figure 4
Typical repetitious sequence of cyclical facies found throughout Gulf Coast Cenozoic strata. The large number of fluctuations and the infrequency of marker beds (ash beds, limestones, etc.) are major problems in correlating by electrical logs or seismic records. The spontaneous potential curve for well A indicates the presence of a thick shale sequence interrupted by five thin silt beds. If well B were not present, it would be extremely difficult to correlate the spontaneous potential curve of well A with that of well C, which shows four thin shale units, nine silt beds, and six sands. Biostratigraphic information from wells A and C would allow correlation without well B.

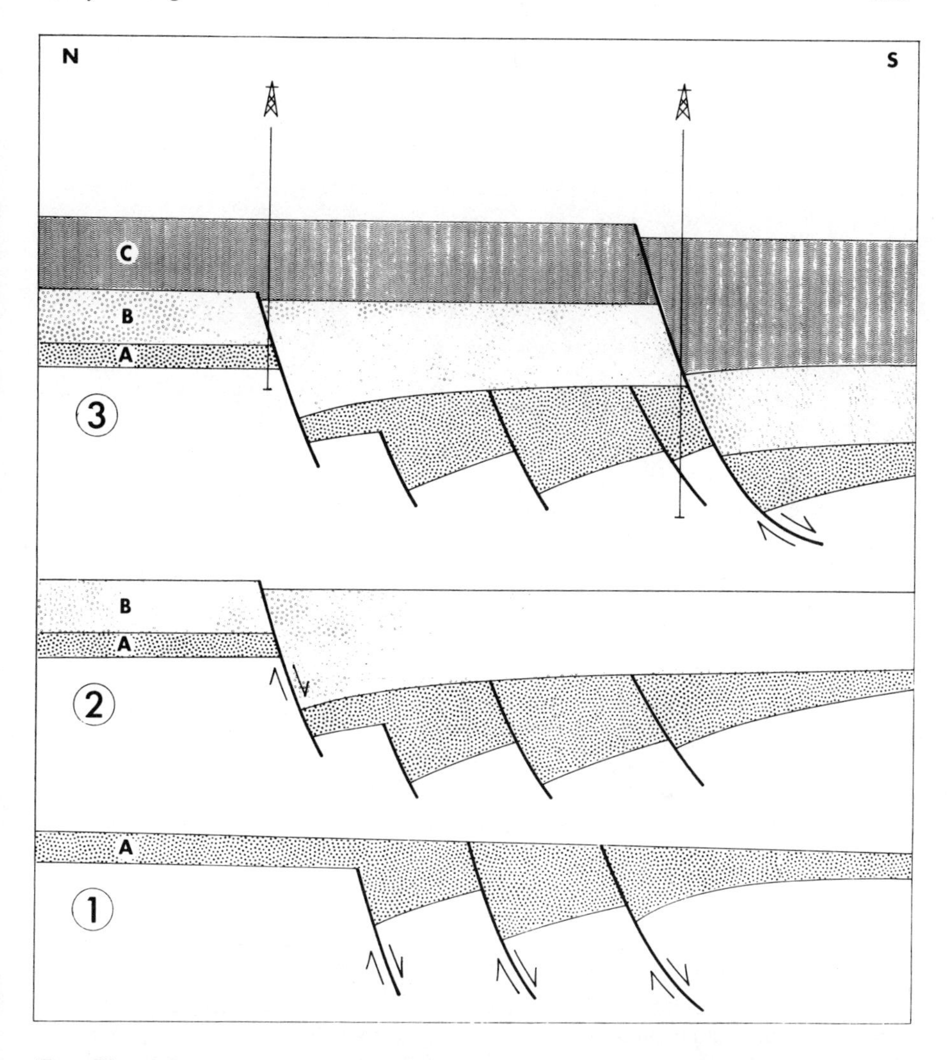

Text-Figure 5
Simplified time series showing sediment accumulation across growth faults and chronologic shift in fault location: (1) bed A cut by three growth faults that allow thicker accumulation on south (downthrown blocks); (2) different growth fault active (arrows) during deposition of bed B; (3) location of growth fault shifts again during deposition of bed C. Sedimentary sequence as recorded in the two wells is complicated as a result of shifting growth faults. Northern well encounters thin part of bed C upthrown to southernmost fault; encounters upper strata of thickened bed B downthrown to northernmost fault; crosses fault into lower strata of thin bed B upthrown to northernmost fault; misses middle strata of bed B; encounters bed A upthrown to all faults and thin. Southern well encounters bed C downthrown to southernmost fault and thick; crosses fault into basalmost strata of bed B, upthrown to southermost fault, but downthrown to northernmost fault; misses upper and middle strata of bed B; encounters bed A downthrown to four northern faults but upthrown to southernmost fault. As the southernmost fault was not active during the deposition of beds A and B, there is no sudden change in the thickness of these two beds across the fault. (After Sloane, 1971.)

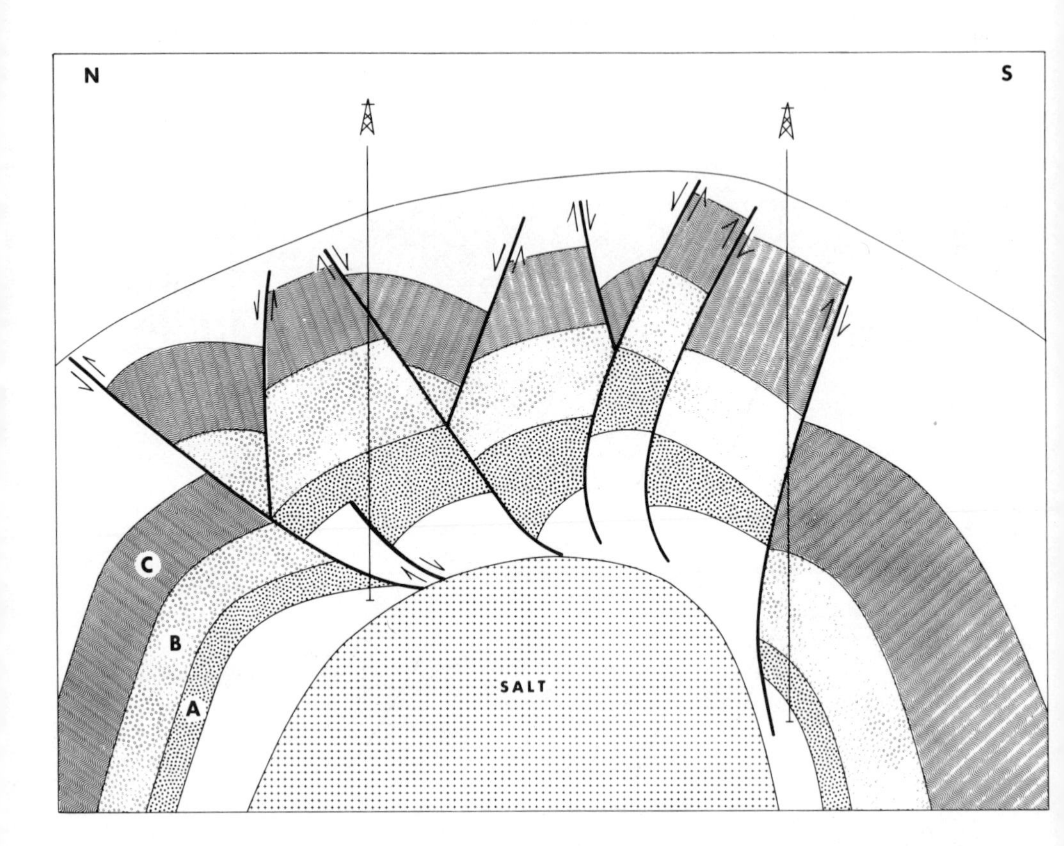

Text-Figure 6
Diagrammatic section through typical piercement salt dome showing complex fault pattern formed when beds A, B, and C of Text-figure 5 are affected by salt uplift after deposition. Northern well encounters upper bed C in a graben; crosses fault into lower bed C; then encounters beds B and A, originally downthrown to growth faults and thickened, but later upthrown in horst; misses the lower bed A; crosses two more faults before the lower bed A is encountered upthrown to the original northernmost growth fault, but downthrown in later movement when the fault was reactivated as a thrust (the thrust is shown just for illustrative purposes; thrust faults are rare in this province). Southern well records an exaggerated apparent thickness for beds C and B, which are downthrown to the original southernmost growth fault (to immediate north of well), but upthrown to younger normal gravity fault (southernmost fault on this section; well misses lower bed B; crosses southernmost fault and reencounters middle and lower bed C, records exaggerated thickness for beds B and A in downthrown position. Actually, salt movement and growth faulting are often contemporaneous and the strata are cyclical, as in Text-figure 4. The resulting confused geometry of fault blocks and the complexity of lithofacies present formidable correlation problems, which usually require biostratigraphic data for solution.

The illustrations herein are simplified and diagrammatic; in reality, these three phenomena are superimposed in incredibly intricate mazes of various fault systems, lithofacies, biofacies, sedimentation rates, and salt movements, all of which have continually shifted through space and time.

BIOSTRATIGRAPHIC ANALYSES

To the great fortune of a society hungry for petroleum energy, the skeletal remains of microorganisms have provided a record of biologic evolution and paleoenvironment that has solved many of these correlation problems. The small size of these organisms, their widespread and abundant occurrence, mineralized tests, moderately rapid evolution, and easily categorized morphology have suited them especially to the type of sampling and analysis required by modern exploration and drilling techniques.

Biostratigraphic Zonation

Benthic zonation. Accurate correlation is the primary requirement for exploration success in this province. Correlation in its strict sense (the tracing of near-isochronous events from place to place) is seldom possible when electrical logs or seismic records are used (see Shaw, 1964). In the search for petroleum traps, however, the term "correlation" is used in a much broader context. The real goal is to map the present and past geometry of sedimentary bodies (formations, facies, zones, etc.), regardless of whether or not isochronous events can be recognized. Knowing the relative order of deposition, alteration, and deformation, is the essential element.

Not even relative chronology can be easily established by using electrical and seismic techniques, but biostratigraphic zones, whose benthic taxa are distributed according to the environmental conditions (biofacies) under which they lived and died, provide just this sort of "correlation." There are, moreover, benthic lineages whose individual morphotypes are distinct stages (chronospecies) in an evolutionary continuum; as such they can be used to document isochronous events in spite of changing facies. For stratigraphic utility, these taxa are usually called species, although the taxonomic problems incorporated in the concept of chronospecies are recognized. An example of a prominent lineage of the Gulf Coast is that of *Brizalina multicostata* in which the evolutionary change in the number and configuration of lengthwise costae is easily detected and several discrete chronospecies can be recognized (Poag and Sweet, 1972).

Ecophenotypic variation is a widespread phenomenon among benthic foraminifers, which are sensitive to changes in a variety of environmental parameters (Murray, 1973). These ecophenotypes are especially useful for detailed correlation in areas of dense well control, where different flanks of a dome and even different fault blocks may contain subtly different facies. Understanding even slight biofacies changes may thus lead to the unraveling of complex fault systems and lithic facies patterns.

As a result of more than 50 years of industrial biostratigraphic study, a regional zonation of benthic foraminifers has been devised for the Gulf Coast (Skinner, 1972). Text-figure 7 lists some of the zones most often cited.

Planktonic zonation. The presence of enormously thick beds containing rich assemblages of planktonic foraminifers has brought this group to the forefront as an incomparable means of establishing regional correlation (Akers and Drooger, 1957).

As with benthic organisms, the easiest method of using planktonic organisms in subsurface exploration is to identify the last occurrences or extinction horizons of individual species; first occurrences generally are harder to recognize because sediment chunks falling from the walls of the well bore may mix young and old specimens. Because final occurrences often are the result of migration or exclusion by unfavorable ecologic conditions, the most reliable groups for establishing correlation datums are those whose lineages can be easily determined and whose members are easily identified. The sphaeroidinellid group is a pertinent example in the Gulf Coast, for despite the several different taxonomic schemes proposed for this group (Bolli, 1957; Blow, 1969; Lamb and Beard, 1972) and much disagreement as to their true biologic relationships (Bé, 1965; Blow, 1969; Bandy et al., 1967), the stratigraphic relationships have clearly been established (Poag and Akers, 1967).

For the last several years, calcareous nannoplankton have begun to play a role in deciphering biostratigraphic sequences in Gulf Coast petroleum exploration. The subject has been only briefly discussed by industrial specialists (Akers, 1965; Wray and Ellis, 1965; Poag, 1971), but several important papers have been published by academicians (Hay et al., 1967; Poag, 1971; Sachs and Skinner, 1973; Smith and Hardenbol, 1973). The biostratigraphic usefulness of this group of organisms in the Gulf Coast Cenozoic beds is largely limited to deep-water marine shales. Their remains are not generally present in sufficient abundance in middle and inner shelf sediments to allow practical application to exploration problems. The discoasters have proved to be the most useful group, because most specimens are relatively large, and taxonomic differentiation at the species level is not complicated by the necessity to examine optical interference figures under crossed nicols. Rate of evolution of the discoasters seems to have been somewhat less rapid than that of the planktonic foraminifers and coccolithophorids, however, resulting in longer species ranges. In conjunction with planktonic and benthic foraminifers, discoasters allow construction of rather detailed biostratigraphic zonations (Text-fig. 8).

The coccolithophorids offer the promise of even finer zonation in purely pelagic facies as demonstrated by the results of the Deep Sea Drilling Project and other publications. Their potential as a practical aid to petroleum exploration is reduced somewhat, however, by the difficulty of defining some taxa without the use of electron microscopy and by their propensity to be redeposited and incorporated into younger beds.

SERIES	STAGE	ZONE
PLEISTOCENE	UPPER	
	LOWER	*Trimosina A* *Angulogerina B*
PLIOCENE	UPPER	*Lenticulina I*
	MIDDLE	*Buliminella I*
	LOWER	
MIOCENE	CLOVELLY	*Bigenerina A* *Buccella mansfieldi* *Textularia L*
	DUCK LAKE	*Bigenerina 2* *Textularia W* *Bigenerina humblei* *Cibicides opima* *Robulus 43*
	NAPOLEONVILLE	*Discorbis bolivarensis* *Marginulina ascensionensis* *Siphonina davisi* *Planulina palmerae*
	ANAHUAC	*Discorbis spp.* *Heterostegina spp.* *Marginulina spp.*
OLIGOCENE	CHICKASAWHAY	*Miogypsinoides sp.* *Cibicides hazzardi* *Nonion struma* *Nodosaria blanpiedi*
	VICKSBURG	*Cibicides mississippiensis*

Text-Figure 7
Standard benthic foraminiferal zonation developed for subsurface strata of Gulf Coast. Generic names have not been corrected to agree with modern taxonomic usage. The large number of nonlatinized trivial names reflects the utilitarian approach used by various industrial micropaleontologists who have contributed to this zonation. Many species have subsequently been properly described and renamed in the literature (see Text fig. 8). Perhaps thousands of additional local zones are used extensively in routine correlation.

GULF COAST BIOSTRATIGRAPHIC DATUMS					
SERIES	AGE (my)	ZONE	BENTHIC FORAMINIFERS	PLANKTONIC FORAMINIFERS	CALCAREOUS NANNOFOSSILS
PLEISTOCENE		N22	AMPHISTEGINA GIBBOSA BULIMINA DENTICULATA ("TRIMOSINA" A) LENTICULINA "SINUATA" HYALINEA BALTHICA TRIFARINA HOLCKI ("ANGULOGERINA" B)	GLOBOROTALIA FLEXUOSA GLOBOROTALIA TRUNCATULINOIDES	GEPHYROCAPSA CARIBBEANICA DISCOASTER BROUWERI
	1.85				
PLIOCENE		N21	BOLIVINA PALANTIA (LENTICULINA I)	GLOBOROTALIA MIOCENICA GLOBOQUADRINA ALTISPIRA	DISCOASTER SURCULUS
		N20		SPHAEROIDINELLOPSIS SEMINULINA	
		N19	TEXTULARIA ARAYENSIS CAUCASINA CURTA ("BULIMINELLA" 1)	GLOBOROTALIA MARGARITAE PULLENIATINA PRIMALIS GLOBIGERINA NEPENTHES SPHAEROIDINELLOPSIS KOCHI	SPHENOLITHUS ABIES CERATOLITHUS TRICORNICULATUS
	5.0	N18	ALVEOVALVULINELLA POZONENSIS	GLOBIGERINOIDES MITRA	
LATE MIOCENE		N17	BRIZALINA "EXMULTICOSTATA" BOLIVINA "FLORIDOIDES" BIGENERINA FLORIDANA (BIGENERINA A)	GLOBOROTALIA MEROTUMIDA	DISCOASTER QUINQUERAMUS DISCOASTER BERGGRENI

Text-Figure 8
Modern biostratigraphic zonation of the Gulf Coast subsurface showing upper limit datums for benthic and planktonic foraminifers and calcareous nannoplankton. Traditional names are shown in parentheses below correct taxonomic citation. Names in quotation marks indicate that name is used loosely or has not yet been published. Stratigraphic resolution provided by this combination is much finer than that of any single group used alone.

Paleoecology

There are often thick sections in a well in which marker species and evolutionary datums are so rare that zonal analysis is ineffective; in such situations, paleoecologic inferences play a predominant role in correlation. Assuming that there are enough marker species to provide a basic zonal framework, the intervening beds may be correlated using Israelsky's principle (1949) of matching maximum points of transgression and regression of biotas to locate isochronous horizons (Text-fig. 9). Because of their sensitivity to conditions on or near the sea bottom, benthic foraminifers lend themselves readily to making the paleoenvironmental assessments needed to produce such oscillation charts. Semiquantitative analysis of certain aspects of community structure, such as species diversity and generic predominance, is especially effective in paleoecological interpretation (Walton, 1964).

Recognizing the ancient taxonomic and ecologic analogs of modern species and genera has always been a major problem for paleoecologists. This problem is somewhat diminished when dealing with foraminiferids no older than Miocene in age, for that is when the benthic and planktonic foraminiferal communities began to acquire their modern aspects. There are, nevertheless, some anomalous associations; among the most intriguing in the Gulf Coast are assemblages composed predominantly of robust, heavy-walled, agglutinated species that are not known in the gulf today but that have been reported from elsewhere as being bathyal and abyssal communities (Bandy, 1967). The ancient agglutinant faunas are quite distinct and have been used as zonal assemblages in Oligocene and Miocene sediments throughout the Gulf and Caribbean region (Renz, 1948). As drilling has progressed farther out from shore, however, nearly identical agglutinated assemblages have been found in younger and younger beds, extending into the upper Pleistocene. Their use in correlation is therefore, largely paleoecological.

Paleoecological analysis is widely used by micropaleontologists at the well site to help determine the course of the drilling operations. Such analysis is used to locate faults, to detect the proximity of high-pressure shale or of salt diapirs, and to pick points at which casing is to be cemented into the well bore. For example, it is nearly axiomatic among many companies that when the benthic foraminiferid *Melonis pompilioides* is found in a well, the drilling should be terminated. This practice stems from the belief that *M. pompilioides* represents such deep-water conditions that the chance of finding hydrocarbon-saturated sand bodies deeper in the hole is negligible. As a result, *M. pompilioides* is jocularly referred to as *M. "pack-your-bag-ensis"*. On-site paleoecologic analyses such as these not only present the loss of millions of dollars in wasted rig time and damaged equipment, but also can safeguard lives and forestall environmental pollution.

Paleotemperature

Pleistocene beds of the Gulf Continental Shelf comprise as much as 10,000 to 15,000 ft (3,000 to 5,000 m) of sands, silts, and shales (Woodbury et al., 1973).

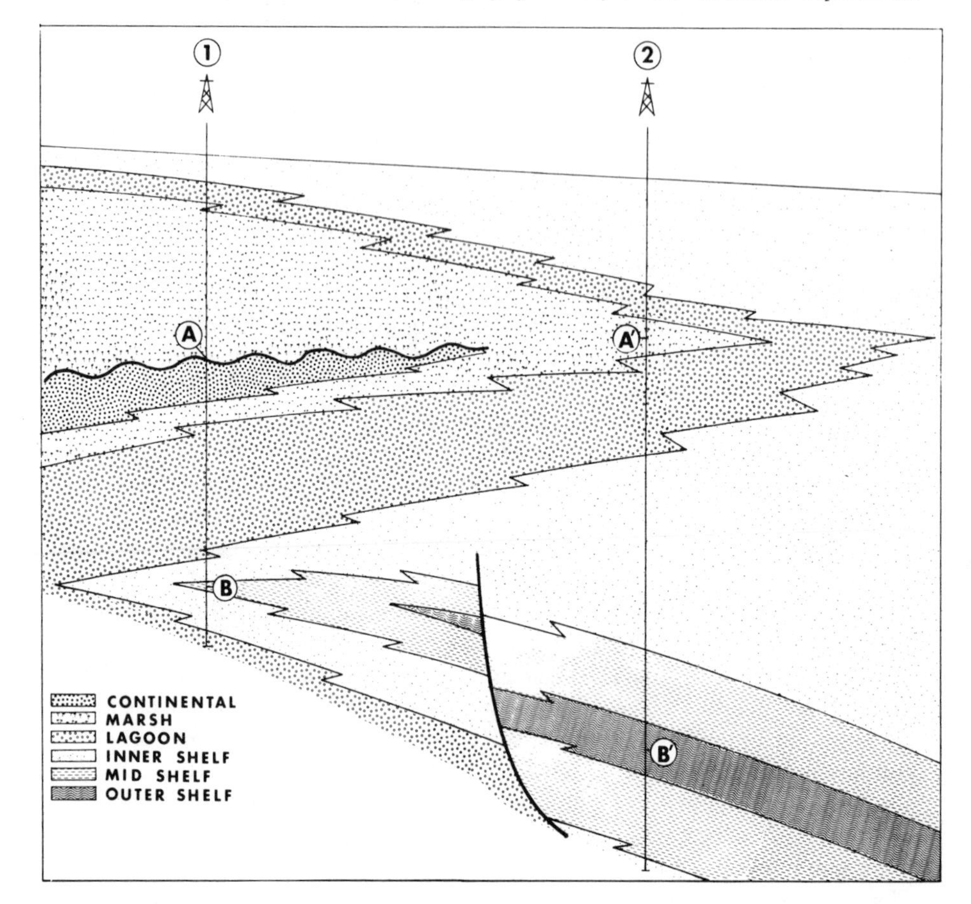

Text-Figure 9
Illustration of the principle behind Israelsky's (1949) oscillation chart and its use in paleoecologic correlation. Maximum regression represented by the unconformity at A is the same maximum regression represented at A′ by marsh beds. A line AA′ therefore would represent an isochronous correlation horizon. Similarly, the maximum transgression at B (mid-shelf fauna) is represented by an outer shelf fauna at B′. A line BB′ would constitute a second isochronous horizon, but the latter example is complicated by an intervening growth fault that offsets this horizon. A third well between 1 and 2 would encounter horizon BB′ at an unexpected depth if the fault were not detected by seismic reflection methods. In the Gulf Coast, the simplistic conditions illustrated here are complicated by the combination of eustatic changes and differential uplift and subsidence within numerous local basins.

In addition to containing benthic and planktonic marker fossils, the Pleistocene foraminiferal remains can be used with facility to infer relative paleotemperature and establish paleotemperature correlations. Very little information on this subject has been published by the industrial paleonotologists, but Beard (1969) demonstrated the principles in cores from the northern Gulf Continental Slope, and

Sidner and Poag (1972) and Poag (1972a) have used this technique on the abyssal plain and the Outer Continental Shelf, respectively (also see Smith, 1965). The basic premise is that during periods of advancing continental glaciers in North America, water depths over the Gulf shelf were reduced, and the tropical pelagic communities of the warm surface waters were replaced by migrant subtropical and temperate forms. Because Pleistocene sea level fluctuated as much as 150 to 200 m (Curray, 1960; McFarlan, 1961; Poag, 1973), glacial intervals are represented in the subsurface by unconformities in the updip areas and by cold-water planktonic faunas in the downdip regions. During the interglacial intervals, deeper-water benthic faunas occupied updip areas and warm-water planktonic faunas were present downdip.

Akers and Holck (1957) traced subsurface Quaternary beds from the lower Mississippi Valley to the nearshore Continental Shelf of Louisiana on the basis of inferred glacial-marine sedimentary and faunal cycles. Later, Poag (1971, 1972b) and Sachs and Skinner (1973) amplified this scheme, demonstrating on the basis of biostratigraphy that glacial-interglacial cycles could be recognized in wells as far as 120 miles offshore (see also Poag and Valentine, 1976).

Certain aspects of foraminiferal community structure may be applied also to the paleotemperature analysis of Tertiary beds in the Gulf Coast. For example, the decrease in planktonic species diversity usually associated with colder waters is well documented among Oligocene planktonic foraminifers (e.g., Cifelli, 1969). Some middle and upper Miocene beds of the Gulf Coast contain unusually rich fossil populations of *Orbulina universa*, a species that reaches maximum abundance today in relatively cool subtropical waters (Bé et al., 1973). Echols and Curtis (1973) have described benthic foraminiferal assemblages that also indicate cool-water conditions in the middle Miocene of Louisiana.

Certain specialized communities of benthic foraminifers are also useful in paleotemperature interpretations. A striking assemblage typical of calcareous banks and shallow reefs and characterized by an *Amphistegina*-miliolid community has been especially useful in assessing the paleotemperature relationships of the inner sublittoral sediments of interglacial intervals (Akers and Dorman, 1964; Poag and Sweet, 1972; Poag and Sidner, 1976).

FUTURE OF BIOSTRATIGRAPHY IN THE GULF COAST

Future Gulf Coast petroleum exploration will be concentrated in the deeper waters of the Outer Continental Shelf and upper Continental Slope and perhaps, given the proper economic incentives, eventually will include the lower Continental Slope and Continental Rise. Tracts were leased in as much as 300 m of water along the upper slope off Louisiana in March 1974 (Text-fig. 10). The biostratigraphy of this area is almost completely unknown except for the work of Lehner (1969) and the brief reports by Lamb and Beard (1972) and Woodbury et al. (1973). These authors reported that most of the sediments penetrated, aside from those on the crests of prominent salt knolls, are hemipelagic beds of Pleistocene age; interspersed among these are displaced shallow-water or deltaic marine strata (Text-fig. 11). At

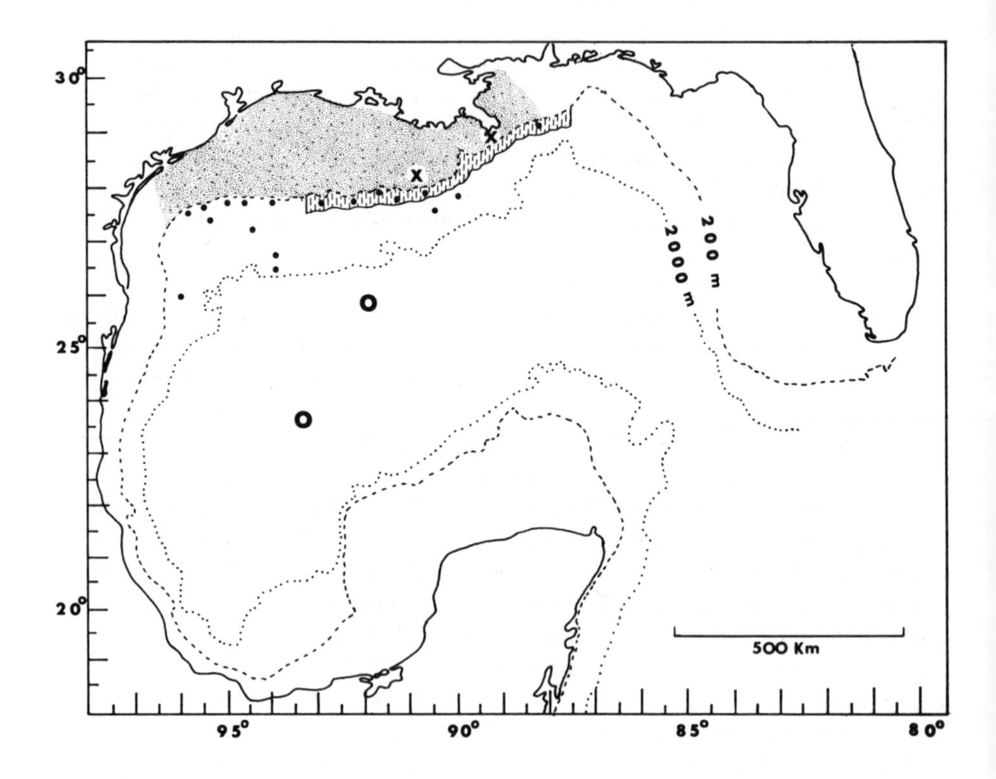

Text-Figure 10
Location of biostratigraphic data points now available in published record from northern gulf: Continental Shelf (X's in stippled area), Continental Slope (solid black dots between 200 and 2,000 m), and abyssal plain (open heavy circles). Two shelf wells are in South Pass Block 41 (Akers and Holck, 1957) and Ship Shoal Block 307 (Poag, 1971). Continental Slope locations are Shell Oil Company boreholes (Lehner, 1969). Abyssal plain locations are DSDP holes 91 and 92 (Worzel et al., 1973). Vertically dashed pattern shows the upper Continental Slope area nominated for commercial leasing in 1974 (see also Poag and Valentine, 1976).

least 400 m of these sediments were identified in basins on the upper slope. The northernmost deep hole drilled by the Deep Sea Drilling Project in the Gulf of Mexico (Leg X, hole 91) penetrated a 282-m sequence of pelagic facies of entirely Pleistocene age (Worzel et al., 1973). The commercial wells farthest out on the Continental Shelf also have penetrated mainly deep-water Pleistocene beds down to 3,000 and 4,000 m (Text-fig. 12; see Hurley et al., 1973). The presence of coarse Cenozoic turbidites on the Continental Slope and Sigsbee Abyssal Plain

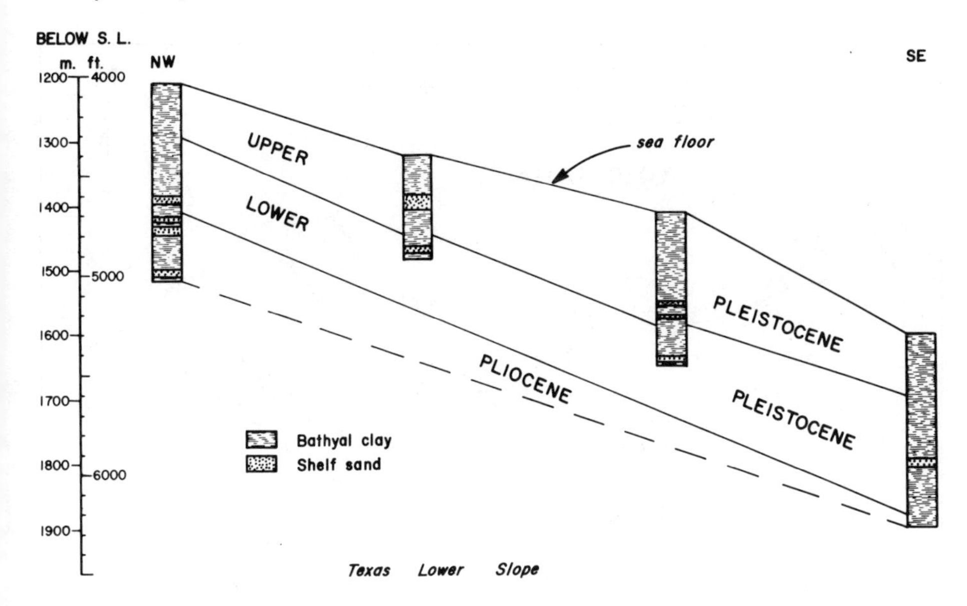

Text-Figure 11
Geologic section through four core holes on the lower Continental Slope off Texas, showing thickness of Pleistocene bathyal clays and displaced shelf sands. Base of Pliocene not reached. (After Lehner, 1969.)

suggests the possibility that displaced shallow-water sands may form significant hydrocarbon traps in deep water. Regardless of how far basinward the exploration effort extends, planktonic microfossils will assume the dominant role in future biostratigraphic interpretation. A thorough knowledge of the benthic sequences established shoreward on the shelf will also be necessary, however, to guide zonation and paleoecologic assessment of the displaced deposits.

The petroleum industry has at its disposal a host of micropaleontologic disciplines with which to attack the problems of biostratigraphy. Dr. Vaughan would, no doubt, be delighted to know that pragmatic foraminiferal biostratigraphy has been a prime element in resolving such problems in the Gulf Coast for more than 50 years and will continue to play an essential role in future exploration and exploitation programs.

SUMMARY AND CONCLUSIONS

Subsurface correlation in the Gulf Coastal province is complicated by the complex arrangement of faulting and salt uplifts combined with differential subsidence, rapid cyclical deposition of sediments, and eustatic fluctuations of sea level. Geophysical methods have not proved successful for accurate correlation. Since the early 1900s, biostratigraphers have helped to decipher the record of Cenozoic history contained in the repetitious strata of sandstone, siltstone, and shale. In shallow-water marine beds, benthic foraminifers provide the primary basis for

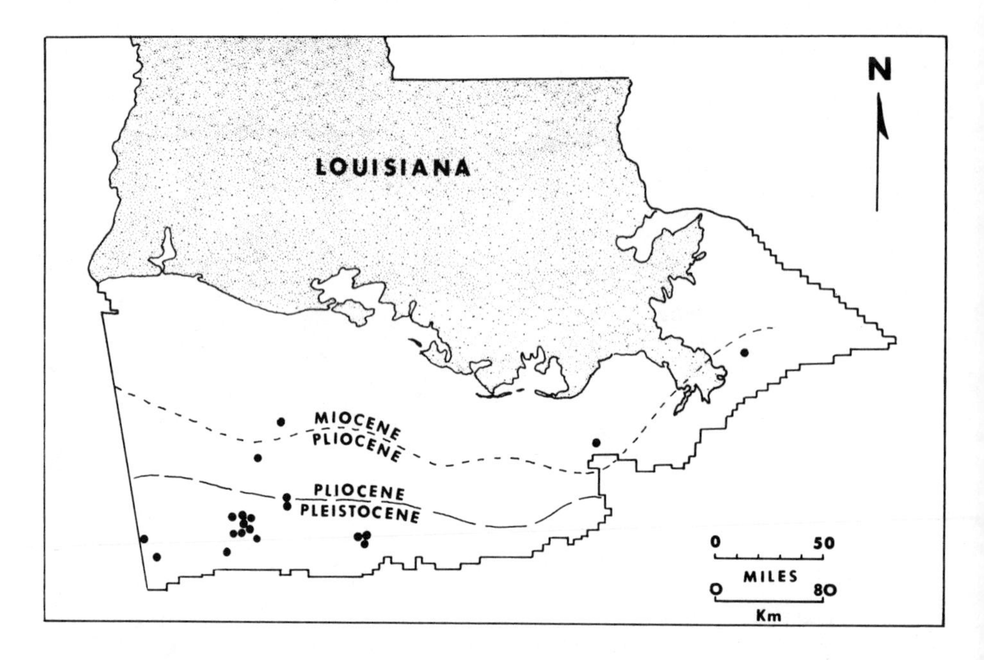

Text-Figure 12
Offshore Louisiana showing discovery wells for 1972 (black dots) and regional disposition of Pleistocene, Pliocene, and Miocene producing trends. Note preponderance of Pleistocene wells and their location on Outer Continental Shelf. (After Hurley et al., 1973.)

biostratigraphic zones. There are, however, frequent hiatuses and estuarine or non-marine beds in the updip areas; in these, ostracodes and palynomorphs provide the essential biostratigraphic ingredients. Paleoecologic and paleotemperature analyses based on all these organisms are also critical elements for correlation of these beds, because guide fossils are often rare or absent.

As drilling has progressed basinward and middle and outer shelf facies have been encountered more frequently, planktonic foraminiferal datums and analyses of calcareous nannofossil assemblages have been integrated with the benthic foraminiferal data to increase the number of correlation horizons. In the most basinward locations, the frequency of pelagic facies has given planktonic forms the dominant role as biostratigraphic indicators, although some benthic forms remain useful.

The vagaries of poorly collected or sandy samples, rapid sedimentation rates, and frequent hiatuses have plagued Gulf Coast biostratigraphers since the beginning, but the immense number of wells now drilled and the practice of exchanging data among companies has greatly alleviated these problems. Correlations are further strengthened by combining biostratigraphic data with the great quantity of other geologic and geophysical information. These data are used to produce the detailed maps, logs, and cross sections that form the basis of sound petroleum exploration.

It is safe to forecast that, as drilling continues to proceed into the deeper-water facies offshore, biostratigraphic analyses of planktonic assemblages will become even more critical for successful programs of petroleum exploration in the Gulf Coastal province.

Acknowledgments

I should like to thank Dr. W. H. Akers of Chevron Oil Co., Mr. C. C. Albers of Amoco Production Co., Mr. D. O. LeRoy of Exxon Co., and Mr. E. B. Picou, Jr., of Shell Oil Co., for their valuable assistance in helping to show here the enormous contribution of the Gulf Coast petroleum industry to the science of biostratigraphy. Special thanks go to Chevron Oil Company in whose employ I gained a practical knowledge of Gulf Coast biostratigraphy and to the editors of this volume for having invited my participation.

Integrated Assemblage-Zone Biostratigraphy at Marine–Nonmarine Boundaries: Examples from the Neogene of Central Europe

Fritz F. Steininger

Institute of Paleontology,
University of Vienna

INTRODUCTION

The modern geologic time scale, which provides a reference for worldwide correlations of Cenozoic strata, is based on marine planktonic organisms such as foraminifera, calcareous nannoplankton, radiolarians and diatoms combined with radiometric dates and (geo-)magnetic events (Berggren, 1972a, 1973; Theyer and Hammond, 1974; Ryan et al., 1974). Historically, the integrated planktonic biostratigraphy (Text-fig. 10) took into account characteristic assemblages, first appearances, extinctions, total ranges, and overlapping ranges of planktonic foraminifers (Bolli, 1959, 1964, 1966; Blow, 1957, 1969). More recent approaches have tried to refine this planktonic zonation by study of morphological changes in evolutionary lineages, especially as data from the Cenozoic of the deep-ocean basins have become available by deep-sea drilling and coring.

Also, in recent years we have become more aware of the environmental factors that limit the distribution of planktonic organisms: temperature, current systems, zonation, and movement of water masses, etc., all of which influence diversity, evolutionary rate, and dispersal. Alternations in plankton faunal compositions are now understood to be caused in part by water depth of sedimentary surfaces (i.e., relation to compensation depth) and fossilization processes. The biogeographic distribution of planktonic organisms has not remained stable through time; it is now well known that there developed a progressive zoogeographic provincialism in plankton during the Neogene, and that the area of major radiation and rapid evolution shifted more and more toward the tropical regions during this time (Berggren and Van Couvering, 1974).

Nevertheless, it is evident that planktonic organisms are the source of primary data for the development of a worldwide chronostratigraphic-geochronologic system during the Cenozoic Era. To date, satisfactory application of this integrated planktonic-geochronologic reference scale within the Neogene has been more or less restricted to sediments that were deposited under open oceanic conditions in low (warm) latitudes. It could not be applied satisfactorily to the widespread and thick Neogene deposits that are found exposed on the continental blocks in facies representing marginal shelf areas, shallow epicontinental seas, and intramontane basins, all of which had mostly limited and only temporary connections with the open ocean, or, in the case of freshwater and terrestrial faunas, none at all. In such areas, sedimentation is not continuous; biotas and facies are complex and varied, and may reflect even minor sea-level changes with complex transgressive and regressive fluctuations. Planktonic forms, if present, are not common, and usually only small portions of the planktonic reference scale can be recognized.

Within the marine to brackish marginal sediments of these areas, however, intercalations of freshwater and terrestrial deposits are commonly preserved that contain the remains of mammals, land snails, and plants. In some cases, these deposits are surrounded by shallow marine sediments and provide a rare opportunity for direct biostratigraphic correlation of "mammal ages," the primary biostratigraphic units of continental deposits for regional and long-distance correlations. These mammal ages are commonly derived from studies of evolutionary lineages of micromammals. Each mammal age is based on a system of faunal zones that represents the evolutionary steps in mammalian lineages and is named after the most typical locality (Thaler, 1966; Cicha et al., 1972; Mein, 1975). Attempts to correlate these vertebrate localities and mammal ages with the marine biostratigraphic zonation and time scale utilizing only radiometric dates have not been particularly successful, considering our present knowledge of Neogene radiometry and the relation of biostratigraphic units to the radiometric time scale (Berggren and Van Couvering, 1974; Van Couvering, 1972; Van Couvering and Miller, 1971). Radiometric dates from seemingly the same biostratigraphic horizon often differ by more than 20 percent. This difference is not only a result of relatively poor biostratigraphic control, but also because of differences in the material used for dating (basalt, glauconite) and the particular technique used (Bandy, 1970; Van Couvering, 1972; Ikebe, 1973; Odin, 1973a, 1973b; Selli, 1970; Vass et al., 1970).

In marginal marine coastal continental areas, it is evident that we have to rely primarily on benthic organisms from various environments and their evolutionary histories to establish biostratigraphic units; these units are for the most part restricted to geographically small bioprovinces. To take into account the many different paleoenvironments from which zonal fossils are drawn, such biostratigraphic units are best defined as "integrated assemblage zones" (composite overlap assemblage zone of Shaw, 1964; Kauffman, 1970; or composite concurrent-range zone of Weller, 1960). The intercalations of planktonic organisms that are recognized in these marginal marine zones are incorporated into the integrated assemblage zones, and offer the possibility of adjusting and correlating such a regional system to a Cenozoic planktonic reference scale. Correlations between such regional systems, using the planktonic biostratigraphic scale as a worldwide reference for the Cenozoic Era, allows the more precise delineation of regional and interregional tectonic and climatic events, major faunal migrations in invertebrates (Berggren and Phillips, 1971) and vertebrates (Fahlbusch, 1970, 1973; Simpson, 1947; Strauch, 1970; Thenius, 1972; Wilson, 1968; Wood et al., 1941), and the determination of differences in evolutionary rates of different biotas.

The Late Oligocene-Neogene biostratigraphy and chronostratigraphy of the Central Paratethys, a relatively shallow epicontinental sea with interspersed basins (Text-fig. 1), provides an opportunity to examine integrated-assemblage-zone biostratigraphy at a marine-nonmarine boundary zone, and provides a test of its utility in regional and interregional correlations.

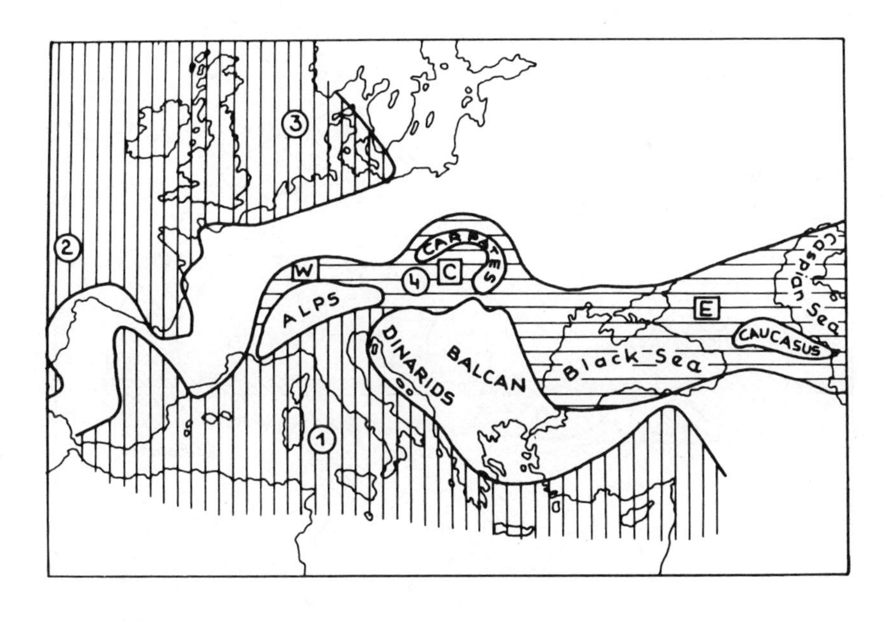

Text-Figure 1
European bioprovincial concept (Upper Tertiary): (1) Mediterranean-Tethyan bioprovince; (2) Celtic-Lusitanian bioprovince; (3) Boreal bioprovince; (4) Transeuropean bioprovince of the Paratethys (W, Western; C, Central; and E, Eastern Paratethys).

UPPER OLIGOCENE-NEOGENE BIOSTRATIGRAPHIC AND CHRONOSTRATIGRAPHIC SYSTEM, CENTRAL PARATETHYS (MIDDLE EUROPE)

During the last 10 years the Upper Oligocene-Neogene sections of the Alpine and Carpathian (Karpatian of European usage; e.g., Text-figs. 1 and 2) foredeep "Molasse zone" and of the inner Alpine and Carpathian basins (Text-fig. 2) have been restudied in considerable detail (see Baldi, 1973; Baldi and Senes, 1975; Buday et al., 1965; Cicha, 1970; Cicha and Senes, 1968, 1973; Cicha et al., 1975; Cicha et al., 1971b; Papp, 1959, 1963a; Papp et al., 1968; Papp and Senes, 1974; Papp and Steininger, 1973, 1975a, 1975b; Papp et al., 1970; Papp et al., 1971; Rögl, 1975; Senes, 1960, 1961, 1967; Senes et al., 1971; Steininger, 1963, 1969, 1975a, 1975b; Steininger et al., 1976; Steininger and Senes, 1971; Thenius, 1959). Two main conclusions have resulted from these studies: (1) beginning in the Late Oligocene and continuing during the Miocene and Pliocene, these troughs and basins developed their own bioprovincial character, best exhibited by benthic marine to brackish water faunas. The areas characterized by such faunas are the west-central and eastern Paratethys, the Transeuropean Bioprovince (Text-fig. 1) (Laskarev, 1924; Senes, 1959, 1960, 1961;

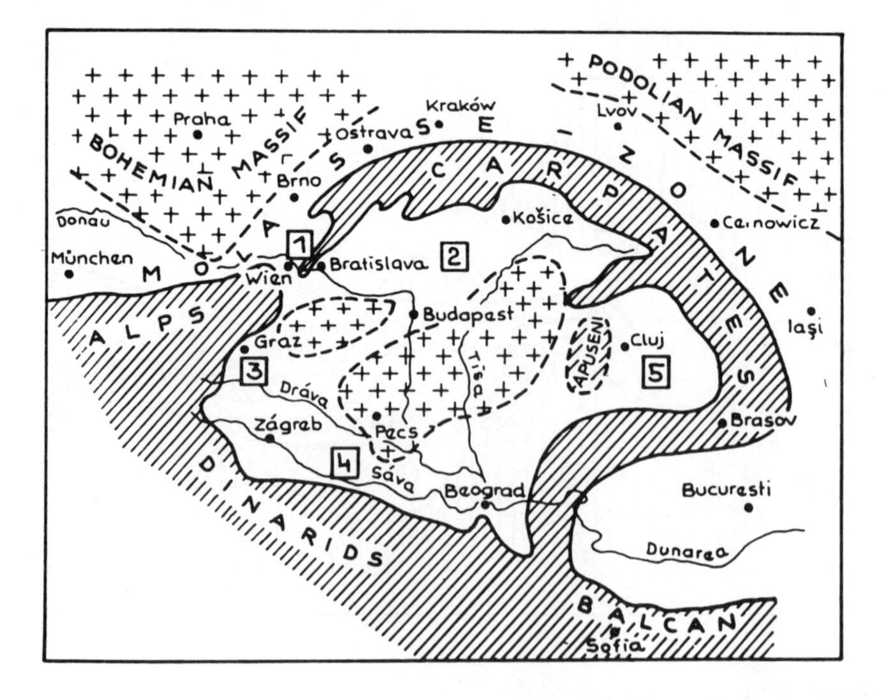

Text-Figure 2
Late Oligocene and Neogene Depositional areas in the Central Paratethys: (1) Vienna Basin; (2) Intra-Carpathian basins; (3) Styrian Basin; (4) Drava-Sava Depression; (5) Transylvanian Basin. (After Senes et al., 1971.)

Senes et al., 1971); and (2) the classic stage terms of the European Neogene that have been applied to the Paratethys cannot be effectively used. Concerning the latter, it can be demonstrated that it is very difficult to correlate the stratotypes of the Chattian, Aquitanian, Burdigalian, Helvetian, and Tortonian stages with the Paratethys sections to which these names have been previously applied (Text-figs. 10 and 11) (Steininger et al., 1976). Therefore, it became necessary to create a new regional chronostratigraphic system, primarily derived from a refined system of biostratigraphic units (Papp et al., 1968; Cicha and Senes, 1968; Baldi, 1969). Because a large group of paleontologists, geologists, sedimentologists, and stratigraphers were working cooperatively in this area, the new chronostratigraphic system could be established for the entire section in question utilizing modern data and techniques.

Paleogeographic Setting and Geodynamics

Beginning in the Late Oligocene, the Paratethys area progressively developed the character of an epicontinental sea with relatively small interspersed rapidly subsiding basins. Connections to oceanic areas were extant but limited (Text-fig. 2). In the upper part of the Egerian stage (Text-fig. 3), a regressive phase terminated the connections and faunal exchange of the Paratethys bioprovince with the Atlantic-Boreal bioprovince. Global eustatic change, probably resulting from plate-tectonic movements, and manifested in the region by tectonic activity in the Alpine-Carpathian and Dinarid-Balcan mountain chains, was followed by a massive transgression in the Early Miocene (M_1, Eggenburgian stage; and M_2, Early Ottnangian stage; Text-fig. 3). Concomitant with this transgression, there was a striking faunal change marked particularly by the introduction of tropical Indo-Pacific faunal elements, which demonstrates a northward shift of the Tropical (Tethyan) faunal realm, as well as the development of an active seaway to the southwest and southeast. A western seaway must have come into existence at least within the Late Eggenburgian and was active throughout the Early Ottnangian (Text-fig. 3). In the later Early Miocene, this western seaway was cut off, and the Western Paratethys started to dry up (M_2, Late Ottnangian, "*Oncophora*"-*Rzehakia* Sea; Text-fig. 3). At about the same time, a seaway again extended from the Mediterranean area (Text-fig. 1) into the Paratethys region (Text-fig. 3; Early Karpatian); such a connection did not exist during the Early Ottnangian. The Middle Miocene (Text-fig. 4; M_3, Karpatian; M_4, Badenian) is characterized by a second peak in the development of faunas with tropical elements, of Mediterranean origin, and by rapid subsidence of the intramontane basins (Text-fig. 2).

Progressive movements and uplift of the Alps and Carpathians and the Dinarides and Balcanic Mountains during the late Middle and Late Miocene subsequently cut off the marine seaways of the Paratethys from the oceanic areas (Mediterranean, and possibly the Indo-Pacific areas), and were followed, from west to east, by a migrating regressive phase of sedimentation. At the same time, a large sea (the isolated remnant of the former marine phase of the Paratethys Sea) with rapidly changing salinity came into existence in the eastern parts of the Central and the

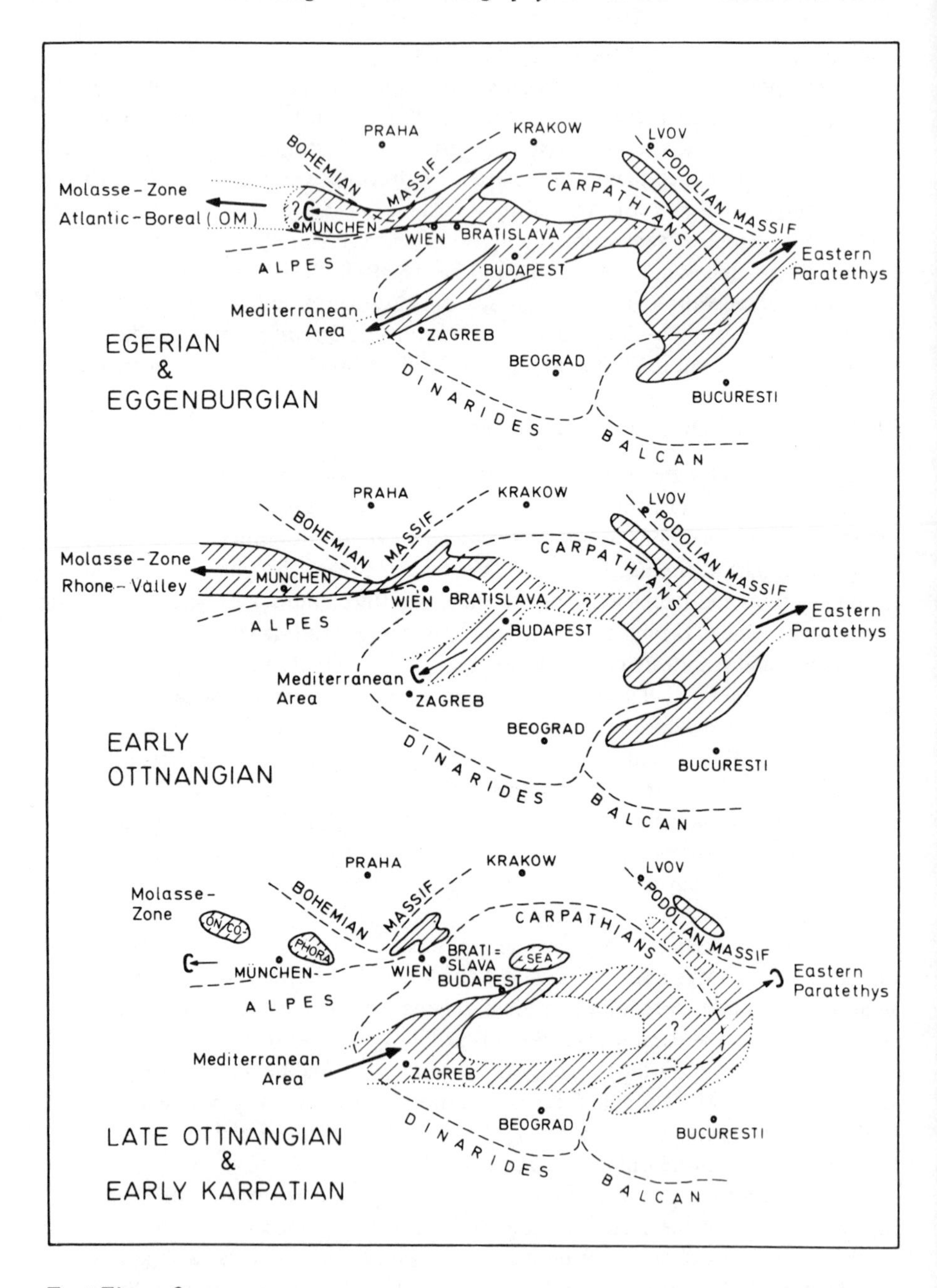

Text-Figure 3
Paleogeographic setting for OM, Egerian; M_1, Eggenburgian; M_2, Early Ottnangian; M_2, Late Ottnangian; M_3, Early Karpatian. Oblique hatching indicates areas where sediments of the given stage are now exposed. Arrows indicate open connections and seaways. Arrows with brackets indicate blocked connections. (After Senes and Cicha, 1973.)

Eastern Paratethys (Text-fig. 1). This produced high-stress environments and seemingly caused accelerated evolution of endemic taxa and communities (Text-fig. 4; M_5, Sarmatian stage). Within the later Late Miocene and Pliocene, this intracontinental sea diminished and small local basins were formed (Text-fig. 4: Pannonian-Rumanian; Upper Bessarab-Akchagyl). These in turn evolved into the "Caspian" brackish lakes and terminated the Paratethys.

Biostratigraphic Concept

In reevaluating the biostratigraphic concepts of the Central Paratethys, it was necessary to develop a model of biostratigraphic units that was suitable for an area of such extreme facies change; it would have to allow interfacies comparisons and correlation, and the correlation of these biostratigraphic units with the planktonic reference scale for the purpose of interregional correlation. In addition to the recent efforts of the Paratethys working group, these studies were based on a continuous tradition of detailed geologic and paleontologic work since about 1840 and the results of intensive oil exploration, especially in the western Molasse zone, the Vienna Basin, and the Intra-Carpathian Basins (Text-fig. 2). This prior geologic and paleontologic work provided extremely valuable stratigraphy for our own work, which began by concentrating on the more or less well defined intervals in the depositional basins of the Central Paratethys.

The new biostratigraphic model developed in this area is composed of units that are probably best characterized by the term "integrated assemblage zone." The methodology and basic concepts of this model and the points made on the definition of an integrated assemblage zone will be outlined and are illustrated by some extremely simplified examples (Text-figs. 5, 6, 7, and 8).

As a first step, the most complete stratigraphic sections (e.g., deep wells) of small uniform depositional areas within the geologically well known Central Paratethys were restudied sedimentologically and paleontologically from the marginal environments to the deeper basins. Detailed composite reference sections were used to build up a reference profile representing all typical environments for such an area. The basic correlations within such small and restricted sedimentation areas could mostly be drawn from previous work or by physical stratigraphic (Text-fig. 5) and/or geophysical methods. An example of a Late Sarmatian sedimentation area illustrates such a physical stratigraphic correlation, resulting in a composite reference section for this area (Text-fig. 5). The composite reference sections of larger regions (e.g., different parts of the Molasse zone, Vienna Basin, etc.) within the Central Paratethys were published by Senes et al. (1971). The research work was done by the Paratethys working group, which concentrated on one after another of the more or less already known time intervals within the composite reference sections.

The next step entailed the collation of all available faunal data from published and unpublished works and museum and private collections. These occurrences were plotted against the composite sections and profiles. Throughout the Central Paratethys area, the first appearances and total ranges of single taxa result in first

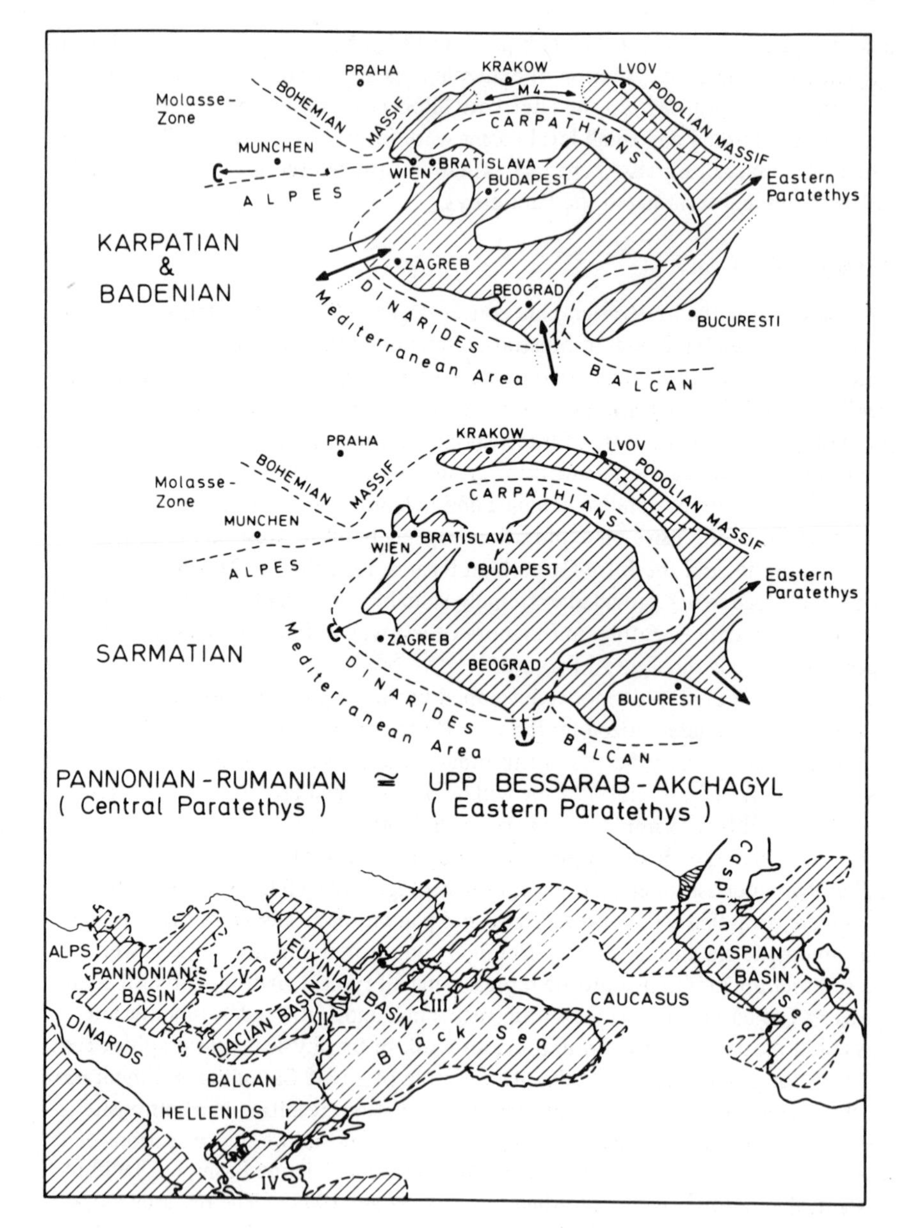

Text-Figure 4

Paleogeographic setting for M_3, Karpatian and M_4, Badenian; M_5, Sarmatian; PM, Pannonian-Rumanian and Upper Bessarab-Akchagyl. Oblique hatching indicates areas where sediments of the given age are now exposed. Arrows indicate open connections and seaways. Arrows with brackets indicate blocked connections. I, Apusen Mountains; II, Dobrodgea Massif; III, Crimea; IV, Agaic Continental Block; V, Transylvanian Depression. (After Senes and Cicha, 1973; Senes et al., unpublished data.)

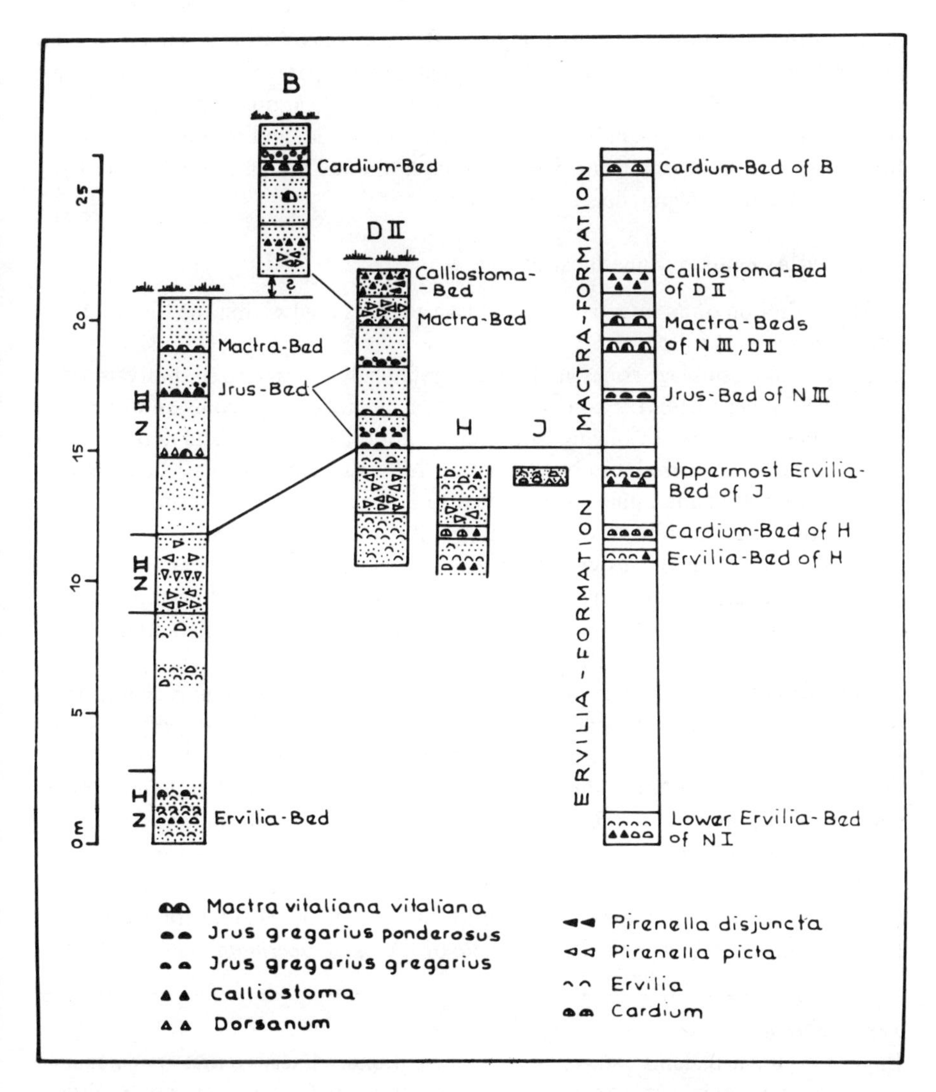

Text-Figure 5
Physical stratigraphic correlation of Upper Sarmatian sections and composition of a composite reference section within the area of Wiesen-Sauerbrunn in Burgenland, southeastern Austria. (After Papp, 1958.)

reference points for zonation. Out of this initial biostratigraphic information, we tried to separate first appearances and ranges of taxa belonging to evolutionary lineages from those dependent on, for example, migrational factors.

Next, known and newly studied evolutionary lineages of different taxa from different environments were identified and plotted against the composite reference sections. Within these lineages the first appearance of taxa were plotted as reference points of primary importance for zonation. Because we constructed such composite reference sections from marginal as well as shelf and deeper-water environments, the biostratigraphic equivalent reference points within these various environmental areas could be ascertained. The most characteristic, widespread, and reliable taxa drawn from these various environments were then used to build up conceptual biostratigraphic units, the integrated assemblage zones.

Integrated Assemblage-Zone Biostratigraphy

The definition of an integrated assemblage zone, as used within the Central Paratethys region, is as follows:

Integrated assemblage zones are biostratigraphically characterized intervals defined by the first appearance, first concurrent appearance, total range zone, and/or partial range zones of (1) taxa belonging to various evolutionary lineages and derived *in situ*, or (2) taxa that are introduced by migration. All taxa characterizing a single integrated assemblage zone cannot be expected to occur all together in a single rock unit, because they are drawn from various time equivalent environments.

Examples of integrated assemblage zones. To illustrate this concept and the composition of integrated assemblage zones, some extremely simplified examples are given in the following sections. The zonal concept with the most characteristic zone fossils indicated is outlined in Text-figure 11.

Late Karpatian-Badenian zones *(Text-fig. 6).* The Late Karpatian is characterized by the first evolutionary appearance and peak zone of several *Uvigerina* species (Text-fig. 6, items 12-18; Papp and Turnovsky, 1953; Steininger et al., 1975). These uvigerinids are concurrent in the Upper Karpatian with the first evolutionary appearance of the planktonic foraminifer species *Globigerinoides sicanus* (Text-fig. 6, items 1 and 2). The evolutionary lineage from *G. sicanus* to *Orbulina suturalis* can be traced across the Karpatian-Badenian boundary. From this lineage, the first appearance of *Praeorbulina glomerosa circularis* (Text-fig. 6, items 3-5) together with the first appearance of *Uvigerina macrocarinata* of the "*macrocarinata*" lineage (Text-fig.

Text-Figure 6

Upper Karpatian-Badenian stage. Selected evolutionary lineages, first appearance and peak zones, planktonic zonal markers, and integrated local vertebrate localities to demonstrate groups used for definition of integrated assemblage zones. 1 and 2, *Globigerinoides sicanus*; 3-5, *Praeorbulina glomerosa circularis*; 6 and 7, *Orbulina suturalis*; 8 and 9, *Globigerina druryi*; 10 and 11, *Velapertina indigena.* Arrows with NN numbers indicate typical first occurrence of calcareous nannoplankton zonal markers. First appearances and peak-zones: 12 and 13, *Uvigerina bononiensis primiformis*; 14 and 15, *U. parkeri breviformis*; 16–18, *U. graciliformis.* Evolutionary lineages: 19-31 represent the *Uvigerina macrocarinata* lineage; 19–21, *U. macrocarinata*; 22-24, *U. grilli*; 25 and 26, *U. v. venusta*; 27 and 28, *U.* cf. *pygmaea*; 29-31, *U. venusta liesingensis*. Items 32-36 represent the *Uvigerina semiornata* lineage: 32, *U. s. semiornata*, 33 and 34, *U. s. urnula*; 35, *U. s. brunensis*; 36, *U. s. karreri.* Items 37-39 represent the *Heterostegina costata* lineage: 37, *H. praecostata;* 38,

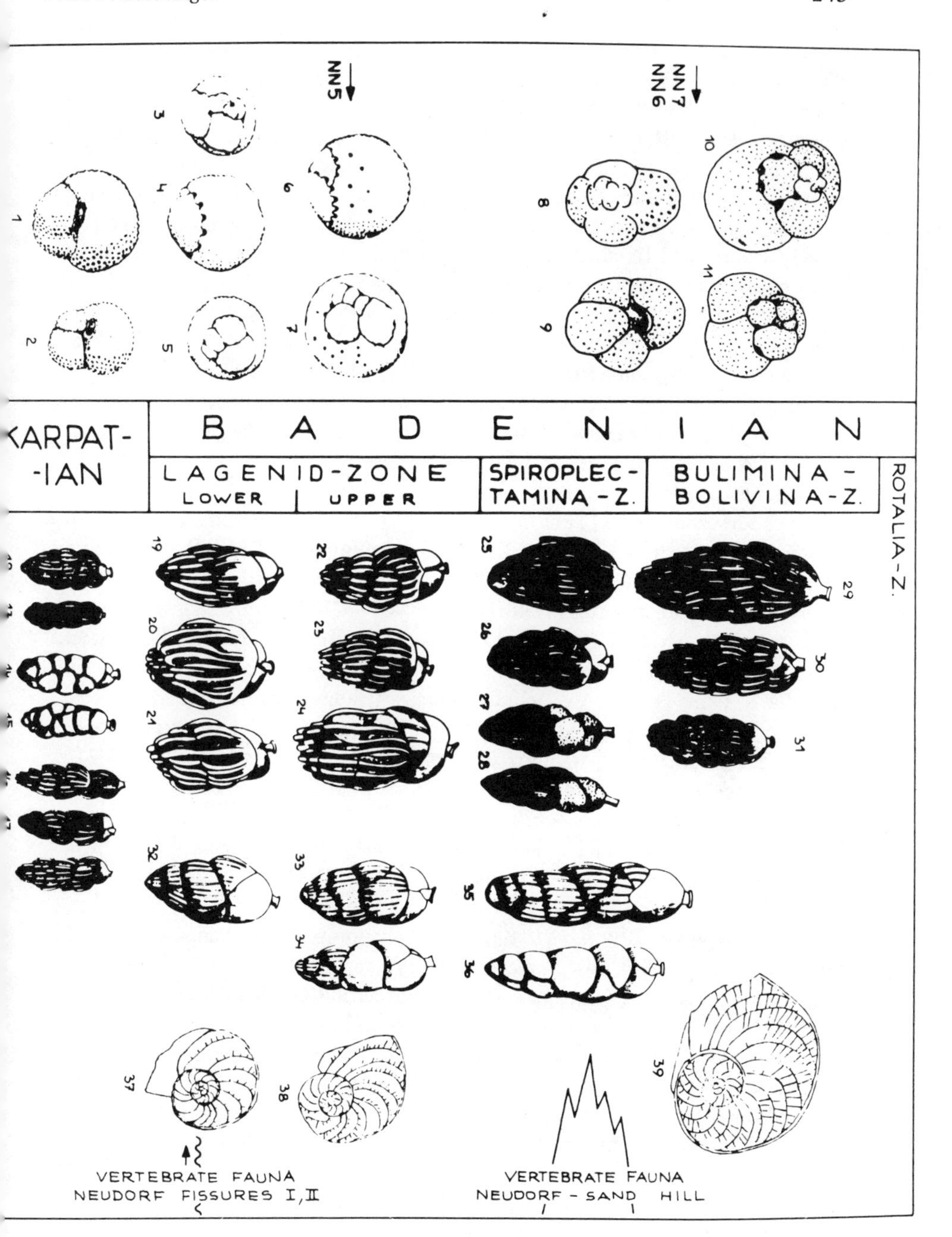

H. c. costata; 39, *H. c. politatesta.* Integrated local vertebrate localities: Neudorf (Devinska Nova Ves, CSSR), fissures I and II; Neudorf (Devinska Nova Ves, CSSR)-Sandhill.

6, items 19-21) and *U. s. semiornata* of the "*semiornata*" lineage (Text-fig. 6, item 32), as well as the first appearance of *Heterostegina praecostata* (Text-fig. 6, item 37) of the "*costata*" lineage, are some of the taxa (also see Text-fig. 11) defining the lowermost integrated assemblage zone and the lower boundary of the Badenian stage. By tracing this zone with concurrent benthonic foraminifers, mollusks, and ostracodes (Text-fig. 11) into the marginal facies (see discussion below), it could be demonstrated that the vertebrate faunas of the Neudorf localities (Devinska Nova Ves, CSSR) fissures I and II, containing a fauna of the "mammal faunal zone" of Sansan, is time equivalent to the lowermost Badenian integrated assemblage zone. This Paratethys integrated assemblage zone correlates with the upper part of planktonic foraminifer zone N8 and lower part of N9. The zonal markers of this interval are *Praeorbulina* and *Orbulina suturalis* (Text-fig. 11). In the next higher zones of the Badenian (Text-fig. 6), we can trace, on the one hand, the evolutionary first appearance of the different lineages mentioned above, and, on the other, the migrational appearances of zonal markers of the calcareous nannoplankton zonation (see Steininger et al., 1975, table 1, pt. 1).

***Upper Sarmatian zones** (Text-figs. 5 and 7).* Two zones defined and named for endemic mollusks characterize this Late Sarmatian interval. The oldest is the *Ervilia* zone (Text-fig. 7) characterized by *Calliostoma poppelacki, Calliostoma podolicum enode,* and transitional forms (Text-fig. 7, items C1-C10); *Mactra eichwaldi* (Text-fig. 7, item M1); *Ervilia d. dissita* and *E. d. podolica* (Text-fig. 7, items E1-E4); and *Irus g. gregarius* and transitional forms between *I. g. gregarius* and *I. g. ponderosus* (Text-fig. 7, items I1-I4). A vertebrate fauna, which contains no *Hipparion*, occurs in the uppermost *Ervilia* zone (Steininger and Thenius, 1965). The next oldest zone is the *Mactra* zone, characterized by *Mactra vitaliana* (Text-fig. 7, item M2), *Calliostoma podolicum enode* together with *C. p. podolicum* (Text-fig. 7, items C11-C18), and *Irus gregarius ponderosus* (Text-fig. 7, items 35-37). In addition to the mollusks, the Lower and Upper Sarmatian zones can be recognized by benthic foraminifera (Grill, 1941) and ostracodes (Cernajsek, 1971; Jiricek, 1973), as indicated in Text-figure 11.

***Pannonian zones** (Text-fig. 8).* Biostratigraphic units for the Pannonian were first defined by Papp (1951), based on evolutionary lineages of endemic mollusk genera (e.g., melanopsids, congerias, and cardiids). For definition of the integrated assemblage zones, ostracodes and mammals are also used (Text-fig. 11). Text-figure 8 demonstrates the lowermost zone A/B (of Papp, 1951), which is characterized by the elements of the "*Melanopsis*" lineage (*M. impressa posterior*, Text-fig. 8, items 1-6) and the "*Congeria*" lineage (*C. praeornithopsis, C. ornithopsis*; Text-fig. 8, items 31, 32), the first migrational appearance of *Hipparion* (Text-fig. 8, item 37), and the evolutionary appearance of *Gomphotherium longirostris* (Text-fig. 8, item 38), of the evolutionary lineage of gomphotheriids of Middle Europe (Zapfe, 1949). The recent discovery of *Hipparion* remains within marine sediments of the Mediterranean Tortonian stage (Text-fig. 10) allows one to relate the Pannonian stage with its endemic invertebrate faunas to the planktonic reference scale (Text-figs. 10 and 11) (de Bruijn et al., 1971; Benda and Meulenkamp, 1972).

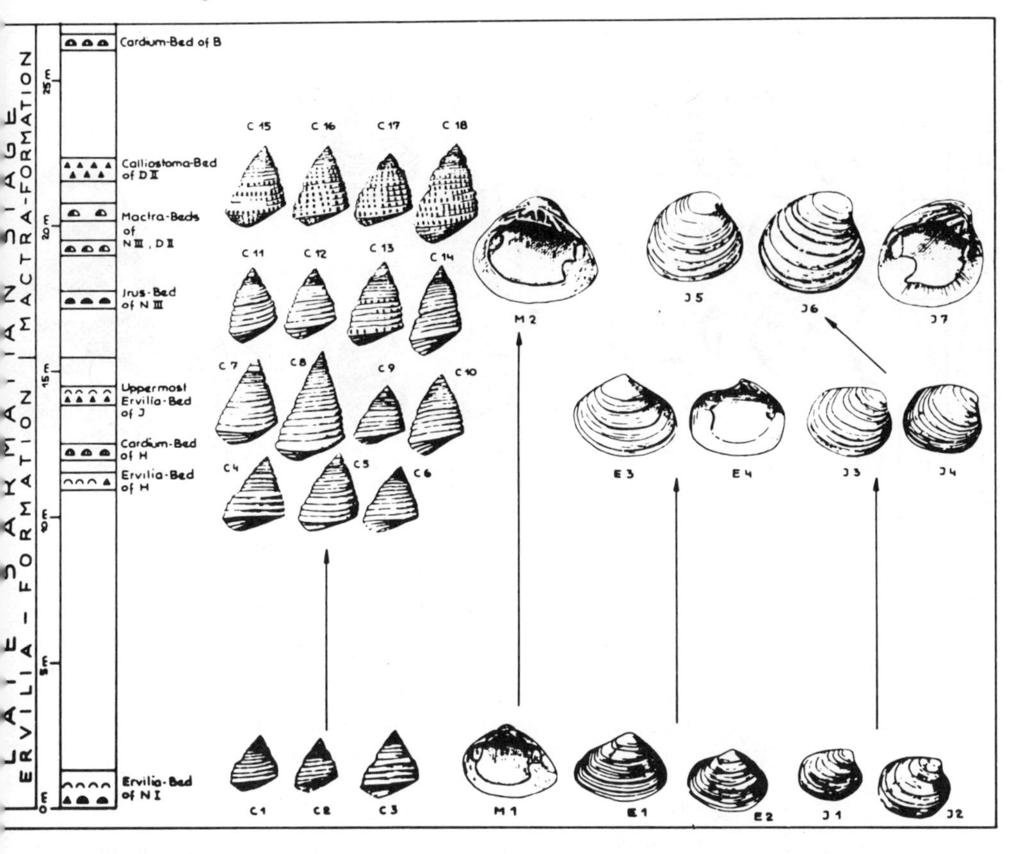

Text-Figure 7
Upper Sarmatian. Selected evolutionary lineages of endemic mollusk genera. Items C1-C18 represent the *Calliostoma* lineage: C1-C3, *C. poppelacki*; C4-C6, transitional forms of *C. poppelacki* to *C. podolicum*; C7-C10, *C. podolicum enode*; C11-C14, *C. podolicum enode*; C15-C18, *C. p. podolicum*. Items M1 and M2 represent the *Mactra* lineage: M1, *M. eichwaldi*; M2, *M. vitaliana*. Items E1 and E4 represent the *Ervilia* lineage: E1 and E2, *E. d. dissita*; E3 and E4, *E. d. podolica*. Items I1-I7 represent the *Irus* lineage: I1 and I2, *I. g. gregarius*; I3 and I4, transitional forms of *I. g. gregarius* to *I. g. ponderosus*; I5-I7, *I. g. ponderosus*.

Integration of local vertebrate faunas and tentative correlations of the European "mammal faunal zones" to the plankton zonation *(Text-figs. 6, 8, 10 and 11).* In depositional regions such as the Central Paratethys, excellent possibilities exist for a direct biostratigraphic correlation of local vertebrate faunas of the integrated-assemblage-zone system with the planktonic reference zonation (Text-figs. 10 and 11) of the Neogene. As outlined above, this integrated-assemblage-zone system attempts to include biostratigraphically valuable taxa from a mosaic of marginal- to open-sea and shallow- to deep-water environments. Because of the rapidly changing

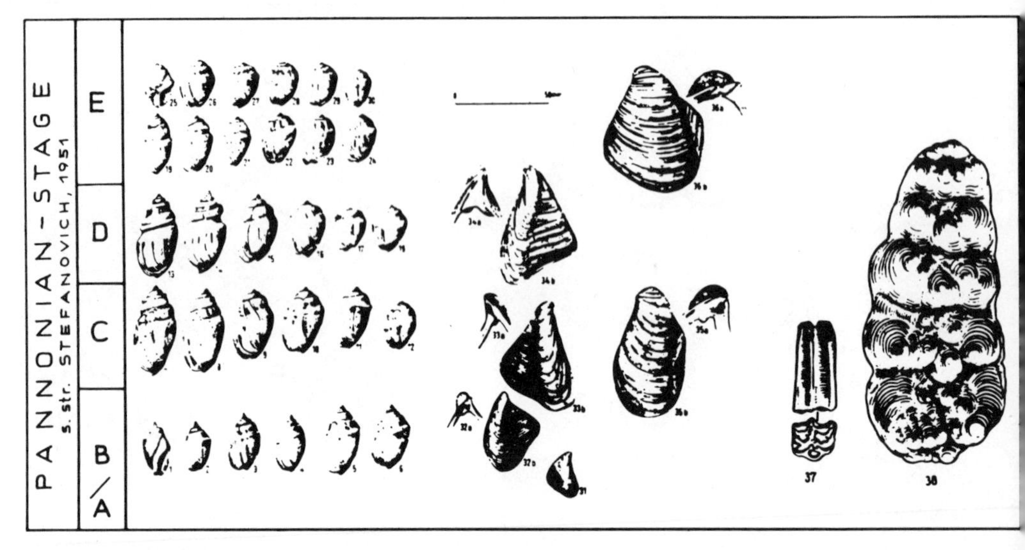

Text-Figure 8
Pannonian: zones (A/B, C, D, E) according to Papp (1951) based on selected evolutionary lineages of endemic mollusk genera, first migrational appearance of *Hipparion* (37), first evolutionary appearance of *Gomphotherium longirostris* (38). Items 1-30 represent the *Melanopsis* lineage: 1-6, *M. impressa posterior*; 7-10 and 13-15, *M. fossilis*; 10, 12, and 16-30, *M. vindobonensis.* Items 31-36 represent the *Congeria* lineage: 31, *C. praeornithopsis*; 32, *C. ornithopsis*; 33, *C. hoernesi*; 34, *C. ungula-caprae*; 35, *C. panici*; 36, *C. subglobosa.*

paleogeographic setting of the Paratethys (Text-figs. 3 and 4), many shallow marine deposits are exposed intercalated or interfingering with beds containing terrestrial vertebrates. Correlation of these marginal deposits by means of mollusks, ostracodes, or benthic and larger foraminifers with time-equivalent deeper-water environments containing planktonic elements has permitted determination of the relationship of these regionally useful units to the open seas (Text-figs. 10 and 11). The standard European continental vertebrate succession was proposed by Thaler (1966) and refined by Cicha et al., (1972) and Mein (1975) as a succession of "mammal faunal

---

Text-Figure 9
Badenian (34-54), Sarmatian (17-33), and Pannonian (1-16) fossils. Stage faunal character based on mollusk genera such as *Congeria* (1, 2, 7, 8, 32, 33), *Limnocardium* (3, 4, 5, 11), *Melanopsis* (6, 9, 12-15), *Viviparus* (10, 16), *Mactra* (17, 22), *Irus* (18), *Cardium* (19, 21), *Solen* (23), *Calliostoma* (24), *Pirenella* (25-27), *Acteocinna* (28), *Dorsanum* (29-31), *Glycymeris* (34), *Megacardita* (35), *Chlamys* (36, 37), *Lucina* (38), *Venus* (39), *Ancilla* (40), *Nassa* (41), *Cerithium* (42, 43), *Turritella* (44, 45), *Natica* (46, 47), *Murex* (48), *Clavatula* (49), *Cancellaria* (50), *Conus* (51), *Cyprea* (52), *Strombus* (53), and *Cassis* (54). (After Papp, 1959.)

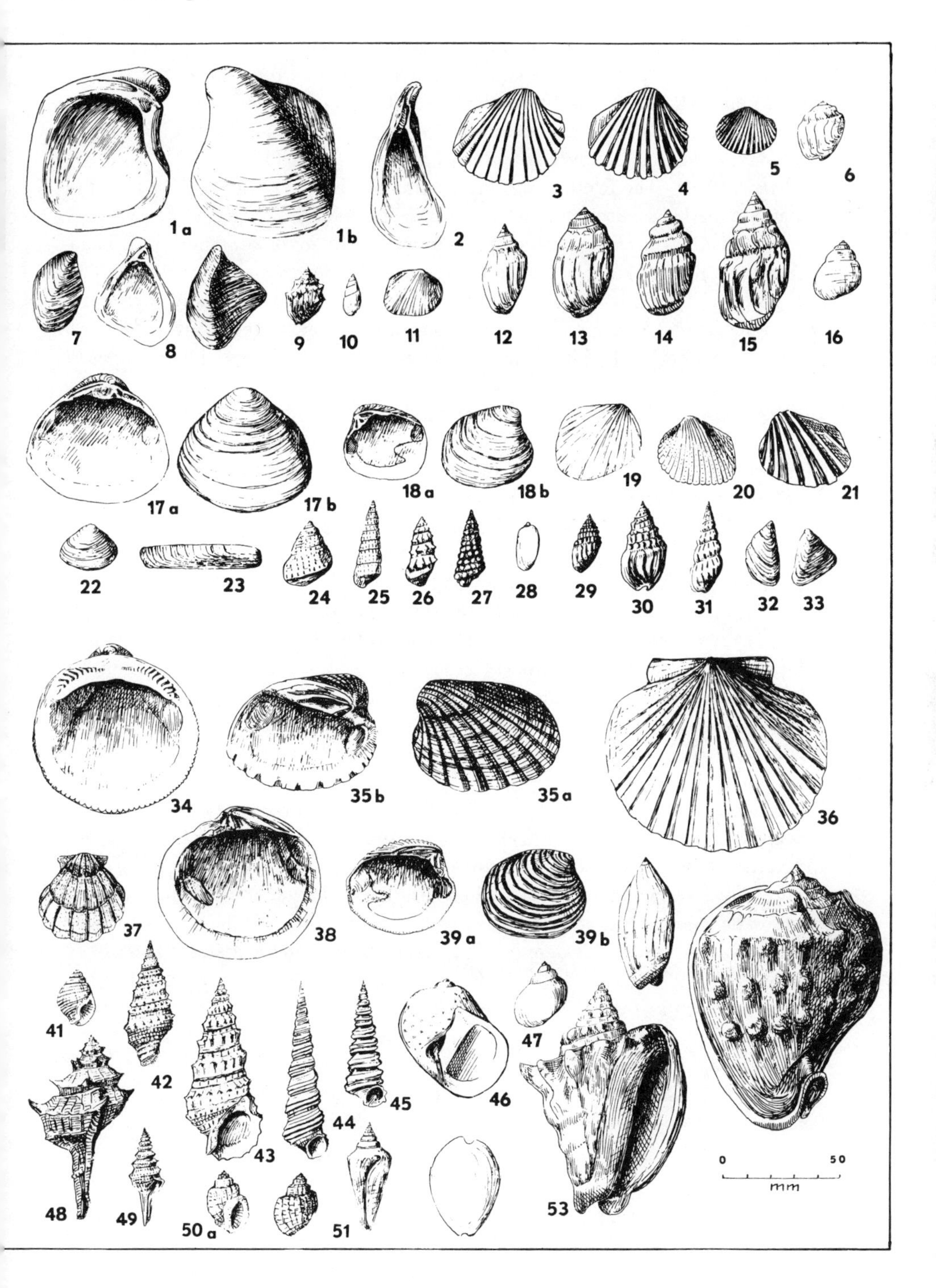
1 a
1 b
2
3
4
5
6
7
8
9
10
11
12
13
14
15
16
17 a
17 b
18 a
18 b
19
20
21
22
23
24
25
26
27
28
29
30
31
32
33
34
35 b
35 a
36
37
38
39 a
39 b
41
42
43
44
45
46
47
48
49
50 a
51
53
0
50
mm

zones," which correspond to evolutionary steps in micromammal lineages and the first concurrent appearance of mammalian taxa. Some of the local vertebrate faunas of the Central Paratethys outlined below (see Text-fig. 11) can be tied very precisely to the marine part of the integrated assemblage zonation, and through this correlated with the planktonic reference zonation (Text-figs. 10 and 11).

Late Egerian. The Egerian vertebrate fauna of Linz (Text-fig. 11) was found in otherwise marine sands with a rich mollusk fauna and *Miogypsina (M.) formosensis.* These sands interfinger and are under- and overlain by marls containing a rich nannoplankton flora of zone NP25 and *Globigerinoides quadrilobatus primordius* (Rögl, 1975; Rabeder and Steininger, 1975; Steininger et al., 1976). Notably absent in this vertebrate fauna are forms known in European vertebrate terminology as "Early Miocene invaders." The Linz fauna correlates with the Paulhiac-Laugnac zones of Thaler (1966).

Eggenburgian. Typical Eggenburgian mollusks are concurrent with uvigerinids and nannoplankton zone NN1-NN2 (Text-fig. 11). With these mollusks are found the first examples of the "Early Miocene invaders." These include Proboscidea, *Anchitherium, Brachyodus*, and the siren *Metaxytherium* (Steininger and Senes, 1971; Thenius, 1959; Zapfe, 1953). This fauna corresponds to Thaler's (1966) zone of Estrepouy.

Ottnangian. The fauna of Orechov and several other time-equivalent faunas (Cicha et al., 1972) are characterized by the first migrational appearance of *Megacricetodons* and *Democricetodons.* The fauna of Orechov is within the Ottnangian "*Oncophora*"-*Rzehakia* beds (Text-fig. 11) and corresponds to Thaler's (1966) zone of La Romieu. A corresponding vertebrate fauna was reported by Mein et al. (1971) from beds containing a marine "Helvetian"-Ottnangian ostracode fauna (Papp and Steininger, 1973); this corroborates the correlation to the marine biozonation of the Central Paratethys.

Karpatian. The Eibiswald fauna interfingers with and is overlain by Lower Karpatian transgressive sediments (Text-fig. 11; Kollmann, 1965). This fauna seems to be equivalent to the Cicha, Fahlbusch, and Fejfar "mammal faunal zone" of Franzensbad-Langenmoosen as based on the evolutionary development of the vertebrates.

Late Karpatian-Early Badenian. The rich fauna of Neudorf fissures I and II (Cicha et al., 1972; Zapfe, 1949, 1951, 1953, 1960) is transgressively overlain by marine sediments of the *Spiroplectammina* zone (Text-figs. 6, 10, and 11). According to the evolutionary level of the fauna (Cicha et al., 1972) and for tectonic and paleogeographic reasons, the stratigraphic position of this fauna can be correlated to uppermost Karpatian-lowermost Badenian. The area of the fissures was first submerged during the Middle Badenian (Kapounek et al., 1965).

Text-Figure 10
Cenozoic planktonic reference zonation and current European stage system. (After Berggren, 1973.)

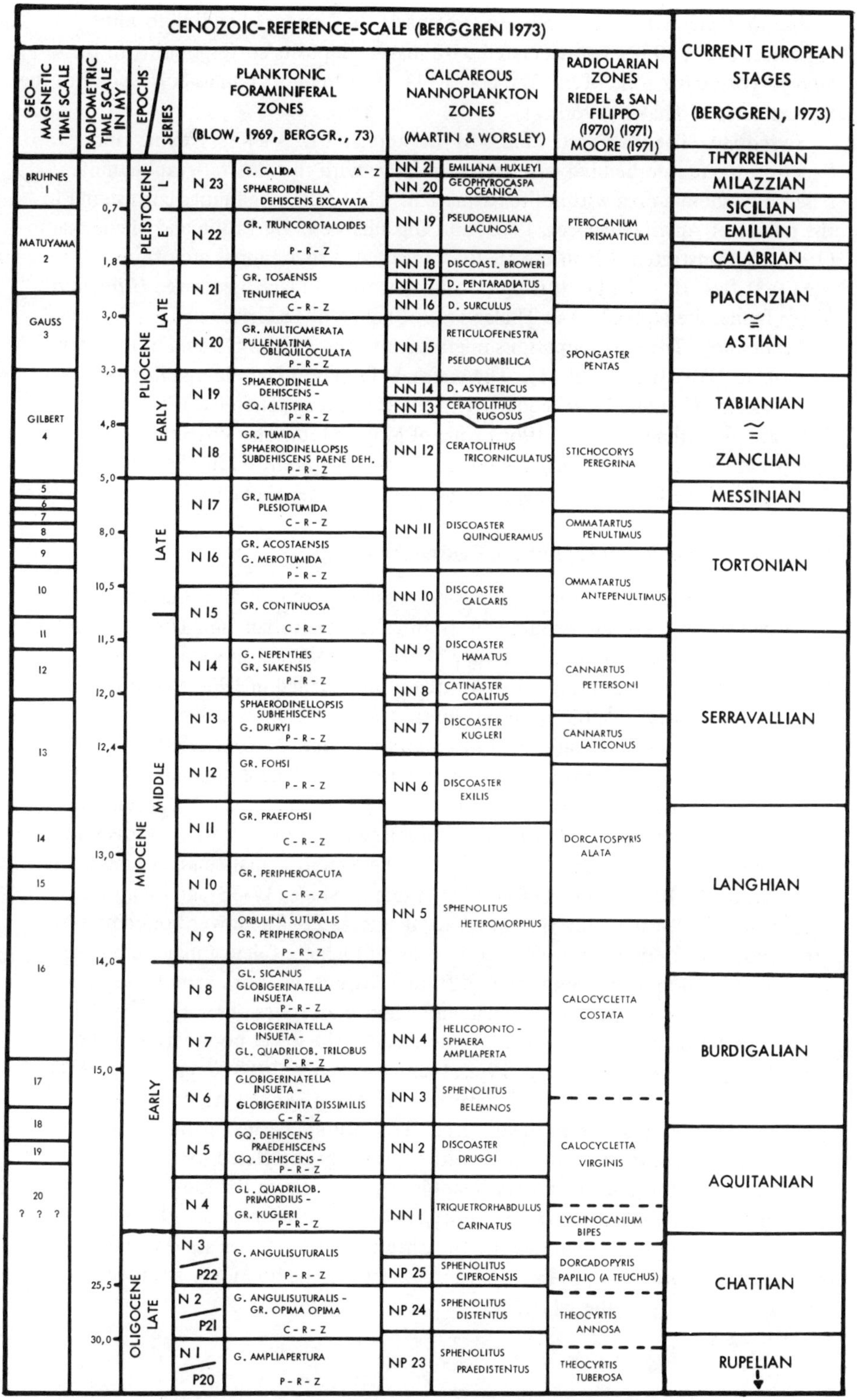

CENOZOIC-REFERENCE-SCALE (BERGGREN 1973)

GEO-MAGNETIC TIME SCALE	RADIOMETRIC TIME SCALE IN MY	EPOCHS / SERIES	PLANKTONIC FORAMINIFERAL ZONES (BLOW, 1969, BERGGR., 73)		CALCAREOUS NANNOPLANKTON ZONES (MARTIN & WORSLEY)		RADIOLARIAN ZONES RIEDEL & SAN FILIPPO (1970) (1971) MOORE (1971)	CURRENT EUROPEAN STAGES (BERGGREN, 1973)
BRUHNES 1	0,7	PLEISTOCENE L	N 23	G. CALIDA A - Z SPHAEROIDINELLA DEHISCENS EXCAVATA	NN 21 NN 20	EMILIANIA HUXLEYI GEOPHYROCASPA OCEANICA	PTEROCANIUM PRISMATICUM	THYRRENIAN MILAZZIAN SICILIAN
MATUYAMA 2	1,8	PLEISTOCENE E	N 22	GR. TRUNCOROTALOIDES P - R - Z	NN 19	PSEUDOEMILIANA LACUNOSA		EMILIAN CALABRIAN
GAUSS 3	3,0	PLIOCENE LATE	N 21	GR. TOSAENSIS TENUITHECA C - R - Z	NN 18 NN 17 NN 16	DISCOAST. BROWERI D. PENTARADIATUS D. SURCULUS		PIACENZIAN ≃ ASTIAN
	3,3		N 20	GR. MULTICAMERATA PULLENIATINA OBLIQUILOCULATA P - R - Z	NN 15	RETICULOFENESTRA PSEUDOUMBILICA	SPONGASTER PENTAS	
GILBERT 4		PLIOCENE EARLY	N 19	SPHAEROIDINELLA DEHISCENS - GQ. ALTISPIRA P - R - Z	NN 14 NN 13	D. ASYMETRICUS CERATOLITHUS RUGOSUS		TABIANIAN ≃ ZANCLIAN
	4,8		N 18	GR. TUMIDA SPHAEROIDINELLOPSIS SUBDEHISCENS PAENE DEH. P - R - Z	NN 12	CERATOLITHUS TRICORNICULATUS	STICHOCORYS PEREGRINA	
5 6 7	5,0	MIOCENE LATE	N 17	GR. TUMIDA PLESIOTUMIDA C - R - Z	NN 11	DISCOASTER QUINQUERAMUS	OMMATARTUS PENULTIMUS	MESSINIAN
8 9	8,0		N 16	GR. ACOSTAENSIS G. MEROTUMIDA P - R - Z			OMMATARTUS ANTEPENULTIMUS	TORTONIAN
10	10,5		N 15	GR. CONTINUOSA C - R - Z	NN 10	DISCOASTER CALCARIS		
11 12	11,5	MIOCENE MIDDLE	N 14	G. NEPENTHES GR. SIAKENSIS P - R - Z	NN 9	DISCOASTER HAMATUS	CANNARTUS PETTERSONI	SERRAVALLIAN
	12,0		N 13	SPHAERODINELLOPSIS SUBHEHISCENS G. DRURYI P - R - Z	NN 8 NN 7	CATINASTER COALITUS DISCOASTER KUGLERI	CANNARTUS LATICONUS	
13	12,4		N 12	GR. FOHSI P - R - Z	NN 6	DISCOASTER EXILIS		
14			N 11	GR. PRAEFOHSI C - R - Z	NN 5	SPHENOLITUS HETEROMORPHUS	DORCATOSPYRIS ALATA	LANGHIAN
15	13,0		N 10	GR. PERIPHEROACUTA C - R - Z				
16			N 9	ORBULINA SUTURALIS GR. PERIPHERORONDA P - R - Z				
	14,0	MIOCENE EARLY	N 8	GL. SICANUS GLOBIGERINATELLA INSUETA P - R - Z			CALOCYCLETTA COSTATA	BURDIGALIAN
			N 7	GLOBIGERINATELLA INSUETA - GL. QUADRILOB. TRILOBUS P - R - Z	NN 4	HELICOPONTO - SPHAERA AMPLIAPERTA		
17	15,0		N 6	GLOBIGERINATELLA INSUETA - GLOBIGERINITA DISSIMILIS C - R - Z	NN 3	SPHENOLITUS BELEMNOS		
18 19			N 5	GQ. DEHISCENS PRAEDEHISCENS GQ. DEHISCENS - P - R - Z	NN 2	DISCOASTER DRUGGI	CALOCYCLETTA VIRGINIS	AQUITANIAN
20 ? ? ?			N 4	GL. QUADRILOB. PRIMORDIUS - GR. KUGLERI P - R - Z	NN 1	TRIQUETRORHABDULUS CARINATUS	LYCHNOCANIUM BIPES	
		OLIGOCENE LATE	N 3 / P22	G. ANGULISUTURALIS P - R - Z	NP 25	SPHENOLITUS CIPEROENSIS	DORCADOPYRIS PAPILIO (A TEUCHUS)	CHATTIAN
	25,5		N 2 / P21	G. ANGULISUTURALIS - GR. OPIMA OPIMA C - R - Z	NP 24	SPHENOLITUS DISTENTUS	THEOCYRTIS ANNOSA	
	30,0		N 1 / P20	G. AMPLIAPERTURA P - R - Z	NP 23	SPHENOLITUS PRAEDISTENTUS	THEOCYRTIS TUBEROSA	RUPELIAN ↓

Middle Badenian. The vertebrate remains of the Neudorf-Sandhill locality (Thenius, 1952, 1959) are associated with marine deposits corresponding to the *Spiroplectammina* zone (Text-figs. 6 and 11). The Neudorf faunas correspond to the Sansan zone of Thaler (1966).

Sarmatian. The vertebrate faunas of St. Stephan i.L., Vienna, Nexing, and Sauerbrunn are interbedded with and correlative with the biostratigraphic units based on endemic taxa within the Sarmatian. The vertebrate faunas correspond to the faunas of Anwil (Engesser, 1972) and Gigenhausen. De Bruijn and Meulenkamp (1972) demonstrated that in the Mediterranean area such faunas are of pre-Tortonian age (Text-figs. 10 and 11). It is significant that they all lack associated *Hipparion.* These faunas are equivalent to Thaler's (1966) zone of La Grive.

Pannonian. The first hipparions migrated into the Paratethys area in the Early Pannonian (Text-figs. 8 and 11). The mammal faunas of Gaiselberg (Zapfe, 1949) and Csakvar (Thenius, 1959) equate with the Höwenegg (Thenius, 1959) fauna. They are older than Thaler's (1966) zone of Montredon. The Vösendorf fauna (Papp and Thenius, 1954) corresponds to the zone of Montredon. See also the discussion of the Pannonian integrated assemblage zones above (p. 246).

Chronostratigraphic Concept of the Central Paratethys

As discussed above (p. 238), there is no possibility of effectively using the classic European stages Chattian, Aquitanian, Burdigalian, Helvetian, and Tortonian in the Central Paratethys (Steininger and Papp, 1973). The Paratethys working group of the International Paleontological Union therefore decided in 1968 to produce a regional chronostratigraphic stage system that would be in accordance with the guidelines of the recently published *International Guide to Stratigraphic Classification, Terminology and Usage* (1972). The definition of these regional stages, as decided by the working group, is as follows: (1) The time interval of a stage should correspond to one or more integrated assemblage zones unified by their overall faunal character. For example (Text-fig. 9), the overall faunal character based on mollusks of the Badenian stage (34-54), Sarmatian stage (17-33), and Pannonian stage (1-16) is distinct from that of adjacent stages. (2) The lower boundaries of a stage should be defined by isochronous levels of high confidence marked by the concurrent first appearance of many different organisms. For example, see the zones of the Late Karpatian-Badenian and Pannonian zones (Text-figs. 6 and 8). The upper boundaries of the stages are defined by the lower boundaries of the next younger stages (Papp et al., 1968; Cicha and Senes, 1968; Papp et al., 1971; Senes et al., 1971; Steininger and Neveskaja, 1975). (3) Each stage is represented by various formations within the different depositional basins of the Central Paratethys. These formations can be rather precisely correlated by using the integrated-assemblage-zone biostratigraphic concept. From the most characteristic exposures of the formation, *Holostratotypus* or unit stratotype should be chosen. The lower and upper boundaries and the different lateral facies are documented by the *Faziostratotypen*; in English, the terms would be boundary stratotype and hypostratotype of Hedberg (1972a).

Every stage of this regional chronostratigraphic system of the Central Paratethys has or will be published, with all the formations and stratotypes and the biotic content indicated, in the editions of the Slovakian Academy of Science entitled *Chronostratigraphie and Neostratotypen–Miozän der zentralen Paratethys* (Senes, 1967, for the Karpatian; Steininger and Senes, 1971, for the Eggenburgian; Papp et al., 1973, for Ottnangian; Papp and Senes, 1974, for the Sarmatian; and Baldi and Senes, 1975, for the Egerian).

CONCLUSIONS

The problem of correlating deposits laid down in open oceanic environments with those of marginal and epicontinental environments (mostly exposed on continental blocks) has necessitated the development and use of an integrated-assemblage-zone model. The model was derived from studies of Late Oligocene and Neogene depositional basins of the Central Paratethys.

The principal features of the model are as follows: (1) a physical stratigraphic framework; (2) evaluation and correlation of time-equivalent composite reference sections representing different environments; (3) plotting of first appearances and stratigraphic ranges of taxa within these composite reference sections; and (4) zonation reference points (origins, extinctions) necessary to build up an integrated assemblage zone. Study of vertebrate faunas demonstrates that by using this model it is possible to correlate the continental "mammal faunal zone" concept across marine-nonmarine boundaries with the Cenozoic marine plankton zonation.

It is necessary and more efficient to develop regional chronostratigraphic stages based on the integrated-assemblage-zone concept for a biogeographically well defined area, such as the Central Paratethys, than it is to use the classic European stage names (e.g., Chattian, Aquitanian, Burdigalian, Helvetian, Tortonian, etc.) in this area. Although used for decades, these classic stages do not correlate well with sections to which these names have been applied previously, because they were originally established within different bioprovincial systems.

Acknowledgments

I am greatly indebted to Professor Dr. A. Papp and to all members of the Paratethys working group of the International Paleontological Union for the valuable unpublished information and advice that made it possible to write this paper. Informal discussions with my colleagues at the Institute of Paleontology, University of Vienna, have contributed to the topics of this contribution. E. G. Kauffman and R. H. Benson of

---

Text-Figure 11
Upper Oligocene-Neogene chronostratigraphic stage system of the Central Paratethys and integrated-assemblage-zone concept. Arrows indicating first (evolutionary, migrational) appearance of taxa used in definition of zones.

LATE OLIGOCENE - NEOGENE BIO-

CENTRAL PARATETHYS BIOSTRATIGRAPHIC - ASSEMBLAGE -

RADIOMETRIC TIME SCALE IN MY	GENERAL FACIES DEVELOPMENT	CURRENT CHRONO-STRATIGRAPH REGIONAL STAGES	BARS CORRELATIVE TO:	PLANKTONIC EVENTS: PLANKTONIC FORAMINIFERA, CALCAREOUS NANNOPLANKTON	BENTHONIC FORAMINIFERA, OSTRACODA, SILICOFLAGELLATA
	BRACKISH TO FRESHWATER TO TERRESTRIAL	RUMANIAN		NO DIRECT CORRELATION TO PLANKTONIC - SCALE	CYPRIA CANDONAEFORMIS; LIMMNOCYTHERE SHARAPOVAS
		DACIAN			CANDONA CANDIDA; CANDONA NEGLECTA; PSEUDOCANDONA MARCHIA
		PONTIAN			CASPIOLLA ACRONASUTA; CASPIOLLA FLECTIMARGINATA; CASPIOLLA, BAKUNELLA
9,5		PANNONIAN			LINEOCYPRIS RETICULATA, CYPRIDEIS SUBLITORALIS; ERPETOCYPRIS RECTA; CYPRIDEIS TUBERCULATA, CYP. SULCATA; HUNGAROCYPRIS AURICULATA, SILICOPLACENTINA; ERPETOCYPRIS., HEMICYTHEREA HUNGARICA
11,2; 12,4; 13,2	BRACKISH TO FRESHWATER TRANSGRESSIVE	SARMATIAN			CIBICIDES BADENSIS; ELPHIDIUM REGINUM; AURILA MEHESI; EL. HAUERI-NUM; EL. ANTONI-NUM; HEMICYTHEREA OMPHALODES; AURILA NOTATA; CYPRIDEIS TUBERCULATA; PROTOELPHIDIUM SUBGRANOSUM
15,1; 17,7	MARINE TRANSGRESSIVE	BADENIAN		NN 7; VELAPERTINA; G. DRURYI, G. DECORAPERTA; NN 6; ORBULINA SUTURALIS; NN 5; PRAEORBULINA GLOMEROSA CIRCULARIS	BORELIS MELO; SPIROPLECTAMINA; HETEROSTEGINA; H. PRAECOSTATA CARINATA; H. COSTATA POLITATESTA; H. C. COSTATA; U. MACROCARINATA; BOLIVINA DILATATA; BULIMINA INTONSA; U. GRILLI; U. V. VENUSTA; U. BON. COMPRESSA; U. S. SEMIORNATA; U.V. LIESINGENSIS; FALUNIA SPINULOSA; LOXOCONCHA CARINATA; MIOCYPRIDEIS ELONGATA; CARINOCYTHEREIS CARINATA; AURILA DIV. SPEC.; ACANTHOCYTH. HYSTRIX.; CYTHERELLA POSIDENTICULATA; MESOCENA ELLIPTICA (75%); CANNOPILUS PICASOL; C. SPHAERICUS; DICTYOCHA SCHAUINSLANDI STRADNERI (SPINES); D. STAURODON; PARADICTYOCHA POLYACTIS MESOCENOIDEA
20,7		KARPATIAN		GL. SICANUS	M. INTERMEDIA; "ROBULUS" EXGR. MELVILLI; U. GRACILIFORMIS; U. ACUMINATA; U. BONONIENSIS-PRIMIFORMIS; U. PARKERI BREVIFORMIS; CYCLAMINA KARPATICA; CYTHERIDEA PRAEACUMINATA; LOXOCONCHA VARIOLATA; NEOMONOCERATINA HELVETICA; AURILA VENTRISULCATA; FAL. PLIC. GLABRA; CORBISEMA TRIACANTHA - ZONE; NAVICULOPSIS NAVICULA ZONE
22,0	BRACKISH / MARINE TRANSGRESSIVE	OTTNANGIAN		GR. ACROSTOMA; GQ. LANGHIANA; GT. DISSIMILIS; NN 4; GT. CF. UNICAVA NN 3 - NN 4; G. CIPEROENSIS OTTNANGIENSIS	HETEROSTEGINA HETEROSTEGINA; SIGMOILOPSIS OTTNANGENSIS; U. CF. UNSERIATA; CF.; CF.; CF.
22,6		EGGENBURGIAN		NN 1 - NN 2; GL. QUADRILOB. TRILOBUS; GQ. DEHISCENS	M. GUNTERI; HAPLOCY. EXGR. HELVETICA; SCHULDERIDEA RHOMBUS; CYTHERIDEA LACUNOSA; MIOCYPRIDEIS FORTISENSIS; FALUNIA PLICATULA; NAVICULOPSIS LATA - ZONE
	BRACKISH / MARINE	EGERIAN		NN 1; GL. QUADRILOBATUS PRIMORDIUS; NP 25; NP 24/25; NP 24; GR. O. OPIMA, GR. O. NANA; FIRST GLOBIGERINOIDES SP., GLOBOQUADRINA SP.	MIOGYPSINA COMPLANATA; M. FORMOSENSIS; M. SEPTENTRIONALIS; M. BANTAMENSIS; OPERCULINA COMPLANATA; LEPIDOCYCLINA (E.) DILATATA; L. (N.) MORGANI; URIGERINA FARINOSA; U. CF. HANTKENI; ALMAENA OSNABURGENSIS; U. POSTHANTKENI; U. PARVIFORMIS; CIBICIDOIDES BUDAYI; MIOCYPRIDEIS RARA; HAPLOCYTHERIDEA DACICA
		KISCELLIAN			

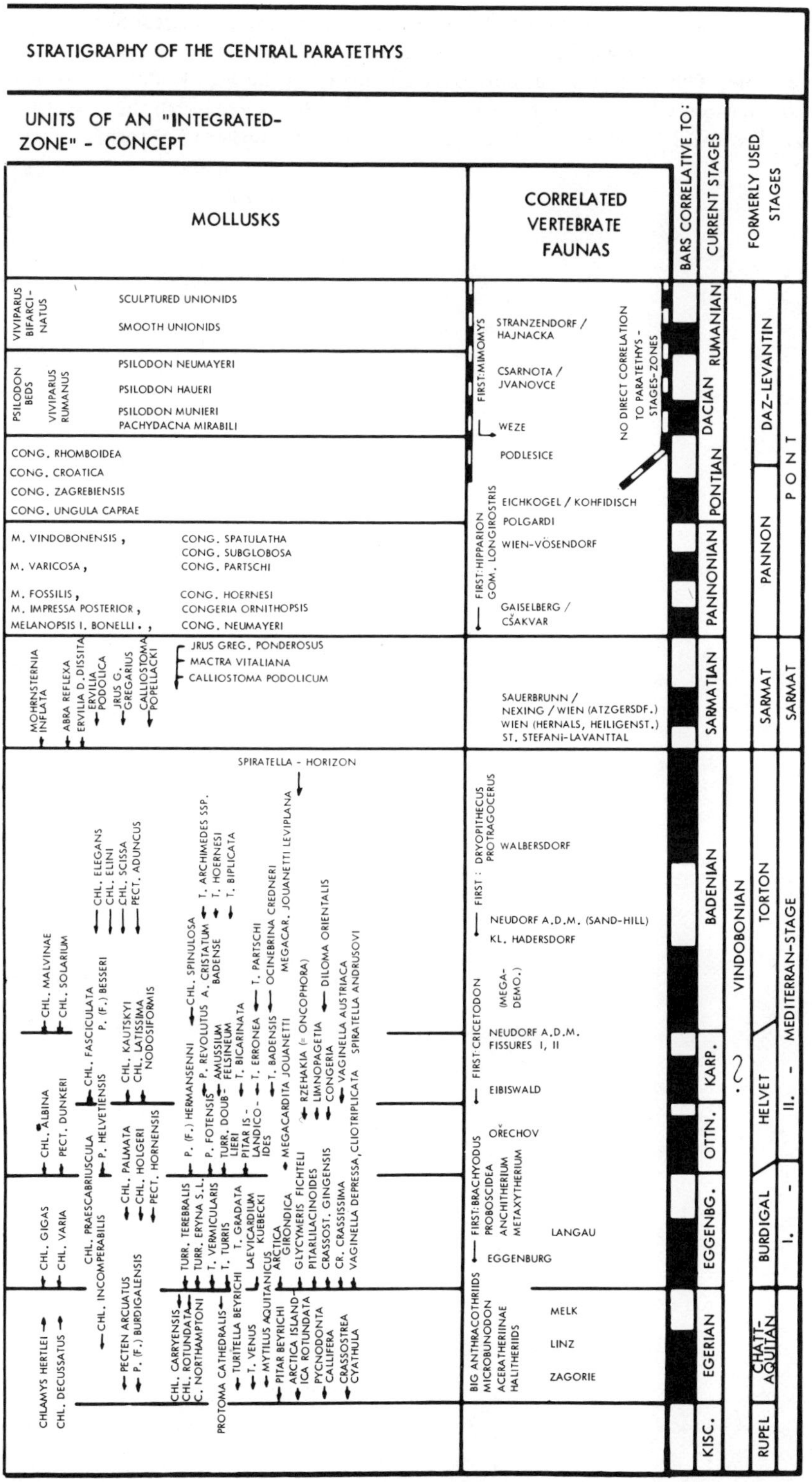

STRATIGRAPHY OF THE CENTRAL PARATETHYS

UNITS OF AN "INTEGRATED-ZONE" - CONCEPT

MOLLUSKS

CORRELATED VERTEBRATE FAUNAS

BARS CORRELATIVE TO:

CURRENT STAGES

FORMERLY USED STAGES

VIVIPARUS BIFARCINATUS
SCULPTURED UNIONIDS
SMOOTH UNIONIDS

PSILODON BEDS
VIVIPARUS RUMANUS
PSILODON NEUMAYERI
PSILODON HAUERI
PSILODON MUNIERI
PACHYDACNA MIRABILI

CONG. RHOMBOIDEA
CONG. CROATICA
CONG. ZAGREBIENSIS
CONG. UNGULA CAPRAE

M. VINDOBONENSIS,
M. VARICOSA,
M. FOSSILIS,
M. IMPRESSA POSTERIOR,
MELANOPSIS I. BONELLI.,
CONG. SPATULATHA
CONG. SUBGLOBOSA
CONG. PARTSCHI
CONG. HOERNESI
CONGERIA ORNITHOPSIS
CONG. NEUMAYERI

MOHRNSTERNIA INFLATA
ABRA REFLEXA
ERVILIA D. DISSITA
ERVILIA PODOLICA
IRUS G. GREGARIUS
CALLIOSTOMA POPELLACKI
IRUS GREG. PONDEROSUS
MACTRA VITALIANA
CALLIOSTOMA PODOLICUM

SPIRATELLA - HORIZON

CHL. MALVINAE
CHL. SOLARIUM
CHL. ALBINA
PECT. DUNKERI
CHL. GIGAS
CHL. VARIA
CHLAMYS HERTLEI
CHL. DECUSSATUS
CHL. FASCICULATA
P. (F.) BESSERI
CHL. PRAESCABRIUSCULA
P. HELVETIENSIS
CHL. INCOMPERABILIS
PECTEN ARCUATUS
P. (F.) BURDIGALENSIS
CHL. ELEGANS
CHL. ELINI
CHL. SCISSA
PECT. ADUNCUS
CHL. KAUTSKYI
CHL. LATISSIMA
NODOSIFORMIS
CHL. PALMATA
CHL. HOLGERI
PECT. HORNENSIS
CHL. SPINULOSA
P. (F.) HERMANSENNI
P. REVOLUTUS
A. CRISTATUM BADENSE
T. ARCHIMEDES SSP.
T. HOERNESI
T. BIPLICATA
P. FOTENSIS
AMUSSIUM FELSINEUM
P. (F.) HERMANSENNI
TURR. DOUBLIERI
T. BICARINATA
PITAR ISLANDICOIDES
T. ERRONEA
T. PARTSCHI
T. BADENSIS
OCINEBRINA CREDNERI
MEGACARDITA JOUANETTI
MEGACAR. JOUANETTI LEVIPLANA
RZEHAKIA (= ONCOPHORA)
LIMNOPAGETIA
CONGERIA
DILOMA ORIENTALIS
VAGINELLA AUSTRIACA
SPIRATELLA ANDRUSOVI
CLIOTRIPLICATA
TURR. TEREBRALIS
TURR. ERYNA S.L.
T. VERMICULARIS
T. TURRIS
T. GRADATA
LAEVICARDIUM KUEBECKI
ARCTICA GIRONDICA
GLYCYMERIS FICHTELI
PITARLILACINOIDES
CRASSOST. GINGENSIS
CR. CRASSISSIMA
VAGINELLA DEPRESSA
CHL. CARRYENSIS
CHL. ROTUNDATA
C. NORTHAMPTONI
PROTOMA CATHEDRALIS
TURITELLA BEYRICHI
T. VENUS
MYTILUS AQUITANICUS
PITAR BEYRICHI
ARCTICA ISLANDICA ROTUNDATA
PYCNODONTA CALLIFERA
CRASSOSTREA CYATHULA

FIRST: MIMOMYS
STRANZENDORF / HAJNACKA
CSARNOTA / JVANOVCE
WEZE
PODLESICE
NO DIRECT CORRELATION TO PARATETHYS - STAGES-ZONES
FIRST: HIPPARION
GOM. LONGIROSTRIS
EICHKOGEL / KOHFIDISCH
POLGARDI
WIEN-VÖSENDORF
GAISELBERG / CŠAKVAR
SAUERBRUNN / NEXING / WIEN (ATZGERSDF.) WIEN (HERNALS, HEILIGENST.) ST. STEFANI-LAVANTTAL
FIRST: DRYOPITHECUS PROTRAGOCERUS
WALBERSDORF
NEUDORF A.D.M. (SAND-HILL)
KL. HADERSDORF
FIRST: CRICETODON (MEGA-DEMO.)
NEUDORF A.D.M. FISSURES I, II
EIBISWALD
OŘECHOV
FIRST: BRACHYODUS PROBOSCIDEA ANCHITHERIUM METAXYTHERIUM
LANGAU
EGGENBURG
BIG ANTHRACOTHRIIDS MICROBUNODON ACERATHERIINAE HALITHERIIDS
MELK
LINZ
ZAGORIE

CURRENT STAGES	FORMERLY USED STAGES		
RUMANIAN		DAZ-LEVANTIN	PONT
DACIAN		DAZ-LEVANTIN	PONT
PONTIAN		PANNON	PONT
PANNONIAN		PANNON	PONT
SARMATIAN		SARMAT	SARMAT
BADENIAN	VINDOBONIAN	TORTON	II. - MEDITERRAN-STAGE
KARP.	?	HELVET	II. - MEDITERRAN-STAGE
OTTN.		HELVET	II. - MEDITERRAN-STAGE
EGGENBG.		BURDIGAL	I. -
EGERIAN		CHATT-AQUITAN	
KISC.		RUPEL	

the Smithsonian Institution and J. E. Hazel of the U.S. Geological Survey critically read the manuscript and made a number of valuable suggestions. The study was supported by a grant from the Max Kade Foundation, which made it possible for me to spend a year with the late Professor Dr. O. L. Bandy at the University of Southern California.

An Integrated Stratigraphical Study of Fossil Assemblages from the Maastrichtian White Chalk of Northwestern Europe

Finn Surlyk — Geologisk Museum, Copenhagen

Tove Birkelund — Institut for Historisk Geologi og Palæontologi, Copenhagen

INTRODUCTION

The Maastrichtian is the youngest Cretaceous stage. Its duration is estimated to be about 5 MY and its boundaries 70 and 65 MYBP. During this interval, a sequence of 250 to 700 m of monotonous white, coccolithic chalk was deposited in the basin now encompassing Denmark, northern Germany, southwest Sweden, the North Sea, and southeast England.

Although the Upper Cretaceous chalk, both in volume and areal extent, is one of the most important Late Mesozoic facies, it represents a depositional environment unknown from other geologic periods and with no recent counterparts. It is peculiar in that it is a pelagic sediment with at least 75 percent planktonic components, deposited in a shallow sea. It thus combines qualities of deep-sea ooze with qualities of shelf sediments, which make it most suitable for detailed biostratigraphical studies. These qualities are (1) continuous sedimentation in the central parts of the basin, (2) fairly high rate of sedimentation (15 cm/1,000 years as a maximum), and (3) rich representation of benthic as well as pelagic faunal and floral elements (more than 1,200 species are now recognized from the Danish Maastrichtian). Finally, the rock can be totally disintegrated, which permits exact quantitative studies to be made.

GEOLOGIC SETTING

Most of the deposits laid down at the margin of the North European Maastrichtian basin have been removed by later erosion; those which remain are poorly exposed. Nevertheless the basin margin is exceptionally well exposed to the northeast, where the coast was situated in Scania, southern Sweden (Text-fig. 1). In that area a shallow archipelago was created by the transgressing Campanian-Maastrichtian sea, the shores, peninsulas, and islands of which were built up of Precambrian crystalline rocks of the Baltic shield. Profuse oyster banks flourished along the coast, and the rocky seashores were dominated by rich epifaunas comprising all kinds of sessile as

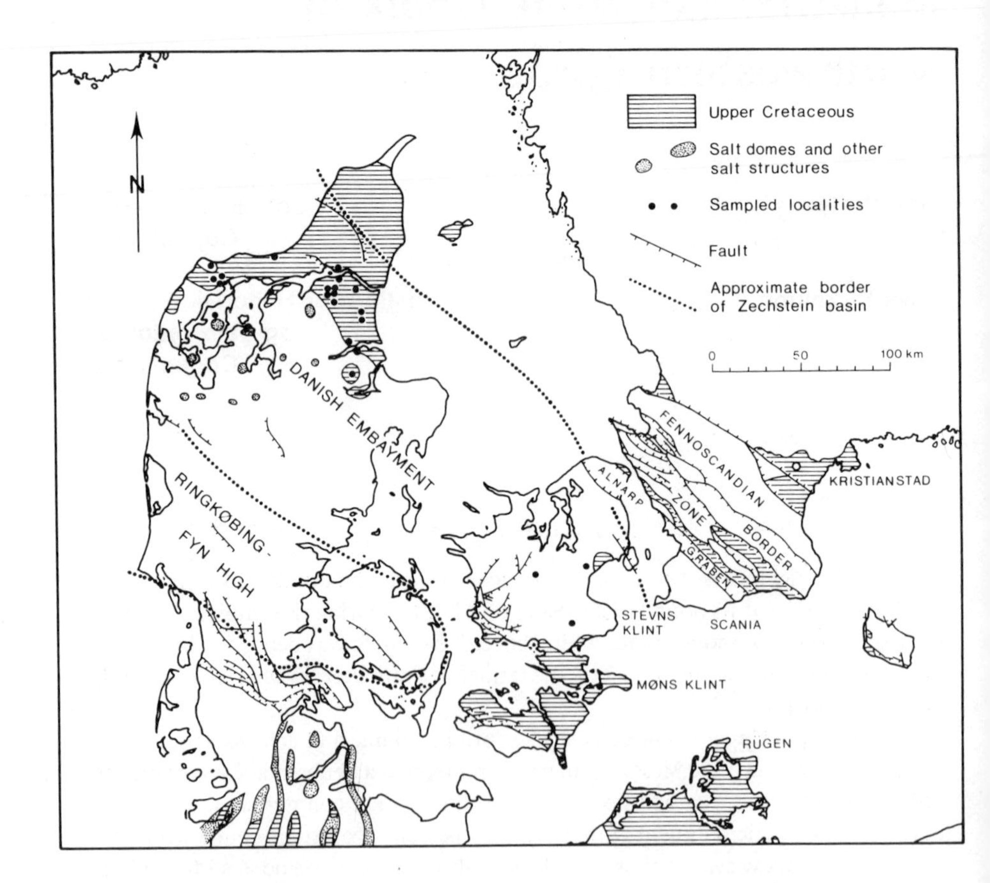

Text-Figure 1
Distribution of Upper Cretaceous sediments on the geologic map of the Pre-Quaternary surface of Denmark and adjacent areas. The most important structural features are shown. The border of the Zechstein Basin gives a fairly good outline of the main structural elements. Black dots show the position of sample localities not mentioned in the text. (Modified from Sorgenfrei, 1966.)

well as vagile animals (Voigt, 1929; Surlyk and Christensen, 1974). The sediment laid down on the sea bottom is almost totally composed of fragmented calcitic skeletons, such as mollusks, echinoids, bryozoans, and calcareous red algae. In an offshore direction, the sediment grain size decreases to fine calcarenites, but increases again to coarse calcarenites, or even calcirudites approaching a system of low NW-SE trending horsts that formed low islands or peninsulas in the late Upper Cretaceous sea (Text-fig. 2). The system of horsts (the "Fennoscandian border zone") was developed between the stable Precambrian shield of Sweden and the Danish embayment (Voigt, 1929, 1963). Along the southwestern flanks of the

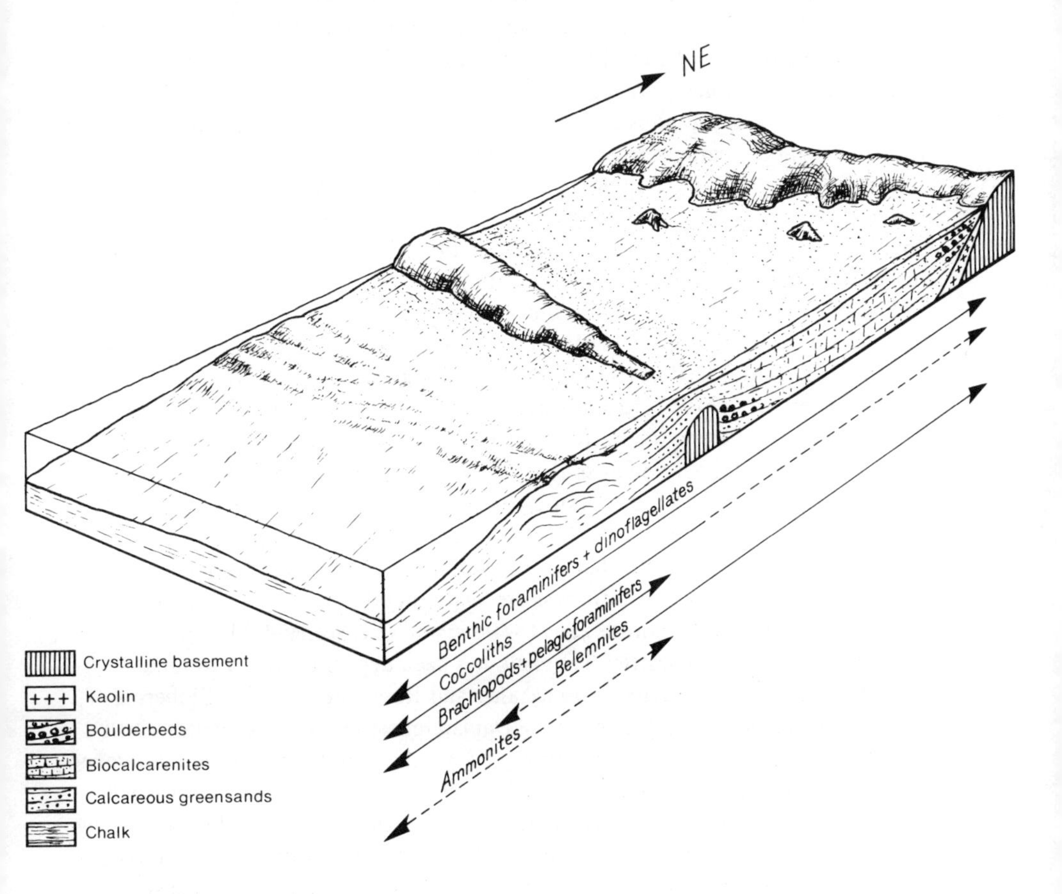

Text-Figure 2
Block diagram showing a highly generalized model of the depositional environments along the northeastern margin of the European chalk sea. Sedimentary environments in which individual fossil groups have been used in Maastrichtian biostratigraphy of the area are indicated below the diagram.

horsts, the thick Campanian-Maastrichtian deposits consist of calcareous greensands, which further offshore pass rapidly into pure soft coccolithic chalk (Text-fig. 2).

THE WHITE CHALK FACIES; ENVIRONMENTAL BACKGROUND

The Maastrichtian white chalk is a pelagic mud with a noncarbonate content usually as low as 0.5 to 10 percent. Clay minerals are dominant in the insoluble residues. Coarse terrigenous detritus is thus practically absent and is restricted to rare grains of eolian quartz. The carbonate of the white chalk is exclusively of biogenic origin (Black, 1953). The fine grain size is due to the fact that the sediment is primarily composed of coccoliths and coccolith detritus. The remaining part consists of skeletal calcite originating from a great number of other fossil groups, among which bryozoans constitute the main part. The exposed chalk is only slightly lithified except for hardgrounds, which are most conspicuous toward the margin of the basin or on local topographic highs. Aragonitic and opaline shell material have been lost in diagenesis, and dissolution also has affected certain species of coccoliths. Opaline silica was presumably the primary source of nodular and sheet flints formed during late diagenesis in most of the sequence.

The chalk is totally bioturbated, but deeper burrows are often well preserved (Bromley, in press; Kennedy, 1971). The profuse occurrence of very small sized epifaunal suspension feeders (Surlyk, 1972) and the common occurrence of cementing animals on even small hard substrates (Ernst, 1969) shows that, except for the topmost few millimeters, the bottom was relatively stable.

On the basis of faunal evidence, the depths of deposition of the white chalk seem to range from the base of the euphotic zone (in restricted areas) to depths down to several hundred meters (Nestler, 1965; Håkansson et al., in press).

In spite of the monotonous character of the white chalk, there is a lateral variation in lithology related to distance from shore, water depth, and biotic composition.

In an area close to the border zone (Sweden and east Sjælland 25 to 60 km from the horsts), the Upper Maastrichtian chalk has been deposited as low mounds or ridges believed to be bioherms (Rosenkrantz and Rasmussen, 1960). Bioherms of greater size, characterizing the overlying Danian limestones, have been described (e.g., Rosenkrantz and Rasmussen, 1960). It is unknown if bioherms are developed in older parts of the Maastrichtian. The heights of the bioherms vary between 3 and 6 m, and they either form low ridges or have circular to elliptical bases. Along the 12-km-long cliff of Stevns, the ridges follow a rather consistent pattern as each ridge overlaps its neighbor, indicating that the ridges become younger in a south to north direction. The mounds and ridges are composed of coccolithic chalk matrix containing an extraordinarily high content (up to 20 percent) of benthic macrofossils, which are dominated by bryozoans. All fossils are extremely well preserved and unsorted, with no signs of wear due to transport, and there has been no winnowing of the fines by current action. Furthermore, the sediments show no signs of scouring or truncations of the inclined bedding on the flanks of the bioherms, the only larger sedimentary irregularities being occasional slumps on the slopes (Surlyk, 1972). There is thus little doubt that the bioherms really are organic build-ups. The sediment contains no frame-building organisms, but it is suggested that the profusely occurring bryozoans, with the possible aid of vegetation, had a baffling effect.

Offshore from the bioherm zone, near the central parts of the basin, the chalk becomes horizontally bedded; in the deepest parts, it is lithologically very uniform and rather poor in benthic fossils.

In the Danish Maastrichtian, there is also a distinct vertical pattern of sedimentation. At two levels, in the mid-Lower and uppermost Maastrichtian, benthic faunas reach their maximum development. These horizons are characterized by extremely diverse faunas of bryozoans, brachiopods, bivalves, serpulids, calcareous sponges, etc. Between these two maxima, a pronounced minimum is found at approximately the boundary between the Lower and Upper Maastrichtian (Surlyk, 1972). Belemnites are also extremely rare in this part of the sequence, but certain ammonites (*Acanthoscaphites, Hoploscaphites,* and *Baculites*) are quite common. The changes in density and diversity of the benthic fauna are accompanied by distinct changes in lithology. The horizons with rich benthic faunas normally contain abundant flint, scattered hardgrounds, and pressure solution seams, and are dominated by the trace fossil *Thalassinoides*; the more sterile parts contain practically no flint and are dominated by the trace fossil *Zoophycos.* The horizontal and vertical fluctuations in the benthic fauna are followed by variations in noncarbonate content from 0.5 to 12 percent.

METHODOLOGY

Traditional macrofossil collecting and microfossil sampling undertaken by individual specialists at different times resulted in only limited progress in biostratigraphical refinement of the Danish chalk sequence. A program was therefore initiated, which involved collection of very large (10 to 30 kg) bulk samples every 2 to 3 m in all major measured sections. It is the intention that as many fossil groups as possible shall be worked out on the basis of these samples. Until now the following groups have been or are at present being studied: Coccolithophorida, Dinoflagellata, Foraminifera, Ostracoda, Bryozoa (Cheilostomata), Brachiopoda, Serpulida, and small-sized representatives of Echinodermata and Bivalvia. In addition, the sparse macrofossils, primarily belemnites and ammonites, have also been collated, with the stratigraphy worked out on the basis of these samples.

Washing of the samples involves repeated boiling in a supersaturated glauber salt solution and deep freezing for about 16 cycles (method described in detail by Surlyk, 1972); after treatment, the chalk breaks down completely into mud and very clean fossils (Text-fig. 3). The wash residues are then picked for all taxa. The method results in immense quantities of perfectly preserved calcitic material. About 70,000 brachiopods have been obtained in this way, each sample of 10 kg containing between 100 and 4,000 specimens, with 500 to 1,000 as the normal amount.

Text-Figure 3
Wash residue >1 mm of two chalk samples from brachiopod zone 8 (lower Upper Maastrichtian). Note the absolute dominance by a very diverse bryozoan fauna. Characteristic are the circular specimens of the *Lunulites* group.

The great density is more easily understood when it is known that the majority of the species has an adult maximum length of a few millimeters. It should be stressed that these fossils are practically impossible to find and collect by normal collecting methods.

After taxonomic determination of the fossils to the species or subspecies level, quantitative graphs are constructed for each fossil group and each section.

ECOLOGY

Community Ecology

Early in the study it became obvious that an ecological evaluation of the fossil groups was not only in itself worthwhile, but also absolutely necessary in order to test the biostratigraphic value and potential of each group. Until now, this analysis has mainly been done on the brachiopods, but work is in progress on bryozoans, ostracods, bivalves, echinoderms, and serpulids.

Nestler (1967) described an important vertical frequency pattern for major groups of organisms having a similar mode of life in the Maastrichtian chalk. Totally unrelated animals with comparable modes of life have similar frequency curves through time, and the curves representing different modes of life have a consistent relationship (Text-fig. 4). The bryozoans (Text-fig. 4a) comprise the dominating benthic fossil group in the chalk. Some chalk bryozoans were initial colonizers on the pure but relatively stable coccolithic mud bottom (Håkansson, 1974). They in turn acted as substrates for the next group of bryozoans; this succession resulted finally in development of a very rich and diverse bryozoan fauna. This stabilized the bottom as a substrate and made colonization possible for other animal groups with stronger demands for harder substrates for attachment. Text-figure 4b shows the groups that on a whole were directly dependent on the bryozoan substrates, such as brachiopods and to a smaller degree serpulids. There is consequently a strong correlation between abundance peaks in Text-figures 4a and 4b.

Text-figure 4c shows the frequency pattern of the cemented animals, such as oysters and other bivalves. They demand larger, hard substrates (e.g., the tests of nonburrowing echinoids). Inasmuch as these larger substrate-forming epifaunal animals first colonized the chalk substrates after the bottom was somewhat stabilized, the abundance peaks of large cementing animals are reached later than those of the initial colonizers.

Text-figure 4d shows the occurrence of nektonic or nektobenthic cephalopods, notably the belemnites. The curve is totally independent of benthic fauna. Belemnites occur most commonly, however, in shallow, nearshore facies, where mixed assemblages of juveniles and mature specimens are usual. In the offshore chalk, the occurrence of belemnites is very scattered and juveniles are rare. It is therefore suggested that burial of belemnite rostra in the Maastrichtian chalk sea bottom merely happened by accident, and that they were far from their normal habitats. Ammonites also show a scattered distribution and are likewise dominated by mature specimens, except for the hardground at the top of the Upper Maastrichtian.

Text-figure 4e shows the free-living forms, such as certain pectinid and inoceramid bivalves. This curve is of fundamental importance for the understanding of the ecological conditions in the chalk sea because of the sudden maximum starting simultaneously with the gradual increase of the bryozoans. This relationship implies that the same environmental factors are responsible for the peaks, but that, when conditions become suitable, the free-living forms are able to colonize the mud bottom directly and not gradually, as in the case of the bryozoans and later benthic invaders. Text-figure 4f shows an actual example from the Lower Maastrichtian of Rügen, East Germany. The curve includes forms loosely attached by a byssus.

The biostratigraphical importance of the curves is obvious: the presence of a group, which is for other reasons stratigraphically useful, may thus be directly dependent on the presence of one of the other four adaptive groups. It is clear, therefore, that, among the benthic animals, early colonizers that then persist in subsequent community succession, especially the free-living forms, have the greatest stratigraphical potential, other things being equal.

Thus, on the basis of quantifying the distribution of major faunal groups in measured sections and developing a knowledge of the ecology of these groups, it is possible to test or even predict the biostratigraphical utility of each group. The present stratigraphical work on the Danish Maastrichtian chalk has therefore concentrated on groups with highest potential, whereas other groups have been left for the time being in stratigraphical oblivion.

Single-Group Ecology

A study of the population dynamics and morphological adaptations of the chalk brachiopods exemplifies the importance of basic ecological studies for each group used in biostratigraphy. A more detailed discussion is given by Surlyk (1972, 1973, 1974).

Text-figure 5 is a quantitative diagram of the numerically important brachiopod species in the Lower Maastrichtian section at Hvidskud, Møns Klint (Text-fig. 1). The first curve shows the weight of the wash residue, which is largely equal or identical to that for the bryozoans—the basic epifaunal substrate. The curve of total number of brachiopod specimens in the samples shows a strong positive correlation with the weight of the wash residue (bryozoans): the number of brachiopods is directly dependent on the number of available substrates. As the bryozoans comprise the only important potential substrate, their number controls the number of brachiopods. In this connection it must be mentioned again that almost all the Maastrichtian chalk brachiopod species are minute and have an adult length of 2 to 4 mm (Surlyk, 1972; and Text-fig. 5), and are thus presumably adaptated to the small size of the common available substrates (Text-fig. 6).

The curve showing the number of brachiopod species parallels the two others in general trend, but it is most important that the diversity decreases in high-density samples. In contrast to the average samples, these samples are numerically dominated by a few species that played a prominent role in competition for space on the minute substrates, and which thus are responsible for the disappearance of some of the rare species as the community reached climax. The next series of curves show the frequency of the more abundant brachiopod species. There is a pronounced positive correlation between the "bryozoan curve" and the majority of the

---

Text-Figure 4
Generalized frequency distributions for major adaptive types represented in the fauna in a chalk section: (a) bryozoans, a group able to colonize the pure coccolithic chalk bottom; (b) animals such as brachiopods and serpulids, which are able to attach themselves to small hard substrates; (c) animals such as oysters and the bivalve *Atreta* restricted to relatively large hard substrates; (d) nektonic animals such as ammonites and belemnites; (e) animals able to lie prone and unattached or loosely attached by byssus, directly on the chalk bottom, including several pectinid bivalves; (f) actual example from the Lower Maastrichtian chalk of Rügen, East Germany. (Modified from Nestler, 1967; not to scale.)

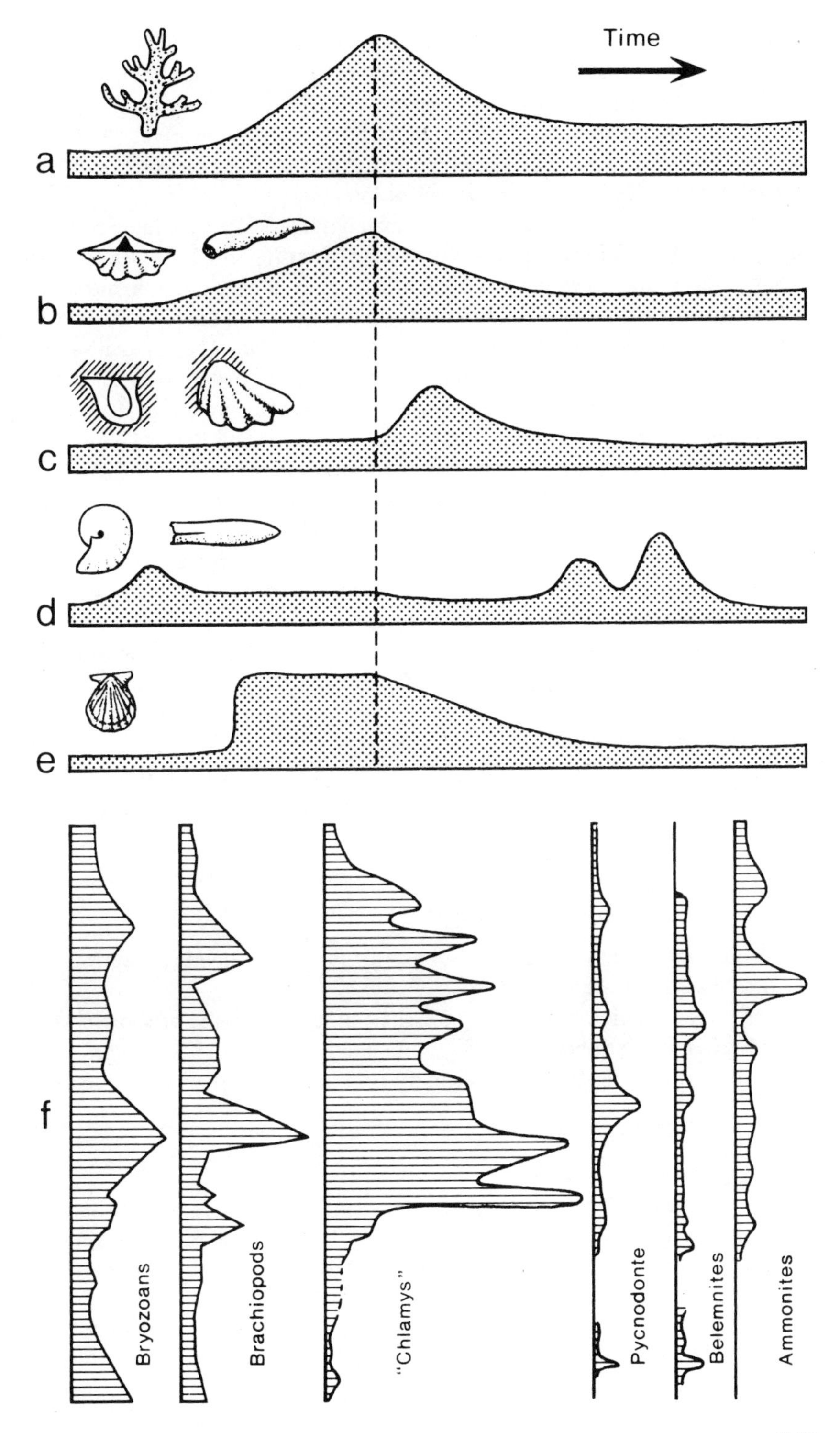
Time
a
b
c
d
e
f
Bryozoans
Brachiopods
"Chlamys"
Pycnodonte
Belemnites
Ammonites

brachiopod curves, and all these species can be considered well-adapted equilibrium species with hard substrates as limiting resources. Only a few species have explosive occurrences at single horizons without any correlation with the main trends of the other curves. These species are typical opportunistic species (Surlyk, 1974), which, however, possess only low correlation potential, as the bursts seem to be of limited lateral extent (on the order of a few tens of kilometers).

Although the brachiopod fauna is dominated numerically by minute pedunculate species, several other life habits are represented (Text-fig. 6). The inset scales show somewhat arbitrarily in which facies the brachiopods of each group in question are most common, ranging from offshore fine-grained coccolith muds to the left, to nearshore skeletal calcarenites and calcirudites to the right. The scales are thus rough measures of the facies tolerance of each group and therefore to some extent of its biostratigraphical potential. It appears from Text-figure 6 that the six free-living hemispherical species, the obese free-living species, and the species attached with a pedicle divided into rootlets have the widest facies tolerance. The two latter species are long-ranging forms showing only slight and very inconspicuous morphological changes through long periods of time; they are therefore of limited biostratigraphical value in refined zonation, but good for general regional correlation. The hemispherical forms, on the contrary,.are very important in biostratigraphical analyses because of their rapid evolutionary rates and characteristic details in exterior ornamentation, which enhance identification. Also, the group of minute species up to 4 mm long (attached to small substrates) is extremely useful in chalk biostratigraphy, because they are relatively easy to determine, normally well preserved, have rapid evolutionary rates, and occur in very large numbers in each sample. As the monotonous chalk facies occupies a very large area through a long period of time, the limited facies tolerance of these species does not prevent them from being very good biostratigraphical tools within this area.

PROBLEMS OF CHALK STRATIGRAPHY

The main stratigraphical problems concerning the Maastrichtian sediments of the area described above are twofold.

First, the correlation of the nearshore skeletal sands with the offshore coccolithic chalks is often difficult because different fossil groups tend to occur in different environments (Text-fig. 2). Thus belemnites, which are standard zonal fossils in the

---

Text-Figure 5
Quantitative diagram of a mid-Lower Maastrichtian brachiopod fauna from the Hvidskud section at Møns Klint. Wash residue is largely equal to the weight distribution of the bryozoans. Diversity is simple species diversity. Boundaries of brachiopod zones 4 and 5 are defined by first appearance in the section of *Terebratulina subtilis* and *Trigonosemus pulchellus.* In addition, zones are characterized by the presence of several other species not shown here.

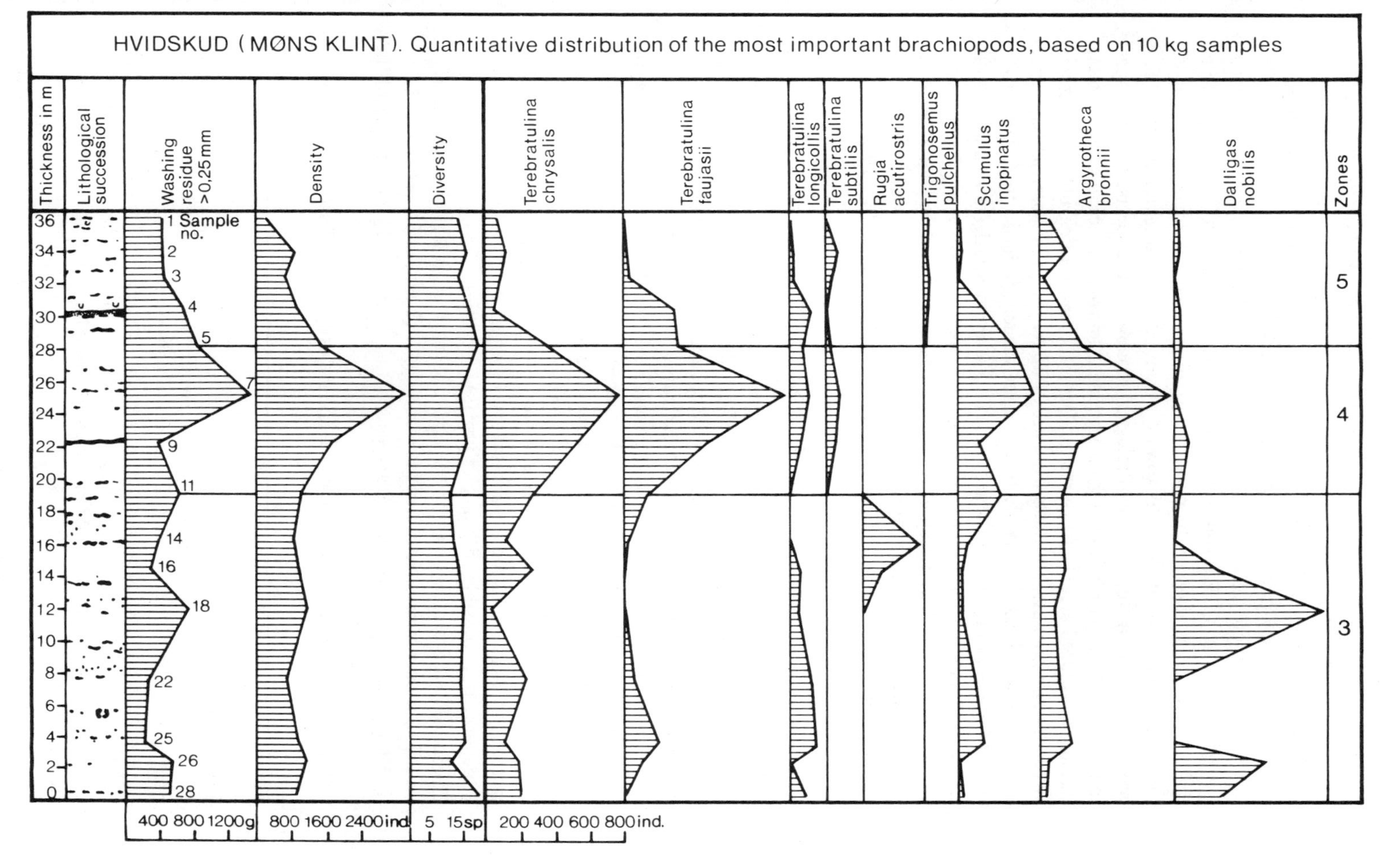

HVIDSKUD (MØNS KLINT). Quantitative distribution of the most important brachiopods, based on 10 kg samples
Thickness in m
Lithological succession
Washing residue >0,25 mm
Density
Diversity
Terebratulina chrysalis
Terebratulina faujasii
Terebratulina longicollis
Terebratulina subtilis
Rugia acutirostris
Trigonosemus pulchellus
Scumulus inopinatus
Argyrotheca bronnii
Dalligas nobilis
Zones
5
4
3
36
34
32
30
28
26
24
22
20
18
16
14
12
10
8
6
4
2
0
1 Sample no.
2
3
4
5
7
9
11
14
16
18
22
25
26
28
400 800 1200g
800 1600 2400ind.
5 15sp
200 400 600 800ind.

European Upper Cretaceous, are very common in the nearshore skeletal sands, whereas they are relatively rare to almost missing in chalk facies.

Second, and more difficult than the correlation of different facies, is the construction of a biostratigraphical zonation of the sequence of up to 700 m of monotonous chalk and the correlation of the numerous small, isolated outcrops. Progress in biostratigraphical studies of the chalk toward a useful detailed zonation has been hampered on account of the extreme lithological monotony, the specialized, very sparse, and apparently rather uniform larger macrofauna, and occurrence in often strongly glacially tectonized areas (East Germany, Denmark, eastern England, and southern Sweden). For example, the thick sequence of up to 700 m of Maastrichtian chalk in Denmark was divided into only two or three vaguely defined zones, until Troelsen (1937) succeeded in dividing the sequence into seven biostratigraphical units on the basis of foraminifers and selected macrofossils, which were for many years the only useful zones for local stratigraphical work. Troelsen's zones were later correlated with more widely applicable zonations, based on belemnites (Jeletzky, 1951; Birkelund, 1957) and foraminifers (Brotzen, 1945; Berggren, 1962, 1964).

STRATIGRAPHICAL METHODOLOGY

Lithostratigraphical correlation of sections measured in detail is possible on a very local scale (Text-fig. 7; and Surlyk, 1971; Steinich, 1972). This type of correlation on the basis of marker beds is very reliable and comes close to a time-stratigraphical ideal because of the pelagic nature of the sediment.

Closely situated localities can also be correlated directly on the basis of quantitative abundance diagrams of one fossil group, such as brachiopods. This technique has been used successfully on several Danish localities (Surlyk, 1969), but the best example is from the Maastrichtian chalk of East Germany (Steinich, 1965, figs. 296-297, modified here as Text-fig. 8). The quantitative variations are often, however, due to more or less local changes in the environment; this kind of correlation is therefore only possible within closely spaced localities.

Quantitative diagrams based on the Lower Maastrichtian section at Hvidskud, Møns Klint, have been constructed for brachiopods (Text-fig. 5), ostracods (Text-fig. 9), foraminifers (Text-fig. 10), and coccoliths (Text-fig. 11) on the basis of the same series of samples.

It appears from all four diagrams that the majority of the illustrated species

---

Text-Figure 6
Different modes of life displayed by the brachiopods of the Maastrichtian chalk. Four main habitat groups occur. From left to right on the (rather arbitrary) scales, the relative abundance of each group is indicated from fine-grained coccolithic muds to the nearshore skeletal sands, giving a rough idea of the facies tolerance of each group and therefore to some extent its biostratigraphical potential. Further details are given by Surlyk (1972, 1973, 1974).

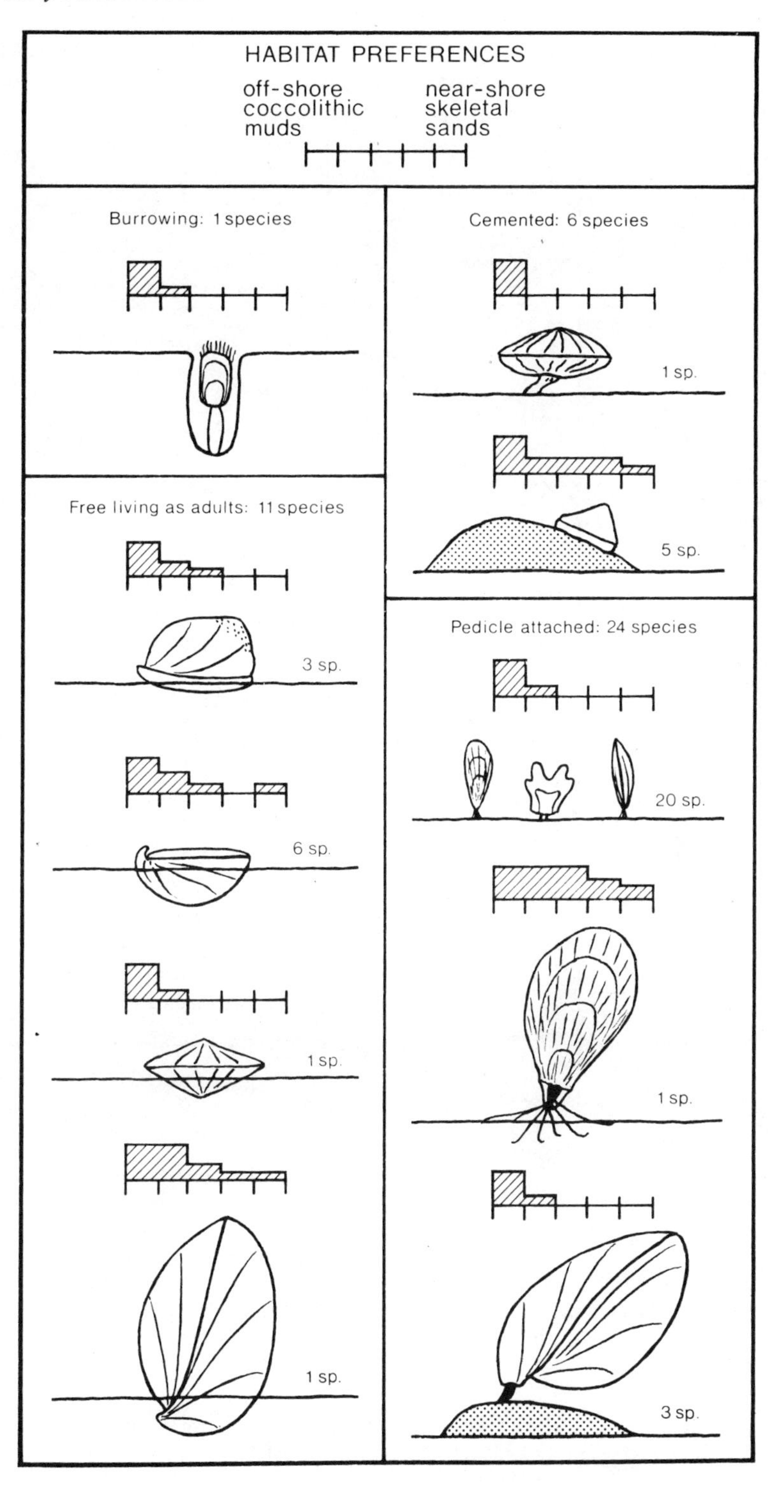
HABITAT PREFERENCES
off-shore coccolithic muds
near-shore skeletal sands
Burrowing: 1 species
Cemented: 6 species
1 sp.
5 sp.
Free living as adults: 11 species
3 sp.
6 sp.
1 sp.
1 sp.
Pedicle attached: 24 species
20 sp.
1 sp.
3 sp.

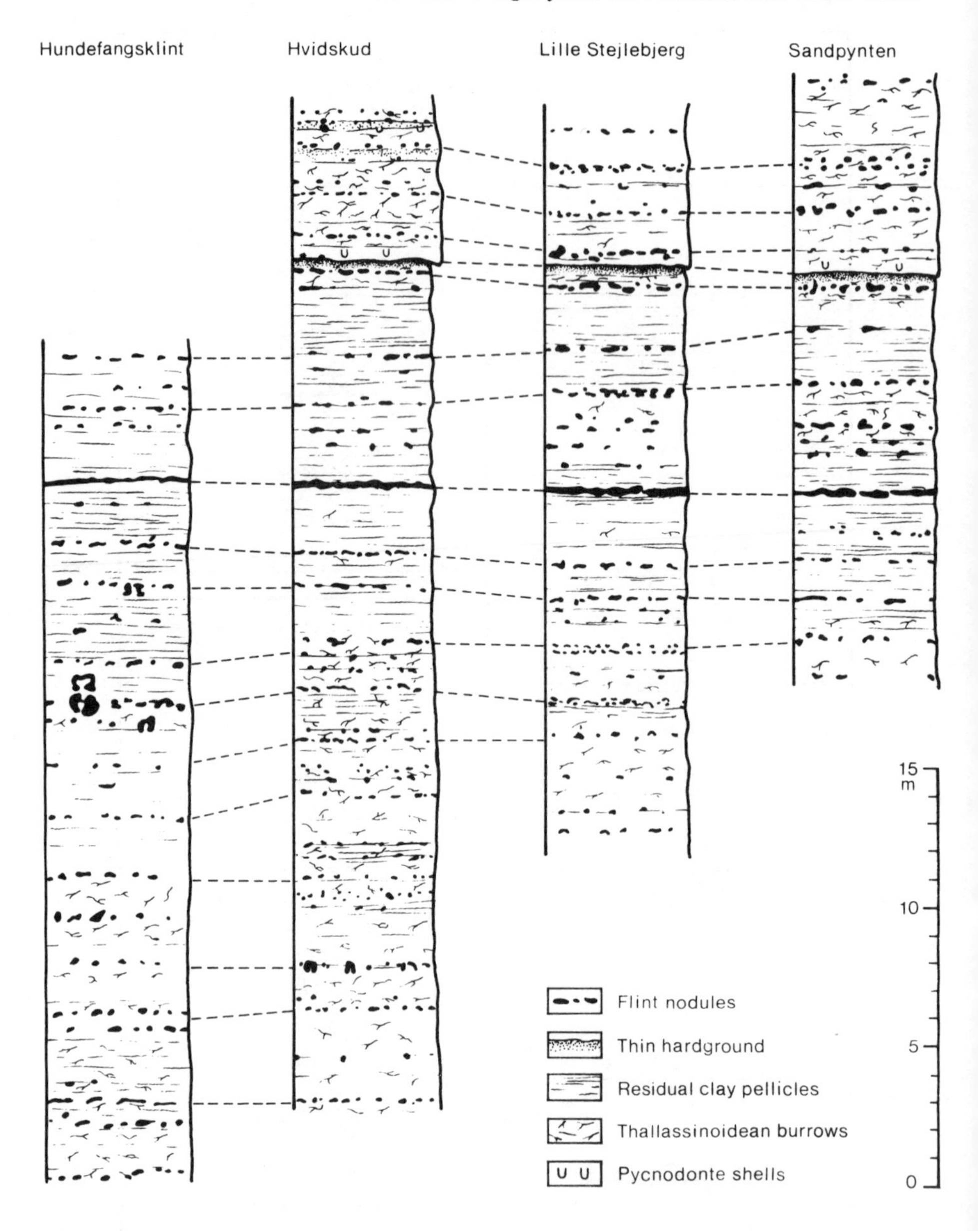

Text-Figure 7
Correlation of different sections from Lower Maastrichtian of Møns Klint, Denmark, on the basis of detailed lithology. Hvidskud section is used as a basis for the quantitative diagrams of Text-figures 5, 9, 10, and 11.

occurs throughout the section. This is a characteristic feature of most chalk localities and impedes biostratigraphical zonation of the whole Maastrichtian sequence.

Among the brachiopods, however, a few species have their first appearance in the Hvidskud section; this makes division into a number of biostratigraphical units possible. The division is not merely local, as the same sequential appearance of the two species in question is found in widely scattered localities within the basin (Rügen, East Germany; as well as in a boring in northern Denmark; Surlyk, 1970).

The ostracods of the Hvidskud section include two stratigraphically important forms, *Bythoceratina acanthoptera* and *B. umbonata.* The first is only found in this part of the Maastrichtian; the second disappears in the upper part of the Upper Maastrichtian. The quantitative variations in the ostracod fauna only rarely show significant trends, and the fluctuations seem to be much influenced by very local factors in the ostracod microhabitats. Nevertheless, two species, *Bythoceratina acanthoptera* and *Cytheropteron* (*A.*) *v-scriptum,* show pronounced positive correlation in the Hvidskud section. These species belong to two widely different genera but show some homeomorphism in that they both have alar prolongation (ostracod data provided by N. O. Jørgensen, 1974).

Although the foraminifers have proved to be most valuable in Maastrichtian biostratigraphy, they show the same type of distribution as the ostracods in the Hvidskud section; all the more common species are found throughout the sequence (Text-fig. 10). This is the case for the main part of the Lower Maastrichtian, whereas the Upper Maastrichtian can be divided into several zones on the basis of the foraminifers (Text-fig. 13) (Stenestad, 1971, and pers. comm., 1974). The relative abundance of some major groups is shown in Text-figure 10. The maximum in calcispheres ("*Oligostegina*") in sample 9 might suggest relatively shallow water conditions (e.g., Adams et al., 1967). The ratio between planktonic and benthic forms varies between 1 and 3, but no definite ecological or biostratigraphical conclusions can be drawn on these fluctuations without detailed investigation of more contemporaneous sections. The Textulariinae, comprising all the agglutinated foraminifers, occur in remarkably constant numbers. The relative variations of the more dominant families, Heterohelicidae, Planomalinidae, Rotaliporidae, and Globotruncanidae, show some significant trends. The open-marine Globotruncanidae show three pronounced maxima in the bottom and top of the section, interrupted by two minima. The maxima presumably correspond to open-marine conditions. The upper minimum corresponds well with a pronounced shallowing of the sea that is well demonstrated by changes in the lithology, which culminate in a thin hardground immediately above sample 4. It is further substantiated by the good negative correlation between the Globotruncanidae and the Heterohelicidae, of which the latter have been used as indicators of nearshore conditions (Eicher, 1969; Sliter, 1972; data on foraminifers were provided by E. Stenestad, 1974).

The coccoliths are somewhat more difficult to treat quantitatively than the above-mentioned groups (Text-fig. 11). Besides technical problems in sample processing and counting, some species are more susceptible to diagenetic solution than others; counts are therefore not always representative of the original composition of the coccolith flora. Some modifications of the floral composition happened even before deposition through preferential destruction while they settled. All the more common species occur throughout the section, as was the case with the ostracods and the foraminifers.

In spite of the above-mentioned reservations, it is most remarkable that some species (e.g., *Cribrosphaerella ehrenbergi*) show a very strong positive correlation with the benthic curve of Text-figure 5 (wash residue), whereas other species seem to show a negative correlation. These relations may be results of different temperature and depth adaptions of the individual coccolith species. *Braarudosphaera bigelowii*, normally considered a characteristic shallow-water form, is thus very sparse in the chalk environment and only found in small numbers in a few samples in the top of the section (data on coccoliths were provided by K. Perch-Nielsen, 1974).

The quantitative diagrams give a good idea of the actual biostratigraphical utility of the different fossil groups, and thus their different dependence on local and more regional environmental variations and their general time range. To illustrate the time range more schematically, some of the major organism groups found in the chalk have been plotted as numbers of species living in either all four, three, two, or only one substage of the Maastrichtian (Text-fig. 12). From this it emerges clearly that the classical macrofossil groups on which most of the Mesozoic stratigraphy traditionally have been based, ammonites and belemnites, comprise a larger number of short-ranging species than other groups here illustrated. Only within other mollusks, such as inoceramids (Kauffman, 1970), is a similar number of short-ranging species seen. New investigations on *Baculites* species of the Maastrichtian chalk have shown that the ranges of these are especially short, like those of the Western Interior of North America (Gill and Cobban, 1966; Kauffman, 1970). Among the macrobenthos shown in Text-figure 12, the brachiopods have a more promising histogram than most other groups, and they are indeed most useful in biostratigraphical zonation (Steinich, 1965; Surlyk, 1970). Among the echinoderms, only the echinoids have a large proportion of potential "guide fossils." This seems to confirm the general evolutionary pattern of Upper Cretaceous irregular echinoids as described by Ernst (1972, p. 150), who stated that in some trends the species are comparable to ammonite species in time range. However, part of the control may be ecological, as the benthic-rich bioherms at the top of the Maastrichtian form good biotopes for several species that are otherwise not found in the chalk.

Among microfossils, almost all the ostracods run through all four substages; those restricted to one or two are in fact very rare forms without stratigraphical value (N. O. Jørgensen, written communication, 1974). Foraminifers are much more useful as a tool in detailed zonation although a considerable number of the species are long ranging (Stenestad, written communication, 1974).

Text-Figure 8

Quantitative distribution of some important brachiopods in two different superimposed upper Lower Maastrichtian sections in Rügen, East Germany. (Redrawn and modified from Steinich, 1965.)

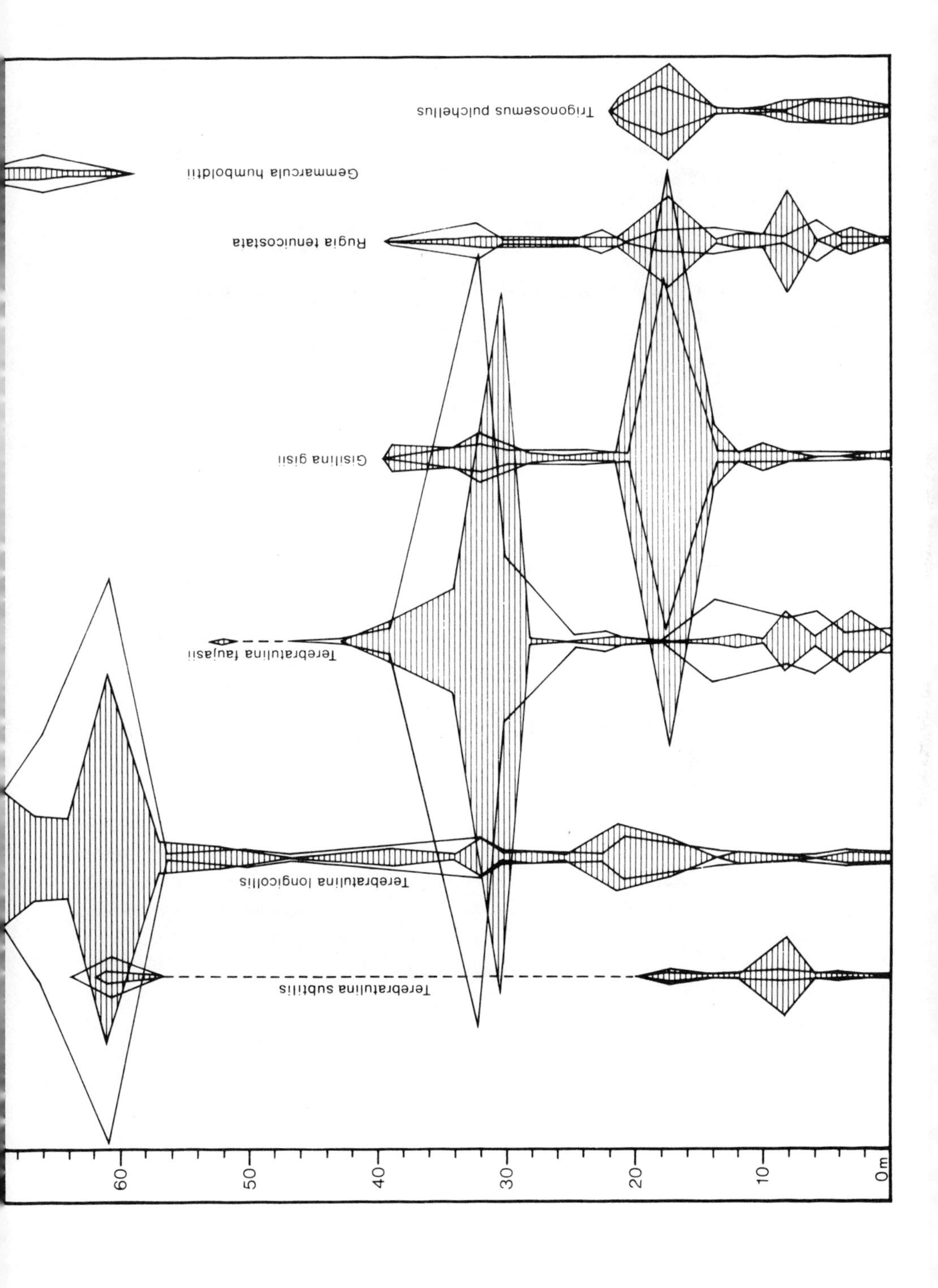

Trigonosemus pulchellus
Gemmarcula humboldtii
Rugia tenuicostata
Gisilina gisii
Terebratulina faujasii
Terebratulina longicollis
Terebratulina subtilis
60
50
40
30
20
10
0m

It is noteworthy that the coccoliths contain mainly long-ranging species and are of no importance for more detailed correlations, whereas dinoflagellates work extremely well (Wilson, 1971, and written communication, 1972; Kjellström, 1973).

It must be emphasized that the patterns described above are characteristic of the Maastrichtian chalk facies of northern Europe and cannot be generalized.

BIOSTRATIGRAPHICAL ZONATION

Quantitative investigation of the fauna shows that the successive assemblages are very much dependent on adaptive ranges of the animals and variations in the physical environment. On this basis it is possible to establish assemblage zones (sensu Hedberg, 1971), but these assemblage zones represent, almost by definition, the spatial and temporal distribution of fossil communities. Therefore, periodically recurrent near-identical environmental relations lead to repetition of the same assemblage zones. As a consequence, assemblage zones are of limited or no time-stratigraphical value.

Thus, in the present study the repeated maxima in the total benthic fauna (notably bryozoans) lead to analogous or almost identical fossil communities in different stratigraphical horizons. Consequently, assemblage zones have not proved useful in the biostratigraphical zonation of the Maastrichtian chalk. In the present study the zonation

> is based on a carefully considered selection (and rejection) of faunal elements, of in part concurrent though not necessarily identical range, with the objective of achieving a biostratigraphical unit which will have optimum time significance and extensibility (Hedberg, 1971, p. 16).

The boundaries of each zone are defined by appearance and, less important, by disappearance of one or two species; but the zone is further characterized by the presence of several other species, which are restricted to the zone in question plus one or two neighboring zones. The species are chosen according to their value in time stratigraphy, evaluated by the aid of continuous control of the relative stratigraphical distribution of the species in other areas.

The stratigraphical distribution of the selected species reflects in one way or another their evolution. Phyletic evolution in the Upper Cretaceous is conspicuous in some planktonic foraminifers (Stenestad, 1969), belemnites (Ernst, 1964), some ammonites (Birkelund, 1966), and some brachiopods (Surlyk, 1973); the pattern of other species seems to reflect allopatric speciation. In accordance with Eldredge and Gould's (1972) hypothesis of punctuated equilibria, these latter species show no integradation with closely related species in spite of the continuous sedimentation.

Text-Figure 9
Quantitative diagram of a mid-Lower Maastrichtian ostracod fauna from the Hvidskud section at Møns Klint. (All data kindly provided by N. O. Jørgensen.)

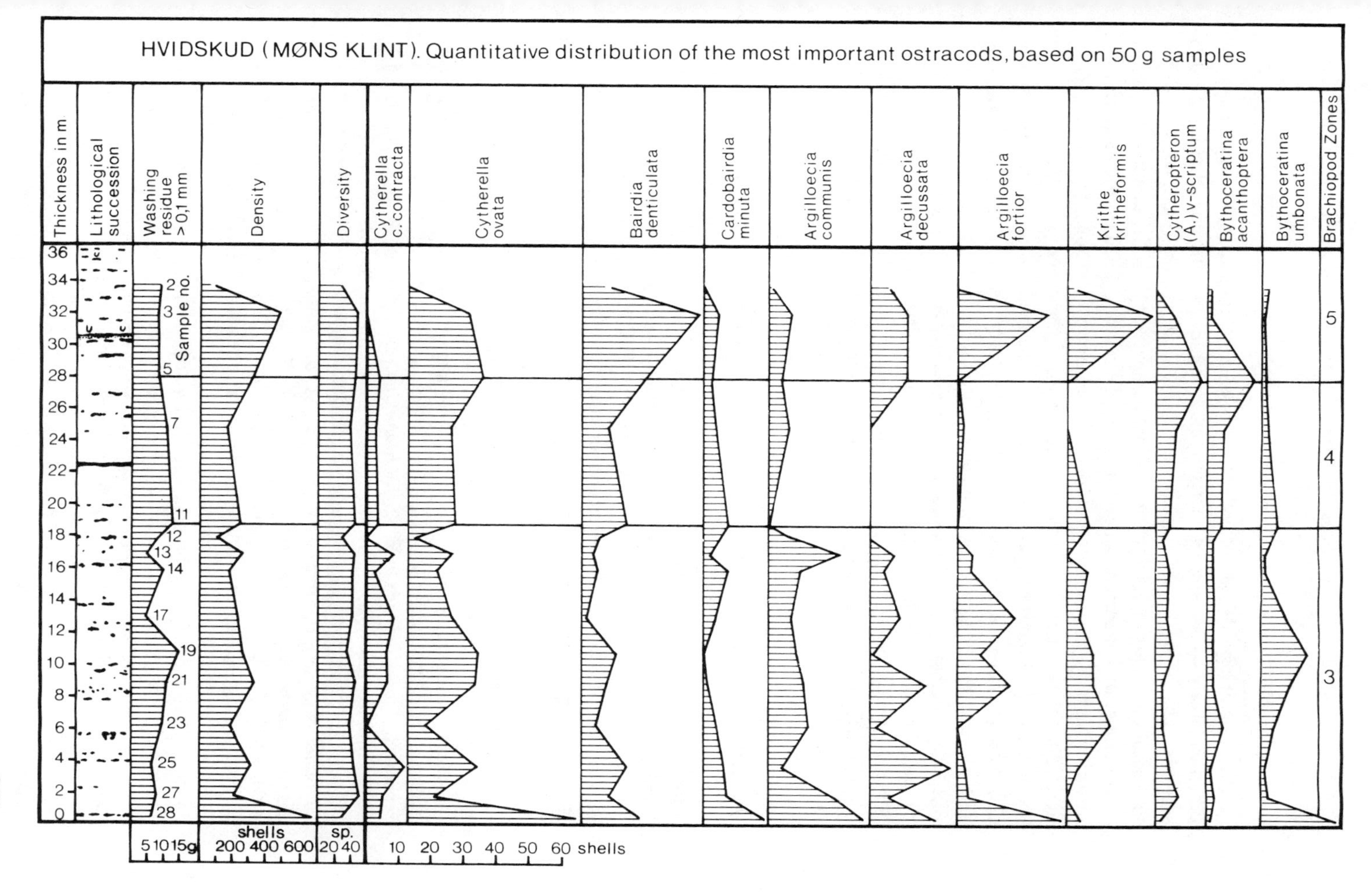
HVIDSKUD (MØNS KLINT). Quantitative distribution of the most important ostracods, based on 50 g samples
Thickness in m
Lithological succession
Washing residue >0,1 mm
Density
Diversity
Cytherella c. contracta
Cytherella ovata
Bairdia denticulata
Cardobairdia minuta
Argilloecia communis
Argilloecia decussata
Argilloecia fortior
Krithe kritheformis
Cytheropteron (A.) v-scriptum
Bythoceratina acanthoptera
Bythoceratina umbonata
Brachiopod Zones
36 34 32 30 28 26 24 22 20 18 16 14 12 10 8 6 4 2 0
Sample no.
2 3 5 7 11 12 13 14 17 19 21 23 25 27 28
5
4
3
5 10 15 g
shells 200 400 600
sp. 20 40
10 20 30 40 50 60 shells

CONCLUSIONS

Rich assemblages of nannofossils, microfossils, and minute macrofossils have been obtained largely by washing of a large number of bulk samples collected in measured sections. The relatively rare, larger macrofossils have been collected, where possible, in the same sections so as to enable a direct comparison with the sample material.

A number of concurrent-range zones have been constructed on the basis of quantitative range diagrams of each major fossil group. The zonal boundaries are defined by appearance or, of less importance, disappearance of one or more common species, and the zone is further characterized by the presence of several other species, which are restricted to that zone plus one or two neighboring zones (Text-fig. 13). The construction of a biostratigraphical zonation for each fossil group can be considered as the first step in the study. The second step involves a comparison and compilation of all the individual zonal schemes, resulting in a composite system of concurrent range zones. This procedure is mainly a reflection of the taxonomic specialization of the biostratigraphers. It would be more logical, although in practice impossible, to produce a single zonal scheme comprising all major fossil groups. This procedure would avoid the initial construction of a zonal scheme based on single fossil groups, which is for logical reasons unnecessary. This drawback is counterbalanced, however, by the use of the same sample series by all workers.

The resulting zonal system has a number of advantages: (1) It is possible to correlate between basinal facies and marginal facies, even though several important fossil groups are restricted to a certain facies (compare groups on Text-fig. 2), and from basin to basin and in some cases between different faunal provinces or realms. Among the fossil groups the ostracods, bryozoans, brachiopods, and echinoderms

---

Text-Figure 10

Quantitative distribution of the foraminifers in the mid-Lower Maastrichtian Hvidskud section at Møns Klint. Right part shows ranges of the biostratigraphically most important species. Majority of the species range throughout the section; many of the more short ranging forms are simply rare forms with only occassional occurrences. 1, *Globotruncana arca.* 2, *Neoflabellina reticulata.* 3, *Eouvigerina gracilis.* 4, *Bolivinoides australis.* 5, *Heterostomella foveolata.* 6, *Angulogavelinella bettenstaedti.* 7, *Pyramidina pseudospinulosa.* 8, *Discopulvinulina* cf. *binkhorsti.* 9, *Eponides beisseli.* 10, *Spirillina subornata.* 11, *Bolivinoides miliaris.* 12, *Heterohelix* cf. *complanata.* 13, *Cibicides complanata.* 14, *Melonis nobilis.* 15, *Cibicides beaumontianus.* 16, *Marssonella oxycona.* 17, *Ataxophragmium variabilis.* 18, *Rugoglobigerina rugosa.* 19, *Bolivina incrassata.* 20, *Cibicides bembix.* 21, *Globigerinelloides multispina.* 22, *Heterohelix striata.* 23, *Osangularia lens.* 24, *Prebulimina laevis.* 25, *Stensioeina pommerana.* 26, *Gavelinella pertusa.* 27, *Gyroidinoides nitida.* 28, *Bolivinopsis suturalis.* 29, *Pseudouvigerina* cf. *rugosa.* 30, *Pullenia americana.* 31, *Bolivinoides decoratus.* 32, *Gaudryina cretacea.* 33, *Praeglobotruncana havanensis.* 34, *Bolivinoides paleocenicus.* 35, *Pseudouvigerina cristata.* 36, *Cibicides plana.* 37, *Rugoglobigerina macrocephala.* 38, *Pullenia cretacea.* 39, *Eouvigerina cretae.* 40, *Heterohelix dentata.* 41, *Gavelinella danica.* 42, *Bolivinoides draco.* 43, *Bolivina decurrens.* (All data kindly provided by E. Stenestad.)

HVIDSKUD (MØNS KLINT). Quantitative distribution and ranges of the most important foraminifers

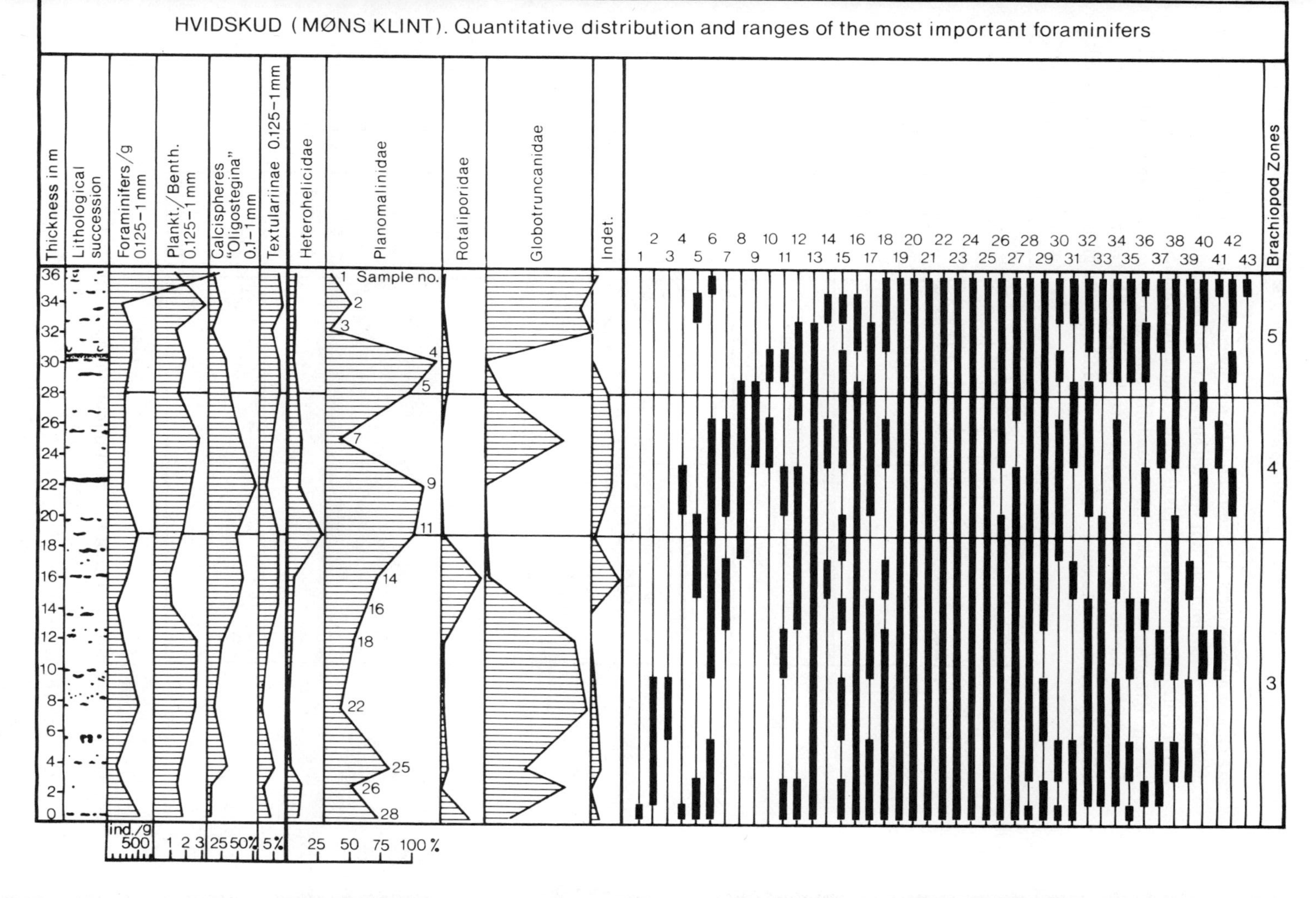

are almost all provincial and restricted to northwest Europe. Some ammonite species have a wider distribution, but also within this group the majority of the species are provincial. The inoceramids, planktonic foraminifers, and coccoliths are spread over large geographical areas. Benthic foraminifers are also surprisingly widespread. (2) The working method outlined here permits the use in biostratigraphy of what are normally considered to be strongly facies dependent fossil groups (e.g., brachiopods)

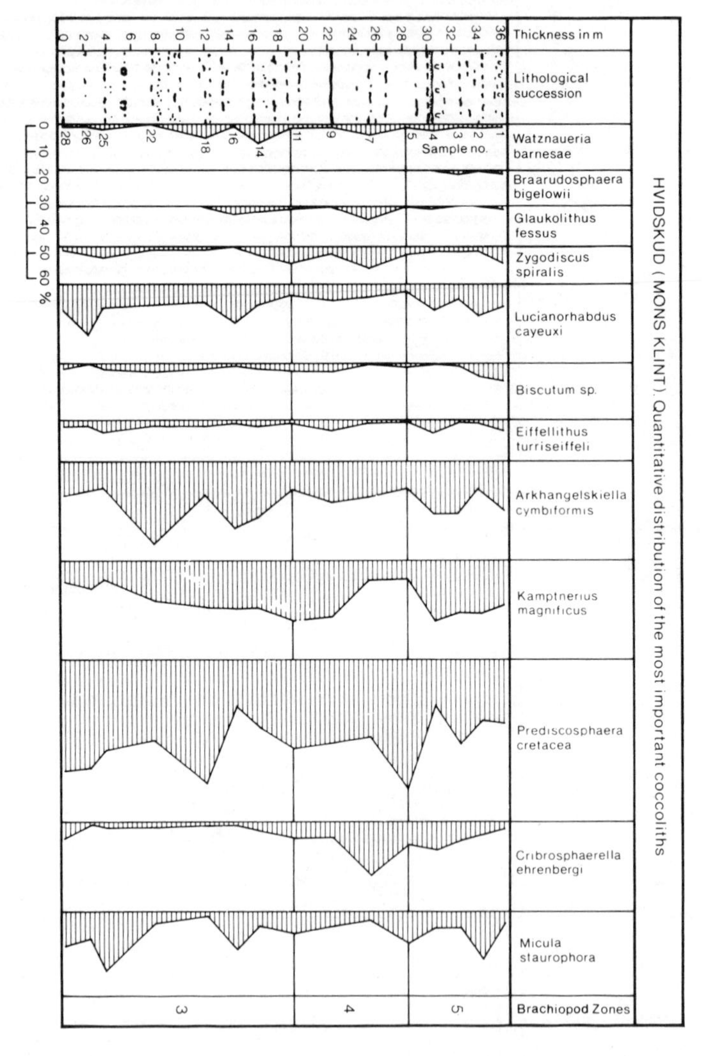

Text-Figure 11

Quantitative distribution of the most important coccoliths in the mid-Lower Maastrichtian Hvidskud section at Møns Klint. (All data kindly provided by K. Perch-Nielsen.)

by continuous control through cross-checking with groups less dependent of facies (e.g., planktonic foraminifers). (3) Classic cephalopod stratigraphy can be correlated with nanno- and microfossil stratigraphy, which is often based on cuttings from well borings. This correlation will aid in avoiding the common problem presented by several separate and uncorrelated zonal systems a problem that invariably results in endless discussions on zone and stage boundaries.

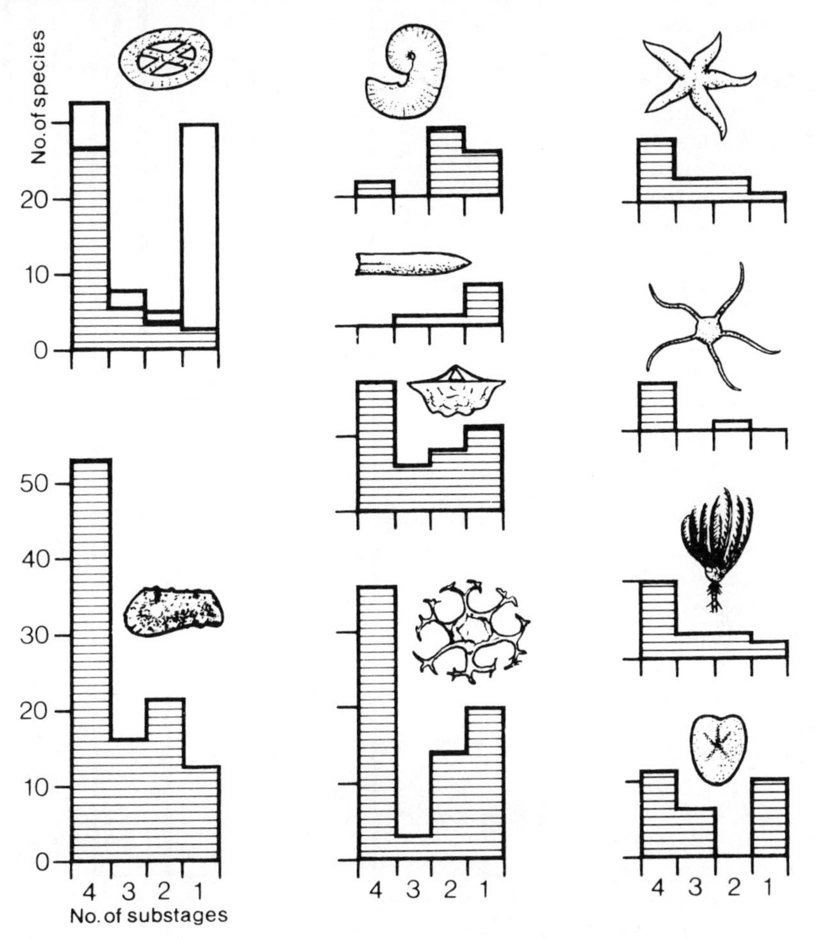

Text-Figure 12
Histograms compiled for the major fossil groups in the Maastrichtian chalk of Denmark showing the number of species found in all four, three, two, or only one of the substages. Coccoliths normally comprise long-ranging forms, but in the Dania locality (uppermost Maastrichtian) a large number of species occur, which are otherwise unknown from the Maastrichtian (unpatterned). (Sources: Rasmussen, 1950, 1961; S. B. Andersen, written communication; Surlyk, 1969; Jørgensen, 1970; G. J. Wilson, written communication; Perch-Nielsen, 1968, and written communication.)

STANDARD SUCCESSION | CHRONOSTRATIGRAPHY | BELEMNITE ZONES | COCCOLITH ZONES

MAASTRICHTIAN
Upper
Lower
UPPER
LOWER
UPPER
LOWER
B. casimirovens.
Bt. junior
B. occidentalis
B. lanceolata
Nephrolithus frequens
Arkhangelskiella cymbiformis
3
4
2
1
6
5

Text-Figure 13 *(above and right)* Zonal scheme of the Danish and German Maastrichtian. 1, *Belemnella lanceolata.* 2, *Belemnella occidentalis.* 3, *Belemnitella junior.* 4, *Belemnella casimirovensis.* 5, *Arkhangelskiella cymbiformis.* 6, *Nephrolithus frequens.* 7, *Rugia spinosa.* 8, *Rugia acutirostris.* 9, *Terebratulina subtilis.* 10, *Trigonosemus pulchellus.* 11, *Rugia tenuicostata.* 12, *Gisilina jasmundi.* 13, *Meonia semiglobularis.* 14, *Gemmarcula humboldtii.* 15, *Argyrotheca stevensis.* 16, *Thecidea recurvirostra.* 17, *Bolivinoides draco miliaris.* 18, *Bolivinoides decorata laevigata.* 19, *Heterohelix dentata.* 20, *Pseudouvigerina cimbrica.* 21, *Pseudouvigerina rugosa.* 22, *Pseudotextularia elegans.* 23, *Stensioeina esnehensis.* 24, *Hoploscaphites constrictus.* 25, *Baculites knorrianus.* 26, *Acanthoscaphites tridens varians.* 27, *Pachydiscus neubergicus.* 28, *Gaudryceras lueneburgense.* 29, *Hoploscaphites tenuistriatus.* 30, *Saghalinites* sp. aff. *wrighti.* 31, *Baculites* sp. 1. 32, *Neophylloceras velledaeforme.* 33, *Baculites* sp. 2. 34, *Saghalinites* n. sp. 35, *Pachydiscus* sp. aff. *colligatus.* 36, *Baculites valognensis.* 37, *Hoploscaphites crassus.* (Sources: Surlyk, 1970, 1972; T. Birkelund, in prep.; K. Perch-Nielsen, pers. comms.; E. Stenestad, pers. comms.)

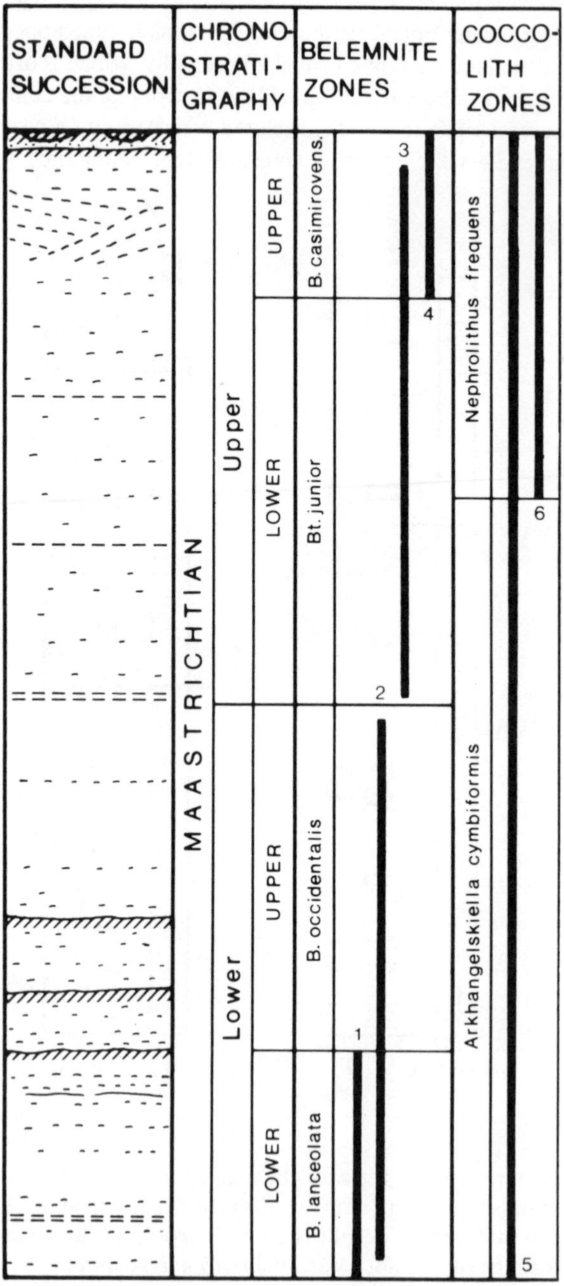

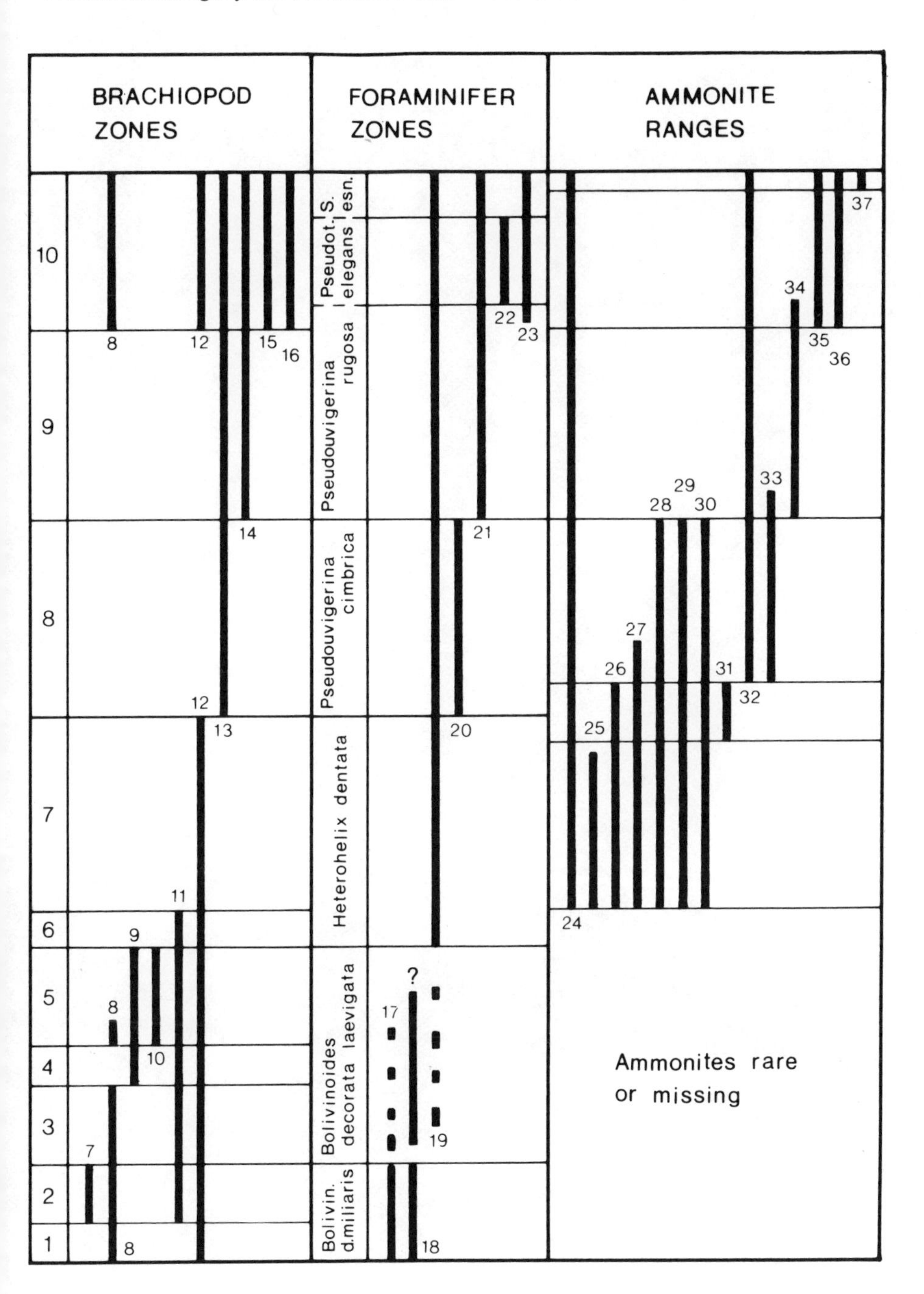
BRACHIOPOD ZONES
FORAMINIFER ZONES
AMMONITE RANGES
10
9
8
7
6
5
4
3
2
1
Pseudot. S. elegans esn.
Pseudouvigerina rugosa
Pseudouvigerina cimbrica
Heterohelix dentata
Bolivinoides decorata laevigata
Bolivin. d.miliaris
Ammonites rare or missing
8
12
15
16
14
12
13
11
9
8
10
7
8
22
23
21
20
17
?
19
18
37
34
35
36
33
29
28
30
27
26
31
32
25
24

Biostratigraphical Basis of the Neogene Time Scale

John A. Van Couvering — **University of Colorado Museum, Boulder**

W. A. Berggren — **Woods Hole Oceanographic Institution**

INTRODUCTION

Interpretations of earth history depend on two different systems of logic, both of which arrange geologic observations into sequences of events. The first and most widely used is the logic of superposition: the ordering of events *iteratively* in a system of invariant properties simply by determining the physical relationship of features in the rocks. This is what is meant by the word *stratigraphy*. The second logical system depends on the recognition of an *ordinal* progression, which links a series of events in a system of irreversibly varying properties. This provides a theoretical basis outside of the preserved geologic record by which the nature and relationship of the events in the progression can be recognized or predicted, and according to which missing parts of the record can be identified. Geology is a historical philosophy, so the ordinal progressions we refer to are progressions in time, just as geologic time is perceived by the progress in one or another ordinal series of events. This is what is meant by the word *geochronology*. In the pages to follow, we discuss the relationship of geochronological systems based on biological evolution (*biochronology*) and on decay rates of unstable isotopes (*radiochronology*) to the stratigraphic record of fossils (*biostratigraphy*) and geomagnetic polarity reversals (*magnetostratigraphy*) during the Neogene. (It should be noted at once that, as yet, no accepted theory specifies the existence and duration of each individual geomagnetic polarity interval, so that a true *magnetochronology* does not exist. It will be shown that the calibrations of magnetostratigraphy upon which pre-Pliocene "paleomagnetic time scales" depend are very imprecise.)

Evidence from paleontology is relevant both to stratigraphy and geochronology. By noting the fossils that marked different sedimentary units and the way that the units succeeded one another, William Smith was able to cogently practice biostratigraphy. It is not important whether or not he speculated about the cause of morphological changes in the fossil shells and tests from one stratum to the next; from a practical point of view, he did not need any explanation. On the other hand, many theories about the process leading to variations in the fossil record were proposed before Darwin and Wallace, years after Smith, could gather together enough information to formulate a useful description of evolution. It is now common practice for biostratigraphical arguments to include some appeal to adaptive selection or other evolutionary principles, but it remains immaterial whether *Bolivina* "B" is descended from *Bolivina* "A" or not for a local biostratigraphy based on key taxa such as these to be valid, and this is probably very fortunate.

Nevertheless, litho-, bio-, and magnetofacies in the rocks are physically discontinuous and are subject to iterative confusion. This means that *most long-distance correlations are geochronological in substance.* Because fossils are more abundant than datable horizons in Phanerozoic sediments, and biological events can be correlated in time more precisely than radiometric dates in all but the youngest levels, long-distance correlations are primarily biochronological in substance. The *biochron*, or unit of geological time measured biochronologically (Tedford, 1970), may take its name and definition from a biostratigraphical stratotype, from a type fauna, or from just an idea—for instance, the Neogene, which is simply the composite of the Miocene, Pliocene, and Pleistocene biochrons (Hoernes, 1856). Biochronological correlations are based, in effect, on three procedures: (1) recognizing the most widespread and distinctive events in biological history, for the most part the first appearance datum (FAD) or last appearance datum (LAD) of key taxa (Van Couvering et al., 1975); (2) locating such events in local biostratigraphies and evaluating their age with respect to as many reinforcing criteria as possible; and (3) stratigraphically relating such events to evidence for other biochronological datum events and to radiometrically dated or calibrated levels, such as a tuff bed or a paleomagnetic boundary.

As observations such as these are synthesized, a time scale of dated events can be built up. But no matter how plausible and internally self-consistent it may be, a geochronological history is still subject to stratigraphic cross-examination. Thus the Neogene, as a biochron, is framed in the biostratigraphy of the Mediterranean region, and it is to *this* biostratigraphy that information from stratigraphic sequences of Neogene age, in the deep sea or in other lands, must be conveyed if such information is to have more than provincial significance. The means of conveyance, as we have pointed out, is geochronology in the form of the Neogene time scale.

Our examples in this paper are drawn from Early Neogene (Miocene) correlations, in part because we see no need to recapitulate our recently published discussions of Late Neogene history (Berggren and Van Couvering, 1974), but mainly because much new information has become available from mid-Cenozoic strata in Europe and elsewhere, which required assimilation into the time scale (cf. Berggren, 1972a). These observations and their correlation suggest new interpretations of Mediterranean Neogene stratigraphy and thus reflect on the definition of the Neogene portion of earth history.

The first part of this paper, therefore, is devoted to a summary of Late Oligocene and Miocene interhemispherical correlations based on calcareous and siliceous planktonic microfossil biochronology and the calibration of this biochronology. The information is synthesized in Text-figure 1, but a complete review of all the pertinent details is beyond the scope of this paper (for more extensive treatment, see Berggren, 1972a, and Berggren and Van Couvering, 1974).

NEOGENE TIME SCALE IN THE DEEP SEA

Deep-sea cores offer excellent biostratigraphic and magnetostratigraphic records, and the results of correlations to other deep-sea cores through microfossil biochronology and models of paleomagnetic polarity sequences are generally very consistent. Deep-sea biochronology has also been linked to Mediterranean Neogene biostratigraphy with very important consequences (e.g., Blow, 1969; Cita and Blow, 1969; Martini, 1971; Van Couvering, 1972; Berggren and Van Couvering, 1974), and has greatly changed the point of view of many European biostratigraphers. It is important, for this reason, to understand the basis of year-ages, which have been used to calibrate the deep-sea record, because these "dates" are widely referred to in discussions of the Neogene time scale.

Deep-sea cores of strata deposited during the last 5 Ma (mega-annum, or years $\times 10^6$) are correlated to the land-based K-Ar calibration of the paleomagnetic polarity intervals without much ambiguity (e.g., Hays et al., 1969; Gartner, 1969, 1973; Opdyke, 1972; Berggren, 1973). The biochronological correlation of Late Neogene circumglobal marine-microfossil datum events to exposed and K-Ar calibrated biostratigraphic sequences is in close agreement with the magnetostratigraphic calibration; thus the time scale of this period appears to us to have a high degree of accuracy and reliability.

For the period older than 5 Ma to the base of the Neogene (and beyond), the deep-sea record is not nearly so well calibrated. Mainly, this is because the K-Ar method loses its resolution with increasing age, so that pre-Pliocene geomagnetic polarity events are increasingly "out of focus" to radiochronology (Cox and Dalrymple, 1967; Dalrymple, 1972; Opdyke, 1972). Thus one may observe and measure the stratigraphic sequence of geomagnetic reversals expressed in paleomagnetic polarity intervals in cores or spread out in bands on the sea floor, but one cannot validly determine the radiometric age difference between adjoining paleomagnetic facies, nor conclusively identify a given polarity interval by any practical number of radiometric age determinations. However, each event in an *ordinal* sequence carries its own identification, so K-Ar ages in Miocene strata can be significant when they are applied to biochronology. Some few K-Ar ages have in fact been related to Miocene planktonic microfossil history (i.e., the FAD of *Orbulina* and of *Globigerinoides*; see below), but most of the year-ages now applied to deep-sea cores of this age have been obtained by one of two roundabout ways: either by extrapolating a local rate of sea-floor spreading, or by extrapolating a local rate of deep-sea sedimentation.

SEA-FLOOR SPREADING TIME SCALE

In this method of year dating Miocene deep-sea cores, the K-Ar calibration of Plio-Pleistocene magnetostratigraphy is used to calculate the rate of formation of sea-floor magnetic anomalies during the last 5 million years, and this "spreading rate" is then extrapolated into older parts of the sea floor. According to this rate, theoretical ages for the boundaries of pre-Pliocene sea-floor magnetic anomalies are computed as a direct linear function of their distance normal to the spreading axes (Heirtzler et al., 1968). However, variations in the relative widths of anomaly bands in different parts of the world indicate that spreading rates are locally inconsistent; the arbitrary decision by Heirtzler et al. (1968) to designate a single South Atlantic magnetic profile as the invariant standard results in incongruous and apparently unreal accelerations in calculated plate motions elsewhere. For this reason, Blakely (1974) has proposed a revised calibration of Miocene sea-floor spreading. Two of the most easily distinguished peaks in the Miocene sections of sea-floor magnetic profiles are designated as anomaly 5 and anomaly 6. Both are intervals with predominantly (if not exclusively) normal polarity. The calculated age limits of these anomalies are as follows: (1) anomaly 5: 8.71 to 9.94 Ma; anomaly 6: 20.19 to 21.31 Ma (Heirtzler et al., 1968, Text-fig. 3; Talwani et al., 1971, Text-fig. 11); (2) anomaly 5: 8.71 to 10.21 Ma; anomaly 6: 20.11 to 21.31 Ma (Blakely, 1974, Table 1 and Text-fig. 8). Ages of the numerous intervening Miocene anomaly boundaries are shifted radically younger, for the most part, in the Blakely (1974) calibration. It should be noted that the number of significant places quoted in these anomaly "ages" reflects the precision with which the stratigraphy can be measured and not the degree of certainty in the chronology.

Whatever its calibration, to apply a "spreading-rate time scale" to cores it is necessary to match up the pattern of sea-floor anomalies to the magnetostratigraphy in the cores, like matching a tree-ring section to a drilled plug. Although the sea-floor anomaly pattern has recently been enhanced by "stacking" reinforcement techniques (Blakely, 1974; Parker, 1974), it remains very difficult to correlate it unambiguously to the core polarity intervals. This is evidently because the quantities involved are stratigraphic measurements in different systems and are *not* the geomagnetic polarity intervals themselves.

In attempting to match the sea-floor magnetic anomaly pattern to Late Miocene paleomagnetic epochs 6 through 11 in deep-sea cores, Foster and Opdyke (1970) and Opdyke (1972) relied heavily on a correlation between anomaly 5 and epoch 11. On the other hand, M. Dreyfus and W. B. F. Ryan, in a verbal presentation at the International Geological Congress, Montreal, 1972, correlated anomaly 5 to epoch 9, the next youngest period of mostly normal polarity seen in cores. At the level of epoch 9 and older, this calibration differs by 2 Ma from that of Foster and Opdyke (1970), as shown in Table 1.

The Dreyfus and Ryan assumption is supported by paleontological arguments. Deep Sea Drilling Project (DSDP) site 16, on the western side of the mid-Atlantic Ridge, is located over a segment of anomaly 5. The oldest part of this core, a few meters above basement, contains calcareous nannofossils and planktonic foraminifera, which indicate a mid-Tortonian age. The *base* of the Tortonian is equivalent to a level within zone N.15 of the planktonic foraminiferal zonation (Blow, 1969; Cita and Blow, 1969) and is also with the upper part of calcareous nannoplankton zone NN.9 (Martini, 1971). Studies of Equatorial Pacific cores by Burckle (1972, p. 223, and pers. comm. to W. A. B.) and by Saito (in Burckle, 1972) indicate that both N.15 and NN.9 are younger than the lower part of epoch 11. Since zone NN.9 extends down into zone N.14 levels (Bronnimann et al., 1971), this suggests to us that the base Tortonian, in zone N.15, must be significantly younger than medial epoch 11. In the Foster and Opdyke (1970) calibration this would have to mean that the base Tortonian was younger than the central age of anomaly 5 (ca. 9.4 Ma); but this is unlikely, because Middle Tortonian sediments rest directly on rocks formed during anomaly 5, which strongly suggests that the base Tortonian should be older than 10.0 Ma. This agrees with calibration evidence from Mediterranean Neogene biostratigraphy (the Foster and Opdyke date does not); in view of the evidence cited, we would tentatively locate the base Tortonian close to the beginning of epoch 10 at ca. 10.6 Ma (Table 1).

SEDIMENTATION-RATE TIME SCALE

General

Recently, papers by Opdyke et al. (1974) and Theyer and Hammond (1974) presented calibrations of the paleomagnetic polarity sequence in Central Pacific deep-sea sediments that extended into Early Miocene time. In both papers the calibration was achieved by interpolation between presumably well established calibration ages and with the assumption that the magnetostratigraphy was recorded in sediments accumulating at relatively constant rates. Opdyke et al. (1974) compared the ages suggested by this method to ages presented by Berggren (1972a) for planktonic foraminiferal datum events with good agreement over the interval from epochs 5 to 15 (5 to 15 Ma). Theyer and Hammond (1974) presented a calibration of the magnetostratigraphy in radiolarian sequences which was almost identical to that of Opdyke et al. (1974) down to epoch 15, but their study suggested slightly older boundaries down to a level close to the base of the Miocene in epoch 20. Both

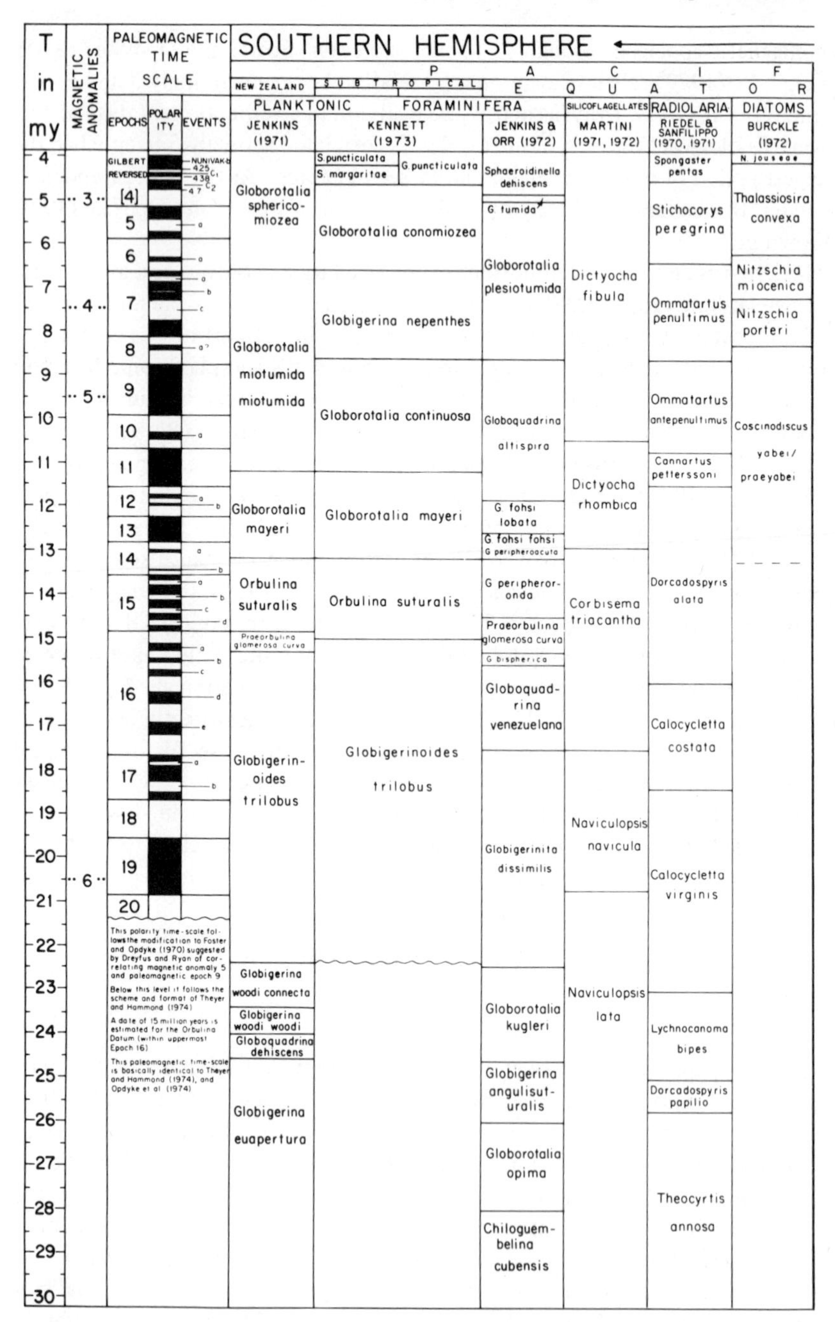

Text-Figure 1 (*above and right*)
Interhemispherical correlation of Miocene planktonic zones and calibration to the paleomagnetic time scale (see text for further explanation).

NORTHERN HEMISPHERE

Region	Group	Authority	Zones (top to bottom, 4–30 my)
ITALIC	PLANKT. FORAM	BLOW (1969)	N19; N18; N17; N16; N 15; N 14; N 13; N 12; N 11; N 10; N9; N8; N 7; N6; N5; N4; P22; P21 b; P21 a; P 19/20
ITALIC	CALCAR. NANNOPL.	MARTINI & WORSLEY (1970)	NN13 / NN14; NN12; NN 11; NN 10; NN 9; NN 8; NN 7; NN 6; NN 5; NN4; NN3; NN2; NN1; NP25; NP24; NP23
N.E.	DIATOMS	SCHRADER (1973 a & b)	NPD 10; NPD 11; NPD 12; NPD 13; NPD 16; NPD 17; NPD 18; NPD 19; NPD 20; NPD 21 ↕ NPD 25
JAPAN	DIATOMS	KOIZUMI (1973)	Denticula seminae-D. kamtshatica; Denticula kamtshatica; Denticula hustedti; Denticula hustedti-Denticula lauta; Denticula lauta
ATLANTIC, NORTH EQUATORIAL	PLANKTONIC FORAMINIFERA	BERGGREN (1972 b)	Globorotalia puncticulata; Globorotalia conoidea; Globorotalia miozea; G. praemenardii; Orbulina; Praeorbulina; Globoquadrina dehiscens; Globoquadrina praedehiscens; Globoquadrina baroemoenensis; Globigerinita dissimilis and G. unicava / Globorotalia munda
ATLANTIC, NORTH EQUATORIAL	PLANKTONIC FORAMINIFERA	BLOW (1969)	N19; N18; N17; N16; N15; N 14; N 13; N 12; N 11; N 10; N 9; N 8; N 7; N6; N 5; N4; P22; P21 b; P21 a; P19/20
ATLANTIC, NORTH EQUATORIAL	CALCAR. NANNOPL.	MARTINI & WORSLEY (1970)	NN13 / NN14; NN 12; NN 11; NN 10; NN 9; NN 8; NN 7; NN 6; NN 5; NN4; NN 3; NN 2; NN 1; NP25; NP24; NP23

Time scale: T in my, 4–30.

EPOCH	SERIES	
PLIOCENE	E	NEOGENE
MIOCENE	LATE	NEOGENE
MIOCENE	MIDDLE	NEOGENE
MIOCENE	EARLY	NEOGENE
OLIGOCENE	LATE	PALEOGENE

Table 1
Calibration of Late Miocene Paleomagnetic Epochs by Correlation to Heirtzler et al. (1968) and Talwani et al. (1971) Sea-Floor Anomaly "Time Scale"

Beginning Paleomagnetic Epoch	*Calibration Age (Ma)*	
	Dreyfus and Ryan	*Foster and Opdyke*
6(–)	6.6	6.6
7(+)	8.1	7.2
8(–)	8.7	7.7
9(+)	10.0[a]	8.1
10(–)	10.6	8.7
11(+)	–	10.0[a]

[a]Base of anomaly 5 is 9.94 Ma in Heirtzler et al., 1968, or 10.21 Ma in Blakely, 1974.

studies adopted the Dreyfus and Ryan correlation of anomaly 5 to epoch 9, and Theyer and Hammond (1974) also matched up epoch 19 to anomaly 6. Epoch 19 falls within the upper *Calocycletta virginis* radiolarian zone, which Riedel and Sanfilippo (1970, Text-fig. 2) show extending just above the base of the Miocene and corresponding to a somewhat vaguely correlated interval in the planktonic foraminiferal zonation centered on zones N.4 and N.5, but which Theyer and Hammond (1974) confirm as extending well up into zone N.6. Theyer and Hammond predict age limits of 19.6 to 20.8 Ma for epoch 19, and thus for anomaly 6, in preference to the spreading-rate "age" limits of Heirtzler et al. (1968) or Blakely (1974), which place the upper boundary of anomaly 6 about 1 Ma older. Land-based dating suggests an age for zone N.6 between the limits of 19 to 17.5 Ma, so an age near 20 Ma for the center of epoch 19 in the upper part of the *Calocycletta virginis* zone is in reasonably close agreement; the spreading-rate "age" for anomaly 6 seems a bit too old. Of course, the correlation of epoch 19 with anomaly 6 may be incorrect, but it should be noted that Blakely's decision (1974) to make the age for the base of anomaly 6 the same as that given by Heirtzler et al. (1968) was arbitrary, for the purpose of evaluating the differences in apparent spreading rate *between* the top of anomaly 5 and the base of anomaly 6. In this paper we have taken the Theyer and Hammond (1974) calibration of the radiolarian sequence as the standard of reference for deep-sea biochronology in the Miocene. We have already discussed some details of correlation between Miocene radiolarian and calcareous microplankton zones in the text and appendix of Berggren and Van Couvering (1974). The main points can be summarized as follows:

1. As noted above, the base of the Tortonian biochron appears to correspond in time to the beginning of epoch 10 (ca. 10.6 Ma). Theyer and Hammond (1974) put this level at the base of the *Ommatatrus antepenultimus* zone, and date it at 10.7 Ma.

2. The FAD of *Globigerina nepenthes*, at the base of zone N.14 and close to the base of zone NN.8, is within the upper part of the *Dorcadospyris alata* zone. It therefore lies within the upper part of epoch 12 (Theyer and Hammond, 1974), which has extrapolated age limits of 11.5 to 12.2 Ma based on sedimentation rates. (Correlation to the sea-floor model of geomagnetic polarity reversals at this level is completely ambiguous; see Blakely, 1974, Text-fig. 8). We estimate the age of this important datum at 11.7 Ma, in part because of the evidence in item 3 below.

3. The zone N.12-N.13 boundary, which lies within the *Discoaster kugleri* nannozone NN.7, is correlated to the macroforaminiferal zone boundary Tf1-2-Tf3 (Blow, 1969) and is dated in New Guinea at 12 to 12.5 Ma by Page and McDougall (1970). It is also correlated to sediments cored in the experimental Mohole, which were dated 11.4 ± 0.6 and 12.3 ± 0.4 by Dymond (1966). L. Burckle (pers. comm. to W. A. B.; see also Opdyke et al., 1974) has determined that this level in the Mohole corresponds to a time close to the epoch 11-epoch 12 boundary based on diatom biochronology. Thus the base of zone N.12 and also the beginning of epoch 12 can be placed at 12.2 Ma, in accordance with the Theyer and Hammond (1974) calibration and the K-Ar dating and correlation noted above.

Geochronology of the *Orbulina* Datum

In tropical microfaunas, the first appearance of *Orbulina suturalis is* a widely noted and important event. Like the fable of the elephant and the blind men, however, a heated controversy has been generated accordingly about the age of this event as observed in different biostratigraphic sequences. Recently, a general consensus has been reached in the biochronology that places the *Orbulina* datum at the base of zone N.9 (Blow, 1969) and in the middle part of the Langhian biochron according to studies of the stratotype fauna (Cita and Blow, 1969). A crucial element in this consensus lies in the recognition that this datum is preceded by the appearance of *Praeorbulina glomerosa curva*, the ancestor of *Orbulina suturalis*, and itself a short-lived guide fossil confined to zone N.8.

The radiochronological argument continues, however. Our own estimates of the year-age of the *Orbulina* datum, based on various lines of evidence, have varied from 18 Ma (Berggren, 1969) to 16 Ma (Van Couvering, 1972; Berggren, 1972a) and, briefly, 14 Ma (Berggren and Van Couvering, 1974). We now advocate an age of 15 Ma for the *Orbulina* datum, with the following points in mind:

1. The estimate of 15 Ma by Ikebe et al., (1972) is based on a number of consistent K-Ar dates correlated to the Japanese biostratigraphy in beds that bracket the observed FAD of *Orbulina.*

2. Opdyke et al. (1974) and Theyer and Hammond (1974) come to impressively close agreement on an age of 15 Ma for levels very close to the *Orbulina* datum in deep-sea cores. In both studies, the beginning of epoch 15 is also related to the

same level, defined as the base of the *Denticula lauta* (base NPD.25) diatom zone and a little above the base of the *Dorcadospyris alata* radiolarian zone, respectively.

3. Biostratigraphic relationships of the marine record and the K-Ar calibrated mammalian sequences of both the North American and Mediterranean regions (discussed below) are in agreement with an age for the *Orbulina* datum of approximately 15 to 16 Ma.

4. Page and McDougall (1970) presented dates from New Guinea, which they stated were in support of an age of 14 Ma for the *Orbulina* datum in California (see Item 5), but the New Guinea volcanics lie stratigraphically *above* beds containing *Orbulina* and suggest instead an age of ca. 15 Ma for the datum (Van Couvering, 1972; Lipps and Kalisky, 1972).

5. Lipps and Kalisky (1972) correlate the base of the Luisian stage in California close to the base of nannoflora zone NN.5, and note that the base of NN.5 corresponds in tropical microfaunas to the base of zone N.9, the *Orbulina* datum. (The datum itself is not observed in California.) Turner (1970) dated the base of the Luisian at about 14 Ma, and the base of the next older stage, the Relizian, at about 15 Ma. Since Lipps and Kalisky (1972) correlated the Relizian to zones NN.3 and NN.4, this would appear to limit the base of NN.5 and thus the *Orbulina* datum to an age no older than 14 Ma. However, R. Z. Poore (written communication, 1974) points out that Bandy (1972) correlated the Lower Relizian with zone N.9, and even found the zone N.10 indicator *Globorotalia peripheroacuta* in the "Relizian" of the experimental Mohole (Bandy and Ingle, 1970). In addition, Poore noted that the correlation of NN.5 by Lipps and Kalisky (1972) may be incorrect because the co-occurrence of *Sphenolithus heteromorpha* and *Discoaster exilis*, which is restricted to NN.5 (Martini, 1971, pp. 745-746), is shown by these authors to extend to the base of the Relizian. Accordingly, the *Orbulina* datum correlates to the base Relizian, not the base Luisian, and can be calibrated at ca. 15 Ma in California in excellent agreement with the other dating itemized above.

6. The FAD of *Orbulina* in a tropical microfossil context is well documented in the Vienna Basin biostratigraphy, where it appears in the lower part of the Badenian deep-water deposits together with a zone NN.5 nannoflora (Steininger, this volume). At Hrušov in Slovakia, two samples (AV-12, AV-13) of andesite lava underlying strata with zone N.9 microfauna have been dated at 15.2 and 17.7 Ma, but the 18 Ma age indicated for the *Orbulina* datum in the Paratethys is based solely on the older of the two (Vass et al., 1971). This age is consistent with other Slovakian dates, particularly those from the Lower Miocene (see Steininger, this volume), but we shall attempt to show in a following section that there is good reason to believe that this part of the Paratethys Neogene calibration is erroneous. (We have already converted the Slovakian ages in this paper to values 6 percent younger than those originally published to make them compatible with others we quote that have been calculated according to "Western" decay constants; see Odin, 1973b, and Vass et al., 1974.)

Paleogene-Neogene Boundary

The earliest planktonic foraminiferal event with which we are concerned is the evolutionary appearance of the genus *Globigerinoides* in the form of *G. primordia* (*G. trilobus primordius* auctt.). In his review of the tropical zonation, Blow (1969) redefined the base of zone N.4 as coincident with this datum. He also concluded that the presence of *G. primordia* with an N.4 microfauna in the stratotype of the Aquitanian stage in France justified equating the base of zone N.4 with the base of the Aquitanian interval, which by present consensus defines the beginning of the Miocene (see George et al., 1969) and thereby the beginning of the Neogene. In this interpretation, however, Blow appears to have been in error, because the stratotype rests on a transgressive unconformity (Eames, 1970); in more complete sequences elsewhere in southern France, the *Globigerinoides* evolutionary datum occurs well below the level of the Aquitanian stratotype, e.g., at Carry (Anglada, 1971a, 1972) and probably at Escornebéau (Scott, 1972; but see Pujol, 1970). It seems probable that Mayer-Eymar (1857-1858) intended that all the marine deposits of the "Aquitanian Cycle" of transgression should be included in the stage (Szöts, 1965) so it is unfortunate that the section subsequently designated as the stratotype should be so abbreviated. It is obvious nevertheless that a transgressive episode is a diachronous process and cannot define a moment in time such as the beginning of the Miocene.

Following the accepted definition of the base of the Miocene, the Paleogene-Neogene boundary is thus younger than the base of zone N.4 (*Globigerinoides* datum) and older than the base of zone N.5, which Anglada (1971b) found near the lower boundary of the Burdigalian neostratotype (Text-fig. 2).

Nannoplankton have not been reported from the Aquitanian stratotype, but recently P. Lohmann (pers. comm. to W. A. B., 1974) found evidence that the *Globigerinoides* datum may occur within nannozone NP.25 in the Codrington College section of Barbados. (This has the effect of greatly reducing the span of zone P.22, restricted in Blow's 1969 definition to the interval between the *Globigerinoides* datum and the extinction of *Globorotalia o. opima* at the top of P.21, which equates with the base of NP.25; see Text-fig. 1). By placing the base of zone N.4 within NP.25, the NP.25-NN.1 boundary closely approaches the base of the Miocene in the middle of zone N.4 (Text-fig. 2).

The radiometric calibration of the Paleogene-Neogene boundary involves the question of a possible faunal and radiometric "gap" between the youngest strata designated by Beyrich (1856) in his definition of Oligocene (represented in the Chattian stage of the North Sea Basin sequence) and the base of the Aquitanian stratotype. Berggren (1972b) noted that Chattian nannofloras did not appear to range younger than zone NP.24, whereas the *Globigerinoides* datum (below the base of the Miocene) was commonly reported to be within zone NN.1, leaving a hiatus spanning NP.25 and much of NN.1 between the two epochs. In apparent confirmation, glauconites from the type Chattian and equivalent strata had been dated by Odin et al. (1969, 1970) between approximately 31 and 28 Ma, whereas Turner (1970) had dated the base of the Saucesian in California, correlated to the base of the Miocene, at ca. 22.5 Ma. This age for the beginning of the Miocene appeared to us (Berggren, 1972a; Van Couvering, 1972) to be substantially in agreement with

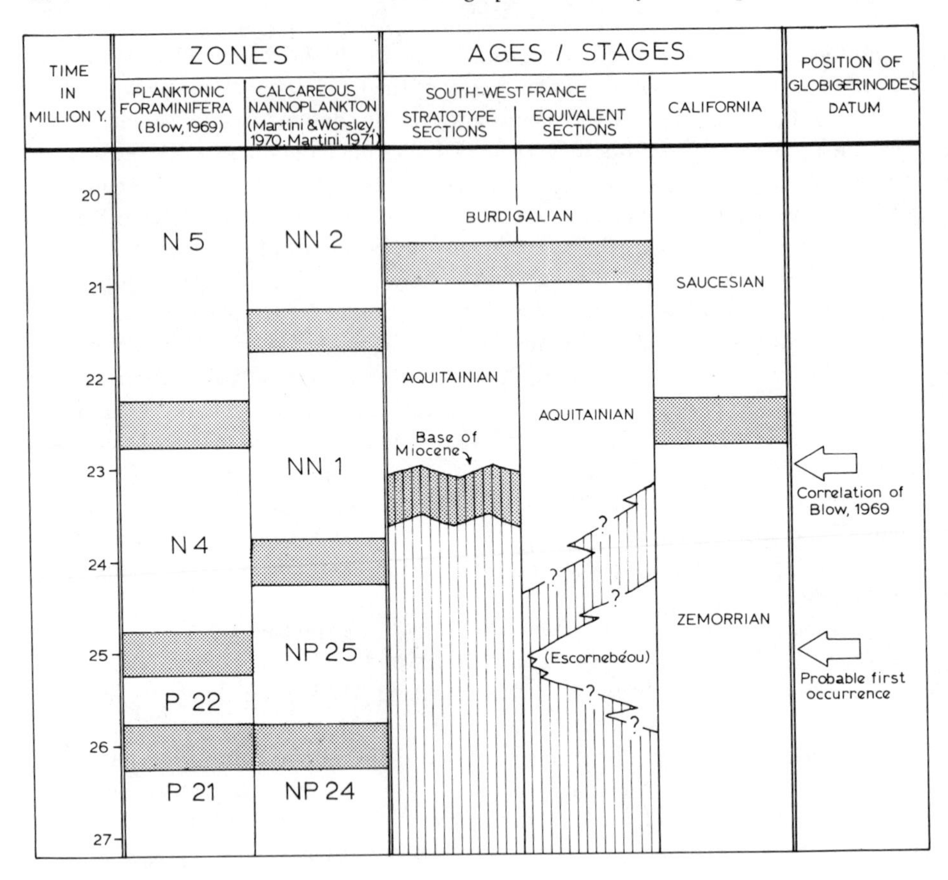

Text-Figure 2
Biostratigraphical relationships and biochronology of the Oligocene-Miocene boundary.

evidence from sea-floor spreading rates (Berggren, 1969), a 22.5 Ma date on "Chattian" small mammals (Lippolt et al., 1963), and a 23.3 Ma date on glass shards close to the *Globigerinoides* datum in Italy (Selli, 1970).

In the time since these studies were published, however, Lohmann's work in Barbados (see above) has indicated that the *Globigerinoides* datum goes down into NP.25, and Müller (1971) identified NP.25 nannofloras in the uppermost Chattian as well. Furthermore, Martini (in Cicha et al., 1971a) assigned Upper Egerian nannofloras of the Paratethys to NP.25 and NN.1, and Cicha and Senes (1973) located the first occurrence of *G. primordia* in the same interval. The faunal and microfloristic gap between Chattian and Aquitanian is thus narrowed considerably, and uppermost Chattian levels extend into, or just up to, zone N.4. As for the radiochronological gap, Kreuzer et al. (1973) have reported glauconite ages on uppermost Oligocene strata in North Germany which they take to indicate an Oligocene-Miocene boundary at 23.5 Ma, and Odin (1973b) has revised his own estimates to agree with the German dates in advocating an age for this boundary at 24.0 Ma. (Both estimates, it should be noted, are based on evidence for the upper age limit of the Oligocene and not on bracketing ages.) Meanwhile, back in California, studies of "Vaqueros" mollusks by Addicott (1973a, 1973b; 1974) and Steininger (1973, and pers. comm. to W. A. B., 1974), and also of Zemorrian foraminifera by Hornaday (1972) and Lamb and Hickernell (1972), have indicated that the Zemorrian-Saucesian boundary is near the base of zone N.5 and not N.4 as previously thought; the 22.5 Ma date on the base of the Saucesian therefore is closer to the beginning of the Burdigalian than to the Aquitanian. The base of the Miocene is bracketed by this date–none of the others mentioned above have much independent validity–and by the age of ca. 24 Ma for uppermost Oligocene levels (above the *Globigerinoides* datum). Estimates of 23.5 Ma for the Oligocene-Miocene (Paleogene-Neogene) boundary, 25 Ma for the *Globigerinoides* datum, and 26 Ma for the base of zone NP.25 appear reasonable (Text-fig. 1). The "gap" *if it exists* is a rather short interval within lower zone NN.1, equivalent to a part of medial zone N.4, between limiting radiochronological values of ca. 23.5 and 24.0 Ma.

THE MEDITERRANEAN NEOGENE

General

The Neogene biochronology of the Mediterranean region is based on a "standard sequence" of fossil beds in France and Italy from which many of the original subdivisions of Cenozoic time and stratigraphy were drawn. The framing of this Mediterranean Neogene "standard sequence" and its correlation to global Neogene history is still in progress, but the literature is already vast almost beyond comprehension. Among numerous recent studies the reader is directed to our review of the Late Neogene (Berggren and Van Couvering, 1974), which includes an extensive bibliography, and to authoritative discussions in Blow and Smout (1968), Blow (1969), Cita and Blow (1969), Martini (1971, 1972) and Cita (1972, 1974).

The Neogene, as defined by Hoernes (1856), includes geologic time from the beginning of the Miocene to the present, and in our version of the "standard sequence" would therefore be composed of the biochrons based on the stage ages given in Table 2. We note that according to the explicit caveat of Lyell (1865, p. 187) the epochs that he framed were based strictly on "conchological data," i.e., mollusks, echinoderms, and other megafossils. This was also true of the original designations of the stages above (see Carloni et al., 1971). Micropaleontological definitions are therefore secondary, but here at least are based on the same stratigraphic sequences used by Lyell and the others. In continental biochronology discussed in a following section, the trend has been to develop an internal system of "mammal ages" rather than to maintain a difficult and often misleading synonymy with the marine biochronology; the mammalian subdivisions are not equivalent to the marine except by accident, and so are not shown here.

Table 2

Estimated Radiometric Age at Base (Ma)	*Standard Stage/Age*	*Epoch*
0.07	–	(Late Pleistocene-Holocene)
0.5	Sicilian	Middle Pleistocene
1.8	Calabrian	Early Pleistocene
3.2	Piacenzian	Late Pliocene
5.0	Zanclian	Early Pliocene
6.5	Messinian	Late Miocene
11.0	Tortonian	
13.5	Serravallian	Middle Miocene
15.6	Langhian	
22.5	Burdigalian	Early Miocene
23.5	Aquitanian	
	(Chattian)	(Late Oligocene)

Early Neogene Mammalian biochronology. Mammalian biochronology is necessarily subdivided and correlated by nonmarine features. In Europe and neighboring parts of Africa and Asia, many new names are now being used in Neogene continental history to avoid the implication, no longer to be taken for granted, that, e.g., Pontian mollusks are of the same age as "Pontian" mammals. We prefer to use the current large-mammal ages, rather than small-mammal "zones," as a basis for our discussion because these ages are more useful in long-distance correlation. However, we recognize that small-mammal biochronology in southwest Europe has become fundamental to detailed study and that parts of it may soon be useful over much larger areas. A correlation of Thaler's (1965) small-mammal "zones" to large-mammal ages is presented in Van Couvering (1972) and Benda and Meulenkamp (1972).

A full characterization of the land-mammal ages will not detain us here, and the reader is referred to the comprehensive presentation of Thenius (1959). Suppressed or still-used age names with marine synonyms are indicated in quotation marks. As noted in the preceding section, the epochal subdivisions of Lyell do not apply directly to mammal biochronology and are not indicated in Table 3.

Datum events and calibration of the Late Neogene are fully described in Berggren and Van Couvering (1974). The most important events in the Early Neogene mammalian history of the Mediterranean region were the "Proboscidean datum" and the "Cricetine datum" in middle "Burdigalian" time, representing the concurrent invasion of western Eurasia by two separate faunal groups (Text-fig. 3). The original homeland of the cricetines and other advanced Cricetidae that appeared at this time is as yet unknown (Mein and Freudenthal, 1971); according to R. Lavocat and L. Thaler (pers. comm., 1973), they were not closely related to the Early Miocene rodents of East Africa. On the other hand, the "Proboscidean datum" marks the sudden appearance in Eurasia of a wide variety of mammals with an African Early Miocene

Table 3

Estimated Radiometric Age at Base (Ma)	*Mammal Age*
	(Late Quaternary-Recent)
0.5	Oldenburgian
1.0	Biharian
3.0	Villafranchian
6.0	Ruscinian ("Astian")
10.0	Turolian ("Pontian s.s.")
12.0	Vallesian ("Early Pontian s.l.")
13.5	Oeningian ("Sarmatian")
~16	"Vindobonian"
~22.5	"Burdigalian"
~25.0	"Aquitanian"

record, and new systematic studies indicate an explicit relationship with the North African fauna exemplified at Gebel Zelten and Wadi Moghara (Van Couvering, 1972; Tobien, 1973; Harris, 1973; Hamilton, 1973a, 1973b).

In East Africa, Early Miocene faunas have K-Ar ages ranging from approximately 23 to 17 Ma (Bishop et al., 1969; Van Couvering et al., 1974). On the other hand, the dating of Eurasian "Burdigalian" mammals is relatively poor (Van Couvering, 1972). Conflicting dates just on the "Late Burdigalian" range from ca. 16 to 22 Ma, and the "Burdigalian" mammal fauna in general is correlated to North American Arikareean and Hemingfordian levels, which range in age between 24 and 18 Ma (Evernden et al., 1964). Van Couvering (1972) assumed that the Early Miocene fauna of Africa was isolated from that of Europe, and that the mid-"Burdigalian" Proboscidean datum, therefore, marked the formation of a trans-Tethyan land bridge in the Middle East sometime close to 18 Ma ago. This was in general agreement with oceanographical and geophysical evidence (see also Maglio, 1973,

Text-Figure 3 (*following two pages*)
Early Neogene Mediterranean geochronology. Significant datum events, identified in the first column, are shown in stratigraphic context or related to selected K-Ar ages in the appropriate regional biostratigraphies. Marine information is shown on the left side of each column and continental information on the right. The Neogene foraminiferal zones of Blow (1969) are calibrated according to the present consensus of paleomagnetic and deep-sea information (see Text-fig. 1), and the marine datum events are positioned accordingly. "Standard" marine and continental ages are shown with bold outlines in the regional biostratigraphies in which they are typified; the "chronostratotypes" of the Paratethys marine sequence are calibrated by paleontological correlations, rather than by Slovakian K-Ar ages, below the Sarmatian level. Inconsistency between the "type Vindobonian" mammal age placement and the Tajo Basin correlation is discussed in the text. Base of the Miocene (beginning of the Neogene) is equivalent to the base of the Aquitanian stratotype.

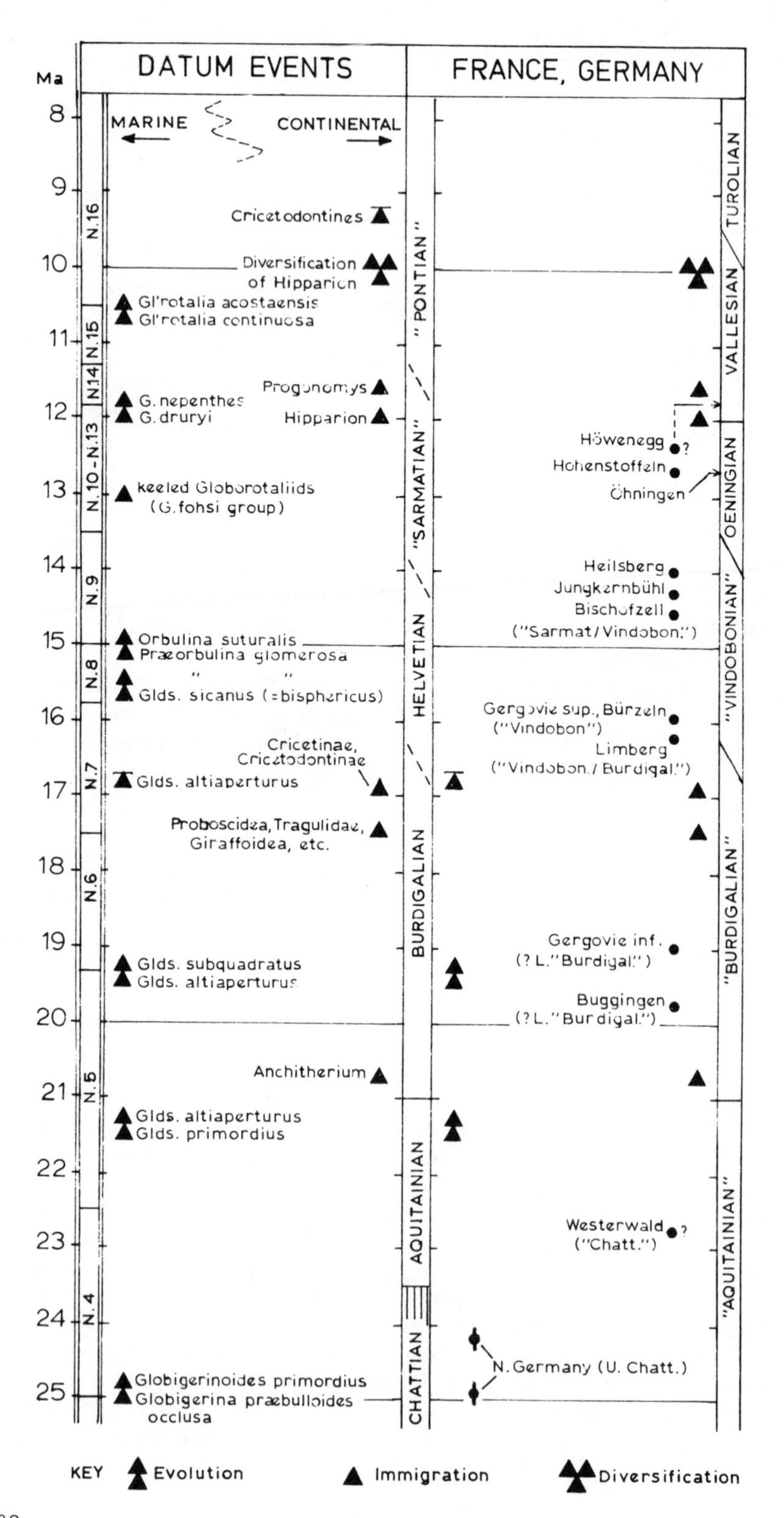

Ma
DATUM EVENTS
FRANCE, GERMANY
MARINE
CONTINENTAL
N.16
N.15
N.14
N.10-N.13
N.9
N.8
N.7
N.6
N.5
N.4
Cricetodontines
Diversification of Hipparion
Gl'rotalia acostaensis
Gl'rotalia continuosa
Progonomys
G. nepenthes
G. druryi
Hipparion
keeled Globorotaliids (G. fohsi group)
Orbulina suturalis
Praeorbulina glomerosa
Glds. sicanus (=bisphericus)
Cricetinae, Cricetodontinae
Glds. altiaperturus
Proboscidea, Tragulidae, Giraffoidea, etc.
Glds. subquadratus
Glds. altiaperturus
Anchitherium
Glds. altiaperturus
Glds. primordius
Globigerinoides primordius
Globigerina praebulloides occlusa
"PONTIAN"
"SARMATIAN"
HELVETIAN
BURDIGALIAN
AQUITAINIAN
CHATTIAN
Höwenegg ?
Hohenstoffeln
Öhningen
Heilsberg
Jungkernbühl
Bischofzell
("Sarmat/Vindobon.")
Gergovie sup., Bürzeln
("Vindobon")
Limberg
("Vindobon./Burdigal.")
Gergovie inf.
(? L."Burdigal.")
Buggingen
(? L."Burdigal.")
Westerwald ?
("Chatt.")
N. Germany (U. Chatt.)
TUROLIAN
VALLESIAN
OENINGIAN
"VINDOBONIAN"
"BURDIGALIAN"
"AQUITAINIAN"
KEY
Evolution
Immigration
Diversification

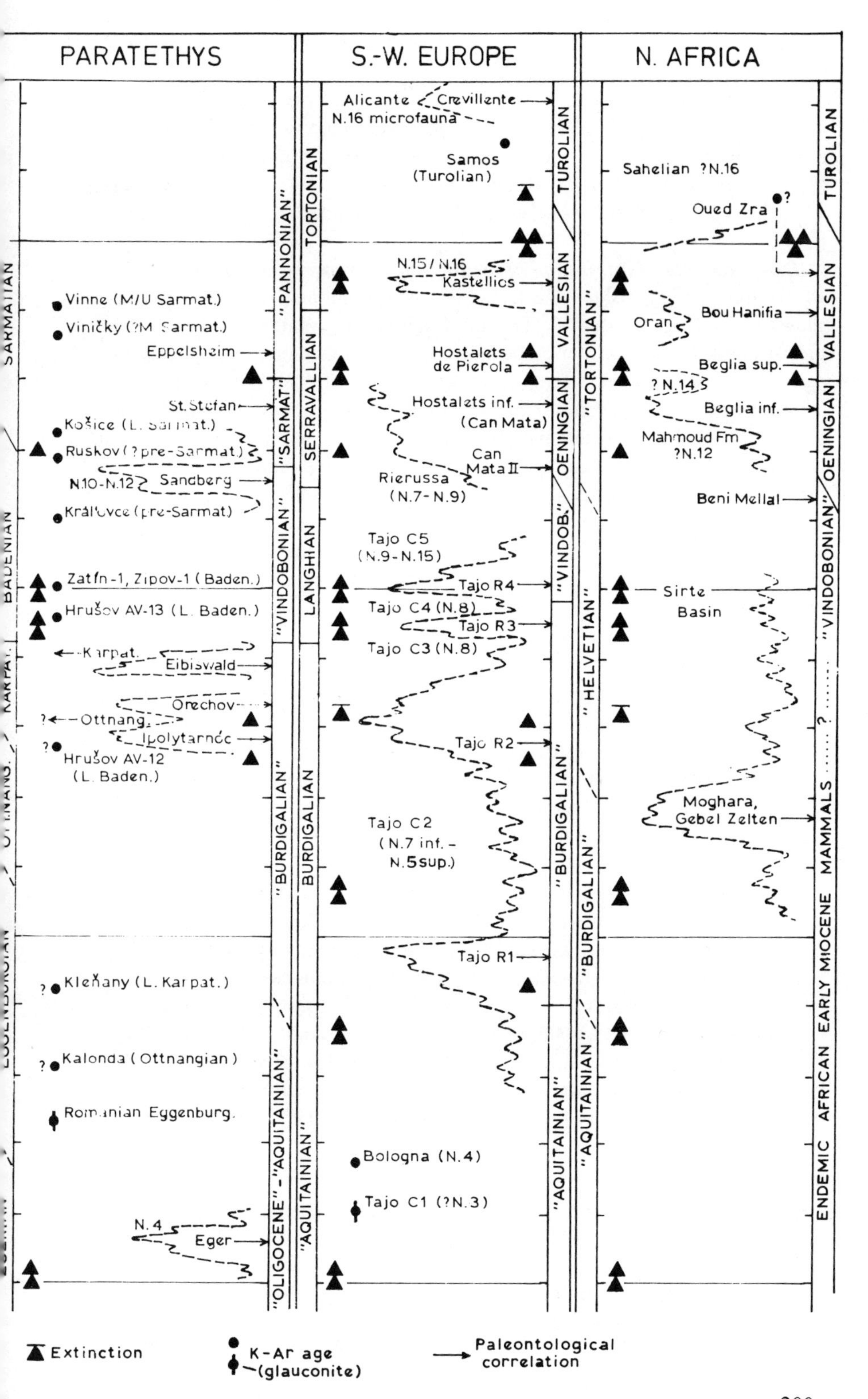
PARATETHYS
S.-W. EUROPE
N. AFRICA
Alicante
Crevillente
N.16 microfauna
Samos
(Turolian)
TUROLIAN
TORTONIAN
"PANNONIAN"
Sahelian ?N.16
Oued Zra
N.15 / N.16
Kastellios
VALLESIAN
"TORTONIAN"
Vinne (M/U Sarmat.)
Viničky (?M Sarmat.)
Eppelsheim
SERRAVALLIAN
Hostalets de Pierola
Oran
Bou Hanifia
Beglia sup.
?N.14
Beglia inf.
St. Stefan
"SARMAT"
Hostalets inf.
(Can Mata)
OENINGIAN
Košice
Ruskov (?pre-Sarmat.)
Can Mata II
Mahmoud Fm
?N.12
N.10-N.12
Sandberg
Rierussa
(N.7-N.9)
Beni Mellal
"VINDOBONIAN"
"VINDOB."
Tajo C5
(N.9-N.15)
LANGHIAN
Zatín-1, Zipov-1 (Baden.)
Tajo R4
Sirte
Basin
Hrušov AV-13 (L. Baden.)
Tajo C4 (N.8)
Tajo R3
Tajo C3 (N.8)
Karpat.
Eibiswald
"HELVETIAN"
Orechov
Ottnang
Ipolytarnóc
Tajo R2
Hrušov AV-12
(L. Baden.)
"BURDIGALIAN"
BURDIGALIAN
Moghara,
Gebel Zelten
Tajo C2
(N.7 inf. -
N.5 sup.)
"VINDOBONIAN" ...... ? ......
ENDEMIC AFRICAN EARLY MIOCENE MAMMALS
Tajo R1
Kleňany (L. Karpat.)
Kalonda (Ottnangian)
Romanian Eggenburg.
"AQUITAINIAN"
Bologna (N.4)
Tajo C1 (?N.3)
N.4
Eger
"OLIGOCENE"-"AQUITAINIAN"
Extinction
K-Ar age
(glauconite)
Paleontological correlation

pp. 111-118). Nevertheless, some paleontological evidence suggests that "Aquitanian" and "Early Burdigalian" mammals might have been able to migrate between Eurasia and Africa; for instance, the newly evolved genera *Hyotherium* and *Amphicyon* appear to be present at the beginning of the Miocene on both sides of the Mediterranean. Thus the indirect evidence for the age of the Proboscidean datum has been inconclusive.

Biostratigraphical integration of Mediterranean Neogene chronologies. The stratigraphical relationships of selected marine and nonmarine faunal horizons are shown in Text-figures 3 and 4. The important datum events are shown at their presently established or inferred position, first on the radiochronological scale, and again in the local stratigraphic contexts which establish their relative age.

One of the most significant papers to appear in the last few years is the short note by Antunes et al. (1973) summarizing the superpositional relationship of Portugese Early Miocene strata containing planktonic foraminiferal faunas (denoted by the prefix "C") and land-mammal faunas (denoted by the prefix "R") in the lower basin of the Rio Tajo (Tagus) near Lisbon. Until this study appeared, the biostratigraphic relationship of Early Miocene marine and continental beds in Europe depended on entirely inadequate occurrences of fragmentary mammal remains in association with marine mollusks (e.g., Ginsburg, 1967), and most conclusions were based on the theory of "sedimentary cycles" (e.g., Richard, 1946; Gigot and Mein, 1973). As noted above, a single K-Ar date of ca. 23 Ma on "Chattian" small mammals at Westerwald (Lippolt et al., 1963) seemed to confirm the age equivalence of mammalian "Aquitanian" with marine Aquitanian faunas, despite the fact that some paleontologists were reporting "Aquitanian" mammals associated with Burdigalian mollusks (Gigot and Mein, 1973), and others were putting the Oligocene-Miocene boundary in mammal biochronology above the "Aquitanian" age (see Van Couvering, 1972).

For the most part, the Tajo Basin sequence (Text-fig. 3) confirms the conclusions about Early Miocene correlations that were deduced from the nonbiostratigraphic evidence given above. The earliest level, Cl, contains no diagnostic foraminifera, but was ascribed by Antunes et al. (1973) to "zone N.3" on the basis of a 24 Ma glauconite date. The overlying continental beds contain the R1 mammal fauna, and this is well correlated to the oldest "Burdigalian" mammal levels of the Loire Valley in France, in what might be called the type fauna. The next higher marine (C2) beds contain, near their base, a planktonic foraminiferal assemblage that includes both *Globigerinoides primordia* and *G. subquadrata*, indicating lower zone N.5. Thus a nearly synchronous correlation of the beginning of the continental "Burdigalian" with its marine counterpart, close to the N.4-N.5 boundary, is strongly indicated;

Text-Figure 4

Later Neogene Mediterranean geochronology. Conventions are the same as for Text-figure 3, except for the open triangle representing a general evolutionary level attained by microtine rodents near the beginning of the Villafranchian age. Existence of planktonic zone N.20 is debatable (Berggren, 1973). Base of the Pliocene is equivalent to the base of zone N.18, approximately at the base of the Zanclian age.

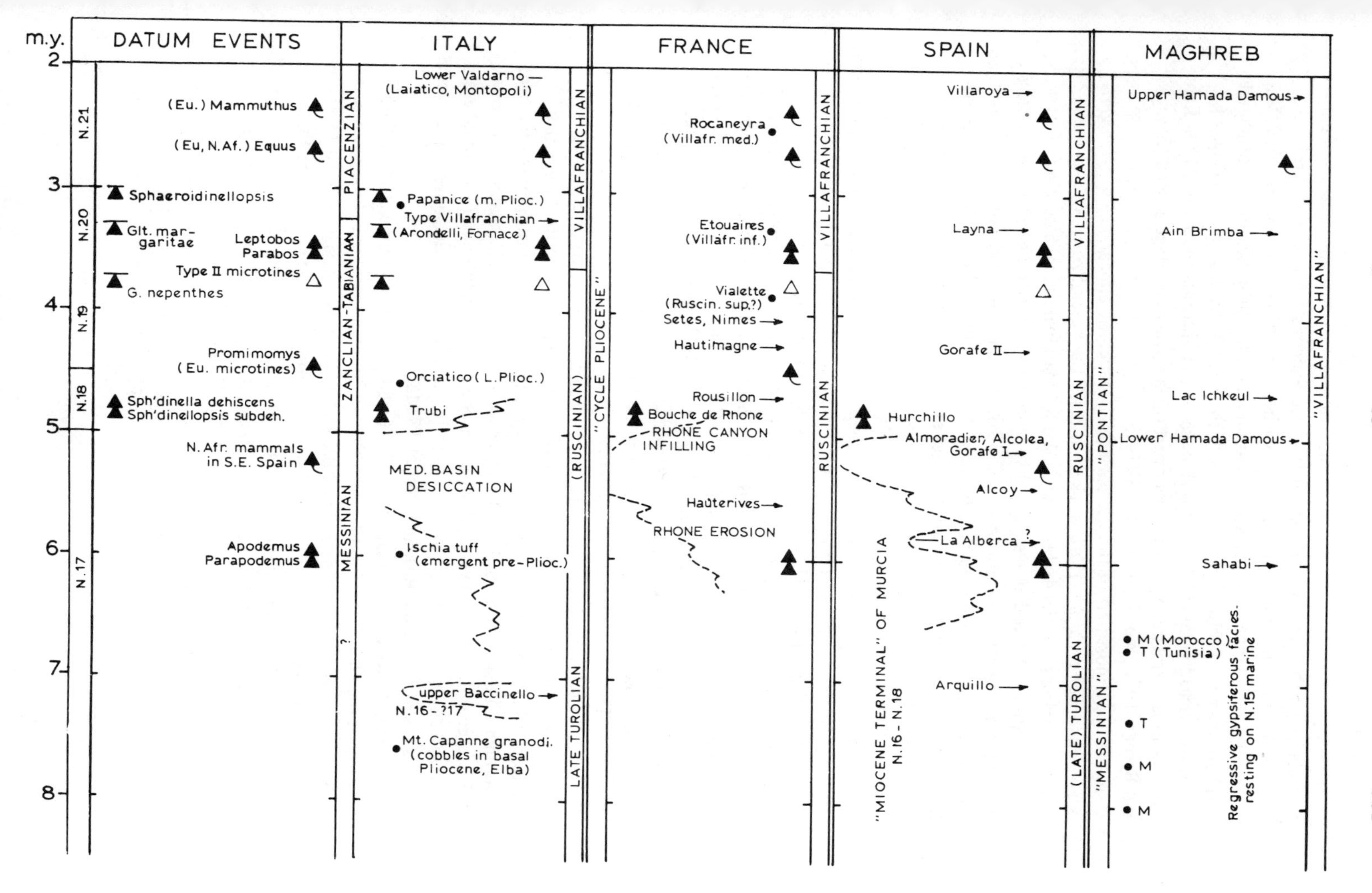
m.y.
2
3
4
5
6
7
8
N.21
N.20
N.19
N.18
N.17
DATUM EVENTS
(Eu.) Mammuthus
(Eu, N.Af.) Equus
Sphaeroidinellopsis
Glt. mar-
garitae
Leptobos
Parabos
Type II microtines
G. nepenthes
Promimomys
(Eu. microtines)
Sph'dinella dehiscens
Sph'dinellopsis subdeh.
N. Afr. mammals
in S.E. Spain
Apodemus
Parapodemus
PIACENZIAN
ZANCLIAN-TABIANIAN
MESSINIAN
?
ITALY
Lower Valdarno —
(Laiatico, Montopoli)
Papanice (m. Plioc.)
Type Villafranchian
(Arondelli, Fornace)
Orciatico (L. Plioc.)
Trubi
MED. BASIN
DESICCATION
Ischia tuff
(emergent pre-Plioc.)
upper Baccinello
N.16-?17
Mt. Capanne granodi.
(cobbles in basal
Pliocene, Elba)
VILLAFRANCHIAN
(RUSCINIAN)
LATE TUROLIAN
"CYCLE PLIOCENE"
FRANCE
Rocaneyra
(Villafr. med.)
Etouaires
(Villafr. inf.)
Vialette
(Ruscin. sup.?)
Setes, Nimes
Hautimagne
Rousillon
Bouche de Rhone
RHONE CANYON
INFILLING
Hauterives
RHONE EROSION
VILLAFRANCHIAN
RUSCINIAN
SPAIN
Villaroya
Layna
Gorafe II
Hurchillo
Almoradier, Alcolea,
Gorafe I
Alcoy
La Alberca ?
Arquillo
"MIOCENE TERMINAL" OF MURCIA
N.16 - N.18
VILLAFRANCHIAN
RUSCINIAN
(LATE) TUROLIAN
"PONTIAN"
"MESSINIAN"
MAGHREB
Upper Hamada Damous
Ain Brimba
Lac Ichkeul
Lower Hamada Damous
Sahabi
M (Morocco)
T (Tunisia)
T
M
M
Regressive gypsiferous facies
resting on N.15 marine
"VILLAFRANCHIAN"

we therefore estimate the beginning of the "Burdigalian" mammal age at roughly 22.5 Ma by correlation with the age of the base Saucesian stage in California (see above). We note that this revision to our earlier estimates (cf. Berggren, 1969; Van Couvering, 1972) quite comfortably accommodates the proposed intercontinental correlation of Early "Burdigalian" levels to the later Arikareean mammal age of North America (Evernden et al., 1964), because in California the 23 to 21 Ma Late Arikareean is correlated with the lower part of the Saucesian stage as well (Savage and Barnes, 1972; Woodburne et al., 1974).

In the uppermost part of the C2 marine sequence, the appearance of *Globigerinoides quadrilobata altiapertura* is taken to mark the base of zone N.7. In the directly overlying continental beds with the R2 mammal fauna, a number of African emigrants are recorded, e.g., *Gomphoterium angustidens, Eotragus, Dorcatherium,* and *Palaeomeryx* (see also Antunes, 1969, for a preliminary faunal list). By comparison with the mammals of the Loire Valley "Burdigalian," the level of evolution in the R2 fauna suggests to Antunes et al. (1973) that the African invaders were newly arrived in the Tajo Basin, and we accept this estimate in placing the Proboscidean datum equivalent to the N.6-N.7 boundary (Text-fig. 3). According to our present evaluation of the marine biochronology and paleomagnetic calibrations, this level is 17.5 Ma old (Text-fig. 1), and the estimate of 18 Ma for the Proboscidean datum (Van Couvering, 1972) is independently confirmed.

The first elements of the Cricetine datum (*Megacricetodon* and *Democricetodon*; see Antunes, 1969) are recorded in the next higher continental level (R3) together with other African invaders not observed in the R2 fauna, such as *Prodeinotherium bavaricum* and *Bunolistriodon.* In terms of the Loire Valley sequence, Antunes et al. (1973) estimate that the R3 mammal assemblage is of Late "Burdigalian" age, and find it to be entirely contained within the time span occupied by foraminiferal zone N.8. This is based on the occurrence of *G. sicanus*, without orbulines, in both the C3 and C4 levels. Mammal fauna R4 is judged, on the basis of evolutionary level in the large mammals, to be very close to the transition to the "Helvetian" (Lower "Vindobonian") of the Loire Valley, and is overlain in turn by marine beds with microfaunas that span the interval between upper zone N.9 and at least zone N.14. According to this evidence, we place the base "Vindobonian" mammal age (in Portugal) at the same age as the N.8-N.9 boundary, or a little above (the same level as the *Orbulina* datum), and thus at 15 Ma.

In a correlative development, R. H. Tedford (written communication, 1974) has correlated the *Orbulina* datum to early Barstovian mammal faunas in the U.S. Gulf Coast. According to Evernden et al. (1964), the Early Barstovian is 16 to 15 Ma, in agreement with our estimate for the *Orbulina* datum, but they correlated this mammal level to middle "Vindobonian" on the basis of the German dating (Lippolt et al., 1963) noted above. In view of the Tajo Basin stratigraphy, it would seem preferable to equate the *Orbulina* datum, the base of the Vindobonian, and the base of the North American Barstovian within the approximate time limits of 16 to 15 Ma.

The Miocene of the western and central Paratethys includes a number of radiometrically dated levels (Vass et al., 1971), mostly in Slovakia. According to our best evidence, however, the mammalian, planktonic foraminiferal, and nannofloral biochronologies of the Early Miocene here are in mutual disagreement; thus it would appear that the biostratigraphy which correlates them is not yet adequately studied. The Ipolytarnóc mammal fauna (Thenius, 1959) contains the giraffoid *Palaeomeryx* and is therefore of post-Proboscidean datum age because giraffes originated in Africa (Hamilton, 1973b). In the open chronstratigraphic scale (OCS) developed by Paratethys stratigraphers, the Ipolytarnóc site is included among the "neostratotypes" of the composite Ottnangian superstage (Senes, in Papp et al., 1973, p. 81). The sparse microfauna attributed to the Ottnangian from various marine "neostratotypes" suggests zone N.6 (Rögl and Cita, in Papp et al., 1973, p. 301), but this would be too old for a post-Proboscidean datum level. Furthermore, the next succeeding Karpatian superstage includes strata with excellent microfauna strictly limited to zone N.7 (Cicha et al., 1967). The reported nannoflora of this sequence is also inconsistent, since Cicha and Senes (1973) attribute Ottnangian nannoflora to NN.4 (N.7; see Text-fig. 1) and those of the Karpatian to NN.5 (N.8 to N.10; see Text-fig. 1), even though the *Praeorbulina-Orbulina* transition (zones N.8-N.9 boundary) is very well represented in strata of the Badenian superstage above the Karpatian. We show (Text-fig. 3) Ipolytarnóc as a continental equivalent of Karpatian, since zone N.7 is as old as it should go, but it might be that Ottnangian strata are (in part) laterally equivalent to Karpatian strata, and that the Karpatian nannoflora is mainly of early NN.5 age.

The radiometric calibration of Early Miocene biochronology in the Paratethys is also in irreconcilable conflict with the time scale based on a synthesis of dating in Western Europe, Africa, North America, and the Pacific Basin. It is impossible to accommodate a Paratethys calibration that places the *Orbulina* datum at 17.5 Ma, a zone N.7 or zone NN.5 (Karpatian) interval at ca. 21 Ma, and a zone ?N.7 to N.6, or zone NN.4, or post-Proboscidean mammal (Ottnangian) interval at ca. 22 Ma. It should be noted, however, that the middle Miocene (Badenian and Sarmatian) and the lowermost Miocene (Eggenburgian and Egerian) calibration is not inconsistent with our time scale (see Steininger, this volume). A reinvestigation of Slovakian calibration should focus on the Lower Badenian dating (Hrušov), since the Ottnangian and Karpatian dates are of doubtful quality without this support (see Vass et al., 1971, p. 322).

We have previously discussed (Berggren and Van Couvering, 1974) the significance of marine microfaunas as young as zone N.12 stratigraphically below the transition to Vallesian mammal faunas (marked by the first appearance, in this area, of *Hipparion*) in the Beglia Formation of central Tunisia, the Valles-Penedés near Barcelona, and in the "Sarmatian" of the Paratethys region (see Text-fig. 3). We have mentioned the date of 12.5 Ma for this datum in the Höwenegg-Deckentuffe sequence of the Hegau (Württemberg). Close upper limits to the datum in Mediterranean marine biostratigraphy are, however, lacking, as are radiometric dates for younger Vallesian levels. The present deep-sea paleomagnetic correlations (Text-fig. 1) indicate an age for zone N.12 that is also centered at 12.5 Ma. Biostratigraphy suggests that N.12 is apparently older than *Hipparion* in Europe, if not older than the entire pre-*Hipparion*

Oeningian age, and new work suggests independently that the age of the *Hipparion* datum should be moved closer to 12.0 Ma (Van Couvering et al., in press).

Relationships of Late Miocene mammalian levels to the marine sequence are depicted in Text-figures 3 and 4. New information is provided by Jaeger et al. (1973), who arranged the known Late Miocene mammal evidence from North Africa to show a sequence of evolutionary levels leading from pre-*Hipparion* rodents of Beni Mellal, through the famous lower fauna at Bou Hanifia of early Vallesian age and the newly discovered Late Vallesian Oued Zra local fauna, up to Early Turolian small-mammal faunas in the upper levels of Bou Hanifia (Sidi Salem l.f.) and at Khendek-el-Ouich. Oued Zra is given an age of 9.7 Ma, based on a single and (under the circumstances) provisional date on underlying basalt. Jaeger et al. (1973) propose an age for the Vallesian-Turolian boundary of 9.5 Ma to fall between the Oued Zra age and that of Samos. Noting that Sidi Salem is overlain by marine levels with N.16 microfauna, and that Khendek-el-Ouich is located in regressive gypsiferous strata commonly termed "Messinian" in the Maghreb, Jaeger et al. (1973) go on to propose a direct chronological correlation between "Turolian typique" and the Messinian.

In our placement of Late Miocene biochronological levels, we have relied instead on biostratigraphic evidence form the northern side of the Mediterranean. The most significant of these, in the context of the North African evidence, is the Kastellios (Crete) locality, where de Bruijn et al. (1971) described Late Vallesian or Early Turolian small mammals directly associated with planktonic foraminifera indicating the lowermost part of zone N.16, which would be ca. 10.5 Ma on present evidence (see Text-fig. 1). The base of the Messinian stage, although not yet firmly defined, has been identified within zone N.17 (Cita and Blow, 1969; Berggren, 1973b) and has an age of about 6.5 Ma. Further to the point, Late Turolian small mammals have been described in association with N.16 planktonic foraminifers both in the upper Baccinello V3 level (Lorenz, 1972) and at Aspe in Alicante (Montenat and Crusafont, 1970). Furthermore, the identification of N.16 (which is wholly within the Tortonian) *above* the Early Turolian of Sidi Salem directly contradicts the proposal (see above) made by Jaeger et al. (1973). Van Couvering (1975) also notes that the regressive gypsiferous facies of the Maghreb, commonly termed "Messinian" by lithologic analogy and dated from ca. 11 to 6.5 Ma in local areas (Choubert et al., 1968; Vass et al., 1974), is biostratigraphically limited only because it overlies beds with Early Tortonian microfaunas and underlies unconformably deposits dated as Pliocene. Considering the Baccinello and Alicante evidence, we believe that the Turolian and Messinian do not overlap in time.

The conception of an isolated Mediterranean Basin emptying itself by evaporation, periodically refilled by deluges, and again baked dry to form a vast desert thousands of meters below sea level is surely the most dramatic vision to enter geological consciousness since plate tectonics or the Ice Ages. The evidence produced by the DSDP (Hsü et al., 1972) explains in this way why the later part of the Messinian age is not well characterized paleontologically, and why so much erosion should have affected circum-Mediterranean areas at the end of the Miocene.

It might also explain the migrations of endemic North African mammals into the Ruscinian faunas of southern Spain. Included among the immigrants are gerbils and the murid *Parethomys* at Gorafe I (Bruijn, 1973) and the distinctive *Hipparion rocinantis* at Almoradier and Alcolea (Alberdi, 1974). None of these have been reported, as yet, from Late Turolian sites such as Alcoy and Arquillo, nor from the La Alberca site, which Montenat and Crusafont (1970) correlated with the lower part of the Andalusian or "Miocène terminal" of southeast Spain. The La Alberca local fauna was originally called Vallesian, but M. Crusafont (pers. comm., 1973) has corrected this to a probable Early Ruscinian or Late Turolian age.

Besides the noted correlation of ?Early Ruscinian, but pre-Maghreb, immigration mammal levels with the latest Miocene of Spain, there has also been a slightly conflicting biostratigraphic correlation, that of earliest Ruscinian (zone de Hauterives) in southern France and the basal Pliocene *Sphaeroidinella* datum in the deposits that filled the Rhone Valley once the Messinian isolation was ended (Ballesio, 1971). The Hauterives site, however, is far upstream and lies below the transgressive wedge of the "cycle Pliocène." Because the local diachronism in this marine Pliocene transgression is not known, the precision of the correlation is open to question. We show (Text-fig. 4) the beginning of the Ruscinian at 6 Ma and the North African incursion, at the height of Messinian isolation, just prior to the flooding that ushered in the Pliocene in the Mediterranean basin. K-Ar dating on Alcolea and on the southeast Spanish "Miocène terminal," now in progress, should add some confidence to our interpretation.

SUMMARY AND CONCLUSIONS

New information, combined with previous calibration and correlation evidence, allows us to make the following revisions and confirmations to previous time scales:

1. The base of the Miocene (Neogene), when correlated with the base of the Aquitanian *stratotype*, has the following attributes: middle or upper zone N.4, lower or middle zone NN.1, and an age of ca. 23.5 Ma. The *Globigerinoides* datum (*G. primordia* first occurrence) at the base of zone N.4 is apparently equivalent to a level in zone NP.25, and would accordingly be the same age as a level within the uppermost Chattian stage of northwest Europe. However, there may still be a short interval in the span of lower zone N.4 that is not included in the time of either the typified Oligocene or Miocene.

2. The Proboscidean datum, representing a time when the Tethys was bridged by plate contact between Africa and Eurasia, is directly correlated to the planktonic foraminiferal sequence at the N.6-N.7 boundary. The age of this boundary, estimated at 17.5 Ma from deep-sea calibrations, is in very close agreement with the calibrated biochronology of African mammals, which suggested the first intercontinental mixing at 18 to 17 Ma ago, and also agrees with the few European mammalian calibration points available for this time.

3. The *Orbulina* datum is best accommodated in deep-sea paleomagnetic stratigraphy, as well as in reevaluated land-based K-Ar dating from the Pacific Basin, if

its age is placed at least as old as 15 Ma. The correlation of the base of the "Vindobonian" mammal age to the *Orbulina* datum in Portugal, and of early Barstovian mammal age to the same datum in Texas, is independent confirmation of its 15-Ma age.

4. Early Neogene K-Ar ages in the Paratethys sequence are evidently erroneous, even when recalculated with western decay constants. Nannoplankton attributions are also seriously at odds with evidence from outside sources. Working from planktonic foraminiferal data, we conclude that the geochronology of the Miocene in the Paratethys could be as shown in Text-figure 3, as follows:

Badenian: N.8 to N.12; NN.5 to middle NN.7; 15.5 to 13.5-13.0 Ma.
Karpatian: (?upper) N.7; ?NN.4 to NN.5; ?16.5 to ?15.5 Ma.
Ottnangian: ?N.6 to N.7; NN.3 to NN.4; ?19.0 to ?16.5 Ma.
Eggenburgian: N.5; upper NN.1 and NN.2; 22.5 to 19.0 Ma.
Egerian: P.21 to N.4; upper NP.23 to upper NN.1; 30.0 to 22.5 Ma.

5. Improved correlation between zones based on planktonic microfossils and their calibration to the paleomagnetic time scale provides a framework for interhemispherical biochronology. The ability to make accurate biochronologic correlations across significant latitudinal distances (ca. 60 N lat. to ca. 60 S lat.) is within the grasp of the micropaleontologist.

Acknowledgments

We would like to thank P. Robinson, P. Lohmann, F. Theyer, N. D. Opdyke, W. B. F. Ryan, L. H. Burckle, and particularly C. Denham for stimulating discussions on the problems of correlation and calibration with the paleomagnetic time scale and access to their unpublished data. We would also like to acknowledge the stimulus that discussions with Dr. Isabella Premoli-Silva on Neogene stratigraphy has furnished. This investigation was supported by grant number GA-30723X to W. A. Berggren from the Branch of Submarine Geology and Geophysics, Oceanography Section of the National Science Foundation. This is Woods Hole Oceanographic Institution Contribution No. 3313.

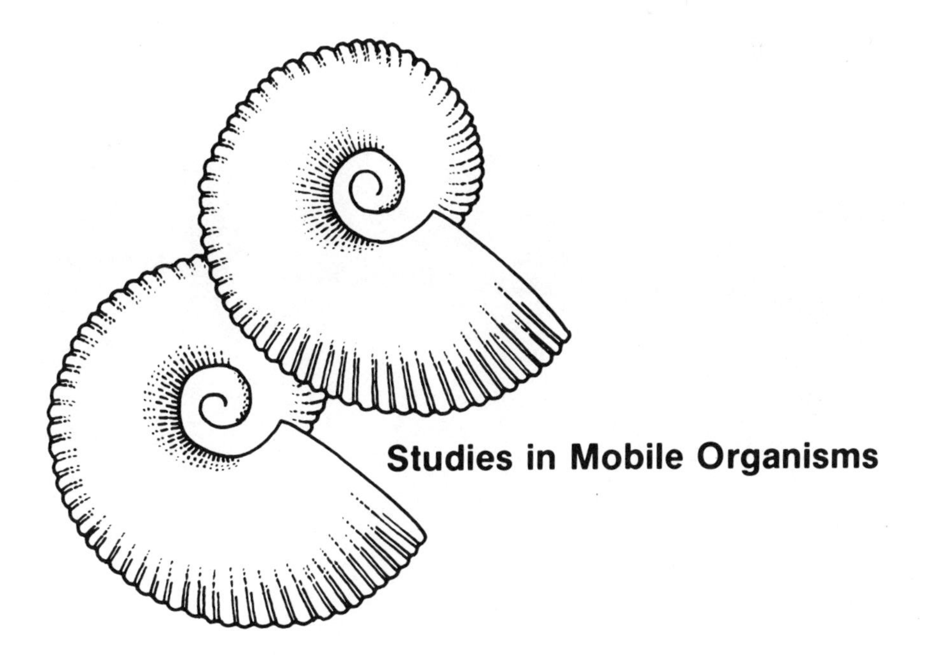

Studies in Mobile Organisms

The Role of Ammonites in Biostratigraphy

W. J. Kennedy — **University of Oxford**

W. A. Cobban — **U.S. Geological Survey, Denver**

INTRODUCTION

Ammonites have been considered prime biostratigraphic indicators in marine sediments from the days of William Smith to the classic works of Oppel, Quenstedt, d'Orbigny, Mojsisovics, Spath, and Arkell. The group is generally held to have possessed many of the characteristics of the ideal index fossil: wide, rapidly attained geographic distribution, high degree of facies independence, rapid evolutionary rates, and high preservation potential. Ammonites are conspicuous and commonly determinable even when fragmentary. These factors allow the erection of fine biostratigraphic subdivisions correlatable over long distances.

Ammonites are, however, a group whose biology, ecology, and functional morphology are poorly understood; our ignorance of these factors brings into doubt many of the classic assumptions made about the group and their biostratigraphic utility. Inasmuch as an understanding of ammonite biology is critical to an evaluation of their biostratigraphic potential, we shall review what is known of their biology and evaluate the degree to which they fulfill their role as key biostratigraphic indicators in terms of their geographic distribution patterns, facies relationships, stratigraphic distribution, and evolutionary rates. Also discussed are problems of classification associated with the widespread sexual dimorphism in the group, problems of inter- and intraspecific variability, chronological and geographical subspeciation, and homeomorphy.

SUMMARY OF AMMONOID PALEOBIOLOGY

The last decade has seen a complete reappraisal of views on the anatomy, functional morphology, and probable ecology of ammonites. The most important new discoveries are as follows. Sexual dimorphism has been widely recognized (e.g., Makowski, 1962; Callomon, 1963, 1969; Cobban, 1969; Davis et al., 1969; and many others). Anaptychi and possibly aptychi are now regarded as jaw apparatus (Closs, 1967; Lehmann, 1967a; Lehmann and Weitschat, 1973), and radular structures are also known (Closs, 1967; Lehmann, 1967a). Egg masses have been recorded in the body chambers of the macroconchs of some species (Müller, 1969) as well as ink sacs (Lehmann, 1967b; Wetzel, 1969) and crop contents (Lehmann, 1971; Kaiser and Lehmann, 1971). Traces of muscle scars, widely described by early authors (e.g., Oppel, 1863; Waagen, 1870; Mojsisovics, 1873; Crick, 1898), have been reinvestigated and interpreted by comparison with living cephalopods (Jones, 1961; Jordan, 1968; Mutvei and Reyment, 1973). Ultrastructure studies have revealed new details of early development and patterns of shell secretion (Erben, 1964, 1966; Erben et al., 1969; Birkelund, 1967; Palframan, 1967; Birkelund and Hansen, 1968). Isotopic studies have shed new light on growth rates and longevity (Stahl and Jordan, 1969; Jordan and Stahl, 1970). Experimental and theoretical approaches to the functional morphology of the shell are making possible the restoration of the life form of an extinct group whose soft parts are unknown (Westermann, 1971; Raup, 1966, 1967; Raup and Chamberlain, 1967; Mutvei and Reyment, 1973; Cowen et al., 1973).

Ammonites were a diverse, variable, and successful group exploiting various habitats within marine ecosystems. The paleobiology of ammonites can be summarized as follows:

1. Ammonites were stenohaline and were absent, scarce, or reduced in generic and specific diversity in quasimarine and marginal marine sediments. Some living cephalopods are tolerant of salinities as low as 30 $^{0}/_{00}$ (Mangold-Wirz, 1963; Hallam, 1969); a similar tolerance was probably attained by certain ammonities.

2. After hatching, many ammonites underwent a planktonic larval stage equivalent to the second phase of shell growth between the protoconch and nepionic constriction. This stage may have lasted from hours to weeks (Ekman, 1953, with references; Thorson, 1961; Scheltema, 1968).

3. Distribution patterns and varying degrees of facies independence suggest that some ammonites may have assumed a benthic mode of life after metamorphosis. Other ammonites were nektobenthic, nektonic, and perhaps planktonic.

4. Differences in juvenile and adult ornamentation and distribution together with apparent segregation of adults and juveniles indicate different life habits according to growth stage.

5. Ammonites, in general, were poorly adapted for swimming; they probably moved slowly by expulsion of water from the mantle by contractions of the funnel.

Rapid jet propulsion, if possible at all, was a seldom-used escape mechanism. Several lines of evidence suggest, however, that many ammonites were capable of slow vertical movement by varying the amount of fluid in their chambers, thus changing their density relative to seawater. This would seem to limit the geographic mobility of adults, classically given as the mechanism of broad species dispersal.

6. Ammonites grew more slowly and lived longer than Holocene cephalopods. Evidence from associated epizoans (Schindewolf, 1934; Seilacher, 1960; Merkt, 1966; Meischner, 1968; Westermann, 1971) and from isotopic studies (Stahl and Jordan, 1969; Jordan and Stahl, 1970) suggests a minimal age at sexual maturity of 11 to 30 years, far greater than the 1- or 2-year period prior to maturity in *Nautilus.*

7. Most ammonites exploited the lower trophic levels of marine ecosystems. Some were plankton feeders, others were benthic and microphagous, and still others may have been vegetarian browsers. A few may have been carnivores and scavengers like Holocene *Nautilus* and coleoids; most probably lacked the ability to capture active prey.

8. Sexual dimorphism, as indicated by size distinction and dissimilar adult form, was widespread, if not universal. Some structural features associated with dimorphism, such as apertural rostra and lappets, may have housed secretory and sensory organs, but other differences between sexes, especially ornamentation, may be better regarded as sexually selected display features. The relatively late discovery of sexual dimorphism in ammonites, *after* a long descriptive phase of research in which dimorphs were treated as distinct taxa, creates serious problems in classification and thus correlation.

9. Predation was apparently not an important limiting factor in adult ammonite dispersal. Few examples of predation have been described, although there are scattered cases of cannibalism (Schwarzbach, 1936; Kaiser and Lehmann, 1971). Ammonites have been recorded from saurian stomachs (Frentzen, 1936), and shell damage has been attributed to predatory decapods (Roll, 1935) and mosasaurs (Kauffman and Kesling, 1960) Masses of entire and fragmentary shells may be fecal accumulations from marine carnivores (Reeside and Cobban, 1960).

POSTMORTEM HISTORY

Necroplanktonic Drifting

Shells of recently dead *Nautilus* are buoyant because of loss of cameral fluid from the chambers (Bidder, 1962; Denton and Gilpin-Brown, 1966; Mutvei and Reyment, 1973). The mechanism of fluid loss is poorly understood, but it results in the shell rising to the surface; until it becomes waterlogged because of puncture, or sinks because of the weight of epizoans, the shell is free to drift on surface ocean currents. Living *Nautilus* distributions cover only a fraction of the area from which dead shells are known (Toriyama et al., 1966; House, 1973). Even if there are living *Nautilus* populations in the western Indian Ocean (Reyment, 1973), dead shells appear to float for hundreds if not thousands of kilometers. This is potentially important to wide dispersal and thus the biostratigraphic utility of ammonites as well. Distribution patterns of some ammonites are compatible with extensive necroplanktonic drifting (Walther, 1897; Scupin, 1912; Reyment, 1958), but other distributions indicate that extensive drift did not occur (Scupin, 1912; Diener, 1912; Kessler, 1923). The abundance of ammonites in offshore or relatively deep water sediments, often without obvious sign of puncture or heavy overgrowth of eopizoans, suggests that many ammonites never floated at all after death or that they rapidly became waterlogged (Gill and Cobban, 1966).

Preburial Dissolution

Ammonite shells are aragonitic and may be dissolved rather than buried if situated below the aragonite compensation depth. Jefferies (1962, 1963) and Hudson (1967) produced convincing evidence that aragonite dissolution can occur at depths of only a few hundred meters. Many sediments lacking ammonites and other originally aragonitic organisms contain zenomorphic oysters and other calcitic epizoans whose attchment areas show traces of the form and ornamentation of diverse aragonitic shells. Absence of ammonites from many open marine sediments, in particular those of the European Chalk sequence and the Niobrara Chalk, could be a result of dissolution.

Postburial Events

Chamber infilling, dissolution of the aragonitic shell, and exhumation and reburial of prefossilized steinkerns is a largely ignored problem, but one that greatly complicates biostratigraphic usage and facies associations of ammonites. The extraordinary range and complexity of these processes have been described by Seilacher (1963, 1966, 1968, 1971). Of particular significance to biostratigraphers is the occurrence of phosphatized remanié ammonites in strata bearing younger autochthonous ammonites, so that dating and correlation are confusing. Such occurrences are particularly important in condensed phosphorites, such as those of the type Vraconian and Clansayesian, and the Cambridge greensand. As many as five ammonite zones may be mixed in one of these condensed sequences, being separable in some places on the degree of phosphatization and wear of the specimens.

TAXONOMIC PROBLEMS IN BIOSTRATIGRAPHIC USAGE

Intraspecific Variation

Ammonites have suffered perhaps more at the hand of the taxonomist than any other group of organisms used widely in biostratigraphy. Failure to recognize and understand the great range of intraspecific variation shown by many forms led to entire works devoted to extensive splitting of a single species or only a few species, such as Buckman's monographic account of the ammonites of the British

Inferior Oolite or Arkell's monograph on the Corallian ammonites. Subsequent authors have reduced many of these accounts to meaningful taxa, usually involving a drastic reduction in numbers of genera and species; Donovan (1958) revised the Early Jurassic ammonite family Echioceratidae from 19 to 5 genera, and Westermann (1966) revised the Bajocian ammonite *Sonninia (Euhoploceras) adicra* (Waagen) (♀)/*Sonninia subdecorata* Buckman (♂) to include *78* previously described "species" and "subspecies."

The biostratigraphic problems raised by failure to recognize high intraspecific variation are obvious. Small numbers of individuals from different areas may be assigned to several "species" and "genera," obscuring synchroneity and similarity of faunas. Differences between successive species are often small by comparison with the wide range of variation within single species. Critical comparison of time-successive ammonite populations within lineages appears to present the most reliable means of defining, identifying, and zoning ammonites.

Homeomorphy

Homeomorphy is widespread in ammonites. Evidence is strong that the classification used in the *Treatise* (Arkell et al., 1957) consists in part of polyphyletic grouping of homeomorphs (Wiedmann, 1963a, 1963b, 1966, 1969, 1970b; Kullmann and Wiedmann, 1970). Many authors have drawn attention to the more striking examples of heterochronous homeomorphy (Schindewolf, 1938, 1940; Haas, 1942; Reyment, 1955). That such homeomorphy presents a trap to the unwary stratigrapher is obvious, especially when available material is poor; there are many records of strata misdated as a result of failure to recognize homeomorphs. Apparent ranges of species and genera have also been unduly extended for similar reasons. For example, Reeside and Weymouth (1931) originally dated the Aspen Shale of southwestern Wyoming as Turonian on the basis of poorly preserved specimens of "*Kanabiceras,*" "*Metoicoceras*," and "*Acompsoceras*," which are known to belong entirely to the homeomorphic late Albian or early Cenomanian hoplitid *Neogastroplites cornutus* (Whiteaves). Recognition of Barremian, Albian, Cenomanian, and Turonian faunas from Salinas, Angola (Douvillé, 1931), were based entirely upon a late Cenomanian assemblage consisting of a series of homeomorphs of well-known Barremian, Albian, and Turonian taxa as well as undoubted Cenomanian forms. As examples of erroneous stratigraphic ranges of species and genera, one may cite many of the "Albian" records of the Cenomanian genus *Schloenbachia*, based on the homeomorphous *Mortoniceras*, or "Aptian" records of the exclusively Turonian genus *Collignoniceras*, which are based upon the homeomorphous Albian genus *Lyelliceras* (Van Hoepen, 1965).

AMMONITES AS FACIES FOSSILS

The inferred life habitats of ammonites suggest that they should show variable facies-linked distributions, as well as independence from facies control. Facies-linked distribution would reflect a benthic habit and limited postmortem transport;

GENUS \ FACIES	CHALKS	GLAUCONITIC CHALKS	PHOSPHATIC FACIES	CALCARENITIC LIMESTONES	CLAYS	NONCALCAREOUS SHALES	CALCAREOUS SHALES	SILTSTONES	SANDSTONES	GREENSANDS
DOUVILLEICERAS	○	○	●	●	●	●	●	●	●	●
SCHLOENBACHIA	●	●	●	●	●	●	●	●	●	●
PLACENTICERAS	●	●	●	●	●	●	●	●	●	●
MORTONICERAS	○	○	●	●	●	●	●	●	●	●
MANTELLICERAS	●	●	●	●	●	●	●	●	●	●
ACANTHOCERAS	●	●	●	●	●	●	●	●	●	●
CALYCOCERAS	●	●	●	●	●	●	●	●	●	●
METOICOCERAS	●	●	●	●	●	●	●	●	●	●
COLLIGNONICERAS	●	●	●	●	●	●	●	●	●	●
HYPOPHYLLOCERAS	●	●	●	●	●	●	●	●	●	●
TURRILITES	●	●	●	●	●	●	●	●	●	●
SCAPHITES	●	●	●	●	●	●	●	●	●	●
SCIPONOCERAS	●	●	●	●	●	●	●	●	●	●
TEXANITES	●	●		●		●		●	●	●
SPHENODISCUS	●			●	●	●		●	●	●
COILOPOCERAS	●	●			●	●			●	●
GAUDRYCERAS	●	●	●		●			●	●	
MAMMITES	●	●		●		●			●	
FORRESTERIA	●		●					●	●	
PRIONOCYCLOIDES					●		●			
BUDAICERAS				●						
NEOSAYNOCERAS					●	●				
FLICKIA					●					
FISCHEURIA					●					

● OCCURRENCE ○ FACIES OF SUITABLE AGE NOT DEVELOPED

Text-Figure 1
Facies occurrences of selected Cretaceous ammonite genera.

facies-independent distribution reflects nektonic or planktonic life habits plus possible postmortem drifting. Text-figure 1 summarizes distribution data for some Cretaceous ammonite genera that we have studied; these distributions (which can be duplicated at specific levels) indicate a high degree of facies independence for many forms. When studied in detail, however, these species and genera commonly show facies-linked variations in relative abundance and/or developmental stage. Broad facies-linked distributions of particular ammonite morphotypes have been frequently sought, as in the examples cited by Scott (1940) and Kauffman (1967a).

Our own view follows that of Ziegler (1967) and others in that no simple predicable patterns are apparent.

Facies-controlled distributions of ammonites are known, however. Several Albian and Cenomanian micromorph genera (e.g., *Flickia, Neosaynoceras,* and *Prionocycloides*) seem to be restricted to offshore clay facies in widely separated geographic regions, such as Algeria, Tunisia, and Madagascar. Some late Campanian nostoceratid heteromorphs seem to have been strongly facies controlled; *Anaklinoceras reflexum* Stephenson is known only from nearshore sandstones in Delaware and Colorado and from shallow-water marl in Texas. Obviously, some knowledge of the degree of facies control is essential in choosing ammonite groups for reliable regional correlation.

GEOGRAPHIC LIMITATIONS ON STRATIGRAPHIC USAGE

Distribution patterns shown by ammonites are the result of a combination of several factors: (1) distribution of living animals by larval dispersion, which depends upon length of larval life, prevailing surface currents, predation rate, and sensitivity of larvae to environmental fluctuations of marine surface layers; (2) postlarval history: settlement leading to a benthic life will reflect the distribution of suitable bottom conditions; a nektonic life will lead to a distribution reflecting tolerance limits of environmental parameters throughout the water column; and a planktonic life will lead to distributions linked to oceanic currents and limited by near-surface environmental parameters; and (3) postmortem distribution, the result of necroplanktonic drift. If minimal, this will scarcely modify the original life distribution; but if comparable to that seen in *Nautilus,* drift will produce a far wider distribution than that shown by the living animals.

Compilations of distribution data have been published in two recent texts (Middlemiss and Rawson, 1971; Hallam, 1973). In these works, various authors have summarized overall distributions during short intervals of time. For the purpose of measuring the biostratigraphic worth of ammonites, we would rather consider the *patterns* of distribution shown by the group. In view of the extensive publications on Jurassic ammonites, we have taken our examples from Cretaceous forms. In general, five life distribution patterns can be recognized: pandemic, latitudinally restricted, endemic, disjunct, and necrotic.

Pandemic Distributions

Arkell (1956b) remarked on the nearly worldwide distribution of many genera and species of Jurassic ammonites; this was later confirmed by Dietl (1973) and Matsumoto (1973). As examples, we have plotted on Text-figure 2 the known distribution of the normally coiled Campanian tetragonitid genus *Pseudophyllites* and the Cenomanian heteromorphic species *Turrilites costatus* Lamarck. *Pseudophyllites* is common in Antarctica, South Africa, Madagascar, southern India, Japan, Greenland, and California. The type species, *P. indra* (Forbes), has almost

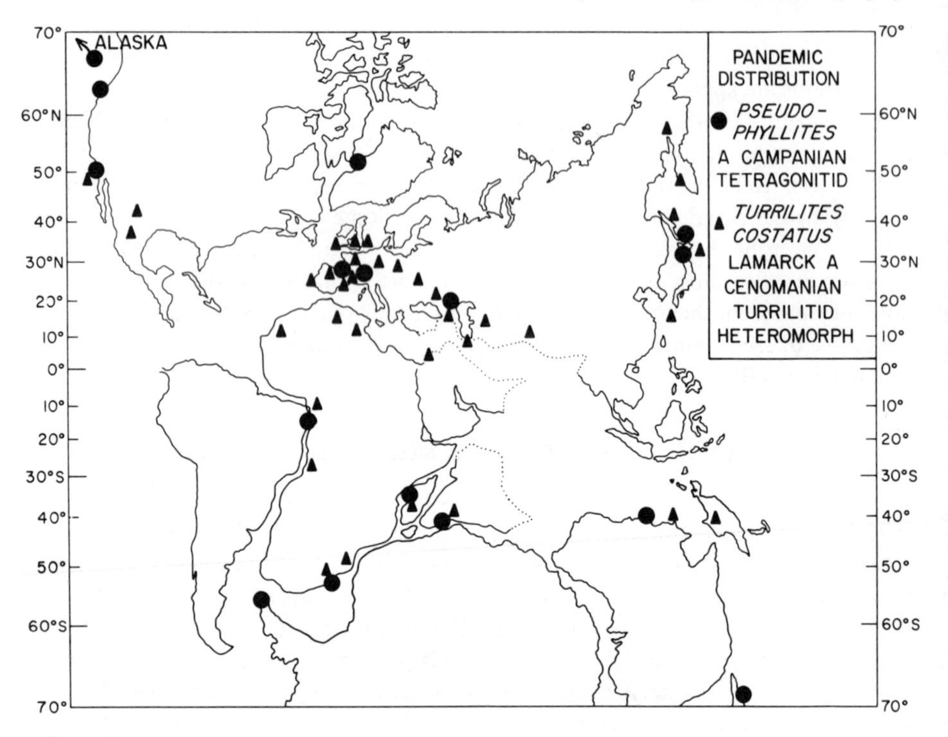

Text-Figure 2
Pandemic distribution of the normally coiled Campanian tetragonitid genus *Pseudophyllites* (circles) and the Cenomanian heteromorphic species *Turrilites costatus* Lamarck (triangles). (Map base modified after Smith et al., 1973, fig. 7.)

as wide a distribution as does *Turrilites costatus,* known from 60° north to almost 60° south. Both species are common throughout their geographic range; their distribution cannot merely be the result of necroplanktonic drift.

Pandemic or near-pandemic distribution presumably reflects the eurytopic nature of the organisms concerned, and these organisms are often marked by facies independence. Potentially prime groups for biostratigraphy, both species are, however, relatively long ranging, as are many other pandemic forms.

Latitudinally Restricted Distributions

Ammonities, in common with other elements of the marine biota, show decreasing diversity when traced poleward. Many groups reveal marked latitudinal restrictions reflecting their relative intolerance of fluctuating environmental conditions. Many distributions of this type have been documented, in particular, elements of Tethyan faunas whose distribution parallels that of many hermatypic corals, rudistid bivalves, and larger Foraminifera. Hallam (1969, 1971), Cariou (1973), Enay (1973), and Howarth (1973a) have documented Jurassic distributions of this type. Text-Figure 3 illustrates a comparable pattern shown by a series of mid-Cretaceous pseudoceratitic hoplitid genera.

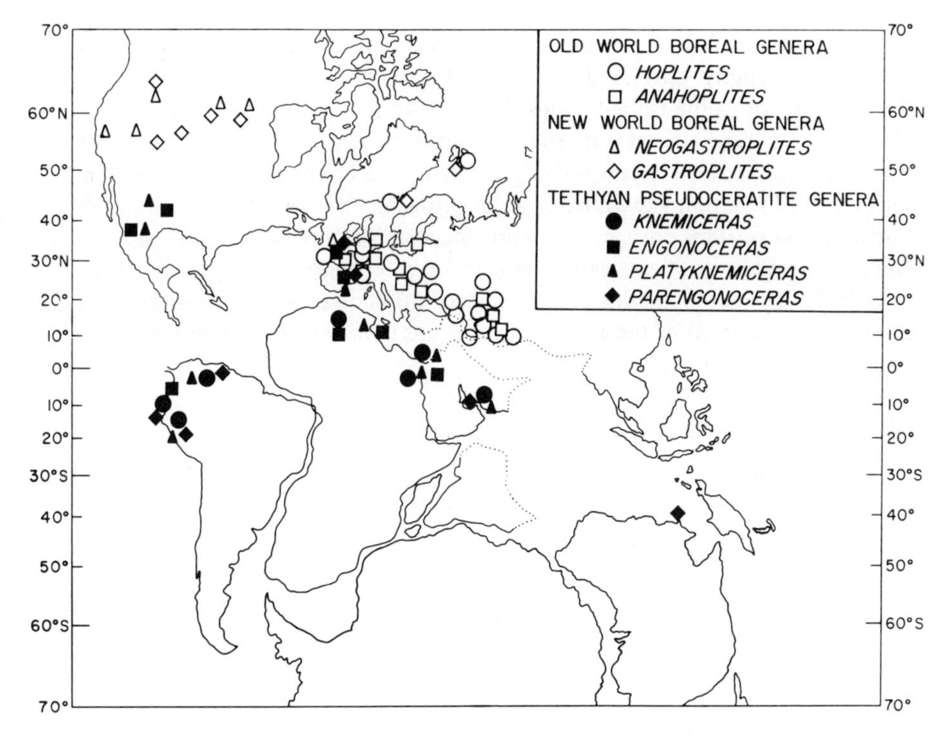

Text-Figure 3
Latitudinally restricted and endemic distribution patterns of some mid-Cretaceous ammonite genera. (Map base modified after Smith et al., 1973, fig. 7.)

Endemic and Provincial Distributions

If longitudinal limitations on distributions (mainly land masses or large ocean basins) are added to latitudinal restrictions, provincial or endemic distribution patterns result. These distributions reflect the presence of barriers to migration. Regional faunal differentiation may also result from salinity gradients (e.g., see Hallam, 1969).

Provincial faunas characterized by many taxa are well known, such as the Old World North Temperate ("Boreal") faunas documented by Hallam (1969, 1971) in Jurassic ammonites, or the various provinces (sub-"Boreal," Mediterranean, Indo-Malagasy, Ethiopian, Himalayan) recognized by Cariou (1973), Enay (1973), and others. Provincialism in Cretaceous ammonites is, likewise, clearly indicated in distributions plotted by Freund and Raab (1969), Matsumoto (1973), and others. The identity of the faunal regions ranges from those that are not persistent and identified by only a few taxa, such as Arkell's (1956b) Pacific Realm, to those characterized by high endemism through long periods of time, such as the "Boreal Realm" of the Late Jurassic and Cretaceous (Hallam, 1969; Juignet and Kennedy, 1977) or the Cretaceous of the Western Interior endemic center. Examples of the latter are two groups of endemic mid-Cretaceous hoplitids, the New World *Gastroplites-Neogastroplites* complex and the Old World *Hoplites* and related forms.

Endemic ammonites are, paradoxically, prime biostratigraphic indicators because they evolved rapidly. Species ranges of Late Cretaceous baculites confined to the Western Interior of the United States allow a precision of correlation as low as 0.2 MY. In general, these species are so restricted to the Western Interior that detailed correlation with an area as near as the Gulf Coast cannot be made by means of them. This type of limitation was probably at its extreme in the Late Jurassic and Early Cretaceous in Western Europe, where provincialism of "Boreal" faunas resulted in a complete breakdown of correlation with Tethyan sequences from late Oxfordian to early Valanginian time; this led to the recognition of an alternative stage nomenclature and placement of the Jurassic-Cretaceous boundary at different points in time in the two regions (Casey, 1963).

Disjunct Distributions

The common occurrence of identical genera and species in now widely separated areas may represent no more than a result of continental displacement. There remain, however, examples of genera and species having disjunct distributions even after continental reassembly. In Text-figure 4, we have plotted the distribution of three

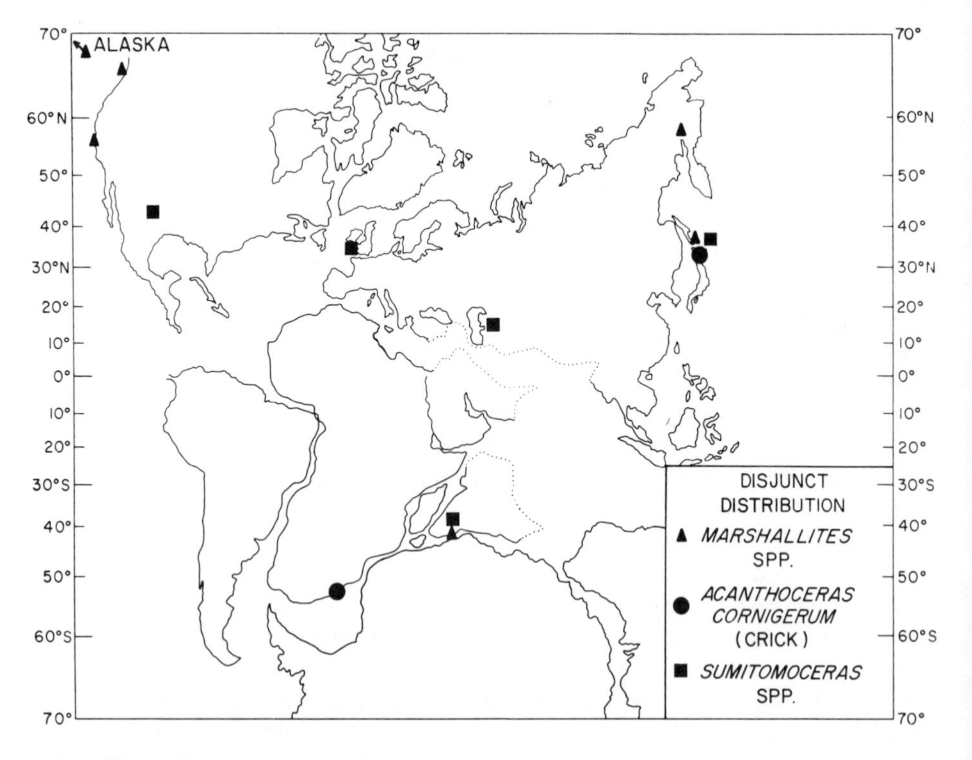

Text-Figure 4
Disjunct distribution patterns of two Cenomanian ammonite genera and a Cenomanian species. (Map base modified after Smith et al., 1973, fig. 7.)

Cenomanian genera from intervals that have been intensively studied over a wide area. These distributions are apparently not an artifact of inadequate data, nor do they reflect a facies-linked distribution. Rather, they may be records of species maintained at very low densities throughout their geographic range; as such, they have considerable potential in correlation of widely separated areas.

Necrotic Distributions

If the possibility of *postmortem* drift of dead shells is accepted, the biostratigraphic potential of some ammonites is greatly increased. All the distribution patterns already outlined may, in part, reflect necroplanktonic distribution, and some occurrences could be interpreted as entirely due to this means. Chief among them are the records of individuals of essentially endemic species far from their centers of abundance, such as the endemic Western Interior Albian gastroplitids from East Greenland, Spitzbergen, and even Kent, England (Spath, 1937), and scarce Tethyan lytoceratids, phylloceratids, and pseudoceratites recorded from the "Boreal Realm" (e.g., Donovan, 1954). A graphic example of such a distribution is shown in Text-figure 5; the genus *Borissiakoceras*, locally common in middle Cenomanian strata in the Western Interior of North America and northern Alaska, occurs sparingly in Texas and Turkestan and is represented by a single individual in western Europe (Kennedy and Juignet, 1973). Such distributions, perhaps the result of westward drift on some proto-Gulf Stream, have no paleobiogeographic significance, but they can be of great value in correlating endemic faunas.

ZONAL DURATION

Ammonites have formed the standard zonal schemes of the Triassic, Jurassic, and Cretaceous periods. Their stratigraphic worth is measured best in terms of zonal duration versus areal scope of the zonal scheme concerned.

As a general indication of stratigraphic value, we have calculated average zonal duration for the Triassic, Jurassic, and Cretaceous periods utilizing the "standard" ammonite zonations used in the *Treatise* (Arkell et al., 1957) and the best available radiometric dates as reviewed by Harland and Francis (1971). The average zonal duration for the Triassic and Jurassic periods is 1.2 MY, and for the Cretaceous, nearly 2.0 MY. These figures are, however, standards for various parts of the world (chiefly Europe) and do not give an accurate measure of the merits of the group. The most precise evaluation of the group available at present is in the Late Cretaceous successions of the Western Interior of North America. There, detailed studies of largely endemic lineages have produced a very fine biostratigraphic framework, which can be dated at many points by virtue of the presence of extensive bentonite beds within the succession. Obradovich and Cobban (1975) have been able to integrate a series of potassium-argon dates very precisely with ammonite successions. Durations of 0.2 to 0.9 MY are revealed; the average duration is 0.6 MY. These figures represent the results of detailed studies of lineages that can be applied over only a limited area. These essentially range zones can, however, be correlated with a global assemblage zonation for the Upper Cretaceous. Thus,

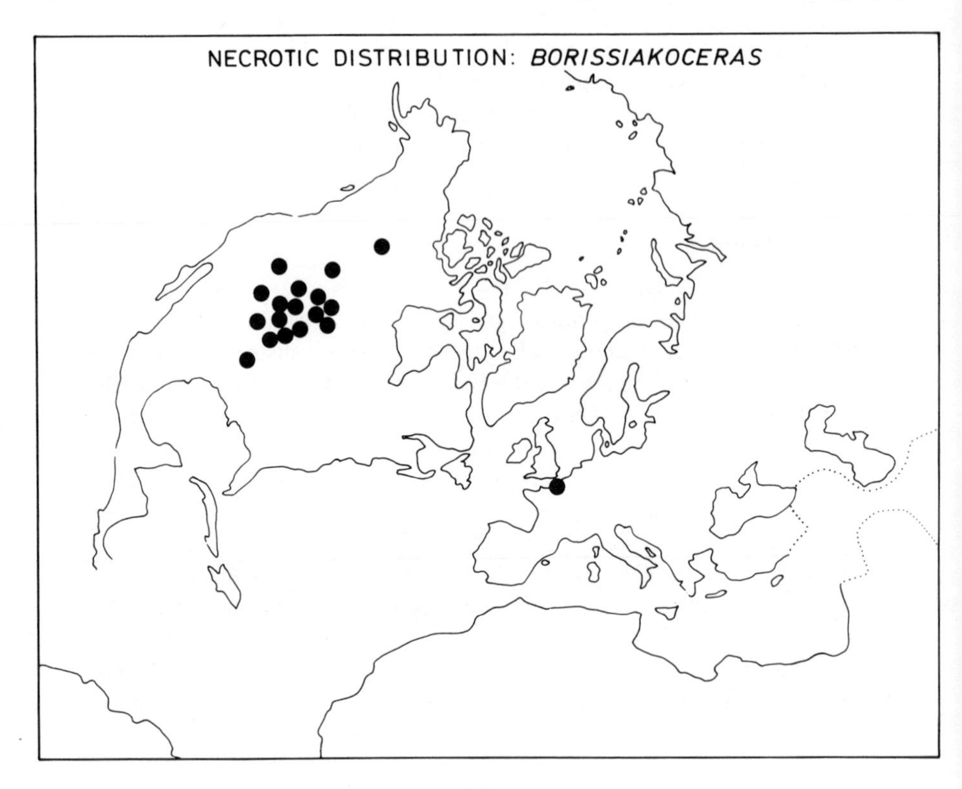

Text-Figure 5
Necrotic distribution of *Borissiakoceras* during middle Cenomanian time. (Map base modified after Smith et al., 1973, fig. 15.)

at this level, a global correlation of the Upper Cretaceous is possible that has an average zonal duration of 1.8 MY and limits of 0.5 to 3.0 MY.

Compilation of similar figures for the pre-Upper Cretaceous is severely limited by the lack of acceptable radiometric dates throughout much of this interval. The best estimates of zonal durations during the Jurassic can be gathered from the Lower Jurassic (Lias) sequences of western Europe, where Dean et al. (1961) have recognized 10 zones of 1.2 MY average duration and 49 subzones of 0.5 MY average duration in the Hettangian and Toarcian interval.

Acknowledgements

This project was supported in part by a Lindemann Visiting Fellowship awarded to W. J. Kennedy, receipt of which is gratefully acknowledged. We thank J. D. Obradovich, J. M. Hancock, and H. C. Klinger for useful discussions and for allowing us to use their unpublished information.

Graptolite Biostratigraphy: A Wedding of Classical Principles and Current Concepts

William B. N. Berry

University of California, Berkeley

INTRODUCTION

Graptolites have long been considered one of the more ideal fossil groups for biostratigraphic use. The class Graptolithina includes the fossil remains of colonial animals; some had a sessile benthic mode of life, and others a planktonic mode of life. The sessile benthic colonies included those that formed encrustations and those that had a bush-like or shrub-like aspect; these have proved to be of little value in biostratigraphy because the fossil remains are commonly highly fragmented and the basic colony form appears to have been relatively stable for long intervals of time. Furthermore, only a few species are widely distributed. The representatives of the class Graptolithina that had a planktonic mode of life (most of the members of the order Graptoloidea), on the other hand, have proved to be extremely valuable to geologists as a tool for documenting short intervals of geologic time and for establishing rock unit correlations. Planktonic graptolites most commonly come to mind when the appellation "graptolite" is used; this discussion will refer hereafter only to them.

Biostratigraphic use of graptolites may be viewed historically in three basic developmental segments. The first included analyses of the relative stratigraphic positions of species and whole faunas in local areas to establish graptolite zones. The analyses were commonly coupled with systematic descriptions. The second encompassed realization that graptolite zonal successions are limited in their applicability to certain areas and rock suites. The third, the current area of interest, involves recognition of specific phylogenies among the graptolites and patterns of graptolite distribution that suggest certain niche breadths and niche changes in some lineages through geologic time.

CLASSICAL BACKGROUND: THE HERITAGE

That graptolites might be used to recognize stratigraphically lower from stratigraphically higher parts of certain graptolite-bearing sequences was documented during the latter part of the nineteenth century and the early years of the twentieth in studies of lower Paleozoic strata in Sweden (Linnarsson, 1876; Tornquist, 1897, 1899, 1901-1904; Tullberg, 1883), Australia (T.S. Hall, 1894, 1896, 1897, 1898, 1899), Great Britain (Lapworth, 1879-1880), North America (Ruedemann, 1904, 1908), and Bohemia (Perner, 1894-1899). Lapworth (1879-1880) summarized the then existing knowledge of the stratigraphic distribution of European graptolites, and suggested recognizing 20 graptolite zones in the latest Cambrian into the Silurian part of the British Early Paleozoic sequence. Lapworth then encouraged E.M.R. Wood and Gertrude Elles to describe and illustrate the British graptolites as fully as possible and to summarize the existing knowledge of them. They carried out this task over several years and produced the now classic *Monograph of British Graptolites* (Elles and Wood, 1901-1918).

Both Lapworth (1879-1880) and Elles and Wood (1901-1918) considered not only the systematic paleontologic, but also the biostratigraphic needs of geologists in their work by first describing and illustrating the graptolite species, and then by trying to summarize the existing vertical stratigraphic range data for each species. Elles and Wood (1914, p. 514) wrote as follows in the concluding pages of their monograph:

> In the preface to this Monograph it was pointed out (Preface, p. 4) that the primary need for a work of this kind was the natural demand of the field geologist and the palaeontologist for figures and descriptions of the British species. Having now completed the figuring and descriptions of the first and most typical section of the British Graptolites—namely, the Graptoloidea—we shall best serve the interests of the field geologist and the palaeontologist if we here summarise in tabular form the main facts bearing upon the vertical range and association in the British Isles of the various species and varieties of the Graptoloidea noticed in the preceding descriptive pages (pp. 1–513) of this Monograph . . . .

Elles and Wood (1914) went on to note that their tabulation of the vertical stratigraphic ranges of the graptolites was an extension of a similar tabulation assembled by Lapworth 33 years earlier.

When Lapworth (1979-1880) and Elles and Wood (1914) examined the vertical stratigraphic ranges of the species, they noted that the overlapping aspect of the ranges of the several species produced unique associations by which each zone could be characterized. Lapworth (1880, p. 202) concluded his summary of the graptolite zones by commenting that "future research will soon fix more definitely the composition and limit of the characteristic (Graptolite) faunas of the zones already recognized, extending the range of some of the forms into neighboring ones, and adding largely to the number of the zones themselves." Elles and Wood (1901-1918) amply documented Lapworth's prediction by recognizing 36 zones in the same stratigraphic interval in which Lapworth had established 20 and by adding substantially to the number of taxa present in British graptolite-bearing sequences. Elles and Wood (1914, p. 515) emphasized that each graptolite zone

is characterized by a *special association* of Graptolites, and that that form in this association which apparently combines restricted vertical range with wide horizontal distribution is most conveniently selected as the *index* of the zone. It is not pretended that each of the Graptolite zones is of equal geological importance. Each zone, however, marks a special stage or horizon in the ascending series of the Lower Palaeozoic rocks. In other words, the Graptolite zones are comparable more or less with the Ammonite zones of the Mesozoic formations. As such their importance as indices of *chronological sequence* can hardly be over estimated; but to what extent they will prove of value as indices of *chronological duration*, only time and future geological research and discovery can be expected to show.

Elles and Wood's (1914) discussion of graptolite zones has been cited at length because the zones they recognized have long served as a basic cornerstone in graptolite biostratigraphy. Their remarks indicate that the graptolite zones classically have been established on the same fundamental principle that Oppel (1856-1858) laid down in his analysis of Jurassic ammonite distributions, i.e., the synthesis of the overlapping vertical stratigraphic ranges of many species through rock sequences over a broad area to recognize certain unique associations of species (the *special associations* of Elles and Wood, 1914, and the *congregations* of Berry, 1966) that characterize each zone. Oppel's principle of zonation permits not only graptolite and ammonite zones, but also zones based upon the overlapping vertical stratigraphic ranges of species of any fossil group to be used as a basis for demarcation of short duration time intervals in the geologic past.

The basic steps in the analysis of overlapping stratigraphic ranges of species include collecting specimens from precisely located stratigraphic positions, from many stratigraphic sections over a broad area. Identification of the taxa is followed by plotting of their ranges in terms of the stratigraphic positions where each was found. The plotting of ranges will reveal overlapping patterns and varying evolutionary histories within and between lineages (Text-fig. 1). These patterns are then studied, and certain congregations (special associations of Elles and Wood, 1914) of taxa are selected as characteristic of certain intervals, termed zones (see Text-fig. 1). The boundaries between zones may be selected at certain first appearances of taxa in the sequence and the disappearance of others (Text-fig. 1). This procedure incorporates many taxa and, in essence, summarizes the evolutionary development of a group as seen in the rock sequence collected. The associations that characterize each zone are unique in time; because they are, they may be used as a basis for delineating chronological units.

As Lapworth (1879-1880) pointed out, the zones originally recognized are subject to division as additional collecting from newly found, as well as initially studied, stratigraphic sections adds new species and new occurrences to the body of data concerning stratigraphic ranges. The zonal range of a species may be extended by means of new collecting if that species is found to occur with special associations or congregations of species that characterize a zone or zones other than those in which it was initially found.

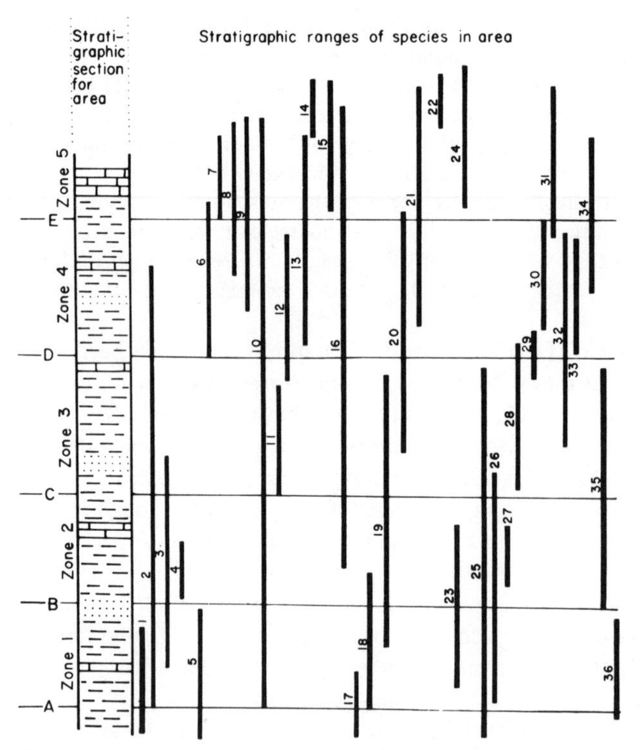

Text-Figure 1
Schematic diagram to indicate Oppel's (1856–1858) principle of zonation. The generalized stratigraphic column is intended to include stratigraphic ranges from several stratigraphic sections throughout a hypothetical broad area. Vertical stratigraphic ranges of taxa identified in collections from many sections are summed up in the section shown. Inspection of the pattern of overlapping vertical stratigraphic ranges suggests positions of zone boundaries (A, B, C, D, E) based on first and last occurrences, as well as restricted ranges of certain taxa. Boundary B, for example, is that thickness of rock in which species 4, 27, and 35 make their first appearance and species 1, 5, and 36 their last. The joint association of species 4, 18, 23, 27, and 35 is unique in the rocks designated zone 2 and is unique in time as well. That association of species comprises the congregation that characterizes zone 2.

Despite Lapworth's (1879-1880) words of caution against making the assumption that, once the range of a species was documented in a published tabulation, the range would never change, the stratigraphic ranges of the graptolite species tabulated by Elles and Wood (1914) commonly have been taken to be absolute the world around by many geologists. Strachan (1971, pp. 2-3 called attention to this lack of understanding of the data behind the Elles and Wood (1913) tabulation by stating:

The stratigraphical range chart which appeared in Part X consisted mainly of the information which had been quoted in the systematic descriptions, modified by Miss Elles's notes on material which she had identified since for the Geological Survey. In many cases the descriptions were based on material collected by Lapworth during his early work in 1870–80 which was not stratigraphically determined with the precision of the zonal scheme put forward in the Monograph, and hence the ranges as published do not have the validity which is assumed in many later published works.

As not only Strachan's (1971) comments on the British graptolite occurrences, but also stratigraphic collecting in British graptolite-bearing sequences in Britain by Toghill (1968, 1970), Rickards (1967, 1970), and P.T. Warren (1971) indicate, new data concerning ranges of species, new species found in the British succession, reevaluation of recognized species, as well as reexamination of originally recognized stratigraphic sections in Britain, has led to certain modifications of the Elles and Wood (1913) zones. The Elles and Wood zones have served, however, as a sort of platform that has been amplified and modified by graptolite biostratigraphic investigations in the years since their recognition, and they are the heritage that graptolite biostratigraphers recognize as the foundation, at least in principle, for their labors.

DEVELOPMENTS SINCE THE ELLES AND WOOD ZONES

Graptolite zonal sequences have been documented through a part or all of the Ordovician and Silurian stratal record in many areas of the world. As the world's graptolite faunas have become increasingly better known, and faunas from many areas are more fully documented, both biogeographic patterns of graptolite distribution and paleogeographic positions of richly fossiliferous graptolite-bearing rock units become increasing clear. The biogeographic patterns include faunal regions and provinces (Bulman, 1958, 1970; Berry, 1959, 1966, 1967; Skevington, 1969, 1973; among others). Examination of the paleogeographic distributions of richly fossiliferous graptolite-bearing rock suites, as well as rock sequences bearing primarily shelly faunas and some graptolites, has led to recognition of the geographic positions and oceanic depths in which the graptolites found optimum and near-optimum envirionmental conditions (Elles, 1939a, 1939b; Ross and Berry, 1963; Ross and Ingham, 1970; Berry, 1972, 1974; Berry and Boucot, 1967, 1970, 1972). In addition, an attempt has been made to document certain phylogenies among the

Silurian graptolites (Sudbury, 1958; Hutt et al., 1972). The patterns of biogeographic and paleogeographic distributions provide certain limits to the use of graptolite zones in chronology and correlation, such as restriction of a zonal sequence to a single zoogeographic province.

Zoogeographic Regions and Provinces

Based upon the distribution of graptolite faunas around the world, Berry (1959, 1966, 1967, 1968a), Bulman (1970), and Skevington (1969, 1973), among others, described two zoogeographic regions during at least the early part of the Ordovician, the Pacific (or Australian-American), and the European (or Atlantic). Skevington (1969) suggested that a degree of faunal provincialism persisted throughout the Ordovician. Berry (1973) reviewed distributions of Silurian and Early Devonian graptolites and indicated that graptolite faunas of that time span were essentially cosmopolitan.

Faunal regions and provinces create limits to the area throughout which a set of graptolite zones (indeed, zones based on the evolutionary history of any organismal group) may be recognized (Berry, 1966, 1967, 1968b; Bulman, 1970; Skevington, 1973). The absence of faunas characteristic of the Ordovician Pacific Faunal Region (e.g., *Oncograptus, Cardiograptus,* and most isograptids of the *I. victoriae* group) in Britain precludes recognition of North American Ordovician graptolite zones in the British provincial part of the European (Atlantic) Faunal Region. Similarly, the absence in North America of *Didymograptus murchisoni* group graptolites and other stocks characteristic of the British Faunal Province prohibits use of the British graptolite zones for the early part of the Ordovician in North America.

Areas of Environmentally Optimum Conditions

A second type of limit to the area in which a graptolite zonal succession may be used is that which results from the geographic spread of the rocks deposited in areas where environmentally optimum and near-optimum conditions for graptolites existed. Marine sediments formed under waters in which graptolites found environmentally most favorable conditions classically have been termed the "graptolite facies" or the "graptolite biofacies." Graptolites are found in greatest numbers for most species and greatest diversity in sedimentary successions deposited upon the continental (or platform) slopes and on the shelf or platform margins of the Early Paleozoic (Elles, 1939a, 1939b; Ross and Berry, 1963; Ross and Ingham, 1970; Berry and Boucot, 1968, 1970, 1972; Berry, 1972, 1974; Whittington and Hughes, 1972).

Elles (1939b) reviewed the relationship of the "graptolite facies" to rocks bearing dominantly trilobites and brachiopods in Britain; she noted that, during the Ordovician, brachiopods dominated faunas from the relatively inshore portions of the continental shelves, whereas trilobites were more prominent in the relatively more offshore parts of the shelves (Text-fig. 2). Berry (1972, 1974) summarized North American Ordovician biofacies relationships and indicated that faunas in the intertidal and shallow subtidal environments on Early Ordovician carbonate

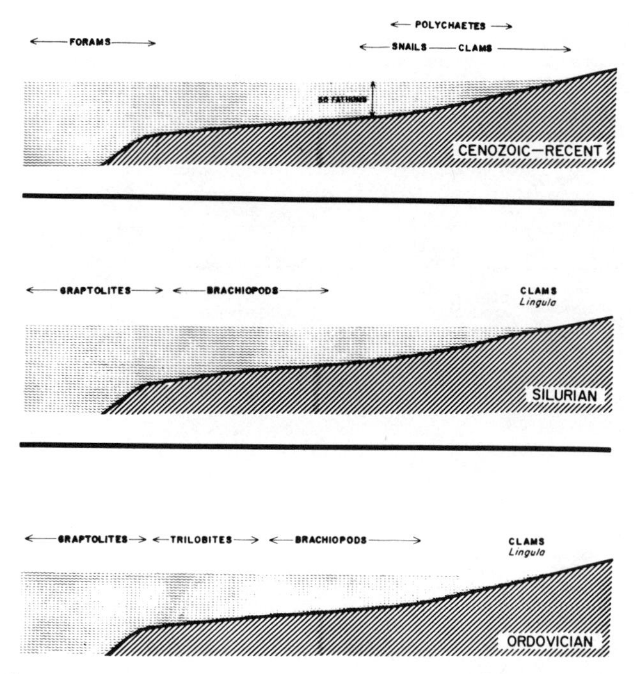

Text-Figure 2
Diagrams indicating the approximate positions of maximum abundance and diversity of selected marine invertebrate taxa in relationship to continental shelves and slopes of the Cenozoic–Recent, Silurian, and Ordovician. Zonal sequences based on the evolutionary history of a particular invertebrate group have their maximum applicability in rock suites deposited in areas in which the animal group was most abundant and diverse. (Data for Cenozoic–Recent taxa, from Bandy, 1961; Keen, 1963; Natland, 1957; Walton, 1964. Data for Ordovician and Silurian are summarized in Berry, 1972, 1974; Berry and Boucot, 1970; Elles, 1939a, 1939b.)

banks were dominated by certain archeogastropods and nautiloid cephalopods (Text-fig. 3). Brachiopods were relatively prominent in somewhat deeper subtidal environments on carbonate platforms. Trilobites dominated faunas near platform margins. Ordovician trilobite-dominated successions are situated where they commonly include interdigitations of the "graptolite facies." Differing tolerances to environmental factors permitted certain genera and species among the trilobites and brachiopods to overlap laterally from the environmentally optimum area of one to the optimum or near-optimum area for the other group (Elles, 1939a, 1939b).

Graptolites are rare in rocks bearing molluscan or brachiopod-dominated faunas, and they occur at infrequent intervals in rock successions bearing trilobite-dominated faunas. Obviously, graptolite zones are inapplicable in biostratigraphic studies in rock suites in which mollusks, brachiopods, or trilobites are most prominent. Similarly, nautiloid, snail, and brachiopod-dominated faunas are rare in graptolite-bearing rock sequences; hence biostratigraphic units based on these

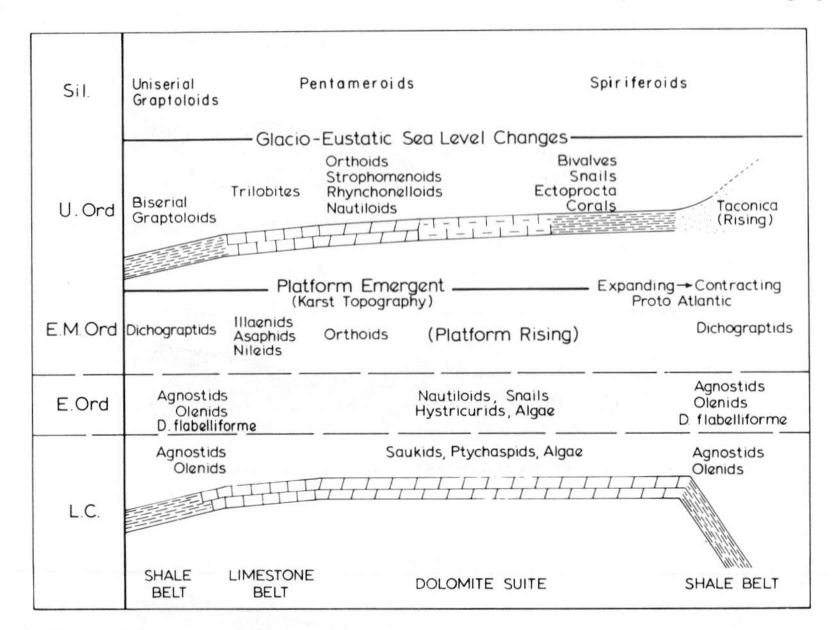

Text-Figure 3
Schematic diagrams drawn essentially west to east across the Early Paleozoic North American platform showing changes in primary rock suites, faunal replacements, tectonic developments, and changes in sea level (both tectonoeustatic and glacio-eustatic). Time intervals: L.C., latest Cambrian; E. Ord., Early Ordovician; E.M. Ord., early part of Middle Ordovician; U. Ord., latter part of the Ordovician; Sil., Silurian. The North American Platform was the site of a plexus of carbonate bank environments during the latest Cambrian-Early Ordovician (Lochman-Balk, 1970, 1971; Berry, 1972). Taconica provided a source for terrigenous sediments that spread across the platform to interdigitate with carbonates in carbonate bank and related environments during the latter part of the Ordovician and the Silurian (Berry and Boucot, 1970). Biochronological sequences (zones, stages) based on the evolution of the graptolites, trilobites, brachiopods, and perhaps other organismal groups in the latter part of the Ordovician and Silurian, for example, are a prerequisite to the patterns depicted. The several biochronological sequences must be equated to provide a network of rock unit correlations basic to documenting lithofacies, paleogeographic, and faunal replacement patterns.

animal groups are clearly inapplicable in biostratigraphic examinations of graptolite successions. Zonal sequences are thus established in rocks formed in areas where the organismal group used to recognize a set of zones is present in greatest abundance and diversity; such areas commonly are those where the organisms found the environmentally most favorable conditions. Because taxa in any organismal group have differing tolerances to environmental conditions, and because some taxa are tolerant of a wide variety of environmental situations, at least some members of any group of organisms on which a zonal sequence is based will be ecologically associated with those organisms on which another zonal sequence is established.

These ecological associations of members of one group with those of another, as well as the interfingering of rock sequences bearing, for example, the "graptolite facies" with the trilobite-rich succession, provide the clues to correlation between zones based upon two different organismal groups. Whittington (1968) suggested some North American Ordovician correlations among graptolite zones, triobite zones, and units based upon nautiloids. Berry and Boucot (1970) indicated correlation between Silurian graptolite zones and chronological units based upon brachipods.

Life Positions of Graptolites

Berry and Boucot (1972) analyzed occurrences of certain Silurian graptolites with brachiopod associations that were indicative of marine benthic life zones. Only a few graptolite taxa were found to occur in the relatively shallow water Silurian marine benthic life zones. Those taxa and a few others occur with brachipod associations indicative of somewhat deeper-water life zones on the shelf. All those graptolites and some others are found with brachiopods indicative of the deepest-water life zone on the shelf. The greatest numbers of graptolite taxa occur in rocks formed from sediment deposited upon the platform slopes. Berry and Boucot (1972) suggested that maximum graptolite diversity was in those waters that had temperatures characteristic of the outer platform margins and their slopes. The Early Silurian shallow shelf seas in northern Africa and southern Europe, for example, were relatively cold as a result of Early Silurian deglaciation (Berry and Boucot, 1973). Oceans in these areas were colder at shallower depths than oceanic waters in the areas of the present British Isles and North America, which were within Warm Temperate to Tropical regions during the early part of the Silurian. Graptolites were abundant and diverse in the apparently relatively shallow but cold waters across North Africa and southern Europe during the Early Silurian.

Because most graptolites floated in waters that were over platform margins and slopes poor in shelly faunas, and because graptolites were distributed at greater depths and lower temperatures than those in which diverse shelly faunas lived, correlations between graptolitic sequences and shelly faunal successions are commonly difficult to document. Those graptolites that floated in relatively near surface and warmer waters are most commonly found with shelly faunal associations and would seem to provide the most direct evidence for correlations between shelly and graptolitic biofacies. The Silurian graptolites that floated highest in the water column and occur with the shallowest brachiopod marine benthic life zone faunas were long-lived stocks, however, in which speciations were infrequent. The great longevity of species in these Silurian stocks makes correlations using them between the graptolite and shelly biofaces relatively broad.

Each graptolite distribution pattern described (zoogeographic provincialism, areas of most favorable environmental conditions, and life positions in the oceans) places limits on the usefulness of any set of graptolite zones. Despite the limits, at least general correlations may be established between graptolite and shelly faunal zones. Careful examination of stratigraphic sequences in which the graptolite and shelly faunas interdigitate and the phylogeny of those graptolites that occur with shelly faunas are requisite to such correlations. The correlations must be attempted, however, to provide a network of chronological units that may be used in making correlations of rock units over broad areas. Such correlations provide a framework basic to lithofacies and paleogeographic reconstructions (e.g., Berry and Boucot, 1970).

Phyletic History

The details of graptolite phylogeny have attracted little attention, although from the point of view of using graptolites in chronology, it is the evolution of the graptolites that lies at the roots of the special associations or congregations that characterize each zone. It is clear that chronological units such as zones, which have the evolution of a group of organisms at their foundation, may be erected and widely used even though the details of phylogenies are not known or understood. Analysis of overlapping vertical stratigraphic ranges provides a mechanism for overcoming lack of knowledge of specific phylogenies because it tends to embrace them.

Some knowledge of evolutionary lineages is helpful in establishing and using chronological units based upon graptolite evolution, because speciations are unique in time and thus are useful events upon which to establish boundaries between chronological units. Other events in evolutionary development, such as invasions of stocks that originated in one zoogeographic province into another province, are also unique events in time useful for drawing boundaries between chronological units. Berry (1967) drew attention to invasions into the Pacific Faunal Region during the Middle Ordovician of certain Ordovician graptolite stocks that had developed in the Atlantic Faunal Region. The appearance of some graptolites with the biserial scandent rhabdosome form (Text-fig. 4) into the Pacific Faunal Region and the North American Ordovician graptolite succession is an example of such an "invasion event."

Text-Figure 4
Schematic sketches indicating some of the evolutionary development of North American Ordovician graptolites. The general progression from many-branched forms (*Anisograptus, Clonograptus,* and *Adelograptus*) to few-branched dichograptids (*Tetragraptus, Didymograptus*), to reclined (*Isograptus*) and essentially biserial (*Cardiograptus*) dichograptids, and to several genera with biserial scandent rhabdosome form (*Glyptograptus, Paraglossograptus, Cryptograptus, Climacograptus, Pseudoclimacograptus, Glossograptus, Orthograptus, Amplexograptus,* and *Diplograptus*) and the reclined *Dicellograptus* and *Dicranograptus* is indicated by the sketches. Discussion of the major evolutionary developments is given in the text. These and certain other evolutionary developments (such as the appearance of *Nemagraptus*) among the North American Ordovician graptolites form the foundation for the zones. Zone numbers are indicated (1 to 15) on the sides of the diagram, and zone names are cited in Table 1. Zonal congregations are based on combinations of species of these genera. (*Orthograptus* q. refers to orthograptids of the *O. quadrimucronatus* group, and *Orthograptus* t. refers to orthograptids of the *O. truncatus* group.)

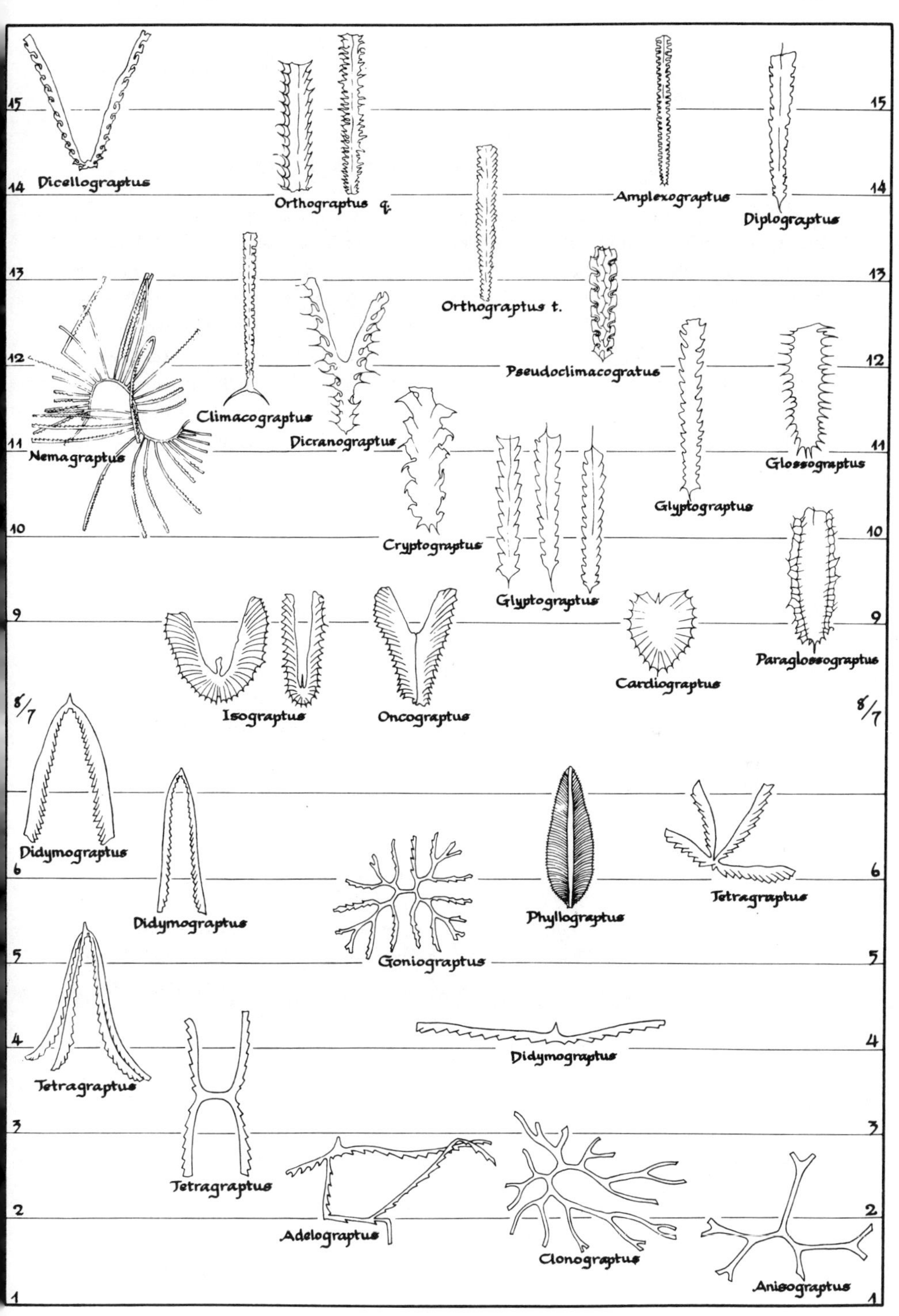
15
14
13
12
11
10
9
8/7
6
5
4
3
2
1
Dicellograptus
Orthograptus q.
Amplexograptus
Diplograptus
Orthograptus t.
Nemagraptus
Climacograptus
Dicranograptus
Pseudoclimacogratus
Glossograptus
Cryptograptus
Glyptograptus
Glyptograptus
Paraglossograptus
Cardiograptus
Isograptus
Oncograptus
Didymograptus
Didymograptus
Goniograptus
Phyllograptus
Tetragraptus
Tetragraptus
Didymograptus
Tetragraptus
Adelograptus
Clonograptus
Anisograptus

Sudbury (1958) and Hutt et al. (1972) suggested some phyletic developments among Silurian graptolites. Phylogenies among Ordovician graptolites are relatively little understood.

CURRENT AND FUTURE INVESTIGATIONS

Having reviewed the classical aspects of graptolite studies that led to recognition of the graptolite zones and then the realization that the zones had certain limits to their widespread usefulness, some future directions of graptolite biostratigraphic investigations may be noted. Clearly, one avenue for future inquiry is refinement of the distribution patterns. Second, the details of graptolite phylogeny must be examined. One aspect of phylogeny includes a consideration of the niche breadths and changes in niche breadths of certain graptolite stocks through time. If the niche breadth of any fossil species may be expressed in terms of the number of animal communities with which it occurs (community used herein in the sense of Petersen animal community of Watkins et al., 1973), then the evolutionary history of the niche of a given species may be investigated by examining its community relationships throughout its history.

Niche breadths among graptolites may be approached by taking into consideration the number of different marine benthic life zones, based upon shelly faunas, with which any species occurs. From such an analysis, the changes in depth positions (if any did take place) may be deduced and the adaptive responses of a species may be determined. Among Silurian graptolite faunas, for example, the ancestors of the pristiograptids may be sought among species that lived at relatively deep levels in the oceanic water column. The pristiograptids (*P. dubius* and relatives), however, lived at comparatively shallow depths in the oceans throughout their life history. They were relatively conservative in their evolution, with few speciation events. They comprise a stock that was relatively long lived in Silurian seas, yet more short-lived stocks (e.g., *Saetograptus*) apparently were derived from them. Another pattern is exhibited by the graptolites whose periderm was essentially reduced to a mesh (retiolitids). They appear to have been surface-water dwellers throughout their history, and a number of different but related taxa, all of which were also surface-water dwellers, appear to have developed regularly among them. D.E. Jackson (1973) described two Late Ordovician graptolite species that lived in shallow waters over carbonate bank environments. He drew attention to the absence of taxa such as the dicellograptids, which are so common in coeval platform marginal and slope sequences in the carbonate bank rock suites and their faunas. Records of graptolite occurrence such as this suggest that certain graptolites may have lived in specific water masses, and they help us to understand the presence or absence of certain graptolites in some areas or associations. Indeed, existing knowledge of graptolite distributions suggests that at least some graptolite species were closely limited to specific water masses or to waters over specific types of depositional environments. An implication for biostratigraphic studies of graptolites that results from these possible distribution patterns is that, as each graptolite fauna is found, it must be examined critically to determine if it is a new association in the general succession of graptolites (and the basis for a new zone, perhaps, or if it is a fauna coeval with others already known, but from a different environment or water mass from which no graptolites had been obtained previously.

NORTH AMERICAN ORDOVICIAN GRAPTOLITE ZONAL SUCCESSION–A CASE HISTORY

The development of a North American Ordovician graptolite zonal succession may be viewed as a case history to illustrate some of the general remarks made concerning graptolite biostratigraphy (see Table 1). A North American Ordovician graptolite zonal succession has developed slowly, starting from a number of essentially descriptive biostratigraphic studies, the results of which were several locally applicable descriptive zonal units. They were summarized by Ruedemann (1947). Berry (1960), following the principle of analysis of overlapping vertical stratigraphic ranges of many species through several rock sections over a broad area, recognized

Table 1
North American Province Ordovician Graptolite Zones and Possible Correlations with Series and Stages Based on Shelly Faunas (Trilobites, Brachiopods)[a]

North American Province Ordovician Graptolite Zones			*Shelly Fossil Series and Stages*
15 *Dicellograptus complanatus ornatus*			Richmond
14 *Orthograptus quadrimucronatus*			Maysville
	13b *Climacograptus spiniferus*		Eden
13 *Orthograptus truncatus intermedius*			Barnveld
	13a *Corynoides calicularis*		
12 *Climacograptus bicornis*			Wilderness
11 *Nemagraptus gracilis*			Porterfield
10 *Glyptograptus* cf. *G. teretiusculus* (or *G. euglyphus*)			Ashby
9 *Paraglossograptus etheridgei*			Marmor
8 } *Isograptus victoriae* group	*Isograptus victoriae*	subzone	Whiterock
7 }	*Didymograptus bifidus*	subzone	
6 *Didymograptus protobifidus*			
5 *Tetragraptus fruticosus,* three and four branched			Canadian Series
4 *Tetragraptus fruticosus,* four branched			
3 *Tetragraptus approximatus*			
2 *Clonograptus–Adelograptus*			
1 *Anisograptus* (with *Dictyonema flabelliforme* as a lower subzone)			

[a]See Whittington, 1968, for shelly fossil correlations.

15 graptolite zones in the Ordovician succession in the Marathon region, west Texas. That zonal succession was a set of working hypotheses to be tested for their general applicability in other areas in North America by analysis of the vertical stratigraphic ranges of species in rock sequences in many other areas. Ruedemann's (1947) review of eastern North America graptolite sequences and reinvestigations of some of the sequences (Berry, 1962, 1963; Osborne and Berry, 1966; Riva, 1969) documented the presence there of many zonal congregations characteristic of the zones recognized in the Marathon region in eastern North American graptolite-bearing sequences. Additional species' stratigraphic ranges and association data came from studies of graptolite-bearing successions in Newfoundland (Kindle and Whittington, 1958; Whittington and Kindle, 1963), the Yukon and adjacent areas in western Canada (Jackson and Lenz, 1962; Jackson et al., 1965; Larson and Jackson, 1966; Jackson, 1964, 1966) the Great Basin (Ross and Berry, 1963), and east-central Alaska (Churkin and Brabb, 1965). The zonal congregations are still being tested and modified (Berry, 1970) as collecting continues in many areas of North America.

The graptolite zonal succession (Text-fig.4) is based on a general knowledge of the evolution of Ordovician graptolites in North America as revealed by their development in many stratigraphic sections. The stratigraphically lowest zones (zones 1 to 3) are characterized by numbers of different graptolite colonies with many branches (anisograptids and dichograptids) and a few dichograptid colonies with two or four branches (Text-fig. 4). Stratigraphically higher (zones 4 to 6) faunas are typified by graptolite colonies with two (didymograptid) and four (tetragraptid) branches (Text-fig. 4). The branches of most of these colonies are declined or pendent from the initial member of the colony (the sicula). The four-branched, leaf-like (when compressed) phyllograptids are common associates of the didymograptids and tetragraptids. These graptolites are replaced in dominance stratigraphically upward (zones 7 to 9) by graptolites with colonies in which the branches are arrayed upward from the sicula (Text-fig. 4). Commonly, two branches are present in these colonies. The generalized upward arrangements of the branches are termed the reclined and the scandent conditions. The first reclined and scandent colonies to appear in the stratigraphic succession are isograptids, cardiograptids, glossograptids, glyptograptids, and cryptograptids (Text-fig. 4). The colonies with scandent form dominate nearly all graptolite faunas from the middle and upper parts of the North American Ordovician sequence (Text-figs. 3 and 4). The scandent colonies include those in which two stipes are pressed closely back to back (e.g., orthograptids, climacograptids, and diplograptids) and those in which two stipes are essentially side by side (glossograptids and cryptograptids) (Text-fig. 4). Essentially reclined graptolites include those in which two stipes begin closely pressed together but then separate to give the colony a Y shape (dicranograptids), and those in which two stipes are separated and give the colony a U- or V-shaped appearance (dicellograptids) (Text-fig. 4). Colonies with these basic shapes characterize the Middle and Late Ordovician graptolite faunas.

Basic colony form, commonly easily seen in the field, is thus a good indication of general position within the overall Ordovician succession in graptolite-bearing strata (Text-fig. 4). Attention to associations of particular species among these basic kinds of graptolites determines the zone.

The basic replacements in kinds of graptolites are indicated in Text-figures 3 and 4. The replacements in general kind of graptolite appear to be linked to coeval changes in the benthic faunas and floras and to major environmental changes across the platform (Text-fig. 3).

The faunas of zones 4 through 6 are particularly rich and diverse in number of different species. That time interval was one in which a major diversification took place among the Ordovician graptolites. The faunas of the zones 4 to 6 interval are more diverse than those of the preceding three zones or of the succeeding two zones.

Graptolite evolutionary development was affected by major Middle Ordovician environmental changes (Text-fig. 3). Many extinctions took place with the result that the faunas of zones 7 and 8 are relatively poor in species.

The North American Faunal Province was invaded in the Middle Ordovician by graptolite stocks that had developed in the Atlantic Faunal Region (Berry, 1967). The invading stocks enriched the taxonomic diversity of the North American Ordovician graptolites, and they apparently underwent extensive adaptive radiations. The faunas of zone 9 are particularly diverse and rich in many new lineages (Text-fig. 4).

A slight decline in diversity took place in the zone 10 interval, but from zone 11 onward, the faunas are relatively rich in those taxa with scandent and reclined colony forms. Radiations probably took place in several lineages, although knowledge of the evolutionary history of the Late Ordovician graptolites is scanty. The orthograptids (Text-fig. 4) appear to have become particularly diverse, apparently radiating such that members lived at nearly all depths in the oceanic water column.

The zones are, of course, most clearly defined in those parts of the succession in which the faunas are richest, as in the zones 4 to 6 interval. The zones are most difficult to recognize and distinguish in those parts of the succession in which only a few stocks are present, as in the zone 7 to 8 interval.

Graptolite collections continue to be made from all parts of the Ordovician succession in several different areas in North America. As more and more collections are made, certain changes in the zones (the working hypotheses) are suggested.

One such change involves the faunas of zone 13 (Berry, 1970). The faunas typical of that zone in western North America differ from those in eastern North America. Some of the same species occur in both western and eastern North America, but certain species are limited to either western or eastern areas. The eastern faunas considered characteristic of zone 13 are more diverse than those in the west, and they may be divided into those typical of a lower and an upper subzone (Berry, 1970).

Waters in which graptolites abounded in the eastern areas of North America were being reduced in extent during the Late Ordovician by deltaic environments encroaching from the east. Because they were, ecologic factors or limited circulation of water masses may have been responsible for the faunal differences. Latitudinal distribution of water masses and their contained faunas across an Ordovician equator that passed across the North American platform may have been responsible also.

Another change that has developed in the initial working hypotheses as collecting brought in new information involves the faunas of the Marathon region *Didymograptus protobifidus, D. bifidus,* and *Isograptus* zones. *Didymograptus protobifidus* zone congregations occur in many places in North America. They are succeeded by congregations typical of first the *D. bifidus* and then the *Isograptus* zone congregations only in the Marathon region sequence. Elsewhere, zonal faunas dominated by either *D. bifidus* or *Isograptus* stratigraphically succeed the *D. protobifidus* zonal congregation. This evidence suggests that for widespread use of the zones, a single zone, designated "*Isograptus* zone," should be recognized stratigraphically above the *D. protobifidus* zone. Where possible, as in the Marathon region, that single zone may be divided to a lower, *D. bifidus*, subzone and an upper, *I. victoriae*, subzone. Inasmuch as a small member of the *I. victoriae* group, *I. victoriae lunatus*, is commonly an associate of *D. bifidus*, the designation "*Isograptus* zone" draws attention to the prominence of isograptids in most faunas stratigraphically above the *D. protobifidus* congregation.

As additional graptolite collections are made throughout North America, the working hypotheses may be modified further. The North American Ordovician zonal succession differs from that in Britain, and it is limited in its applicability to the graptolitic biofaces. Dissimilar yet approximately coeval faunas are known to occur within some of the zones (especially zone 9). Aspects that limit the use of the North American Ordovician zones as well as links between evolution of the graptolites and shelly faunas with environmental changes will be considered next (Text-fig. 3) to emphasize the need for understanding graptolite distribution patterns as well as evolutionary history in biostratigraphic studies.

Comparison of the North American Ordovician graptolite faunas with those from Australia, Britain, China, Scandinavia, and areas in Russia (Berry, 1959, 1966, 1967; Bulman, 1970; Skevington, 1973; among others) indicated that Ordovician graptolite faunas were distributed in discrete faunal provinces and regions. The American, Australian, and Chinese faunas appear to have been within one faunal region, the Pacific (Mu, 1963). British and Scandinavian Ordovician faunas were within a second, the European or Atlantic Faunal Region. Provinces were present within the regions. For example, the Chinese Ordovician faunas differ to a degree from coeval North American and Australian faunas; hence China lay within a faunal province distinct from either North America or Australia. As Berry (1966, 1967) and Bulman (1970) have pointed out, the North American Ordovician graptolite zones are based upon special associations or congregations that do not occur in the British Ordovician graptolite faunal province of the European faunal region. The North American zones are confined in their applicability to North America and small areas in western Ireland and western Norway that lay in the North American Ordovician graptolite faunal province.

Correlation of the Early and early Middle Ordovician North American graptolite zones with approximately coeval graptolite zones in Britain has led to heated debate (Berry, 1966, 1967, 1968a; Skevington, 1969; Williams, in Williams et al., 1972). A resolution of that debate has been hindered because shelly faunas in North America and Britain appear to have comprised faunal provinces and regions that had an areal extent similar to that of the graptolites (Whittington, 1963, 1966, 1973; Jaanusson, 1973). British-American correlations, summarized by Williams (in Williams et al., 1972) in a review of British Ordovician correlations are closely similar to British-American correlations proposed by Berry (1960, 1966, 1967, 1968a).

North American Ordovician graptolite faunas are mostly richly developed in rock suites that were along the platform margins of the Ordovician and in the apparently deeper-water trough that lay west of a rising Taconica land mass during the latter part of the Middle Ordovician (Text-fig. 3). Analysis (Berry, 1972, and Text-fig. 3) of distributions of North American Early Paleozoic invertebrate faunas indicates that trilobites dominated faunas from all rock suites and nearly all platform and platform slope environments during the latter part of the Cambrian. Trilobites that inhabited platform environments were different from those (such as the olenids and agnostids) found in sedimentary successions deposited on the platform margins and slopes (Text-fig. 3), later to become the graptolite facies. As the platform became the site of a broad expanse of tidal flats in which stromatolites were common (Lochman-Balk, 1970, 1971) in the latest Cambrian-earliest Ordovician, nautiloids and archeogastropods became dominant members of the faunas in the shallow marine environments of the platform interior. (Flower, 1964, 1968, discussed evolutionary developments among nautiloids from these environments.) Ordovician trilobites are relatively more prominent in faunas from areas that lay near the platform margins than those of the platform interior. Early Ordovician trilobite-dominated faunas occur primarily in limestone suites that rim coeval dolomitic successions in which molluscan faunas dominate (Berry, 1972). Graptolites (primarily dichograptids and members of the Dendroidea group, the anisograptids) replaced olenid and agnostid trilobite-dominated faunas in rock suites (the Shale Belts of Berry, 1972) that formed from sediments deposited on the platform slopes. As platform uplift took place during the Middle Ordovician, the terrain of subtidal environments between those that had richly molluscan faunas and the slopes became divided in terms of faunal dominance such that brachiopods are relatively more prominent in faunas from the shallower subtidal environments, and trilobites are the more common in faunas from the relatively deeper subtidal environments (Text-fig. 3).

Most of the North American platform was above sea level during the Middle Ordovician, and a karst topography developed over much of it (Lochman-Balk, 1970, 1971; Berry, 1972). A land area, Taconica, formed along the eastern side of the platform from an area that had formerly been continental or platform slope and continental rise (Text-fig. 3). Taconica may have formed as a response to a change from an expanding to a contracting proto-Atlantic Ocean (Bird and Dewey, 1970).

Marine environments returned across the platform during the latter part of the Middle Ordovician and faunas in many of the environments are characterized by entirely new phyletic stocks (Text-fig. 3). Graptolites in the slope terrain are dominated by rhabdosomes with biserial scandent rhabdosome form and include diverse dicellograptids and dicranograptids as well, whereas they had been primarily dichograptids before platform emergence. Intertidal and subtidal environments across the eastern part of the platform, those within the influence of the spread of terrigenous clastic sediments, were populated by faunas dominated by orthoid, strophomenoid, and rhynchonelloid brachiopods and, locally, by certain bivalves. Many other invertebrates radiated markedly in these environments (Walker and Laporte, 1970). Intertidal and shallow subtidal environments across carbonate banks remained the sites of archeogastropod and nautiloid cephalopod-dominated faunas, but they did include some corals. The replacements among the graptolites thus paralleled replacements among the benthic marine invertebrates, a relationship which suggests the significance of nutrient supplies as an important factor behind certain major faunal replacements.

Glacioeustatic sea-level lowering, which exposed most of the platform in the Late Ordovician (Berry and Boucot, 1973), and subsequent sea-level rise in the Early Silurian were accompanied by other faunal replacements (Text-fig. 3). Pentameroid and certain spiriferoid brachiopods replaced orthoid-rhynchonelloid-strophomenoid brachiopod-dominated faunas across most of the platform. Certain pentameroids became common in some carbonate bank environments as well. Graptolite colonies with uniserial scandent rhabdosome form became the dominant graptolite type in the Silurian, replacing graptolites with biserial scandent rhadbosomes.

CONCLUSIONS

The paleogeographic positions of the richest graptolite faunas indicate those areas in which graptolite zones may be established and provide greatest use in correlations. Because most platform rock suites bear few graptolites, chronological units must be developed from the faunas available in them, i.e., the trilobites in faunas from near the platform margins, and the brachiopods and nautiloids and archeogastropods from faunas in rock suites from the platform interior. Through careful collecting in areas where graptolitic sequences interdigitate with sequences in which trilobites are abundant, certain correlations between the graptolite zones and shelly faunal chronological units may be documented. A net of correlations must be developed to provide a basis for precise rock unit correlations that will include rocks and faunas formed in all environments that were present. Whittington (1968) summarized correlations between North American Early Ordovician trilobite and graptolite zonal faunas.

Future developments in graptolite biostratigraphy will include more precise comprehension of graptolite geographic and stratigraphic distribution patterns and the limitations they impose upon applicability of the graptolite zones. An investigation into graptolite phyletic history will indicate certain speciation events that may be used for greater precision in recognizing zone boundaries. These developments will have the effect of sharpening the basic tool of graptolite biostratigraphers, the graptolite zones, to continue to provide geologists with a method of making precise, specific geologic correlations in studies involving Early Paleozoic rock units.

Spores and Pollen: The Potomac Group (Cretaceous) Angiosperm Sequence

James A. Doyle **University of Michigan**

INTRODUCTION

From its beginnings early in this century as a tool in studies of Quaternary vegetational and climatic changes recorded in northwest European bogs (von Post, 1916, translation 1967; Faegri and Iversen, 1964), paleopalynology, or the study of fossil spores and pollen grains, has grown explosively to become a major source of stratigraphic information throughout the geological column, thanks in large part to its usefulness in petroleum geology (see Kuyl et al., 1955; Hopping, 1967; Tschudy and Scott, 1969). Spores and pollen, represented by their highly resistant walls or exines (composed of sporopollenin, a carotenoid polymer; Brooks et al., 1971), are the most abundant identifiable remains of land plant life in the geological record. Wide distribution in both continental and marine sediments and highly varied morphology, perhaps as indicative of systematic relationships as any plant structure (see Erdtman, 1952; Muller, 1970), have added a whole new dimension to studies of both the stratigraphic relations of continental sediments and land plant evolution. In conjunction with the study of organic-walled microplankton (chitinozoans, acritarchs, dinoflagellates, etc.; see Evitt, 1964), now frequently included under the term palynology, spores and pollen are also important in facies analysis of transitional continental-marine sequences, one of the main applications of palynology in the oil industry (Muller, 1959). Furthermore, their very abundance and consequently nearly continuous records point up the theoretical limitations of conventional biostratigraphic methods based on species and zones, and have prompted the search for new approaches capable of taking fuller advantage of the fact of biological evolution (see Hughes, 1970, and the following discussion).

After a general introduction to the biological, morphological, and geological peculiarities of spores and pollen, specific examples of the use of palynology in stratigraphy will be drawn from the mid-Cretaceous Potomac Group and Raritan Formation of the Atlantic Coastal Plain of the United States. Studies of this sequence have dealt with both the whole flora and the angiosperms, or flowering plants, which appear and begin their initial rapid diversification in this interval, and have been simultaneously directed toward intra- and interregional correlation and conclusions on angiosperm evolution (Brenner, 1963, 1967; Doyle, 1969a, 1969b, 1973; Doyle and Hickey, 1972, 1976; Wolfe and Pakiser, 1971; Wolfe et al., 1975). The studies therefore illustrate both special problems of palynology and the general theoretical and practical links between studies of evolution of rapidly radiating groups and stratigraphic correlation.

BIOLOGY AND MORPHOLOGY OF SPORES AND POLLEN

Although their behavior in sedimentary environments is essentially identical, spores and pollen differ considerably in their biological roles (see Faegri and van der Pijl, 1966). These differences are closely tied to differences in the life cycle of the groups they represent: (1) bryophytes (mosses and liverworts) and pteridophytes (ferns and "fern allies"), and (2) seed plants (gymnosperms, such as conifers and cycads, and angiosperms).

Both spores and pollen grains are haploid cells produced by meiosis in the sporangia or pollen sacs of the adult diploid plant (sporophyte); they are the first stage in the haploid (gametophyte) phase of the life cycle, which is the dominant generation in bryophytes, but relatively inconspicuous in the other groups, or vascular plants. In the presumably primitive homosporous pteridophytes and in many bryophytes, the spores are of one kind and produce bisexual gametophytes, so that a single spore is capable of establishing a new population of the species. Hence the spores act as the primary mechanism for dispersal as well as gene flow, like the planktonic larvae of sessile marine benthos. In fact, one may speculate that the exine, permitting air dispersal of the meiotic products from one pool to another, was the first terrestrial adaptation of the hypothetical freshwater algal ancestors of land plants (Jeffrey, 1962). In contrast, the more advanced heterosporous pteridophytes produce male microspores and female megaspores, so that two spores are required to establish a new population; the larger megaspores set the limits on dispersal. Seed plants represent a still more specialized version of the heterosporous condition: the microspores are the pollen grains; one megaspore is retained within each megasporangium, which together with its protective integument(s) constitutes the ovule or immature seed. In gymnosperms, the pollen grains are transported by wind to the micropylar opening of the naked ovule; in angiosperms, they are carried by wind, insects, or other animals to the sticky stigmatic surface of the carpel, which encloses and protects the ovules. Hence pollen grains still function as agents of gene flow, but the dispersal function is taken over by the seed and its associated structures, a fact that should not be overlooked in discussing the dispersal capacity of seed plant groups.

Major variations in spore and pollen morphology, especially symmetry and the shape and position of germination apertures, are correlated with these successive levels of specialization in the life cycle (see Chaloner, 1970a; and Text-fig. 1). The majority of bryophytes and pteridophytes, like the first land plants of the Silurian (Hoffmeister, 1959; Chaloner, 1970b; Gray and Boucot, 1971), have radially symmetrical trilete spores, characterized by a triradiate crest, representing the lines of juncture on the proximal faces of the four spores in a tetrahedral meiotic tetrad, and functioning in germination. Certain more advanced groups (e.g., polypodiaceous ferns) have bilaterally symmetrical monolete spores, with a simple linear tetrad scar, reflecting a shift from tetrahedral to tetragonal (square) tetrad form, or alete spores with no tetrad scar at all.

The earliest gymnosperms of the latest Devonian and Carboniferous (cordaites and seed ferns) retained trilete or monolete microspores or "pre-pollen" (see Chaloner, 1970a). However, within the Carboniferous, the differentiation of an elongate thin area or sulcus on the distal face and the reduction of the tetrad scar resulted in bilaterally symmetrical monosulcate pollen, as in modern *Ginkgo,* cycads, and some conifers. This step is presumably connected with the origin of a pollen tube, first functioning to anchor the grain to the roof of the pollen chamber in the ovule, as in *Ginkgo* and cycads, and later for sperm transfer, as in conifers, other gymnosperms, and angiosperms. Other noteworthy morphological specializations among gymnosperms are (1) the elaboration of air bladders or sacs (usually two, the bisaccate condition), as in the modern conifer families Pinaceae (pines, spruces, etc.) and Podocarpaceae, presumably an adaptation for flotation of the grain up the micropylar canal (Doyle, 1945), and (2) loss of the aperture, as in other conifer families (Wodehouse, 1935; Florin, 1951).

Just as the first gymnosperms retained "pteridophytic" microspores, "gymnospermous" monosulcate pollen is retained by a variety of modern monocotyledons (e.g., lilies), a few primitive dicotyledons (e.g., *Magnolia,* some waterlilies), and the first recognizable angiosperms of the mid-Lower Cretaceous. Such primitive angiosperm pollen may generally be distinguished from that of gymnosperms on the basis of its tectate-columellate exine structure, with two layers connected by well-defined rods or columellae (Couper, 1958; Van Campo, 1971; Doyle, 1969a, 1973). Much of the subsequent origin and diversification of pollen types with more than one aperture and with radial or global symmetry is documented by fossil as well as comparative evidence (Wodehouse, 1935; Takhtajan, 1969; Doyle, 1969a, 1973; Muller, 1970; see Text-fig. 2). The most important transformations are (1) the origin of radially symmetrical tricolpate grains, with three elongate apertures or

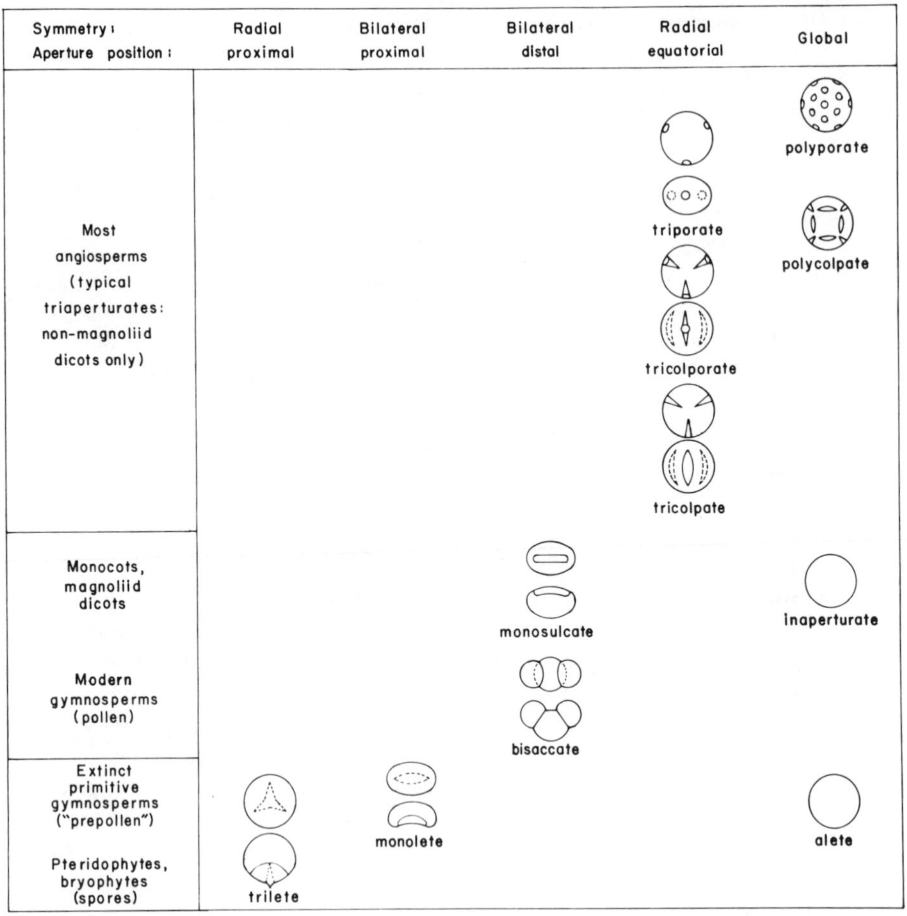

Symmetry: Aperture position:	Radial proximal	Bilateral proximal	Bilateral distal	Radial equatorial	Global
Most angiosperms (typical triaperturates: non-magnoliid dicots only)				triporate tricolporate tricolpate	polyporate polycolpate
Monocots, magnoliid dicots			monosulcate		inaperturate
Modern gymnosperms (pollen)			bisaccate		
Extinct primitive gymnosperms ("prepollen") Pteridophytes, bryophytes (spores)	trilete	monolete			alete

Text-figure 1
Major evolutionary levels in form, symmetry, number, character, and position of germination apertures in spores and pollen. Systematic groups in which each type is represented are indicated to the left. Proximal, distal, polar, and equatorial are in relation to an axis defined by the center of the original meiotic tetrad and the center of the individual grain. Upper member of each pair of sketches represents a distal-polar view; the lower, an equatorial view. Dashed lines indicate features on the opposite side of the spore or pollen grain.

colpi arranged along lines of longitude (as defined by an axis passing from the center of the original meiotic tetrad through the center of the individual grain), (2) the origin of compound-aperturate tricolporate grains, the most common type in modern dicots, with specialized thin areas or ora at the centers of the colpi, (3) reduction of the colpi to round pores (triporate), and (4) shifts from radial to global symmetry

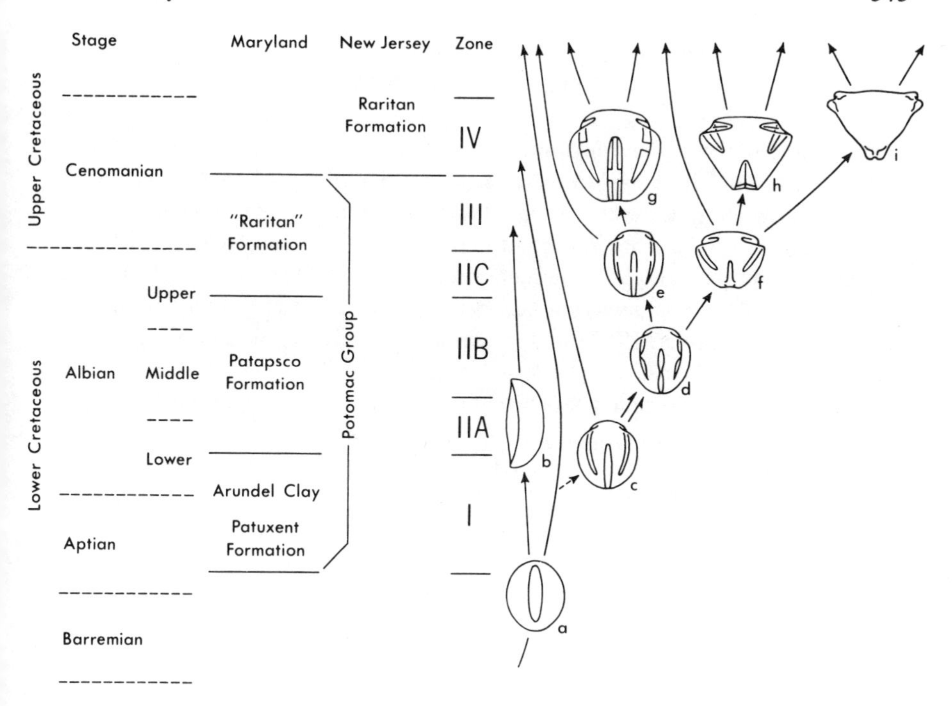

Text-figure 2
Postulated evolutionary relationships of major angiosperm pollen types and stratigraphic units in the mid-Cretaceous section of the Atlantic Coastal Plain (Maryland, northern New Jersey). Roman numerals in the center refer to biostratigraphic units (zones and subzones of Brenner and Doyle) based on the spore and pollen flora. Suggested correlations with standard faunally defined subdivisions of the Cretaceous (4 of 12 stages) are indicated to the left (see text for discussion). Pollen types shown are (a) generalized tectate-columellate monosulcates (*Clavatipollenites,* etc.); (b) monocot-like reticulate monosulcates (*Liliacidites* spp.); (c) generally reticulate-tectate tricolpates (nonmagnoliid dicots); (d) tricolporoidates; (e) small, generally smooth-walled, prolate tricolporoidates; (f) small generally smooth-walled, oblate-triangular tricolporoidates; (g) larger, often more sculptured, prolate tricolpor(oid)ates; (h) large, often more sculptured, oblate-triangular tricolpor(oid)ates; (i) early members of the triporate Normapolles complex (*Complexiopollis,* etc.).

and numerous scattered apertures (polycolpate, polyporate) (see Erdtman, 1952; Kuyl et al., 1955; or Faegri and Iversen, 1964, for further classes and terminology). These trends presumably reflect in various ways the new selective pressures imposed on pollen morphology in angiosperms, such as the diversity of pollination agents and the necessity for germination on the exposed surface of the stigma, and particularly the advantage of having points for pollen tube germination on more than one hemisphere (see Smart and Hughes, 1972; Doyle and Hickey, 1976).

GEOLOGICAL AND STRATIGRAPHIC ROLE OF SPORES AND POLLEN

Spores and pollen grains, including the pollen of insect-pollinated species (either as free grains or in whole shed flowers), generally first enter sedimentary environments from the air; but, except in special environments such as peat bogs, water transport appears to be more important in controlling their subsequent sedimentary distribution (Muller, 1959; Germeraad et al., 1968). Because of their small size (on the order of 10 to 100 μm) and low density, spores and pollen tend to sort as fine silt particles (Muller, 1959), and are abundant in clays, silts, and fine-grained sands, as well as in the peats, lignites, and coals on which the earliest studies tended to concentrate. The extreme resistance of exines to most geological processes except metamorphism and strong or prolonged oxidation not only ensures their preservation in a variety of sediments containing no other fossils (the most significant exceptions being redbeds and many limestones), but also facilitates their extraction by such techniques as maceration in HF or HCl, separation from silicates by heavy liquid flotation, removal of humic materials with dilute KOH, mild oxidation, and acetolysis (boiling in a 9 : 1 mixture of acetic anhydride and sulfuric acid) (see Faegri and Iversen, 1964; Kummel and Raup, 1965; Tschudy and Scott, 1969). As with other microfossils, special care must be taken to avoid laboratory and sample contamination (e.g., by drilling mud, Traverse et al., 1961). Because spores and pollen can be isolated from small core samples, it is possible to establish long, closely sampled, vertically controlled subsurface sequences covering a variety of alternating facies, generally an impossible task with plant megafossils.

As with all thanatocenoses, the composition of fossil spore and pollen assemblages reflects an interplay of ecological variation in the original plant communities and differential transport and sorting factors, complicating both correlation of time-equivalent assemblages and paleoecological reconstructions. For example, although the transportability of spores and pollen might theoretically permit direct correlation of continental and marine sediments, this is often rendered difficult by the fact that many continental facies (especially coals) are dominated by spores and pollen of lowland swamp plants, whereas equivalent marine beds are dominated by a paradoxical mixture of upland pollen and marine plankton (see Chaloner and Muir, 1968; Hopping, 1967; Clapham, 1970; Hughes and Pacltová, 1972). In post-Eocene sediments, the facies sensitivity and stratigraphic reliability of forms can often be evaluated from the ecology of related modern plants, but in older deposits dominated by extinct or now relictual taxa, such conclusions must be drawn from empirical and geological evidence (see Kuyl et al., 1955). As a generalization, the greater abundance, transportability, and extractability of spores and pollen make them more valuable in stratigraphy and studies of regional vegetation changes than plant megafossils, but more problematical in local ecological reconstruction (see Muller, 1970; Doyle and Hickey, 1976). Hence, for stratigraphic purposes, they may perhaps be classed with high-mobility organisms; however, it should not be forgotten that they are only one stage in the life cycle of large, rooted, "benthic" plants whose distribution is largely controlled by different factors (climate, soil, biotic associations, etc.).

The same sorts of correlation and zonation methods used elsewhere in biostratigraphy have been applied in stratigraphic palynology: "qualitative" (presence-absence) methods using ranges of individual "index" forms and assemblages, and overlaps of first and last occurrences of different groups (concurrent-range zones), and "quantitative" methods, based on either relative (percentage) or absolute frequencies (grains per unit sediment or time), often presented in terms of "sawtooth" diagrams (see Kuyl et al., 1955; Jekhowsky, 1958; Tschudy and Scott, 1969; Christopher and Hart, 1971). In general, quantitative fluctuations may serve as a basis for accurate and rapid correlations when they reflect tectonically or climatically controlled changes in the regional vegetation, as they often do within relatively homogeneous lacustrine or nearshore marine environments or within a single climatic belt. Hence quantitative correlation methods are widely used in the oil industry, where studies often focus on sedimentary sequences that accumulated in just such situations. However, quantitative methods tend to break down in fluviatile and other heterogeneous environments, where local facies shifts often obscure regional trends, and when attempts are made to correlate from one climatic zone, sedimentary basin, or continent to another. In such cases, correlations must be based more on long-term qualitative, evolutionary changes in the composition of the spore and pollen flora, although the problem remains of distinguishing appearances resulting from immigration, and hence useful only within one region, from truly evolutionary originations (see Kuyl et al., 1955, and Germeraad et al., 1968, for further discussion). Rather than attempt an exhaustive review of these methods and their strengths and weaknesses, I prefer to evaluate them in the context of a discussion of the Cretaceous Potomac Group sequence of the Atlantic Coastal Plain, and to concentrate here on certain proposals prompted by palynology for improvement of evolution-based qualitative correlation methods.

As has been emphasized in a series of papers by Hughes and his collaborators (Hughes and Moody-Stuart, 1967, 1969; Hughes, 1970, 1973a; Hughes and Croxton, 1973), the very abundance of spores and pollen points up well-known but often disregarded theoretical contradictions between the fact that the geological record adds a continuous time dimension to the record of evolving groups and the conventional use in paleontology of the concepts of species and biostratigraphic zones. Although the existence of reproductive isolating mechanisms may (at least in theory) provide objective criteria for defining species at one time plane, the "vertical" definition of species must be completely arbitrary, or at best based on population-derived means or other statistics (see Simpson, 1953b), or else each segment or internode of the phyletic tree must be recognized as a separate species (see Hennig, 1966), regardless of its length or practical biostratigraphic utility. In practice, apparently discrete "species" are often recognizable because of breaks in the record and the abrupt immigration of allopatrically evolved species. However, both cases cause notorious correlation problems when rocks of intermediate age or adjacent geographic areas are investigated. As stratigraphic and geographic sampling becomes nearly enough continuous, as it often does in palynology (see Germeraad et al., 1968), the evolutionary changes within zones based on species become greater than those between them, and their boundaries become blurred. Rather than being frustrated by these phenomena, biostratigraphers should seek new methodologies which might allow exploitation of the increase in information that they represent.

Hughes's proposal, which perhaps does not represent such a radical departure from current concepts as his terminology might suggest, is to abandon species and zonal concepts as the formal basis of correlation, and to substitute a system of informal "biorecords," essentially detailed and semiquantitative descriptions of populations from single horizons in a reference section, and "reference point" correlations, based on "graded comparisons" rather than identifications of populations in the sample being studied with biorecords in the reference section, and expressed as brackets between detectably older and younger reference samples. The length of the correlation brackets varies inversely with the amount and quality of information; they can thus express differing degrees of confidence and can be progressively refined (shortened) with a simplicity and flexibility that is impossible with hierarchical zonal schemes. Hughes recognizes four grades of comparison, most recently defined as follows (Hughes, 1970, 1973a): cfA for populations with no qualitative and only minor quantitative differences from the biorecord; cfB, with one minor qualitative difference; cfC, with greater differences; cf., material believed related but inadequate for grading. It is possible that in practice cfA will often correspond to very critical identification (by a "splitter"), and cfC to the conventional qualification of identification with "aff.," which implies a distinct but related species (compare the definition of cfC in Hughes and Moody-Stuart, 1967). However, the other conventional qualification, "cf.," is an ambiguous combination of the concepts denoted by Hughes's cfB and cf. Besides computer-assisted quantitative analysis of such data, perhaps the stratigraphically most useful refinements of this system would be means of expressing (1) the temporal "direction" of morphological "deviation" from the biorecord (i.e., whether toward older or younger populations in the reference section) and (2) the relative probabilities of different-length correlation brackets (see Hughes, 1973a). As Hughes emphasizes, the sort of study involved in these methods may permit not only finer stratigraphic correlations but more secure evolutionary and paleoecological conclusions.

THE POTOMAC POLLEN SEQUENCE

Introduction

Palynological studies of the Potomac Group and Raritan Formation of the Atlantic Coastal Plain of the United States have been concerned with both stratigraphic correlation and the evolutionary interpretation of the early record of the angiosperms and its bearing on the controversial question of the timing of their origin and adaptive radiation. Until recently, most authors (e.g., Seward, 1931; Axelrod, 1952, 1959, 1960, 1970; Takhtajan, 1969) interpreted the rapid rise to dominance of angiosperms in the mid-Cretaceous in terms of immigration of already highly diversified forms from some ecologically or geographically restricted area, largely on the basis of identifications of the earliest angiosperm leaf fossils with advanced and distantly related modern taxa. In contrast, the pollen record (and more recently the leaf record: Wolfe, 1972; Doyle and Hickey, 1972, 1976) has been interpreted as direct evidence of a Cretaceous adaptive radiation (Hughes, 1961, 1973b; Doyle, 1969a, 1973; Muller, 1970; Pacltová, 1961, 1971; Wolfe et al., 1975). The Atlantic Coastal Plain section shows a sequence of progressively more differentiated morphological types of angiosperm pollen (Text-fig. 2), as would be expected in an evolutionary radiation, rather than a random sequence of unrelated advanced types, as would be expected with immigration of previously differentiated taxa. Furthermore, contemporaneous sections from other geographic areas reveal angiosperm sequences generally consistent with the evolutionary scheme inferred from the Potomac Group. One exception, although not an inconsistency, is the later appearance and rise of angiosperms in higher-latitude areas, which seem to have acted as refuges for gymnosperm-dominated floras (Axelrod, 1959; Hughes, 1961; Doyle, 1969a). A more significant exception is the apparently more advanced character of early angiosperm floras in South America, Africa, and Israel, where tricolpates appear in the Aptian rather than the Albian, and endemic polyporates and tricolpodiorates occur in the upper Albian (Jardiné and Magloire, 1965; Müller, 1966; Brenner, 1968, 1976; personal observation). Although this indicates that at least some angiosperm appearances in the Potomac Group represent immigration of types which originated to the south rather than evolution in place, the fact that the order of appearance of major types is the same suggests that the migrational lag was not sufficiently great to distort the underlying evolutionary pattern.

Although the following sections emphasize the development of a palynological reference section based on subsurface material from the Potomac Group, the specific examples discussed have been chosen for their relation to evolutionary problems as well. The sample chosen to illustrate correlation methods in detail correlates with the same interval in the reference section as an independently dated flora described from Oklahoma (Hedlund and Norris, 1968), and hence serves to demonstrate both the potential of palynology in long-distance correlation of continental sequences with the standard marine stages and the uniformity of early angiosperm floras over wide areas. The immediately overlying beds illustrate problems encountered in correlation with western Canada (Norris, 1967), an area where angiosperms appear later than they do to the south, and between continental and marine facies, and hence methods for unraveling the factors of environmental restriction, migration, and evolution.

Geological Background

The Potomac Group is the oldest stratigraphic unit in the portion of the Atlantic Coastal Plain sedimentary sequence exposed from Virginia through Maryland and Delaware into southern New Jersey. Further north, from the Raritan Bay area of northern New Jersey through Long Island to Martha's Vineyard, the younger Raritan Formation occupies an analogous position in the sequence. Both units are composed of fluvial-deltaic sediments, and they exemplify the difficulties of stratigraphic subdivision and correlation in such deposits, i.e., the characteristically limited lateral extent of beds, rapid facies changes, and sporadic distribution of megafossils usable for time correlation within the sequence and with other areas.

Between Washington, D.C., and Baltimore, Maryland, it is possible to subdivide the Potomac Group into three formations: in ascending order, the Patuxent Formation (predominantly sands), the Arundel Clay, and the Patapsco Formation (predominantly sands and variegated clays) (see Clark et al., 1911; Weaver et al., 1968; and Text-fig. 2). However, to the north and south, where the Arundel Clay is not recognizable, there are no reliable lithological criteria for subdividing the Potomac Group, and it is hence often considered a single formation (see Jordan, 1962; Owens, 1969). Furthermore, although lithological differences can now be recognized between the Potomac Group and Raritan Formation (Owens and Sohl, 1969), they are so subtle as to have led in the past to misidentification of the sandy upper Potomac beds of Maryland and Delaware ("Raritan" in Text-fig. 2) with the Raritan Formation (see below), and even to such greater errors as Spangler and Peterson's (1950) correlation of the entire Potomac Group of Maryland with the Raritan Formation.

Below the Woodbridge Clay Member of the Raritan Formation, which contains marine mollusks indicating a mid-Cenomanian age (Sohl, cited in Wolfe and Pakiser, 1971), the fauna consists of rare and seemingly stratigraphically undiagnostic vertebrates, freshwater mollusks (Clark et al., 1911), and insect wings (personal observation). Plant megafossils, mostly leaves, are more common, but before the application of palynology to the relations of the leaf localities (Doyle and Hickey, 1972, 1976; Doyle, 1974), facies variations and difficulties in establishing superposition meant that only three broad floristic units could be recognized: lower Potomac, Patapsco, and "Raritan" (see Fontaine, 1889; Berry, 1911a, 1911b).

Understanding of the stratigraphic relations in the Potomac-Raritan sequence has been markedly improved by palynological studies, thanks especially to the ubiquity of spores and pollen and the availability of extensive well sections. Although Potomac palynology began with the brief reconnaissance studies of Groot and Penny (1960), Groot et al. (1961), and Stover (1964), the first comprehensive study was that of Brenner (1963), who established a palynological zonation based on two wells drilled through the Potomac Group of Maryland, and demonstrated its utility in outcrop stratigraphy. Subsequent work has clarified the relations between Potomac and younger beds (Brenner, 1967; Wolfe and Pakiser, 1971; Doyle, 1969a), and has led to proposals to extend the palynological zonation upward (Doyle, 1969b, 1973). At the same time, more detailed studies of angiosperm pollen from Delaware well sections (Doyle, 1970) have revealed not only new forms and range extensions, but also the sort of theoretically expected intergradations between complexes, gradual intrazonal changes, and blurring of zonal boundaries discussed above. To some extent this new information has led to a breakdown of the zonational framework; but with the application of the concepts of graded comparison and bracket correlation of samples with a reference section, it has permitted significantly finer correlations than were previously possible.

Palynological Zonation in the Potomac-Raritan Section

On the basis of his study of the total flora (pteridophyte spores and gymnosperm and angiosperm pollen) in two Maryland wells, Brenner (1963) recognized two broad biostratigraphic units, termed zone I and zone II, with zone II subdivided into subzones A and B, the latter occasionally further divisible into subzones B1 and B2. The lithological sequence in the wells and palynological correlations with outcrop samples indicated that zone I corresponds to the Patuxent Formation and Arundel Clay, whereas zone II corresponds to the Patapsco Formation. Brenner's zones are defined primarily by the simultaneous entry of several spore and pollen "index species" at the base of zone II, and of several additional types at the base of subzone IIB. Among the most important zone II index forms were several species of tricolpate angiosperm pollen. Brenner in fact recognized no angiosperm pollen in zone I; however, two or three very rare tricolpate species have subsequently been found in the upper part of intervals defined on other palynological criteria as zone I (Wolfe et al., 1975; and discussed below). Furthermore, recent comparative studies have strengthened the interpretation of several zone I monosulcates with tectate-columellate exine structure (*Clavatipollenites, Retimonocolpites, Liliacidites* spp.) as angiospermous (Couper, 1958; Van Campo, 1971; Doyle, 1973; Wolfe et al., 1975, and above). Brenner also recognized certain gradual quantitative changes within the Potomac sequence, such as the decline of schizaeaceous fern spores and "Jurassic" gymnosperm pollen types (e.g., the extinct conifer *Classopollis*), and the rise of typically Upper Cretaceous conifers and angiosperms. However, he considered these trends to be of secondary importance in his zonal definition because the well data showed that they were frequently masked by short-term, presumably facies-related fluctuations, a judgment borne out by subsequent studies.

Addition of a new subzone of zone II, subzone C, and of zones III and IV, has been based on palynological studies of outcrop material from the type Raritan Formation of New Jersey and the "Raritan" of Maryland (Wolfe and Pakiser, 1971; Doyle, 1969a; Doyle and Hickey, 1976), and of well sequences in Maryland (Waldorf), Delaware (Delaware City, discussed in detail below), and New Jersey (Toms River, Butler Place). At the outcrop, subzone IIC and zone III together (the "Patapsco-Raritan transition zone" of Doyle, 1969a) are represented in the Maryland "Raritan"; zone IV corresponds to the lower part of the true Raritan Formation (Farrington Sand and Woodbridge Clay members). Very small, smooth-walled tricolporoidate pollen (i.e., with weakly developed ora, such as *Tricolporoidites* cf. *subtilis*) and the conifer *Rugubivesiculites rugosus* Pierce enter at the base of subzone IIC, and several additional tricolpate and tricolporoidate species, especially larger triangular forms, enter in zone III (See Wolfe et al., 1975, and below). The base of zone IV is marked by the first appearance of the triangular triporate Normapolles group and several larger tricolporates (see Wolfe and Pakiser, 1971, fig. 2; Doyle, 1969a, fig. 3).

An important feature of the Potomac-Raritan palynological sequence, which is clearly reflected in the definition of both Brenner's and the new zones, is that, even though new forms tend to be relatively rare when they first appear, first occurrences are of generally greater stratigraphic utility than last occurrences, because older elements (especially among the gymnosperms and spores) tend to continue sporadically well above their level of maximum abundance. Some of this phenomenon is undoubtedly due to reworking, a common problem in palynology, but I would suggest that it is also a reflection of the fact that angiosperms were rapidly radiating into new niches, some of them essentially unoccupied (e.g., aquatic, Doyle, 1973; Doyle and Hickey, 1976), and only gradually replacing older elements, which might persist for considerable time as relicts in progressively more restricted niches before becoming extinct. Whatever its biological explanation, this phenomenon means that zones are more readily defined by the entry of new forms than by the overlap of the upper and lower portions of the ranges of different forms, as in true concurrent-range zones; although assignments to higher zones can be based on positive occurrences, assignments to lower zones tend to be based on absence of younger index forms or on relative proportions. Hence, especially near zonal boundaries where index forms are rare, samples that are too poor, represent peculiar facies, or are not studied in sufficient detail may be incorrectly assigned to too low a zone (see the example of subzone IIA in Delaware well D13, discussed below). "Sufficient detail" almost always means more intensive study than the count of 200 grains common in Quaternary quantitative studies, often a complete scan of at least one slide. These difficulties are in part overcome by the use of graded comparison and reference point correlation concepts, as discussed below.

Even without further refinement, palynological zonation has helped elucidate broad patterns in mid-Cretaceous continental sedimentation in the Atlantic Coastal Plain, especially a gradual northward migration of the locus of maximum deposition (Doyle, 1969b). In Virginia, the outcrop Potomac Group consists of sediments ranging in age from zone I through subzone IIB. In Maryland, sediments of subzone IIC and zone III age are added at the top, while in Delaware zone I appears to be pinching out. Zone IV age beds appear in New Jersey; zones II and III are still present in southern New Jersey and in downdip wells, but from the Raritan Bay area northward they too have dropped out and the sequence begins with zone IV. Finally, in Long Island wells (personal observations on slides from Brookhaven studied by Steeves, 1959), zone IV is overlain by a great thickness of sediments apparently intermediate in age between zone IV and the younger Magothy Formation of Santonian-Campanian age (Wolfe and Pakiser, 1971), and at least in part equivalent to the upper part of the Raritan Formation of New Jersey, believed to be of late Turonian-Coniacian age (Doyle, 1969a).

Delaware Well Sequence

The two wells that have proved most useful in refinement of the Potomac palynological sequence were drilled about 2.5 miles (4 km) apart through the Potomac Formation near Delaware City, Delaware (Delaware Geological Survey well numbers Dc53-7 and Ec14-1, here designated D12 and D13, respectively). Results of a preliminary palynological study of well D13 were reported by Brenner (1967), who kindly provided me with most of the samples; departures from his stratigraphic interpretation are noted below. The existence of two wells and the close sampling interval (every 5 ft or 1.5 m) have together proved invaluable in recognition of facies-related fluctuations and as controls on the ranges and stratigraphic reliability of morphological types.

The angiosperm pollen sequence in the Delaware City wells and its stratigraphic significance may be summarized with reference to Text-figure 3. The lithological sequence is presented in the form of two gamma-ray logs, where peaks (to the left) indicate lower radioactivity, and hence coarser-grained sediments (sands). Samples yielding spores and pollen are indicated by horizontal lines and depths in feet; notice that sampling is more complete in the finer-grained intervals. The stratigraphic distributions of the more important angiosperm pollen morphological complexes (right) are generalized from well D12 for the lower part of the section (zone I, not represented in well D13), and from D13 for the remainder (zones II and III). The distinction between solid and dotted lines concerns the degree of comparison with the taxa indicated, and is related to the convention used in the text and to Hughes's system (cfA, cfB, cfC, cf., discussed above) in the following manner. Solid lines represent two grades of comparison used in the text: simple identification, which corresponds to stringent cfA-level comparison, where there are no more than minor quantitative differences from the named taxon; and identifications qualified with cf., used here both for populations with minor cfB-level qualitative differences and for cases where present or original material is inadequate to determine whether cfA- or cfB-level comparisons are involved (Hughes's cf.). Dotted lines, represented in the following text by aff., indicate members of the same morphological complex as the named taxon, but with greater cfC-level differences, roughly comparable to those in contemporaneous species (operationally, consistently distinguishable populations). Hughes might include some of these under his cf.; however, in practice, differences of this order seem to be recognizable even when material is very meager. This system is admittedly far cruder and more subjective than that of Hughes, especially because of the use of comparisons with often poorly defined and illustrated published species, but I feel that at the present state of knowledge of the morphological variation in Potomac angiosperm pollen it is less premature than complete adoption of his system would be. In the future, careful biorecord-like characterizations of populations in the reference section may allow comparisons with previously published species to be dispensed with (at least in intraregional correlations), and distinctions to be made within the cf. category.

Correlations between the two wells (mostly of samples in well D12, using well D13 as the reference section) are indicated by double-headed arrows, corresponding to Hughes's brackets. Each arrow terminates immediately above and below the first samples in the reference section, which on the basis of the sorts of arguments discussed below can be recognized as definitely older and younger than the sample in question. The lengths of these arrows and the degree in which they overlap vary greatly depending on the number of elements, their ranges, their degree of facies restriction in both the reference section and the sample under consideration, and

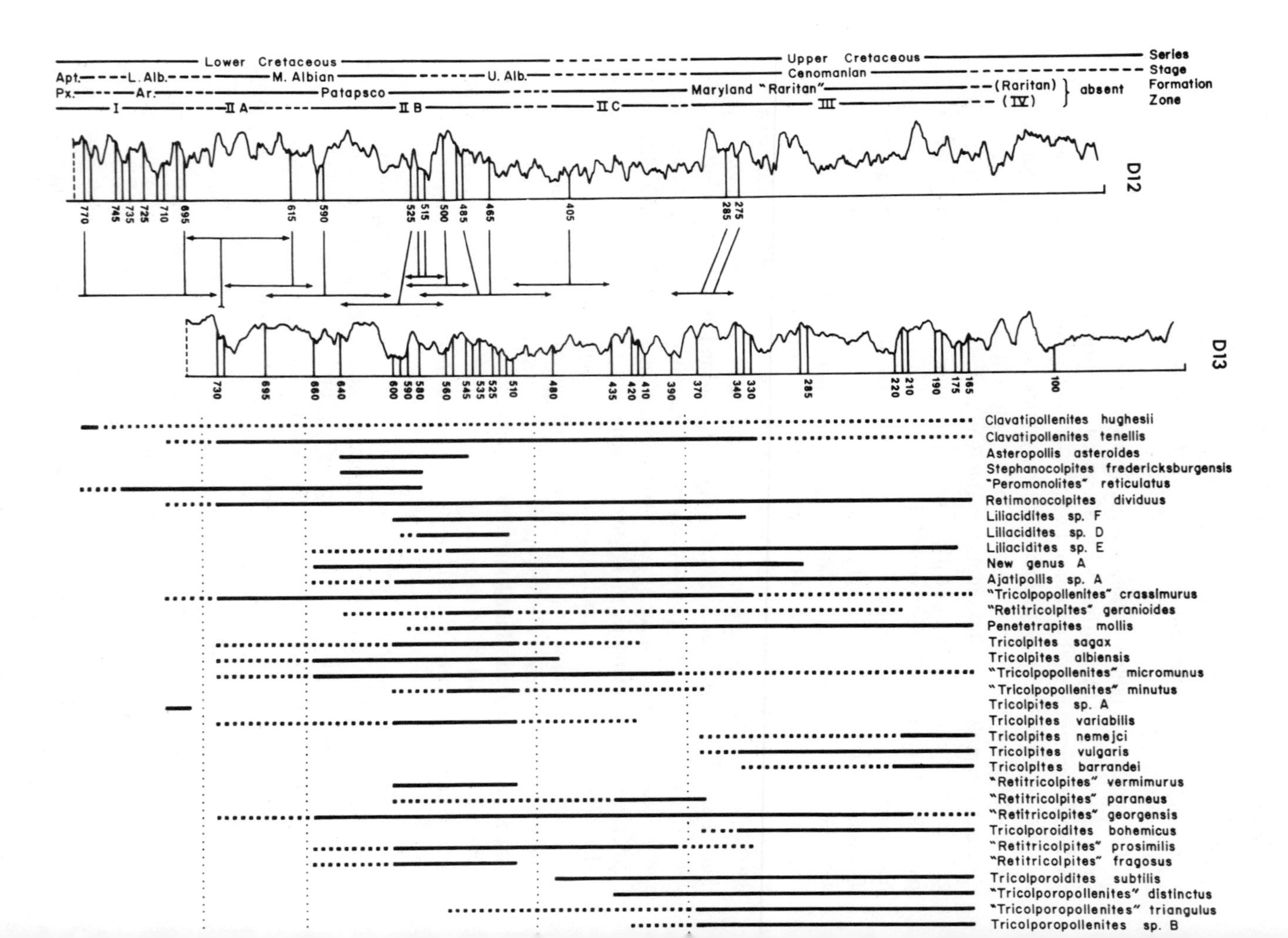
Series
Stage
Formation
Zone
Lower Cretaceous
Upper Cretaceous
Apt.
L. Alb.
M. Albian
U. Alb.
Cenomanian
Px.
Ar.
Patapsco
Maryland "Raritan"
(Raritan)
absent
I
II A
II B
II C
III
(IV)
D12
770
745
735
725
710
695
615
590
525
515
500
485
465
405
285
275
D13
730
695
660
640
600
590
580
560
545
535
525
510
480
435
420
410
390
370
340
330
285
220
210
190
175
165
100
Clavatipollenites hughesii
Clavatipollenites tenellis
Asteropollis asteroides
Stephanocolpites fredericksburgensis
"Peromonolites" reticulatus
Retimonocolpites dividuus
Liliacidites sp. F
Liliacidites sp. D
Liliacidites sp. E
New genus A
Ajatipollis sp. A
"Tricolpopollenites" crassimurus
"Retitricolpites" geranioides
Penetetrapites mollis
Tricolpites sagax
Tricolpites albiensis
"Tricolpopollenites" micromunus
"Tricolpopollenites" minutus
Tricolpites sp. A
Tricolpites variabilis
Tricolpites nemejci
Tricolpites vulgaris
Tricolpites barrandei
"Retitricolpites" vermimurus
"Retitricolpites" paraneus
"Retitricolpites" georgensis
Tricolporoidites bohemicus
"Retitricolpites" prosimilis
"Retitricolpites" fragosus
Tricolporoidites subtilis
"Tricolporopollenites" distinctus
"Tricolporopollenites" triangulus
Tricolporopollenites sp. B

Text-figure 3

Reference-point correlations of Delaware City wells D12 and D13 and stratigraphic distribution of principal angiosperm pollen types. The lithological sequence is represented by gamma-ray logs (peaks are coarser-grained sediments); sample depths are in feet. Correlation brackets (double-headed arrows) terminate just above and below the first samples in the other well that can be demonstrated on palynological grounds to be older and younger than the sample in question. To the left are suggested correlations with the standard stage sequence, Maryland and New Jersey lithological units, and palynological zones and subzones; dashed lines indicate intervals that cannot be correlated with one unit rather than the next. Ranges of pollen types (right) are based on well D12 for zone I and well D13 for zones II and III. Solid lines indicate ranges of forms strictly identifiable with or closely comparable to (cf.) the named taxon; dotted lines, forms that appear related but which have more than one qualitative difference from the named taxon (aff.). See text for discussion. Angiosperm pollen taxa indicated and their principal diagnostic characters: *Clavatipollenites hughesii* Couper: small, finely tectate-columellate monosulcate (aff.: rounder contour, granular sulcus); *Clavatipollenites tenellis* Paden-Phillips and Felix: like *C.* aff. *hughesii* but larger, coarser (aff.: intermediate); *Asteropollis asteroides* Hedlund and Norris: like *C. tenellis* but tri-, tetra-, pentachotomosulcate; *Stephanocolpites fredericksburgensis* Hedlund and Norris: like *C. tenellis* but polycolpoidate; *"Peromonolites" reticulatus* Brenner: medium-sized, reticulate (aff.: transitional to *Clavatipollenites; "P." peroreticulatus* Brenner is coarser, with no visible pila); *Retimonocolpites dividuus* Pierce (*Clavatipollenites rotundus* Kemp?): like *"P." reticulatus* but with loose reticulum, infolded sulcus margins (aff.: intermediate); *Liliacidites* sp. F (*Retimonocolpites* sp. A of Doyle, 1973): elongate monocotyledonoid monosulcate, reticulate, finer at ends; *Liliacidites* sp. D (*Retimonocolpites* sp. B of Doyle, 1973): like *L.* sp. F but larger, coarser (aff.: intermediate); *Liliacidites* sp. E (*Retimonocolpites* sp. D of Doyle, 1973): trichotomosulcate, reticulate, finer at sulcus margins and proximal pole (aff.: unbranched sulcus); new genus A: large monosulcate, "crotonoid" sculpture; *Ajatipollis* Krutzsch sp. A: tetrads with poroid apertures arranged according to "Garside's law" (aff.: less tightly fused); *"Tricolpopollenites" crassimurus* Groot and Penny: medium-sized, tectate-columellate tricolpate, thickest at poles (aff.: more reticulate, Wolfe et al., figs. 11-13, or smoother); *"Retitricolpites" geranioides* of Brenner: large, coarse pila (aff.: finer or more open sculpture); *Penetetrapites mollis* Hedlund and Norris: oblate, three colpoid apertures, polar thin area (aff.: more clearly colpate); *Tricolpites sagax* Norris: medium-sized, spheroidal, tectate (aff.: some tectal perforations); *Tricolpites albiensis* Kemp: like *T. sagax* but smaller; *"Tricolpopollenites" micromunus* Groot and Penny (same as most figured specimens of *T. albiensis* Kemp): small, finely reticulate, often tricolporoidate above D13-640'; *"Tricolpopollenites" minutus* Brenner: like *"T." micromunus* but very small (9 to 13 μm) (aff.: intermediate); *Tricolpites* sp. A: small (18 to 22 μm), open reticulate; *Tricolpites variabilis* Burger (*"Retitricolpites"* sp. B of Hedlund and Norris, *"R." virgeus* of Brenner, in part): like *"T." micromunus* but larger (aff.: intermediate); *Tricolpites nemejci* Pacltová: medium-sized, prolate, tall pila (aff.: thinner, finer sculpture); *Tricolpites vulgaris* (Pierce) Pacltová: spheroidal, reticulate, thin nexine; *Tricolpites barrandei* Pacltová: like *T. vulgaris* but smaller, coarser pila the density of sampling in the comparable interval in the reference section. A related convention is used (left) to indicate proposed correlations with the palynological zonation, Maryland and New Jersey lithological units, and the standard European stage sequence. Here the solid portions of the lines are placed next to those samples which it is believed can be definitely assigned to or correlated with the unit designated, whereas dashed portions indicate samples that, because of the limitations of stratigraphic resolution, cannot be correlated with one unit rather than the next (e.g., D13-510' and D13-370' are considered definitely upper Albian and Cenomanian, respectively, but intermediate samples could be of either age).

In several ways, these conventions and the correlation methods used help overcome the principal difficulties of the zonational framework discussed above, i.e., the rarity of "index" forms at their first appearance and the poor definition of upper age limits. First, with more critical morphological study and graded comparisons, it is often possible to discriminate between older and younger samples on the basis of stratigraphic trends within individual morphological complexes and the degree of differentiation between related complexes. Furthermore, differing degrees of confidence in correlation can be indicated by brackets of different length, and cases where it would be uncertain whether the sample lies near the top of one zone or the base of the next can be indicated by brackets straddling the zonal boundary.

In the oldest samples in Delaware well D12, from 770 and 765 ft, the angiosperm pollen flora consists solely of highly subordinate monosulcate forms. Among these are grains closely comparable to *Clavatipollenites hughesii,* the oldest definite angiosperm pollen yet described, from the Barremian of England (Couper, 1958; Kemp, 1968; see Doyle, 1973, figs. 2a, 2b). However, the presence of several additional angiospermous types, including aff. *"Peromonolites" reticulatus, "P." peroreticulatus,* and monocot-like *Liliacidites* species, and of the conifer *Parvisaccites rugulatus,* which first occurs at the base of the marine Aptian of England (Kemp, 1970), suggests a post-Barremian age (Aptian?).

Several important new complexes appear in samples D12-715' through D12-695'.

Text-Figure 3 (continued)
(aff.: intermediate); *"Retitricolpites" vermimurus* Brenner: rugulate-reticulate; *"Retitricolpites" paraneus* Norris: striate-reticulate (aff.: more reticulate); *"Retitricolpites" georgensis* Brenner: medium-sized, reticulate, finer at poles (aff.: finer); *Tricolporoidites bohemicus* Pacltová: like *"R." georgensis* but smaller, spheroidal, tricolporoidate (aff.: weaker colpus margins, ora; Wolfe et al., figs. 33, 34); *"Retitricolpites" prosimilis* Norris: like *"R." georgensis* but smaller, prolate (aff.: intermediate); *"Retitricolpites" fragosus* Hedlund and Norris: small, tectate, reticulate at poles (aff.: reticulate, less differentiation at poles); *Tricolporoidites subtilis* Pacltová: very small (9 to 13 μm), psilate, tricolporoidate; *"Tricolporopollenites" distinctus* Groot and Penny: like *T. subtilis* but larger; *"Tricolporopollenites" triangulus* Groot, Penny and Groot: very small, triangular tricolporoidate (aff.: rounder, larger; Wolfe et al., figs. 37, 38); *Tricolporopollenites* sp. B. (Wolfe et al., figs. 39, 40): medium-sized, triangular tricolporate (aff.: smaller).

These include two distinctive monosulcates (*Clavatipollenites* aff. *tenellis* and *Retimonocolpites* aff. *dividuus*) and the first rare reticulate tricolpates (*Tricolpites* sp. A and aff. *"Tricolpopollenites" crassimurus*). Brenner (1963) considered both *R. dividuus* and tricolpate angiosperm pollen to be zone II index forms, but the absence of any other zone II index species and the fact that the angiosperms in question are more comparable to forms found in Arundel localities which Brenner assigned to zone I (Allen, United Clay Mine) than to their relatives in D13-730' indicate that we are dealing with genuine range extensions (see Wolfe et al., 1975).

The oldest samples in well D13, from 730 and 725 ft, yield both zone II index spores and a variety of small- to medium-sized, reticulate to nearly smooth walled tricolpates, which show tendencies toward the younger forms with which they are compared in Text-figure 3, but which are more generalized and poorly differentiated from each other. All these forms are relatively rare, and they were apparently missed by Brenner (1967), who assigned the 730- to 695-ft interval to zone I. One of the zone II index spores is *Apiculatisporites babsae,* which first occurs at the base of the marine middle Albian in England, associated with the first small tricolpates (Kemp, 1970). This, together with the presence in the British lower Albian of monosulcates comparable to *R. dividuus* and possibly of medium-sized reticulate tricolpates, suggests that the lower-middle Albian boundary falls near the boundary between zones I and II; i.e., above D12-695' and below D13-730' (see Kemp, 1970).

These results indicate that, although zone I sediments are present in well D12, the section in well D13 begins with subzone IIA. This appears to be an effect of basement relief, since well D12 is updip from but deeper than well D13 (compare the correlated sample depths in Text-fig. 3).

The samples from 660 and 640 ft in well D13 and from 590 ft in D12 contain the first of Brenner's (1963) subzone IIB index spores and several new angiosperms, including monosulcates and tricolpates with more specialized sculpture patterns (e.g., *Liliacidites* aff. sp. E, new genus A, and cf. *"Retitricolpites" georgensis*), apparent *Clavatipollenites* derivatives with an irregularly branched sulcus or several blotchy colpus-like apertures (cf. *Asteropollis asteroides* and cf. *Stephanocolpites fredericksburgensis*), and peculiar permanent tetrads (aff. *Ajatipollis* sp. A). Many of the tricolpates are strikingly intermediate between the generalized forms in D13-730' and more differentiated younger groups (e.g., *Tricolpites* aff. *sagax* and a complex of small reticulate forms whose extremes suggest *"Retitricolpites" fragosus* and *"R." prosimilis*). Finally, among other small tricolpates we see the first rare tricolporoidate variants, with weak areas in the centers of their colpi, a tendency that becomes progressively more common higher in the section (see Doyle, 1969a).

At 600 and 595 ft in well D13, cf. *"Retitricolpites" fragosus* and cf. *"R." prosimilis* have become quite distinct; new elements include tricolpates with distinctly rugulate-reticulate and striate-reticulate sculpture (cf. *"Retitricolpites" vermimurus* and aff. *"R." paraneus*) and large monocot-like monosulcates (*Liliacidites* sp. F and *L.* aff. sp. D). This interval is discussed later in connection with the correlation of the 515 ft sample from well D12 and the Oklahoma middle Albian flora of Hedlund and Norris (1968).

Several complexes appear or attain their typical morphology at 560 ft in well D13: monocot-like monosulcates with a three-armed sulcus (*Liliacidites* sp. E), large, coarsely sculptured tricolpates (*"Retitricolpites" geranioides*), smooth tricolporoidates with the first indications of the trend toward triangular shape that becomes conspicuous in zone III (aff. *"Tricolporopollenites" triangulus*), and abundant very small, finely reticulate tricolporoidates (*"Tricolpopollenites" minutus*). Finally, tricolporoidate and often oblate grains constitute a majority of the *"T." micromunus* complex (see Doyle, 1969a, figs. 2p–t).

The pollen flora from 480 ft and higher in well D13 and from 405 ft in well D12 has features not seen in the typical outcrop Patapsco Formation, but which are observed in the overlying Maryland "Raritan" beds (see Wolfe and Pakiser, 1971; Doyle, 1969a, 1969b; Doyle and Hickey, 1976). These include the appearance of small, smooth-walled tricolporoidates (*Tricolporoidites* cf. *subtilis* and cf. *"Tricolporopollenites" distinctus*), the disappearance of cf. *"Retitricolpites" vermimurus,* and the common occurrence of two species of the conifer *Rugubivesiculites, R. reductus* (which may occur sporadically lower and was used by Brenner to define subzone B2) and *R. rugosus.* The similarity of this flora to that of the latest Albian of western Canada (Norris, 1967) is discussed later. The fact that similar changes occur at the outcrop (e.g., Elk Neck) at an abrupt lithological contact between typical Patapsco variegated clays and "Raritan" alternating sands and silts might have suggested that they were simply facies effects, but this is ruled out by the fact that they occur in both Delaware wells in the middle of a thick clay unit (see Text-fig. 3).

Except for the features cited, the spore and pollen flora from D13-480' through D13-390' is rather difficult to distinguish from that of the Patapsco Formation. Brenner (1967) in fact included this interval in subzone IIB, but because of its interest in Maryland lithostratigraphy, it is here designated a new subzone, subzone IIC. In comparison, samples from 370 through 165 ft in well D13 and 285 and 275 ft in well D12 are obviously younger than the typical Patapsco Formation. Brenner (1967) correlated them with the type New Jersey Raritan Formation, from which, however, they differ in the absence of triporate Normapolles and several distinctive tricolporate species. They are therefore placed in another new biostratigraphic unit, zone III.

Among the elements that first occur in D13-370' are several species of tricolpates (e.g., *Tricolpites* cf. *vulgaris* and *T.* aff. *nemejci*) and small tricolpor(oid)ates, including both smooth-walled, triangular and reticulate, spheroidal forms (cf. *"Tricolporopollenites" triangulus, Tricolporopollenites* sp. B, and *Tricolporoidites* aff. *bohemicus*). More typical *Tricolporoidites* cf. *bohemicus* and larger smooth tricolporates occur in D13-340' and higher samples, while *"Retitricolpites"* cf. *paraneus* drops out. Toward the top of zone III, from 220 through 165 ft in well D13, the tricolpates include typical *Tricolpites* cf. *nemejci* and *T.* cf. *barrandei,* while many of the tricolporates are intermediate in size and sculpture between the small lower zone III forms and their larger relatives from the Raritan Formation (see Doyle, 1969a, fig. 3i; Wolfe et al., 1975, figs. 41–43). Most zone III angiosperm pollen types are reported from dated lower to middle Cenomanian rocks elsewhere: the Dakota Formation of Minnesota (Pierce, 1961), the Woodbine Formation of Oklahoma (Hedlund, 1966), and especially the Peruc Formation of Czechoslovakia (Pacltová, 1971), at the top of which are found Normapolles similar to those in the lower Raritan Formation (zone IV) of New Jersey (see Wolfe and Pakiser, 1971; Doyle, 1969a).

Correlation of D12-515'

A detailed comparison of the angiosperm flora of the rich sample from 515 ft in well D12 with the D13 reference section and with an independently dated late-middle-Albian (Fredericksburgian) flora described from the Antlers Sand and "Walnut" Clay of Oklahoma (Hedlund and Norris, 1968) illustrates both the sorts of arguments used in the "reference point" correlations shown in Text-figure 3 and one of the best connections with the standard stage sequence, since essentially all the angiosperms in the Oklahoma flora have close analogs in D12-515' or the correlative interval in D13. Correlative floras also occur at several important outcrop angiosperm megafossil localities (Mt. Vernon, White House Bluff: Doyle and Hickey, 1972, 1976; Doyle, 1974).

Many angiosperms in D12-515', and even more so the spores and gymnosperm pollen, are so long ranging or difficult to differentiate from related forms that they would serve only to bracket the flora as late zone I or zone II: e.g., *Clavatipollenites* aff. *hughesii* (Pl. 1, figs. 1 and 2), *"Peromonolites" peroreticulatus, C.* cf. *tenellis,* cf. *"C." minutus, Retimonocolpites* cf. *dividuus,* cf. *"Tricolpopollenites" crassimurus* (Pl. 1, figs. 6 and 7), *Tricolpites* cf. *variabilis, T.* cf. *albiensis* (Pl. 1, figs. 8 and 9), and cf. *"Tricolpopollenites" micromunus.* All these, except *C.* aff. *hughesii, C.* cf. *tenellis,* and *T.* cf. *albiensis,* have close analogs in Oklahoma. Future careful study of these complexes may well prove them to be more useful in correlation.

Other angiosperms enter or attain a morphology comparable to that in D12-515' at higher levels in well D13, raising the lower age bracket to 590 ft: cf. *Asteropollis asteroides* (Pl. 1, fig. 3), cf. *Stephanocolpites fredericksburgensis* (Pl. 1, figs. 4 and 5), new genus A (Pl. 1, fig. 16), cf. *"Retitricolpites" vermimurus* (Pl. 1, figs. 17-19), aff. *"R." paraneus, Tricolpites* cf. *sagax* (Pl. 1, figs. 10 and 11), *Liliacidites* sp. D (Pl. 1, fig. 20), *L.* sp. F (Pl. 1, fig. 21), *Ajatipollis* sp. A (Pl. 1, figs. 22 and 23), and cf. *"Retitricolpites" prosimilis* (Pl. 1, figs. 12 and 13). All but the last four species are either reported by Hedlund and Norris or seen in their type slides at the U.S. National Museum (e.g. new genus A). Three additional Oklahoma forms, *"Retitricolpites" georgensis, "R." fragosus,* and the grain figured as *"R." vulgaris,* are absent or only questionably present in D12-515'; however, since they occur above or below (500 and 525 ft), this is presumably an example of a facies effect (and of the advantages of close sampling in establishing the ranges of facies-restricted forms). Finally, a fairly large proportion of the small, reticulate tricolpates in D12-515' are actually tricolporoidate (see Pl. 1, figs. 14 and 15), indicating a post-640-ft-age.

Two of the most typical Oklahoma forms, *Asteropollis asteroides* and

Stephanocolpites fredericksburgensis, help provide an upper age bracket for the two floras as well; they last occur at 500 ft in well D12 and at 580 or 545 ft in well D13. The absence of several forms that enter at 560 ft in well D13 also tends to indicate a pre-560-ft age: *Liliacidites* sp. E, aff. *"Tricolporopollenites" triangulus,* and *"Tricolpopollenites" minutus* (perhaps the most significant, in view of its abundance in D13-560' through D13-510').

All the forms considered so far have been consistent with a post-590-ft, pre-560-ft age bracket for D12-515'. However, both D12-515' and the Oklahoma flora contain two conspicuous additional elements which first occur in D13-560', *Penetetrapites mollis* (Pl. 1, figs. 26 and 27) and cf. *"Retitricolpites" geranioides* (Pl. 1, figs. 24 and 25). Here we have reached the limit of resolution of the available data: either D12-515' is intermediate in age between D13-580' and D13-560', or we are dealing with facies-related absences in D12-515', D13-580', or D13-560'.

Using the data presented, the Oklahoma flora may be correlated with the same post-590-ft, pre-560-ft interval in well D13, and with the post-525-ft, pre-490-ft interval in well D12. Since the Oklahoma flora can be dated almost directly by comparison with the nearby Texas ammonite sequence (Young, 1966) as late middle Albian, this represents an important palynological tie between the continental Potomac Group sequence and the faunally defined subdivisions of the Cretaceous.

Plate 1

Angiosperm pollen types used in correlation of sample D12-515'. In three cases (figs. 10, 11; 20; 26, 27), the grain figured is from the immediately underlying and very similar sample D12-520'. Coordinates: UMMP Zeiss RA microscope No. 4767359. All photographs ×1,000. Figs. 1, 2: *Clavatipollenites* aff. *hughesii,* monosulcate, two focal levels, surface view and optical section (D12-515-1d, 8.7 × 86.3); 3: cf. *Asteropollis asteroides,* pentachotomosulcate (D12-515-1a, 20.0 × 99.1); 4, 5: cf. *Stephanocolpites fredericksburgensis,* polycolpoidate, two focal levels (D12-515-1d, 14.8 × 89.0); 6, 7: cf. *"Tricolpopollenites" crassimurus,* tricolpate, two focal levels (D12-515-1d, 8.5 × 96.2); 8, 9: *Tricolpites* cf. *albiensis,* tricolpate, two focal levels (D12-515-1d, 13.7 × 99.5); 10, 11: *Tricolpites* cf. *sagax,* tricolpate, two focal levels (D12-520-1a, 12.2 × 88.8); 12, 13: cf. *"Retitricolpites" prosimilis,* tricolpate, two focal levels (D12-515-1a, 3.1 × 98.2); 14, 15: cf. *"Tricolpopollenites" micromunus,* tricolporoidate, two focal levels (D12-515-1d, 10.8 × 97.9); 16: new genus A, monosulcate, optical section with inset of surface sculpture (D12-515-1d, 22.3 × 84.8); 17–19: *"Retitricolpites" vermimurus,* tricolpate, three focal levels (D12-515-1c, 13.0 × 91.7); 20: *Liliacidites* sp. D, monosulcate (D12-520-2a, 4.5 × 83.5); 21: *Liliacidites* sp. F, monosulcate (D12-515-1a, 13.3 × 84.8); 22, 23: *Ajatipollis* sp. A, tetrad of triaperturate grains, two focal levels (D12-515-1a, 19.6 × 88.9); 24, 25: cf. *"Retitricolpites" geranioides,* tricolpate, two focal levels (D12-515-1a, 9.4 × 89.9); 26, 27: *Penetetrapites mollis,* tricolpoidate, two focal levels (D12-520-1a, 20.4 × 96.3). See text for discussion.

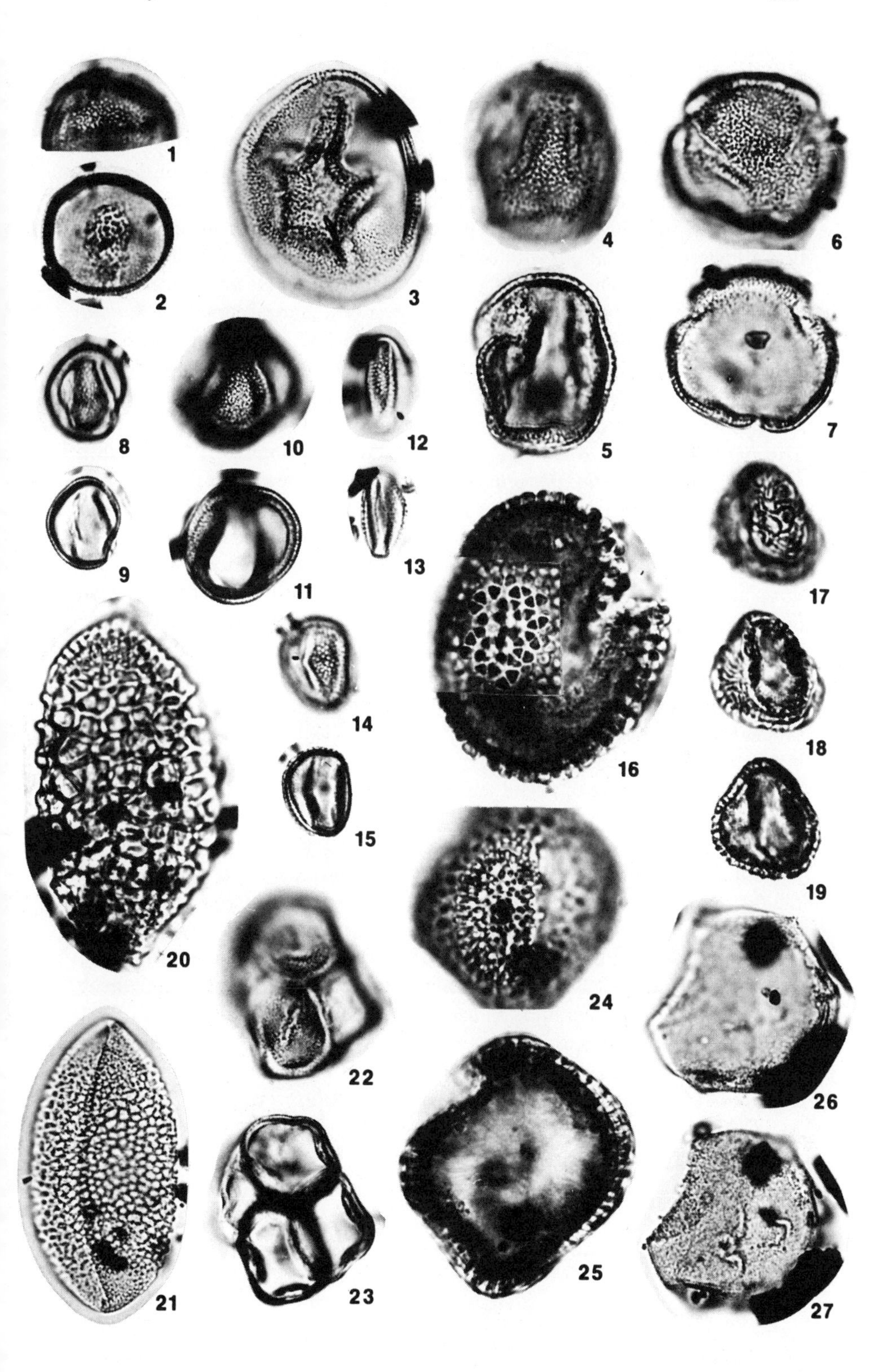
1
2
3
4
5
6
7
8
9
10
11
12
13
14
15
16
17
18
19
20
21
22
23
24
25
26
27

Problems in Interregional Correlation

As discussed above, the relative role of migrational and evolutionary phenomena is of central importance in both "qualitative" interregional correlations and discussions of early angiosperm evolution. The lack of anomalous associations of elements in the Oklahoma flora of Hedlund and Norris (1968) in terms of the Delaware sequence is consistent with the assumption that there were no significant barriers or lags in angiosperm migration between the two areas, at least for the interval in question. However, correlations with certain other areas do in fact appear to be complicated by migration. One example, noted by Hedlund and Norris, concerns the flora of the upper Albian (-basal Cenomanian?) lower Colorado Group of the Western Canada plains (Norris, 1967). Marine evidence indicates that this sequence is just younger than the Oklahoma flora (and, by extrapolation, than some point in the 590- to 560-ft interval in well D13). In agreement with this, most of the angiosperms reported by Norris have close analogs in the 560- to 390-ft interval (.e.g., *Tricolpites sagax, "Retitricolpites" georgensis, "R." prosimilis, "R." paraneus, "Tricolpopollenites" minutus, Clavatipollenites* spp.), while several of the types most important in the definition of subzone IIC enter somewhat above the base of the Canadian section (*"Psilatricolpites" parvulus,=Tricolporoidites* cf. *subtilis?, Rugubivesiculites reductus,* and *R. rugosus*). Some of the differences between Oklahoma and Canadian floras that puzzled Hedlund and Norris (1968) are what would be expected from the Potomac sequence, e.g., the absence of the *Asteropollis–Stephanocolpites* complex. What is more striking, however, is the fact that no angiosperm pollen has been described from the Mannville beds which directly underlie the Colorado Group, apparently without a major unconformity, and which should therefore be at least in part equivalent to Oklahoma and Potomac Group beds with abundant angiosperm pollen. This suggests that there was a significant delay in the appearance of angiosperms at higher latitudes, as in Axelrod's (1959) poleward migration theory (see Hughes, 1961; Doyle, 1969a). At the same time, a more detailed examination shows that this cannot have represented a simple wave of migration, since the first angiosperm pollen in western Canada is not comparable to the first in the Potomac, but rather to pollen higher in the section. These data can be explained by a model which postulates that the first angiosperms, although highly mobile, were initially ill adapted to colder climates, so that outside a broad low- to middle-latitude belt, where evolutionary phenomena are relatively unobscured by migration, their invasion had to await the evolution of species with broader climatic tolerances (based on discussions with G. J. Brenner and a suggestion by J. Muller, pers. comm.). The practical lesson which can be drawn is that the absence of angiosperms, especially in higher-latitude areas, is an unreliable criterion for correlation, but that careful consideration of the specific composition of the first angiosperms can be useful in correlation with lower latitudes.

Facies Problems

Another important pitfall in mid-Cretaceous stratigraphic palynology is the wide variation in the percentage of angiosperms in the total spore and pollen flora, both from one continental facies to another (noted by Brenner, 1963, and Doyle, 1969a) and between equivalent continental and marine beds. This phenomenon appears to be responsible for many of the apparent contradictions cited by Axelrod (1970) between megafossil and palynological evidence on the quantitative role of angiosperms in Albian-Cenomanian vegetation. An example is provided by samples D12-500', D12-485', and D13-555', which on present palynological evidence are indistinguishable in age (early upper Albian?). While D12-485' yields only 2 percent angiosperm pollen, being dominated by spores and gymnosperm pollen, D12-500' contains 45 percent. Some outcrop localities of similar age yield even higher percentages and are dominated by angiosperm leaves, e.g., the Severn and West Brothers localities of Brenner (1963), Doyle (1969a), and Doyle and Hickey (1976). D 13-555' resembles D12-485' in being dominated by gymnosperms and ferns, although the percentage of angiosperms is somewhat higher at 14 percent. The key to understanding these variations may be the abundance in D13-555' of planktonic cysts (acritarchs and some dinoflagellates), indicating at least marginally marine conditions; this is consistent with the downdip location of well D13.

This suggests we are dealing with an example of the "Neves effect" of Chaloner and Muir (1968): some forms (typically pteridophyte spores in the Late Paleozoic and Early Mesozoic) are abundant in certain continental facies, especially coals, whereas others (typically cordaite or conifer pollen) dominate in equivalent marine sediments and some stream-laid continental sands and silts. Chaloner and Muir (1968) and Clapham (1970) independently interpret this phenomenon on a simple geological model, postulating that the spore-bearing plants tended to be restricted to certain lowland habitats (e.g., swamps), while the others were dominant in the "upland" vegetation and hence contributed most to the total pollen load entering the sea from streams (see also Hopping, 1967). If the analogy can be drawn, this means that in the late Albian conifers were still the regional vegetational dominants, but that angiosperms were already locally abundant in the lowlands, particularly in disturbed and aquatic habitats (Doyle and Hickey, 1976), a conclusion also drawn by Pierce (1961) from a study of still younger (Cenomanian) deposits. This leads to another cautionary principle in using early angiosperm pollen in stratigraphy: correlations should be based not on the percentage of angiosperms in the total flora, but rather on the qualitative composition of the angiosperm element. This in turn may require more intensive study than is often done in reconnaissance work.

SUMMARY AND CONCLUSIONS

Spores and pollen, by virtue of their small size, transportability, chemical resistance, and distinctive morphology, offer unique potential in stratigraphic subdivision of otherwise unfossiliferous continental deposits, their correlation with faunally dated marine sequences, and the study of land plant evolution. Because they can be extracted in great numbers from subsurface core material, they allow closely sampled vertical sequences to be established, which point up the limitations of neontological species concepts and zones in biostratigraphy. This has led to proposals of alternative stratigraphic methodologies based on graded comparisons of populations and "reference-point" (bracket) correlations. Illustrations are drawn from palynological studies of the rapidly evolving angiosperm element in the Potomac Group (mid-Cretaceous) of the Atlantic Coastal Plain. Because of rapid facies alternations in this fluvial-deltaic sequence, correlations must be based more on qualitative evolutionary changes than on relative proportions. A zonal scheme based largely on the progressive entry of new elements has clarified time relations in the Potomac sequence, but finer correlation brackets have been made possible by intensive study of temporal-morphological variation within the angiosperm element in two well sections from Delaware. Close correlations are possible with dated marine sequences in other areas, but careful study of the qualitative character of the angiosperm element is necessary to overcome quantitative differences between continental and marine facies and the effects of late migration of angiosperms into higher-latitude areas. Suggestions for continued progress in stratigraphic palynology include (1) practical and conceptual refinements of reference-point stratigraphy (including computer-assisted quantitative analysis), (2) detailed analysis of geographic and facies variation in spore and pollen assemblages, aided by recent and sedimentological models, and (3) the development of techniques making practical the use of scanning electron microscopy in stratigraphic studies of fossil pollen.

Acknowledgments

I wish to thank Gilbert J. Brenner, State University College, New Paltz, N.Y., and the Delaware Geological Survey (Robert R. Jordan, director) for the Delaware well samples used in this study, and Harold E. Gill, U.S. Geological Survey, Trenton, N.J., for the gamma-ray logs. This paper was completed while I was an associate of the Laboratoire de Palynologie du C.N.R.S., Montpellier, France (Mme. Van Campo, director); I am most grateful to Mme. Van Campo for her encouragement and the use of the facilities of the laboratory.

Addendum

Since completion of this paper, "new genus A" has been formally described as *Stellatopollis barghoornii* Doyle, and *"Peromonolites" reticulatus* and *"P." peroreticulatus* have been transferred to the genus *Retimonocolpites* on the basis of transmission electron microscopic studies confirming their angiospermous affinities (J. A. Doyle, M. Van Campo, and B. Lugardon. 1975. Observations on exine structure of *Eucommiidites* and Lower Cretaceous angiosperm pollen. *Pollen et Spores 17*(3): 429–486).

Discovery of an assemblage of monosulcate angiosperm pollen in the upper Wealden (Barremian) of England which shows much the same range of morphological variation seen in lower Zone I (unpublished data of Hughes and Laing, presented by Hughes at the XII International Botanical Congress, Leningrad, 1975) indicates that the base of the Potomac Group may well be Barremian in age (as originally proposed by Brenner, 1963) rather than Aptian, as argued above. Previously, *Clavatipollenites hughesii* was the only angiosperm reported from these beds. This illustrates the dangers of using angiosperm pollen to correlate between sequences in which the angiosperm element has been studied with differing degrees of intensiveness.

Despite recognition of the fact that abrupt appearances of species should be expected as a result of allopatric speciation, the discussion above of the theoretical bases of reference point correlations and graded comparisons (and even more so Hughes's discussions on which it is based) takes too much for granted many of the assumptions of "phyletic gradualism" criticized by Eldredge and Gould (1972). In fact, many phenomena observed in the Potomac Group angiosperm record may be equally consistent with Eldredge and Gould's alternative model of "punctuated equilibria." For example, because of the difficulty in recognizing species limits in such small parts of the whole plant as pollen grains (so that pollen "species" may often represent genera), it is possible that some of the series of stratigraphically and morphologically "transitional" forms cited actually represent immigration of successive, slightly differentiated, allopatrically derived species (i.e., Stufenreihen rather than Ahnenreihen: cf. Simpson, 1953). In addition, the attention given to taxa (whether species or species groups) which show stratigraphically useful changes should not obscure the fact that many other "species" persist over long stratigraphic intervals with no appreciable morphological changes (e.g., *Retimonocolpites peroreticulatus* throughout the Delaware Potomac Group, *R. dividuus* from Subzone IIA onward). However, I strongly suspect that more quantitative study will show that some of the observed changes are more consistent with gradual phyletic evolution, particularly readjustment to new physical and biotic factors following entry of former peripheral isolates into new environments (including competitive interaction and resulting character divergence of sister species, as emphasized by Gingerich, 1974. Stratigraphic record of Early Eocene *Hyopsodus* and the geometry of mammalian phylogeny. *Nature 248:* 107–109). The replacement of the variety of poorly differentiated tricolpates of Subzone IIA and lower Subzone IIB by related upper IIB forms which are morphologically more distinct from each other is especially suggestive of such a process. I would suggest that such cases of "punctuated gradualism" should be more common in groups actively radiating into new adaptive zones, where more "ecological space" should be available for competitive displacement of sister species after reestablishment of sympatry, than in groups (such as Devonian trilobites) already occupying stable or even contracting adaptive zones, where because of the more closed ecological situation extinction of one lineage would be the more likely result of renewed sympatry.

Ecologic and Zoogeographic Factors in the Biostratigraphic Utilization of Conodonts

Frank H. T. Rhodes University of Michigan

R. L. Austin University of Southampton

INTRODUCTION

Conodonts were first described by C.H. Pander in 1856, but they remained an obscure and little studied group until 70 years later, when Ulrich and Bassler (1926) published a description of Devonian and Mississippian forms, and proposed a comprehensive scheme for their classification. This provided the impetus for further studies, and led to three decades of sporadic studies. These publications, in which Branson and Mehl and their students were preeminent, consisted chiefly of descriptions of isolated faunas, but provided a general understanding of the broad stratigraphic sequence of conodont faunas, except for their earliest (Cambrian) and latest (Permo-Triassic) history. This period also saw the independent discovery by Schmidt (1934) and Scott (1934) of natural conodont assemblages, composed of a variety of individual conodont elements. In the last 20 years, there has been an explosion of conodont publication, with particular emphasis upon their use in refined biostratigraphic studies. More than 600 articles were published in the period from 1961 to 1969 (Sweet and Bergström, 1971b, p. 1). The use of conodonts in detailed biostratigraphy, first demonstrated by Sannemann, Ziegler, and Bischoff in the Upper Devonian of Germany, has been extended by intensive studies in Europe and North America and preliminary studies in Asia and Australia, which have indicated the unique value of these microfossils in Paleozoic biostratigraphy. The recent publication by Sweet and Bergström (1971b) provides an important summary of the present general state of the art of conodont biostratigraphy.

COMPOSITION AND STRUCTURE

Conodonts are extinct, marine microfossils, ranging from about 0.1 to 1 mm in length. They appear to be a relatively homogeneous group, composed of calcium phosphate resembling francolite, having the general formula $Ca_5 Na_{0.14} (PO_4)_{3.01}$-$(CO_3)_{0.16} F_{0.73}(H_2O)_{0.85}$, and having grown by the centrifugal addition of broadly similar lamellae deposited one above another (Pietzner et al., 1968; Armstrong and Tarlo, 1966). Some conodont elements exhibit breakage and repair, which can be seen to have involved regeneration and reconstruction, presumably within soft tissues, thus suggesting that they were internal structures. Conodonts have a basal cavity, which in well-preserved specimens of many ages occasionally contains a basal filling of phosphatic material. This material, although it shares the lamellar structure of conodont elements proper, has a somewhat different fabric, which is, however, quite distinct in structure from vertebrate bone.

The use of the scanning electron microscope in recent years has allowed a comprehensive study of conodont ultrastructure, fabric, and microarchitecture, and has contributed significantly to systematic and evolutionary studies. Barnes et al. (1973a), who studied the ultrastructure of 24 form species of Ordovician age, concluded that these conodonts represented two major groups. The first, the hyaline group, includes a subgroup of robust, massive elements, known chiefly from the Midcontinent region (the neurodonts). These forms, apparently restricted to Ordovician nearshore, hypersaline, carbonate environments, have shallow basal cavities, distinctive brittle basal fillings, and a cone-in-cone structure made up of long acicular crystallites. The related nonneurodont hyaline elements, although widely distributed, are found chiefly in shelf and miogeosynclinal sediments. They have a lamellar structure of cone-in-cone form, but display more fusion between the lamellae than do the neurodonts. These forms, in contrast to the neurodonts, develop limited white matter in the denticles, apparently forming this by resorption, and thus perhaps achieving a strengthened structure.

The other major group of conodonts are the cancellate forms. These contain an appreciable amount of white matter, which tends to be concentrated in the denticles. Although it seems probable that this white matter served a strengthening function (Barnes et al., 1973b), it has also been suggested that it may have been related to element weight and phosphate availability.

Function and Affinities

The function of conodonts has long exercised the attention of conodont workers, and several alternative explanations have been offered (see Rhodes, 1954; Lindström, 1973). It now seems clear that the irregular form, delicate construction, divergent axes, and variable denticulate profile of conodonts preclude the possibility of their having acted as jaws or teeth. In particular the position of the "anterior cusp" makes it almost inconceivable that conodonts could have been used in either grasping or chewing. Furthermore, the patterns of conodont growth, repair, and resorption suggest that they were covered with soft tissue during the life of the organism.

Melton and Scott (1973) have described specimens that are interpreted as complete conodont animals from Carboniferous strata in Central Montana (Text-Fig. 1). These specimens contain conodont assemblages identified as *Lochriea* and

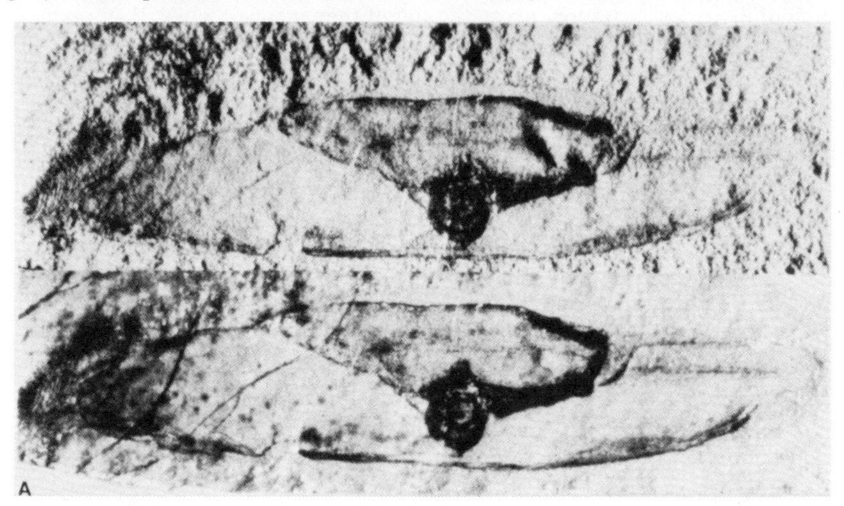

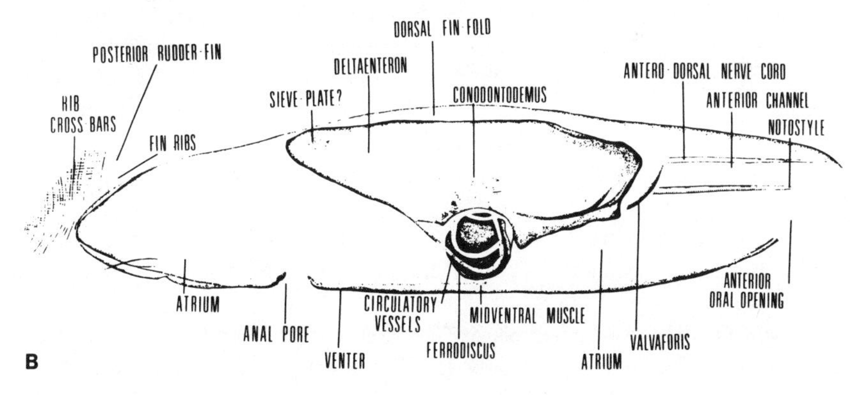

Text-Figure 1
(A) *Lochriea wellsi* Melton and Scott. A conodont-bearing organism from the Carboniferous Bear Gulch Limestone of Montana. Top figure photographed in air, bottom figure photographed under water. Note detailed wall structure of notostyle and faint oblique lines subjacent to notostyle. ×1.70. Specimen 6027, University of Montana. (After Melton and Scott.) (B) Sketch of reconstruction of *L. wellsi* showing names given to various anatomical features. (After Melton and Scott.)

Scottognathus. The conodont-bearing fossil organisms are about 7 cm long, elongated, and bilaterally symmetrical, with an anterior oral opening and structures that are interpreted as a dorsal nerve cord and a notostyle. A gut-like structure, the deltaenteron, and what is interpreted to be a circulatory system are also present. The former is believed to have functioned as a food-filtering system, and a posteroventral anal pore is also identified.

Lindström (1973), offering an alternative functional interpretation to that of Melton and Scott, has suggested that the conodont animal was tentaculate, with the conodont denticles corresponding to the site of the tentacles, which they partly supported. He believes that the pitting commonly found on the platforms of conodonts probably corresponds to the site of muscle attachments, being similar in form to that found in living brachiopods, to which he suggests they may be distantly related.

Taxonomy

Conodont taxonomic studies are at present in a state of transition. Pander (1856), in the first description of conodonts, based his study on several thousand isolated specimens, and erected a system of nomenclature based on variation in individual elements. He established 14 genera and 56 species, which he recognized as being utilitarian and arbitrary, since the individual conodont animals may have borne several different types ("genera") of elements.

Although this latter supposition proved correct, an expanded form of this "form classification," as it came to be called, founded upon variations in "cones, bars, blades and platforms" formed the basis of almost all taxonomic studies until the mid-1960s. Employing simple descriptive morphologic terms for the conodont elements ("cusp, blade, lateral faces, basal cavity," etc.), his system was readily applicable, and it has formed the basis of the spectacularly successful biostratigraphic utilization of conodonts during the last two decades. Suprageneric classification has, on the whole, been less successful. It is little used and of limited value.

The recognition of natural assemblages of conodonts, containing several elements of diverse forms, has been based upon several distinct kinds of data. These include the following:

1. *In situ* bedding-plane assemblages. The first tentative identification of such an assemblage was made by Hinde as early as 1879, but it was not until 50 years later that Scott (1934) and Schmidt (1934) provided more convincing evidence, which has been supported by subsequent studies (e.g., Scott, 1942; Rhodes, 1952, 1962; Schmidt and Müller, 1964; Collinson et al., 1972; Mashkova, 1972; Melton and Scott, 1973) (Text-fig. 2).

2. Associations of conodonts in bituminous material of presumed organic origin have been described as original associations by Lange (1968) and Scott (1969).

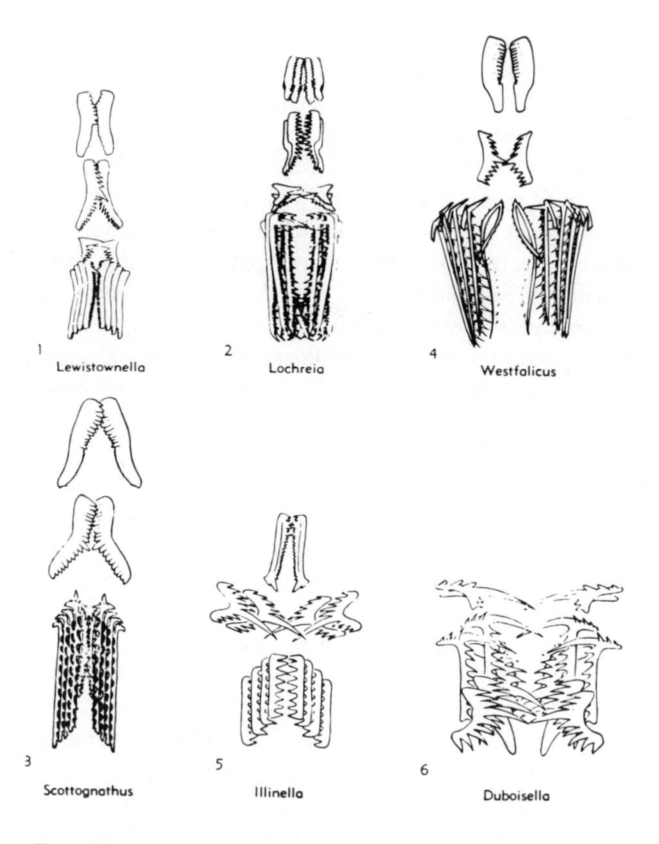

Text-Figure 2
Diagrammatic illustrations of natural multielement conodont assemblages from the Carboniferous: (1) *Lewistownella agnewi* Scott; (2) *Lochriea montanaenis* Scott; *(3) Scottognathus typica* (Rhodes); (4) *Westfalicus integer* (H. Schmidt); (5) *Illinella typica* Rhodes; (6) *Duboisella typica* Rhodes; all X15 (approx.). (After Rhodes.)

3. Fused element assemblages are sometimes found in phosphatic residues after solution of the enclosing rock matrix. Barnes (1967), Pollock (1968), and Austin and Rhodes (1969) have described examples of this.

4. The occurrence of distinctive surface microarchitecture has been used by Lindström and Ziegler (1965) as an indication of original associaton.

5. Similarity of form, color, denticulation, ornamentation, and basal attachment have been used by Bergström and Sweet (e.g., 1966) as evidence of original association.

6. Similarity of occurrence, involving geographic, stratigraphic, and ecologic ranges, and similarity of relative frequency of isolated elements are the criteria of association most frequently used in recent studies. This method has involved both simple empirical counts and numerical comparison (e.g., Walliser, 1964; Bergström and Sweet, 1966; Webers, 1966; Schopf, 1966; Jeppsson, 1971) and sophisticated statistical analysis (e.g. Kohut, 1969, who used rank correlation techniques to confirm the validity of Bergström and Sweet's groupings, and Druce et al., 1971, who used polythetic cluster analysis, as well as other methods, to establish multielement assemblages).

None of these methods provides an infallible guide to original association, and many "form species and genera" are as yet unassigned to multielement assemblages. Furthermore, at least some conodont assemblages appear to have consisted of only a single element, and some described as multielement assemblages are almost certainly partial and incomplete; a few others may include two or more natural species having identical ranges and tolerances. Many common elements, especially "bars and blades," are found in several different types of assemblage. In spite of these problems, the overwhelming majority of conodont workers accept the existence of multielement associations, and most now seek to employ a taxonomy and nomenclature established upon them. Prevailing conodont taxonomy, based on individual element variation, is seen as arbitrary and artificial, in spite of the basis it has provided for biostratigraphic studies. Its replacement by a "natural" taxonomy presents a number of problems, however; these are discussed next.

Recognition of the need to provide a nomenclature for natural multielement assemblages, based on the assumption that these represent the remains of individual organisms (conodontifers), has involved the use of a number of different schemes of nomenclature. Four more or less distinct systems have been employed.

1. Erection of a new Linnean genus and species for the assemblage, leaving existing "form classification" undisturbed (e.g., Hinde, 1879; Scott, 1942; Rhodes, 1952, 1957; Schmidt and Müller, 1964).

2. Application of the oldest available name for one of the elements included in the assemblage (e.g., Bergström and Sweet, 1966; Webers, 1966).

3. Designation of natural assemblages by open nomenclature and symbols (e.g., Walliser, 1964; Pollock, 1968; Druce et al., 1971).

4. The use of "parataxa" to accommodate form classification categories (e.g., Moore and Sylvester-Bradley, 1957).

The application of any one of these schemes involves substantial problems, both legal (under the international Rules of Zoological Nomenclature) and practical, but it is now generally accepted by most workers that method 2, the application of the oldest available element name, is the most acceptable. The most serious practical problem is that the overwhelming majority of conodonts studied are extracted as discrete elements from insoluble residues, and that conodont nomenclature and zonation are largely based on these isolated elements and must be changed nomenclaturally at a time they are so well established—even committed to memory. Although bedding-plane assemblages are very rare and will probably continue to be, those few that are known provide models against which one can evaluate whether or not associations established by empirical means make "biological sense."

Two examples will illustrate the application of this method to systematic and biostratigraphic studies. Walliser (1964) established nine tentative, and not necessarily complete, groupings of conodont elements from the Silurian and Early Devonian, some of which were later named by Jeppsson (1971, and references therein).

Subsequently, Pollock (1968) described fused elements from the Silurian of Indiana, which agree well with some of the associations established by Walliser. More recently, Mashkova (1972) has described a bedding plane conodont assemblage, which she names *Ozarkodina steinhornensis,* from Lower Devonian rocks of the Fana Mountains in Central Asia (Text-fig. 3). This agrees in detail with the empirical assemblage J of Walliser, containing one pair of *Ozarkodina,* one pair of *Spathognathodus,* together with six or more hindeodelliform, two or more neoprioniodiform, and one or more plectosphathodiform elements. Mashkova has also been able to recognize subspecies of the assemblage species *Ozarkodina steinhornensis* by tracing the stratigraphic succession of the component elements, comprising successive subspecies *O. s. interposita, eosteinhornesis, remscheidensis, praeoptima, optima,* and *steinhornensis.* The stratigraphic succession of these forms is shown in Text-figure 3, which provides an excellent example of the evolutionary relationships to the stratigraphic sequence.

Provincialism and Ecologic Restrictions

The spectacular biostratigraphic success of conodonts in local zonal schemes, established by studies in Europe and North America, has encouraged intensive worldwide study of successions and correlations. Inevitably, and quite properly, these studies tended to be concentrated in strata in which conodonts were abundant and from which they could be readily extracted by the use of standard reagents and techniques. In almost all cases, these were carbonate strata, representing shelf or miogeosynclinal sediments. Zonal schemes, many of unusual refinement, have been established, applied, tested, and duplicated in both local and regional correlation. So impressive have the results of these studies been that the global similarity of conodont faunas has received more emphasis than regional differences, so that subtle provincial and ecologic differences have sometimes been regarded as anomalous aberrations and imperfections in what had come to be wishfully regarded as a pattern of worldwide uniformity. In the euphoria of the early 1960s, it almost appeared that conodonts were facies breakers of such cosmopolitan occurrence that they were somehow exempt from the limitations of other groups. Conodonts do, in fact, occur in a great many sediments, from geosynclinal to relatively shallow water, most genera and many species having a worldwide occurrence. Some forms, however, are far more restricted, and it is now becoming clear that there are subtle but significant geographic and ecological variations which are of importance in the use

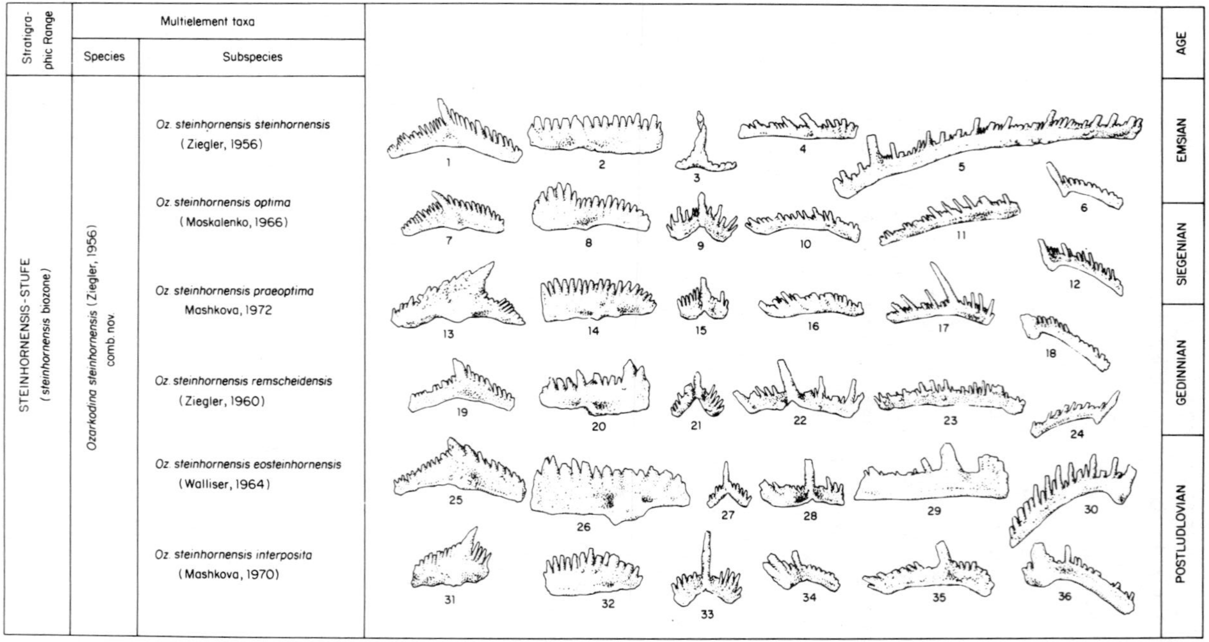

Stratigraphic Range	Multielement taxa: Species	Multielement taxa: Subspecies	Figures	AGE
STEINHORNENSIS-STUFE (*steinhornensis* biozone)	*Ozarkodina steinhornensis* (Ziegler, 1956) comb. nov.	*Oz. steinhornensis steinhornensis* (Ziegler, 1956)	1 2 3 4 5 6	EMSIAN
		Oz. steinhornensis optima (Moskalenko, 1966)	7 8 9 10 11 12	SIEGENIAN
		Oz. steinhornensis praeoptima Mashkova, 1972	13 14 15 16 17 18	GEDINNIAN
		Oz. steinhornensis remscheidensis (Ziegler, 1960)	19 20 21 22 23 24	
		Oz. steinhornensis eosteinhornensis (Walliser, 1964)	25 26 27 28 29 30	POSTLUDLOVIAN
		Oz. steinhornensis interposita (Mashkova, 1970)	31 32 33 34 35 36	

Text-Figure 3
Evolution of subspecies of the multielement conodont assemblage *Ozarkodina steinhornensis* (Ziegler) in Late Silurian-Early Devonian. (After Mashkova.)

of conodonts in biostratigraphic correlation. Thus the Ordovician chirognathids (p. 374), for example, may have been extreme shallow water forms. These faunal differences and restrictions, once they are adequately understood, may ultimately provide an even more refined basis for conodont biostratigraphy, because they will enable what are often now regarded as local "problems" or "anomalies" in zonation to be understood in the light of the original distribution and abundance of the conodont animals.

The following case histories, typical of many others, will illustrate some of the more general characteristics of conodont provincialism.

Ordovician provincialism and ecologic partioning. Since biostratigraphic correlation depends on comparisons of separated fossil sequences, an understanding of the original distribution of the organisms is vital to our conclusions. Cambrian conodont faunas seem to have been cosmopolitan during the whole Period and into the succeeding Tremadocian interval of the Ordovician, although Bergström (1973, p. 49) has suggested that Upper Tremadocian faunas displayed provincialism. Throughout the rest of the Ordovician and into Middle Silurian (Early Wenlockian) times, however, conodonts were characterized by provincial diversification. Later Silurian faunas are increasingly cosmopolitan. Although our knowledge of South American, Asian, Tethyan, and African faunas is inadequate, two major realms ("provinces" of many authors), the European and the North American Midcontinent, first identified by Sweet et al. (1959), are now generally recognized, the latter having two or three "provinces" that may be distinguished at different times. It may also be that Australian faunas represent a third distinctive realm (Bergström, 1971). The presence of provinces is still imperfectly understood, although they were clearly present, and equally clearly fluctuated in time and extent. Thus, in the late Middle Ordovician and Late Ordovician faunas of the eastern Midcontinent, two provinces, the northern and the southern, are recognized; their faunas were intermixed at times with migrant European forms, and differed in some respects from the faunas of a third Western Interior Province, centered around the Upper Mississippi Valley (Bergström, 1973), which extended into the Canadian Arctic and the Western United States. Similarly, a Welsh-Irish Province can be recognized within the European Realm in Middle Ordovician times.

Bergström has provided a provocative map (Text-fig. 4) showing the relation of Middle Ordovician conodont faunal realms to reconstructed continental and paleoequatorial positions in the Northern Hemisphere. On the reconstruction, the Midcontinent Realm appears as an equatorial, presumably warm water type, agreeing well in this respect with the megafaunas of the same age and localities; these are interpreted by Spjeldnaes (1961) as tropical or subtropical. The European Faunal Realm would then presumably be temperate in character. It is noteworthy that such a distribution, although it accords with that of some other invertebrate faunas (Jaanusson, 1972), does not accord with the tentative pattern of oceanic circulation proposed by Williams (1969). Other problems also exist, and it has been recently suggested that some Ordovician conodonts were benthic, rather than pelagic. One curious, and as yet unexplained, phenomenon is the very sharp boundary between the two provinces in the Appalachians (Bergström, 1971, 1973).

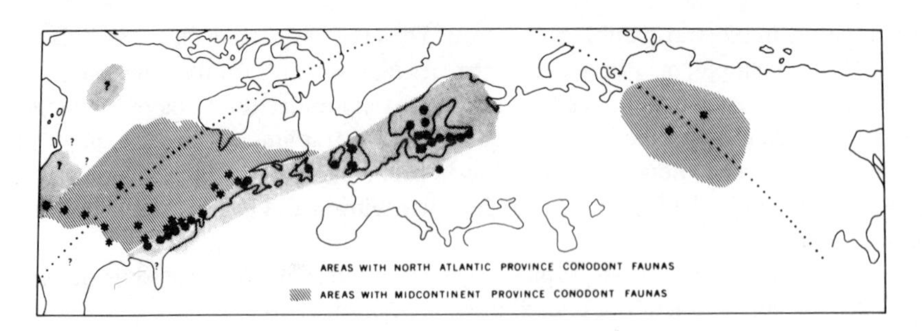

Text-Figure 4
Sketch map showing the distribution of conodont faunal provinces in the Northern Hemisphere during mid-Middle Ordovician time. European continent has been moved to a position relative to the North American continent that is similar to that proposed in several recent reconstructions of the Appalachian–Caledonian geosynclinal area in "pre-drift" time. Location of the equator (dotted line) in North America and Siberia is slightly modified from that in Williams (1969) and Spjeldnaes (1961), respectively. Note the relation between the distribution of the conodont faunal provinces and the location of the equator in the North American and Euro-Asiatic continents. (After Bergström.)

Bergström (1973, p. 56) concludes that "except in possibly a few cases, it is difficult to discern any relation whatsoever between type of Ordovician conodont faunal assemblage, and type of sediment; indeed, very few, if any, animal groups are less facies-dependent that conodonts." However, studies by Barnes et al. (1973c) have shown that a distinctive pattern of ecosystem partioning may be identified in conodonts. They have shown that Ordovician communities are differentiated in the littoral and nearshore sublittoral shelf, shallow offshore shelf, deep shelf, and miogeosyncline, continental margin, and eugeosyncline. These authors suggest that temperature and oceanic circulation were the chief controlling factors in conodont distribution, with depth, salinity, lithofacies, and physical barriers being factors of secondary importance.

The importance of the study by Barnes et al. (1973c) is that by careful identification of conodonts with respect to paleotectonic setting (Text-fig. 5), as well as stratigraphic succession (Text-fig. 6), it provides the first comprehensive model of the evolution of Midcontinent conodont communities during the Ordovician. The potential stratigraphic usefulness of the model lies in clarifying existing biostratigraphic anomalies of distribution. It also reveals striking evolutionary diversification, involving the successive partioning of ecosystems into three broad divisions, the inshore shallow shelf, the offshore shallow shelf, and the deep shelf and miogeosynclinal areas. The first of these, dominated by chirognathids, and characteristically low in species diversity, grades laterally into the higher-diversity shallow-shelf fauna of *Plectodina, Bryantodina* and *Ozarkodina,* which grades seaward into the deep shelf-miogeosynclinal, lower-diversity, *Phragmodus*-dominated community.

ENVIRONMENT / SERIES	INCREASING DEPTH →				
	LITTORAL				
ASHGILLIAN	?	*Amorphognathus* *Baltoniodus* *Belodina* *Plectodina*	*Belodella*	?	?
CARADOCIAN	NEURODONTS	*Amorphognathus* *Baltoniodus* *Eoplacognathus* *Icriodella* *Rhodesognathus*	*Belodella* *Cordylodus*	*Periodon*	?
LLANDEILIAN	NEURODONTS	*Amorphognathus* *Baltoniodus* *Eoplacognathus* *Icriodella*	*Belodella* *Cordylodus*	*Periodon*	?
LLANVIRNIAN	NEURODONTS HYALINE PRIONIODIDS	*Baltoniodus* *Eoplacognathus*	*Belodella* *Cordylodus*	*Amorphognathus* *Oistodus* *Periodon*	?
ARENIGIAN	NEURODONTS HYALINE PRIONIODIDS	?	*Belodella* *Cordylodus*	*Amorphognathus* *Baltoniodus* *Microzarkodina* *Oistodus* *Periodon*	*Prioniodus*
TREMADOCIAN	?	?	*Cordylodus*	?	?

Text-Figure 5
Lateral segregation of the main conodontophorid communities in the Ordovician North Atlantic Province. (After Barnes and Fahraeus.)

Bergström's zoogeographic conclusions are not exactly coincident with those of Barnes et al. (1973c). They argue that the general evidence derived from paleomagnetic data, and the distribution of such distinctive rock types as evaporites, carbonates, tillites, and thick-shelled invertebrate faunas, indicate a pole lying in or near West Africa during Ordovician times and a generally north-south equator with respect to North America; related evidence suggests a very narrow Atlantic Ocean.

Against this paleogeographic background, the east-west or northwest-southeast migrations of Ordovician communities (Text-fig. 6) are regarded by them as chiefly climatic in origin, corresponding to those displayed by other contemporary marine invertebrates (Sweet et al., 1959), although the European fauna, which circumscribes the North American craton, would then cut across these paleolatitudinal belts. Barnes et al. (1973c, p. 178) suggest this may reflect the preference of the latter faunas for deeper and therefore cooler waters of the Tropical zones or the influence of oceanic currents on distribution (Williams, 1969; Lindström, 1970).

Upper Paleozoic provincialism. A similar pattern of faunal restriction now begins to emerge from recent studies of Upper Paleozoic conodont faunas. At the generic level, Upper Paleozoic faunas show virtually no provincialism, but they do show significant ecological differences. In the Late Devonian, for example, conodont faunas consisting of *Icriodus, Polygnathus,* and simple cones generally represent relatively shallow paleoenvironments; the more intensively studied *Palmatolepis-*

AREA	SER.	STAGES	DIRECTION OF IMMIGRATION ROUTES: MAJOR	MINOR	MAJOR IMMIGRATING ELEMENTS	MAJOR EMIGRATING ELEMENTS
CINCINNATI REGION	CINCINNATIAN	Richmondian	•		*Panderodus, Belodina, Plegagnathus, Oulodus*	*Phragmodus*
				•	*Rhipidognathus*	European elements
				•	*Rhipidognathus*	
		Maysvillian		•		
				•	*Rhipidognathus*	
		Edenian		•		
NEW YORK STATE	CHAMPLAINIAN	Trentonian		•	*Panderodus, Belodina*	
				•	European elements	*Polyplacognathus, Bryantodina*
				•	*Polyplacognathus, Bryantodina*	
			•		*Phragmodus*, European elements	*Erismodus*, etc.
		Blackriveran		•	*Panderodus, Belodina*	
			•		*Erismodus, Polycaulodus, Cardiodella*, etc.	

Text-Figure 6
Pattern of conodont migrations in the eastern Midcontinent Province of North America during the late Middle and Upper Ordovician. Arrows mark approximate direction of immigration routes (top of diagram as representing north). (After Barnes, Sass, and Monroe.)

and *Ancyrodella*-dominated faunas apparently represent deeper-water deposits. These grade laterally into still-deeper-water deposits, characterized by faunas containing *Ancyrognathus, Playfordia,* and *Polylophodonta.* Such ecological assemblages are not entirely mutually exclusive, however, since small numbers of characteristically shallow water elements are found in deep-water faunas and, occasionally, deep-water elements in shallow-water faunas (Druce, 1973).

The same situation exists in the Early Carboniferous, where the shallow-water assemblages are represented by *Spathognathodus, Polygnathus,* and *Clydagnathus,* and the deeper-water faunas by *Siphonodella* and *Pseudopolygnathus.* The "exotic" genera, such as *Dinodus, Dollymae, Scaliognathus,* and *Staurognathus,* appear to be confined to deep-water faunas. The problems that arise in these restrictions are of some significance, because many zonal schemes are based on deeper-water faunas and therefore require supplementation for application to shallow-water deposits. It therefore becomes all the more important, as Druce has pointed out, that the typical icriodid, polygnathid, and spathognathid elements which are present in deeper-water faunas should be accurately reported and described.

Within a given province there are also significant ecologic differences in Upper Paleozoic faunas. One of the best documented examples of this is provided by Druce (1969, 1973, and references therein) in his descriptions of Upper Devonian and Lower Carboniferous faunas from Australia. Druce's studies confirm the independent recognition by Seddon (1970) of conspicuous ecologic partitioning in various shelf carbonate environments. The reef complex of the Upper Devonian Ningbing Limestone of the Bonaparte Gulf Basin provides a typical example.

Druce identifies four distinct facies, one of which, the reef itself, is virtually barren of conodonts. Biofacies I, representing the shallow-water zone around the margin of the reef, is characterized by single cone conodonts, especially *Belodella,* and the less common *Acodina.* Biofacies II, the backreef shallow-water (<50 m) zone, is characterized by the predominance of *Icriodus* and *Pelekysgnathus,* and the presence of *Polygnathus* and *Spathognathodus.* The deeper-water (>50 m) zone of the forereef and interreef areas is characterized by *Palmatolepis,* accompanied by *Ancyrodella* and *Ancyrognathus* in the Frasnian, and *Polygnathus* and *Polylophodonta* in the Famennian. Undenticulate spathognathodids and *Spathognathodus ziegleri* are also present.

Two original alternative explanations (Text-fig. 7) were offered to account for this faunal differentiation. Seddon (1970) and Seddon and Sweet (1971) suggested bathymetric restriction as the underlying cause. But lateral and vertical biofacies changes, occurring in short distances and within single outcrops, and the complex patterns of mutual biofacies associations do not suggest that simple stratified bathymetric restriction is a likely general explanation. Druce (1970) offered an alternative explanation which implied that distance from shore may be the controlling factor in producing "deep and shallow" water faunas. Such a model, however, does not satisfactorily explain the relative intermixing found between the several biofacies, which seems to imply some selective filter connections between them; e.g., some elements of the *Icriodus* biofacies are found to be associated with the typical *Palmatolepis* biofacies, but not vice versa. A third, undeveloped alternative is that they were not pelagic but benthic saltators, although the weight of other evidence favors a pelagic life for most conodonts.

Because all biofacies have a worldwide distribution, some form of depth stratification appears a more likely explanation for their differences than does total water depth, but the relative scarcity of the shallow-water elements in deeper-water faunas suggests that other factors, perhaps biological in character, are also important. Thus Druce (1973) (Text-fig. 7) has suggested a higher abundance of conodont organisms in the nearshore areas of each depth zone.

Druce (Text-fig. 8) has extended this distributional model to the remainder of the Upper Paleozoic, basing his extrapolations on published descriptions, analagous morphology of particular elements, and "intuition" (1973, p. 211). The model is important as a cautionary guide in biostratigraphic correlation.

Merrill (1973) has shown in Pennsylvanian faunas that nearshore shale and siltstone biofacies, characterized by *Cavusgnathus,* can be distinguished from the more widespread ("ubiquitous") offshore carbonate biofacies, dominated either by *Idiognathodus* or *Streptognathodus,* depending on the age of the cycles, with only rare cavusgnathids (Text-fig. 9). Of these two faunas, the *Idiognathodus-Streptognathodus* is much the more common, the *Cavusgnathus* fauna being localized and rare, both stratigraphically and geographically.

In addition, two other distinctive Pennsylvanian faunas may be recognized, a Midcontinent and an Appalachian fauna. In spite of the fact that the terms "Appalachian" and "Midcontinent" are used to describe these two faunas, it is

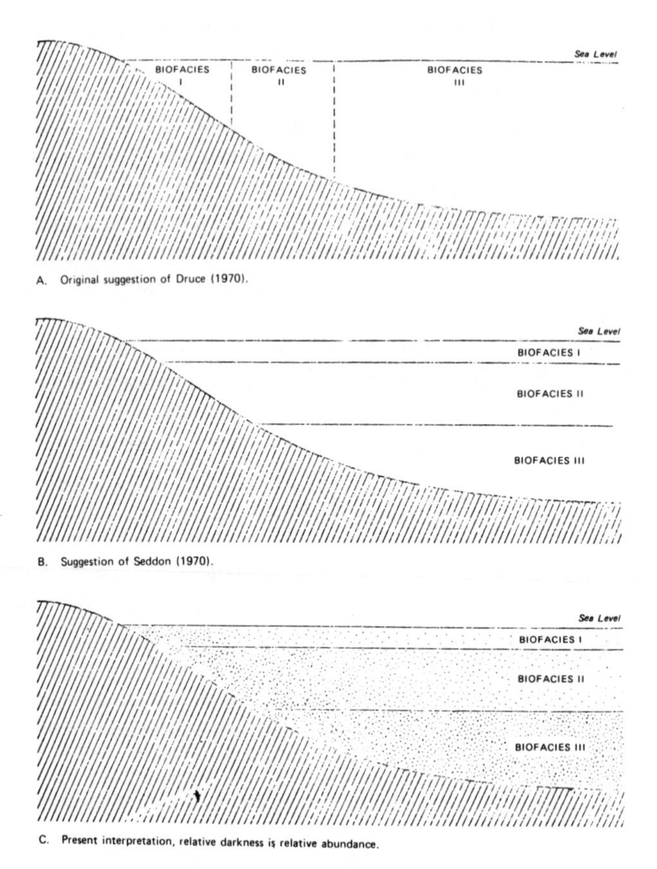

Text-Figure 7
Upper Paleozoic conodont biofacies and distribution. For explanation see text. (After Druce.)

now clear that they reflect environmental, rather than provincial, differences (see below). These faunas are not wholly mutually exclusive, and a fifth fauna of *Gondolella* elements is also known; it is of very limited distribution, but is usually associated with the Midcontinent fauna. The weight of evidence tends to suggest that the environmental factors responsible for the differences between the "Appalachian" and "Midcontinent" faunas differ from those which were responsible for *Cavusgnathus* and *Gondolella* faunas. None of the more conspicuous factors, such as depth, salinity, energy, substrate, or faunal association seem sufficient alone to explain these differences. Text-figure 10 represents a model in which rock type is related to conodont biofacies. The conodont groups also show distinctive faunal associations with elements of other phyla.

In spite of these variations, it is interesting that Merrill, in establishing a biostratigraphic zonation for the Central Appalachians, writes (1971, p. 410), "The

Stratigraphic Divisions	Biofacies I	Biofacies II	Biofacies III	Biofacies IV
Triassic	Bar and blades? Celsigondolella I or II	Celsigondolella? I or II Tardogondolella II	Neogondolella II	Gladigondolella II Paragondolella II Furnishius? II or IIIb Platyvillosus I or IIIb
Permian	Spathognathodus? I or II Celsigondolella I or II	Streptognathodus II	Gondolella II	
Late upper Carboniferous		Streptognathodus II Idiognathoides II or IIIB Spathognathodus I or II	Gondolella II Idiognathodus II	
Early upper Carboniferous	Cavusgnathus? IV	Ganthodus (simple) II Spathognathodus I or II Cavusgnathus? IV	Gnathodus (complex) II	
Viséan	Cavusgnathus? IV	Cavusgnathus Gnathodus (simple) II Mestognathus IV Spathognathodus I or II Taphrognathus? I	Gnathodus (complex) II	
Tournaisian	Clydagnathus? IV	Clydagnathus IV Polygnathus (inornatus grp.) Pseudopolygnathus (ex. tr. grp.) IIIa Spathognathodus (nodose) IIIa Patrognathus II Gnathodus (simple) II	Dinodus II Doliognathus? II Dollymae? I --- Staurognathus? IV? Siphonodella II Bactrognathus? II Gnathodus (complex) II Pseudopolygnathus triangulus grp. IIIb	Doliognathus? Dollymae? --- Staurognathus?
Famennian	Simple cones (Acodina)	Icriodus II Pelekysgnathus I or II Spathognathodus (nodose) IIIa Scaphignathus? IV	Palmatolepis II Polygnathus (germanus grp.) II Polylophodonta II	
Frasnian	Simple cones (Belodella)	Icriodus II Pelekysgnathus I or II Polygnathus (simple) II	Ancyrodella II Ancyrognathus II Palmatolepis II	
Middle Devonian	Simple cones (Belodella)	Icriodus II Spathognathodus I or II Polygnathus (simple) II	Polygnathus (complex) II	
Lower Devonian	Simple cones (Panderodus)	Icriodus (simple) II Pelekysgnathus I or II Spathognathodus (simple) I or II	Icriodus (complex) II Ancyrodelloides? II Spathognathodus (complex)	Ancyrodelloides?

Text-Figure 8
Distribution of Upper Paleozoic conodont genera with respect to biofacies. (After Druce.)

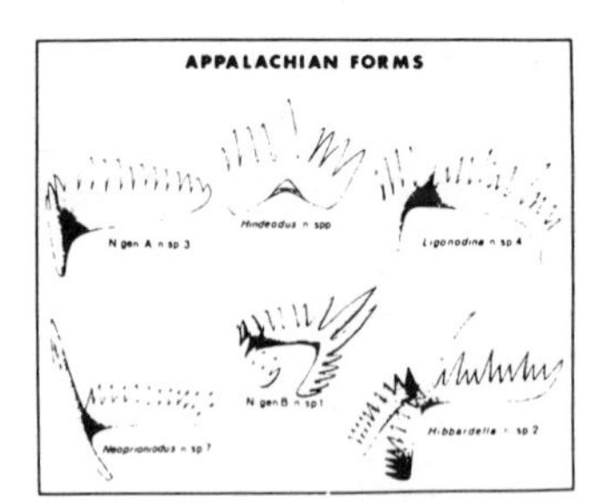

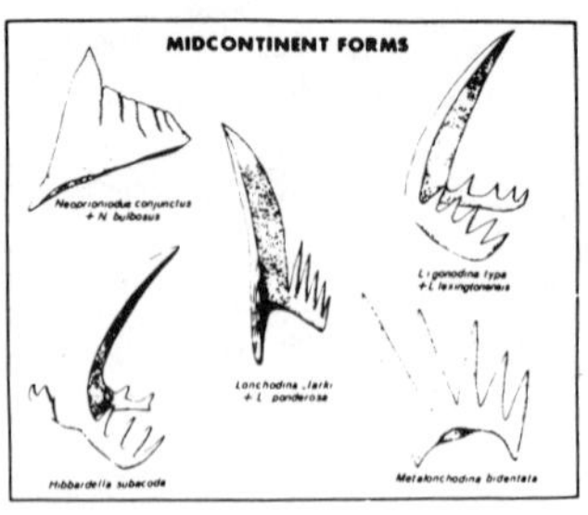

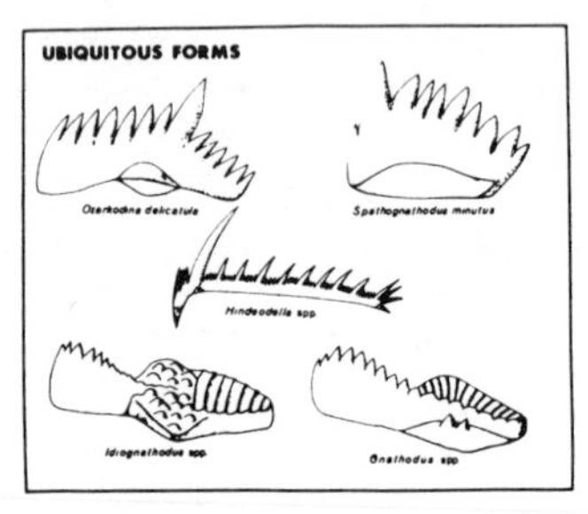

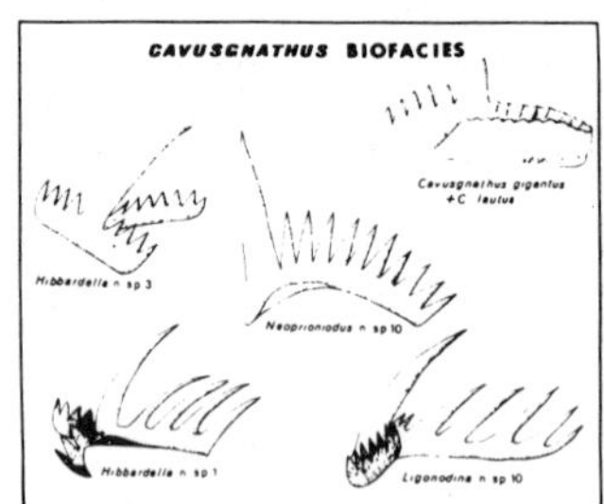

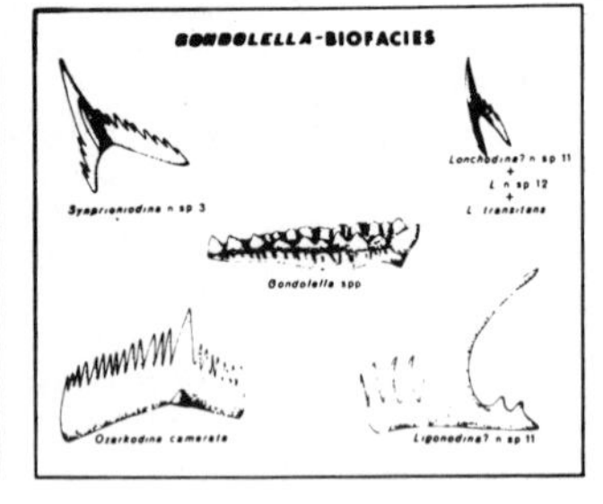

Text-Figure 9

Major environmental groupings among Pennsylvanian conodonts. Four distinct, but less than totally independent, groups can be distinguished as well as a central or "normal" one. The normal or "ubiquitous" grouping is represented in nearly every Pennsylvanian conodont sample and usually waxes or wanes with fluctuations in the *Cavusgnathus*-biofaces (nearshore) elements. Most of the ubiquitous grouping can be thought of as an age-dependent *Streptognathodus*- or *Idiognathodus*-dominated offshore biofacies plus some elements *(Ozarkodina, Hindeodella)* that probably were anatomical associates of distinctive platform taxa, plus a few others truly ubiquitous within their ranges *(Spathognathodus, Neognathodus)*. Confirmation and minor amendment of these groups have been provided by von Bitter (1972) (see this article, p. 382). (After Merrill.)

biostratigraphically important genera cross biofacies and provincial boundaries without specific change, permitting recognition and identification of zones wherever

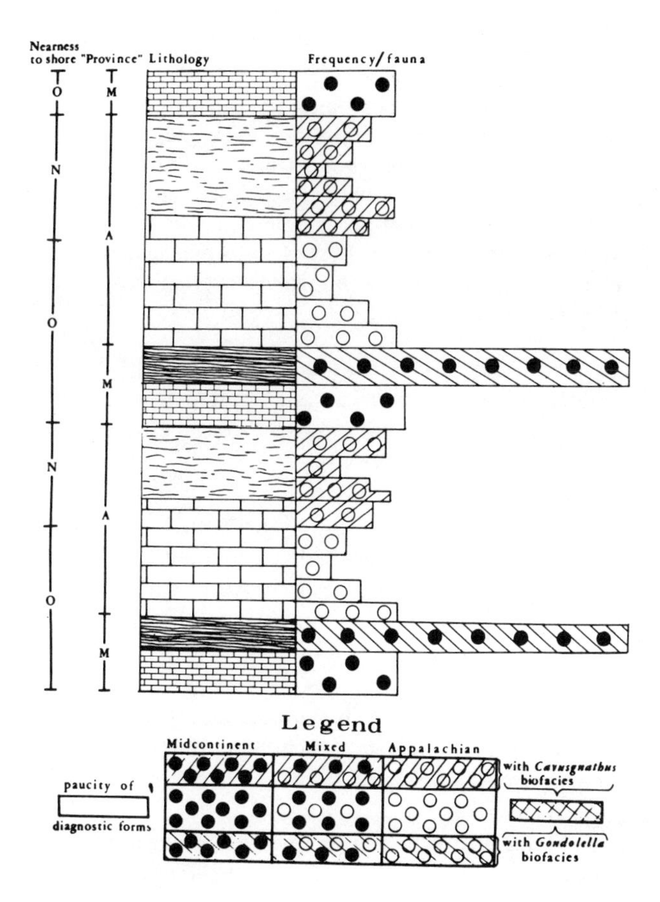

Text-Figure 10
Model synthesized from Midcontinent conodont occurrences where the rock succession ideally consists of, in ascending order, thin limestone, thin shale, thick limestone, thick shale. This regularity in lithic repetition is uncommon, and the overall pattern may actually be obscured. The model is constructed to reflect these ideal conditions, and departures from them are not considered; therefore, the model is overly simple. Little of the section is dominated by the Midcontinent Fauna, despite its location within that region. No "mixed" faunas are shown, although the locations where they are most likely to occur, the meeting zones between Appalachian and Midcontinent faunas, are indicated. In practice, the transition between these faunas at the lithic boundary between thin black shale and thick limestone is as sharp as the lithic changes, and the gradual transition is more likely to be found at the other lithic change (thick shale to thin limestone). Normally *Gondolella* only occurs in the thin, often black,

Midcontinent Fauna-dominated shale, and not in all of them. *Cavusgnathus* may occur through a fairly thick interval, upper part of thick limestone through the thick shale, and even into the thin limestone, but most commonly is not present throughout. Unlike the Appalachian and Midcontinent faunas, the offshore and nearshore faunal limits do not necessarily coincide with lithic boundaries, nor with those of the Appalachian and Midcontinent faunas, although the overlap may not be large. (After Merrill.)

adequate material is available." The zonation is based on only three genera, *Gnathodus, Idiognathodus,* and *Streptognathodus,* and Merrill has not yet provided the supporting evidence for it, or related it to the ecologic differences he has described.

Merrill's general conclusions have been confirmed by von Bitter (1972) in a careful study of conodont distribution in the Upper Pennsylvanian of Eastern Kansas. *R*-mode cluster analysis confirms the five conodont biofacies recognized by Merrill, and identifies a sixth. Von Bitter has provided a new taxonomic identification of the conodont elements, based on multielement taxonomy, and has also provided initial paleoecological descriptions of the five major biotopes. These may be summarized as follows:

1. *Streptognathodus* biofacies, which is typically an offshore carbonate facies.
2. *Streptognathodus gracilis* biofacies, probably representing a transitional deepening environment from reducing to "normal" marine, presumably of nearshore or even lagoonal character.
3. *Cavusgnathus* biofacies, apparently representing the nearshore shale biotope.
4. *Lonchodina* biofacies, characteristic of clear, shallow-water (<20 m) carbonate deposits.
5. *Neoprioniodus conjunctus* biofacies.
6. *Gonolella* biofacies, which is rare and restricted, probably representing shallow nearshore, and possibly lagoonal, deposits.

The foregoing discussions of environmental control are of great importance in the growing utilization and refinement of conodont biostratigraphy. The correlation of the Lower Carboniferous of Europe and North America provides a typical example (see Rhodes et al., 1969; Rhodes and Austin, 1971; Collinson et al., 1971; Austin and Barnes, in press). The recent zonal scheme for the British Lower Carboniferous developed by Rhodes, Austin, and Druce was characterized by an oldest zone containing *Patrognathodus variabilis, Spathognathodus plumulus, Clydagnathus gilwernensis,* and *Pseudopolygnathus vogesi.* The first three of these species are not recorded in the German succession (Ziegler, 1971), which consists of bathyal deposits dominated by *Protognathodus,* and *Siphonodella.* In Britain this oldest Carboniferous zone is replaced in vertical succession by oolitic limestones, yielding *Clydagnathus, Spathognathodus,* and *Pseudopolygnathus,* and bioclastic limestones, which contain the two latter genera and rare polygnathids, and, higher in the succession, abundant polygnathids and early *Siphonodella*-like forms. This difference is regarded as both chronological and environmental in origin, the standard British zonal sequence of Rhodes et al. (1969) representing a shallower-water shelf regime than the German. It is noteworthy that the intervening Belgian succession at Huy (Austin et al., 1970), presumably representing a deeper shelf environment, contains elements of both faunas and also yields its oldest siphonodellids at a point of distinctive facies change.

Austin and Barnes (in press) have suggested a tentative model for the transgressive sequence that reflects the present environmental interpretation of these faunas (Text-fig. 11). The biostratigraphic significance of this is that although the base of the Carboniferous can be adequately recognized in many areas by the first appearance of *Siphonodella sulcata,* an additional complementary scheme is required for carbonate shelf areas, such as those of England and Wales. It is noteworthy that the revised North American Mississippian scheme of Collinson et al. (1971) contains elements suggestive of both shelf and bathyal facies.

The implication of these studies is of major importance in biostratigraphy. The increasing detail in conodont biostratigraphy demands not only recognition of regional zones based on non-facies-controlled forms, but recognition of the degree of facies control on other forms, development of zonations within each facies based on typical facies species, and thus the ability to interpret biostratigraphic succession in mixed facies situations. Only thus may we hope to separate what is essentially time equivalent, but facies restricted, from that which is really time successive.

Spathognathodus remscheidensis: the significance of variation. Although environmental and geographic variation may appear at first as handicaps in the biostratigraphic utilization of conodonts, they may, when fully evaluated, provide the basis for more refined stratigraphic and environmental delineation. What is not yet available for most conodont faunas is an evaluation of the differing significance of geographic, ecologic, and evolutionary changes. Thus chronologic succession and

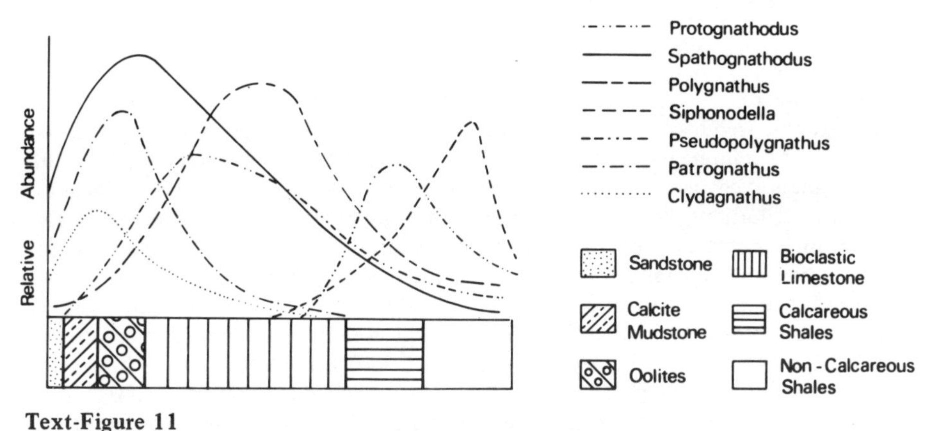

Text-Figure 11
Relative abundance of seven platform genera in the Lower Carboniferous K zone of the British Southwestern province. (After Austin and Barnes.)

morphological change, for example, may reflect the results of any one or all of these underlying cases. The most satisfactory zonal scheme will presumably be that which recognizes the significance of each.

A recent attempt to evaluate these various influences is that of Barnett (1971), who made a study of the environmental distribution of the cosmopolitan conodont species *Spathognathodus remscheidensis* in a composite reference section established in northern New Jersey and southeastern New York. Barnett based his study on 1,000 individuals of the species from 11 successive Upper Silurian, Upper Cayugan, and Helderbergian horizons, representing 200 ft of strata; he used simple statistical studies to show the presence of distinct changes in the eight characters that he analyzed, in spite of the fact that these changes were not observable by visual inspection.

He selected various univariate characters (see Text-fig. 12) that appeared essentially constant throughout the size range of each sample. Bivariate characters studied included the dimensions of the margins of the basal cavity and its position. Barnett used the overall lithologic faunal characteristics of the succession to establish an environmental profile, demonstrating transgression, regression, and inferred salinity variations, against which (Text-fig. 12) he compared the morphologic trends he had determined.

Barnett's comparison of character trends against inferred environmental changes in the reference section was supplemented by a study of geographic variation along three independently determined time horizons (Text-fig. 13). He was thus able to show that certain characters of conodont elements were markedly affected by environmental conditions, while other characters were interpreted as ecophenotypic only under particular environmental conditions. The remaining characters (basal flare shape, basal flare length-width, angle made by conjunction of anterior and posterior basal margins, and angular relationship between two denticles) exhibited no detectable geographic or ecophenotypic tendencies, and were therefore regarded as evolutionary in character, although each was stable for considerable periods

---

Text-Figure 12

Stratigraphic trends of the eight measured characters of *Spathognathodus remscheidensis* at the reference section (composite of sections 14 and 16) in comparison with environmental change. For univariate characters (1, 3–6, 8), a line connects means of sample distributions, heavy horizontal lines indicate 95 percent confidence intervals for the mean, and vertical cross bars represent standard deviations. For bivariate characters (2, 7), a line connects ratio equation slopes of each sample population. Heavy horizonal lines indicate the 95 percent confidence interval for each slope value. Sample means, ratio equation slopes for all characters, and environmental interpretation of samples 4-57′(b), 4-115′(d), 10-29′(e), 10-33′(f), and 15-37′(c) are plotted at that horizon to which each of these samples best correlates. The same data for sample 6-270′(a) are plotted at its independently established correlative position in order to demonstrate geographic variation between the reference section and locality 6 (Longwood Lake, N.J.). (After Barnett.)

of the stratigraphic interval studied. When change did occur, it took place at different times and rates for each of the characters studied.

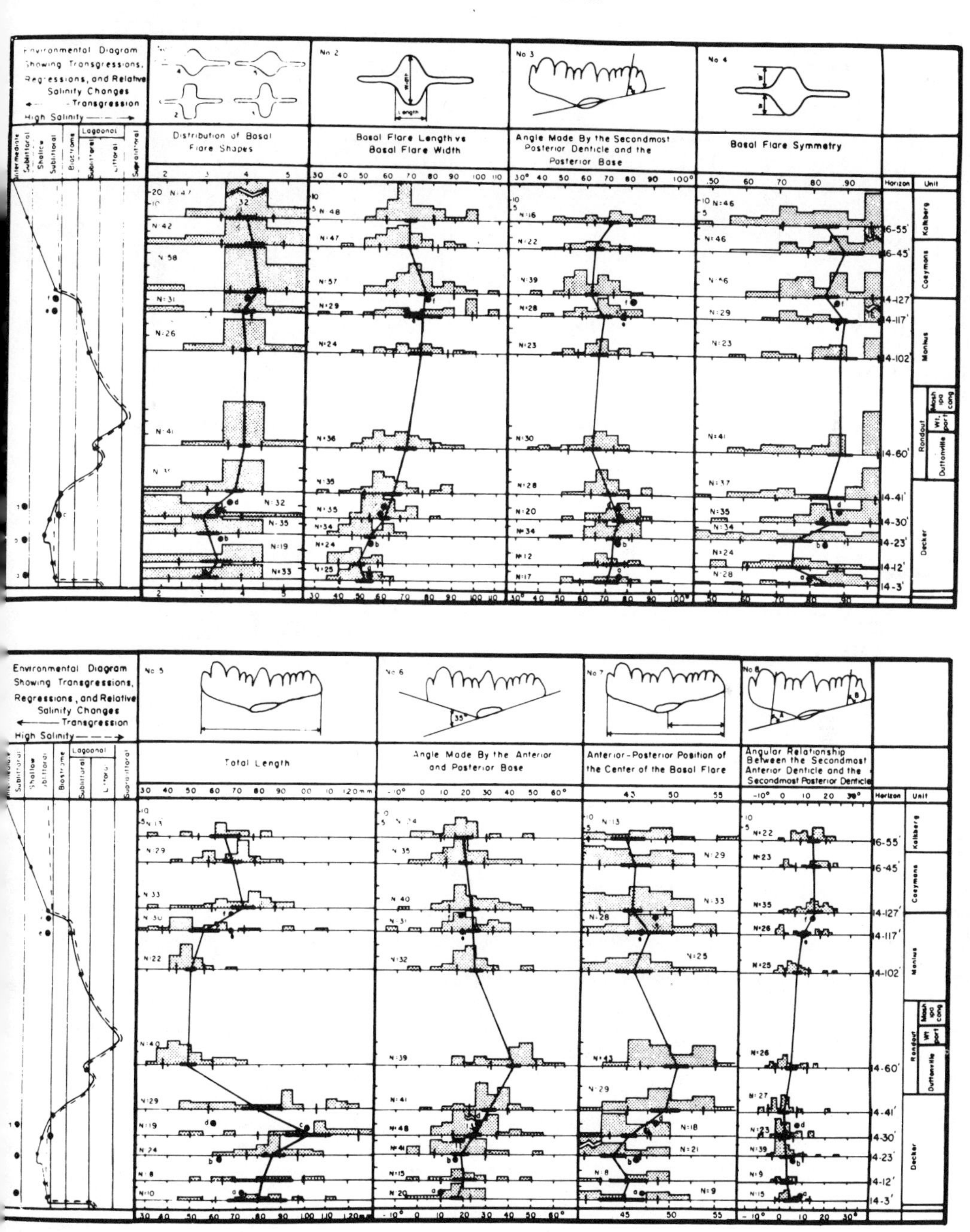

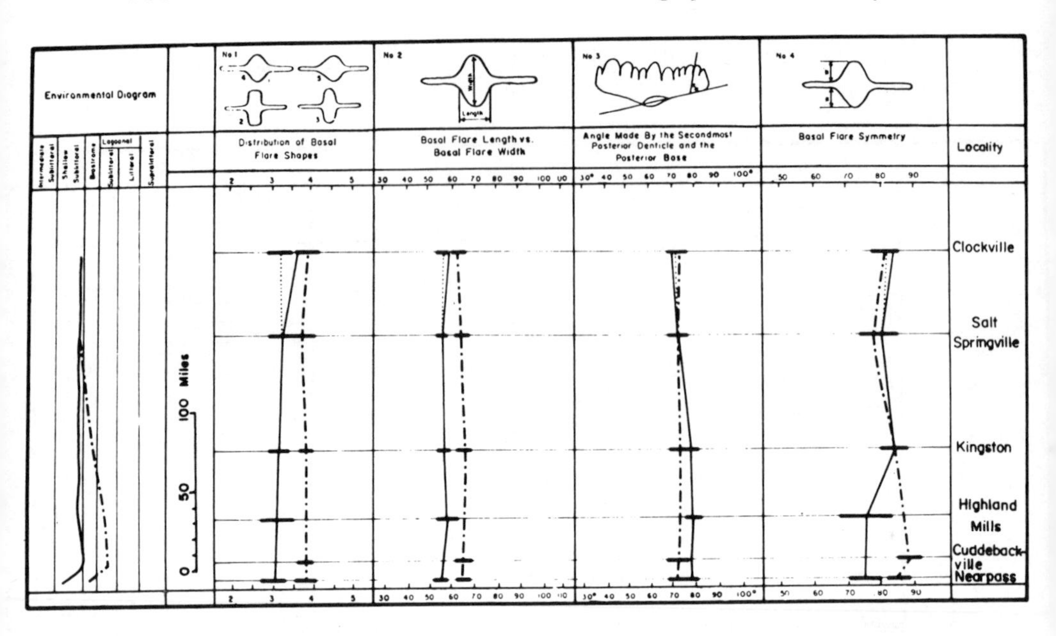

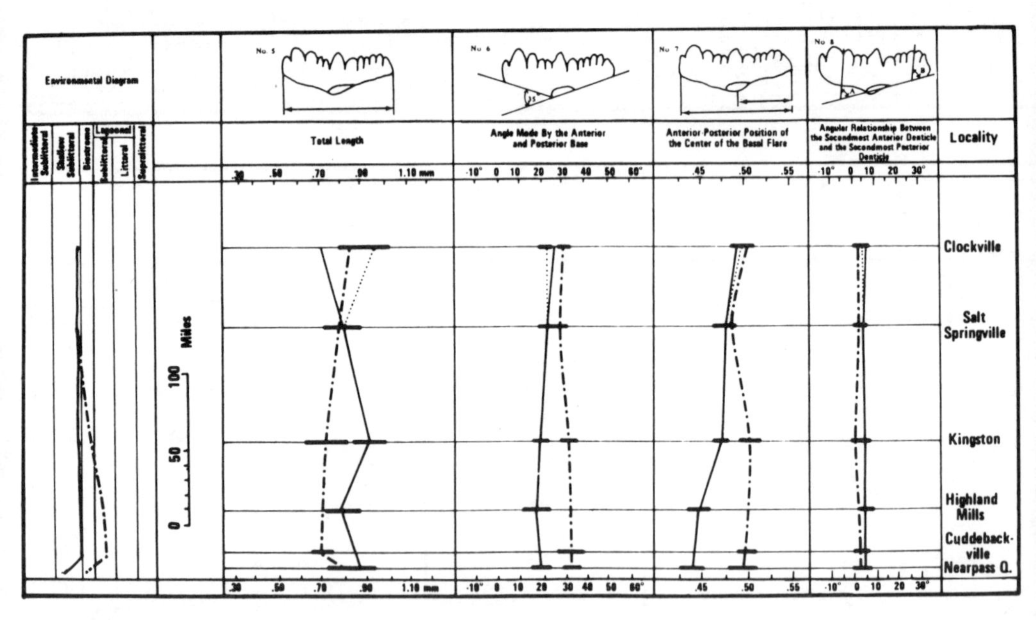

Text-Figure 13

Geographic variation with respect to environment of the eight measured characters of *Spathognathodus remscheidensis* from Nearpass Quarry, N. J. (Loc. 14, bottom of diagram) to Clockvi N.Y. (Loc. 2, top of diagram) along two independently established time horizons correlative with 14-23′ (solid lines) and 14-41′ (dashed lines) at Nearpass Quarry. Lines connect means of univariate characters (1, 3–6, 8) and ratio equation slopes for bivariate characters (2, 7). Heavy horizontal lines indicate 95 percent confidence intervals. Dotted lines connect means and slope values of samples 2-64′ at Clockville with the sample from the lower time horizon (sample 4-1′) at Salt Springville. (After Barnett.)

Barnett also undertook a study of the overall distribution of the species in various rock types. His studies indicate that the *S. remscheidensis* animal was abundant in sublittoral lagoons, some biostromal reefs, and crinoidal meadows, but decreased in abundance further seaward, being generally not abundant even in the deeper-water portions of the intermediate sublittoral environment. The abundance of the species was variable in reef (biostromal) samples, being greatest in coral-dominated reefs and least in stromatoporoid-dominated reefs. The significance of this discovery is of some importance, because *Icriodus woschmidti,* which is used to identify the base of the Lower Devonian (Gedinnian), is not present in New York and New Jersey in lagoonal and biostromal deposits, and its appearance may not everywhere be a contemporaneous event, but may rather reflect local environmental conditions.

Barnett (1972), followed his biometric analysis of the species *Sphathognathodus remscheidensis* from Upper Cayugan and Helderbergian strata of New Jersey and New York with similar studies in Nevada and Czechoslovakia. These confirmed his earlier conclusions. The potential value of such detailed studies is great. They promise a high degree of refinement by the use of carefully identified evolutionary characteristics. What is less clear is whether they will ultimately provide more precision than is already available from more traditional biostratigraphic comparisons.

USE OF CONODONTS IN BIOSTRATIGRAPHY

The recent increasingly successful use of conodonts in both local and intercontinental correlation suggests that these fossils now provide the most comprehensive and widely applicable basis for Paleozoic biostratigraphy. The study by Sweet and Bergström (1971a) provides an important introduction to the general topic, but the following examples will illustrate the degree of refinement that is now available in various systems and will illustrate the methodology involved.

Biostratigraphy of the Middle and Upper Ordovician

Bergström (1971) has reviewed the conodont biostratigraphy of the Middle and Upper Ordovician of Europe and eastern North America. He was able to use the relatively condensed calcareous sequence of the Balto-Scandinavian area as a basis for establishing a scheme of conodont zonation, and he has correlated this with the standard graptolite sequence and zonation. To avoid problems that arise when correlations are established on species that may have different local ranges, Bergström has established a zonation based largely on characteristic stages of rapidly evolving conodont lineages, especially of *Amorphognathus* and *Eoplacognathus,* as well as other genera (Text-fig. 14). On the basis of the stratigraphic ranges of these, the

Middle and Upper Ordovician sequence is divided into five zones and ten subzones, which provide a degree of refinement greater than that established on the basis of any other group, including graptolites.

When these Baltic-Scandinavian faunas are compared with those in the Appalachians, many species are common to both areas, in addition to numbers of genera and species that are not. Bergström notes that the conodont faunal sequence "is basically the same on both sides of the Atlantic" (1971, p. 115), and the conodont zonation established on the basis of the Swedish sections can be applied with little difficulty to the much more structurally complex sequence in the Appalachian region. Bergström claims that the degree of precision in this correlation is such that in several cases it is possible to correlate a thin Appalachian unit with one of less than 1.5 m thickness in the condensed Balto-Scandic sequence (Text-fig. 15).

In spite of the similarities between the Balto-Scandic-Scottish faunas with those of the Appalachians, they show very little similarity to faunas of comparable age from the Midcontinent area of North America (see p. 374), where the division into 12 "faunas" is based on overlapping associations and sequential range determinations. The extent of this difference is striking. Bergström observes that, although it is almost impossible at present to correlate the well-known Middle Ordovician sequence of the Tazewell area of Virginia with the presumably equivalent section of the Saltville–Marion area, only 20 km to the south, the latter can be correlated with remarkable precision with the Scandinavian faunas, which are more than 10,000 km away. The boundary separating the eastern and western Appalachian faunas coincides with an extensive fault system running from Alabama into the St. Lawrence Valley. This tectonic boundary also marks a striking lithological boundary and a change in the megafaunas. Such differences presumably reflect factors other than purely ecological controls, probably a radically different paleogeography.

During Upper Middle and Lower Upper Ordovician time, there was an influx of Atlantic forms into the North American Midcontinent region. The 12 faunas that characterize the Middle and Upper Ordovician conodont faunas of the Mid-

---

Text-Figure 14

Conodont zones and subzones and the stratigraphic ranges of some important multielement conodont species in the Middle and Upper Ordovician of Sweden. Illustrated specimens represent the most characteristic elements among those present in the apparatus of each of the following species: (1) *Pygodus serrus* (Hadding); (2) *P. anserinus* (Lamont and Lindstrom); (3) *Amorphognathus tvaerensis* Bergström; (4) *A. superbus* (Rhodes); (5) *Eoplacognathus suecicus* n. sp.; (6) *E. foliaceus* (Fahraeus); (7) *E. reclinatus* (Fahraeus); (8) *E. robustus* n. sp.; (9) *E. lindstroemi* (Hamar); (10) *E. elongatus* (Bergström); (11) *"Bryantodina"* n. sp.; (12) *"Scolopodus" varicostatus* Sweet and Bergström; (13) *"S."* n. sp. cf. *"S." insculptus* (Branson and Mehl); (14) *Polyplacognathus sweeti* Bergström; (15) *"Distomodus" europaeus* Serpagli; (16) *Prioniodus variabilis* Bergström; (17) *P. gerdi* Bergström; (18) *P. alobatus* Bergström; (19) *Amorphognathus complicatus* Rhodes; (20) *Icriodella* sp.; (21) *Amorphognathus superbus* (Rhodes); (22) *A. ordovicicus* Branson and Mehl. (After Bergström)

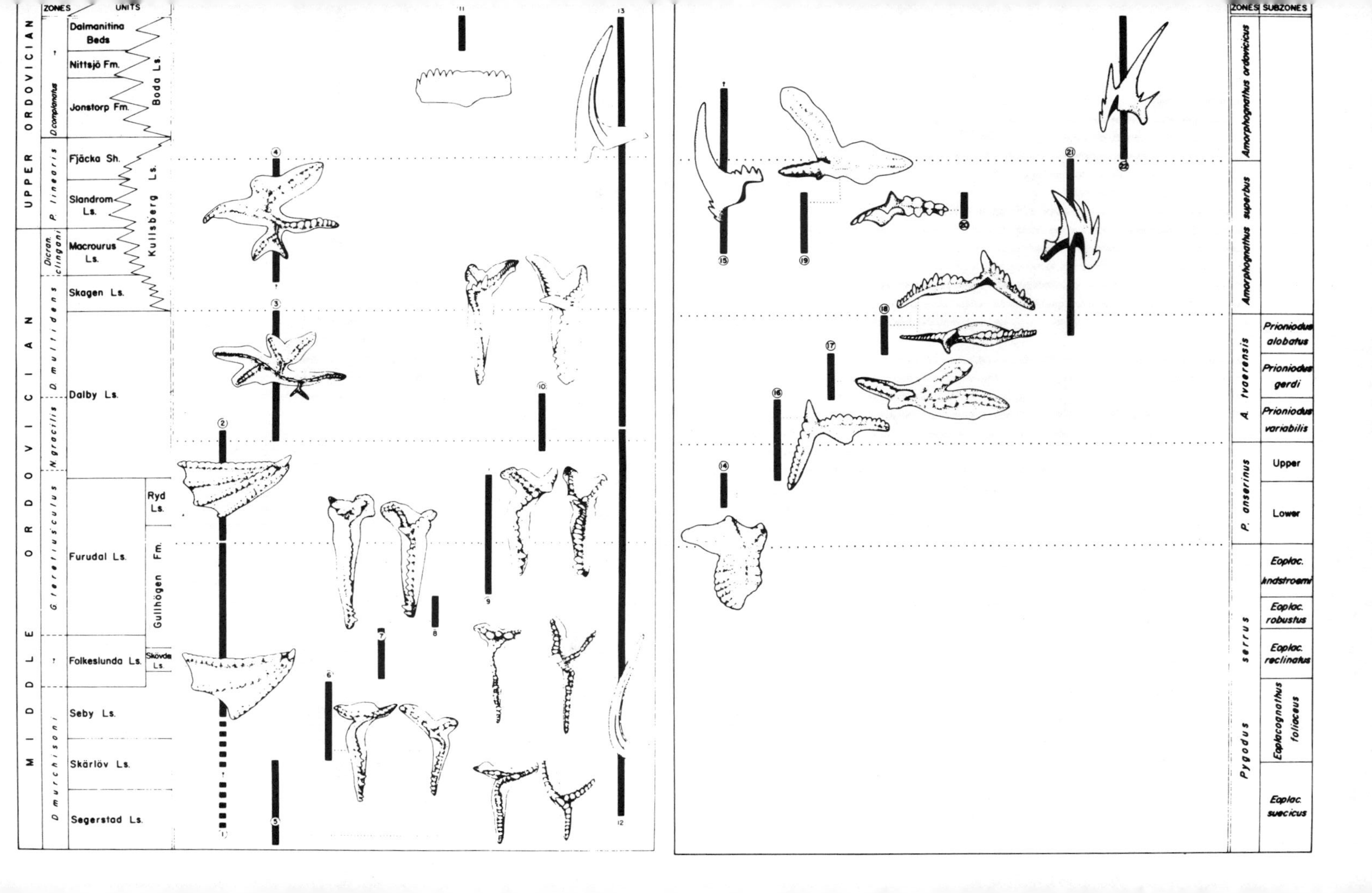
ZONES
UNITS
UPPER ORDOVICIAN
MIDDLE ORDOVICIAN
D. complanatus
P. linearis
Dicran. clingani
D. multidens
N. gracilis
G. teretiusculus
D. murchisoni
Dalmanitina Beds
Nittsjö Fm.
Jonstorp Fm.
Boda Ls.
Fjäcka Sh.
Slandrom Ls.
Kullsberg Ls.
Macrourus Ls.
Skagen Ls.
Dalby Ls.
Ryd Ls.
Furudal Ls.
Gullhögen Fm.
Folkeslunda Ls.
Skövde Ls.
Seby Ls.
Skärlöv Ls.
Segerstad Ls.
ZONES
SUBZONES
Amorphognathus ordovicicus
Amorphognathus superbus
A. tvaerensis
Prioniodus alobatus
Prioniodus gerdi
Prioniodus variabilis
P. anserinus
Upper
Lower
serrus
Eoplac. lindstroemi
Eoplac. robustus
Eoplac. reclinatus
Pygodus
Eoplacognathus foliaceus
Eoplac. suecicus

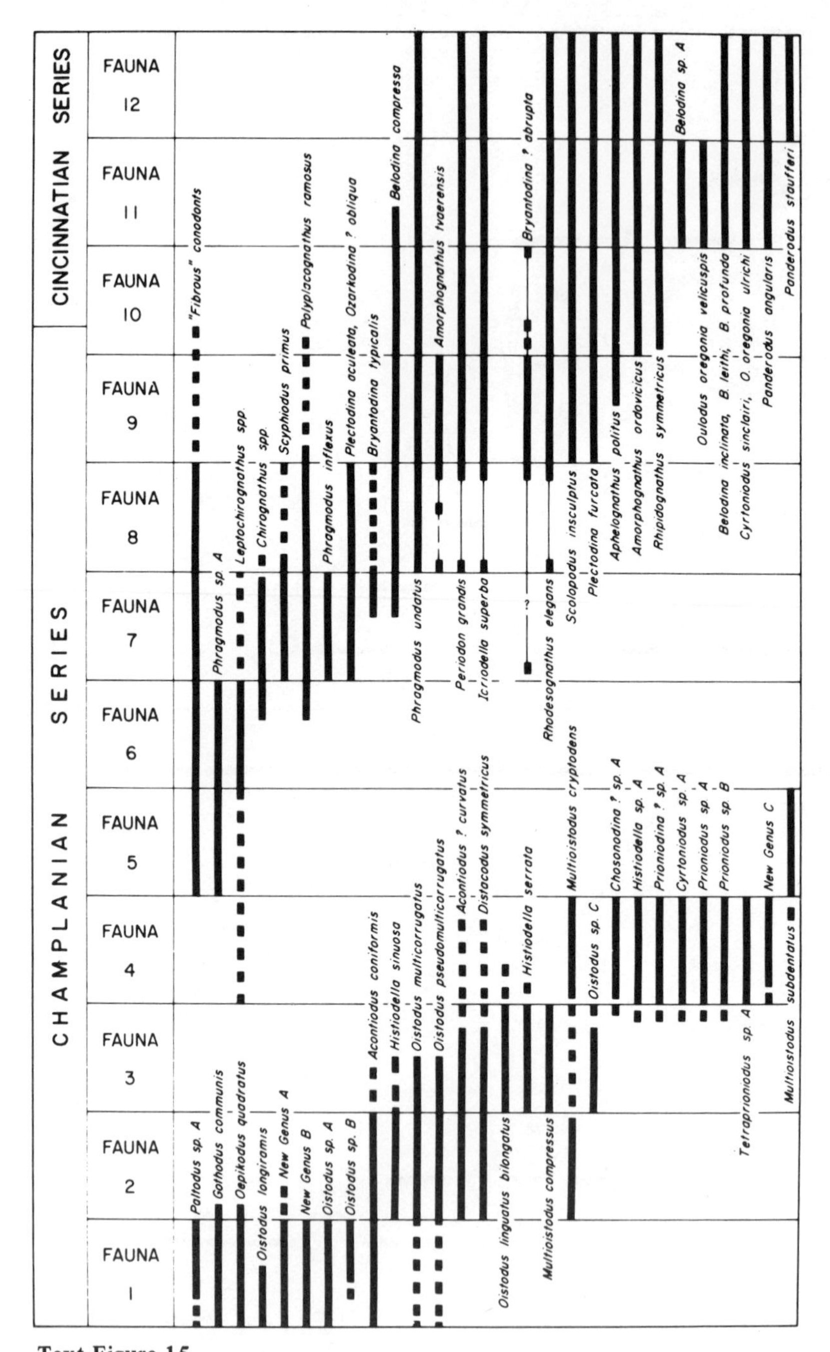

Text-Figure 15

Middle and Upper Ordovician conodont faunas of North America and ranges of stratigraphically important species. (After Sweet, Ethington, and Barnes.)

continent region have been described by Sweet et al. (1971, p. 163) and are summarized in Text-figure 15. Comparison of this figure with Text-figure 14, representing Bergström's zonation of the Scandinavian area, shows that there is virtually no similarity between the two conodont successions.

The methods employed in these Ordovician correlations are based on recorded ranges of associated species (assemblage zones), using a direct comparison of association and sequence to compare zones. One potentially useful alternative method of correlation is the use by Bergström and Sweet (1966) of the relative abundance of the species *Phragmodus undatus* in the correlation of the Middle Ordovician successions of Ohio, Indiana, and Kentucky (Text-fig. 16). These

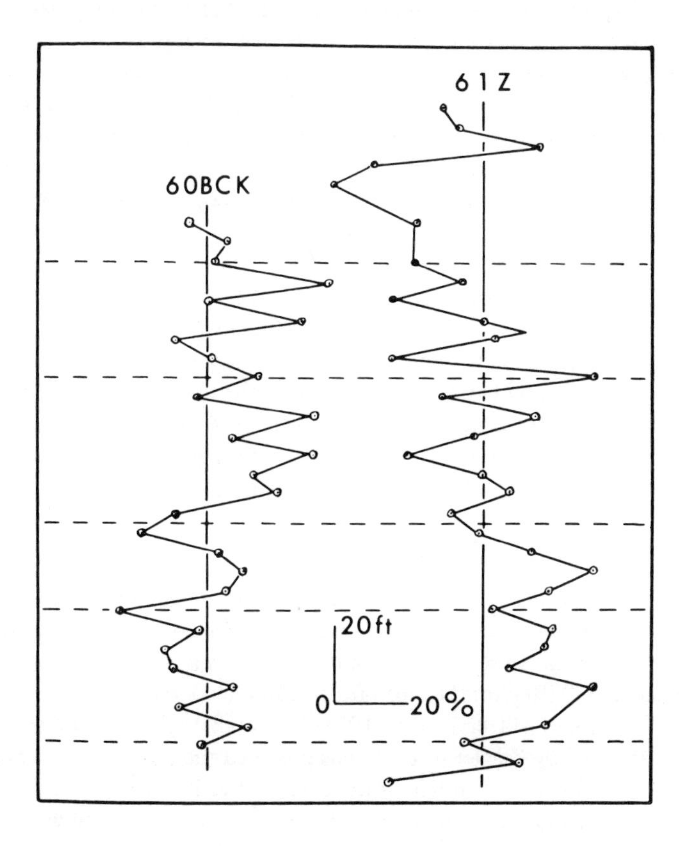

Text-Figure 16
Relative-abundance logs of *Phragmodus undatus* for two sections of the Lexington Limestone in southern Ohio. The center line of each log is 50 percent and dashed horizontal lines bracket comparable intervals in each section. (After Bergström and Sweet.)

authors noted (1966, p. 285) that virtually all the conodont species represented had ranges which exceeded the thickness of their sampled sections in the Lexington Limestone, and that correlation was also made more difficult by fluctuation and repeated migrations within the area of at least two distinct faunas. Under such conditions the "assemblage"-zone basis of correlation is clearly inapplicable, and Bergström and Sweet employed the graphic analysis of vertical changes in relative abundance of the multielement species *P. undatus,* the dominant species in their collections. Correlation by this method has been successfully employed in palynology, based upon certain assumptions, the most important of which is that such variation is time dependent and depends upon distributional or environmental factors other than those directly reflected in the particular lithologic character of the strata. The authors concluded that abundance variation did "appear to be generally unrelated to the gross lithological character of the rocks in any section" and concluded that it was "reasonable to assume that the environmental changes, whatever their nature, were basin-wide and that at least the major fluctuations in relative abundance of *Phragmodus undatus* were essentially contemporaneous events in all parts of the Cincinnati Region" (1966, pp. 286-287).

On the basis of these assumptions Bergström and Sweet compared logs of relative abundance for each section, and showed a close correspondence in the general profiles represented by two control sections of essentially similar thickness. The comparison, applied to other sections, yielded the basis for a local correlation and provided a composite section against which the distribution of other species was plotted. Application of the same techniques to Trenton Group sections in New York, Minnesota, and Ontario yielded consistent results, after allowance for the difference in relative thickness of the sections. This method has been little used by subsequent workers. Although its assumptions are not easily proved, it would seem to have some potential value in intrabasinal correlation.

Biostratigraphy of the Devonian

The European Devonian represents one of the best documented of all conodont sequences. Although our knowledge of Lower Devonian conodonts and stratigraphy is incomplete, a reasonable sequence of faunas has been described. Seven zones were established by Wittenkindt (1965) for the Middle Devonian, although the lower boundary is problematical (see Ziegler, 1971, for details). The Upper Devonian zonation, established by Ziegler in 1962, has provided the basis for a worldwide correlation, and is the best example of the degree of refinement that conodonts now provide. The sequence of faunas, which has also been integrated with the standard ammonoid sequence, has enabled Ziegler to divide the Upper Devonian into 12 major conodont zones and 24 subzonal divisions. These were based on the limited vertical ranges of a large number of associated species in assemblage zones, and information was also gathered concerning the phylogeny of a number of the conodont genera (Ziegler, 1962; Helms, 1963). Text-figures 17, 18, and 19 illustrate the zonation of the Upper Devonian.

Ammonoid Stufe	Ammonoid zone	Conodont zone	Subzone
		Lower Carboniferous	
Gonioclymenia-Stufe / ?Wocklumeria-Stufe	do VI	Protognathodus	
		?	?
	do VI	costatus	Upper
		costatus	Middle
	do V/VI ?	costatus	Lower
		styriacus	Upper
	do V	styriacus	Middle
	do IV/V ?	styriacus	Lower
Platyclymenia-Stufe	do IV	velifer	Upper
	do III β	velifer	Middle
	do III α	velifer	Lower
		quadrantinodosa	Upper
Cheiloceras-Stufe	do II β	quadrantinodosa	Lower
	do II β	rhomboidea	
	do II α	crepida	Upper
	do II α	crepida	Middle
	do II α	crepida	Lower
Manticoceras-Stufe	post-do I δ	triangular.	Upper
		triangular.	Middle
	do I δ	triangular.	Lower
	do I δ	gigas	Uppermost
	do I γ/δ	gigas	Upper
	do I γ	gigas	Lower
	do I γ	Ancyrognathus triangularis	
	do I (β) γ	asymmetric.	Upper
	do I α	asymmetric.	Middle
		asymmetric.	Lower
		asymmetric.	Lowermost
no ammonoids but probably do I α		hermanni-cristatus	Upper
Maenioceras-Stufe		hermanni-cristatus	Lower

Text-Figure 17
Correlation of Upper Devonian conodont zones with ammonoid stratigraphy. The post-doI δ represents a period with *Manticoceras* but without *Crickites holzapfeli.* (After Ziegler.)

Ziegler's Upper Devonian conodont zonation provides a classic example of the value of conodont studies, for it represents a proved and tested basis for worldwide correlation of impressive precision. First established more than a decade ago, the fundamental framework has undergone no major change, although Ziegler himself has introduced some minor refinement, supplementation, and taxonomic revision (e.g., see Ziegler, 1971; Sandberg and Ziegler, 1973).

The original succession and zonation, first established largely on a series of condensed limestone sequences in West Germany, is now supplemented by compilation of stratigraphic ranges of additional conodont species, now totaling some 120, or more than twice the number used in the original zonation (see Ziegler, 1971). Virtually all these represent a relatively small number of platform genera, especially *Polygnathus, Palmatolepis, Ancyrodella, Ancyrognathus,* and *Spathognathodus,* plus a few species of *Ancyrolepis, Icriodus, Polylophodonta, Protognathodus,*

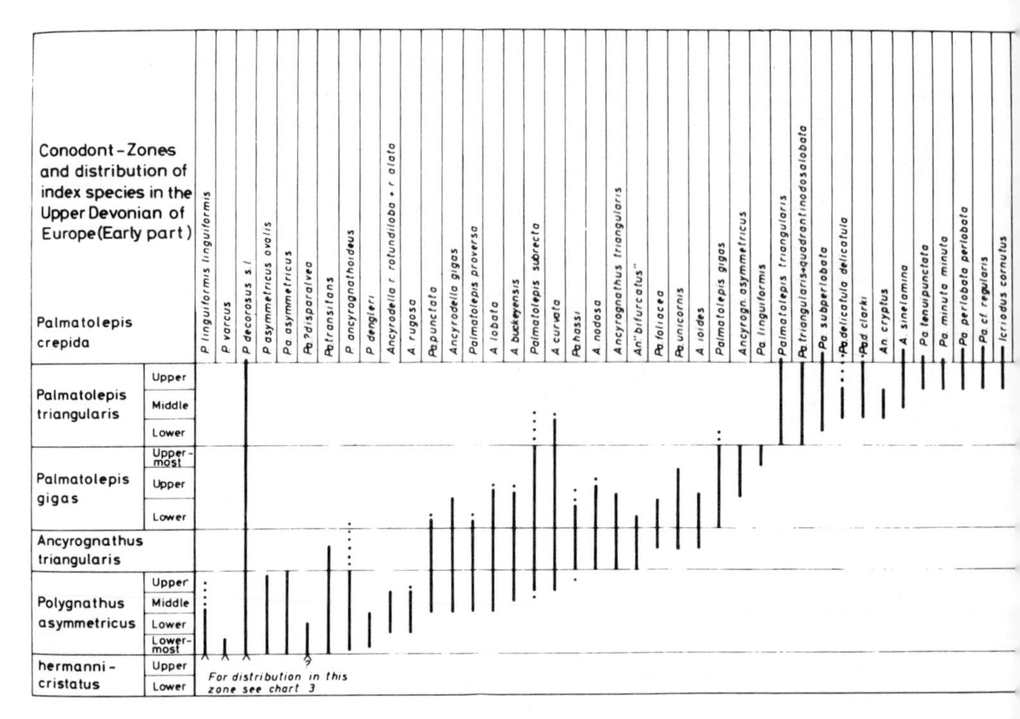

Text-Figure 18
Conodont zones and distribution of index species in the Upper Devonian of Europe (early part): *P., Polygnathus; Pa., Palmatolepis; A., Ancyrodella; An., Ancyrognathus; I. Icriodus.* Solid vertical lines indicate frequent occurrence; dotted lines indicate single specimens. (After Ziegler.)

Pseudopolygnathus, and *Scaphignathus.* Little dependence has been placed, and little comparably detailed knowledge is yet available, on the distribution of bars and blades.

Even for the Upper Devonian conodont zonation, however, some ambiguities remain. These result from morphological transition between a few zonal species, limited problems of correlation with parts of the standard ammonoid chronology, and some still unresolved geographic differences, such as those between the higher Devonian faunas of North America and Europe.

FUTURE DEVELOPMENT OF CONODONT BIOSTRATIGRAPHICS

The biostratigraphic basis of this zonation is relatively simple. The boundaries of the zones themselves are generally identified by the first or last appearance of the name-giving species, or some clearly identifiable segment of its range, but their recognition and subdivision are further based on the detailed distribution and association of numbers of other species and subspecies. Its success, however, depends not only on its simplicity, and therefore the relative lack of ambiguity in its application by specialists other than its authors, but the fact that it rests on

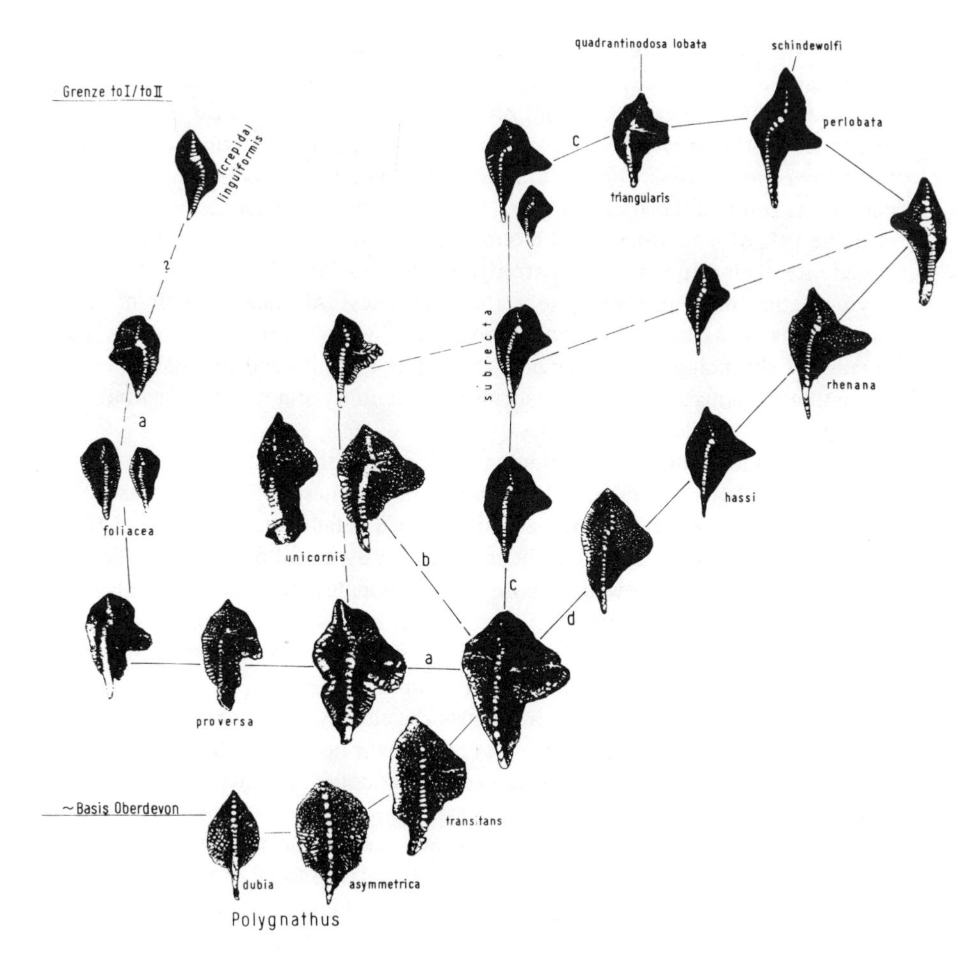

Text-Figure 19
Origin and development of *Palmatolepis* species in the *Mantiococeras* Stage of the Upper Devonian of Germany. The specimen from which the lines a-d originate is *P. marthenbergensis.* All the specimens on the horizontal line belong to *P. proversa* (giving rise to *P. linguiformis* and perhaps to *P. unicornis*) and those on the vertical line to *P. subrecta*, which leads in turn to several younger species. A fourth trend leads to *P. hassi*. Note the general decrease in time in platform size, and sculpture in each of the major groups. (After Ziegler, 1962).

a meticulously careful stratigraphic approach. The parachronological zonation was established in a relatively well exposed, richly fossiliferous, intensively studied, previously subdivided succession of strata, from which modern techniques allowed the extraction of very large numbers of isolated specimens, all collected with scrupulous care on a virtually layer by layer basis. Furthermore, the paratype strata, although not uniform, are relatively free of major facies variation and tectonic complication and provide a reasonably continuous fossiliferous succession.

Some of the success of conodont biostratigraphy also reflects the characteristics of conodonts themselves. In spite of the various limitations described above, and the many other subtleties of distribution that will undoubtedly come to light, conodonts are, to a remarkable degree, unaffected by the major facies restrictions characteristic of most other fossil groups. The relative abundance and diversity of conodonts, their frequently short stratigraphic ranges, and their suitability for phylogenetic studies all contribute to their usefulness. Although they do not provide any easy panacea for Paleozoic biostratigraphy, no other group exceeds them in range of chronologic and ecologic distribution, diversity and abundance, ease of extraction from many strata, and morphologic subtlety and variety of a kind appropriate for phylogenetic studies.

There is much promise in continuing phylogenetic studies of both single-element forms and multielement taxa, and the comprehensive description of faunas from type and stratotype sections. We particularly need detailed studies of the distribution of bars and blades, as well as the more obviously useful platforms, and of the faunas of clastic strata, as well as those from carbonates. Bars and blades may provide the only common elements of conodont faunas from carbonate and noncarbonate strata. There is also a need to link these studies to exhaustive lithological studies and census studies of other faunal groups.

Such studies as these promise to enhance the already formidable success of conodont biostratigraphy. The degree of refinement now attainable in some periods, such as the Late Devonian, also suggests that conodonts may provide a unique opportunity to test many of the implicit assumptions of paleontological correlation, such as the degrees of similarity required over a variety of distances and environments to justify correlation, migration rates, the relative value of the various existing methods of correlation, and so on. These studies, although they may well modify some of our present basic assumptions, could also provide a new basis of confidence in the fundamental practice of biostratigraphy.

Acknowledgments

We wish to thank Ruth Frazier for her patient and skillful typing of the manuscript.

Late Cambrian of Western North America: Trilobite Biofacies, Environmental Significance, and Biostratigraphic Implications

Michael E. Taylor

**U.S. Geological Survey,
Washington, D.C.**

INTRODUCTION

Cambrian stratigraphic paleontology of the western United States has been studied for approximately 100 years (see Palmer, 1971, p. 3). During that time, research has concentrated mainly on the systematics of the biota, principally trilobites, and the construction and refinement of biostratigraphic zonations for stratigraphic correlation. Such studies usually emphasize the "vertical" element of faunal distribution and the extension of biostratigraphic schemes into new areas. A new element was added when faunas began to be analyzed for their paleozoogeographic importance. Such studies are important to biostratigraphy because they provide understanding of the geographic limits that can be expected for the recognition of zonal schemes. They also can focus attention on those areas where faunal provinces are in juxtaposition, and thus might provide data for interprovince stratigraphic correlations.

Efforts to synthesize geographical characteristics of North American Cambrian trilobites were first made by Wilson (1957) and Lochman-Balk and Wilson (1958). They recognized three trilobite biofaces belts surrounding the North American craton, which were considered to be related to tectonic regions (Text-fig. 1). They delineated a Cratonic Faunal Realm characterized by trilobites having restricted

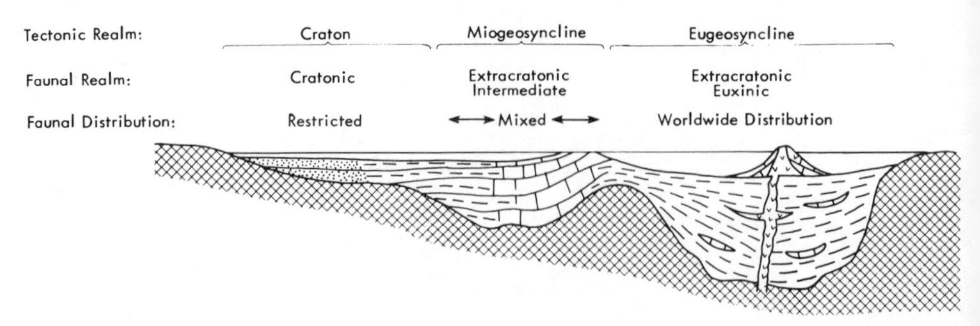

Text-Figure 1
Model of trilobite biofacies distribution in North America. Not to scale. (Based on Wilso
1957; and Lochman-Balk and Wilson, 1958.)

geographic distribution, an Extracratonic–Intermediate Realm characterized by mixed trilobite faunas in miogeosynclinal sites, and an Extracratonic Euxinic Realm, which occupied eugeosynclinal areas and contained more strongly cosmopolitan faunas.

Palmer (1960, 1965b, 1973) refined this biofacies model and suggested that trilobite distribution patterns were more closely related to degree of access to open-ocean conditions than to tectonic areas. Palmer's model (Text-fig. 2) shows relatively endemic Inner Detrital Belt faunas characterized by restricted access to the open ocean. These faunas are separated from more cosmopolitan Outer Detrital Belt faunas, which are thought to have had free access to open-ocean conditions. A shallow-water Outer Carbonate Belt was thought to serve as a major physical barrier to migration between inner- and outer-belt faunas.

The Lochman-Wilson and Palmer models have been discussed in more detail elsewhere (Taylor and Halley, 1974, p. 12).

The purpose of this paper is to reexamine the geographic distribution of polymerid trilobites for part of the Late Cambrian across the western United States,

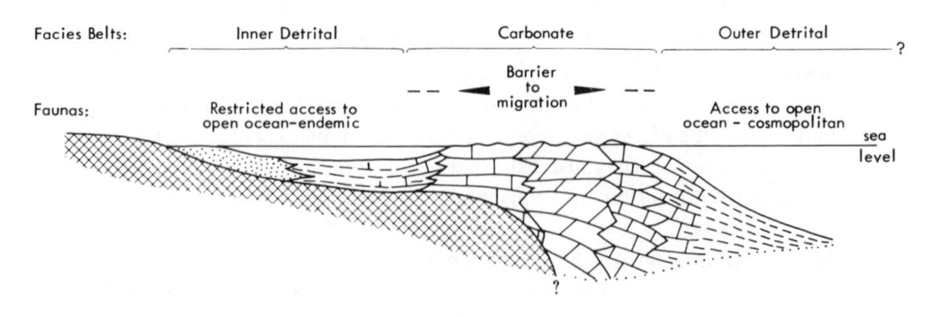

Text-Figure 2
Model of Cambrian trilobite biofacies distribution in North America. Not to scale.
(Based mainly on Palmer, 1960, 1965b, 1973.)

using published data and new faunal and sedimentological information (Taylor, 1971, 1976; Cook and Taylor, 1975; Taylor and Cook, 1976).

To better understand the environmental implications of the trilobite biofacies, characteristics are compared with some zoogeographic data for modern marine isopod crustaceans. This comparison provides better understanding of the temporal, environmental, and paleogeographic relationships between North American and some other faunal provinces during part of the Late Cambrian, and allows construction of a model of trilobite faunal distribution, which may be further tested and compared with biofacies patterns for other parts of the Phanerozoic record. To meet these objectives I have drawn freely from the work of others, especially Christina Lochman-Balk, A. R. Palmer, and James H. Stitt. The work of R. J. Menzies, R. Y. George, and G. T. Rowe (1973) on Holocene isopod distribution serves as a basis of comparison with trilobite distribution. A recent paper by O. G. Kussakin (1973) contains an important discussion of isopod distribution, but unfortunately the work was received too late to incorporate here.

CAMBRIAN CHRONOSTRATIGRAPHY

The interpretation of dynamic paleozoogeography of trilobites requires synthesis of data mainly from three sources: (1) systematic information on the faunas, (2) an established chronostratigraphic (mainly biochronologic) framework within which natural paleozoogeographic patterns can be recognized and delimited, and (3) sedimentologic data on the environments associated with the paleozoogeographic patterns. Paleozoogeography (and closely related paleosynecology) is additive to these more basic studies and in turn contributes to a better understanding of regional stratigraphic correlations based on trilobite zones. It is therefore necessary to briefly examine some characteristics and the current status of Cambrian chronostratigraphy for North America.

Biochronology

Currently recognized Cambrian and Lower Ordovician biostratigraphic units for North America are shown in Text-fig. 3. The biostratigraphic scale is a composite of units that can be recognized from place to place around North America, although all units have not been recognized in any one area. Background information for most parts of the scale can be found in Bell et al. (1956), Fritz (1972), Hintze (1952), Howell et al. (1944), Lochman-Balk and Wilson (1958), Lochman-Balk (1974), Longacre (1970), Palmer (1965b), Robison (1964a, 1964b), Ross (1951), Stitt (1971b), and Winston and Nicholls (1967).

Biostratigraphic assemblage zones have been traditionally defined in the Cambrian of the United States on the basis of stratigraphic associations of trilobite genera. Assemblage zones have a definite homotaxial position relative to other such units, and they are temporally nonrepetitive. The boundaries between assemblage zones may be gradational or abrupt, and they are usually placed at horizons

MYBP	SYSTEM	SERIES	STAGE	BIOSTRATIGRAPHIC UNITS	
	ORDOV.	CANAD.		ZONES C – J	
				Symphysurina	
				Missisquoia	
500	CAMBRIAN	UPPER	TREMP.	Saukia	Corbinia apopsis
					Saukiella serotina
					Saukiella junia
					Saukiella pyrene
			FRANCONIAN	Ellipsocephaloides	
				Idahoia	Drumaspis walcotti
					Idahoia lirae
				Taenicephalus	Orygmaspis llanoensis
					Parabolinoides contractus
				Elvinia	UPPER
					LOWER
			DRESBACHIAN	Dunderbergia	
				Prehousia	
				Dicanthopyge	
				Aphelaspis	
				Crepicephalus	
				Cedaria	
515		MIDDLE		Bolaspidella	
				Bathyuriscus – Elrathina	
				Glossopleura	
				Albertella	
540		LOWER		Bonnia – Olenellus	
				Nevadella	
				Fallotaspis	
570				"TOMMOT BIOTA"	

Text-Figure 3
Composite biostratigraphic scale for the Cambrian and Lower Ordovician of North Ame ica based on trilobites. Time before present in millions of years (MYBP) is considered t best available approximation following a review of radiometric dates by Cowie (1964) and Cowie et al. (1972). Assignment of lettered zones to the Canadian Series follows R. J. Ross, Jr. (pers. comm., March 1974).

where a relatively high proportion of local ranges of genera begin and end (Robison, 1964b). Assemblage subzones are based primarily on associations of trilobite species and are similar to assemblage zones in concept. Subzones have been established where local conditions permit detailed collecting of fossils from carefully measured stratigraphic sections and the interpretation of local ranges of many species under a relatively high degree of stratigraphic control.

Cambrian assemblage zones and subzones given in Text-fig. 3 are considered the best available approximations of biochronologic units from which regional chronostratigraphic units can be deduced.

Radiometric Dates

Data bearing on the temporal duration of Cambrian biostratigraphic units are poor. However, Cowie (1964) and Cowie et al. (1972) have evaluated radioisotope dates published for various parts of the Cambrian System around the world. On the basis of their best estimates for the boundaries between Cambrian epochs (Text-fig. 3), an approximate average duration for the biostratigraphic subdivisions of each of those epochs can be interpolated (see Table 1).

Biostratigraphic units in both the Early and Middle Cambrian average 7.5 million years duration. In contrast, biostratigraphic refinement of the Late Cambrian is approximately 1 million years average duration for each unit. Whether refinement is greater in the Late Cambrian as a result of more intensive collecting and study or because of more rapid evolutionary diversification in trilobite stocks is not yet known.

UPPER CAMBRIAN FACIES

The biostratigraphic interval analyzed in this paper extends from the top of the *Elvinia* Zone to the base of the *Missisquoia* Zone (Text-fig. 3). This interval corresponds to the Conaspid Biomere of Palmer (1965a, 1965b), and was later renamed the Ptychaspid Biomere by Longacre (1970). The unit has subsequently been studied by Stitt (1971a, 1971b, 1973, 1975).

Table 1
Average Duration of North American Cambrian Biostratigraphic Units[a]

Epoch	*Estimated Years Duration*	*Biostratigraphic Units*	*Average Years/Unit*
Late Cambrian	$15. \times 10^6$	16	0.94×10^6
Middle Cambrian	$30. \times 10^6$	4	7.5×10^6
Early Cambrian	$30. \times 10^6$	4	7.5×10^6

[a]Based on radiometric ages suggested by Cowie et al. (1972, p. 12). Averages should be considered only gross approximations.

Data are examined along a stratigraphic profile from basinal deposits in the western Great Basin to near-shoreline deposits in the upper Mississippi Valley. Text-figure 4 shows the distribution of control sections for this analysis. My own fieldwork has been concentrated in the Great Basin area of Utah and Nevada. Data for other sections are from studies in the literature (see recent syntheses by Lochman-Balk, 1970, 1971, 1972) and from examination of museum collections of trilobites from those areas.

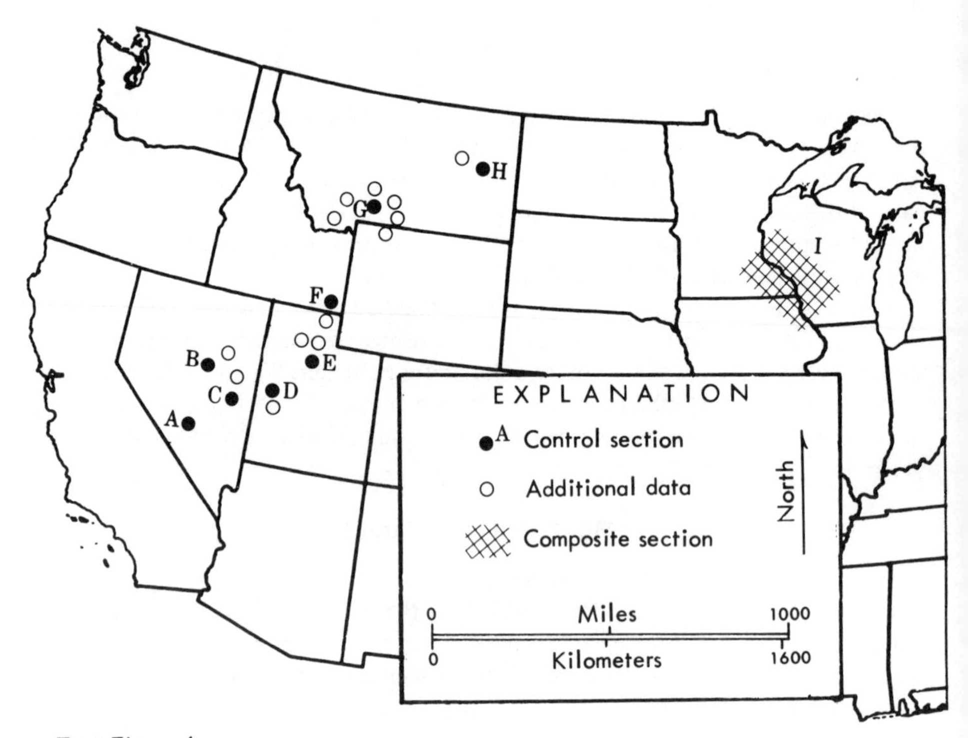

Text-Figure 4
Index map showing location and source of data for control sections discussed in text: A, Hales Limestone, Hot Creek Range, Nev. (Cook and Taylor, 1975b; Taylor, 1976; Taylor and Cook, 1976); B, Windfall and lower part of Goodwin formations, Eureka Mining District, Nev. (Taylor, 1971); C, Whipple Cave Formation, central Egan Range, Nev. (Taylor, 1971; Cook and Taylor, 1975b; Taylor and Cook, 1976); D, Notch Peak Limestone, House Range, Utah (Taylor, 1971; L. F. Hintze and J. F. Miller, unpublished data); E, Ajax Dolomite, East Tintic Mountains, Utah (Morris and Lovering, 1961 p. 50); F, St. Charles Formation, Bear River Range, Idaho (Lochman and Hu, 1959; Armstrong, 1969; Haynie, 1957; M. E. Taylor, unpublished data); G, Snowy Range Formation, southwestern Montana and northwestern Wyoming (Grant, 1965); H, Deadwood Formation, east-central Montana (Lochman, 1950, 1964; Lochman-Balk and Wilson, 1967); I, Franconia, St. Lawrence, and Jordan formations, Minnesota and Wisconsin (Austin, 1972; Bell et al., 1952; Grant, 1962; Howe et al., 1972; Nelson, 1951; Ostrom, 1970; Raasch, 1951; Ulrich and Resser, 1930, 1933).

Shelf-to-basin biofacies changes are so strong during the Late Cambrian that refined correlation along the profile is difficult. Therefore, I have combined some of the zones and subzones that can be recognized locally to produce the smallest biostratigraphic units that can be correlated across the entire profile. Two units can be so recognized, the upper Franconian Stage (*Taenicephalus* through *Ellipsocephaloides* zones) and the Trempealeauan Stage (*Saukia* Zone). The time represented may be on the order of 4 or 5 million years for each unit (Table 1). Shifting of boundaries between subprovinces undoubtedly occurred during these 4 or 5 million year intervals, resulting in some degree of time averaging of the faunas.

The shelf-to-basin lithofacies pattern (Text-fig. 5) is characterized by an inner shelf area with shoreline terrigenous sandstones and siltstones in the upper Mississippi Valley (I) that grade seaward to subtidal siltstones, shales, and limestones, which are commonly glauconitic (G, H). Outer shelf areas of Utah and eastern Nevada (C–E) are characterized by light-colored lime mudstones and lime wackestone, high-relief stromatolites, and skeletal and lithoclastic lime grainstones. The outer shelf has a dolomitized limestone facies (C-F) composed of laminar and digitate stromatolites, oncolites, and cross-bedded grain-supported fabrics similar to somewhat older Upper Cambrian rocks described in detail by Kepper (1972) from the same general area. These outer-shelf carbonates are thought to represent near-sea-level depositional environments with some low-lying supratidal mudflats. Shallow-water carbonates are replaced westward (A, B) by dark-colored thinly bedded to laminated lime mudstones and dark-gray shales. These rocks show many of the characteristics of "deeper-water" limes mudstones described by Wilson (1969).

The regional facies reconstruction suggests that changes in stratigraphic thickness, but not in predominant sedimentary environments, characterized the transition from craton to miogeocline during the Late Cambrian.

Regional palinspastic relationships of lower Paleozoic rocks in the Great Basin, and their inferred paleogeographic significance, have been discussed by Stewart and Poole (1974). Their conclusions concerning the presence of an early Paleozoic shallow shelf, slope, and deeper ocean basin to the west are consistent with the conclusions reached here.

Because of the importance of the off-shelf facies in the interpretations presented here, an example is discussed in some detail in the following paragraphs. Examples of stratigraphic sections typical of other facies can be found in the references cited in the caption to Text-figure 4. Regional characteristics of the outer-shelf carbonate facies were discussed previously (Taylor and Halley, 1974, pp. 12–15).

The Hales Limestone of the Hot Creek Range in south-central Nevada (Text-fig. 6) is an example of the off-shelf or slope facies. The Hales Limestone is composed of dark-gray to black thin-bedded argillaceous lime mudstone and some interbedded black laminated chert. Interbedded conglomerates, breccias, and folded beds are common. The clastic rocks and folded beds are thought to represent allochthonous limestone debris flows and slumps generated from a shelf edge, or in an off-shelf slope environment (Cook and Taylor, 1975; Taylor and Cook, 1976). These limestone conglomerates and breccias show many of the characteristics of debris flows described by Cook et al. (1972) from the Devonian of Alberta, Canada.

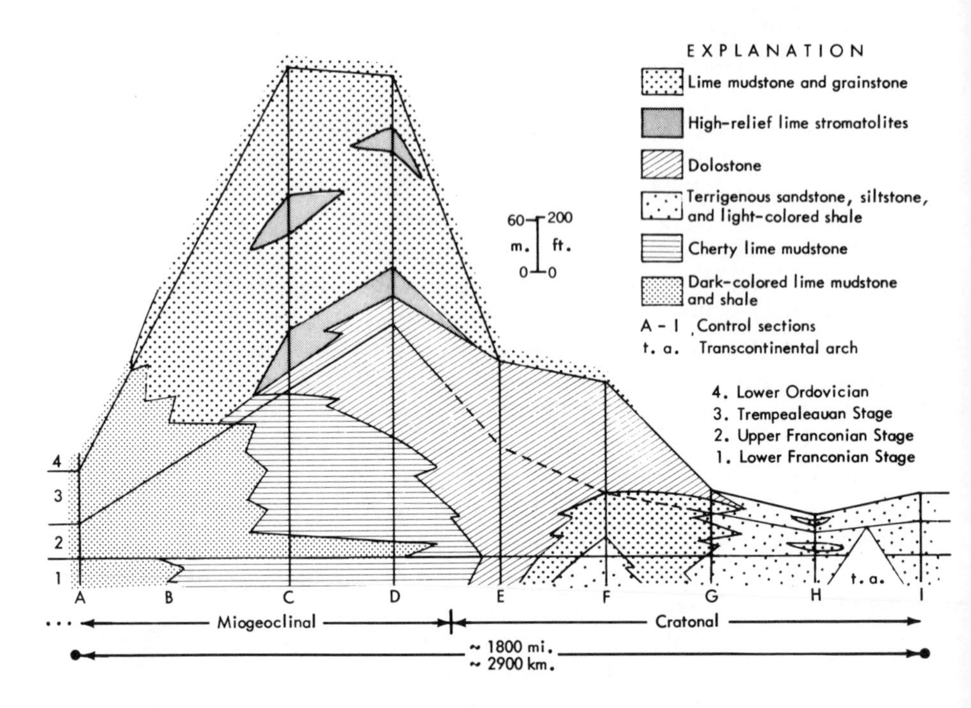

Text-Figure 5
Panel diagram showing generalized lithofacies distribution from central Nevada to upper Mississippi Valley. Time boundaries dashed where poorly known. Location of control sections and source of data given in Text-figure 4: 1, *Elvinia* Zone; 2, *Taenicephalus, Idahoia,* and *Ellipsocephaloides* zones; 3, *Saukia* Zone; 4, *Missisquoia* and *Symphysurina* zones. Stratigraphic thickness to scale except at B (Eureka Mining District, Nev.), which is expanded to facilitate showing facies relationships. Geographic distance not to scale; t.a., Transcontinental Arch.

Study of fossils from the Hales Limestone (Taylor, 1976) suggests that two mutually exclusive types of assemblages can be recognized. The first is in the matrix of allochthonous debris flows where individual fossils are disarticulated, broken and abraded, usually well sorted, and often concentrated in graded beds. The second type is in dark-colored lime mudstones similar to the "deeper-water" limestones of Wilson (1969). These latter occurrences of fossils are interpreted as autochthonous or parautochthonous faunules. They are characterized by a high number of articulated individuals that are usually oriented parallel to bedding and tend to be unabraded and unbroken, except by breakage through compaction.

The allochthonous assemblages contain genera typical of the *Hungaia* fauna, an association of Late Cambrian trilobite genera that generally occurs in a narrow but widespread band in shelf-edge and basinal sites around most of North America (see review in Taylor and Halley, 1974). Some elements of the *Hungaia*

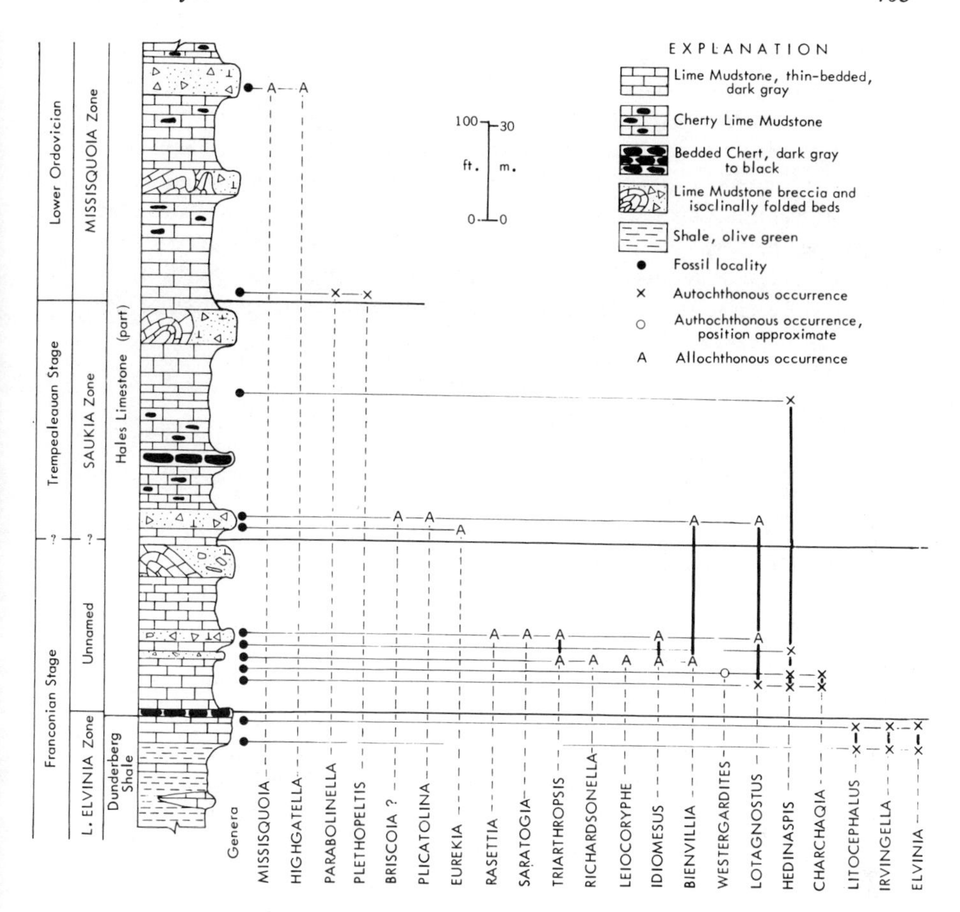

Text-Figure 6
Columnar section of Hales Limestone (lower part) at Tybo Canyon, Hot Creek Range, central Nevada (control section A in Text-fig. 4). Known occurrences and local ranges of trilobite taxa are shown. Lime mudstone breccia and isoclinally folded beds are interpreted as allochthonous debris flows and slump deposits. Allochthonous fossils shown occur in breccia matrix, not in breccia clasts. (Section based on Taylor, 1976; additional data are given in Cook and Taylor, 1975 and Taylor and Cook, 1976.)

fauna are known from recent work in shallow-water sites coeval with the Hales Limestone in eastern Nevada (Taylor, 1971; Taylor and Cook, 1976). Thus, occurrence of the *Hungaia* fauna in allochthonous debris flows in central Nevada suggests that the fossils were either transported from the shelf edge located to the east, or displaced from higher slope environments and redeposited in deeper-water sites.

The autochthonous *Hedinaspis* faunule is characterized by taxa distinctly different from the *Hungaia* assemblage in allochthonous debris flows. The association consists of *Hedinaspis, Charchaqia, Westergaardites,* and *Lotagnostus.* Except for *Lotagnostus*, the genera are unknown from the North America Faunal Province, except possibly for some *Hedinaspis* fragments reported by Palmer (1968) from east-central Alaska. However, the *Hedinaspis* fauna is widely distributed in Asia (see p. 424). The occurrence of the *Hedinaspis* faunule in deep-water deposits in Nevada and its absence from shallow shelf sites of the North American Province suggests the faunule lived in deeper-water environments and lacked ecological tolerance for shallower-water conditions.

LATE CAMBRIAN PALEOZOOGEOGRAPHY

Methods and Definitions

The recognition of both shallow- and deep-water faunas in the Upper Cambrian of the western United States provides an opportunity to examine some of the broader characteristics and implications of regional biofacies patterns.

In developing an effective analysis of faunal distribution, it was found important to distinguish between "cosmopolitan" taxa that are widespread because of tolerance to a wide range of environmental conditions, and stenotopic taxa that are widespread because their preferred habitat happened to be widespread. To meet these analytical needs and to provide a basis for simple quantitative analysis of the faunas, I have classified the trilobite genera according to distributional characteristics by using the following operational definitions: *Eurytopic trilobite,* a genus that is known to occur coevally in two or more different habitats; *stenotopic trilobite*, a genus that is only known to occur in one habitat; *habitat*, either inner shelf, outer shelf, or off-shelf slope as recognized for the Cambrian on lithologic, sedimentologic, and paleogeographic criteria; *eurygeographic trilobite*, a genus known to occur on the North American continent and at least one other continent; *stenogeographic trilobite*, a genus known to occur on only the North American continent.

Geographic occurrences of polymerid trilobite genera along the shelf-to-basin profile from the upper Mississippi Valley to the Great Basin are compiled in Tables 2 and 3. Major faunal discontinuities are evident from inspection of these tables. To quantify the shelf-to-basin faunal changes, and to provide a basis for defining paleozoogeographical units, I have adapted the simple methods used by Menzies et al. (1973, pp. 74–78) in studies of the Holocene distribution of isopod crustaceans. The percentage of genera not in common between two samples is taken as a measure of distinctiveness (D). However, when tabulating occurrence data to calculate distinctiveness, a "range-through" method was used. For example, the procedure was to count a taxon present at stations F and G if it occurs at stations E and H but not F and G (see Tables 2 and 3). This tends to decrease the influence of incomplete collecting and local ecological factors operating at the community level, and to increase emphasis on the regional dispersal potential

Table 2
Trilobite Genera and Geographic Distributional Data for the Upper Franconian Stage in the Western United States[a]

Habitat:	*Slope*		*Outer Shelf*				*Inner Shelf*			
FSP:	*"Lower" Slope*	*"Upper" Slope*	*Outer Shelf*				*Inner Shelf*			
Control Section: / *Taxa*	*A*	*B*	*C*	*D*	*E*	*F*	*G*	*H*	*I*	*Geographic Distribution*
Charchaqia	X									E
Hedinaspis	X									E
Rasettia	a									S
Richardsonella	a									E
Triarthropsis	a									S
Bienvillia	a	X								E
Idiomesus	a	X								S
Westergaardites	X	X								E
Leiocoryphe	a	–	X							S
Saratogia	a	–	X	–	–	X	X	–	X	S
Loganellus		X								E?
Parabolina		X								E
Plicatolina		X								E
Simulolenus		X								E
Sulcocephalus		X								S
Orygmaspis		X	X	X	–	–	X			S
Drumaspis		X	X	–	–	X	X	–	X	S
Parabolinoides		X	X	X	–	–	X	X	X	E
Hungaia			X							E
Idahoia			X	–	–	X	X	X	X	E
Taenicephalus			X	–	–	–	X	–	X	E?
Wilbernia			X	X	–	X	X	X	X	S
Bellaspis						X				S
Taenicephalina						X				S
Ptychaspis						X	X	X	X	E?
Irvingella							X			E
Kendallina							X			S
Maladia							X			S
Pinctus							X			S
Stigmacephaloides							X			S
Maustonia							X	X		S
Croixana							X	–	X	S
Ellipsocephaloides							X	–	X	S
Monocheilus							X	–	X	S
Prosaukia							X	–	X	E
Camaraspis								X	X	E?
Stigmacephalus								X	X	S
Chariocephalus									X	S
Conaspis									X	S
Dartonaspis									X	S
Eotychaspis									X	S
Psalaspis									X	S
Stigmaspis									X	S

[a]Location of control sections shown in Text-figure 4; X, verified occurrence; a, known only from matrix of allochthonous debris flow deposits; E, eurygeographic distribution; S, stenogeographic distribution (see text); FSP, trilobite faunal subprovince.

Table 3
Trilobite Genera and Geographic Distributional Data for the Trempealeauan Stage in the Western United States[a]

Habitat:	*Slope*		*Outer Shelf*				*Inner Shelf*			
FSP:	*"Lower" Slope*	*"Upper" Slope*	*Shelf Undifferentiated*							
Control Section: / *Taxa*	*A*	*B*	*C*	*D*	*E*	*F*	*G*	*H*	*I*	*Geographic Distribution*
Bienvillia	a	X								E
Hedinaspis	X									E
Plicatolina	a									E
Briscoia?	a	–	X	–	–	–	X	–	X	E?
Eurekia	a	X	X	X	X	–	X	–	X	S
Apatokephaloides		X	X							S
Magnacephalus		X	X							S
Richardsonella		X	X							E
Bayfieldia		X	X	X						S
Leiocoryphe		X	X	X						S
Acheilops		X	X	–	–	–	–	–	X	S
Bowmania		X	X	X	–	–	–	–	X	S
Euptychaspis		X	X	X	–	–	–	–	X	S
Plethometopus		X	X	X	–	–	–	X	X	S
Saukiella		X	X	X	–	–	–	–	X	S
Triarthropsis		X	X	–	–	–	–	–	X	S
Keithiella				X						S
"Leiobienvillia"				X						S
Theodenisia				X						S
Yukonaspis				X						S
"Acidaspis"			X	X						S
Heterocaryon			X	X						S
Idiomesus			X	X	–	–	X			S
Calvinella			X	X	–	–	–	–	X	S?
Corbinia			X	X	–	–	–	–	X	S
Dikelocephalus			X	–	–	–	X	X	X	S
Entomaspis			X	–	–	–	–	–	X	S
Illaenurus			X	X	–	–	X	–	X	S
Macronoda			X	X	–	–	–	X		S
Prosaukia			X	–	–	–	–	–	X	E
Rasettia			X	–	–	–	X	–	X	S
Saukia			X	–	–	–	–	X	X	E
Stenopilus			X	X	–	–	X	–	X	S
Plethopeltis				X	–	–	–	–	X	S
Bynumina							X			S
Bynumiella							X			S
Monocheilus			X	X	–	–	X	X	X	S
Promesus								X		S
Osceolia									X	S
Tellerina									X	E?
Walcottaspis									X	S

[a]Location of control sections shown in Text-figure 4. "*Leiobienvillia*" refers to forms assigned to *Leiobienvillia leonensis* Winston and Nicholls (1967). Unpublished data suggest that this form probably represents an undescribed genus. "*Acidaspis*" refers to the form taxon "*Acidaspis*" *ulrichi* Bassler (see Rassetti, 1959, p. 393) and is probably not *Acidaspis sensu stricto*. Symbols the same as in Table 2.

of the faunas. Unfortunately, the method eliminates the possibility of recognizing disjunct endemics and patchiness of preferred environments if such were present.

Data on faunal distinctiveness, stenotopy, and generic richness are given for both the upper Franconian and Trempealeauan stages in Tables 4 and 5.

The Late Cambrian faunas from North American cratonal and miogeoclinal rocks constitute an unique paleozoogeographic province (Lochman-Balk and Wilson, 1958; Palmer, 1973; Rowell et al., 1973) characterized by predominance of the trilobite families Dikelocephalidae, Ptychaspididae, Saukiidae, and Parabolinoididae. The North American Faunal Province is here divided into subprovinces. The term "faunal subprovince" (FSP) is used for a relatively homogeneous assemblage of genera, trilobites in the case of the Late Cambrian, that occupied specific geographic space at any given time, or persisted with some evolutionary change through an interval of time, and which are separated from adjacent coeval subprovinces by saltations in generic faunal distinctiveness values (D) of greater than 50 percent. Location of boundaries between specific FSP's may change through time, and they may merge, divide, or terminate in time. FSP's are named according to the habitat in which they are most commonly found, but the boundaries between FSP's need not necessarily coincide with lithofacies boundaries. It should be emphasized that the assemblage of particular genera operationally defines

Table 4
Data on Late Franconian Trilobite Stenotopy, Generic Richness, and Faunal Distinctiveness (D; see p. 406)[a]

Control Section	*Stenotopy* N_s	*%*	*N* *RT*	*Ob*	*Distinctiveness (D)*	*Faunal Subprovince (FSP)*
A**	5	50.	10	10		
A*	2	67.	3	3		"Lower" slope
					92.	
B	7	64.	11	11		"Upper" slope
					82.	
C	1	11.	9	9		Outershelf
					22.	
D	0	0.	7	3		Outershelf
					0.	
E	0	0.	7	0		Outershelf
					30.	
F	2	20.	10	7		Outershelf
					60.	
G	5	28.	18	18		Innershelf
					40.	
H	0	0.	14	7		Innershelf
					35.	
I	6	32.	19	19		Innershelf

[a]High distinctiveness values define boundaries of trilobite faunal subprovinces (FSP). Location of control sections is shown in Text-figure 4: A*, data from section A with allochthonous taxa excluded from calculations; A**, data from section A with allochthonous taxa included in calculations; N_s, stenotopic genera; %, percent stenotopic genera; N, total genera; RT, number of genera by range through method of tabulation; Ob, number of observed genera.

Table 5
Data on Trempealeauan Trilobite Stenotopy, Generic Richness, and Faunal Distinctiveness[a]

Control Section	*Stenotopy* N_s	*%*	*N* *RT*	*Ob*	*Distinctiveness (D)*	*Faunal Subprovince (FSP)*
A**	2	40.	5	5		
A*	1	100.	1	1		"Lower" slope
					100.	
B	1	8.	13	13		"Upper" slope
					57.	
C	0	0.	27	27		Shelf
					25.	
D	4	14.	29	21		Shelf
					28.	
E	0	0.	21	1		Shelf
					0.	
F	0	0.	21	0		Shelf
					9.	
G	2	9.	23.	10.		Shelf
					17.	
H	1	5.	21	6		Shelf
					21.	
I	3	14.	22	22		Shelf

[a]High distinctiveness values define boundaries of trilobite faunal subprovinces (FSP). Location of control sections is shown in Text-figure 4. Symbols same as in Table 4.

the subprovince, not the lithofacies in which it is found. Upper Cambrian lithofacies are independently interpreted here using sedimentologic and paleogeographic criteria, and include inner shelf, outer shelf, and off-shelf slope habitats.

The faunal subprovinces recognized here are similar in concept to "communities" of some authors and "benthic marine life zones" of others (see Watkins et al., 1973). The term community is considered inappropriate here because the Late Cambrian FSP's persisted long enough through time that genera living at the beginning of the units became extinct or evolved into other forms by the end of the units. A more basic difficulty is that the presence-or-absence data upon which the trilobite analysis is based are not sufficient to provide information on species dominance needed for community-level analysis.

Objections to the use of "benthic marine life zones" are mostly semantic. The term "zone" has been given many different meanings in stratigraphy (see discussion in Berry, 1966, p. 1488) so that yet another use seems poorly advised. In addition, Stitt (1975) has suggested the possibility that many catillicephalid trilobites, a group common to Late Cambrian "Upper" Slope and Outershelf FSP's, lived attached to floating algae rather than as part of the benthos. General problems related to the application of differing concepts and nomenclature in biogeographic classifications have been discussed by Sylvester-Bradley (1971) and Middlemiss and Rawson (1971).

Late Franconian Age

Distinctiveness values (D) for the late Franconian polymerid trilobites show three major peaks above 50 percent (Table 4). A D value of 82 percent clearly establishes a major discontinuity between faunas occupying shelf and slope habitats. This faunal change occurs between the outer-shelf shallow-water carbonate lithofacies and the deeper-water dark-colored thin-bedded lime mudstone lithofacies. Faunas within the slope habitat are also highly differentiated, with a D value of 92 percent. The differentiation separates an "Upper" Slope FSP characterized by olenids and eurytopic parabolinoidids from a "Lower" Slope FSP characterized by *Hedinaspis* and *Charchaqia*. The "Upper" Slope FSP contains 64 percent stenotopic trilobite genera, whereas the "Lower" Slope FSP contains 67 percent. The terms "upper" and "lower" slope are used here reservedly because upper- and lower-slope environments have not yet been documented on independent sedimentologic criteria. In addition, it is not certain whether the faunas occupying the "upper"- and "lower"-slope FSP's were actually parallel bands or a mosaic of communities on the slope and/or perhaps partly in the water mass above the slope.

The next major faunal discontinuity is represented by a D value of 60 percent between control sections F (Bear River Range, Idaho) and G (southwestern Montana-northwestern Wyoming). This faunal difference may be enhanced by differences in intensity of collecting in the two areas. However, additional collecting in the Bear River Range (the author, unpublished data) has not yielded evidence for extension of geographic ranges.

A D value of 60 percent separates an Innershelf FSP dominated by early ptychaspidinids and drumaspidinids from an Outershelf FSP dominated by parabolinoidids. The distinctiveness of the Outershelf FSP during the late Franconian is mainly a result of exclusion of inner-shelf stenotopes from the outer-shelf habitat. The Innershelf FSP contains 69 percent stenotopic trilobite genera, whereas the Outershelf FSP contains only 33 percent.

Control sections H (east-central Montana) and I (upper Mississippi Valley) were probably separated by the Transcontinental Arch during at least part of Late Cambrian time (Lochman-Balk, 1971, figs. 19-21). Faunal distinctiveness values suggest that trilobite dispersal was less influenced by the Transcontinental Arch than by factors associated with inner- and outer-shelf lithofacies differentiation. However, the apparent relatively high incidence of stenotopic genera in inner-shelf sites during the late Franconian (see p. 421) could have resulted in part from existence of the Arch during that time.

The proportional distribution of genera for each FSP of the late Franconian is shown in Text-figure 7.

In summary, the late Franconian trilobite faunas in the western United States can be differentiated into four subprovinces: Innershelf, Outershelf, "Upper" Slope, and "Lower" Slope. Both shelf FSP's show association with predominant sediment types. The "Upper" and "Lower" Slope FSP's collectively occur in

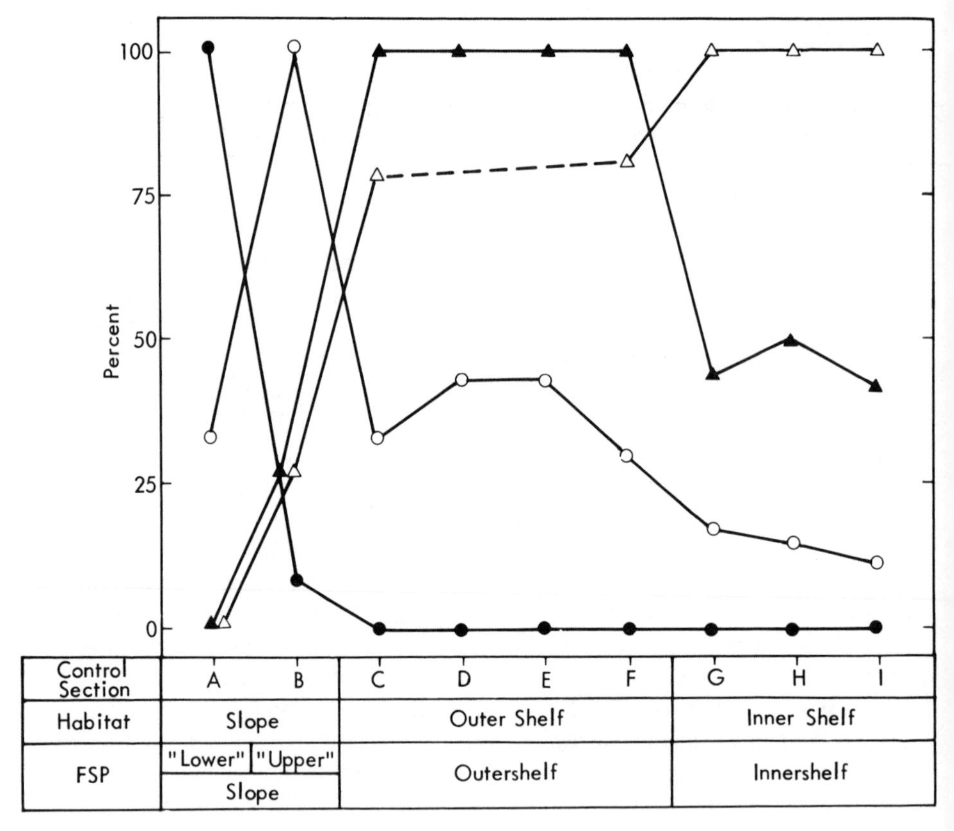

Text-Figure 7
Percentage of each late Franconian subprovincial trilobite fauna that occurs at each control section. Location of control sections shown in Text-figure 4: open triangle, inner-shelf FSP; solid triangle, outer-shelf FSP; open circle, "upper"-slope FSP; solid circle, "lower"-slope FSP. Line dashed where strongly influenced by the dolostone facies.

grossly similar lithic types that differ markedly from the shelf lithofacies. The two slope FSP's are differentiated on the basis of high faunal distinctiveness, biostratinomic characteristics, and inferred paleogeographic position.

Trempealeauan Age

Distinctiveness values (D) for Trempealeauan polymerid trilobites show a change of 100 percent between autochthonous fossils at control sections A (south-central Nevada) and B (Eureka District, Nevada) (Table 5). The calculation is probably biased by the small number of taxa present, and more data are needed to confirm a "lower"-slope FSP for the Trempealeauan Age. The occurrence of the stenotope *Hedinaspis* hints at persistence of the "Lower" Slope FSP from the late Franconian, however.

The next major faunal change is reflected by a distinctiveness value of 57 percent between the "Upper" Slope FSP and Undifferentiated Shelf FSP. This faunal change is less sharp than the shelf-to-slope change of the late Franconian. Stenotopy within the Trempealeauan "Upper" Slope FSP is 8 percent, whereas the Undifferentiated Shelf FSP shows 68 percent stenotopic genera. The differences are accounted for by the appearance of several taxa characteristic of the *Hungaia* fauna, which seem to range across the shelf-to-slope break during Trempealeauan time.

Distinctiveness values are below 30 percent across the outer-to-inner shelf part of the profile. These low values preclude subdivision of an Undifferentiated Shelf FSP and suggest widespread equitable environmental conditions for the North American Trempealeauan shelf. Proportional distribution of genera in each Trempealeauan FSP is shown in Text-figure 8.

HOLOCENE ZOOGEOGRAPHIC MODEL

A better understanding of the significance of trilobite distributional patterns may be gained by examining Holocene faunas for possible analogues. The living isopod crustaceans are a group that may serve this purpose. Many benthic isopods have no free-living larval stage and are found at almost all depths in modern oceans (Menzies et al., 1973, p. 76). Furthermore, some genera of isopods, such as *Serolis* (Text-fig. 9), show many morphological features convergent with trilobites (Menzies and George, 1969; Krebs, 1974).

The discussion that follows is based principally on the work of Menzies et al. (1973, pp. 79-128, Table 4-2), who compiled data on bathymetric distribution, zonation, and physical environments associated with genera of marine isopod crustaceans along a profile from near Cape Lookout, North Carolina, to the abyssal plain north of Bermuda.

Rowe and Menzies (1969) and Menzies et al. (1973) divided the northwestern Atlantic isopod fauna into several "faunal provinces" and "faunal zones" based on associations of genera and species. Their methods used for quantitatively determining the positions between faunal units are the same as those discussed on page 406.

The term "faunal province" is used as a major division of a faunal realm that can be distinguished from other provinces by high rates of faunal change among assemblages of isopod genera along a geographic (bathymetric) gradient. The assemblage of genera within a province defines its fauna, rather than bathymetric limits occupied by the fauna along any given topographic profile. This concept of isopod faunal provinces (IFP) is considered comparable to that of trilobite faunal subprovinces (FSP) adopted here. A principal difference is that isopod provinces represent a geological "instant" in time, even though some seasonal time averaging may occur, whereas the Late Cambrian trilobite data are agglomerative for time intervals of perhaps 4 or 5 million years duration.

Menzies et al. (1973) were able to subdivide their isopod faunal provinces into "faunal zones" based on assemblages of isopod species. The systematics of Late Cambrian trilobites are not sufficiently refined that regional paleogeographic

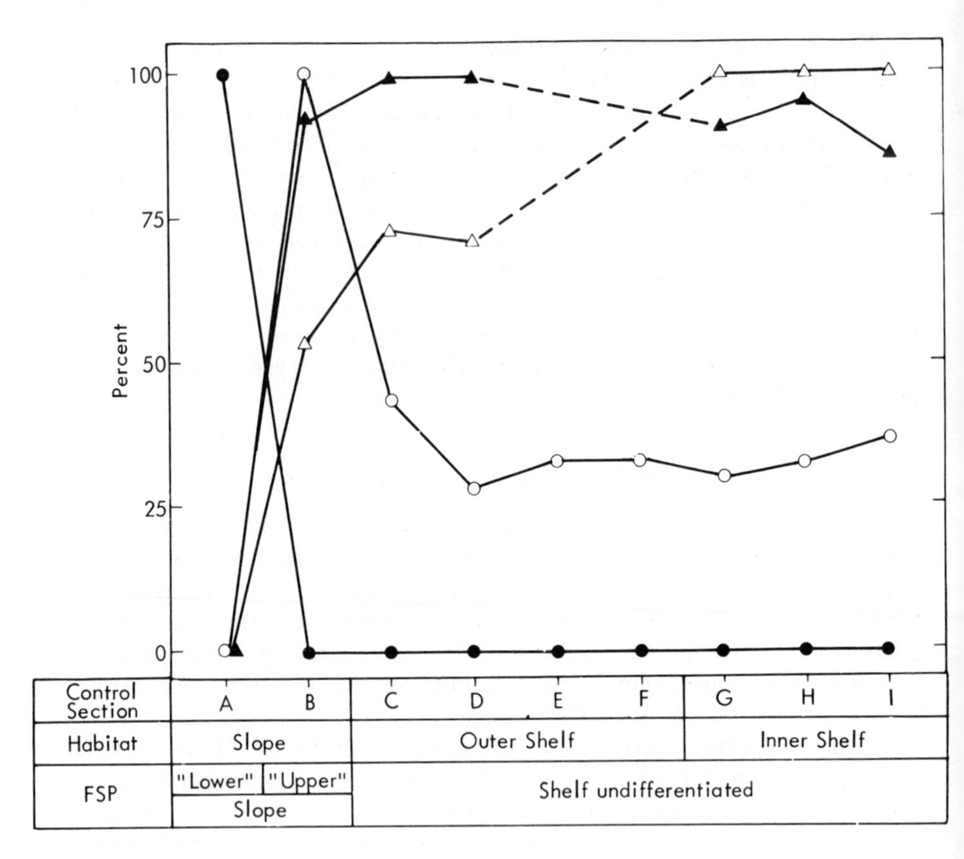

Text-Figure 8
Percentage of each Trempealeauan subprovincial trilobite fauna that occurs at each control section, except inner- and outer-shelf habitats are differentiated to facilitate comparison with late Franconian data in Text-figure 7. Inner- and outer-shelf sites contain an Undifferentiated Shelf FSP. Location of control sections is shown in Text-figure 4. Open triangle, Inner-shelf; solid triangle, Outershelf; open circle, "Upper" Slope FSP; solid circle, "Lower" Slope FSP. Lines dotted where strongly influenced by dolostone facies.

analyses of species are meaningful, except in special instances (Taylor and Halley, 1974). Therefore, review here is restricted to the analysis of isopod genera and the zoogeographic units defined by them.

To allow comparison of data on isopod and trilobite distribution, the Holocene marine isopod genera listed by Menzies et al. (1973, p. 98) are here classified as follows: *Eurytopic isopod,* a genus that is known to occur in two or more isopod faunal provinces off the northwestern Atlantic coast of the United States; *stenotopic isopod,* a genus that is known to occur in only one isopod faunal province off the northwestern Atlantic coast of the United States; *eurygeographic isopod,* a genus

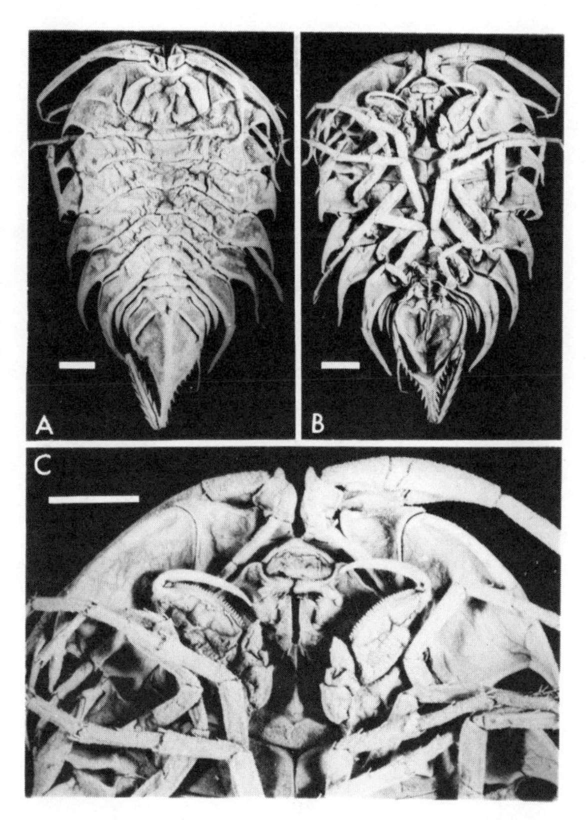

Text-Figure 9
Serolis meridionalis Hodgson, a Holocene trilobitiform isopod crustacean from 1,437 m deep in Drake Passage, Antarctic Ocean (USNM Invert. Zool. coll. 123963): (A) dorsal view; (B) ventral view; (C) closeup of oral region. Tail broken in (A) and (B). Specimen wrinkled during freeze-dry process; highlighted with ammonium chloride sublimate. Bar scale, 5.0 mm.

that is known to occur off the northwestern Atlantic coast of the United States and within the Antarctic Faunal Region; *stenogeographic isopod,* a genus that is known to occur off the northwestern Atlantic coast of the United States, but not within the Antarctic Faunal Region. Menzies et al. (1973, pp. 301-327) used broader definitions in a more comprehensive review of isopod distribution than is presented here.

Data on faunal distinctiveness, stenotopy, generic richness, and isopod faunal provinces are given in Table 6. Proportional distribution of isopod genera from each province is shown in Text-figure 10.

The Intertidal IFP (depth 0 to 3 m) contains a variety of habitats consisting of terrigenous clastic substrates in estuaries, bays, beaches, and offshore sandbars.

Table 6
Data on Northwestern Atlantic Holocene Isopod Stenotopy, Generic Richness, and Faunal Distinctiveness (D)[a]

Depth Interval (m)	*Stenotopy*		*N*		*Distinctiveness (D)*	*Province (IFP)*
	N_s	%	*RT*	*Ob*		
1. 0–3	7	47.	15	15		Intertidal
					65.	
2. 5–20	2	13.	16	16		Shelf
					29.	
3. 21–70	0	0.	13	10		Shelf
					15.	
4. 80–90	0	0.	11	4		Shelf
					8.	
5. 91–150	0	0.	12	11		Shelf
					31.	
6. 200–250	1	10.	10	7		Shelf
					67.	
7. 450–550	0	0.	6	4		Archibenthal
					17.	
8. 600–650	0	0.	5	3		Archibenthal
					74.	
9. 900–1,500	5	26.	19	16		Abyssal
					38.	
10. 2,000–2,400	1	7.	15	5		Abyssal
					7.	
11. 2,400–2,635	0	0.	14	2		Abyssal
					22.	
12. 2,640–3,100	3	17.	18	16		Abyssal
					42.	
13. 3,300–3,600	1	8.	12	7		Abyssal
					15.	
14. 3,800–4,100	1	8.	12	10		Abyssal
					50.	
15. 4,800–5,100	1	11.	9	8		Abyssal
					30.	
16. 5,200–5,400	1	13.	8	8		Abyssal

[a]High distinctiveness values define boundaries between isopod faunal provinces (IFP). N_s, stenotopic genera; %, percent stenotopic genera; *N*, total genera; RT, number of genera by range through method of tabulation; Ob, number of observed genera. Data from Menzies et al. (1973, p. 90, 98).

Natural rocky shore environments are absent. The general habitat contains a high stress environment subject to high fluctuations in salinity resulting from surface runoff and annual ranges in water temperature from approximately 0 to 30°C. Stenotopic genera compose 47 percent of the Intertidal IFP isopod fauna.

The Shelf IFP (depth 5 to 246 m) occupies the Continental Shelf off the Carolinas and the upper part of the Continental Slope below the shelf break. The shelf sediments are rich in skeletal carbonate debris, terrigenous sand, and rock debris. The lower boundary of the Shelf IFP, although imprecisely known, seems to correspond with the beginning of foraminifera-rich glauconitic sands. The Shelf IFP is separated from the Intertidal IFP by distinctiveness of 65 percent, and from the Archibenthal IFP by 67 percent. These values are higher by a factor of 2 than changes within the province (Table 6). Thirty-seven percent of the Shelf IFP isopod genera are stenotopic.

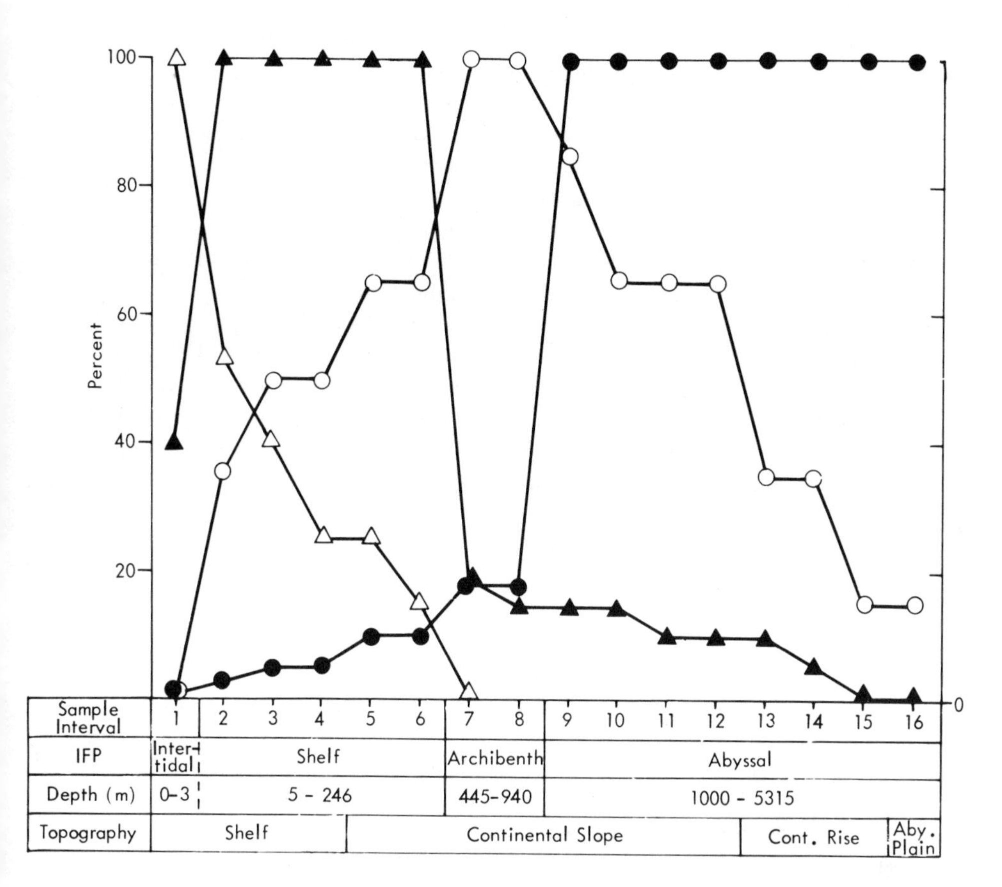

Text-Figure 10
Percentage of each northwestern Atlantic provincial isopod fauna that occurs within each depth interval: Open triangle, Intertidal IFP; solid triangle, Shelf IFP; open circle, Archibenthal IFP; solid circle, Abyssal IFP. Depth ranges for depth intervals given in Table 6. (Redrawn from Menzies et al., 1973, p. 93.)

Some data suggest (Blanton, 1971) that relatively cool Atlantic shelf waters are seasonally displaced by onshore flow of the warmer waters of the Gulf Stream. Such fluctuations would presumably increase environmental stress within the Shelf IFP or cause the boundaries of the province to shift seasonally.

The Archibenthal IFP is a transition zone beginning between 246 and 445 m and extending to a depth of about 940 m. Oxygen content and temperature of the Archibenthal IFP are seasonally variable and related to a transitional interval between the northerly flowing warm Gulf Stream above and the southerly flowing cold Western Boundary Undercurrent below. Temperature within the province ranges from 5 to 10°C and is subject to seasonal fluctuations of at least 4°C.

Sediments characterizing the archibenthal IFP consist of glauconite-rich foraminiferal and pteropod oozes deposited under the influence of the Gulf Stream.

The isopod fauna of the Archibenthal IFP is characterized by low generic richness, an absence of stenotopic genera, and predominance of taxa most closely related to the Abyssal IFP. The province is distinguished from the Shelf IFP by distinctiveness values of 67 percent, and from the Abyssal IFP by 74 percent.

The Abyssal IFP occupies the lower Continental Slope, Continental Rise, and Abyssal Plain from about 1,000 to 5,340 m depth. Associated sediments consist of fine-grained terrigenous silt, clay, and foraminiferal ooze that grades to red clay and pebble-sized manganese nodules on the Abyssal Plain. The province corresponds to cold waters with seasonally stable temperatures below the 4°C isotherm, and includes the Western Boundary Undercurrent and the northerly flowing Antarctic bottom waters. Faunal distinctiveness at the transition between the Abyssal and Archibenthal IFP's is 74 percent, which represents the most abrupt change along the shelf-to-basin transect. Stenotopic genera compose 83 percent of the Abyssal IFP isopod fauna.

In summary, environmental stability, and especially temperature-related factors, have an important influence on marine isopod distribution. The northwestern Atlantic shelf faunas are associated with warm and seasonally variable temperatures, whereas the Abyssal IFP is associated with water masses that are cold and temperature stable throughout the year. Location of the upper boundary of the Abyssal IFP is apparently regulated by the lower limits of the permanent thermocline and may migrate up or down seasonally as temperature and current circulation patterns vary, as well as emerge into relatively shallow water in high latitudes (Menzies et al., 1973, pp. 249-252).

The greatest degree of faunal change in the northwestern Atlantic occurs between the Shelf and Abyssal IFP's. Only 17 percent of the isopod genera found in the Abyssal IFP extend into the Shelf IFP. In contrast, 63 percent of the northwestern Atlantic abyssal genera are found among the diverse isopod faunas of the Antarctic Faunal Region, where water temperatures range from 0 to 2-4°C (Menzies et al., 1973, p. 201) and remain relatively stable seasonally. The northwestern Atlantic shelf contains only 29 percent of genera in common with the Antarctic Faunal Region.

CONCLUSIONS

Zoogeographic Implications

The habitat and geographic distributional relationships of each depth interval for northwestern Atlantic isopods are shown in Text-figure 11. The shelf faunas as a whole tend to show less than half the habitat restriction that abyssal faunas show, even though few taxa range into the Abyssal IFP (Text-fig. 10). In contrast, shelf faunas are more restricted geographically than abyssal faunas. These and other data discussed above show that the Abyssal IFP of the northwestern Atlantic Realm

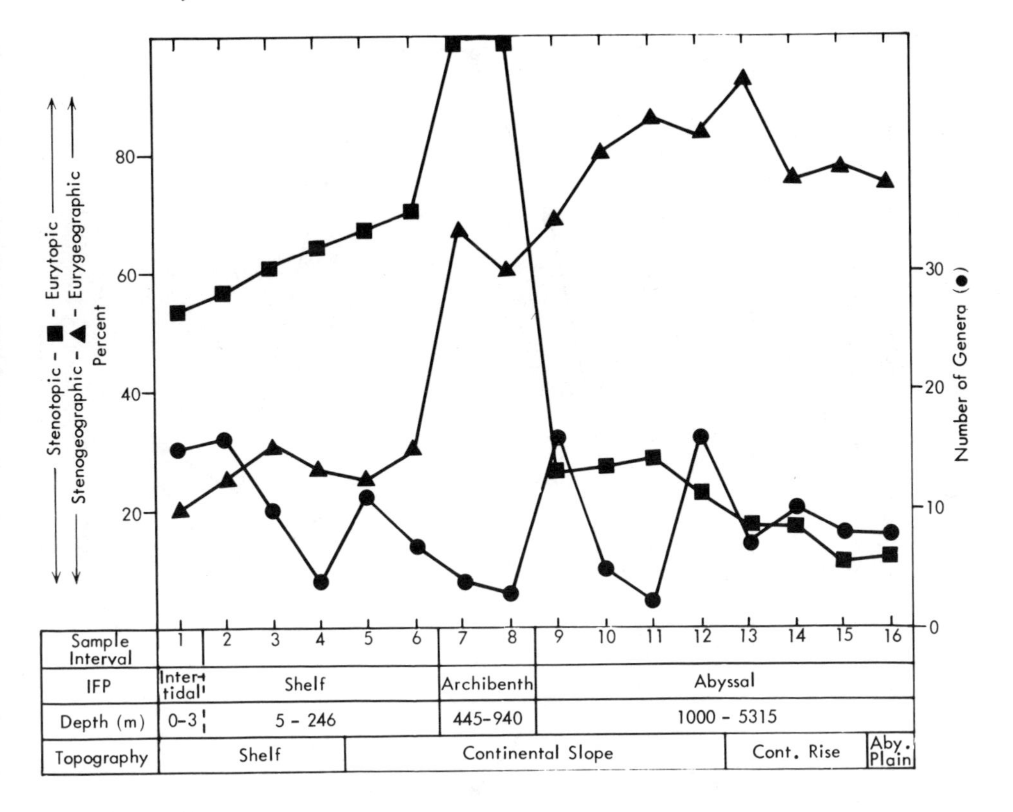

Text-Figure 11
Data on geographic distribution, habitat distribution, and generic richness of Holocene isopod crustacean faunas from the northwestern Atlantic: Solid square, percent tolerant (eurytopic) genera; solid triangle, percent widespread (eurygeographic) genera; solid circle, observed generic richness; IFP, isopod faunal province. See Table 6 for depth range of sample intervals. (Dervied from data given in Menzies et al., 1973, pp. 79-126, Tables 4-2 and 7-8.)

contains a predominantly cold water (cryophilic) and widely distributed (eurygeographic) isopod fauna, in comparison with the environmentally variable, predominantly warm water (thermophilic), and geographically restricted (stenogeographic) faunas of the Shelf IFP. The reversal point of habitat and geographic distribution curves (Text-fig. 11) seems to correspond with the top of the Abyssal IFP and lower limits of the permanent thermocline (see Menzies et al., 1973, fig. 4-1). This relationship has important paleontological implications and provides a basis for deducing probable temperature-related faunal discontinuities in the fossil record, particularly for the early Palezoic where cryophilic or thermophilic characteristics of faunas cannot be directly observed.

Graphs showing the habitat and geographic distributional characteristics of late Franconian and Trempealeauan trilobites are given in Text-figures 12 and 13. Both

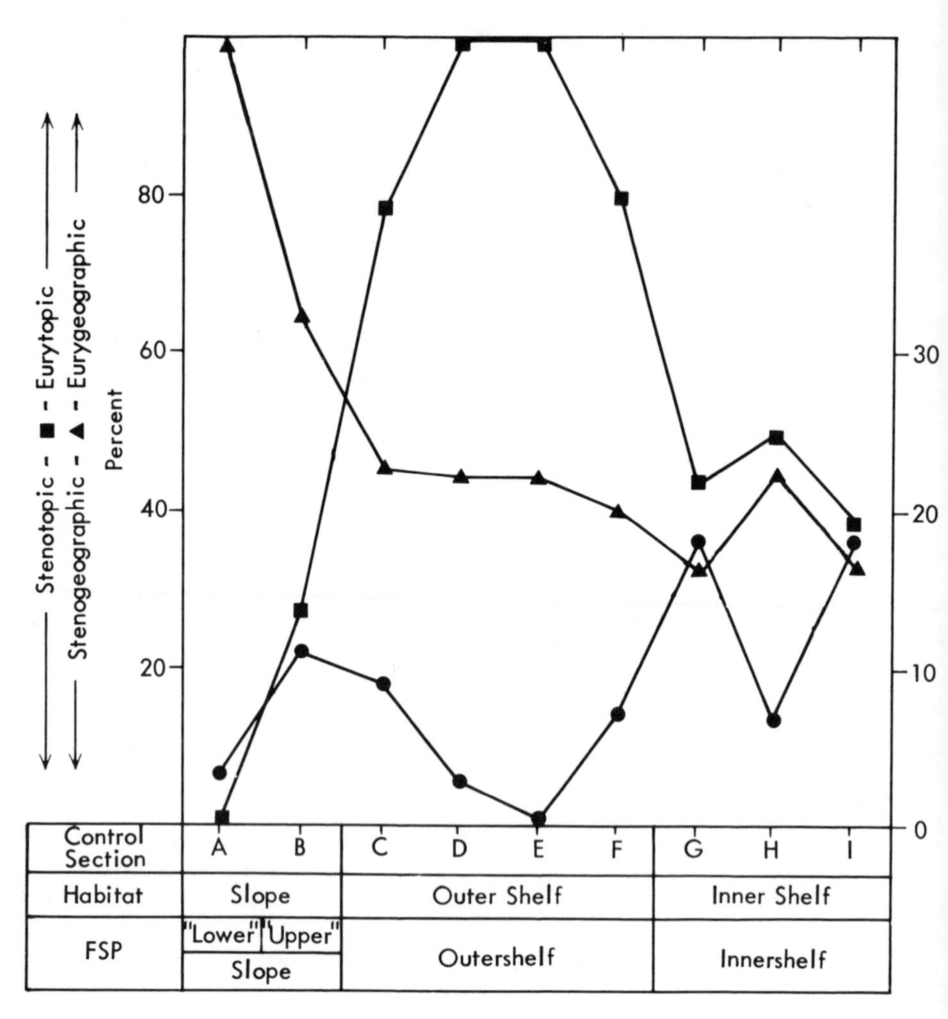

Text-Figure 12
Data on geographic distribution, habitat distribution, and generic richness of late Franconian trilobite faunas from the western United States: Solid square, percent tolerant (eurytopic) genera; solid triangle, percent widespread (eurygeographic) genera; solid circle, observed generic richness, excluding allochthonous occurrences; FSP, trilobite faunal subprovince. Location of faunal control sections is shown in Text-figure 4.

figures show that shelf faunas are generally eurytopic and stenogeographic, whereas slope faunas are stenotopic and eurygeographic. These general patterns are similar to that shown for the northwestern Atlantic isopods, suggesting that temperature barriers and differences in environmental stability may account for strong shelf-to-slope biofacies discontinuities among the Late Cambrian trilobite faunas of the western United States.

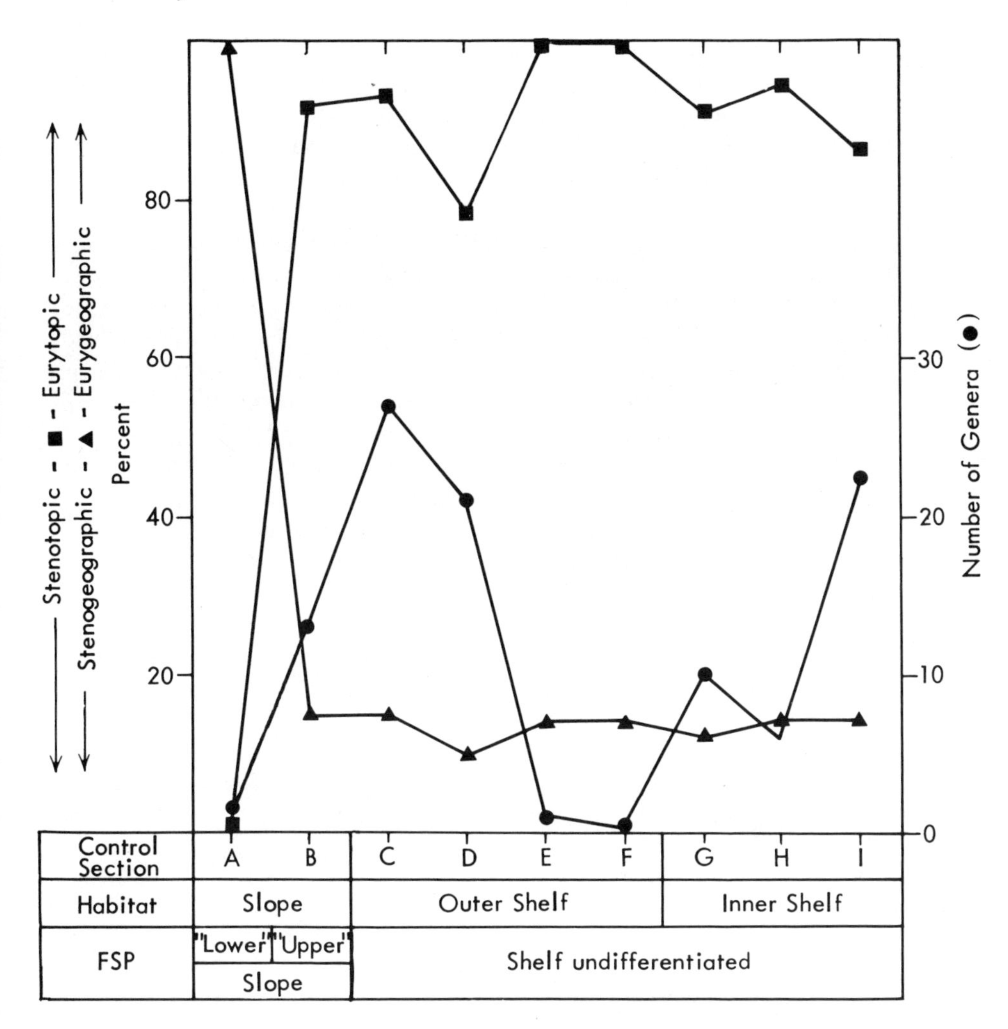

Text-Figure 13
Data on geographic distribution, habitat distribution, and generic richness of Trempealeauan trilobite faunas from the western United States. Symbols same as in Text-figure 12.

Some time-related changes in distributional characteristics of the Late Cambrian trilobites are also evident. During the late Franconian, faunal provincialism was greater across the shelf-to-basin profile than during the Trempealeauan. The late Franconian Innershelf FSP is stenotopic relative to the more oceanward FSP's and to the Trempealeauan Undifferentiated Shelf FSP. Also, the position of the cross-over from eurygeograph-stenotope dominated biofacies shifts from between control sections B (Eureka District) and C (Egan Range) during the late Franconian to between B and A (Hot Creek Range) during the Trempealeauan. These observations are consistent with an interpretation that the shelf was cooler in the late Franconian than during the Trempealeauan. The Trempealeauan probably represented a time of more equitable environmental conditions (a lower temperature gradient) across the shelf. Some possible faunal restriction during the late Franconian resulting from emergence of the Transcontinental Arch cannot, however, be entirely ruled out.

The hypothetical relationships between inferred thermospheric and psychrospheric faunas for the Late Cambrian of the United States are shown in Text-figure 14. The terms "thermosphere" and "psychrosphere" are used in a modified sense after Bruun (1957, p. 641, following Wüst, 1950) for the major shallow and deep temperature divisions of the world oceans. Bruun thought that the boundary between divisions coincided with a major ecological boundary at 10°C. However, more recent data suggest that the boundary is more closely related to colder or latitudinally variable temperatures in the Holocene (Benson, 1972, 1975; Menzies et al., 1973, p. 78). Therefore, the terms are used here in a general sense, without specific temperature or depth connotations, for tropical and other warm marine waters subject to high seasonal temperature instability (thermosphere) and colder, generally deeper or high-latitude waters that are relatively temperature stable throughout the year (psychrosphere).

The inferred differentiation of Late Cambrian world oceans into two thermal realms implies the presence somewhere of a cold surface marine climate that provided a source of psychrospheric waters (see Warren, 1971; Menzies et al., 1973, p. 354). Late Cambrian trilobite occurrences in the Acado-Baltic Faunal Province of Scandinavia, western Europe, Poland, and maritime Canada (Henningsmoen, 1957, pp. 31-57, and references cited therein; Orlowski, 1967, 1968; Tomczykowa,

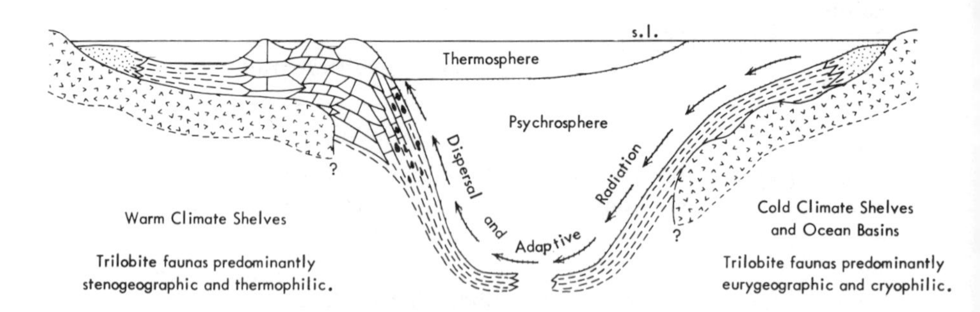

Text-Figure 14
General model of Late Cambrian trilobite dispersal based on presumed marine climatic differentiation. Some examples of warm-climate shelves are North America and eastern Asia. The cold-climate shelf is hypothetical but may be represented by the western Baltic region (see text). Not to scale.

1968a, 1968b) show some features that suggest a cold-climate shelf habitat. Characteristics include (1) trilobite faunas that are dominated by few families (i.e., the Olenidae), (2) relatively low generic richness, (3) common occurrences of high numbers of individuals of dominant species, (4) relatively uniform faunal dispersal throughout the province (Henningsmoen, 1957, p. 52), and (5) stratigraphic sequences composed of fine-grained terrigenous sediment that lack appreciable accumulations of lime mudstone.

Perhaps of further significance is the fact that although olenid trilobites predominate in the Acado-Baltic shelf sites, they commonly occur only in off-shelf sites in the North American Faunal Province. This distributional pattern may be analogous to that shown for the generally cryophilic asellote isopods, a group that predominates at all depths in the Antarctic Region, and in the cold waters below the permanent thermocline in the northwestern Atlantic and southeastern Pacific off Peru (compare Menzies et al., 1973, Figs. 4-11, 5-8, and 7-12). These analogies suggest that more detailed ecological and zoogeographical studies comparing trilobites and living marine crustaceans might prove fruitful.

A marine climate model, based in major part on analogy with marine isopod distribution, explains the observed distributional characteristics of trilobite faunas and associated lithofacies for the Upper Cambrian of the western United States. These conclusions are consistent with earlier ones based on other lines of evidence (see especially Lochman-Balk, 1970, 1971; Stitt, 1971a, 1973, 1975). However, study is needed of other faunal provinces in both space and time to test the validity and general practicability of the model. In addition, independent evidence for ambient temperatures at the time Upper Cambrian sediments were deposited could provide a rigorous test of the hypothesis.

Biostratigraphic Implications

The Late Cambrian North American Faunal Province contains a relatively endemic trilobite fauna developed on a Subtropical to Tropical, shallow-water, broad marine shelf that surrounded a low-relief land area. A warm marine climate is suggested independently by the abundance of lime mudstone and by paleomagnetic data (Smith et al., 1973, text-figs. 13, 21). Analysis of trilobite distribution shows that taxonomic content of biofacies changes most abruptly between the shelf and deeper-water basinal deposits which surrounded the shelf. These biofacies changes are apparently related to water temperature differences that separate faunas typical of the North American Faunal Province from cosmopolitan deeper-water faunas with Asiatic affinities. These facies relationships suggest that, in general, use of nominal biostratigraphic zones based on polymerid trilobites should be restricted to single paleozoogeographic provinces. Attempts to apply the same zonal nomenclature to two or more warm-climate trilobite provinces obscures the biofacies differences between them and implies precision in correlation that exceeds what is justified by the data. This conclusion also applies to stage-level nomenclature, and suggests that attempts to apply the terms Franconian and Trempealeauan stage should be abandoned in those regions that have only a few taxa in common with the North American Faunal Province, or in regions that merely exhibit similar grades of trilobite development.

These conclusions are consistent with those of Valentine (1963) and Berry (1966, pp. 1494-1498), both of whom have previously discussed in a general context the relationship between biostratigraphic and biogeographic concepts.

Sequences of basinal strata can play a key role in biostratigraphic correlation. The demonstration that redeposited shallow-shelf provincial assemblages occur interbedded with autochthonous cosmopolitan assemblages provides a basins for intercontinental correlation between faunally distinct warm-climate provinces (Taylor, 1976). For example, the *Hedinaspis* fauna has been described and illustrated from western China (Troedsson, 1937) and southern China (Lu, 1954). Kobayashi (1967) has reviewed occurrences of the fauna in other parts of Asia. Asian occurrences are associated with the Machari lithofacies (Kobayashi, 1967, pp. 458-466, 478-479; Palmer, 1973, p. 6), dark gray to black carbonaceous shale, "impure" dark-colored thin-bedded limestone, and cherty shale that is thought to have been deposited in "off-shore" open ocean sites (Kobayashi, 1967, pp. 459, 479). The available lithic descriptions, although meager, suggest similarity between the Machari lithofacies and the deep-water deposits of the Hales Limestone that contain the *Hedinaspis* fauna in central Nevada.

The Machari lithofacies contains faunas assigned to the Chiangnan Faunal Province. The Chiangnan Province of eastern Asia borders the Hwangho Faunal Province, which contains a predominantly Asiatic endemic fauna associated with carbonate rocks of a warm shallow-water shelf in southern Asia (Kobayashi, 1967, pp. 475-478). The details of Upper Cambrian correlations between North America and Asia are yet to be worked out. The important point here is that the potential solution of interprovince correlations are best sought in the basinal sediments surrounding carbonate platforms. Biofacies analysis suggests that the reason this is so is that some cosmopolitan trilobite faunas are geographically widespread and environmentally restricted in apparent response to the widespread, cold, and environmentally stable conditions encountered in the deep sea.

Acknowledgments

Discussions or correspondence with Richard H. Benson, Harry E. Cook, Robert R. Hessler, and Robert J. Menzies were helpful in formulating some of the ideas expressed here. James H. Stitt is especially thanked for making available a preprint of his paper on the Ptychaspid Biomere (Stitt, 1975). Thomas Phelan brought to my attention some trilobitiform isopods in the zoological collections of the National Museum of Natural History. Laurie Joseph assisted with compilation of isopod data. I thank the following for reviewing the typescript:

R. H. Benson, H. E. Cook, J. E. Hazel, R. R. Hessler, E. G. Kauffman, R. J. Menzies, A. R. Palmer, J. Pojeta, Jr., R. A. Robison, R. J. Ross, Jr., P. M. Sheehan, and J. H. Stewart.

Julia E. Taylor prepared several versions of the typescript. Fieldwork was supported during 1966 and 1969 by Geological Society of America research grants 1082-66 and 1221-69, and by the U.S. Geological Survey during 1971 to 1973.

This article is dedicated in appreciation to Professor Emeritus J. Stewart Williams of Utah State University.

Aspects of Vertebrate Paleontological Stratigraphy and Geochronology

Donald E. Savage **University of California Museum of Paleontology**

INTRODUCTION

Animals here called *vertebrates* include early bony water-dwelling fish-like forms, shark-like fishes, bony fishes, amphibians, reptiles, mammals, and birds. Although a few soft tissues and molds or casts of soft tissues have been preserved in rocks, the fossilized bony skeletons and hard dental parts of these animals have given us most of our information about the biology of ancient vertebrates. Also, morphologic character analysis of the fossils led to their assignment to taxa, the named species, genera, etc. and these taxa have become the tools for stratigraphic and geochronologic studies.

The record of fossil vertebrates begins in Ordovician time, 500 million years ago, and increases in general stratigraphic-geochronologic significance through the Late Paleozoic, Mesozoic, and Cenozoic as the preserved fossils and fossiliferous rocks become progressively more voluminous and widespread (see Gilluly, 1949). Specific information about the broad aspects of the fossil record of vertebrates may be gained from the very useful summary by Romer (1966, pp. 311-346). Vertebrate paleontologists who have specialized in studies on Late Mesozoic and Cenozoic fossils, particularly in continental sedimentary sequences where stratigraphic tie-ins with fossiliferous marine strata are nonexistent or tenuous, have been deeply concerned with the concepts, methods, and problems of zonation, dating, and correlation, and with the general relationships between vertebrate paleontology and stratigraphy. The excellent review by Tedford (1970) is essential reading for those who wish to attain historical perspective on the development of principles and practices in mammalian stratigraphy and geochronology in North America. The work of C. W. Hibbard in the Late Cenozoic of the southern Great Plains region of the United States, published in a 30-year series of papers beginning around 1940, serves as a standard for meticulous collecting and analysis of the total (paleo-) organic content of strata. Hibbard's methods led to synthesis of data not only from the vertebrate fossils, but also from the other fossil organisms that may be associated.

During the past decade an explosive increase of discoveries of fossil terrestrial vertebrates associated with marine fossils in coastal districts has stimulated vertebrate paleontologists to a greater awareness of classic concepts in stratigraphy, concepts that were first exemplified by data from marine strata. There is nothing unique in the processes of using fossil vertebrates for stratigraphic-geochronologic conclusions: *vertebrate paleontologic stratigraphy is governed by the same principles that apply to marine microorganism and invertebrate stratigraphy.*

Vertebrate paleontologists are forced to work with more geographically restricted fossiliferous rocks and with more sporadically distributed and generally smaller samples of fossils than are invertebrate paleontologists. On the other hand, vertebrate paleontologists work with rapidly evolving, rapidly dispersing, complex organisms, which compensates strikingly for the sporadic nature of the fossil vertebrate record.

An integration of data and disciplines is necessary for optimum development of stratigraphic-geochronologic controls. It has been maintained that little is to be gained by trying to construct a vertebrate (or other) zonation in a stratal succession where a microorganism or invertebrate zonation has been established. Nevertheless, the stratigraphies of all areas of the earth, even those calibrated by foraminiferal or ammonite trellises, have not yet reached the level of refinement needed for answering many questions pertaining to short-duration evolutionary or geologic events. Independently derived cross-checks from the fossil successions of various organisms have the clear potential for giving us less biased and ultimately more refined paleontostratigraphic controls. Thus analysis and definition from *one* specialty as well as collation and definitions from *all* specialties ought to be continued simultaneously, even though each procedure strives for the same goal in stratigraphy and geochronology.

One must remember, also, that there are many districts on continents and islands where large bodies of nonmarine strata are exposed and in which only a greatly restricted sample of fossils can be obtained. These districts may be important to understanding the dynamics of earth and life history. In such districts the vertebrate or plant paleontologic stratigrapher must frequently develop an independent provincial zonation with the hope of corroboration or refinement eventually from geochronometric datings and from interregional correlations.

PRINCIPLES

Character and Use of the Fossil Record

The fossil record is known to be an inadequate depiction of total life of the past. Durham (1967) estimated that only 1 percent of the number of Phanerozoic species is known, and Raup (1972) has detailed many of the reasons for biases in the samples of fossils. Is the admittedly poor general record of ancient life worth the conclusions it has fostered? Alan Shaw (1964, p. 103) said,

> Preoccupation with the unattainable is a stultifying approach to any problem. Practical paleontology cannot be concerned with any of the fossils we *cannot* find .... Geologically, we can only be interested in finding the total stratigraphic range through which a species *is preserved.* While the life and death of the millions of unrepresented individuals is of theoretical interest, we cannot gain practically useful information from them.

Eldredge and Gould (1972, p. 96) are even less concerned with the sample inadequacies recognized by Durham and Raup:

> Many breaks in the fossil record are real; they express the way in which evolution occurs, not the fragments of an imperfect record. The sharp break in a local column accurately records what happened in that area through time. Acceptance of this point would release us from a self-imposed status of inferiority among the evolutionary sciences. The paleontologist's gut-reaction is to view almost any anomaly as an artifact imposed by our institutional millstone—an imperfect fossil record.

Clearly, Eldredge and Gould and Shaw were combating sophistic stagnation. They were urging paleontologists to exercise cerebrally with the sample available, to hypothesize, to construct "paleobiological models," and the like. The practicality of Shaw's declaration that unfound fossils are of no practical concern is questionable, however. Insofar as the fossil vertebrate record from the Late Mesozoic and Cenozoic is concerned, the samples used by earlier workers are being improved to an astounding degree. Hibbard (1949), McKenna (1962), D. E. Russell (1964), W. Kühne (1971), Lillegraven (1969), Rensberger (1971), Lindsay (1972), and many other vertebrate paleontologists have been demonstrating convincingly that improved and intensive collecting techniques can produce divers organisms in the death assemblage with ancient vertebrates—from strata that at first may appear to be almost nonfossiliferous—and the complete assemblage may have utility in evolutionary, ecological, and chronological interpretations. For example, in the 1961-1962 field seasons, D. E. Russell and his colleagues doubled the number of taxa of fossil vertebrates known from the lower Eocene strata of the Paris Basin, an area frequently claimed to be one of the most thoroughly studied in the world.

Better syntheses of facts and development of conclusions supported by modern concepts in biology, ecology, and unifying geologic theory are the accomplishments that all aspire to attain. But all too often, in the compulsive drive toward erudition, we may become little more than article synthesizers and book reviewers, slighting the alpha phase of paleontologic stratigraphy and paleontologic geochronology: meticulous collecting, preparing, curating, collating field data, section measuring, mapping, and the like. *Fieldwork methodology in vertebrate paleontologic stratigraphy has not yet reached maturity.* There is an enormous amount of fieldwork yet to be accomplished.

Paleontological Discipline in Stratigraphy and Geochronology

Two natural processes of past and present are, in gross, unidirectional and nonrepetitive with respect to flow of time. Each process has evidences, thus documentation, in the rocks of the earth's crust. These two processes are evolution among organisms and progressive chemical-physical alteration and marking of certain substances. Metaphorically, organic evolution is a chain reaction, and the larger links in this chain are witnessed as a sequence of individually peculiar platforms of morphologic development. The preservable components of each platform may become fossils, which therefore give record for the platform. Paleontologists and other students of evolution have observed that, although evolution progressed in a myriad of directions, it never followed a circular path. From experience we judge that the odds are overwhelmingly against a brontosaur stage in reptilian evolutionary radiation or an eohippus stage in the phylogeny of horses and tapirs ever emerging again on this planet. Thus *the superpositionally controlled fossil record of continual and nonrepetitive developmental change among organisms is the sequence of unique events that has been primarily responsible for the conceiving and constructing of our earth-life calendar = geologic time table, and most of our more detailed zonations as well.* This generalization is valid even though we must admit that the rhythm and flow of evolution has been stagnated for thousands or even millions of years in certain lineages with respect to detectable change in preservable parts of some organisms. Fossils, then, gave our forebears the concept of quantum changes in life through time.

Phylogeny and Stage of Evolution

Fossils, stratigraphically arranged, provide historical documentation of the course of morphological evolution of the hard parts and, indirectly, soft parts in a lineage of organisms; this is a paleontological phylogeny. Even the best paleontological phylogenies are probably only approximations to the actual biologic-genetic story. Such phylogenies inculcated us with the notion of a continuum of almost imperceptible developmental changes in lineages through time, which might be recorded by a comparably gradual stratigraphic-paleontologic succession if preservation were complete. This is the concept of phyletic gradualism. Eldredge and Gould (1972), however, question this concept and offer an alternative hypothesis. They assert that, with phylogeny of species, and following the theory of allopatric speciation, the phenomenon of speciation may be peripherally localized with respect to areas of abundance and areas of most frequent fossil preservation of organisms within a lineage. Thus, according to these authors, breaks or gaps in the fossil record in areas of abundant fossils are to be expected, even though sedimentation there may have been essentially continuous. Schaeffer et al. (1972), Cracraft (1973), and others believe that one can approach actuality more closely by using cladistic analyses in interpretation of phylogeny. The phylogeny, by whatever means conceived, is still a powerful tool for use by vertebrate paleontologists in geochronological assignments. We assert empirically that a platform of evolutionary development—the *stage of evolution*—usually signified by a particular genus or species, may be used as indicator for a discrete segment in a time scale. Fossils, therefore, and the taxonomic names we apply to them, remain our indexes for saying "older than," "younger than," or "contemporaneous with."

Although my colleagues might not admit it, vertebrate paleontostratigraphers forge ahead in their phyletic and stratigraphic-geochronologic interpretations with a simplistic dictum: *lacking evidence to the contrary, organisms appearing most alike are most closely related, and organisms which are alike lived during the same geochronological interval, no matter how extensive their geographic ranges.* Every paleontologist realizes immediately that this dictum opens Pandora's box for an explosion of diverse qualifications and questions. T. H. Huxley (1862) presented the phenomenon of homotaxy as an antidote for such paleontologic pragmatism. Nevertheless, index taxa have been and continue to be used in the manner prescribed by the dictum for preliminary dating and correlation. Nowadays, vertebrate paleontologists are expanding their paleontostratigraphic base by determination and the plotting of local and total stratigraphic ranges for each consensus-recognized species and genus. Examples of these procedures are given later. The collation and comparing of stratigraphic ranges of species and genera and the use of selected clusters of species ranges for zonation and correlation are the predictable sequel to these modern procedures.

Premises of the Paleontologic Stratigrapher Who Works with Fossil Mammals

The fossil-mammal worker accepts that many mammals, marine or nonmarine, contributed fossils which are admirable tools for paleontologic stratigraphy and geochronology, and especially for age-magnitude correlations from continent to continent. Reasons for this belief may be summarized under three headings:

1. *Many genera of mammals* (consensus) *and probably many species* (my opinion) *have had transoceanic and transworld geographic distribution in the northern hemisphere during the Mesozoic and Cenozoic.* Fossils of these taxa may be found around the world and are thus available for geochronologic correlation. Among the extant species of Carnivora and Cetacea, *Felis (Puma) concolor* is found from British Columbia to Patagonia, *Mustela erminea,* the ermine, and *Gulo gulo*, the wolverine, are circumboreal species. Even the tiny *Reithrodontomys*,

harvest mouse, ranges from southwestern Canada to Colombia and Ecuador, and its species, *R. megalotis*, extends from southwestern Canada to Oaxaca, Mexico. In the marine sphere, *Physeter catodon*, sperm whale, is found in tropical to polar water; *Orcinus orca*, killer whale, lives in all oceans. *Delphinus delphis*, common dolphin, is worldwide in warm and temperate seas, but is found also at times in cooler seas. *Zalophus californianus*, California sea lion, ranges from the west coast of North America to the Galapagos and into the southern part of the Sea of Japan. These range data are from Walker (1968). Three of the 13 genera known from the Late Jurassic Purbeck Formation of southern England are known also from the Morrison Formation at Como Bluff, Wyoming. Thirty to thirty-five percent of the 140 or more nonvolant nonmarine mammalian genera of the North American Pleistocene are also known from Pleistocene rocks of greater Europe, and a comparable percentage is known from Pleistocene rocks in eastern Asia. An almost incredible 50 to 60 percent of the 50 genera of mammals from the Sparnacian beds (Lower Eocene) of the Paris Basin, France, are found also in the Wasatch strata of Wyoming. Thus genera of land mammals are potentially a valuable asset to Mesozoic and Cenozoic correlations between continents of the northern hemisphere. The continental areas of the southern hemisphere, however, have been partly (South America) or completely (Australia) unavailable to land-mammal dispersal through much of the Cenozoic, and fossils of marine organisms as well as other means of correlation are being employed for these areas.

2. *The taxa of mammals used in zonation and correlation have restricted stratigraphic and geochronologic ranges.* Text-figure 1 shows ranges of a spectrum of mammalian genera, including some with longest known ranges as well as some of the most frequently used genera with shortest known ranges. I have concentrated on comparing genera of transoceanic geographic range, and the year dates are based on general radiometric tie-in to approximate epoch boundaries (e.g., the Cretaceous-Paleocene boundary is placed at about 65 megennia BP). Many fossil-mammal specialists will hold that I have overextended the time range on some of these genera. There are insurmountable biological, genetic, and taxonomic uncertainties involved with any comparison between taxa within or between classes of organisms, but Text-figure 1 supports the assertion that mammalian genera are relatively short ranging in time. This is not a new idea. Sir Charles Lyell (1833, p. 253) noted, "We have more than once adverted to the fact that extinct mammalia are often found associated with assemblages of *recent* shells, a fact from which we have inferred the inferior duration of species in mammalia as compared to the testacea...." Genera are compared simply because there is agreement among vertebrate paleontologists as to identification only at this level of the taxonomic hierarchy. Although Text-figure 1 may persuade that a recognized genus was extant for a relatively short time and is thus most useful for preliminary interregional and intercontinental correlation of geochronologic intervals in the magnitude of about 2 to 6 million years, it must be remembered that today joint occurrence and zones of overlapped stratigraphic ranges of taxa are being used to define smaller subdivisions of the geochronologic scale. For

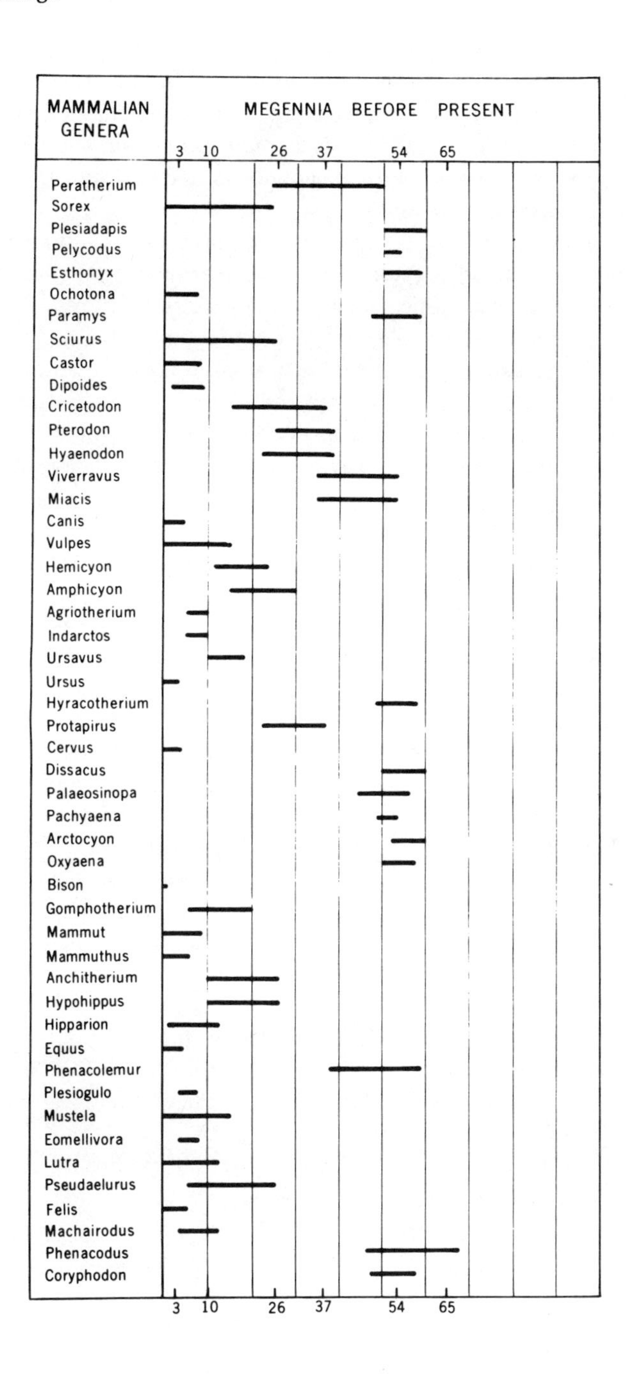

Text-Figure 1

example, among the Foraminifera an interval defined and recognized by the potential joint occurrence of fossils of *Cyclammina, Miliola, Alveolina*, and *Siphogenerina*, and among the Mammalia, an interval defined and recognized by the joint occurrence of *Hypolagus, Agriotherium, Hipparion*, and *Machairodus* is of less duration than the geochronologic interval indexed by the occurrence of *Miliola* or *Hypolagus*, respectively.

Intrabasin and interbasin zonation by concurrent ranges of species of vertebrates seldom has been attempted because of disagreement on the morphologic scope of the vertebrate paleospecies. Such zonations have been, of course, the special contribution of microfossil paleontology and, in certain instances, megafossil paleontology (mollusks and graptolites, for example). Data on stratigraphic and chronologic ranges of fossil-vertebrate species whose geographic ranges are greater than a single basin of deposition or a single faunal province are embryonic, but this type of synthesis is now being attempted and the published results will be most useful.

3. *The genera and species of mammals disperse rapidly*; they walk, run, climb, swim, and fly. They are free moving. Relative vagility of different classes of organisms is as difficult to assess as relative duration in geologic time. In asserting the mobility of marine invertebrates, Woodford (1965, p. 96) said, "Modern marine clams and snails have spread through scores or even hundreds of miles of shallow water in a few decades (Elton, 1958). If the ammonites got around as quickly as the clams and snails do now, which seems likely, the spread of a new European zonal fauna must have been geologically instantaneous." Erle Kauffman has reemphasized to me (written communication, 1974) that each generation of mollusks, echinoderms, worms, etc., in the marine realm normally disperses hundreds or thousands of miles by planktonic larval drift. I can only affirm that certain populations and/or individuals of marine and land mammals and birds are known to travel hundreds, even thousands, of miles in migration over their home range and in permanent extension of their geographic range. Some species of mammals evidently have dispersed across a continental or ocean-basin surface in a few thousand years.

Tyndale-Biscoe (1973, pp. 225-226) summarizes the dispersal record of the opossum, *Didelphis marsupialis*, which is known from the United States, Mexico, Central America, and into South America. This species was introduced in California during the period 1870 until 1915. It spread north into British Columbia and eastward to elevation of 1,500 m by 1958 and is now established along the Pacific Coast from southern Canada to northern Mexico. It was recorded to have spread northward at a rate of 50 km/year during a 26-year interval, although the influence of man's activities on this dispersal is not completely certain. Thus, as a conservative projection and if there were no ecological barriers or other restraints, *D. marsupialis* might extend its range 20,000 km (to Europe and Africa via Asia) in 1,000 to 2,000 years. This would be an insignificant interval as compared to presently recognized durations of chronozones or the like. Such evidences strongly suggest that a species of placental mammal, known from far-flung provinces of the world, dispersed to these provinces at speeds too fast to be calibrated by the present-day methods of geochronology.

Kurtén (1957) compiled significant information on the rate of dispersal of mammalian species in the faunas of the Pliocene, Pleistocene, and Recent of Europe and China. He concluded (p. 217) from the data amassed that "an unchecked spread of some 1,000 kilometers in a century would seem a moderate estimate for most larger land mammals." Thus, following his criteria and assuming no special restraints, a mammalian species might extend its geographic range from the Bering Straits to western Europe in about 1,000 years. Combined evidence indicates that all conceivable barriers to mammal dispersal over vast distances on land were less effective during the first 175 million years of mammalian history than during the Pleistocene and Recent. Thus we can agree with Kurtén and other workers that the Tertiary species of mammals, and perhaps the Mesozoic species, dispersed as quickly or more quickly and were as far-ranging as Pleistocene and Recent species.

The preceding premises of the mammalian paleontologist comprise two of the three main requirements for the traditional index fossil—actually index taxon: it should be wide-ranging geographically and it should have restricted stratigraphic and temporal range. These two attributes indicate that populations representing the taxon dispersed rapidly. The third traditional requirement, that the index fossil should be usefully abundant throughout its geographic range, unfortunately is not often fulfilled by the fossil sample of mammals. But as stated previously, great improvements in collecting techniques are underway, and the recent increase in number and diversity of samples has been remarkable.

METHODS AND SOME RESULTS

I am happy to concur with Tedford (1970) in urging that vertebrate paleontologic stratigraphers should now push forward energetically into an era of principled and meticulous collection, section measurement, collation, and synthesis of the stratigraphic, sedimentologic, and paleontologic facts in their respective areas and specialties. The results obtained from such an approach already make it evident that markedly improved vertebrate paleontostratigraphy is possible. As an endeavor in this direction, J. Howard Hutchison and Barbara T. Waters and I are undertaking a paleontologic-stratigraphic and paleoecologic study of a 3,000-ft continental Lower Eocene succession, Wasatch and Green River formations, near Bitter Creek in the northwestern segment of the Washakie Basin, southern Wyoming. This succession includes diverse sedimentary rock types in a series of imbricating tongues of the two formations. Five seasons of work here have been organized as follows:

1. Prospecting by the usual bed-by-bed surface crawling, and anthill searching for all kinds of fossils.

2. Standard quarrying operations for salvageable fossils exposed by weathering, as well as quarrying in sedimentary pockets where obvious concentrations of fossils are entombed.

3. Not only are the concentrations of fossils quarried and collected by the established methods, but also the fossiliferous strata in these pockets, some of which are only a few centimeters thick, are being collected, dried, kerosened, soaked in water, sieved through fine-mesh screens by gentle spraying, dried, and picked for fossil content using binocular microscopes. These procedures obviously must be followed by hundreds of people-hours of laborious preparing, identifying, mounting for study, organizing, cataloging, tagging, and systematically conserving the collected specimens in the museum during non-fieldwork intervals of the year (Waters and Hutchison, 1975).

4. In conjunction with steps 1, 2, and 3, we are measuring many control sections through the stratigraphic succession in order to be provided with accurate spatial control, vertically (stratigraphically) as well as laterally, for future detailed evaluation of lithofacies and paleontologic facies.

5. Oriented samples of the fine-grained clastic strata, suitable for paleomagnetic studies, must be collected also.

In general, the above operations are well-known standard procedures for all paleontologic stratigraphers, and our techniques are innovative only in minor details and in aspects of the discipline of performance. The result of the alpha-stratigraphy outlined is our knowledge that there are more than 40 important and successively superposed fossiliferous stratigraphic intervals (zonules) in this 3,000-ft succession, representing about 5 million years of earth and life history. As is customary with fossil vertebrates, some of these zonules extend for only a few feet laterally; others, however, have been traced for several miles. Many of the zonules are yielding voluminous samples of populations of divers nonmarine vertebrates (amphibians, fishes, reptiles, birds, and mammals) associated with other fossils, such as pollen, charophytes, ostracodes, and mollusks. The Bitter Creek section is producing microvertebrate taxa that have not been found in other Lower Eocene collecting areas of North America, and it offers a stratigraphic continuum in one district, which, when properly sampled, can provide a paleontologic zonation in one faunal province, based on large mammals, small mammals, turtles, and associated nonmammalian vertebrates. Such a zonation has not been derived from the previously worked nonmarine districts of the world where correlative rocks and fauna are known. This kind of study is not accomplished quickly. Those who feel pressured to publish quickly should not undertake a project of this sort.

Lindsay (1972) has published a detailed stratigraphic ordering of a large sample of small-mammal taxa within the Barstow Formation of southeastern California. He attempted to obtain a bed-by-bed bulk sample, continuously distributed through the columnar section in various transects through the formation, employing the general methods that we are using at Bitter Creek, described above. He was hampered, however, by the sporadic concentration of fossils, the ubiquitous problem for vertebrate paleontostratigraphers. As a first step in synthesis of a provincial paleontologic stratigraphy and geochronology, he recognized four *assemblage zones* in the formation, each bearing the name of a conspicuous fossil rodent species. He chose to use the term assemblage zone because (p. 6),

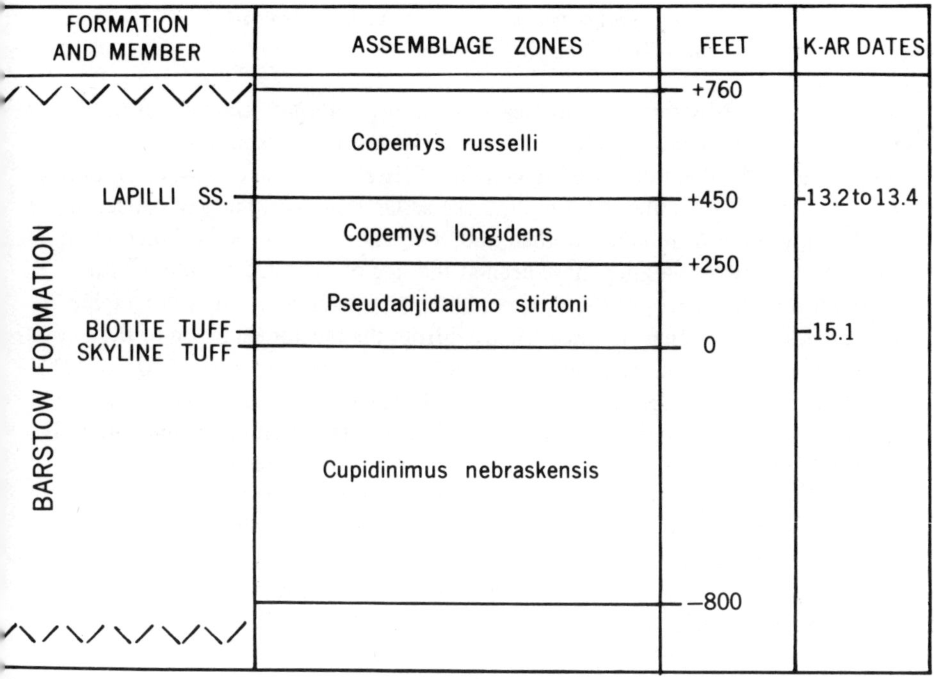

FORMATION AND MEMBER	ASSEMBLAGE ZONES	FEET	K-AR DATES
		+760	
	Copemys russelli		
LAPILLI SS.		+450	13.2 to 13.4
	Copemys longidens		
		+250	
	Pseudadjidaumo stirtoni		
BIOTITE TUFF			15.1
SKYLINE TUFF		0	
BARSTOW FORMATION	Cupidinimus nebraskensis		
		–800	

Text-Figure 2

Each ... is comprised of a distinct assemblage of fossils and, according to present sampling, is stratigraphically discrete. Characterization of each assemblage zone is based upon stratigraphic range of individual taxa .... These are descriptive biostratigraphic units and may someday be expanded into interpretive Oppelian zones by further sampling, e.g., comparable sampling in the Mojave Desert or elsewhere in southern California.

A composite diagram, showing his arrangement of zones, is given in Text-figure 2. The *Copemys longidens* assemblage zone, for example, is characterized by the assemblage *Hypolagus parviplicatus, Perognathoides halli*, and *Copemys longidens.* Its fauna comprises the following:

Lanthanotherium sawini
Paradomnina cf. *relictus*
Limnoecus tricuspis
Chiroptera
Hypolagus parviplicatus
Tamias ateles
Miospermophilus sp.
Sciuropterus jamesi
Sciuropterus minimus
Perognathoides minutus
P. furlongi
Cupidinimus nebraskensis
Perognathoides halli
Mojavemys lophatus
Parapliosaccomys sp.
Copemys pagei
C. tenuis
C. longidens

Monosaulax pansus	*C. barstowensis*
Mookomys cf. *formicorum*	*C. russelli*

The stratigraphically restricted species is *Mookomys subtilis.* The bulk of the *Copemys longidens* and *Pseudadjidaumo stirtoni* assemblage zones is included in an interval dated by the potassium-argon method (Evernden et al., 1964; G. H. Curtis, pers. comm., 1974) at about 1.7 million years. On this basis, Lindsay's zones in the Barstow are probably no less than 800,000 years in duration. Sedimentation through these two zones may have been at the rate of only 235 ft/million years.

In plotting the ranges of the species and the zones and in noting stratigraphic discreteness of the zones, Lindsay departed from the recommendation of the Code of the American Commission on Stratigraphic Nomenclature (1961-1970, Art. 21), which defines *assemblage zone* as a body of strata characterized by a certain assemblage of fossils *without regard to their ranges.* His assemblage zones are well characterized as concurrent-range zones, although he chose not to use this latter designation, evidently because of the restricted geographic extent of his zonation.

J. M. Rensberger and associates have been intensively re-collecting paleontological and stratigraphical information from the John Day Formation in north-central Oregon through a 10-year period, endeavoring to accumulate all data that may be obtainable from each stratum in the succession. Fisher and Rensberger (1972) defined three concurrent-range zones in the John Day based on stratigraphic ranges of 10 genera of mammals and extended these zones to the Great Plains area. For example,

> The base of the *Meniscomys* concurrent-range zone is defined by the first appearance of *Meniscomys* and *Allomys* .... The upper boundary of this zone is defined by the upper limit of the range of *Pleurolicus* and the lower limit of the range of *Entoptychus*. The zone is also characterized by the first appearances of *Palaeocastor*, in its upper portion, and *Promerycochoerus* near its base (Fisher and Rensberger, 1972, p. 23).

The tie-ins that Fisher and Rensberger had between the strata they studied and potassium-argon dates suggest that one of their concurrent-range zones had a duration of 2+ million years. This is not a refined zonation as compared to the asserted durations of recognizable zones in the marine Cretaceous of the Rocky Mountain region (Kauffman, 1970); but it is a step in the right direction, and there is promise of much greater chronostratigraphic resolution when zonation based on paleospecies can be effected.

A final example of current methods and results in paleontological chronostratigraphy based upon the vertebrate record is a circumglobal mammalian zonation of the northern hemisphere (see Text-fig. 3). By use of concurrent stratigraphic ranges of land-mammal genera, we may recognize the *Wasatchian stage* in North America, Europe, and China. Wasatchian was first designated a North American Provincial Age by Wood et al. (1941), but can now be substantiated as a stage in the sense of the Code of the American Commission of Stratigraphic Nomenclature and extended transglobally. Subdivision of the Wasatchian by means of analysis of the stratigraphic ranges of land-mammal species is feasible also in North America and in Europe, but it appears that such subdivisions may never be worldwide in application. For example and as shown in Text-figure 3, a *Lambdotherium* concurrent-range zone (equaling approximately the duration of the Lostcabinian land-mammal subage of informal usage) can be recognized easily in North America, but a subdivision of different magnitude will probably be more useful in the European section.

The lower limit of the Wasatchian stage is defined by the lowest joint occurrence of a majority of the following genera: *Hyracotherium, Coryphodon, Haplomylus, Pelycodus, Apatemys, Didelphodus, Palaeosinopa, Prototomus, Viverravus, Miacis, Pachyaena, Hyopsodus, Homogalax*, and *Diacodexis.* The upper limit of the Wasatchian is subjacent to the lowest joint occurrence of *Anaptomorphus, Smilodectes, Uintanius, Washakius, Hemiacodon, Mesonyx, Uintatherium, Palaeosyops, Orohippus, Helaletes, Trogosus, Leptotomus*, and *Homacodon* (North American recognition), and is the highest occurrence of *Hyracotherium, Esthonyx, Palaeosinopa, Meniscotherium*, and *Coryphodon* (North American and Eurasian recognition). Each of these suites of mammalian genera represents a wide spectrum of ecological adaptation on land. Recognition of the upper limit intercontinentally is based on the recorded highest stratigraphic placement of the genera characterizing the Wasatchian, inasmuch as the taxa superjacent to this stage in Eurasia are quite dissimilar to those superjacent to it in North America. Land dispersal routes were evidently flooded at about the end of the Wasatchian.

The seemingly abrupt appearance of so many genera at 54 megennia BP, for example (Text-figure 3), is an artifact of lack of refined tie-in to radiometric dates. These appearances could have been scattered artistically through the interval 54.5 to 53.5, for example, and might thus have given a more persuasive presentation. It is axiomatic to remind ourselves, also, that these respective *lowests* and *highests* of genera *do not demark planes*; the boundaries thus defined are most likely to be fuzzy and must be picked probably in a paleontostratigraphic interval of disconcerting thickness!

The lower limit of the *Lambdotherium* concurrent-range zone is defined by the lowest stratigraphic location of a majority of the following: *Shoshonius, Patriofelis, Hyrachyus, Eotitanops, Lambdotherium, Bathyopsis*, and *Antiacodon.* The upper limit coincides with the upper limit of the Wasatchian.

Several of my interpretations, potentially, give bias to the depiction of the Wasatchian stage:

1. The Sparnacian beds, typified in the outskirts of Epernay, Paris Basin, France, and employed as a stage name for earliest Eocene by many, represent only a segment (middle part) of the Wasatchian in my opinion. For this reason, I have chosen to use Wasatchian rather than Sparnacian for the transglobal name.
2. The sudden appearance of many genera in Europe at the beginning of the time represented by the Sparnacian beds (about 52 million years BP) is the result of a dispersal flood from North America.

If these interpretations are validated by further data, we will be assured that

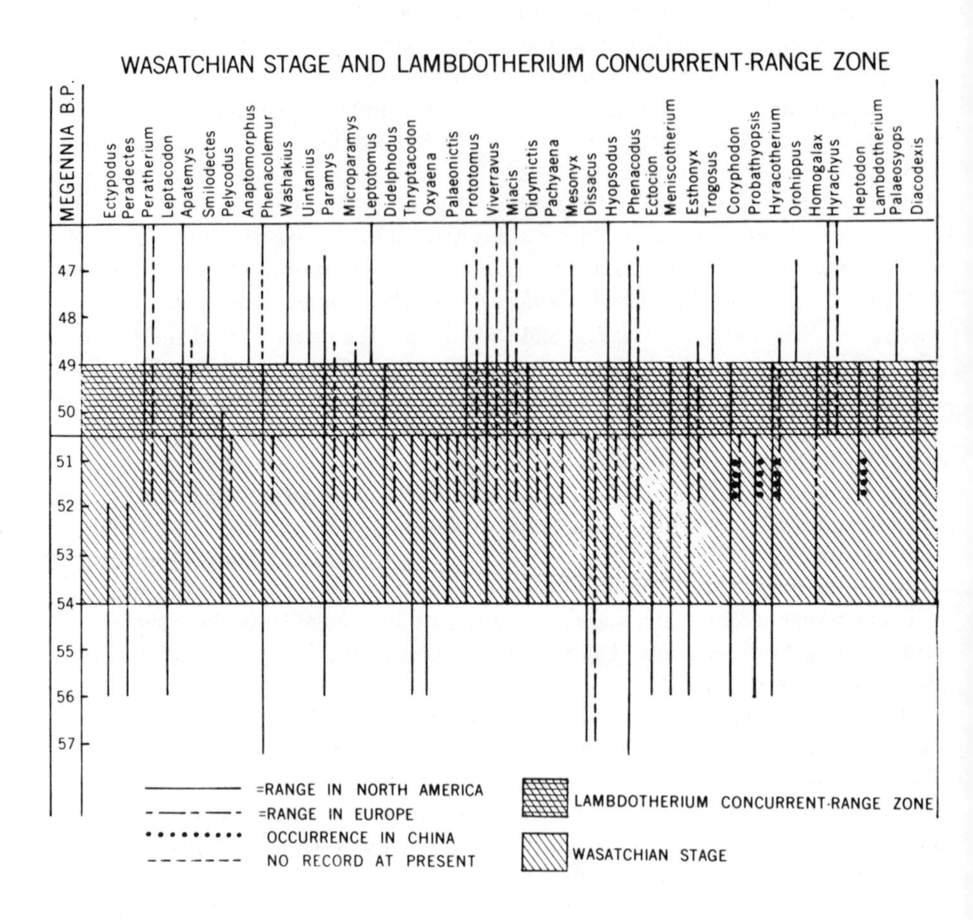

Text-Figure 3 *(above and right)*

Wasatchian had a duration of about 5 million years and that the *Lambdotherium* concurrent-range zone had a duration of less than 2 million years. My presentation of the Wasatchian stage and the *Lambdotherium* concurrent-range zone is little more than a reasserting of long-established correlations based on land mammals, using the terms and methodology of paleontological stratigraphers who work with fossil microorganisms and with invertebrates. For previous consideration of these datings and correlations, see Wood et al. (1941), Van Houten (1945), Robinson (1966), Jepsen (1963), Jepsen and Woodburne (1969), Tedford (1970), and McGrew and Sullivan (1970).

WASATCHIAN STAGE AND LAMBDOTHERIUM CONCURRENT-RANGE ZONE

FORMATIONS AND MEMBERS				STAGES, FORMATIONS AND MEMBERS			
COLORADO	NEW MEXICO	CENTRAL AND NORTHERN WYOMING	SOUTH-CENTRAL WYOMING	FRANCE	ENGLAND	BELGIUM	CHINA
HUERFANO	GALISTEO		CATHEDRAL BLUFFS	CALCAIRE GROSSIER AND LUTETIAN LAKE BEDS			
HUERFANO	GALISTEO; SAN JOSE	LOST CABIN; WILLWOOD; WIND RIVER	WILKINS PEAK; TIPTON; NILAND	TEREDINE SANDS	WOOLWICH-BLACKHEATH BEDS; LONDON CLAY	YPRESIAN BEDS	FORMATIONS UNPUBLISHED
"HIAWATHA"	SAN JOSE	WILLWOOD; WIND RIVER; INDIAN MEADOWS; LYSITE	LUMAN; MAIN BODY OF WASATCH	SPARNACIAN BEDS	WOOLWICH-BLACKHEATH BEDS	YPRESIAN BEDS; LANDENIAN BEDS	FORMATIONS UNPUBLISHED
TIFFANY BEDS		SILVER COULEE; FORT UNION	FORT UNION		THANET SAND	LANDENIAN BEDS	

SUMMARY

Vertebrate paleontological stratigraphy is governed by the same principles governing the procedures and interpretations of paleontological stratigraphers who work with marine microorganisms and marine invertebrates. With respect to geographic range, vertebrate paleontologists are forced to base conclusions on relatively restricted fossiliferous rocks and more sporadically distributed and generally smaller samples of fossils, but they work with rapidly evolving, rapidly dispersing, complex organisms. Current research in vertebrate paleontological stratigraphy trends toward more intensive bulk sampling of potentially fossiliferous stratal successions and more detailed analyses and plotting of ranges of genera and species. These procedures constitute the necessary alpha phase of refined zonation.

Land-vertebrate fossils, along with many plants and a few representatives of other phyla of organisms, may be the only sample obtainable from certain geologically important paleobasins of deposition, and thus form the basis for relative datings, zonations, and correlations in those areas. In other areas, particularly in districts where marine fossils are abundant and where geophysical-geochemical data for aging and correlation are available, the fossil vertebrate record may be a minor but nevertheless valuable contributor to integrated systems of paleontologic stratigraphy and geochronology. Greater refinement in vertebrate paleontostratigraphy must come from consensus definition and utilization of paleospecies.

Studies in Benthic Organisms

Foraminiferal Zonation of the Lower Carboniferous: Methods and Stratigraphic Implications

Bernard Mamet **Université de Montréal**

INTRODUCTION

Although fusulines have been widely used in Paleozoic biostratigraphy, they represent only 20 percent of the 400 known Paleozoic foraminiferal genera. The remaining diverse microfauna is composed of siliceous agglutinated forms (e.g., saccamminids, ammodiscids, textulariids, hyperamminids) and of calcareous secreted forms [e.g., semitextulariids, "endothyrids" (s. 1.), nodosariids, and nodosinellids]. There is therefore a great variety of Paleozoic foraminifers that can be used for stratigraphic correlations at the points of faunal radiation. The following is a brief summary of their development from the Cambrian to the Permian (Text-fig. 1).

The existence of Cambrian foraminifers remains poorly documented and often unconvincing. Most of the reputedly calcareous secreted forms, such as *Reitlingerella, Lukashevella, Syniella,* and *Chabakovia,* ought to be transferred to the botanical realm. However, primitive agglutinated taxa, such as *Thuramminoides*, are certainly present in rocks as old as Middle Cambrian (Alexandrowicz, 1969).

Evolution was slow during the Ordovician, and Silurian populations were rather small and diversity very low. The first major radiation occurred in the Middle

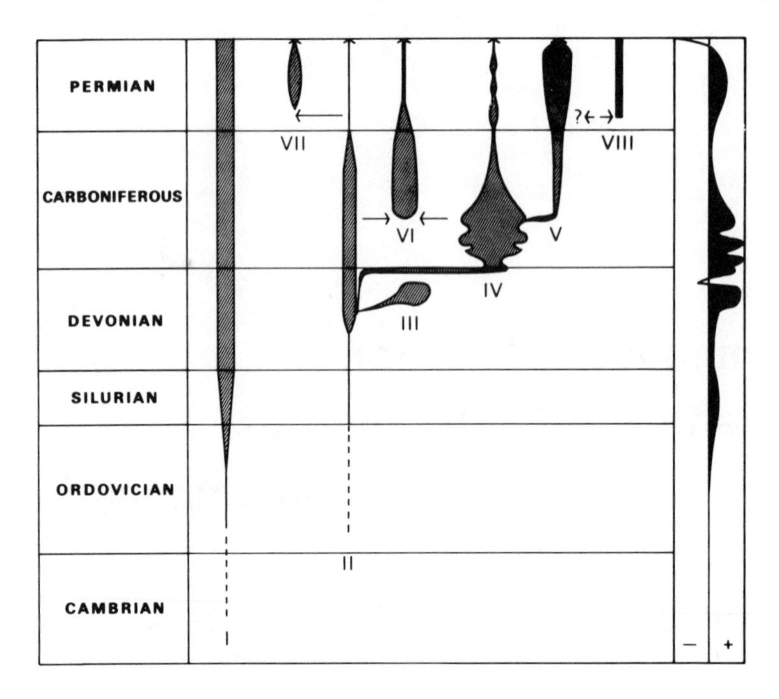

Text-Figure 1
Stratigraphic distribution, main groups of Paleozoic Foraminifera. Key: I, siliceous agglutinated (saccamminids, ammodiscids, textulariids, hyperamminids, etc.); II, primitive calcareous secreted (earlandiids, pseudoammodiscids etc.); III, eogeinitzinid-nanicellid fauna; IV, "endothyrids" (*sensu lato*) (endothyrids, endothyranopsids, tournayellids, bradyinids, eostaffellids, pseudoendothyrids, etc.); V, fusulinids (ozawainellids, schwagerinids, verbeekinids, etc.); VI, fischerinids, apterrinellids, etc; VII, nodosinellids; VIII, nodosariids. Last column shows the net profit and loss balance of the fauna. Middle Famennian and Late Permian are obvious times of crisis.

Devonian and culminated in the Late Frasnian. The brief and worldwide outburst of the *Eogeinitzina-Nanicella* fauna occurred at that time and led to morphologically advanced forms such as *Multiseptida.*

Although many Devonian taxa are morphologically convergent with Permian nodosinellids, nodosariids, or colaniellids, they have different wall structure, are phylogenetically unrelated, and should not be confused with these Late Paleozoic groups. Indeed, this Devonian microfauna became nearly extinct in Early Famennian, and was gradually replaced by the tournayellid stocks from which the endothyrids were derived in the Late Famennian. Both families had a major outburst in Tournaisian and Viséan time, the eostaffelids and the pseudoendothyrids being the most advanced of the endothyrid group.

The first occurrence of fusulines is a matter of semantics, because the passage from the endothyrids is gradational. Contrary to the Russian school usage (Rozovskaia,

1963), the first occurrence of a true tunnel is used here to characterize the first fusuline. If this concept is correct, *Millerella* is the first fusuline, and its first occurrence marks the base of the Pennsylvanian of North America.

Fischerinids and apterrinellids are abundant from the Middle Carboniferous upward. Nodosinellids *sensu stricto* occur for the first time in the Permian, associated with nodosariids.

The development rhythm of the fauna is quite uneven (Tappan and Loeblich, 1972). If one plots the succession of first and last occurrences (Text-fig. 1), one observes a strong outburst in the Frasnian, a regression in the Famennian, a Lower Carboniferous radiation (with minor oscillations at the base of the Viséan and at the Viséan-Namurian boundary), a regression in the Upper Carboniferous, a Middle Permian outburst, and a profound crisis at the Permian-Trias boundary. This net profit and loss diagram is quite similar to that outlined in the "fossil record" (Harland et al., 1967), although the latter treatment appears to suffer from some monographic fluctuations. It shows that small foraminifers can be used effectively in biostratigraphy at least three times in the Paleozoic: the Upper Devonian, the Lower Middle Carboniferous and the Permian.

This paper will be restricted to the study of the evolution of Lower Carboniferous calcareous secreted "endothyrids" in time and space. As an example of the concepts, methodology, and types of organisms that can be used in constructing biostratigraphic zones, we shall (1) first review some observations concerning their habitat and mode of life, (2) discuss their phylogenetic relationships and establish a model for their dispersal patterns, and (3) define global zones and evaluate their validity for biostratigraphic correlations.

MODE OF LIFE

Lower Carboniferous "endothyrids" were exclusively benthic. They thrived mostly in shallow water and are associated with calcareous grainstones, packstones, and wackestones or calcareous silts and shales. On a Viséan carbonate platform for instance, they are usually found with green or blue-green algae, mostly dasycladaceans (Mamet, 1972). All over the Tethys, a *Koninckopora* grainstone contains up to 300 to 500 individuals/cm^3 of sediment. Thus, as much as 3 to 5 percent of the rock is composed of foraminiferal tests; in very exceptional cases, this figure reaches up to 20 percent. However, quantitatively, crinoids, bryozoans, and algae normally play an even greater role in carbonate accumulation.

Foraminifers are found in order of decreasing importance in the following facies: (1) pelmatozoan-bryozoan-brachiopod-dasycladaceae-ungdarellaceae grainstone-packstone-wackestone; (2) pelletoidal bothrolithic and pseudo-oolitic grainstone, (both facies are encountered in the neighborhood of banks, at low to very low bathymetry, in agitated, well-aerated waters); (3) barrier boundstones (such as brachiopod "knolls," or brachiopod-pelmatozoan-algae banks); (4) pelletoidal lagoonal wackestones and mudstones associated with kamaenid algae; and

(5) kamaenid packstone-wackestone with bryozoan, pelmatozoan, and spongiostromid algae.

Foraminifers are scarce in the following facies: (1) deeper-water pelmatozoan-bryozoan-brachiopod-echinid mudstone-packstone (the microfauna is usually reduced to Earlandiidae); (2) waulsortian reefs (fenestellid wackestones-boundstone); (3) true oolitic grainstone (not to be confused with pseudo-oolitic bothrolithic grainstone, which are usually quite rich); (4) crinoid-bryozoan-brachiopod mudstone-packstone; and (5) evaporitic penesaline mudstone, loferite mudstone, dolomite, and dedolomite.

The following facies are at best very poor to completely devoid of fauna: (1) fenestellid-pelmatozoan packstone-"pseudograinstone"; (2) pelmatozoan packstone-wackestone-"pseudograinstone"; (3) spiculitic mudstone-packstone-wackestone; (4) spiculite-radiolarian mudstone-wackestone; (5) radiolarian packstone-mudstone; and (6) freshwater mudstone.

Carboniferous foraminifers thrived nearly exclusively within the euphotic zone, in water of normal marine salinity. This is particularly true for the Eostaffellidae, Pseudoendothyridae, Tetrataxidae, and Archaediscidae. Moreover, the last family was particularly sensitive to salinity changes and was readily eliminated when freshwater entered the platform. Endothyridae (notably *Latiendothyra* and *Priscella*) and Earlandiidae were more tolerant to salinity changes and lived even in slightly hyper- or hyposaline lagoons.

There is apparently no well-defined community succession in the sense of Berry and Boucot (1972) (marine benthic life zones) although an Earlandiidae versus Earlandiidae-Endothyridae versus Earlandiidae-Endothyridae-Archaediscidae-Tetrataxidae-Eostaffellidae-Pseudoendothyridae bathymetric succession could tentatively be proposed.

The Paleozoic foraminifers differ mainly from their modern counterpart in the absence of pelagic forms. This lack of pelagic life is well substantiated by studies of Carboniferous rhythmically layered carbonate turbidites. In such graded sequences, foraminifers are sorted by size. At the base of a Late Viséan turbidite, *Bradyina rotula* d'Eichwald (1,500 to 2,000 μm), *Climacammina, Haplophragmella tetraloculi* Rauzer-Chernoussova, and *Globoendothyra globulus* (d'Eichwald) (the three taxa ranging from 1,000 to 2,000 μm) abound in the coarse-grained grainstones.

Medium-grained calcarenites contain *Endothyranopsis crassa* (Brady), *Tetrataxis, Eostaffella, Pseudoendothyra, Omphalotis circumplicata* (Howchin) (in the 700 to 1,000 μm range). Finer grained carbonates yield abundant *Endothyra* of the group *E. bowmani* Phillips *emend* Brady, or *Endothyra similis* Rauzer-Chernoussova and Reitlinger, *Mediocris, Archaediscus* of the group *A. krestovnikovi* Rauzer-Chernoussova, and *Valvulinella*, which range from 400 to 750 μm. Fine-grained carbonates have minute *Neoarchaediscus* and *Planospirodiscus* (150 to 400 μm) associated with calcispheres. Finally, the pelagic fine-grained laminations yield no foraminifer at all. Thus foraminifers form a rather homogeneous entity in platform sediments, but in turbidites, where they are segregated by size, the pelagic niche is unoccupied, in contrast to what would be observed in the Cretaceous or Cenozoic.

In conclusion, Paleozoic small foraminifers were restricted to the shallow areas of the platform, usually associated with euphotic algae. Depending on the clay content and agitation of the water, this euphotic zone may range from a dozen to several dozen fathoms. The deeper parts of the platform and all the continental slope were barren. This clearly defines the main facies in which calcareous secreted benthonic foraminiferal biostratigraphy can be applied in the Paleozoic.

BIOGEOGRAPHY AND ORIGIN OF THE MICROFAUNA

Contrary to the opinion of many authors (Nalivkin, 1957; Vdovenko, 1961; Conil and Lys, 1964), no well-defined *stratigraphic* provincialism can be detected in the Lower Carboniferous, at least in the northern hemisphere. Unfortunately, little is known from the southern hemisphere, but Australia has a microfauna and microflora very similar to those of Indochina-Malaya. All 19 extant families are known in American, Europe, Asia, and Africa. Of the known 181 valid genera extant in 1972, at least 120 appear to be circumhemispheric (Tethys-North America), 36 are restricted to the Tethys, and the paleogeographic dispersal of 25 remains to be checked. Hence three freely interconnected realms can be detected: (1) a rich Tethyan Realm covering Ireland, Great Britain, Belgium, Germany, Poland, France, Spain, Morocco, Algeria, Libya, Egypt, the Russian platform, the central and southern Urals, the Donbass, Kazakhstan, Iran, the Tian-Shian, Malaya, Vietnam, and northwestern and southeastern Australia; (2) a poor Taimyr-Alaska Realm (Taimyr, Lena, Kolyma, Omolonsk Massif, Arctic Canada); and (3) an intermediate Kuznets-North American Realm (covering the Kuznets Basin, the Primori, Japan?, and all North America, with the exception of the Arctic part of Alaska). The place of China in this scheme remains doubtful, as we have little information, concerning its faunal affinities. Southern China is certainly Tethyan. This scheme (Mamet, 1962) is rather similar to the paleogeographical reconstruction recently proposed by Juferev (1973) and Lipina (1973). It is in reasonable agreement with the conclusions of Bogush and Juferev (1966) and Ustritz (1967).

The origin of the microfauna is not well established. However, a number of observations suggest that the Tethys was the main breeding ground and the center of development from which the majority of the foraminifers were derived. Considerations supporting this thesis are discussed below.

Abundance pattern. The Tethys has a prolific microfauna. As previously mentioned, 300 to 500 "endothyrids"/cm^3 for the *Koninckopora* grainstone facies is common. This number is only about a third as large in the North American Realm in the same *Koninckopora* facies. In the Alaska-Taimyr Realm, 50 to 100 individuals/cm^3 is considered an exceptionally rich fauna. Therefore, abundance consistently decreases away from Tethys.

Specific diversity. The total number of valid species is approximately 800 for the Tethys, 400 for the North American Realm, and 100 for the Taimyr-Alaska Realm. These figures are highly debatable and would vary from author to author. They are used here only for comparison, the author using the same species concept within the three realms. (In 1972, there were more than 1,800 published taxa for the European Viséan, most of them being junior synonyms based on poorly oriented sections. For instance, in 20 years, 220 species have been attributed to *Archaediscus,* but only 82 fit the generic diagnosis. No more than 22 of these are satisfactorily defined and, in our opinion, they represent only 19 species.)

Theoretical interpretations of diversity patterns remain inconclusive (Stehli, 1968; Dunbar, 1968; Valentine, 1967). However, diversity distribution of Paleozoic foraminifers coupled with stratigraphic criteria strongly suggests that cold-water microfauna have low specific diversity (Text-fig. 2).

Species-genus ratio. The total number of genera does not vary appreciably from one realm to another; therefore the species-genus ratio drops rapidly from the rich, diverse Tethyan assemblages to the monospecific assemblages of Alaska (Text-fig. 3).

Morphology. The genera occurring in the Tethys, but excluded from the Taimyr-Alaska Realm, are all morphologically advanced forms such as the curiously partitioned tetrataxid *Valvulinella,* the aberrant *Klubovella,* the shield bearing *Cribrospira,* or the uncoiled evolute, although pseudoseptated, *Lituotubella.* In particular, uncoiled forms, cribrate forms, and especially uncoiled cribrate forms are abundant in the Tethys, scarce in North America, and absent in the Arctic (Ganelina et al., 1972).

The influence of size is less obvious, and the average size for each genus does not vary appreciably from realm to realm. However, gigantic forms are restricted to the Tethyan Realm. For instance, in the latest Viséan, Zone 16_{inf}, enormous *Bradyina rotula* d'Eichwald, *Archaediscus karreri* Brady *non auct.,* and *Omphalotis omphalota* (Rauzer-Chernoussova and Reitlinger) proliferate in the Tethys, but are absent or very scarce in the two other realms.

Endemism. The percentage of endemic species is much higher in the Tethys than in the Arctic. This leads to a stratigraphic paradox; the Arctic microfauna is poor and undiversified but lends itself well to zonation, because most of the faunal indexes are known from other basins.

Heterochronism of the first occurrence. Numerous widespread genera permit the establishment of a succession of assemblage zones in an order that appears constant in the Northern Hemisphere (Text-fig. 4). However, some advanced forms, mostly among the Bradyinidae and the Palaeotextulariidae occur for the first time much later in North America than in the Tethys. *Bradyina*, for instance, is a good case of heterochronic dispersion (Text-fig. 5). The genus occurs for the first time in the latest Viséan (Zone 16_{inf}) of Eurasia, and its sudden radiation is a good stratigraphic marker, associated with ammonoids of the IIIβ zone. Because there is no foraminiferal keriotheca known in Early or Middle Viséan time, this first occurrence can be recognized in thin section, even on a very small wall fragment. *Bradyina* is not known in the North American Midcontinent until the very latest

Namurian, a full stage later than in Eurasia. This "late" arrival is further underlined by the fact that the earliest known Midcontinent *Bradyina* belong to the *cribrostomata* group while the whole evolutionary sequence of the *rotula* group is missing.

When such heterochronic dispersal is known, the earliest appearance usually occurs in the Tethys, the latest in North America and the Arctic.

Role of island arcs. Island arcs are crucial for correlation because they contain, in unusual associations, faunas that are normally endemic to the two adjacent realms. The Late Tournaisian-Early Viséan *Eotextularia* is basically Tethyan and has only been encountered once in North America. The same applies to *Mediocris* and *Omphalotis,* which are very abundant in the Middle Viséan of Eurasia, but are still doubtful or scarce outside the Tethys. *Valvulinella* and *Howchinia* are not known in the late Viséan Chester Group of the Midcontinent. *Endothyranopsis sphaerica* (Rauzer-Chernoussova and Reitlinger) is widespread in the Tethys, but unrecorded in the North American Midcontinent. Elements of such fauna, however, are known in Carboniferous island arcs bordering the North American Cordillera (Pacific British Columbia, Pacific islands of Alaska, and Oregon). The assemblages are strikingly different from those of surrounding formations of British Columbia, Yukon Territory, or the remainder of Alaska. But on the island arcs, Tethyan "endemic" genera are observed mixed with North American "endemic" taxa, allowing an important integration of regionally distinct biostratigraphic systems.

Moreover, in terms of zones, the first occurrence of diachronic genera is intermediate between the Tethyan and the North American Realms. For instance (see Text-fig. 5), *Bradyina* is found in the Peratrovich Formation for the first time in Zone 18, three zones younger than in the Tethys, but two zones earlier than in the North American Realm.

Heterochronism of the base of the acme. First occurrence heterochronism, as mentioned above, is rather uncommon. Usually the earliest occurrence is observed at about the same level in the different realms, while the acme is strongly diachronic (Mamet and Skipp, 1970). This is obvious in the case of *Eostaffella* (see Text-fig. 6), a very prolific genus occurring for the first time in the Early Viséan of the northern hemisphere. It is very poorly represented in North America until the latest Viséan, where it has a sudden outburst, underlying the Chester Group. This late acme is due here to the earlier proliferation of the morphologically, ecologically similar and apparently competitive North American genus *Eoendothyranopsis.* The extinction of the latter at the Meramec-Chester boundary permits the Eostaffellidae to blossom from Zone 16_{inf} upward.

Disjunct phylogenies. Evolutionary sequences are complete in the Tethys; transitions from one genus to another can be slow or rapid. They are, however, morphologically progressive. Let us examine an example among the Tetrataxidae. The first known *Tetrataxis* are Late Tournaisian in Eurasia; at the beginning of the Viséan, the four-folded opening becomes two-folded and finally circular. The coiling symmetry passes from four, to two, and finally one, giving *Howchinia.* In the Late Viséan, this genus flattens, still keeping its inner pseudofibrous lining on the ventral side of the test. It finally becomes planispiral (*Monotaxinoides*). In

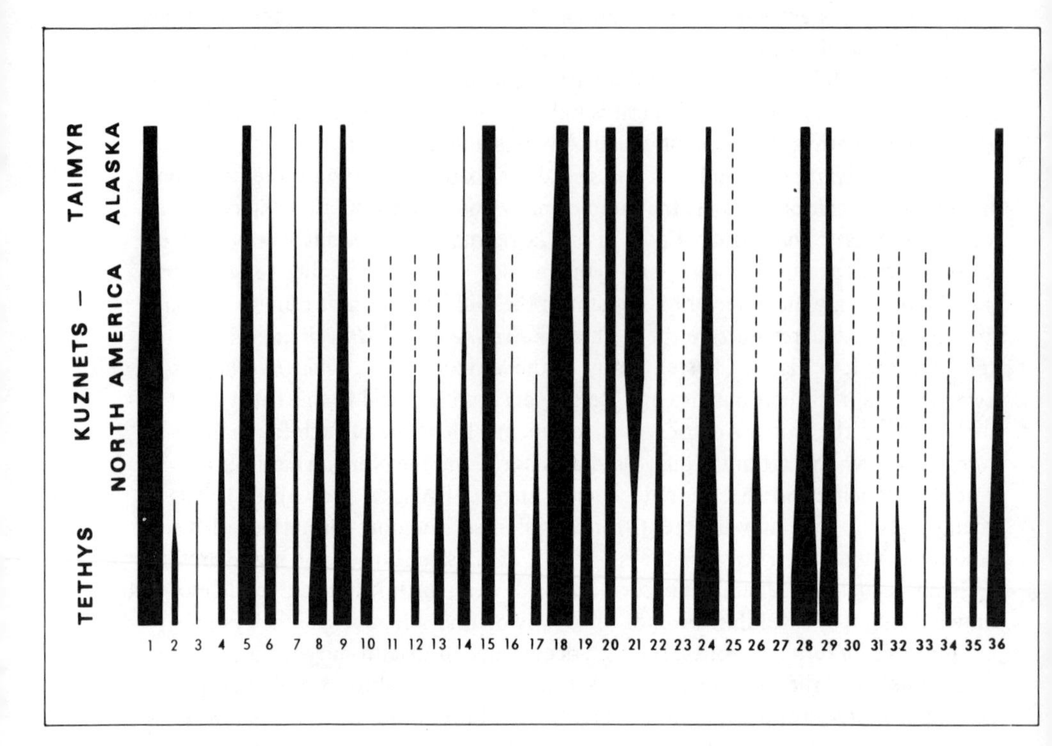

Text-Figure 2 *(above and right)*
Specific diversity of 71 common Lower Carboniferous genera in the Tethys, the Kuznets-North America, and the Taimyr-Alaska Realms. Key: 1, *Archaediscus* sp.; 2, *Birectochernyshinella* sp.; 3, *Biseriammina* sp.; 4, *Bradyina* sp.; 5, *Brunsia* sp.; 6, *Brunsiina* sp.; 7, *Carbonella* sp.; 8, *Chernyshinella* sp.; 9, *Pseudoammodiscus* sp.; 10, *Climacammina* sp.; 11, *Cribroendothyra* sp.; 12, *Cribrospira* sp.; 13, *Cribrostomum* sp.; 14, *Dainella* sp.; 15, *Earlandia* sp.; 16, *Endospiroplectammina* sp.; 17, *Endostaffella* sp.; 18, *Endothyra* sp.; 19, *Endothyranella* sp.; 20, *Endothyranopsis* sp.; 21, *Eoendothyranopsis* sp.; 22, *Eoforschia* sp.; 23, *Eoparastaffella* sp.; 24, *Eostaffella* sp.; 25, *Eotextularia* sp.; 26, *Forschia* sp.; 27, *Forschiella* sp.; 28, *Globoendothyra* sp.; 29, *Glomospiranella* sp.; 30, *Haplophragmella* sp.; 31, *Howchinia* sp.; 32, *Janichewskina* sp.; 33, *Klubovella* sp.; 34, *Koskinobigenerina* sp.; 35, *Koskinotextularia* sp.; 36, *Latiendothyra* sp.; 37, *Lituotubella* sp.; 38, *Lugtonia* sp.; 39, *Mediocris* sp.; 40, *Mikhailovella* sp.; 41, *Monotaxinoides* sp.; 42, *Neoarchaediscus* sp.; 43, *Omphalotis* sp.; 44, *Palaeospiroplectammina* sp.; 45, *"Palaeotextularia"* (group *P. consobrina*); 46, *Palaeotextularia sensu stricto,* 47, *Paraendothyra* sp.; 48, *Permodiscus* sp.; 49, *Planoarchaediscus* sp.; 50, *Planoendothyra* sp.; 51, *Planospirodiscus* sp.; 52, *Propermodiscus* sp.; 53, *Pseudoendothyra* sp.; 54, *Pseudoglomospira* sp.; 55, *Quasiendothyra* sp.; 56, *Rectocornuspira* sp.; 57, *Rectoseptabrunsiina* and *Rectoseptatournayella* sp.; 58, *Rectoseptaglomospiranella* sp.; 59, *Septabrunsiina* sp.; 60, *Septaglomospiranella* sp.; 61, *Septatournayella* sp.; 62, *Spinoendothyra* sp.; 63, *Tetrataxis* sp.; 64, *Tournayella* sp.; 65, *Tuberendothyra* sp.; 66, *Urbanella*

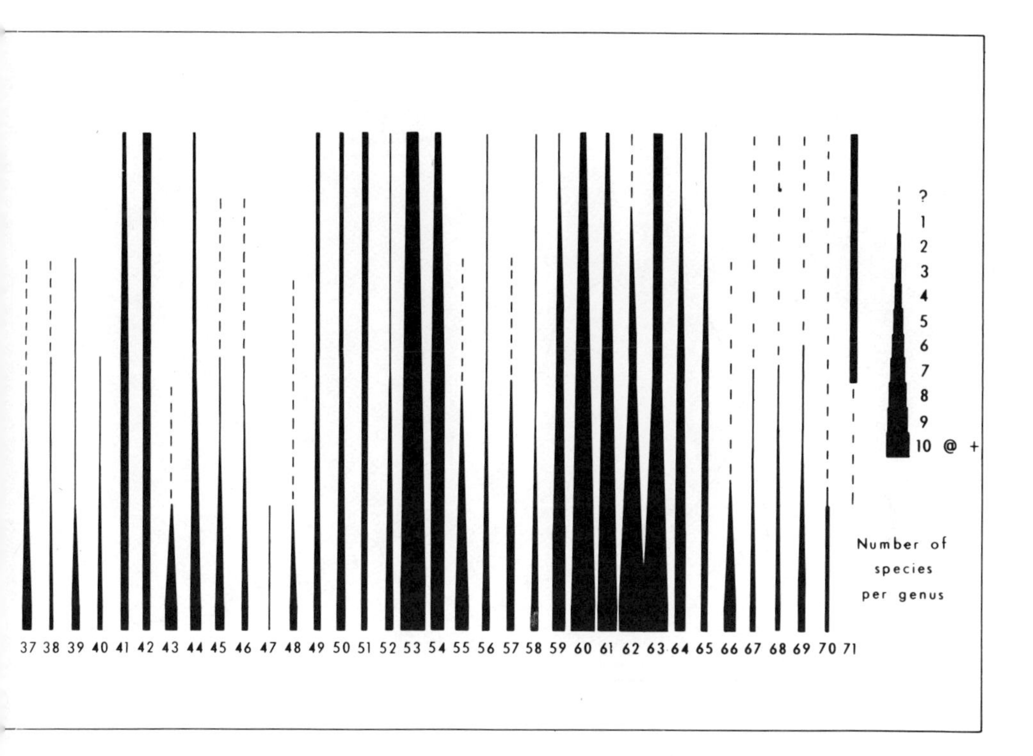

sp.; 67, *Uvatournayella* sp.; 68, *Uviella* sp.; 69, *Valvulinella* sp.; 70, *Vissariotaxis* sp.; 71, *Zellerina* sp.

Namurian time, secondary partitions and small fissures develop in *Eolasiodiscus,* the earliest representative of the Lasiodiscidae.

Howchinia has not been observed in the North American Midcontinent, in the Cordillera or in the Appalachians, although the ancestral *Tetrataxis* and its derivative *Monotaxinoides* are abundant and widespread. There is apparently no link between the two genera, and the Tetrataxidae and Lasiodiscidae appear unrelated. The phylogeny appears disjunct here.

Many phylogenies are incomplete in North America. This is particularly striking among the earliest endothyrids, where the "*Eoendothyra*"-*Quasiendothyra*-*Cribroendothyra*-*Klubovella* fauna is missing.

In the Midcontinent and in the Cordillera, archaediscids appear to have no link with their ancestors *Brunsia*-*Pseudoammodiscus*. No "*Parapermodiscus*-*Eodiscus*" fauna has ever been recorded. Even *Permodiscus* is absent, and the most primitive archaediscid is the fairly advanced *Propermodiscus*. As a result, any attempt to reconstruct phylogenic lines or establish cladogenic-phylogenic comparison in North America or Alaska is bound to fail because the breeding ground is mainly the Tethys.

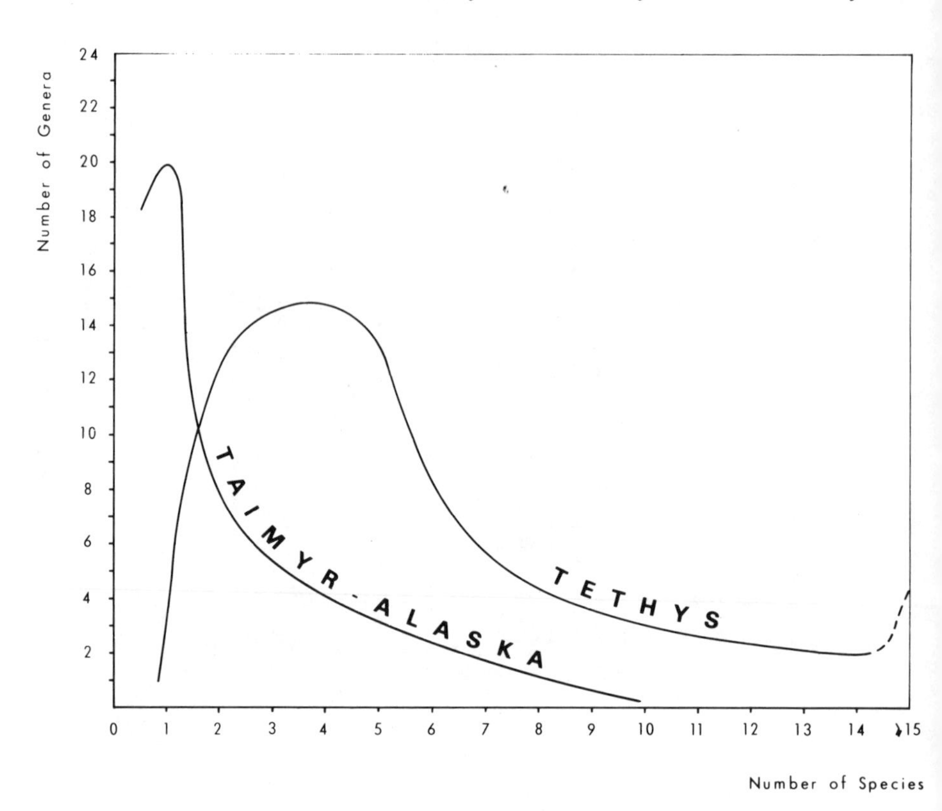

Text-Figure 3
Species-genus ratio in the Tethys and the Taimyr-Alaska Realms.

A paleontologist working in Alaska would observe a succession of apparently unrelated taxa, "occurring from nowhere," certainly useful for zonation, with clear zonal boundaries, but in truth lagging behind the first occurrence in the Tethys. The same paleontologist working in the Tethys would observe highly diversified and rich fauna, with transitional taxa, each transition representing a possible new species, new genus, or even new family. Hence the temptation to erect a multitude of taxa and zones.

ZONATION

Global Zones

As previously mentioned, 1,800 Viséan "species" are known in the literature, hence the possibility of erecting 1,800 first occurrence zones, 1,800 last occurrence zones, and 1,800 acme zones. This is common practice and gives a maze of

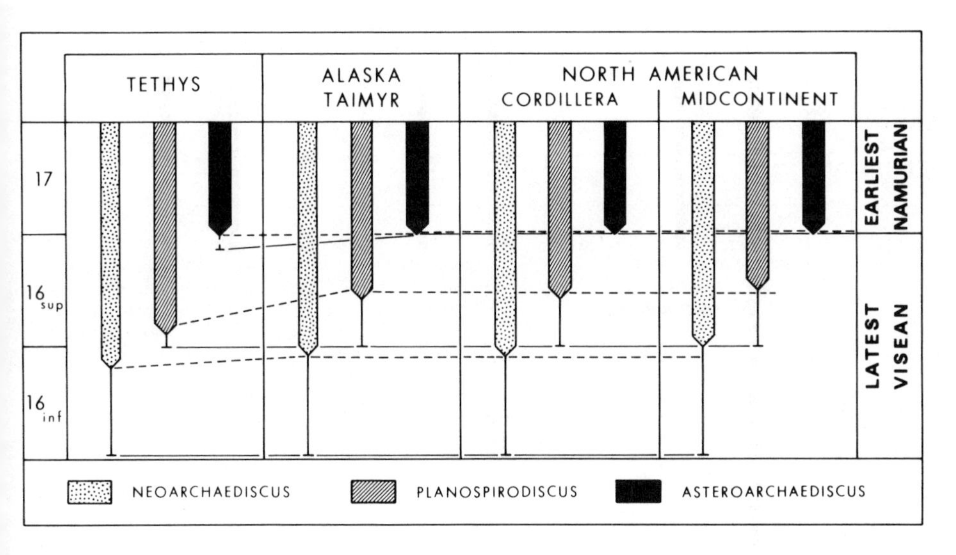

Text-Figure 4
Neoarchaediscus-Planospirodiscus-Asteroarchaediscus succession in the northern hemisphere. Note the remarkable similitude of the base of the range zone and of the peak zone. (Correlations on first occurrences in solid lines. Correlations on the base of the acme in dashed lines.)

incomplete ranges inducing pseudocorrelations. Moreover, as could be expected, total ranges found in the literature are too short, and the base of the acme is usually confused for the first true occurrence. Some selection must be made and some screening done in this accumulation of data in order to keep a common language among stratigraphers (Text-fig. 7).

Global zones used herein are first occurrence clusters of assemblages observed among different phylogenies that have wide lateral dispersal. The base of the zone is chosen at the first occurrence of a number of indexes if the ancestral forms, from which the indexes are derived, are present in the underlying strata (principle of phylogenic continuity).

Zonal "boundaries" are transitional, have in some case a measurable thickness, and cannot be underlined by the "golden spike" method (unless the spike is rather thick). Should first occurrences of multiple phylogenies occur exactly at the same level, they would indicate a paraconformable hiatus, and not a true first occurrence.

As an example of this concept, let us consider how to underline the passage of Zone 9 to 10, the classical Tournaisian-Viséan boundary (Text-fig. 8).

In the Tethys, the Late Tournaisian is marked by the end of the Tournayellidae reign (the last *Septatournayella, Tournayella, Septabrunsiina, Spinotournayella,* and *Carbonella,* will promptly be eliminated in the Viséan) and the widespread

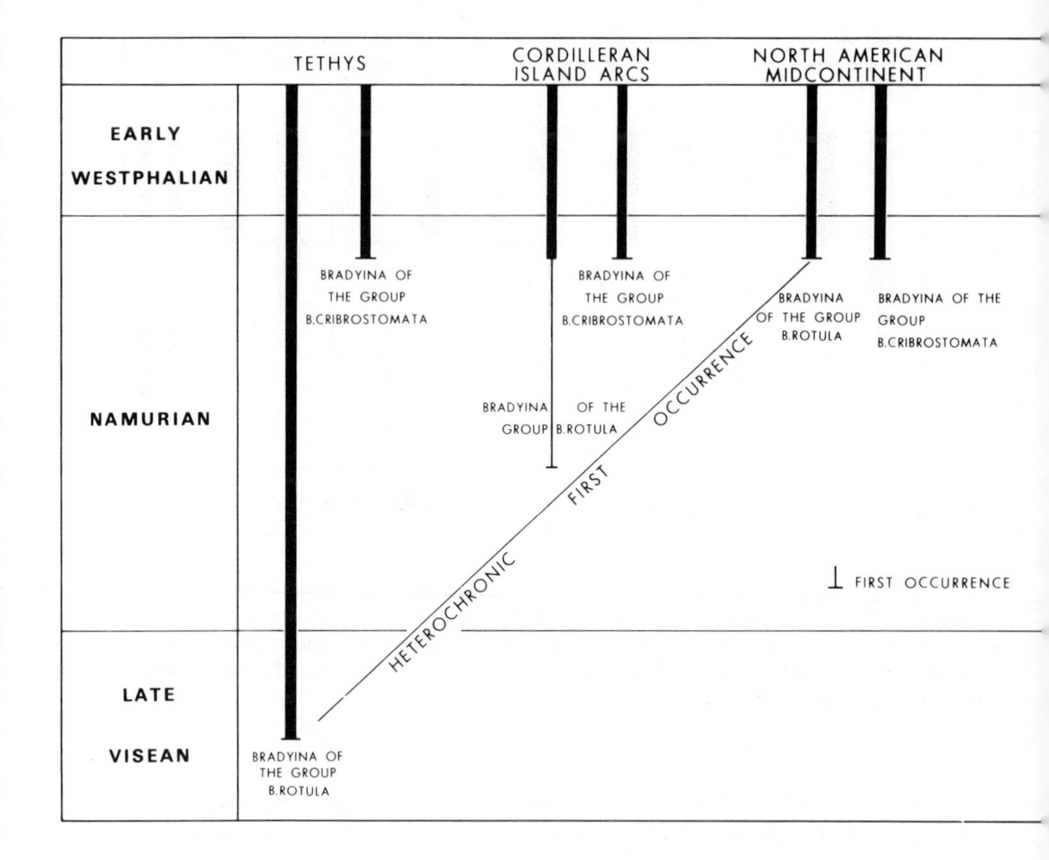

Text-Figure 5
Heterochronic first occurrence of *Bradyina* in the Tethys, the Cordilleran island arcs, and the North American Midcontinent.

occurrence of *Latiendothyra* and *Priscella; Endothyra sensu stricto* occurs for the first time, mixed with numerous *Spinoendothyra*. In the very latest Tournaisian, *Inflatoendothyra* derives from the spinose endothyrids by resorbtion of the secondary deposits. Tetrataxidae also occur for the first time and are derived, as the first Biseriamminidae, from an endothyrid stock. The first pseudoagglutinated calcareous forms, such as the Palaeotextulariidae (*Eotextularia*) and the Forschiidae (*Eoforschia*) also occur for the first time. Thus not less than four families find their roots in the Late Tournaisian.

The conventional Tournaisian-Viséan boundary has been placed for a century at the base of the Black Marbles of Dinant (V I a), a rhythmic sequence of lagoonal and marine shallow-water euxinic limestones in the Dinant region (Belgium). This boundary is underlined by an outburst of Globoendothyridae, notably *Globoendothyra* and *Dainella.* Dainellids are derived from the *Inflatoendothyra*

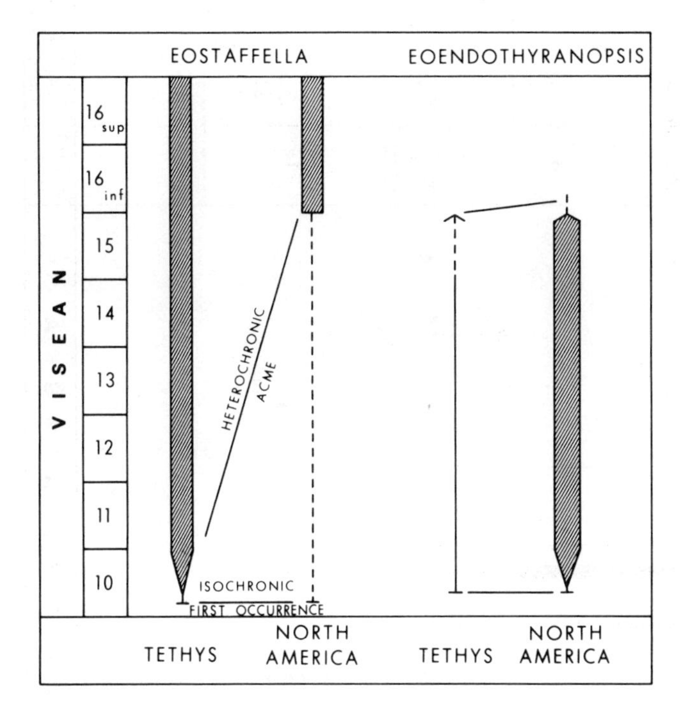

Text-Figure 6
Isochronic first occurrence and heterochronic peak zone of *Eostaffella.*

stock, which was widespread in the latest Tournaisian strata. *Globoendothyra* are recognized on the first occurrence of a well-defined diaphonetheca among endothyrid coiled forms. At about the same level, a fibrous radial layer occurs for the first time among the Archaediscidae. These form the *"Parapermodiscus"* OBJ-*"Eodiscus"* OBJ assemblage, which rapidly leads to the *Permodiscus-Propermodiscus* assemblage (the *Uralodiscus-"Glomodiscus"* OBJ assemblage of Malakhova, 1973).

A little higher in the Viséan, *Eostaffella* and *Pseudoendothyra* are recognized for the first time. *Pseudoendothyra* is derived from *Eoparastaffella*, which occurs in the very latest Tournaisian. *Eostaffella* is probably derived from the Late Tournaisian-Early Viséan *"Endothyra" transita* group. Scarce *Eoendothyranopsis* are encountered at that level for the first time.

We may therefore underline Zone 10 in the Tethys on not less than four new families. What is left of this assemblage in North America?

As previously mentioned, the roots of the Archaediscidae has not been observed in North America. The *"Parapermodiscus"* OBJ-*"Eodiscus"* OBJ

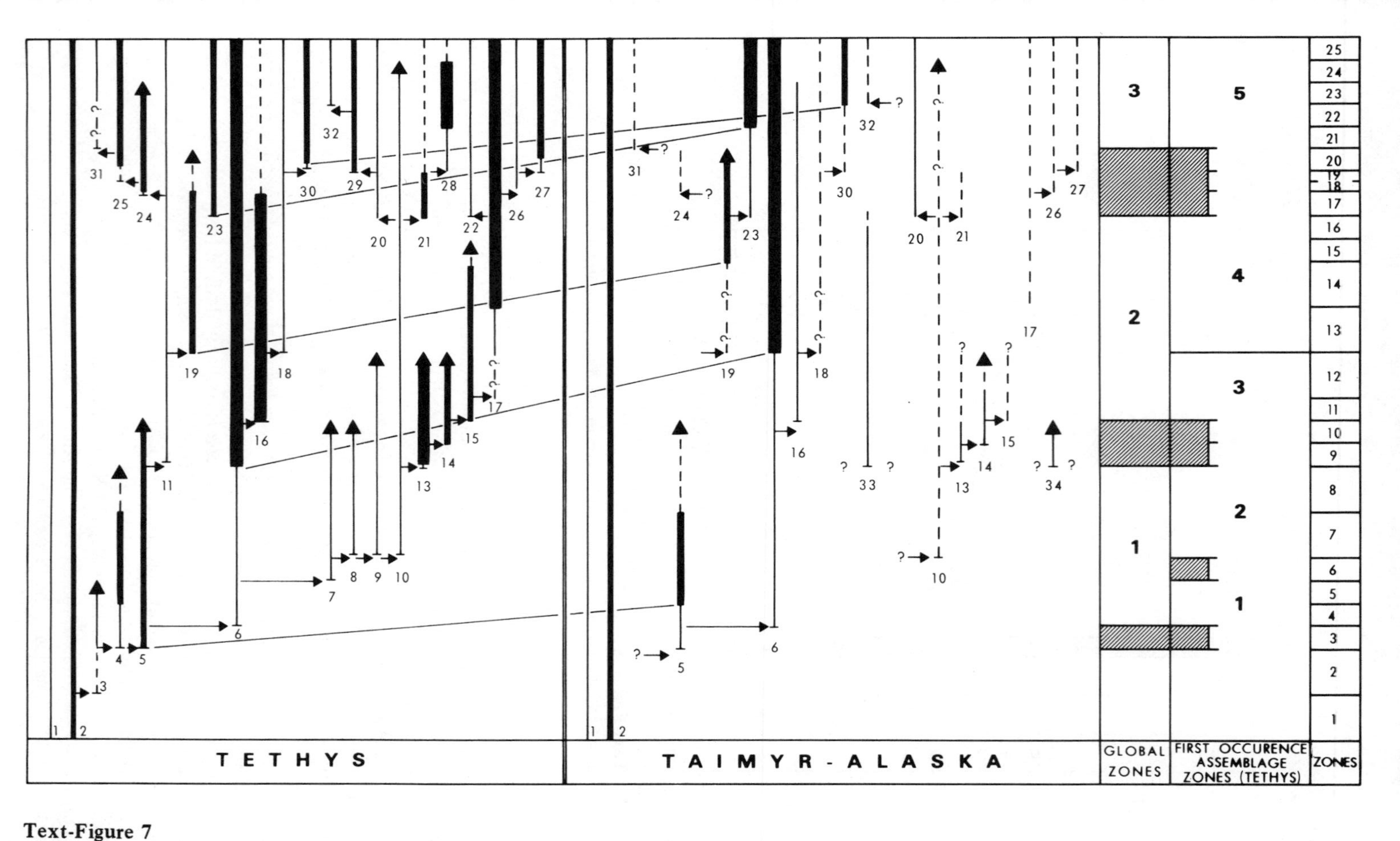

Text-Figure 7

Diagrammatic representation of faunal distribution in the Tethys and in the Taimyr-Alaska Realms. The distribution of 32 taxa enables recognition of 25 "zones" (first occurrence, last occurrence, and acme). In the Tethys, five clusters of first occurrences are recognized. Of these, only three are common to the Tethys and the Taimyr-Alaska Realm; these are global zones.

assemblage and even the fairly advanced *Permodiscus* have not been encountered yet. The earliest known archaediscid of North America are *Propermodiscus, Planoarchaediscus,* or *Archaediscus.* These are scarce and cannot be used for "daily stratigraphy."

We have also mentioned that Eostaffellidae and Pseudoendothyridae are not widespread before the Late Viséan Chester. Their scarcity below that level precludes any stratigraphic use in the Early Viséan.

In the Madison Group of Wyoming and Utah, as well as in the Shunda and the Turner Valley Formations of Alberta and British Columbia, earliest *Dainella* are encountered at the Tournaisian-Viséan boundary, and they occur above their ancestral *Inflatoendothyra.* This certainly indicates that the genus spread in the North American Realm and that it can be used in the Cordillera, as well as in the Tethys. However, the genus is scarce and has not been found yet in the American Midcontinent. On the other hand, *Globoendothyra,* the second representative of the Globoendothyridae, is represented in all known basins by the *baileyi* group, and is as abundant in North America as it is in the Tethys; thus this taxon is a good index, especially when associated with the earliest *Eoendothyranopsis.*

Zone 10 is therefore ideally defined on a number of indexes observed in the Tethys, where the phylogenies are complete and where the earliest occurrences are observed. When Zone 10 is extended into other paleobiological realms, a number of indexes are lost, but the zone is still recognizable on part of the fauna.

Zonal Boundaries and Validity of the Approach

"First occurrence" of a fauna should truly be called "first occurrence of a facies favorable to a given fauna." This applies to benthic as well as pelagic life. Hence our zonal boundaries are by definition diachronous and, as a rule, must be too young in all but the place of origin of the taxon.

Moreover, global zones are based on multiple phylogenies forming new appearance clusters. We may place golden spikes at the base, in the middle, or at the top of such clusters, but the chronostratigraphic value of this approach is debatable.

Aside from the criterion of superposition, which appears to be respected in all northern hemisphere basins, no good check has been made of the validity of the approach. At the present time, the only attempt to verify the validity of foraminiferal global zones versus other biostratigraphic zones is at the basin scale (see Text-fig. 9). In the northern Cordillera of the United States, Sando and Dutro (Sando et al., 1969) have established a sequence of 12 megafaunal zones based on corals and brachiopods, a sequence that can be applied successfully to the zonation of the Mississippian part of the Carboniferous. The samples used for megafaunal determinations were then reprocessed for foraminiferal identifications and 14 global zones could be observed. Although based on completely different criteria, global foraminiferal zones and megafaunal assemblages were found complementary and gave consistently reliable ages; for instance, megafaunal zone C_I equals Zone 7 and the lowermost part of Zone 8. Zone C_2 equals Zones 8, 9, and 10?. Zone D equals Zones 10, 11, and 12; etc.

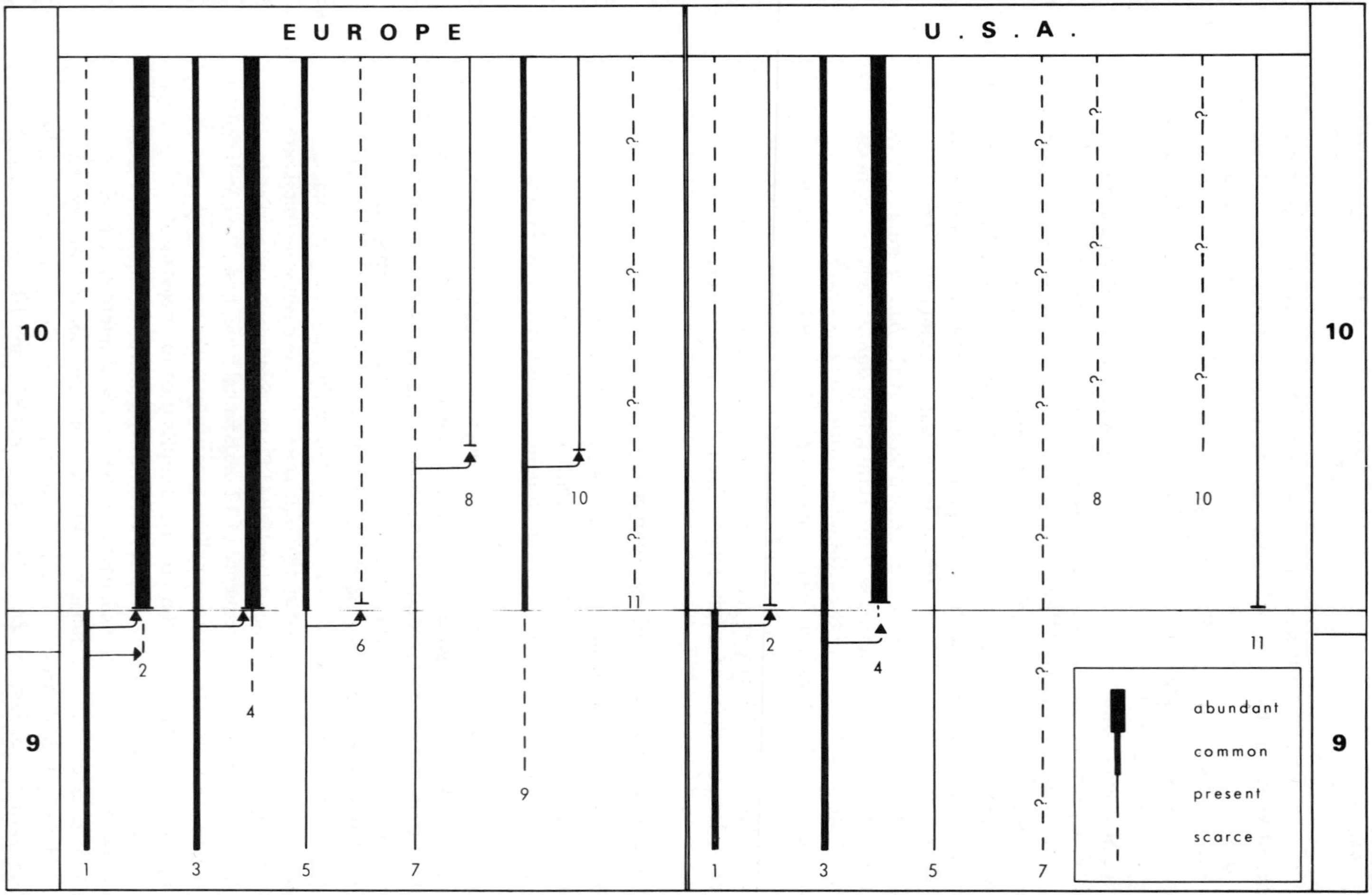

Text-Figure 8

Definition of the Tournaisian-Viséan "boundary" in Europe and the United States. Key: 1, *Inflatoendothyra:* 2, *Dainella;*

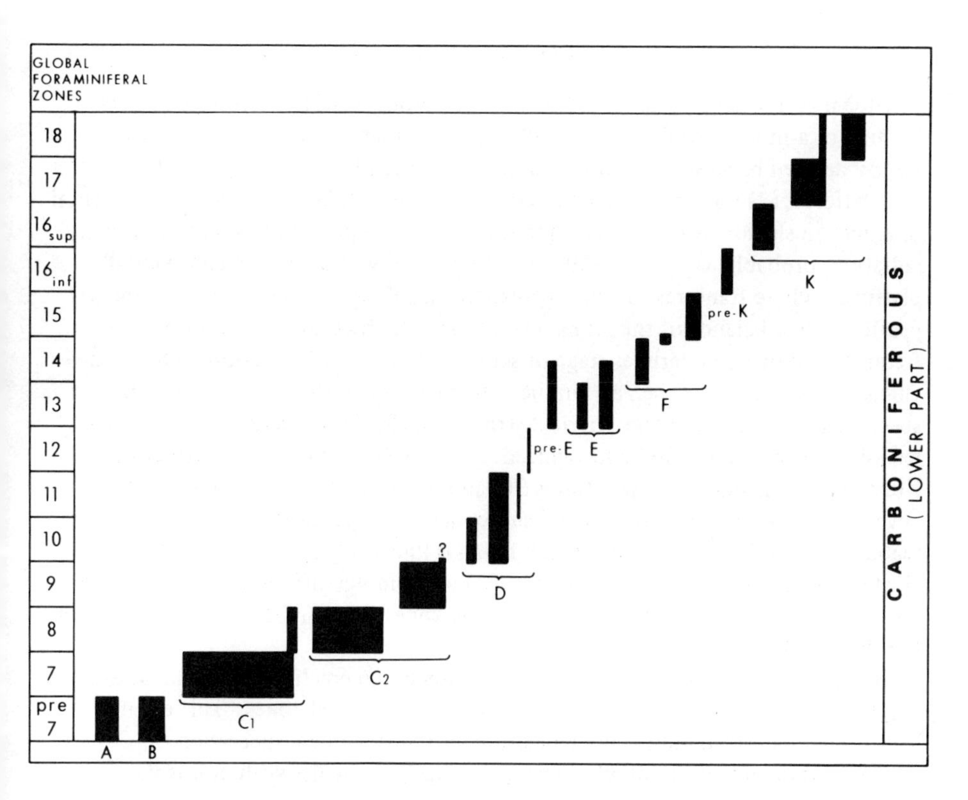

Text-Figure 9
Equivalence of megafaunal zones (A, B, . . .) of Sando and Dutro and global foraminiferal zones (7, 8, . . .). (Slightly modified after Sando et al., 1969.)

In the Lower Tournaisian part of the column, megafaunal zones are certainly more refined than the global foraminiferal zones. In the Late Viséan-Early Namurian, the opposite is true.

Moreover, none of the megafaunal zone boundaries coincides with those of global zones. If they had coincided, one would have suspected nonsequence in the sedimentation, or facies-controlled assemblages.

Global zones can be quite long, and there is a temptation to "refine" them furthermore. For instance, Zone 13 can easily be split into two units in Belgium, northern France, and southern England, but this division appears questionable in Poland and in the Russian Platform and is lost in Asia. It is certainly not recognizable in the North American Realm. Although such refinement would be useful for the study of a given basin, the "refined" zones would only be local.

The basic advantage of global zones versus other types of zones is that they enable recognition of discrete chronostratigraphic sequences in widely dispersed basins. They are not a refinement of the stratigraphic scale.

CONCLUSIONS

Global zones can be established on Carboniferous benthic small foraminifers because of a number of favorable conditions: (1) marine benthic life zones are poorly defined because most of the fauna was restricted to the uppermost part of the platform; (2) the fauna spread laterally very far, displaying broad environmental tolerance in shallow waters; (3) foraminifers were in a period of rapid evolutionary radiation, probably associated with the existence of widespread epicontinental platforms where transgression and regressions had a profound influence on the net profit and loss balance of the fauna; cladogenesis with period of eruption was frequent and often underlines stage or series boundaries; (4) no *stratigraphic* provincialism is recognized; i.e., all families, the majority of the genera, and many species have a northern hemispheric distribution; (5) three freely interconnected paleobiological realms can be recognized: a Tethyan Realm, a North American-Kuznets Realm, and an Alaska-Taimyr Realm; (6) most of the fauna originally developed in the Tethyan Realm and subsequently migrated into the two other realms; little endemic North American fauna is known; (7) zones are established on clusters of first occurrences of assemblages taken among different phylogenic (or cladogenic) lines; and (8) global zones are the zones that can be recognized through the three realms.

It has been a tradition in stratigraphy to pursue the erection of new taxa, to emphasize faunal differences, and to create numerous local zones. This usually leads to parallel stratigraphic nomenclature with unfortunate succession of "chronostratigraphic" terms whose usage strangely coincides with political borders.

It is, however, equally interesting to emphasize faunal similarities of phylogenic or cladogenic evolution, and it is quite rewarding to follow the lateral extent of the zones and to correlate from basin to basin. It is finally worthwhile to screen the gigantic amount of faunal zones and to select not those with a deceivingly short range, but those with wide dispersal.

The Development of Fusulinid Biostratigraphy

Raymond C. Douglass

U.S. Geological Survey, Washington, D.C.

INTRODUCTION

Fusulinids are among the more useful fossils for correlation in marine rocks of late Paleozoic age. They are widespread geographically and are commonly found in such abundance that many rocks appear to be made of fusulinids with a minimum amount of matrix holding them together. Consequently, they attract much attention; but many who are attracted so easily lose interest in further study when confronted by the problems of making systematic and biostratigraphic sense out of this varied group. The initial problems of obtaining the required oriented thin sections from minute specimens embedded in hard matrix, or even of loose specimens washed from a silt, are only the beginning. The seemingly endless variability of characters developed in their shells is baffling, as no two sections through a single individual produce the same view, and variation within a sample population is considerable.

The purposes of this paper are to provide background for the study of fusulinids, to discuss the development of concepts and methods in fusulinid biostratigraphy, and to indicate fields needing further study.

Before proceeding, it would be useful to have a definition of the term "fusulinid." Unfortunately, fusulinids are not easily categorized, as they include a wide variety of late Paleozoic benthic marine protozoans (foraminifera) judged to be closely related to the spindle-shaped or fusiform genus *Fusulina.* Size varies from minute forms of less than 0.5 mm in the largest dimension to forms as much as 140 mm long. The shape of many fusulinids is fusiform, but shape varies from discoidal through rounded to subcylindrical (Text-fig. 1). All fusulinids are coiled septate forms, but the septa assume complex shapes, and other structures form within and between the chambers. Many of the changes in size, shape, and internal complexity are time related.

HISTORICAL BACKGROUND

Significant discoveries in the first 100 years of fusulinid study help place our problems and procedures in perspective. The first described fusulinids, *Miliolites secalicus* (Say, 1823, p. 51), came from the banks of the Missouri River near the Platte. This species was later designated the type for the genus *Triticites* (Girty, 1904, p. 239). The name *Fusulina* was proposed for the Carboniferous "fossil wheat" of the Moscow area (Fischer, 1829, p. 330). Fischer named two species, *Fusulina cylindrica* and *F. depressa*, published in 1837. By 1846, E. de Verneuil (1846a) had recognized fusulinids in the Carboniferous of Ohio as his "good Russian friend" *F. cylindrica.* This was the first published correlation using fusulinids.

As more occurrences of fusulinids were reported, it became obvious that they had stratigraphic usefulness, not only because of their wide distribution, but also because they evolved rapidly. The original discoveries in Nebraska and the Moscow Basin were followed by initial identifications in Ohio (de Verneuil, 1846a, 1846b), Iowa (Owen, 1852), the Arctic (Albert Land) (Salter, 1855), the Caucasus (Abich, 1858), Kansas (Meek and Hayden, 1858), Texas and New Mexico (Shumard, 1858, 1859), California (Meek, 1864), the Alps (Suess, 1870), Illinois (Meek and Worthen, 1873), Japan (Gümbel, 1874), Peru (Gabb, 1874-1881), Sumatra (Verbeek, 1875), British Columbia (Dawson, 1878), Persia (Möller, 1880), Spitzbergen (Göes, 1883), China and Japan (Schwager, 1883), the Salt Range of Pakistan (Schwager, 1887), India (Noetling, 1893), Guatemala and Mexico (Sapper, 1894a, 1894b), and Greece (Renz and Reichel, 1900). By 1900, fusulinids had been recognized worldwide. Their taxonomy was still in a primitive state, but a number of species had been described in several genera, and some stratigraphic distribution within the Carboniferous and Permian was recognized. The first extensive taxonomic study, with good illustrations of axial and equatorial sections and wall structure, was published by Schwager (1883) on fusulinids from China and Japan.

Schellwien (1898) described and illustrated beautifully 15 fusulinids from the Permian of the Carnic Alps, and provided a table of occurrence of the taxa in various beds. Schellwien also grouped the taxa into four sets he called "levels" but used as zones, and he correlated the four named levels with other parts of the world. Subsequently, Schellwien and co-workers began to monograph the fusulinids of the world; but Schellwien died before the first volume was published, so the series was limited to three parts: the Russian arctic faunas, completed by Staff (Schellwien, 1908); the Darwas faunas completed by Dyhrenfurth (Schellwien and Dyhrenfurth, 1909); and the North American faunas completed by Staff (Schellwien and Staff, 1912). The papers are illustrated with excellent photographs of well-oriented fusulinid thin sections, and in the first two parts the stratigraphy of the fusulinid-bearing beds is discussed and the associated fauna listed. The third part includes a summary of all the fusulinids reported from Alaska to Guatemala and from Ohio to California. No attempt at a summary zonation is made in these papers.

The first major compilation of global stratigraphic and geographic distribution of Carboniferous and Permian fusulinids was presented by Deprat (1912), accompanying a description of fusulinids from China and Indochina. He showed the stratigraphic ranges of the taxa and suggested a phylogeny for the fusulinids. The tables and diagrams were updated in 1913 (Deprat, 1913, pp. 62-71), and additional information from other parts of Indochina and from Japan was added in 1914 and 1915. A monumental summary of these studies by Deprat (1912-1915) was prepared by Colani (1924) in which she updated and collated the work of Deprat and described and illustrated many new taxa. Colani made extensive use of graphic representation of numerical data based on measurements and counts taken from thin sections.

A major step forward in the study of North American fusulinids was presented by Dunbar and Condra (1927) in which they systematically outlined the morphology, methods of study, classification, and distribution of Pennsylvanian and some Permian fusulinids. A distribution table was provided showing the occurrence and range of 21 species in 54 stratigraphic units occurring in seven states. They proposed three zones based on genera, and indicated that these could be subdivided into more limited faunal zones based on species. They listed nine possible zones from the Middle Pennsylvanian into the Lower Permian.

The early success of correlations based on fusulinids led to intensification of studies on faunas from all parts of the world. Although most of the reports continued to be on small collections from isolated samples, more comprehensive studies from stratigraphically controlled samples were also undertaken. Studies in the United States include Dunbar and Skinner (1937) on the Permian fusulinids of Texas, Dunbar and Henbest (1942) on the Pennsylvanian fusulinids of Illinois, Thompson (1954) on Wolfcampian fusulinids, Ross (1963) on the Wolfcampian of Texas, Skinner and Wilde (1965) on the Permian biostratigraphy and fusulinids in California, Petocz (1970) on biostratigraphy and Early Permian fusulinids in Alaska, and King (1973) on Pennsylvanian fusulinids from Texas and New Mexico.

Studies from other areas include Rauzer-Chernousova (1940) on the Upper Carboniferous and Lower Permian of the Urals, and (1961) on biostratigraphical subdivisions of the Middle Carboniferous of the Samara Bend area; Shcherbovich (1969) on fusulinids of Gzhelian and Asselian time of the pre-Caspian area; Kahler and Kahler (1937b, 1938, 1941) on fusulinids from the Carnic Alps; Toriyama (1958, 1967) on fusulinids and fusulinid zones of the Akiyoshi Plateau; Leven (1967) on the fusulinids of the Pamirs; and Ginkel (1965, 1973) on Carboniferous fusulinids from Spain.

Recent refinement of fusulinid taxonomy is found in papers such as those by Ishii (1958) on *Fusulina, Beedeina,* and allied genera, where morphology, stratigraphy, and geography were all used to work out the lineages of Carboniferous fusulinids; Stewart (1968) on *Eowaeringella* and related forms; Sanderson and Verville (1970) on the morphologic variability of the genus *Schwagerina*, who showed that variation within one sample population encompasses as many as six named species; and Douglass (1970) on *Monodiexodina* from West Pakistan, who gave methods for discerning relationships between superficially dissimilar forms.

Among papers that interpret the phylogeny of fusulinids, Ishii (1958) gave a careful analysis of the lineages in the Upper Carboniferous; Miklukho-Maklai (1959a) discussed the systematics and phylogeny of the fusulinids with special reference to *Triticites* and related genera; Ross (1962) presented the evolution and dispersal of the Permian fusulinid genera *Pseudoschwagerina* and *Paraschwagerina*; Leven (1963) presented a phylogeney of the higher fusulinids in the Upper Permian; Kochansky-Devidé (1969) suggested parallel tendencies in the evolution of the fusulinids; Rosovskaya (1969) revised the taxonomy presenting her views on the phylogeny of the fusulinids; Ozawa (1967) presented a phylogeny for the species groups in *Pseudofusulinella* and the phylogeny and classification of the superfamily Verbeekinoidea (Ozawa, 1970); and Wilde (1971) presented another view on the phylogeny of *Pseudofusulinella*. These refinements of the phylogeny have resulted in greater precision in the definition of the biostratigraphic zones.

The influence of local environment and of geography on the development and distribution of fusulinids has received attention from Rauzer-Chernousova and Kulik (1949); from Ross, who discussed fusulinids as paleoecological indicators (1961, 1969) and the development of fusulinid faunal realms (1967); and from Gobbett, who presented the paleozoogeography of the Verbeekinidae (1967). Recognition of fusulinid provincialism and ecological restrictions set new, more conservative limits on their application to biostratigraphy.

EVOLUTIONARY TRENDS AND THEIR STRATIGRAPHIC SIGNIFICANCE

Because the taxonomy and phylogeny of the fusulinids are closely interwoven and because fusulinids tend to evolve rapidly, stages of development are useful for stratigraphic correlation.

Some general trends in the development run parallel, having evolved concurrently in several species groups or genera (see Kochansky-Devidé 1969); other trends are repetitive, occurring again and again at different stratigraphic levels. Where sequences of closely spaced collections containing abundant fusulinids are available, the changes can be seen as a continuous progression. More commonly, fusulinids are found concentrated in certain beds separated by barren beds, and the changes in morphology appear to be more abrupt. Several trends having stratigraphic significance are discussed below; for further discussion the reader is referred to Thompson (1964, p. C381-C387).

Size and Shape

The earliest forms included among the fusulinids are small lenticular forms (Text-fig. 1) less than 0.5 mm in diameter, and some of the latest are subcylindrical forms as much as 140 mm long. Increase in size, however, is not a regularly progressive attribute. Within any one species group, the trend to greater size through time is apparent, but different groups enlarge at different rates and originate at different times, so the size must be observed in the context of the specific group to have more than general utility in biostratigraphy.

Size increase occurs in some groups by elongation of the axis, developing a rounded or fusiform shape from a discoidal shape. Increase also takes place in diameter, either separately or in combination with the increase in length. One dimension may increase more or less rapidly than the other, ontogenetically in a specimen, or phylogenetically through a series of populations.

Shape varies with the relative dimensions but is also influenced by the curvature of the outer wall, which may tend to be straight, sunken or concave, or inflated and rounded. Most forms are involute, so that the last whorl completely encloses the preceding whorls; but some of the more discoidal forms are evolute, showing some or all of the inner whorls along the axis. A few forms develop an uncoiled series of chambers before growth ends. The uncoiled part may be a flare on elongate forms or a simple uniserial extension on discoidal forms. Aberrant shapes, probably not genetically controlled, are also found occasionally (Wilde, 1965). These may form because of multiple proloculi, damage during growth, inclusion of foreign objects, or other causes. Little stratigraphic value can be attributed to forms with abnormal growth, but they do tend to develop at some horizons more than others, perhaps because of environmental influence.

Proloculus

Growth of the fusulinid shell starts with an initial chamber, the proloculus, that tends to be spherical. Sexual and asexual (microspheric and megalospheric) generations are recognized in several species. Very little information is available on the size of the proloculus in microspheric forms. A general trend toward increase in size of the megalospheric proloculus through time is apparent. Pennsylvanian prolocular diameters seldom exceed 200 μm, whereas diameters of 500 μm are not uncommon in Permian fusulinids, and some of almost 1,000 μm occur. The usual caution should be noted, as many Permian fusulinids also have small proloculi. The particular lineage must be considered within the general trend.

Spirotheca

The wall that creates the external form of the fusulinid is called the spirotheca. Each added chamber is formed by an extension of the wall as spirotheca and a downward extension, which temporarily forms the antetheca but which, when the next chamber is added, is referred to as a septum. The structure of the wall is not uniform throughout the shell and is generally simplest in the last added chamber. Although the wall structure of all genera has not been thoroughly studied, there appears to be a regular progressive development in structural complexity through time. The development takes place at different rates in different lineages, and some forms of specialization tend to mask the overall trend.

The initial or primitive wall (Text-fig. 1, item 2a) is composed of a thin dense layer (tectum) and a thicker inner layer (diaphanotheca) that appears to be structureless or finely granular in most specimens. Some specimens with exceptional preser-

Text-Figure 1

Evolutionary trends in the fusulinids; specimens 1 and 2 enlarged. Specimens 3 to 24 show a general increase in size with time but vary considerably. Specimen 1 and all the wall-structure details are at the same magnification. 1, *Millerella* has lenticular shape, simple septa, small but obvious chomata, and thin wall with little structure apparent. 2-2a, *Profusulinella* shows inflated fusiform shape, nearly plane septa, small but obvious chomata, and a thicker wall composed of a tectum, diaphanotheca, and outer tectorium visible in the inner volutions. 3, *Fusulinella* shows typical fusiform shape, nearly straight septa, well-developed chomata; wall (3a) has a tectum, diaphanotheca, and tectoria. 4, *Wedekindellina* shows elongate fusiform shape, nearly straight septa, well-developed chomata, axial filling, and wall structure as in 3a. 5-6, Axial and tangential sections of *Beedeina* showing inflated fusiform shape, well-developed chomata, tightly but irregularly fluted septa forming some chamberlets visible in 6; wall in this form is similar to 7a-b. 7, *Beedeina*, an advanced species showing elongate shape, intensely fluted septa, and less massive chomata; wall commonly resembles that shown in 7a, but unusual preservation (7b) shows the perforate nature of the wall. 8, *Triticites*, an early form showing an elongate subcylindrical shape, well-developed chomata, irregularly fluted septa, and wall structure resembling 12a but much finer pores. 9, *Triticites*, somewhat similar to 8 but smaller; septa show less fluting. 10, *Triticites*, a large fusiform species with massive chomata and irregular fluting. 11, *Triticites*, an elliptical form with more open coiling and irregular fluting. 12, *Triticites*, one of the largest forms, fusiform with prominent chomata and irregular fluting of the septa; keriothecal wall (12a) is relatively coarse. 13, *Pseudofusulinella*, with fusiform shape, prominent chomata, irregularly fluted septa, and wall similar to 3a. 14, *Pseudoschwagerina*, showing a subrounded shape, inner volutions similar to *Triticites* with obvious chomata, and irregular septal fluting; the outer volutions greatly expanded with fluting only near the base of the septa; wall similar to 12a. 15, *Paraschwagerina*, showing inflated fusiform shape, inner volutions tightly coiled resembling *Schwagerina* with no chomata and with tightly fluted septa; outer volutions expand rapidly and continue without chomata and with tightly fluted septa; wall is similar to 17a. 16, *Schwagerina* showing an inflated fusiform shape, a regular expansion of the shell, no chomata, and tightly fluted septa; wall is similar to 17a. 17, *Pseudofusulina* showing rapid expansion from a large proloculus to a large elongate fusiform shape; septa are tightly fluted, and wall (17a) is thick and coarsely alveolar; 18, *Parafusulina*, tangential section showing the tight septal fluting producing chamberlets and cuniculi. 19, *Parafusulina* with elongate-fusiform shape, no chomata, and regularly tightly fluted septa. 20, *Verbeekina*, showing rounded shape, loosely coiled form, nearly plane septa. 21, *Polydiexodina*, part of an axial section of an elongate subcylindrical specimen showing tightly and regularly fluted septa, multiple tunnels, and axial filling. 22, *Polydiexodina,* tangential section showing the fluted septa forming cuniculi and the presence of multiple tunnels. 23, *Sumatrina*, showing the inflated fusiform shape and the development of parachomata, transverse septa, and septula. 23a-b, Details of the wall in equatorial and axial section showing the parachomata, septa, and septula. 24, *Lepidolina*, showing the inflated fusiform shape, tight coiling, and the development of parachomata, septa, and septula, details of which are shown in 24a.

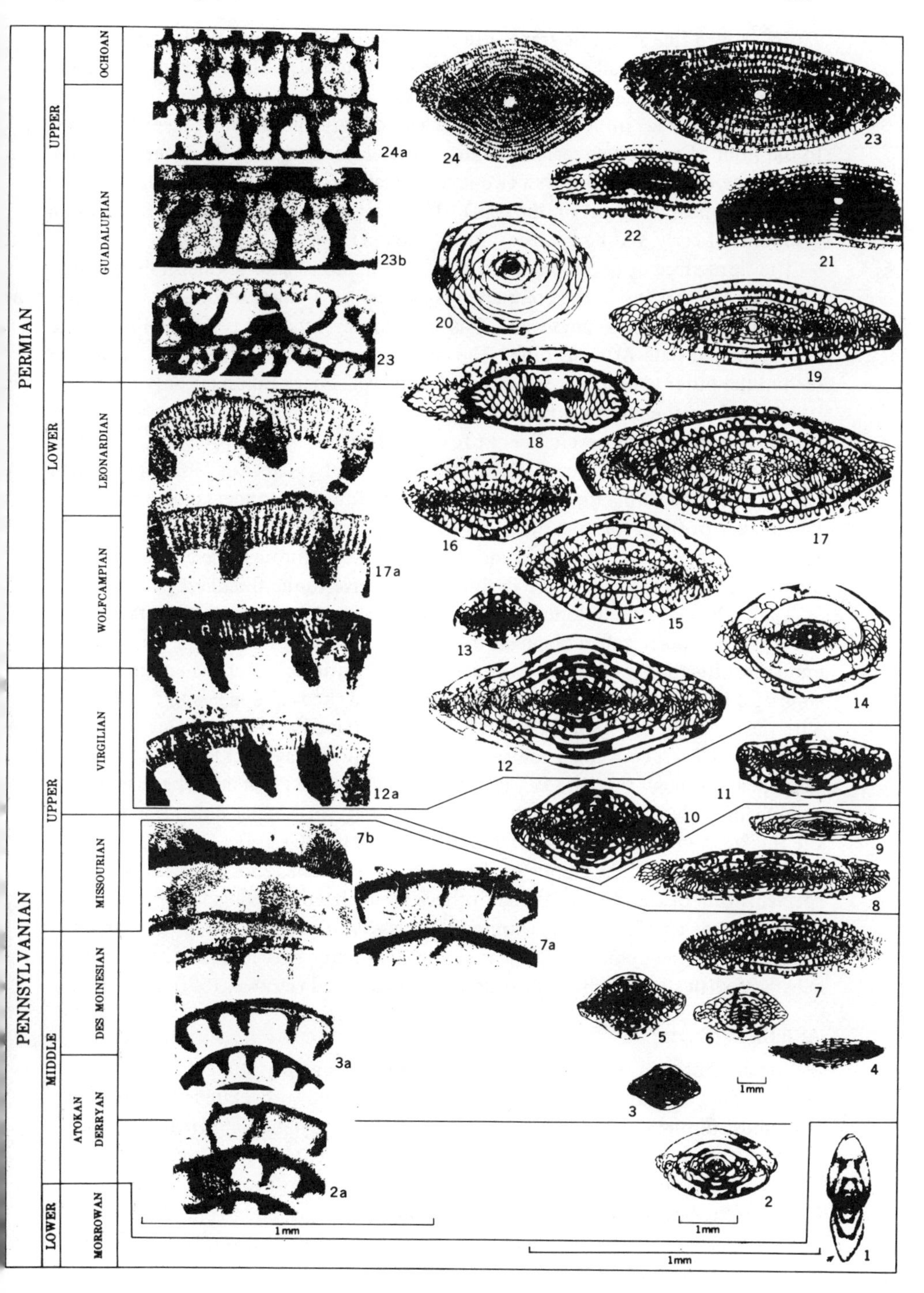
PERMIAN
UPPER
LOWER
OCHOAN
GUADALUPIAN
LEONARDIAN
WOLFCAMPIAN
PENNSYLVANIAN
UPPER
MIDDLE
LOWER
VIRGILIAN
MISSOURIAN
DES MOINESIAN
ATOKAN DERRYAN
MORROWAN
24a
23b
23
17a
12a
7b
7a
3a
2a
24
23
22
21
20
19
18
17
16
15
14
13
12
11
10
9
8
7
6
5
4
3
2
1
1mm
1mm
1mm
1mm

vation show a fine structure transverse to the wall (Skinner and Wilde, 1954a, pl. 49, fig. 3). This kind of wall is seen in the simplest fusulinids of Early Pennsylvanian and persists in some of the more conservative taxa.

The second stage of development adds a secondary fine granular layer (tectorium) to the floor of each chamber, producing a three-layered wall in the inner volutions but leaving the last whorl with a two-layered wall. The tectorium is referred to as an epithecal deposit (deposit on the wall). This stage of development is best shown in fusulinids of early Middle Pennsylvanian (Atokan) age.

The third stage of development (Text-fig. 1, item 3a) adds epithecal deposits on the other surfaces inside the completed chambers, forming a second or inner tectorium and producing a four-layered wall. This stage of development is attained in Middle Pennsylvanian (late Atokan) time. The development of a four-layered wall can be seen in the outer volution or volutions of some specimens that have only a three-layered wall in the inner volutions.

The fourth stage of development (Text-fig. 1, item 7a) is a coarsening of the structure transverse to the wall. Late Middle Pennsylvanian (Des Moinesian) time is characterized by forms that have a wall with the four-layer appearance. An increasing number of these forms show an alveolar or porous transverse structure (Text-fig. 1, item 7b) in the diaphanotheca and, to some extent, through the entire wall. The diaphanotheca tends to increase in thickness relative to the thickness of the tectoria.

The fifth stage of development (Text-fig. 1, item 12a) is a further coarsening of the layer previously called diaphanotheca. The alveoli are now visible under moderate magnification, and the layer is called the keriotheca. Tectoria are diminished to insignificant thickness or are not deposited, and the wall appears to be two layered with only tectum and keriotheca. This stage is attained at the beginning of the Late Pennsylvanian. A continued coarsening of the keriotheca (Text-fig. 1, item 17a) takes place through the Late Pennsylvanian and into the Permian. The diameters of individual alveoli in some Early Permian Schwagerinas are as much as one fourth the thickness of the wall (Dunbar and Skinner, 1937, pl. 44, figs. 1 and 2).

The sixth stage of development (Text-fig. 1, items 23a, 23b, and 24a) is a specialization of elements of the keriotheca which become elongated, extending down into the chamber space, forming septula (Skinner and Wilde, 1954a, p. 449). This stage of development is known only in the family Verbeekinidae, and several substages with stratigraphic significance are recognized (Ozawa, 1970, fig. 10).

Antetheca (Septa)

The antetheca in the most primitive forms are essentially straight or slightly curved. They remain unfluted in the ozawainellids and verbeekinids, but in most fusulinids, fluting is formed progressively with time. Dunbar and Henbest (1942, p. 44) outlined six grades of septal fluting arbitrarily chosen from a continuously evolving series. Fluting is generally most intense (1) near the poles of a specimen, (2) along the base of the septa, (3) in the outer volutions, and (4) in geologically younger specimens. Several lineages develop intensive fluting at different times: (1) late in the Middle Pennsylvanian (Text-fig. 1, item 7) for the fusulinids;

(2) late in the Early Permian (Text-fig. 1, item 19) for the schwagerinids; and (3) early in the Late Permian for the schubertellids.

Chomata

Chomata are secondary deposits formed lateral to the tunnel, producing levee-like ridges along the base of the chamber. A general trend is the reduction of the chomata from relatively large deposits in the Middle Pennsylvanian (Text-fig. 1, item 3) to smaller deposits in the Late Pennsylvanian (Text-fig. 1, item 11). Many Early Permian forms have chomata formed only in the innermost volutions (Text-fig. 1, item 14) and pseudochomata, or discontinuous chomata-like deposits at the base of the septa, beside the tunnel, but not forming a continuous band. Chomata are not present in most of the later Permian forms (Text-fig. 1, item 19).

Axial Filling

Secondary or epithecal deposits along the axis and coating the septa in the polar areas are referred to as axial filling. Axial filling occurs in several genera distributed in time for Middle Pennsylvanian (Text-fig. 1, item 4) to Late Permian (Text-fig. 1, item 21). No clear trends have been recognized in the occurrence of axial filling.

THE INFLUENCE OF ENVIRONMENT ON FUSULINID DISTRIBUTION

Fusulinids are widely distributed in rocks of Late Paleozoic age, but they are not found in all parts of the world nor in all kinds of rocks. Furthermore, some kinds of fusulinids that are common in some parts of the world are unknown where other kinds of fuslinids are abundant. The distribution of fusulinids was apparently controlled by local environments and also by larger-scale geographic features.

Local Environments

Fusulinids are marine organisms that lived on shallow clastic or carbonate shelves marginal to the Pennsylvanian and Permian seaways. Although they are found in varied sediments from silica sands to reef limestones, they are most commonly found in limestones and calcareous sandstones. Ross (1961, p. 398) showed that some fusulinid species are closely associated with particular rock types. In a study of species of *Schwagerina* and *Parafusulina* in the Leonard Formation of the Glass Mountains in Texas, he plotted fusulinid occurrence on a grid with calcium carbonate grain size plotted against percentage of silt and clay impurities. The species proved to be in nearly discrete groups. Ozawa (1970, pp. 22-26) discussed the paleoecology of fusulinids and diagrammed the relationship between several taxa and the environment of sedimentation. Another study by Ross (1969) related species of *Triticites* and *Dunbarinella* to the sedimentary facies in the Upper Pennsylvanian deposits of Texas. He found that the fusulinids were adapted only to the shelf environments from shallow water out to near wave base. Eleven lineages of *Triticites* and one of *Dunbarinella* were related to subenvironments on the shelf, and tended to adapt in shape, wall thickness, and spacing and folding of the septa to the local physical environment.

Faunal Provinces and Fusulinid Distribution

Some fusulinids are cosmopolitan, occurring almost everywhere that fusulinids of the proper age are found. Other fusulinids are restricted to more limited parts of the total fusulinid range. Ross (1967) summarized information on fusulinid distribution and recognized three faunal provinces: Eurasian-Arctic, Tethyan, (in the Permian only), and Midcontinent (USA)-Andean. Some fusulinid lineages are restricted to one province, such as the verbeekinids in the Tethyan. Gobbett (1967) described the biogeography of this group. Most lineages seem to be generalized in their early history before becoming localized in one of the more restricted provinces.

The provinces may not be independent of environment as interpreted on the local level. Most fusulinids attributed to the Tethyan province are not in the normal shelf facies. Instead, they are found in massive limestones commonly associated with volcanic rocks and other sediments generally referred to as eugeosynclinal facies. A generalization that the Midcontinent-Andean province is miogeosynclinal and the Tethyan realm, eugeosynclinal has exceptions, but is reasonably accurate. One principal exception is the presence of normal shelf facies with Tethyan fusulinids in the heart of Tethys at Djebel Tebaga, Tunisia.

WORLD DISTRIBUTION OF FUSULINIDS

Fusulinids are known from most areas of the world where there are marine rocks of Late Paleozoic age with the exceptions of Australia, Tasmania, Antarctica, and Madagascar. The deposits in Antarctica are poorly known and may yet include fusulinid-bearing rocks, because fusulinids are known as far as 52°S along the Andean geosyncline. Fusulinids are known well within the Arctic Circle in several areas, from the Yukon in western Canada through the Canadian Arctic Islands, and Greenland, Spitzbergen, Novaya Zemlya, and northern Siberia. The distribution is primarily northern hemisphere on maps of the present world, with exceptions in South America, New Zealand, and the East Indies (Text-fig. 2).

REPRESENTATIVE ZONATION OF FUSULINIDS

The fusulinid zones based on genera are recognized worldwide. Local species zones

---

Text-Figure 2
Distribution of known fusulinid faunas. Localities are generalized in areas of common occurrence. Note that both Pennsylvanian and Permian faunas are reported from within the Arctic Circle and as far south as southern Chile.

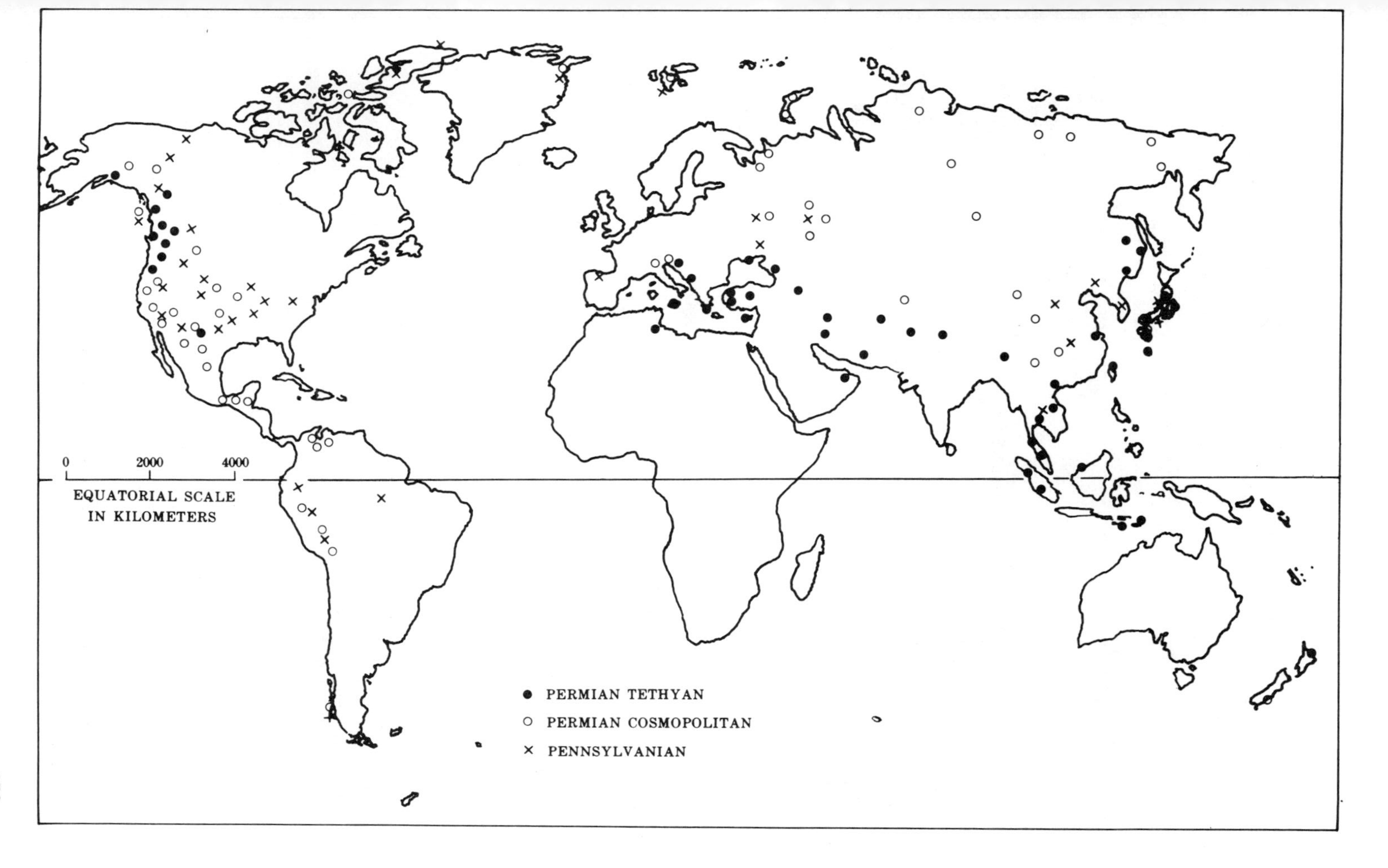
0
2000
4000
EQUATORIAL SCALE
IN KILOMETERS
PERMIAN TETHYAN
PERMIAN COSMOPOLITAN
PENNSYLVANIAN

are recognized in areas where more intensive studies were made. The general zones are represented in Text-figure 3 and are discussed below, followed by more specific

		Eurasia		North America		Japan	Fusulinid Zones: North America	Fusulinid Zones: Cosmopolitan	Fusulinid Zones: Tethys
PERMIAN	PERMIAN	DZHULFIAN	PAMIRAN	OCHOAN		KUMAN			PALAEOFUSULINA LEPIDOLINA
		KAZANIAN	MURGABIAN	GUADALUPIAN	CAPITANIAN	KUMAN	LEELLA POLYDIEXODINA	RAUSERELLA REICHELINA CODONOFUSIELLA	SUMATRINA YABEINA
		KAZANIAN	MURGABIAN	GUADALUPIAN	WORDIAN	AKASAKAN			PSEUDODOLIOLINA AFGHANELLA
		ARTINSKIAN	DARVASIAN	LEONARDIAN		NABEYAMAN		PARAFUSULINA	VERBEEKINA NEOSCHWAGERINA NAGATOELLA CANCELLINA
		KARACATIRAN	SAKMARIAN ASSELIAN	WOLFCAMPIAN		SAKAMOTOZAWAN	PSEUDOSCHWAGERINA PARASCHWAGERINA	SCHWAGERINA PSEUDOFUSULINA	BREVAXINA MISELLINA ORIENTOSCHWAGERINA ACERVOSCHWAGERINA
PENNSYLVANIAN	CARBONIFEROUS	GZHELIAN	ORENBURGIAN KASIMOVIAN	MISSOURIAN	VIRGILIAN	HIKAWAN		TRITICITES	
		MOSCOVIAN	UPPER LOWER	DES MOINESIAN		KURIKIAN AKIYOSHIAN		BEEDEINA FUSULINA WEDEKINDELLINA FUSULINELLA	
		BASHKIRIAN		DERRYAN	ATOKAN	ATETSUAN		PROFUSULINELLA	
		NAMURIAN	BASHKIRIAN	MORROWAN		KAMITAKARAN		PSEUDOSTAFFELLA MILLERELLA EOSTAFFELLA	

Text-Figure 3
Generalized fusulinid zones represented in the three faunal provinces. Greater detail is given in Text-figures 4 to 6.

zonations for three regions: the Soviet Union (Text-fig. 4), the United States (Text-fig. 5), and Japan (Text-fig. 6). Throughout the discussion of zones, it should be kept in mind that the genus or species for which a zone is named may not occur throughout the zone or be confined to the zone. The name is used only because the taxon is present and perhaps most characteristic or most striking in the bed or group of beds assigned to the zone. The generic zones also tend to be defined in terms of

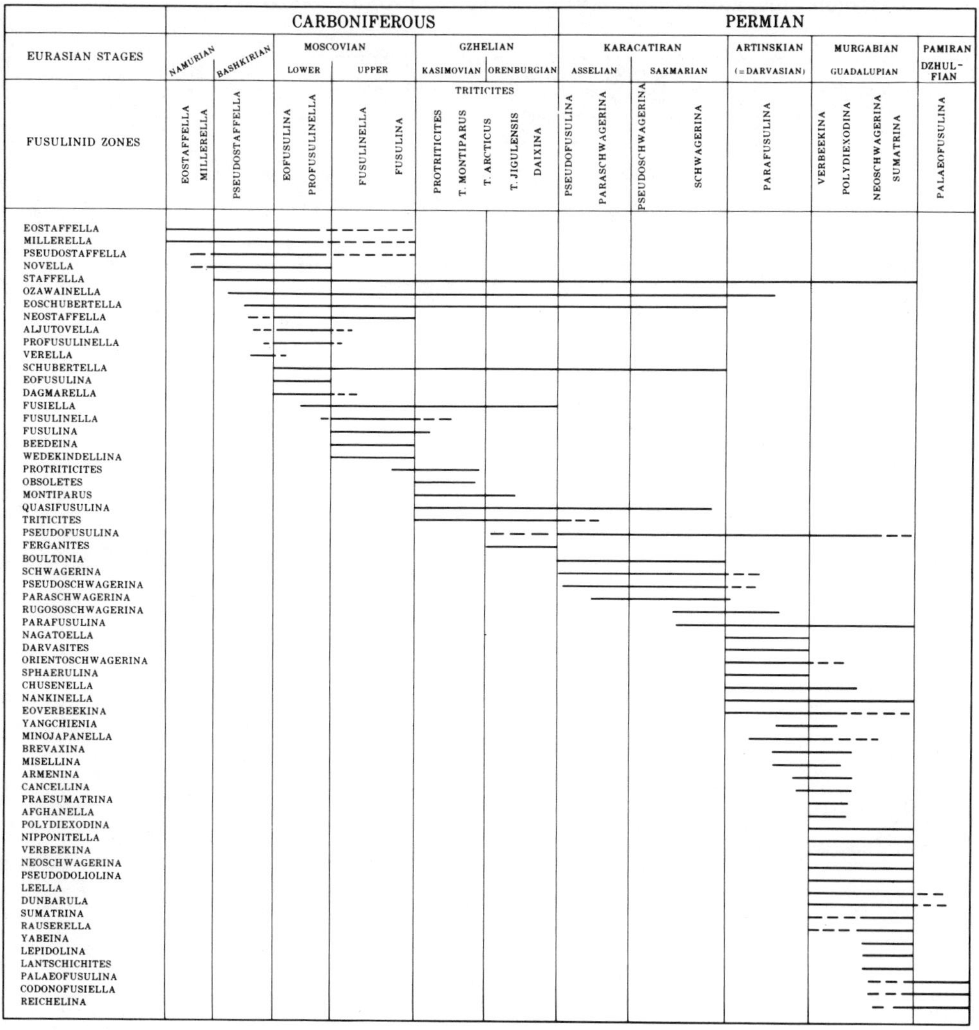

CARBONIFEROUS PERMIAN

EURASIAN STAGES: NAMURIAN; BASHKIRIAN; MOSCOVIAN (LOWER, UPPER); GZHELIAN (KASIMOVIAN, ORENBURGIAN); KARACATIRAN (ASSELIAN, SAKMARIAN); ARTINSKIAN (= DARVASIAN); MURGABIAN (GUADALUPIAN); PAMIRAN (DZHUL-FIAN)

FUSULINID ZONES: EOSTAFFELLA; MILLERELLA; PSEUDOSTAFFELLA; EOFUSULINA; PROFUSULINELLA; FUSULINELLA; FUSULINA; TRITICITES: PROTRITICITES, T. MONTIPARUS, T. ARCTICUS, T. JIGULENSIS, DAIXINA; PSEUDOFUSULINA; PARASCHWAGERINA; PSEUDOSCHWAGERINA; SCHWAGERINA; PARAFUSULINA; VERBEEKINA; POLYDIEXODINA; NEOSCHWAGERINA; SUMATRINA; PALAEOFUSULINA

EOSTAFFELLA
MILLERELLA
PSEUDOSTAFFELLA
NOVELLA
STAFFELLA
OZAWAINELLA
EOSCHUBERTELLA
NEOSTAFFELLA
ALJUTOVELLA
PROFUSULINELLA
VERELLA
SCHUBERTELLA
EOFUSULINA
DAGMARELLA
FUSIELLA
FUSULINELLA
FUSULINA
BEEDEINA
WEDEKINDELLINA
PROTRITICITES
OBSOLETES
MONTIPARUS
QUASIFUSULINA
TRITICITES
PSEUDOFUSULINA
FERGANITES
BOULTONIA
SCHWAGERINA
PSEUDOSCHWAGERINA
PARASCHWAGERINA
RUGOSOSCHWAGERINA
PARAFUSULINA
NAGATOELLA
DARVASITES
ORIENTOSCHWAGERINA
SPHAERULINA
CHUSENELLA
NANKINELLA
EOVERBEEKINA
YANGCHIENIA
MINOJAPANELLA
BREVAXINA
MISELLINA
ARMENINA
CANCELLINA
PRAESUMATRINA
AFGHANELLA
POLYDIEXODINA
NIPPONITELLA
VERBEEKINA
NEOSCHWAGERINA
PSEUDODOLIOLINA
LEELLA
DUNBARULA
SUMATRINA
RAUSERELLA
YABEINA
LEPIDOLINA
LANTSCHICHITES
PALAEOFUSULINA
CODONOFUSIELLA
REICHELINA

Text-Figure 4
Fusulinid zonation in the Soviet Union. (After Miklukho-Maklai, 1963a, 1963b, Ruzhentsev and Sarycheva, 1965, Rosovskaya, 1969, and Shcherbovich, 1969; more details of the zonation can be found in these papers.)

the stratigraphic series or stages and are therefore subject to all the problems encountered in the definitions of series and stage boundaries. Circular reasoning involving series based on zones and zones based on series continues to be a problem. The zones are discussed in ascending order.

Zone of *Millerella*

The Lower Pennsylvanian (Morrowan) rocks and their equivalent parts of the Carboniferous (mostly Upper Namurian) are called the zone of *Millerella*. *Millerella*

	PENNSYLVANIAN						PERMIAN				
NORTH AMERICAN SERIES	LOWER MORROWAN	MIDDLE DERRYAN ATOKAN	MIDDLE DES MOINESIAN	UPPER MISSOURIAN	UPPER VIRGILIAN		WOLFCAMPIAN	LEONARDIAN	GUADALUPIAN WORDIAN	GUADALUPIAN CAPITANIAN	OCHOAN
FUSULINID ZONES	MILLERELLA-EOSTAFFELLA	PROFUSULINELLA	WEDEKINDELLINA, BEEDEINA	T. OHIOENSIS	TRITICITES	T. CULLOMENSIS	PSEUDOSCHWAGERINA	NEOSCHWAGERINA, PARAFUSULINA	FUSULINELLA	YABEINA, POLYDIEXODINA	LEPIDOLINA

OZAWAINELLA
NANKINELLA
STAFFELLA
PSEUDOSTAFFELLA
EOSTAFFELLA
MILLERELLA
EOSCHUBERTELLA
PROFUSULINELLA
FUSULINELLA
WEDEKINDELLINA
BEEDEINA
BARTRAMELLA
KANSANELLA
TRITICITES
DUNBARINELLA
PSEUDOFUSULINELLA
SCHUBERTELLA
PSEUDOFUSULINA
CHUSENELLA
SCHWAGERINA
PARASCHWAGERINA
PSEUDOSCHWAGERINA
NAGATOELLA
CHALAROSCHWAGERINA
CUNICULINELLA
EOPARAFUSULINA
KLAMATHINA
PSEUDOREICHELINA
BOULTONIA
MONODIEXODINA
PARAFUSULINA
MISELLINA
CANCELLINA
NEOSCHWAGERINA
SKINNERINA
VERBEEKINA
YANCHIENIA
AFGHANELLA
LEELLA
DUNBARULA
PALAEOFUSULINA
REICHELINA
RAUSERELLA
CODONOFUSIELLA
POLYDIEXODINA
YABEINA
LEPIDOLINA
SUMATRINA
PARADOXIELLA
PARABOULTONIA

Text-Figure 5
Fusulinid zonation in the United States. (After Thompson, 1948, 1964, Skinner and Wilde, 1965, and Ross, 1967, 1970; more details are available in these papers.)

SYSTEM	STAGE	FUSULINID ZONE	TYPICAL FUSULINIDS
PERMIAN	KUMAN	PALAEOFUSULINA CODONOFUSIELLA LEPIDOLINA	LEPIDOLINA KUMAENSIS L. MULTISEPTATA
		YABEINA	YABEINA GLOBOSA, Y. COLUMBIANA, SUMATRINA ANNAE LEPIDOLINA TORIYAMAI
		POLYDIEXODINA	YABEINA GUBLERI, Y. SHIRAIWENSIS, POLYDIEXODINA
	AKASAKAN	NEOSCHWAGERINA: N. MARGARITAE	NEOSCHWAGERINA MARGARITAE, VERBEEKINA COLANIA DOUVILLEI N. COLANIAE, N. DOUVILLEI
		NEOSCHWAGERINA: N. CRATICULIFERA	SUMATRINA ANNAE PSEUDODOLIOLINA OZAWAI NEOSCHWAGERINA CRATICULIFERA, VERBEEKINA VERBEEKI CANCELLINA NIPPONICA
	NABEYAMAN	NEOSCHWAGERINA: N. SIMPLEX	NEOSCHWAGERINA SIMPLEX, PARAFUSULINA KAERIMIZENSIS
	SAKAMOTOZAWAN	PSEUDOFUSULINA AMBIGUA MISELLINA	PSEUDOFUSULINA AMBIGUA, PSF. KRAFFTI, PARAFUSULINA JAPONICA MISELLINA CLAUDIAE, CANCELLINA PRIMIGENA
		PSEUDOFUSULINA VULGARIS PSEUDOSCHWAGERINA MORIKAWAI	PSEUDOFUSULINA VULGARIS, PSF. DONGVANENSIS, PSF. JAPONICA PSEUDOSCHWAGERINA SCHELLWIENI, PSS. HIDENSIS PSS. MUONGTHENSIS, PSS. UDDENI, PSS. ORIENTALE RUGOSOFUSULINA ARCTICA, QUASIFUSULINA LONGISSIMA TRITICITES SIMPLEX, T. OZAWAI
CARBONIFEROUS	HIKAWAN	TRITICITES: T. YAYAMADAKENSIS	TRITICITES YAYAMADAKENSIS, SCHUBERTELLA SP., STAFFELLA SP. T. EXSCULPTUS, T. HIDENSIS, T. SAURINI, T. SAKAGAMII
		TRITICITES: T. MATSUMOTOI	T. MATSUMOTOI, T. NAKATSUGAWENSIS, T. OPPARENSIS
	KURIKIAN	FUSULINELLA: F. OHTANII	FUSULINA OHTANII FUSULINELLA GRACILIS
		BEEDEINA: B. HIGOENSIS	BEEDEINA HIGOENSIS WEDEKINDELLINA PROLIFICA
		STAFFELLA PSEUDOSPHAEROIDEA	STAFFELLA PSEUDOSPHAEROIDEA, FUSULINELLA HIROKOAE
	AKIYOSHIAN	FUSULINELLA: F. BICONICA	FUSULINELLA BICONICA, F. ITOI, F. SUBSPHAERICA, F. ASIATICA FUSIELLA TYPICA
		FUSULINELLA: F. SIMPLICATA	FUSULINELLA BOCKI, F. ITADORIGAWENSIS FUSULINELLA SIMPLICATA, F. KAMITAKAENSIS, F. JAMESENSIS
	ATETSUAN	PROFUSULINELLA	PROFUSULINELLA BEPPENSIS, P. RHOMBOIDES, P. TORIYAMAI P. FUKUJIENSIS NANKINELLA PLUMMERI STAFFELLA POWWOWENSIS EOSCHUBERTELLA LATA
	KAMITAKARAN	EOSTAFFELLA MILLERELLA	PSEUDOSTAFFELIA ANTIQUA, P. KANUMAI EOSTAFFELLA KANMERAI MILLERELLA JAPONICA, M. GIGANTEA, M. KOMATUI

Text-Figure 6

Fusulinid zonation in Japan. (After Toriyama, 1958, 1967, 1973, and Ozawa, 1970; more details are available in these papers.)

originates earlier, possibly during Visean time, and continues on into the Upper Pennsylvanian and possibly into the Permian. The zone is commonly recognized as those Lower Pennsylvanian beds below the first occurrence of *Profusulinella*. Thus the zone is poorly defined and difficult to recognize in most sections. King's (1973) study of the taxonomy and stratigraphic distribution of the primitive fusulinids of this zone may lead to a more precise understanding and correlation of the zone. Other fusulinid genera that are reported from this zone include *Eostaffella* (a form closely related to *Millerella* and perhaps congeneric) and some primitive forms of *Pseudostaffella*. In Japan, an upper subzone has been recognized in some areas (Toriyama, 1967, p. 38) and named for the *Pseudostaffella* present.

Zone of *Profusulinella*

The lower part of the Atokan (or Derryan) series of early Middle Pennsylvanian age is assigned to the zone of *Profusulinella*. Equivalents include the Bashkirian stage and the Atetsuan series of Japan. The zone includes the beds between the zone of *Millerella* and the first occurrence of *Fusulinella*. Genera represented include *Profusulinella, Millerella, Eostaffella, Pseudostaffella, Eoschubertella, Nankinella,* and possibly *Staffella* and *Ozawainella*. Where more or less complete sequences are available, it is not uncommon to find species that are transitional in morphology between *Profusulinella* and *Fusulinella*. The fusulinellid wall commonly forms first in the outer volutions. The zone boundary is no less precise than it would be if the transitional form were missing.

Zone of *Fusulinella*

The upper part of the Atokan (or Derryan) series of early Middle Pennsylvanian age is assigned to the zone of *Fusulinella*. Equivalents include the lower Moscovian stage and the Akiyoshian series. The zone includes the beds between the top of the *Profusulinella* zone and the first occurrence of *Beedeina* or *Fusulina*. The genus *Fusulinella* occurs from the base of the zone and continues into the overlying zone. This is one of the most widely recognized fusulinid zones in the world and includes representatives of the following genera: *Fusulinella, Millerella, Eostaffella, Schubertina, Taitzehoella, Wedekindellina, Pseudowedekindellina, Fusiella, Eoschubertella, Pseudostaffella, Ozawainella,* and *Staffella*.

Zone of *Beedeina*

The Des Moinesian series of late Middle Pennsylvanian age is assigned to the zone of *Beedeina* (*Fusulina* of author). Equivalents include the upper Moscovian stage and the Kurikian stage of Japan. The zone includes the beds of Middle Pennsylvanian age above the zone of *Fusulinella*. *Beedeina* is restricted to the zone but is not recognized worldwide. The zone has been called the zone of *Fusulina*, but reinterpretation of that genus has placed all the U.S. forms of Des Moinesian age in *Beedeina*, so the name of the zone was changed. *Fusulina* has a range through the zone and into younger beds in Eurasia and the Far East. *Fusulina* is known from rocks of early

Missourian age in the United States (Thompson et al., 1956). The genera found in the zone of *Beedeina* include *Millerella, Eostaffella, Schubertina, Fusiella, Beedeina, Fusulina, Wedekindellina, Pseudowedekindellina, Bartramella, Taitzehoella, Putrella, Hidaella, Plectofusulina, Ozawainella, Staffella,* and *Pseudostaffella.*

Zone of *Triticites*

The Upper Pennsylvanian (Missourian and Virgilian) is assigned to the zone of *Triticites.* Equivalents include the Kasimovian and Gzhelian (probably synonymous in part), the Orenburgian, and the Hikawan series of Japan. The zone includes the beds from the first occurrence of *Triticites* to the base of the Permian. *Triticites* occurs throughout the zone but continues into the Permian. Other genera represented in the zone include *Millerella, Fusiella, Oketaella, Fusulina, Quasifusulina, Pseudofusulinella, Dunbarinella, Kansanella, Eowaeringella, Ozawainella,* and *Staffella.*

Zone of *Pseudoschwagerina*

The Wolfcampian series of Early Permian age is assigned to the zone of *Pseudoschwagerina.* Equivalents include the Asselian and most of the Sakmarian stage, and all but the upper part of the Sakamotozawan stage of Japan. The zone includes the beds from the base of the Permian to base of the Leonardian. *Pseudoschwagerina* occurs from near the base of the zone to the top of the zone in most areas, and some specialized forms continue into the overlying beds. The zone is the most widely recognized of the fusulinid zones and contains one of the most diversified faunas. The genera include *Pseudoschwagerina, Zellia, Occidentoschwagerina, Robustoschwagerina, Paraschwagerina, Chalaroschwagerina, Cuniculinella, Klamathina, Schwagerina, Pseudofusulina, Stewartina, Leptotriticites, Triticites, Nagatoella, Pseudofusulinella, Quasifusulina, Toriyamaia, Rugochusenella, Eoparafusulina, Alaskanella, Monodiexodina, Staffella, Pseudoreichelina, Ozawainella, Nankinella, Millerella, Eoschubertella, Biwaella, Schubertella, Boultonia, Brevaxina,* and *Misellina.*

Zone of *Parafusulina-Neoschwagerina*

The Leonardian and lower Guadalupian series (Wordian) of Early Permian age are assigned to the zone of *Parafusulina-Neoschwagerina.* The double name is necessary because of the two faunal realms that evolve different faunas. Equivalents include the Artinskian (or Darvasian) and the lower part of the Kazanian (or lower Murgabian), and the Nabeyaman and lower Akasakan of Japan. *Parafusulina* occurs from above the base of the zone into the overlying zone. *Neoschwagerina* occurs through most of the zone and continues into the overlying zone. The zone is widespread geographically, but the faunas are more provincial than in the older zones. The genera include *Pseudoschwagerina, Acervoschwagerina, Schwagerina, Pseudofusulina, Parafusulina, Monodiexodina, Chusenella, Skinnerina, Dunbarula, Rauserella,*

Boultonia, Toriyamaia, Yangchienia, Fujimotoella, Hayasakaina, Gallowayinella, Chenella, Eostaffelloides, Eoverbeekina, Pisolina, Sphaerulina, Armenina, Minojopanella, Nipponitella, Nagatoella, Misellina, Brevaxina, Cancellina, Colania, Neoschwagerina, Pseudodoliolina, Verbeekina, and *Afghanella.*

Zone of *Polydiexodina-Yabeina*

The upper Guadalupian (Capitanian) series of Late Permian age is assigned to the zone of *Polydiexodina-Yabeina.* Equivalents include the upper Kazanian or upper Murgabian, and the upper Akasakan and lower Kuman of Japan. *Polydiexodina* is a prominent form in the Texas section but is not widespread in its distribution. *Yabeina,* widespread in the Tethyan realm, is limited to the zone. The genera represented in the zone include *Polydiexodina, Parafusulina, Condonofusiella, Reichelina, Paradoxiella, Rusiella, Rauserella, Leëlla, Lantschichites, Sichotenella, Metadoliolina, Neoschwagerina, Sumatrina, Yabeina,* and *Lepidolina.*

Zone of *Lepidolina-Palaeofusulina*

The Ochoan series of Late Permian age is assigned to the zone of *Lepidolina-Palaeofusulina.* Equivalents include the Dzhulfian or Pamiran, the Lopingian in China, and the upper Kuman. The zone includes the beds above *Yabeina* and below the Triassic boundary. This zone is not recognized in most areas of the world but is found in the area of the trans-Caucasus, south China and Southeast Asia, and Japan. The fusulinid fauna is greatly diminished and includes *Lepidolina* in Japan and Southeast Asia, *Chenia* in south China, and the smaller fusulinids *Palaeofusulina, Parareichelina, Reichelina, Leëlla, Dunbarula, Nankinella,* and *Codonofusiella.*

SUBJECTS FOR FURTHER STUDY

The study of fusulinids is in an advanced state in many ways, but the number of places in which a continuous sequence of faunas is known are few. There is a real need to undertake a series of studies that would fill in the many gaps in the record. Studies are needed of continuous fossiliferous sequences that can be collected in detail at closely spaced intervals. The faunas of these collections would be studied to determine the range of variation represented for each of the taxa encountered. The local variation present in single samples could then be compared with the variation observed in samples of the same age but from differing environments. This, in turn, could be contrasted with the variation observed through successive beds in the stratigraphic section. The evolutionary changes are generally rapid, and the taxa commonly change within a few feet of section, even if the environment is not significantly different. Morphologic studies of the limits of variation, and of variation trends in the species, will help to stabilize the taxonomy. Continuing studies of individual taxa and their relationship to the environment will help to identify those taxa likely to be more sensitive to time than to environment. Integration of the data from the fusulinids with those from associated faunas and with the sediment characteristics should be included.

Systems capable of handling the masses of data should be devised, but caution should be observed in the selection of data. Studies based on the available literature would be erroneous, or at best misleading, because the taxonomy is unstable and the stratigraphic data are incomplete in many publications. Meaningful summaries can be prepared only from detailed studies of the kinds suggested above.

North American Mississippian Coral Biostratigraphy

William J. Sando

U.S. Geological Survey,
Washington, D.C.

INTRODUCTION

Corals have been widely used in biostratigraphic studies of Upper Paleozoic rocks. In North America, more than half the described Late Paleozoic coral species are from the Mississippian System, which shows a greater diversity and abundance of corals than either the Pennsylvanian or the Permian. The abundance and widespread distribution of corals in the Mississippian has led to their considerably greater use for zonation than in the other two Upper Paleozoic systems.

In this paper, attention is focused on the Mississippian as an example of coral biostratigraphy. After examining the kinds of coral zonations and the places in North America where they have been used, we shall examine the effect of zoogeography, and ultimately ecology, on the areal extent over which coral zones can be reliably recognized. The geochronometric resolution of coral zones is also compared with that of other zonal fossils as a measure of biostratigraphic utility.

DEVELOPMENT OF CORAL BIOSTRATIGRAPHY

History of Early Usage in Biostratigraphy

Corals have played an important role in biostratigraphic studies of North American Mississippian rocks since the earliest days of North American stratigraphy. The first Late Paleozoic coral to be described from North America was *Turbinolia cynodon* Rafinesque and Clifford 1820 (*Cyathaxonia*) from the New Providence Shale (Lower Mississippian) at Buttonmold Knob near Louisville, Kentucky (Bassler, 1950, p. 214).

The earliest uses of corals for correlation of North American Upper Paleozoic strata centered in the classical Carboniferous sequence of the Mississippi Valley area. Nuttall (1821) and Nicollet (1843) compared the American Mississippian rocks with the Mountain Limestone of England and included corals in their lists of fossils common to both sequences. Buckland (1839, pp. 171-173) noted the same similarities in his study of fossils from Cape Thompson and Cape Lisburne, Alaska. Corals were included in the paleontological characterization of the earliest stratigraphic units to be established in the type Mississippian (Owen, 1852; Hall, 1858; Worthen, 1866). Hall (1852) was the first to describe Mississippian corals from the western territories. As geologic exploration of the United States spread westward, many Mississippian species were described from collections made by expeditions of the latter half of the nineteenth century.

By the second decade of the twentieth century, the use of individual species or genera as stratigraphic markers was common practice, and this "index-fossil" concept has continued to the present. Studies by Butts (1917, 1922, 1926, 1940) of the Mississippian rocks of the Appalachian region are particularly noteworthy examples of the use of corals as index or guide fossils. Two lithostrotionoid species, *Lithostrotion mammillare* (Castelnau) and *L. proliferum* Hall, had long been regarded as diagnostic guides to the St. Louis Limestone of the Mississippi Valley area (Owen, 1852, p. 96; Hall, 1858, pp. 667-668; Worthen, 1866, p. 85). Butts extended the St. Louis Limestone into the southeastern states and defined its stratigraphic limits on the occurrence of these corals. Other examples of the use of corals as index fossils are in studies by Girty (1923), Van Tuyl (1925), J. M. Weller (1931), and Cooper (1948), all of which deal with the Mississippian of the Mississippi Valley and the southeastern United States.

Local and Regional Zonation

The use of corals as index fossils led the way to more sophisticated zonation concepts based on ranges of individual species or genera or on assemblages of taxa. As early as 1880, C. A. White (1883, p. 159) called attention to the widespread occurrence of a distinctive coral assemblage at the top of the Kinderhook division of the type Mississippian in Iowa, Illinois, and Missouri. Bowsher (1961) later gave this assemblage the name *Cleistopora typa typa* faunule, and pointed out that it can be recognized in Missouri, Iowa, Illinois, Ohio, and at many localities in the western United States.

Stuart Weller (1926) was the first to use corals in a formal zonation of Upper Paleozoic rocks in North America. He divided the type Mississippian sequence of Iowa, Illinois, and Missouri into 14 zones based largely on assemblages of brachiopods and echinoderms but also on corals and other marine invertebrates. One of these zones was named for corals, the *Lithostrotion canadensis* Zone, characteristic of the St. Louis Limestone, and was defined on the stratigraphic range of the name bearer.

Since Weller's work, a number of local and regional zonations of Mississippian rocks have been proposed in North America. Some of these zonations are based entirely on corals, but many have integrated coral data with that from other fossils on a regional scale. Most of the schemes are a combination of range- and assemblage-zone concepts. The various zonation schemes are discussed next and are summarized in Text-figure 1.

Mississippi Valley. Most of the described species are from the Mississippi Valley region (Text-fig. 2), yet only a few coral zones have been established in that area, largely because of greater emphasis placed on other fossils, particularly brachiopods and crinoids. Local Lower Mississippian assemblage zones of Laudon (1931, 1933) and the Upper Mississippian range zone of Stuart Weller (1926) utilizing corals were eventually meshed with biologically more heterogeneous regional zonations proposed by Moore (1948) and J. M. Weller et al. (1948). A later zonation of the Kinderhookian-Osagean transition interval proposed by Bowsher (1961) overlaps the earlier established Lower Mississippian zones.

The taxonomy of coral faunas described by earlier workers in the Mississippi Valley has been largely revised, providing a sound base for more detailed biostratigraphic work. However, the complex stratigraphy of the Mississippi Valley area requires a complete restudy of coral distribution in the sequence before a reliable zonation can be established for the type Mississippian sequence.

Eastern North America. A moderately large coral fauna has been described and cataloged stratigraphically in the southeastern United States, but only two "index-fossil" zones, both in the Upper Mississippian, have been formally recognized (Butts, 1929; S. Weller, 1926). Like the Mississippi Valley, the southeastern states are a fertile area for more detailed biostratigraphic study. Very little is known about the coral faunas of the northeastern United States, where the Mississippian is largely nonmarine.

Corals have been used extensively in analyzing the Upper Mississippian stratigraphy of Nova Scotia, where a fauna with strong affinities to the Avonian of Great Britain was described by Bell (1929) and Lewis (1935). Two coral-brachiopod assemblage zones established by Bell have been used in subsequent work by Stacy (1953) and Sage (1954).

Western North America. The Mississippian of the western United States is characterized by a large and varied coral assemblage, but few faunas have been described and illustrated. Nevertheless, a great deal has been done on zonation, especially in the northern Rocky Mountain states of Utah, Wyoming, Montana, and Idaho.

In the northern Rockies, local zonations of Parks (1951), Sando (1960), Sando and Dutro (1960), and Dutro and Sando (1963) have been organized into a comprehensive zonal scheme for the entire area (Sando, 1967a). Although most of the zones are defined on assemblages and ranges of coral genera, they include brachiopods and other fossils. This megafossil zonation has also been correlated with intercontinental microfossil zones established by B. L. Mamet (Sando et al.,

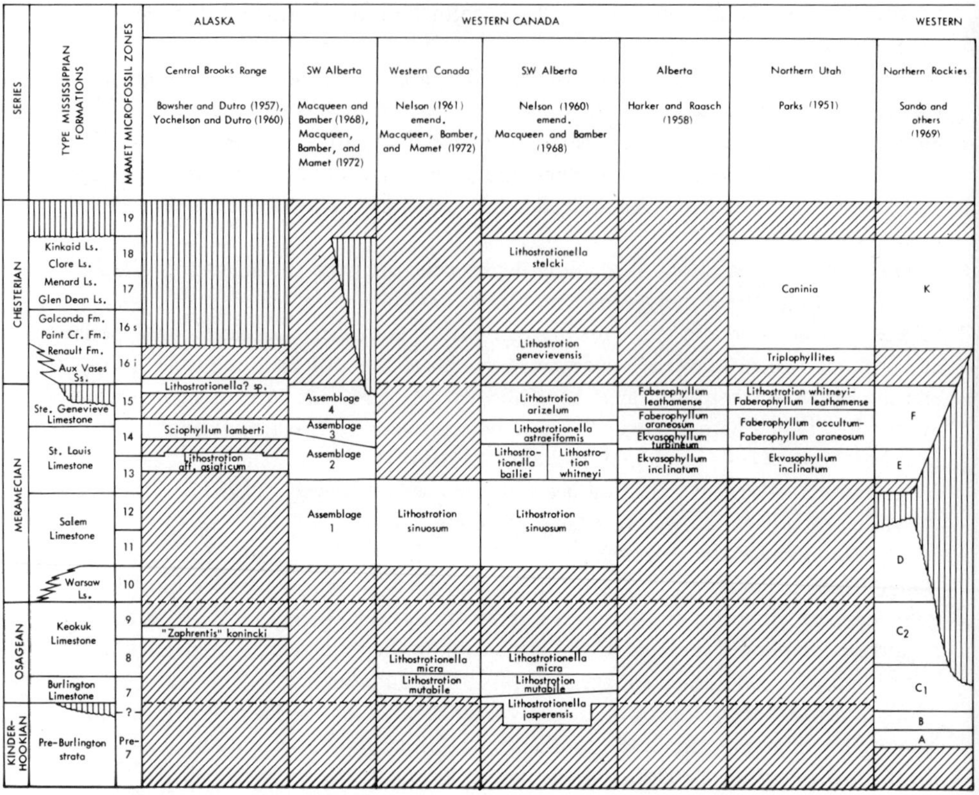

SERIES	TYPE MISSISSIPPIAN FORMATIONS	MAMET MICROFOSSIL ZONES	ALASKA: Central Brooks Range, Bowsher and Dutro (1957), Yochelson and Dutro (1960)	WESTERN CANADA: SW Alberta, Macqueen and Bamber (1968), Macqueen, Bamber, and Mamet (1972)	WESTERN CANADA: Western Canada, Nelson (1961) emend., Macqueen, Bamber, and Mamet (1972)	WESTERN CANADA: SW Alberta, Nelson (1960) emend., Macqueen and Bamber (1968)	WESTERN CANADA: Alberta, Harker and Raasch (1958)	WESTERN: Northern Utah, Parks (1951)	WESTERN: Northern Rockies, Sando and others (1969)
CHESTERIAN		19							
CHESTERIAN	Kinkaid Ls., Clore Ls.	18				Lithostrotionella stelcki		Caninia	K
CHESTERIAN	Menard Ls., Glen Dean Ls.	17						Caninia	K
CHESTERIAN	Golconda Fm., Paint Cr. Fm.	16 s				Lithostrotion genevievensis			
CHESTERIAN	Renault Fm., Aux Vases Ss.	16 i				Lithostrotion genevievensis		Triplophyllites	
MERAMECIAN	Ste. Genevieve Limestone	15	Lithostrotionella? sp.	Assemblage 4		Lithostrotion arizelum	Faberophyllum leathamense; Faberophyllum araneosum	Lithostrotion whitneyi–Faberophyllum leathamense	F
MERAMECIAN	St. Louis Limestone	14	Sciophyllum lamberti	Assemblage 3		Lithostrotionella astraeiformis	Ekvasophyllum turbineum	Faberophyllum occultum–Faberophyllum araneosum	F
MERAMECIAN	St. Louis Limestone	13	Lithostrotion aff. asiaticum	Assemblage 2		Lithostrotionella bailiei; Lithostrotion whitneyi	Ekvasophyllum inclinatum	Ekvasophyllum inclinatum	E
MERAMECIAN	Salem Limestone	12		Assemblage 1	Lithostrotion sinuosum	Lithostrotion sinuosum			
MERAMECIAN	Salem Limestone	11		Assemblage 1	Lithostrotion sinuosum	Lithostrotion sinuosum			D
MERAMECIAN	Warsaw Ls.	10							D
OSAGEAN	Keokuk Limestone	9	"Zaphrentis" konincki						C_2
OSAGEAN	Keokuk Limestone	8			Lithostrotionella micra	Lithostrotionella micra			C_2
OSAGEAN	Burlington Limestone	7			Lithostrotion mutabile	Lithostrotion mutabile			C_1
		?				Lithostrotionella jasperensis			B
KINDERHOOKIAN	Pre-Burlington strata	Pre-7				Lithostrotionella jasperensis			A

Text-Figure 1 *(above and right)*
Coral zonations of the Mississippian System in North America. Correlations based mainly on foraminiferal studies by B. L. Mamet and his colleagues. Vertical lines denote hiatus. Diagonal pattern denotes strata not zoned principally on corals or not discussed.

1969). Unpublished studies indicate that many coral zones are recognizable over a much broader area in the western states, although unfavorable paleoenvironments in some parts of the area make recognition of coral zones difficult (e.g., dolomitic facies in central Wyoming; Sando, 1967b).

In the southern Rocky Mountains, a few local zones based on ranges of narrowly restricted coral species have been recognized in northern Arizona (Sando, 1964, 1969). Although studies by Laudon and Bowsher (1941) and Armstrong (1962) in southwestern New Mexico and southeastern Arizona have documented the ranges of many species, Bowsher's (1961) Lower Mississippian assemblage zones are the only ones formally recognized in that area.

No zonations have been proposed elsewhere in the western United States, although small to moderately diverse faunas have been described from Oklahoma, California, Oregon, Nevada, and Sonora, Mexico.

UNITED STATES			MISSISSIPPI VALLEY				SE U.S.	NOVA SCOTIA	MAMET MICROFOSSIL ZONES	TYPE MISSISSIPPIAN FORMATIONS	SERIES
Central Wyoming Sando (1967b)	Northern Arizona Sando (1964, 1969)	Southern New Mexico Bowsher (1961)	Missouri Bowsher (1961)	Iowa Laudon (1931, 1933)	Mississippi Valley Weller, S. (1926), Weller, J.M., and others (1948)	Mississippi Valley Moore (1948)	Kentucky, Tennessee, Alabama, Georgia, Virginia Weller, S. (1926), Butts (1929)	Bell (1929), Stacy (1953), Sage (1954)			
									19		CHESTERIAN
									18	Kinkaid Ls. Clore Ls.	
								E (Caninia dawsoni)	17	Menard Ls. Glen Dean Ls.	
							Campophyllum		16 s	Golconda Fm. Point Cr. Fm.	
								C (Dibunophyllum lambarti)	16 i	Renault Fm. Aux Vases Ss.	
									15	Ste. Genevieve Limestone	MERAMECIAN
					Lithostrotion canadensis = Lithostrotionella castelnaui	Lithostrotion canadense	Lithostrotion canadensis		14	St. Louis Limestone	
									13		
									12	Salem Limestone	
D Diphyphyllum	Dorlodotia inconstans								11		
	Lithostrotion oculinum								10	Warsaw Ls.	
C									9	Keokuk Limestone	OSAGEAN
									8		
	Michelinia expansa	Cleistopora typa typa	Cleistopora typa typa	Cyathophyllum		Cyathaxonia arcuata Cyathophyllum			7 ?	Burlington Limestone	
? A-B?		Cleistopora typa gorbyi	Cleistopora typa gorbyi	Cyathaxonia arcuata					Pre-7	Pre-Burlington-strata	KINDER-HOOKIAN

The richly coralliferous Mississippian sequence in the Rocky Mountains of western Canada has been the focus of considerable attention from the standpoint of biostratigraphy. Although much descriptive work remains to be done, taxonomic studies of the western Canadian coral faunas have provided a good basis for biostratigraphic analysis. An assemblage zonation for Alberta proposed by Harker and Raasch (1958) was based mainly on brachiopods, but included four zones in the Upper Mississippian named for species of solitary corals originally described from Utah. Colonial rugose corals formed the basis for a system of 10 assemblage and range zones in Alberta and British Columbia (Nelson, 1960), the only zonation of a complete Mississippian sequence in North America based entirely on corals. These coral zones were later integrated with other fossil zones to form a 15-assemblage-zone scheme for the Mississippian of Alberta, British Columbia, Yukon Territory, and southwestern Northwest Territories (Nelson, 1961). The use of syringoporoid species was investigated by Nelson (1959, 1962) and incorporated into the zonation system, but no zones were established on syringoporoids alone. Subsequent work by Bamber (1966), Macqueen and Bamber (1967, 1968), and Macqueen et al. (1972) produced revisions in some of the previous zonation

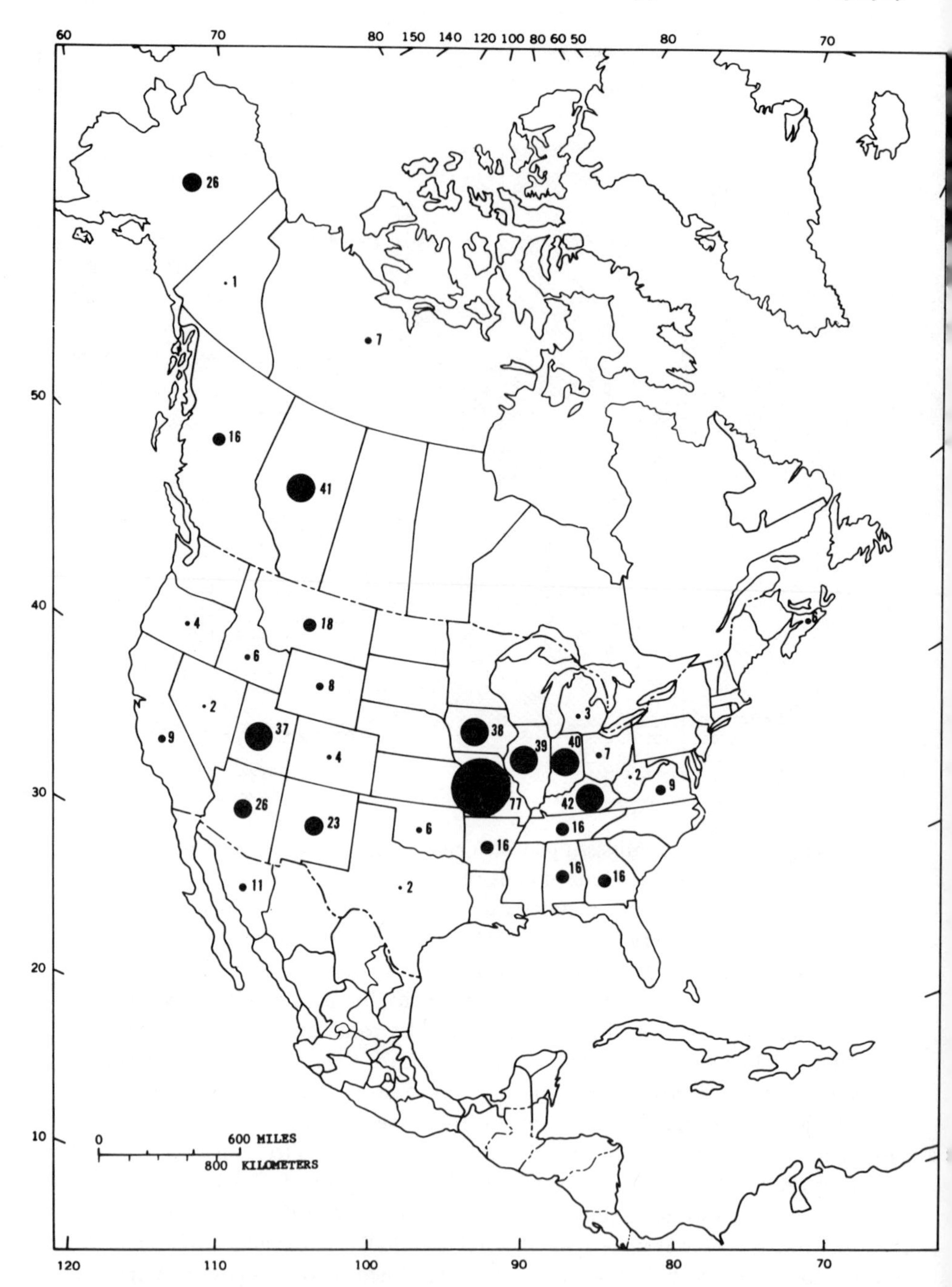

Text-Figure 2
Map of North America showing number of Mississippian coral species described and illustrated from each state and province as of mid-1973. Diameters of circles are approximately proportional to numbers of species described from each geographic subdivision.

concepts based on corals, and resulted in the recognition of four assemblage zones in the Upper Mississippian of Alberta.

The only published zonation established for Alaska is the local zonation of Bowsher and Dutro (1957) in the Brooks Range, where four local occurrence zones named for corals occur in a sequence of 13 zones based on various fossils. A new regional zonation for northern Alaska has been devised by A. K. Armstrong but is not yet published.

Analysis

Coral zonation of the North American Mississippian has been more extensive than in the other Upper Paleozoic systems largely because the Mississippian faunas are better known and the outcrop area of coralliferous marine rocks is greater than that of the Pennsylvanian or Permian. Moreover, the Mississippian rocks are predominantly relatively clean carbonate sediments deposited on shallow open marine shelves that provided optimum environments for coral growth. Although corals are locally abundant in the Pennsylvanian and Permian, the profusion of terrigenous sediments in the Pennsylvanian and large areas characterized by highly saline conditions in the Permian were factors preventing more widespread distribution of corals.

Although more species have been described from southeastern North America than from western North America (Text-fig. 2), the greatest use of corals for zonation has been in the western United States and Canada (Text-fig. 1). This is mainly because most of the more recent work has been concentrated in the western areas and because these recent studies have emphasized biostratigraphic application rather than pure taxonomy.

One of the more fruitful recent developments in coral biostratigraphy has been the coordination of coral and microfossil zones (calcareous foraminifers and algae) in the Mississippian of the western United States and Canada (Text-fig. 1). In Sando et al. (1969), microfossil samples were taken from numerous stratigraphic sections in the Rocky Mountains that had been collected in detail and zoned by means of corals and brachiopods. The microfossil assemblages were then placed by B. L. Mamet in his previously developed microfossil zonation for the northern hemisphere (Mamet and Skipp, 1970). Prior to the coordinated microfossil study, corals had proved valuable for correlations within the Rocky Mountain area, but correlations with the Mississippi Valley area were difficult (Sando, 1969, pp. 268-271). The wider geographic distribution of key microfossil assemblages permitted more precise correlations to the type Mississippian and to standard Carboniferous sequences in western Europe. Subsequent testing of the coordinated system at many additional localities has revealed only a few places where there is any disagreement on the age of the rocks, and where disagreement exists it is not more than one zone off.

There are many advantages to this coordinated approach to biostratigraphy. Corals, brachiopods, and conodonts offer a finer zonation than foraminifers and algae in the Kinderhookian part of the Rocky Mountain sequence, where environments were generally unfavorable to calcareous microfossils. In the post-Kinderhookian Mississippian, the foraminifers and algae are abundant, widespread, and afford a zonation as fine or finer than the other fossils. Both megafossils and microfossils are characterized by shoreward reduction in abundance, but corals and other megafossils survive the effects of the postdepositional reflux dolomitization that characterizes a large part of the Mississippian carbonate shelf much more readily than do the calcareous microfossils. Any recrystallization of the calcareous test of foraminifers makes identification difficult, whereas identification of corals is dependent mainly on skeletal elements less affected by recrystallization. Thus in many ways the megafossil and microfossil zonations are complementary.

BIOSTRATIGRAPHY, ECOLOGY, AND ZOOGEOGRAPHY

Inasmuch as the application of corals to biostratigraphy is significantly affected by their geographic distribution, it is important to bring into this discussion some remarks about zoogeography, which is in turn partly dependent on ecology. Although much can be learned by uniformitarian comparisons with modern corals (Scleractinia), these are so temporally and taxonomically distant from the Mississippian corals that many conclusions about Mississippian distribution rest largely on empirical relationships in the ancient record.

Ecology

North American Mississippian corals are found predominantly in carbonate rocks whose textures and structures suggest environments found today on tropical and subtropical shallow-water carbonate shelves (e.g., Bahamas Banks and Florida Bay). Corals are rare or absent in rocks of high terrigenous content, evaporitic beds, and carbonate beds (predominantly dolomitic) that have the characteristics of modern infratidal and supratidal sediments.

Mississippian corals were sessile benthic animals that lived predominantly on the substrate, although many forms undoubtedly lived partly submerged in bottom sediment. A requisite for optimum growth was the absence of rapid influxes of fine sediment that might cover the oral surfaces of the polyps and interfere with food gathering.

Modern hermatypic or reef-building corals are restricted to the photic zone by virtue of their symbiotic dependence on dinoflagellate algae (zooxanthellae), but modern ahermatypic corals, which have a depth range from the surface to 6,000 m, lack zooxanthellae and are not phototropic (Wells, 1957, pp. 1087-1089). Illumination may have been an environmental factor in some Mississippian corals, but no evidence of photic control has been found.

By analogy with living representatives, dispersal of Mississippian corals is assumed to have been accomplished predominantly by a free-swimming larval stage. Although little is know about the actual migration time of modern coral larvae, laboratory experiments indicate that most larvae become attached during the first two days of their existence, although some remain swimming for as long as two months (Connell, 1973, p. 209). The effects of temperature, salinity, currents, and predators on dispersal of larvae are virtually unknown. But presumably corals with the most long lived and environmentally tolerant larvae were those that achieved the most rapid, widespread dispersal, and thus those which would be most useful in regional correlations.

Zoogeography

Despite evidence of low dispersal rates in modern larvae, the distribution patterns of Mississippian corals suggest geologically rapid dispersal within ecologically favorable areas. In a study of North American Mississippian zoogeography, Sando et al. (1975) recognized five zoogeographic provinces and five subprovinces based on the distribution of coralliferous facies and the degrees of endemism and generic similarities of the coral faunas. An example drawn from this study (Text-fig. 3) shows coral zoogeography during Chesterian time. Areas of coralliferous facies (shelf carbonate rocks) are separated by areas of land, shallow-water noncoralliferous facies (predominantly terrigenous rocks), and deep-water noncoralliferous facies (terrigenous and volcanoclastic rocks). Analysis of 33 Chesterian coral genera and subgenera by means of the Otsuka similarity index (Cheetham and Hazel, 1969) and an endemism index (number of genera and subgenera known only in a given zoogeographic region divided by total number of genera and subgenera in the region, multiplied by 100) (Text-fig. 4) yields the following:

1. Endemism values of zero for zoogeographic regions on the periphery of the North American continent (Alaskan, Pacific Coast, and Maritime provinces) indicate that they had favorable connections to other coralliferous areas of the world, which permitted maximum gene flow. These areas are important for intercontinental correlation.
2. Zoogeographic regions in the interior of the North American continent (Western Interior and Southeastern provinces) were more isolated genetically and were characterized by high endemism. Zonal and correlation schemes are correspondingly more localized.
3. Gene flow was highest along continuous shallow-water carbonate shelves and was impeded by areas of terrigenous sedimentation and areas of deeper water. This reflects the normally short mobile larval stage of corals. Organisms with long larval duration should have been able to establish better gene flow across these barriers.
4. Similarities between faunas of different zoogeographic regions tend to vary inversely with the migration-route distance between these regions.

Limitations for Correlation

The analysis cited above lays open to question attempts such as those by Hill (1957, p. 58) and Ross (1970, fig. 7) to base worldwide correlations on corals, and lends support to the conclusion of Vasilyuk et al. (1970, pp. 45, 59-60) that the

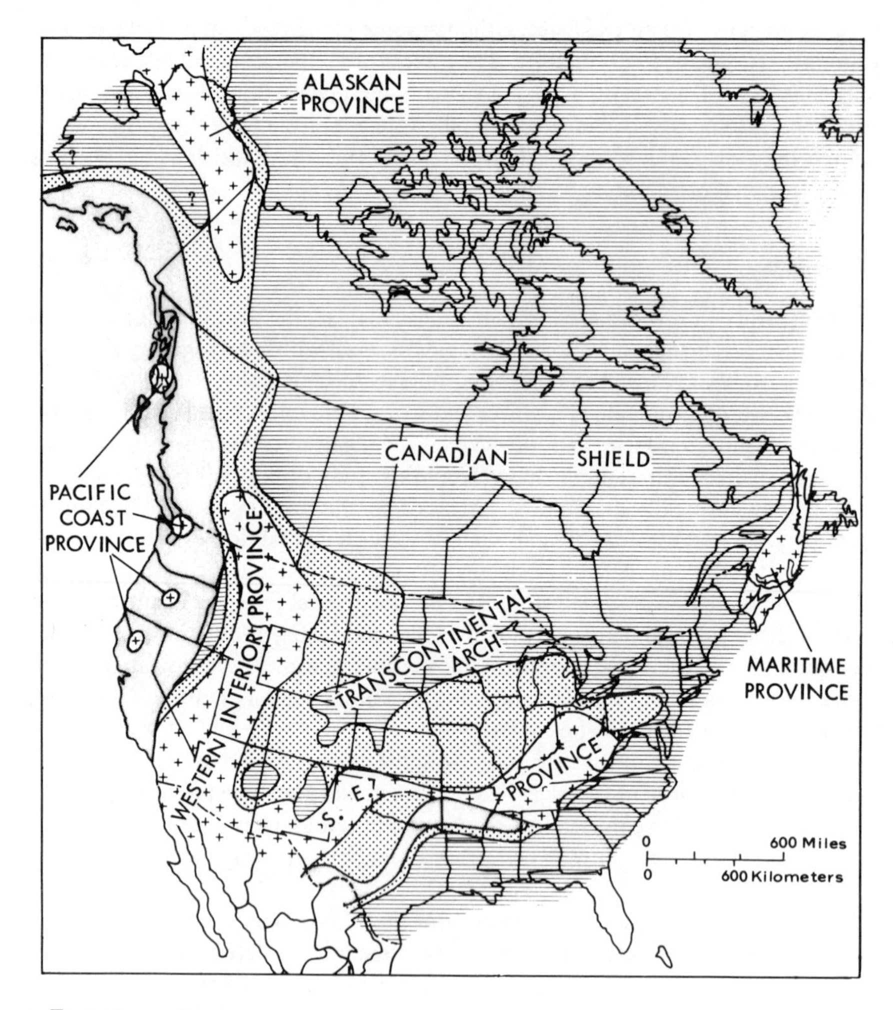

Text-Figure 3

Paleogeographic map of North America showing distribution of land (horizontal lines), deep-water noncoralliferous facies (shaded), shallow-water noncoralliferous facies (stippled), shallow-water coralliferous facies (plus pattern), and coral provinces during Chesterian time.

chief value of corals is in delineating zoogeographic provinces. This conclusion is not meant to discourage the use of corals as zonal fossils within zoogeographic regions, some of which are of considerable areal extent (see Text-fig. 1). It simply means that one cannot intelligently use corals to correlate between ecologically separated and genetically diverse areas of the world with the same precision attained by more rapidly and widely dispersed planktonic and mobile benthic organisms (foraminifers,

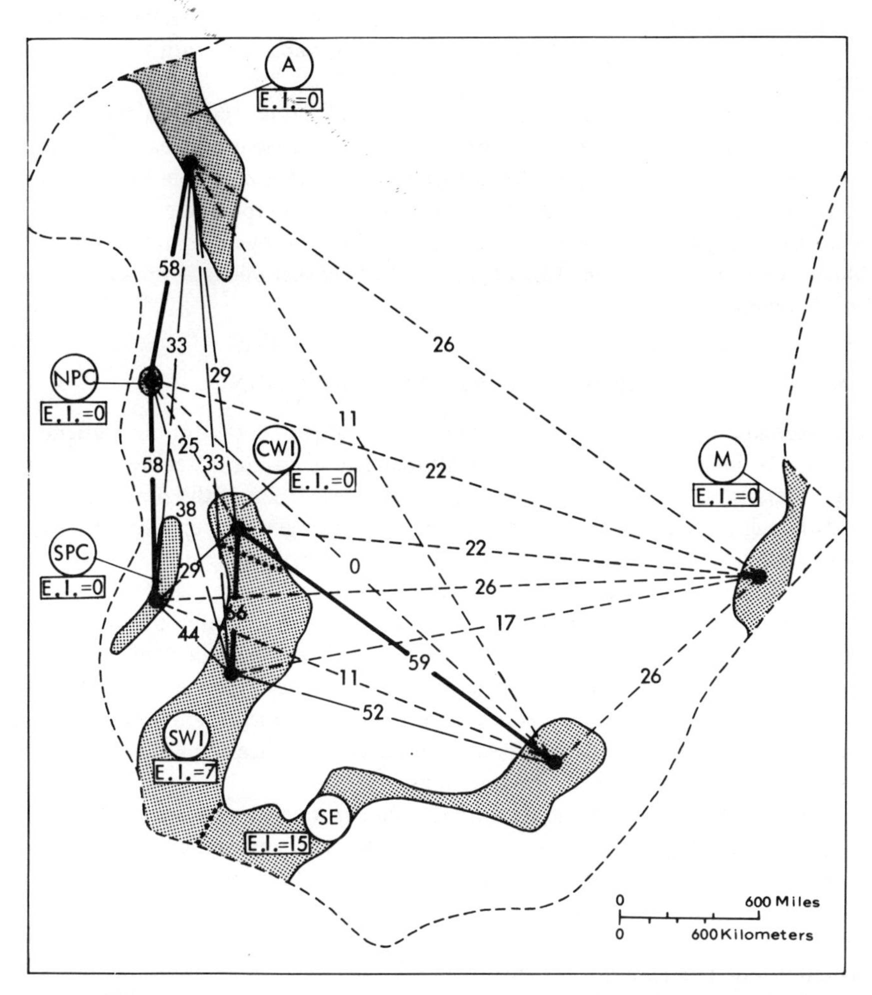

Text-Figure 4

Outline map of coral facies (shaded) in North America during Chesterian time showing Otsuka similarity indexes and endemism indexes (E.I.). Solid lines between zoogeographic regions indicate high similarity; long dashed lines indicate moderate similarity; short dashed lines indicate low similarity. Circled letters indicate coral provinces and subprovinces: A, Alaskan province; NPC, Northern Pacific Coast subprovince; SPC, Southern Pacific Coast subprovince; CWI, Central Western Interior subprovince; SWI, Southern Western Interior subprovince; SE, Southeastern province; M, Maritime province.

conodonts, and cephalopods). Correlation by means of corals may be difficult or impossible even over short distances, depending on paleogeographic circumstances. An example is the difficulty in correlating the Mississippian of the Western Interior province with that of the Southeastern province. It is easier to correlate the Mississippian sequence of northern Arizona with that of Alberta than with that of Missouri.

Attention should be focused on establishing local and regional coral zonations for each zoogeographic region and correlating these zonations by means of species common to adjacent regions and by integration with other zonations based on more widespread taxa. In this way, analysis of the complex interplay of evolutionary centers and migrations can bring about a correlation of zoogeographic regions which will lead to a world picture of geologic history that might be obscured by a broad zonation.

GEOCHRONOMETRIC RESOLUTION

A second criterion for evaluating the practical value of corals for biostratigraphic zonation is the length of the time intervals that can be discriminated by coral zones. This criterion may be called the geochronometric resolution of the zonation.

Although precise radiometric calibration of biostratigraphic zonation schemes in the Mississippian is limited by a paucity of radiometric dates, approximate values in absolute years can be assigned to the Lower and Upper Mississippian series using the data compiled by Francis and Woodland (1964) for the Carboniferous. A rough estimate of the average time span of each zone, expressed in millions of years, may then be calculated by dividing the time span of each series by the number of zones and unnamed intervals between zones. In this manner, geochronometric resolutions may be calculated for zonations based on different biologic groups in order to compare the practical value of the zonations.

Table 1 presents data on the resolution of North American zonations based on corals and other fossils that have been regarded traditionally as the leading biostratigraphic groups in the Mississippian. The compiled data indicate that zonation schemes based on corals compare very favorably in time resolution with those based on foraminifers, algae, conodonts, and cephalopods. In the Lower Mississippian, the conodonts are the leading group, but the corals are exceeded only slightly by the cephalopods, and they are superior to the foraminifers and algae. In the Upper Mississippian, the corals are slightly better than the conodonts and are only slightly exceeded by the foraminifers and cephalopods.

FUTURE PERSPECTIVES

In a summary of the status of Late Paleozoic correlation and chronometry, Ross (1970) concluded that improvement in the accuracy of correlations requires more information on phylogeny, paleoecology, paleozoogeography, and paleoclimatology coupled with more data on factors that affect the accuracy of radiometric dating and the integration of absolute and relative time scales.

Insofar as corals are concerned, the greatest needs are in the following areas:

1. Systematic studies. Species are the basic building blocks of more refined

Table 1
Geochronometric Resolution of Leading Biostratigraphic Groups in Mississippian Zonal Schemes for North America

Series	*Area*	*Reference*	*Average Timespan of Zones (MY)*
Corals			
Upper Mississippian	Rocky Mountains, USA	Sando et al. (1969)	1.7
Upper Mississippian	Western Canada	Nelson (1960) emend. Macqueen and Bamber (1968)	1.3
Upper Mississippian (Chesterian)	Nova Scotia	Bell (1929)	1.4
Lower Mississippian	Rocky Mountains, USA	Sando et al. (1969)	2.4
Lower Mississippian	Western Canada	Nelson (1960) emend. Macqueen and Bamber (1968)	2.4
Foraminifera and algae			
Upper Mississippian	North America	Mamet and Skipp (1970)	1.1
Lower Mississippian	North America	Mamet and Skipp (1970)	3.0
Conodonts			
Upper Mississippian	North America	Collinson et al. (1971)	1.5
Lower Mississippian	North America	Collinson et al. (1971)	1.3
Cephalopods			
Upper Mississippian	Ozark Plateau	Gordon (1970)	1.2
Lower Mississippian	Ozark Plateau	Gordon (1970)	2.0

zonation schemes and better paleozoogeographic systems. Much tedious taxonomic work remains to be done on the North American faunas, many of which are largely undescribed and uncatalogued. A growing tendency to describe only certain types of corals in a fauna is not nearly as rewarding biostratigraphically as evaluating the total coral fauna. Tabulates, solitary Rugosa, and colonial Rugosa are all important zonal fossils in some areas and some parts of the geologic column. One can seldom predict in advance of careful studies which corals will be the most valuable in a given area or stratigraphic level.

Better international communication among coral workers will enhance immeasurably the opportunities to solve basic questions of taxonomy and relationships with zoogeographic regions outside of North America. Solutions to many local problems require a knowledge of faunas in distant areas. The recent international meeting of coral specialists (at Novosibirsk, USSR, 1971) and the consequent establishment of an international organization of coral specialists are excellent steps in the right direction.

2. Correlation of coral zonations with other relative chronometric scales. Ultimate biostratigraphic correlation requires a synthesis of as many different biologic groups as possible. Recent studies indicate that corals as a group are move provincial than foraminifers, conodonts, and cephalopods. It is therefore important to relate coral zonations to zonations based on more widely distributed fossils in order to correlate different zoogeographic regions. Relation to nonbiologic relative time scales is also an important subject for future studies. Work is just beginning on a paleomagnetic time scale, and attempts are now being made to compare this scale with biostratigraphic scales.

3. Better correlation with the absolute radiometric scale. Ross (1970, p. 9) pointed out that radiometrically dated Upper Paleozoic rocks associated closely with fossiliferous strata are extremely rare. Only about a dozen such rocks were known in 1970, and most of these were related to the biostratigraphic scale by means of difficult intercontinental faunal or floral correlations. Francis and Woodland (1964, fig. 1) recorded only one radiometric date from North America in their construction of a Carboniferous time scale. Obviously, a great need exists for more radiometric data on the Mississippian rocks of North America. A better absolute chronology would permit an evaluation of the influence of ecology and migration on relative biostratigraphic scales.

Chronologic, Ecologic, and Evolutionary Significance of the Phylum Brachiopoda

J.B. Waterhouse **University of Queensland**

INTRODUCTION

In paleontology the hazard of correlation arises from the practice of ascribing the presence or absence of fossil species to ecological constraints, on the one hand, and temporal constraints, on the other. To many workers, especially those not actively engaged in correlative paleontology, the theoretical difficulties against disentangling these two constraints are overwhelming, as well exemplified by Huxley (1862), who was so overawed by time and correlation that he, no páleontologist, asserted that the thoughtful geologist would not consider the Lias (Middle Jurassic) of England synchronous with the Lias of Germany, and would not be prepared to guess whether it was deposited in 100 years or 10 million years. This destructive scepticism based on unsubstantiated theorizing is still widespread, and prevails even among paleontologists who are more biologically than geologically orientated. It is as well to recall that Western oil companies employ numerous paleontologists for one purpose, to ensure good correlation essential to their industry. No amount of theoretical objection can deny the fact that correlation works, and pays.

The aim of this essay is to demonstrate from chiefly Late Paleozoic marine sequences and brachiopod faunas that the ecological and temporal constraints are not alternative choices, and do not work in opposition to each other. Rather, the evolutionary record on which we base correlation through delineation of species, genera, and communities resulted from a sequence of chronecological development, consequent on oscillating climatic changes affecting the environment through time.

USE OF BRACHIOPODA FOR CORRELATION

At first sight, members of the phylum Brachiopoda would appear to provide fossils rather unsuited for accurate correlation. They belong to the benthos and are essentially sessile in habit, after a brief larval phase, growing like plants or colonial animals such as corals and bryozoa, at the interface between sea and sea floor. They move principally to open and close the valves, to allow the intake and exit of seawater bearing oxygen and organic matter, as reviewed by Suchanek and Devington (1974).

A second apparent disadvantage for correlation lies in the longevity of the Brachiopoda, which commenced in the Cambrian, peaked in the Late Paleozoic Era, and slowly declined through the Mesozoic and Tertiary to the present day; they still occupy a variety of marine niches, including cryptic habitats in coral reefs (Jackson et al., 1971). The phylum stands in contrast to other fossil groups such as graptolites, trilobites, and goniatites, which evolved and died out more spectacularly and, moreover, potentially achieved wide dispersion by assuming a less passive role in the environment, moving actively, or at least drifting, even as dead shells. These inadequacies in the Brachiopoda are compensated by the very great numerical preponderance of the phylum in terms of individuals, species, and genera. Although no figures are available, they were almost certainly the most common macroinvertebrate fossil of the Paleozoic Era, and somehow penetrated or were preserved in more environments than other Paleozoic macroinvertebrates. A number of species, in spite of their sessile, benthic habit, ranged widely and were short-lived, thus enabling good and consistent correlation at a number of marine horizons.

Biozones

Correlation of rocks and faunas involving brachiopods involves two steps: (1) subdividing the vertical sequence, and (2) correlating the resultant units laterally with units of other sequences. Normally, faunas are divided and assembled into biozones, recognized chiefly by the "faunal assemblage," or key species of that assemblage, deemed to have been significant from abundance, wide distribution, or short range. The faunal assemblage incorporates various fossil communities, which, within one region or sequence, are correlated by the following:

1. Sharing of one or more short-lived species, the range of which can only be determined from observation, and remains open to subsequent correction, extension, or contraction by both systematic revision and further discovery of material.
2. Lateral equivalence, established by field relationships, or by relationships to intermediate communities.

3. Interposition between two identical (recurrent) communities, established by field relationships or by relationships to intermediate communities.

4. Relationship to lithologic feature, or short-lived event, such as ash shower or tillite, usually restricted to one sedimentary basin.

Geochemical and geophysical techniques have not yet proved reliable, at least for Paleozoic problems. Paleomagnetic signature provides some promise, but the precise magnetic sequence remains to be determined. Radiometric values for the Paleozoic Era have not yet provided correlation as refined as even stages, let alone biozones, and most values have low confidence limits for even the period. There is a *minimal* spread of 40 MYBP for the base of the Permian period and 20 MY for the top of the Permian, for example, and these figures are only obtained by omitting, with no objective reason, the much broader deviations. Older periods are still more difficult to delineate objectively through radiometry, because the time constant is overprinted by temperature and pressure effects.

The various means by which biozones are recognized are outlined formally by Hedberg (1972a): acme zones, range zones, assemblage zones, etc. But it must be appreciated that for one sequence a monograph on a genus will refer to key species in an evolutionary sequence, at the same time as a monograph on the total fauna refers to assemblages and communities–the various types of zones are complementary, not opposed. Thus Cooper and Grant (1973), although rightly emphasizing the value of assemblages, were able to enumerate a number of key species and genera in their thorough survey of the Permian faunas of west Texas.

A second misunderstanding may arise from the perhaps unintentional implication that zonation is arbitrary, and that there may be considerable choice over what kind of zone is used. The confusion is in part due to the use of species biozones and generic range biozones as though they were interchangeable; they are not, and biozones are here restricted to species; in my view generic biozones should be prefaced accordingly. That upper Paleozoic biozones based on Brachiopoda may be subdivided with reasonable objectivity is illustrated in Text-figures 1 and 2, where both the range of species and the faunal assemblages yield a number of well-defined segments, or biozones, through two Permian columns. A careful study without prejudgement of the species ranges shown through the stratigraphic columns must conclude that it is undesirable to arrange the segments in any other way. For example, the extensive, well-exposed, and richly fossiliferous Permian brachiopod sequences of north Yukon Territory (Text-fig. 1) display a succession of biozones over a region of 60,000 square miles, and extending into the Canadian Arctic Archipelago and southward far into the Canadian Rocky Mountains. Each biozone is named for one or more short-lived "key species" ranging for all or much of the duration of the zone, and each biozone is also recognized by the general assemblage of species, so that boundaries between faunas can be drawn to within the limits of fossiliferous rock and time available to discover it. Each zone generally includes a number of fossil communities and a variety of sedimentary types that do not always change at the zonal boundary. Thus all the "E" faunas are found in the

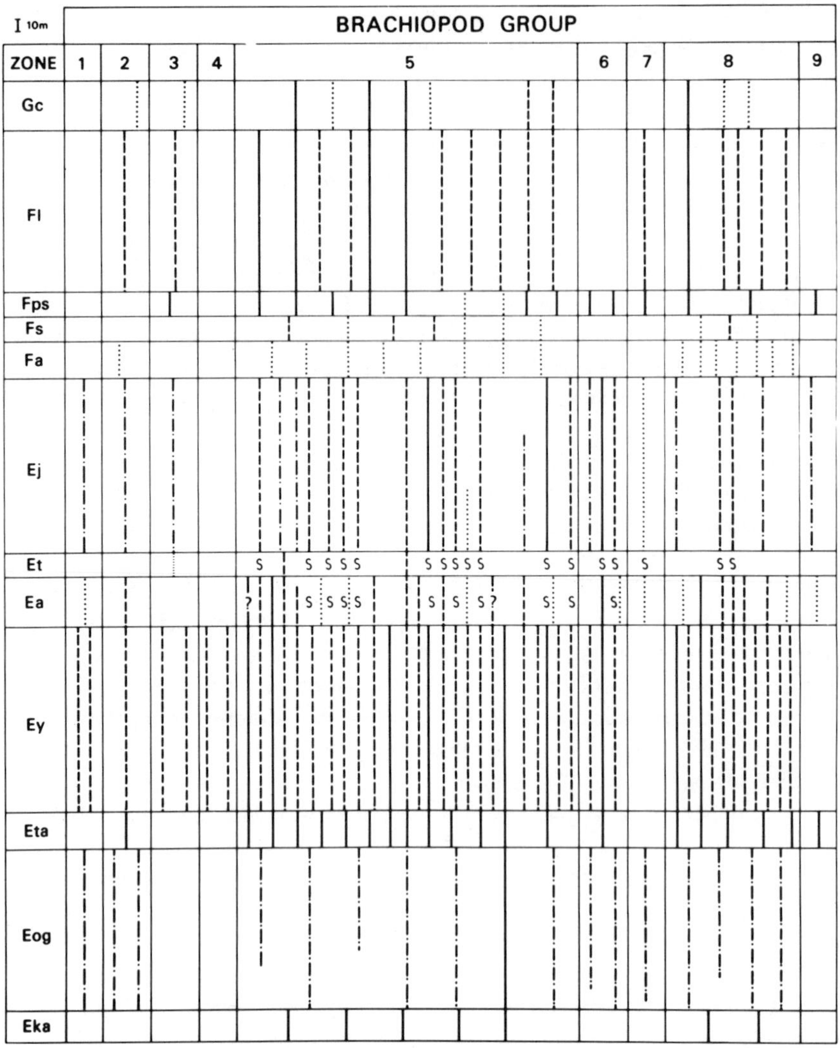

I 10m

BRACHIOPOD GROUP

ZONE 1 2 3 4 5 6 7 8 9

Gc

Fl

Fps

Fs

Fa

Ej

Et

Ea

Ey

Eta

Eog

Eka

Text-Figure 1

Range chart of brachiopod species in Permian beds of the north Yukon Territory, Arcti Canada. Species that commence in a zone are assigned a distinct kind of line, and many are replaced from zone to zone, although a few persist. The species are grouped as follow 1, Inarticulata (zone Ea), Orthida; 2, Davidsoniacea; 3, Chonetidina; 4, Strophalosiacea; 5, Productacea; 6, Rhynchonellida; 7, Atrypida; 8, Spiriferida; 9, Terebratulida. Zones match the type stages and substages, based also on biozones, of the Urals as follows: Gc, Lower Kazanian; Fl, Nevolin; Fps, Filippovian; Fs, Krasnoufimian; Fa, Sarginian; Ej, Aktastinian; Et and Ea, Sterlitamakian; Ey, Tastubian; Eta, Kurmaian; Eog, Uskalikia Eka, Surenan. From preliminary studies (Bamber and Waterhouse, 1971), understating t number of species. The symbol S links species absent from an intervening zone. Thickness approximate; exaggerated for narrow zones.

Jungle Creek Formation, which consists chiefly of recessive fine sandstone and siltstone, but also contains conglomerate, breccia, grits, and limestone. Perhaps the Et zone should be treated as a subzone or community, for it is restricted to black mudstone or fine sandstone. The Ea, Eta, and Eka zones include conspicuous *Attenuatella*-dominated communities, and the remainder include a number of different fossil communities linked by a few shared species, and, thanks to excellent exposures, often mappably correlative.

There is a lithological change higher in the column at the start of the Tahkandit Formation (zone Fa), including more chert, breccia, and limestone. This might suggest that the changes in species were controlled by lithological change, and clearly brachiopods must have been affected by such change. But episodic zonal changes occurred through the formation, not accompanied in some instances by any lithologic change. Moreover, farther north in the Arctic islands the formations differ in lithology, but contain the same sequential changes to biozones, including some of the fossil communities, and sharing some of the short-lived species that are found in the Yukon Territory.

A New Zealand example (Text-fig. 2) shows similar abrupt sequential changes to brachiopods in the complex and rapidly changing environments at the edge of a leading plate margin, in a milieu totally different from that of the Yukon shelf. The same pattern of abrupt sequential changes occurred everywhere in the Carboniferous, Permian, and Triassic periods, exemplifying, to judge from numerous studies, the behavior of brachiopods for all periods. In short, after a variable thickness of section of variable or monotonous lithology, the species and genera and most of all the communities in which they resided change. Because of these changes, the brachiopod faunas may be readily arranged in a succession of biozones on the basis of species assemblage and entry or disappearance of certain species. Few of the boundaries are arbitrary, although discussion is warranted over the nomenclatural ranks of subzone, zone, and superzone, for we still await their formal definition. Although theoretical objections may be offered on the basis of ascribing the change to facies, in fact many of the changes occurred without visible change in facies; elsewhere, lithology changed without visible change to the biozones.

Moreover, Foraminifera, Anthozoa, Bryozoa, and Mollusca reinforce the subdivisions. Space precludes a lengthy exposition of this observation, but it is well demonstrated by the detailed studies of various invertebrate groups in the Permian System as summarized by Likharev (1966) and Waterhouse (1972, 1973a). For instance, Table 1 summarizes the early Permian succession in the Ural Mountains, Soviet Union. The zones based on Fusulinacea agree very closely with the zones based on Ammonoidea, both groups having been intensively monographed. These zones may be extended widely, as for instance to the correlative beds of northern Canada of Text-figure 1, where brachiopods fall into an almost identical number of biozones correlated with those of the Urals by means of various fossil groups. The sequence tabulated for New Zealand in Text-figure 2 could be reinforced by a number of Bryozoa, Anthozoa, Gastropoda, and Bivalvia species. It would seem that the benthos, and some plankton, all changed more or less

ZONE	THICK-NESS	BRACHIOPOD GROUP 1–9
Mwr	200m	
Wan	150 - 1000m	
–	1000 - 5000m	
Pss	100 - 1200m	
Ppm	100m	
Pmw	65m	
	170m	Barren beds
Btb	85m	
Beo	150m	
Bns	60m	
Bem	20m	
Btc	10m	
–	100m	
Mep	5500m	
Mma	4000m	
–	1000m	Igneous
Tnh	550m	
Tnz	600m	

Text-Figure 2
Range chart for brachiopod species of Permian beds in New Zealand; style and species groups as in Text-figure 1, including an Inarticulate in zone Pss. Stratigraphic columns not fully to scale, and including, as shown, intervals without brachiopods. Zones match the type stages and substages of Urals, Russian Platform, and Greater and Lesser Caucasus in Armenia as follows: Mwr, Dienerian stage (of early Triassic age, but with Permian brachiopod genera); Wan, Vedian; Pss, Urushtenian; Ppm, Chhidruan; Pmw, Kalabaghian; Btb, Beo, Kazanian; Bns, Ufimian; Bem, Elkin; Btc, Nevolin; –, ?Filippovian; Mep, Krasnoufimian; Mma, Sarginian; Tnh, Tnz, Aktastinian. (Based on Waterhouse, 1973b.)

Table 1
Outline of Early Permian Zones Recognized in the Ural Mountains (see Likharev, 1966) Compared with Brachiopod Zones in Yukon Territory, Canada (Text-fig. 1). Correlated with the Russian Sequence by Means of Fusulinacea, Brachiopoda, and Ammonoidea[a]

	Biozones		
World Substage	*Ammonoidea, Urals*	*Fusulinacea, Urals*	*Brachiopoda, Canada*
Krasnoufimian	No ammonoids	*Pseudofusulina solidissima*	*Sowerbina* (Fs)
Sarginian	*Popanoceras polypetale*	*Pseudofusulina makarovi,* etc.	*Antiquatonia* (Fa)
Aktastinian	*Aktubinskia* *Agathiceras* *Neoshumardites*	*Pseudofusulina paraconcessa* *Parafusulina*	*Jakutoproductus* (Ej)
Sterlitamakian	New species of genera in Tastubian	*Pseudofusulina devexa*	*Tornquistia* (Et) *Attenuatella* (Ea)
Tastubian	*Synartinskia* *Medlicottia* *Metalegoceras* *Juresanites kharakhorum*	*Rugosofusulina serrata*	*Yakovlevia* (Ey)
Kurmainian	*Sakmarites* *Protopopanoceras*	*Schwagerina sphaerica*	*Tomiopsis* *Attenuatella* (Eta)
Uskalikian	*Juresanites* *Paragastrioceras*	*Schwagerina moelleri*	*Orthotichia* (Eo)
Surenan	*Glaphyrites*	*Schwagerina vulgaris*	*Kochiproductus* *Attenuatella* (Eka)

[a]See Bamber and Waterhouse (1971) and Sarytcheva and Waterhouse (1972). Zones are based primarily on species of the genera listed in the columns, and the table demonstrates that (1) biozones were contemporaneous within the limits of correlation between the Urals and Canada, and (2) the different faunal elements changed more or less simultaneously, within the limits of correlation. Note the presence of only two exceptions, one within the Ammonoidea, the other within the Brachiopoda, where it appears probable that one extensive *Tornquisitia* community was incorrectly treated as a full zone.

simultaneously. We must however recognize one very major exception to this generalization, offered in particular by Mesozoic species of Bivalvia, most notably the Pterioida (including *Halobia, Daonella,* and *Monotis*), and a number of Mesozoic ammonoid species. Many of these species were both widespread and lived for intervals much shorter than the typical benthos, including Brachiopoda. This type of highly successful genus and species that was clearly able to penetrate and dominate numerous communities for very brief intervals, or dwelt in a nektonic or planktonic community but was buried with the benthos, is seldom if ever found among Brachiopoda.

Correlation of Biozones

Biozones are unlikely to extend over the entire globe–the Canadian examples for a few thousand miles, the New Zealand examples for a few hundred miles, in part reflecting the original extent of the fossil communities, in part reflecting tectonic disruption or elision, burial by younger sediments, or loss through erosion. For these reasons, wide-ranging correlation depends heavily on the entry or time range of certain genera and species found on a trial and error basis to retain validity as guides. There are many such instances at generic and specific interval level among brachiopods. The entry of the genus *Kochiproductus* at the base of the Permian period, the entry of certain species of *Streptorhynchus, Cleiothyridina, Spiriferella,* and *Dielasma* in the Kungurian stage, and of species of *Cancrinelloides* at the base of the Kazanian stage enable extensive correlation (Likharev, 1966). The Spiriferid genus *Attenuatella* displayed three episodic and very wide ranging, short-lived "bursts" in the early Permian of wide distribution, and in various lithotopes (Table 2). *Licharewia* and *Permospirifer* did the same in the early part of the Middle Permian (Bamber and Waterhouse, 1971; Waterhouse, 1973a). The ad hoc nature of using these species and genera is obvious, and each is emphatically open to revision and verification through testing from additional sequences. But there is little alternative until we can apply statistical and predictive procedures to the discipline, procedures that are just barely commencing to show promise. It is almost inevitable that mistakes will be made and useful keys overlooked, but the continuing process of correlation is self-correcting, provided that correlations are (1) related to stratigraphic sequence; (2) related to all elements of the faunas, two points well reiterated by Cooper and Grant (1973); and (3) tested for consistency on a worldwide scale. The latter point has to be stressed, because intracontinental correlation may be based on widespread and identical communities of much the same paleolatitude, and by enabling relatively easy, or even superficial, correlation may fail to come to grips with the real problem of correlation.

Alternation of Genera

In some instances, two genera alternate through a succession. For instance, Permian biozones with the genus *Neochonetes* in various communities alternate with biozones containing communities with smooth or variously costate chonetids such as *Lissochonetes, Dyoros,* and *Chonetina* (Table 3). There is no obvious correspondence with the type of sediment; the alternation through time must have been connected with some other ecologic parameter, but I can offer no explanation.

Table 2

Occurrence (A) of Early Permian *Attenuatella* in a Number of Correlative Horizons of Asselian and Sakmarian Stages, and Absence (–) from Intervening Horizons[a]

World Standard Substages	*Yukon*	*West Texas*	*Kolyma R*	*Spitz-bergen*	*Kazhak-stan*	*Mongolia*	*Tibet*	*Carnian Alps*	*Australia*	*Brazil*	*Spain*
Sterlitamak	A	A		–					?A	?A	
Tastub	–	?–		–				–	?–		
Kurmain	A	?–	A				A	A	?–	–	
Uskalik	–	?–		–				–			
Suren	A	?–		A	A	A		A			A

[a]Occurrences poorly controlled for east Australia, Brazil, and west Texas. From studies by the author, summarized in Waterhouse (1973a).

Table 3
Correlation of Early Middle Permian Units of Kungurian Stage with Alternating Costate and Smooth Chonetids[a]

World Standard	Russian Platform	Petchora	Greenland (G); Novaya Zemlya (NZ)	Canadian: Arctic Archipelago (A); Yukon (Y)	West Americas	California	Sikhote Alim, Siberia	Sumatra	Queensland (Q), New Zealand
Ufimian (costate)	*?Neochonetes*								*Neochonetes*
Elkin (smooth)	*Chonetinella*	*Lissochonetes*							
Nevolin (smooth and variably costate)	*Chonetinella*	*Lissochonetes*	*Chonetina (G)*	*Lissochonetes (AY)*	Meade Peak	*Lissochonetes*	*Chonetina*	*Chonetina*	*Lissochonetes (Q)*
	Chonetina	*Tornquistia* *?Dyoros* *Arctochonetes* *Tornquistia* *Lissochonetes* *Chonetina* *Neochonetes*	*Lissochonetes (G, NZ)*	*Chonetinella (A)*	*Lissochonetes*	*Chonetinella*	*Lissochonetes*		
Filippovian (costate)	*Neochonetes*			*Neochonetes (AY)*	Chochal *Neochonetes*				

[a]Based on studies by Russian writers and the author (see Waterhouse, 1973a).

Faunal Signature

Throughout a succession, biozones vary in the number of taxa present, and by plotting even the numbers of species a zigzag curve is derived (Text-fig. 3). This may be called a faunal signature. The relative diversity, Di, for collection i may be assessed by the formula

$$Di = \frac{Pi - Si}{Si}$$

where Pi = number of species at locality i
Si = number of species shared with other localities

As a control for collections from a longer time interval, the number of shared species may be replaced by the number present in each collection of the 10 most common genera present in the entire succession. Williams (in Moore et al., 1965, p. 243) has demonstrated the possibilities of deriving "correlation profiles" from the coefficients of association of Ordovician brachiopod genera. These simple statistical techniques deserve to be applied more widely in correlation. Obviously, they are susceptible to interference from facies, but the interference could be examined by applying the same techniques to collections from particular lithotopes or communities.

It may be well to interpose an observation here on the probability that correlation coefficients of actual species lists are not likely to achieve much in the way of convincing correlation, perhaps because they at best would indicate close links only for identical communities, instead of correlative but different communities. However, one attempt has been made to cluster Permian brachiopod genera to achieve worldwide correlation (see Grant and Cooper, 1973). The results are far from convincing, because, although a very few correct *lateral* correlations were obtained, the data showed even stronger links *vertically* through time, and we know that the Early and Middle Permian suites *cannot* be correlative, but successional. Clearly, the method was grouping similar ecological faunas, as well understood by Williams (1973) in his superb application of clustering to Ordovician brachiopod genera.

Recurrent Generic Assemblages

Over lengthy stratigraphic columns at least within the Upper Paleozoic Era, it is found that particular assemblages, separated by quite different faunas, recur not at specific but at generic or familial level (Waterhouse, 1964, 1974). This overall resemblance at widely different ages may first lead to ready miscorrelation, but ultimately the repetition may provide a key or confirmation for correlation if sequences elsewhere in the world show similar recurrent patterns. It is of particular efficacy in wide-ranging correlation through a substantial succession of different biozones, once the patterns of change are established. As discussed below, the cause probably lay in the episodic recurrence of similar climatic conditions.

Overall Correlation

In making final correlations, all these lines of evidence are used, together with the

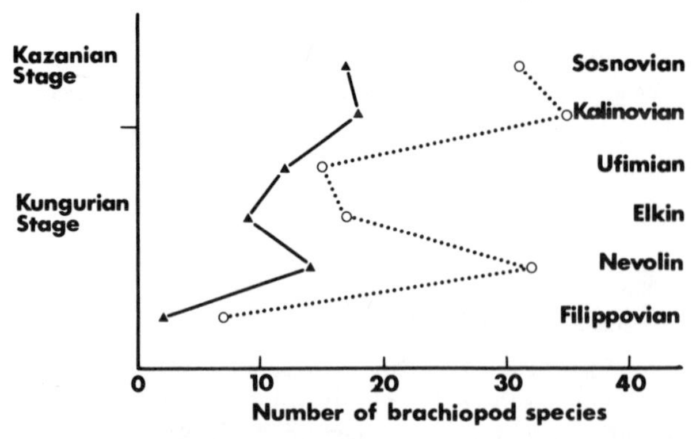

Text-Figure 3
Diversity graph of the number of brachiopod species from early Middle Permian sequences of the Russian Platform and Urals (world standard) and of New Zealand (solid), showing similar faunal signatures. Russian data (dots) based on studies summarized in Branson (1949), Likharev (1966), Chalishev (1966), and Zolotova et al. (1966). New Zealand data summarized from Text-figure 2. Correlations are supported by entry of key species and succession.

significant evidence contributed by other elements of the faunas. Again the stress lies on the value of full sequences, and this is perhaps where Paleozoic and Triassic brachiopods prove so useful; they are frequently found throughout lengthy successions, whereas other fossils tend to be restricted to certain horizons, especially among the short-lived and well-studied Fusulinacea and Ammonoidea that afford such valuable data for correlation, *when* they are present. Once the first correct fit is made to a standard or type sequence, the second and subsequent fits become progressively easier and help confirm or falsify the first fit, like fitting teeth into the sockets of a ratchet. The first tentative correlation severely reduces the remaining possibilities, and the choice is still further reduced by matching major rhythms in the sequence of zones.

SIGNIFICANCE OF PALEOECOLOGY IN CORRELATION

There is no question about the significance of ecological constraints on the distribution of brachiopod species, genera, and families. As with other fossils, brachiopods were limited in distribution by numerous parameters of which some are amenable to study from the geologic record. In the context of biostratigraphy, these parameters may be subdivided into several major categories according to their temporal significance, for paleoecology itself is intimately but variably connected with time.

1. Essentially unchanging through time, e.g., bottom sediment, depth of water, perhaps salinity.
2. Continental movement and changing oceans, distinguished by long-term and erratic periodicity.
3. Large-scale major climatic oscillations, with perhaps 30 million year periodicity and shorter episodicity.
4. Steady and slow cumulative changes to environment, such as day length and tides.
5. Progressive and cumulatively changing biota (despite some regression and repetition), presumably affecting competition for food, nature of food, predator-prey relationships, living space, and other aspects.

Bottom Sediment

The influence of bottom sediment has been emphasized by Ager (1971) and many other authors. Ager considered bottom sediment to be more significant than depth of water, which Ziegler et al. (1968) regarded as a prime parameter in the distribution of Late Silurian brachiopod species and communities. Of course it is possible that brachiopods gradually became less affected by water depth, and more adapted to and therefore affected by bottom sediment through time (Waterhouse, 1973b). To some extent the history of the diversification of the phylum through time involved a morphological response to environment through the development of different modes of attachment to the substrata and a concomitant change in feeding apparatus, so that the substrate was always a significant parameter intimately related to the life of Brachiopoda. The various modes of attachment include the following basic types:

1. Attached by pedicle:
   a. Passing between valves (Lingulida), preferred inshore often plant-rich sediment.
   b. Through foramen of ventral umbo (e.g., some Acrotretidina, Rhynchonellida, Atrypida, and Terebratulida). Preferred fine sediment or attached to algae, often outer shelf, but varied.
   c. Through ventral delthyrium (e.g., some Orthida and Spiriferida). Some species restricted to either siltstone or sandstone, whereas most widely tolerant.
2. Attached by cementation (Craniidina, Strophomenida, and especially Davidsoniacea): Apparent preference for firm substrate, either on rock as indicated by breccia or other shells.
3. Attached by cementation and spines (e.g., Strophalosiacea, Richthofeniacea): Tolerant of wide range of sediment, and established shell banks where sediment fine. Some Aulostegidae and Waagenoconchinae preferred coarse and medium sandstone.
4. Attached by spines: Chonetidina widely tolerant, but apparently preferred inshore siltstone-sandstone, often with infaunal bivalves, though ascribed a swimming role by Rudwick (1970). Productacean families ranged widely, but many

genera and especially species appear to have been restricted to one type of sediment (Table 5).

5. Unattached: A number of Orthida and Spiriferida lost their pedicle and were relatively independent of sediment type (see Table 4); some, including members of Ambocoeliidae, preferred plant-rich, presumably inshore sediment.

Much more study needs to be made along these lines, but available data strongly suggest that the presence or absence *and* the number of individuals of brachiopod species are *variably* affected by bottom sediment (Waterhouse, 1973b). Many species, especially those attached by cementation by umbo, or attached by umbonal pedicle, or in many instances by spines, were strongly controlled in distribution as "specialized" shells adapted to particular bottoms. Others were relatively independent of lithotopes, particularly those attached by pedicle through the delthyrium or by cementation and spines, although the latter were influenced in numbers of individuals. Of course, such generalizations are extremely crude and must be treated with great caution; but it appears likely that a prime factor in the presence of many but not all brachiopod species and the nature of communities in which they participated lay in the sediment type. The relationship between facies and brachiopods, if close, must clearly impose a nontemporal control on the distribution of some brachiopods, varying from family level for some forms to generic and species level for others. This does not mean that only those species which were tolerant of a wide variety of lithotopes can be used for correlation. But it does mean that short-lived, lithologically controlled species have to be sought from only favorable lithotopes, as instanced from *Spiriferella* and *Terrakea*, discussed in Tables 4 and 5. For such species, ecological situations of roughly identical constraints that supported identical species must have been essentially correlative to within one biozone. And if the species of identical genera differed in one kind of lithotope within one region, the possibility increases that they were not correlative.

The second kind of species is that belonging to a genus and family which frequently ranged through various lithotopes, and was unusual in that it differed from its related species by being very short lived, such as species of *Attenuatella* (see Table 2) and *Licharewia* among the Spiriferida. Occasionally species of either group were very successful, and entered a wide range of habitats in many communities.

Different species within the one genus of brachiopod sometimes displayed either of these two attributes. Members of the Spiriferellinae, a Late Paleozoic subfamily, may be used to summarize the overall difficulties and possibilities inherent in brachiopods (Text-fig. 4). In the genus *Timaniella,* species were short lived, two of Filippovian-Irenian age in various lithologies, and one of Kazanian age; but geographic extent, although intercontinental, was limited to the northern hemisphere. *Elivina* ranged throughout much of the period, with a number of short lived species that were rare except during the Kalabaghian and Chhidruan substages.

The companion genus *Spiriferella* was a Carboniferous and Permian genus that

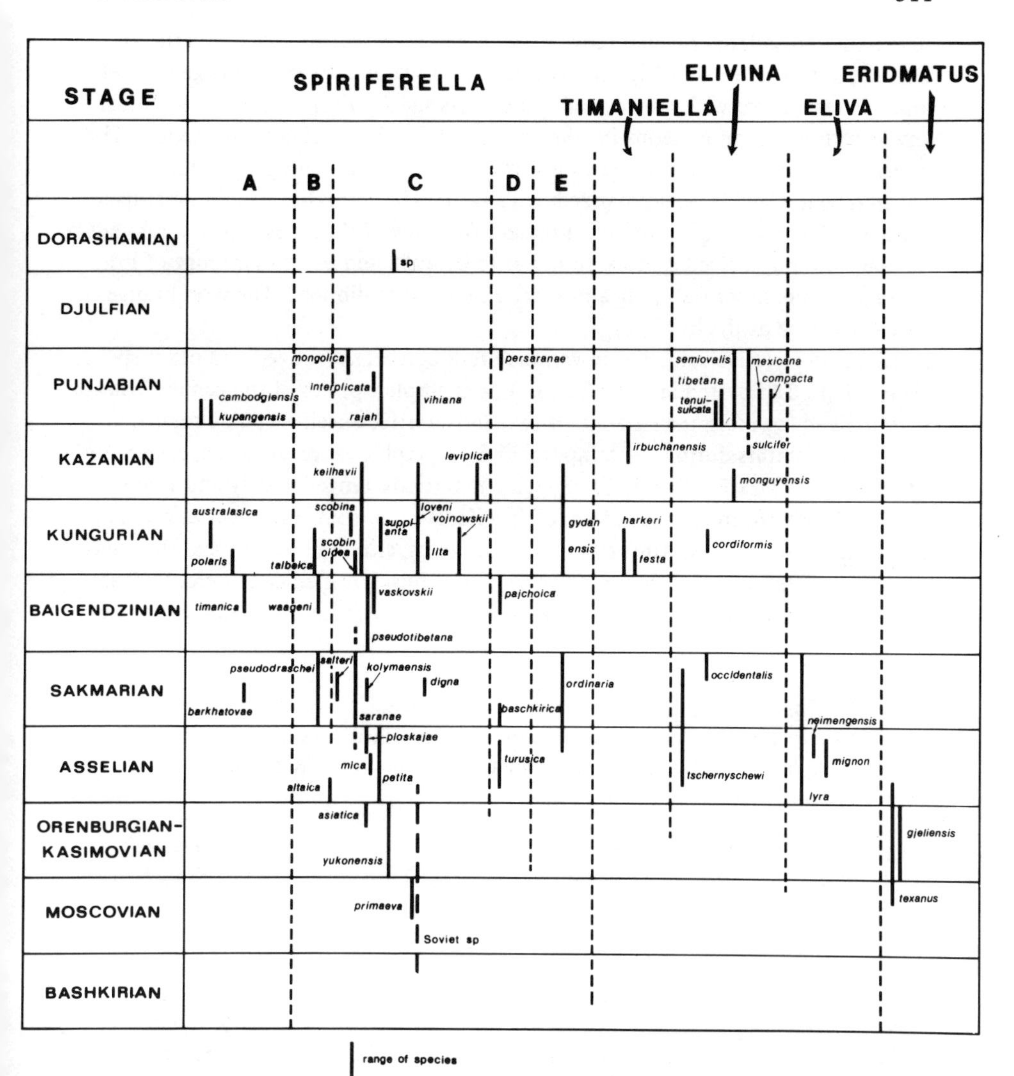

Text-Figure 4
Range chart and interrelationships for the genera of Spiriferellinae, based on examination of material and wide survey of literature. Some species of *Spiriferella* appear to have tolerated a range of lithofacies and to have been unusually long lived.

typified many brachiopod genera. There are some 50 species known in several lineages that have required a great deal of study to clarify. Evolutionary trends provide time subdivisions between one substage and two stages in length, with a number of widely dispersed species in a variety of facies, some covering much of the northern (especially paleotemperate) hemisphere, some ranging into the southern hemisphere. Species ranged widely across North America, Siberia, and Europe, but never into highly saline or very cold waters. Even so, the genus was more diverse, more widespread, and some of its species shorter lived than almost any known contemporaneous fusulinacean, ammonoid, or conodont species. The interplay between species, range of lithotopes, time range, and geographic range of *Spiriferella* in Canada is summarized in Table 4. The genus has several short-lived species rather restricted in distribution and from few lithotopes, and several wider-ranging and longer-lived species, found as to be expected in a wider range of lithotopes and sometimes assuming a primary role in communities. The world range is much the same as the local range.

Higher correlative value lies in various Productacea, short lived and often specially adapted to particular lithofacies. For example, species of the genus *Terrakea* seldom lived for more than a zone or two and so offer excellent guide fossils, provided that habitats during the temporal and geographic range of each species were suited for colonization (Table 5). Species apparently ranged widely and freely for the length of eastern Australia and to New Zealand, in some instances as primary species of communities. Unfortunately, the genus, although widespread throughout much of the Permian, has been neglected or misidentified elsewhere, so that it is premature to chart its world range, geographic range, and duration of species;

Table 4

Species of *Spiriferella* in the Late Paleozoic of Canada, (Yukon Territory and Arctic Archipelago), Summarizing Lithology, Time Biozones, and Geographic range[a] brec, breccia; calc, Calcareous; lmst, limestone; ms, mudstone; sls, siltstone; ss, sandstone.

Species	*Lithology*	*Biozone*	*Occurrence Elsewhere*
Spiriferella			
gydanensis	Ss, sls	Fn, Gc	Kolyma
leviplica	Calc sls, lmst	Gc	
vojnowski	Calc grit, ss	Gc	Kolyma
keilhavii	Ss, sls, calc ms	Fps–Gc	Arctic
loveni	Ss, sls, lmst, brec	Fps–Gc	Arctic
vaskovskii	Sls	?Fs	Kolyma
pseudotibetana	Grit, ss, brec, lmst	Fa–Fs	Russia
saranae	Fine ss, sls	Ey, Ej	NZ, Kolyma
pseudodraschei	Ss, fine ss, sls, lmst	Ey, Ej	Russia, Oregon
ordinaria	Sls	Eta, Ey, Ea	NZ
petita	Ss, sls, lmst	Eka, Eo, Eta	
yukonensis	Ss, sls, ms, lmst	Ck, Dos, D	Aff. Asia
gjeliensis	Sls	Cp	Urals
primaeva	Calc sls	Cp, Cgb	

[a]Zones as in Text-figure 1: C, Carboniferous zones; NZ, Novaya Zemlya. Summarized from Waterhouse and Waddington (in press). See Text-figure 4.

Table 5
Succession of *Terrakea* in New Zealand Showing Short Range and Some Degree of Lithological Control, Correlative with Succession of *Terrakea* in Eastern Australia[a]

Species	*Lithology*	*New Zealand Zone*	*Australia Zone*
Terrakea sp.	Calc ms (coquina)	Wan	–
multispinosa	Calc sls	Ppm	Qld, IV–V
elongatum	Calc sls–ss	Beo	Qld, NSW, IV
brachythaerum	Ss	Btb	Qld, NSW, Tas IV
concavum	Ms, coquina	Bem, Btc	NSW, IIIc
pollex gunjum	Breccia	Tnh, Tnz	Qld, high II

[a]Dear (1971), Waterhouse (1971). calc, calcareous; ms, mudstone; sls, siltstone; ss, sandstone; Qld, Queensland; NSW, New South Wales; Tas, Tasmania. Australian Biozones numbered II-V, correlative with those of New Zealand, which are shown in sequence in Text-figure 2 and elaborated in Waterhouse (1973b).

but it is clearly represented in North America, the Arctic, Russia, and Pamirs with two lineages, species of which alternate in appearance for parts of the column (Dear, 1971; Waterhouse, 1971).

Continental Displacement

The gradual movement and growth of continents and birth-to-death cycle of oceans provided a slowly changing background environment, which affected lithofacies and discriminated between epicontinental seas or volcanic arcs and trenchs that played a large role in bottom facies (Valentine, 1971; Waterhouse, 1973b), as well as changing sea level (Burke and Waterhouse, 1973) and perhaps salinity. Attention has been drawn also to the effect on animal migration (Hallam, 1967b) and the slowly changing paleolatitude (Waterhouse and Bonham-Carter, 1972).

Climate

The chief environmental control over the distribution of brachiopods within the marine environment is thought to have been climate (Williams, 1973; Middlemiss, 1973; Waterhouse, 1974), although the influence must have been very complex, involving both temperature and seasonality, and profoundly affecting many other parameters significant to brachiopods, such as food chains, breeding seasons, geochemistry, and sediment type. A comprehensive analysis of the distribution of Permian brachiopod families over the world by Waterhouse and Bonham-Carter (1975) recognized three major associations, one found close to ice caps, and therefore presumably of polar origin; one found along the Tethys, Texas, and China, suggestive of a tropical association, as supported by the presence of major rugose coral reefs and paleomagnetic evidence; and the third an intervening association, presumably of temperate climate. These associations were largely independent of sediment type (although not completely because there is some correlation between climate and sediment). Because the associations are at family level, they are independent of time divisions within the Permian period; but they do shift in geographic position from substage to substage, suggesting cyclic changes in climate, as discussed later, or shift of continents toward or away from the poles.

Earth Rotation and Tide

Length of day, tides, and other environmental parameters related to the rotation of the earth and the proximity and effect of the moon and sun provided a pervasive and slowly changing background to brachiopods, which, although affecting feeding habits and growth of shell, presumably did not affect distribution or duration of species to any significant degree.

Biota and Communities

Biotic influence on the presence or absence and number of brachiopod species was probably significant in terms of competition for living space and food. This is expressed by the occurrence of brachiopod species in a finite number of fossil communities, in which various life forms were organized to exploit and live in harmony with the environment. Ziegler et al. (1968) analyzed communities of Late Silurian age in Wales and showed stable recurrent compositions with apparently consistent proportions of various life forms, including brachiopods. Permian communities have been outlined for New Zealand by Waterhouse (1973b). Each fossil community, more or less (but not rigidly) unique in terms of its predominant or primary species, was governed in its distribution by the ecologic parameters outlined above, including the time-independent factors such as depth of water and type of bottom sediment, the time-recurrent factors of climate and temperature, and the time-unique factors of other biota, with trophic resources and food chains. It is of course the time-unique factors that provide a record of the evolution of communities within similar bottom facies and similar water depth and, being shared to some extent with contemporaneous communities adapted to different parameters, enable intercorrelation to assemble biozones.

Spasmodic or Episodic Evolution

To judge from their common recurrence through a faunal zone, fossil benthic communities were established, self-protecting, and self-sustaining. There is little evidence of major change within communities apart from a few incoming and outgoing species. Even more emphatically, it is very difficult to perceive consistent morphological trends in brachiopod species within a zone. Brachiopod species, such as those summarized in Tables 4 and 5, seldom showed any morphological cline, variation being random. Clines can be produced of course, but chiefly between species rather than within species, to suggest that gene pools during the life of species were stable and that homeostasis prevailed, as also observed by Eldredge and Gould (1974) in a wider context. It is this observation, based on actual sequences rather than neontological theorizing (as Huxley, 1862, was well aware in launching his attack on paleontology because he knew it was incompatible with the newly enunciated dogma on gradual incremental change) that partly explains why correlation works, and how ecology and chronology are so closely intertwined. The second aspect of actual geologic sequences completes the basic data to be evaluated—that faunal changes were abrupt, geologically speaking, instantaneous, and independent of facies. Entire communities appear all to have changed in the interval between one bed and the next, so that each successive biozone may be characterized by any one of a number of species or genera, for often the entire faunal assemblage has changed. This agrees with what Eldredge and Gould have established—evolutionary jumps independent of facies. Potentially, there is a significant interval of time unrepresented by sediment between any two beds, and obviously the chances are quite strong of this being so between two very different faunas. But when the same faunal change occurs widely over the world, as shown above, at what appears to be the same interval, it appears that abrupt changes are real. Any appeal to the contrary must rely on special pleading.

Biozones Reflect Climatic Change

What is the underlying cause to zonal changes that so widely and within the limits of evidence synchronously affect faunas? Clearly, the explanation does not lie in depth of water or in lithofacies, because these remain much the same from zone to zone, even though the facies and depth belts may have shifted position. The parameters of tide and light were changing too slowly and steadily, and tectonic plates moved too erratically, converging here, diverging there. Nor does the answer appear to lie within life itself, through steady evolution under random genetic mutation, because morphological change was, for the life of species, negligible, and erratic—fully in keeping with genetic theory of course, but not producing new species through gradual cumulative change. Some outside interference seems necessary to upset the homeostasis intrinsic in communities and gene pools if the changes were worldwide and synchronous. The obvious factor that is amenable to change is climate. Climate played a vital role in distribution and succession of faunas (Spjeldnaes, 1961; Williams, 1969; Waterhouse 1973b) expressed in modern ecological terms by the concept of biomes, which are the major faunal and floral subdivisions of the globe today, numbering eight or nine, fewer in the seas, and paired each side of the equator. Each biome includes many communities. If climate were to have changed drastically, fossil biomes must have moved toward or away from the poles, and regions previously occupied by one biome are likely to be colonized by another of quite different appearance, thus leaving in the fossil record two very different sets of communities, or two biomes, or two biozones. In any prolonged stratigraphic succession, climatic conditions will recur several times, giving rise to several zones of rather similar appearance with the same genera, but different species.

Migration Catastrophe

If the climate deteriorated, and especially if glaciation developed, the environment must have changed drastically in many ways and severely affected communities. Over the world, the sea level would have dropped by at least 100 m, to judge from the Pleistocene record, and thus affected sediments, which is likely to be revealed in the record as slump conglomerates (Burke and Waterhouse, 1973). In glaciated areas, the weight of ice led by isostatic adjustment to subsidence and marine incursions. Salinity increased, at least during the Permian period, with the extraction of water as ice, and under unusually widespread continentality shallow epicontinental seas developed extensive saline deposits, again affecting communities, even at family level according to analyses by Waterhouse and Bonham-Carter (1975). Food chains as discussed by Valentine (1971) and geochemical conditions were especially responsive to temperature changes. The depressed temperatures discouraged the growth of tropical flora and fauna and permitted the territorial expansion of cold-water species. Just as during the Pleistocene, there must have been extensive migration of life forms over short intervals of time to form brief times of flux, in the sense of migration of species, and change in gene pools, and in community structure. It is likely that during migration community structure became difficult to sustain because species migrated at different rates, and the homeostatic mechanisms protecting the stability of individual species as well as communities broke down. This was the time when evolution proceeded rapidly under stress of changing conditions of habitat, food, and climate, resulting in rapid changes in genera, with the appearance of new forms and the loss of old species.

In recent years stress has been laid on the significance of geographic isolation as a means of effecting change in the gene pool, and no doubt this may have applied, although the present most isolated parts of the world often seem more like havens for old life forms, rather than breeding grounds for new species. It would appear that reproductive isolation and change would be affected just as readily, and more pervasively, by large-scale migration, which moreover would deprive species briefly of communal protection.

Confirmation from Geologic Record

Judged from the geologic record, there is good reason to believe that the climate has oscillated periodically and severely since the beginning of earth history, and frequently since the origin of life. Tillites and other rock indicative of glacial conditions alternate with sediments deposited in warm conditions through the Paleozoic Era for the lower Paleozoic in the Sahara, in the early and Late Devonian in Brazil, in the early Pennsylvanian in Brazil, Argentina, South Africa, and Australia, and very widely indeed in the early Permian over Gondwana and northeast Siberia, with more restricted glacigene rock in east Australia, New Zealand, and Siberia for the Middle Permian, and New Zealand for the late Permian (see Burke and Waterhouse, 1973; Waterhouse, 1974).

Independent confirmation that temperature changed during the Permian period is offered by oxygen isotope values (as reviewed by Waterhouse and Bonham-Carter, 1972), which show that cool episodes in Australia had temperatures as low as 5 to 7°C, compared with 17 to 18°C for warm intervals. For the later record, oxygen isotope data set out by Bowen (1966) suggest 30 million year periodicity for depressions in temperature, one in the Jurassic, three in the Cretaceous, one in early mid-Tertiary, and another in the Pleistocene. Admittedly, these Mesozoic and especially Cenozoic cool episodes may not have affected life so drastically as in the Paleozoic, because Paleozoic marine life, notably amongst Brachiopoda and Fusulinacea, was conceivably more sensitive to temperature than the Mollusca now forming the predominant benthos.

Permian Example

Glaciations, by chilling the seas, must have had profound effects on marine faunas by developing distinct temperature belts, more or less latitudinally controlled, and life adapted to those distinct temperature belts. The different communities within each particular temperature belt were each controlled by different lithofacies, depth, and salinity, and biotic interplay and the species represented the stage of evolution then attained. The distribution is particularly well established for the Permian period when brachiopods such as *Wyndhamia, Terrakea, Notospirifer, Tomiopsis, Fletcherithyris,* and *Maorielasma*, the bivalves *Vacunella, Eurydesma, Myonia,* and *Astartila,* and the gastropods *Mourlonopsis, Walnichollsia,* and others were clearly adapted for dwelling in cold waters. In some cases, species have been driven out by competition from more Temperate and Tropical latitudes and have managed to survive by tolerating cold conditions. Other species possibly evolved to exploit the lengthy summer growing season and rich nutrients of polar regions, and yet were able to tolerate the severe winter, unfavorable for growth. Such faunas indicate cold conditions even when tillites are not found, either because of later destruction or nonexposure. The entry of Australian-related brachiopod species into the Canadian and Siberian Arctic during the Filippovian and Kurmain horizons strongly suggests that cool conditions obtained there also, at the same time as ice prevailed in east Australia.

By contrast, other zones in the Arctic and New Zealand are characterized by an array of Fusulinacea and compound rugose corals with locally abundant Ammonoidea, suggestive of warm waters, and during these intervals no tillites are known. Accompanying brachiopods may include warm-water indexes and few or no allies of east Australian species.

SUMMARY

The geologic record of evolution, paleoecology, and chronology underlines the importance of the climatic regime through time and space to the distribution and evolution of life. Episodic changes in climate brought about large-scale migration of life forms toward and away from the poles, changing species, genera, and communities morphologically, functionally, and geographically. Although the climate itself was cyclic and repetitious, it induced a cumulative affect on life and communities, recording progressive change which can be used to construct a linear time scale that summarizes the response of life both to the unchanging, the cyclic, and the changing environment and to surrounding life. Such a model fits well with what we know of the geologic record of evolution, and appears to be confirmed by what is known of the stratigraphic record and geochemical and isotopic data. It would explain the alternating associations of generic communities discussed above, and the episodic invasions of species of certain genera, favored by either particular climatic regime (e.g., *Attenuatella*), the demise of species, or the incoming of others after the migrationary flux. The hypothesis would explain the observations that biozones changed simultaneously over much of the globe, but it must be noted that a number of factors require either verification or qualification. Of these, the most critical lies in the actual connection between biozones and climate. The connection appears likely and has had some substantiation from isotopically determined temperature values and lithologic record of glaciation and nonglaciation, but much research is required to verify the correlation. The supposition that biozonal changes were largely synchronous also requires a great deal of study, and is not yet proved. Moreover, considerable confusion has been caused by the use of very extremely short lived fossils as zonal indicators, such as Mesozoic ammonoids. Since the accompanying benthos and plankton appear to have changed less rapidly, such ammonoids might be better regarded as indexes for subzones, related perhaps to genetic change or brief ecological parameters other than climate. The lack of change within zones, or failure of species to exhibit clines, has exceptions, but appears to be a general rule. The better known supposed exceptions have occasioned wide debate, and many examples are clearly clines between species, not clines within species. There is no question that species evolved. But it is my contention that they evolved episodically for brief intervals during times of exceptional stress.

Utility of Gastropods in Biostratigraphy

Norman F. Sohl

U.S. Geological Survey, Washington, D.C.

INTRODUCTION

The Gastropoda are a long-lived group of mollusks with a fossil record dating back to the Late Cambrian. They represent a plastic and successful group, which constantly increases in importance through its long evolutionary history, exhibiting successive periods of adaptive radiation and diversification. The various estimates of living species of from 105,600 (Jaeckel, 1958) to 37,500 (Boss, 1971) are a measure of their success. They represent about 80 percent of the taxonomic diversity of the Phylum Mollusca and rank second only to the Insecta in diversity as a class. Both in terms of feeding adaptations and habitats occupied, the Gastropoda must also be counted among the most successful of animals. They range from the deep sea to intertidal marine environments, and large numbers have accomplished the transition to freshwater and a wide variety of terrestrial environments. Feeding types include browsers and grazers on algae, deposit and filter feeders, scavengers, carnivores, and full parasites. Many crawl on the surface; others burrow, swim, or may be sedentary. This combination of diversity, abundance, and wide distribution in almost all environments should make the Gastropoda potentially one of most utilized groups for biostratigraphic studies.

For many years gastropods have been successfully utilized in the biostratigraphic zonation of Cenozoic rocks throughout the world. They were early used as an integral part of the Lyellian scheme of classification of Tertiary strata that was based upon percentages of extant species which occur in fossil assemblages. In this way they were used for intercontinental correlation. More commonly, they have proved useful especially in provincial correlations. For a variety of reasons they have seldom been ulitized in pre-Cenozoic rocks, but they have a similar potential for use. This potential can best be gauged by an understanding of their biological characteristics, disperal potential, and evolutionary history.

Distribution Potential

Wide and rapid dispersal of species are prerequisite to their successful use in biostratigraphy. Although fairly large numbers of prosobranch gastropod species may hatch in the crawling stage, the great majority possess a pelagic larval phase during their early development. The ratio of forms with a pelagic larval stage to those without varies with the latitudinal temperature gradient. Thorson (1946), for example, calculated that 85 percent of the warmer-water mollusk species of Bermuda undergo a free-swimming larval stage. In contrast, not a single prosobranch having a pelagic larval stage was found by Thorson (1936) in the Arctic seas.

Length of larval life varies greatly within the class. Among the prosobranchs, the Mesogastroda and Neogastropoda gènerally have a longer larval stage than the Archeogastropoda and may extend to periods of several months. Thus, given the proper set of circumstances, the potential for long-distance transport of larvae is great. Ager has provided some measure of relative rate of gastropod planktonic larval spread (1963, p. 167). After its accidental introduction into the Thames estuary in 1891, the sedentary calyptraeacean *Crepidula fornicata* (Linné) spread southward and then westward along the south coast of Britain at a rate of about 13 km/year. Ager points out that, providing suitable habitats were available, this snail could be dispersed around the world in about 3,000 years. This is a rate of dispersal that would appear virtually instantaneous relative to our available degree of resolution in biostratigraphy. Similarly, Robertson (1964) has documented the trans-Atlantic distribution of planktonic larvae of the architectonicid *Philippia krebsii.* In addition to sheer longevity of the planktonic stage, many larvae possess the ability to test the substrate for suitability of settlement. If conditions are not suitable, they rise back into the plankton and larval life is extended (Thorson, 1946).

The presence or absence of a *planktonic larval stage* in either living or fossil mollusks may generally be discerned from the shell. This provides an initial guideline for selection of groups with greatest biostratigraphic potential. Prosobranch gastropods without a larval stage possess a proportionally larger protoconch, with a smooth surface and inflated whorl profile that approaches a spherical shape. In addition, the protoconch grades to the teleconch without abrupt change. Thus, as Thorson (1946, p. 246) has pointed out, a large clumsy apex is a safe criterion for distinguishing a species having a nonpelagic development. In contrast, pelagic gastropod larvae undergo a relatively abrupt change of life at metamorphosis when bottom dwelling is assumed. This change is recorded in the shell by a distinct separation of the protoconch from the teloconch. Robertson (1971) has provided exceptional illustrations of such protoconch-to-teloconch changes. Usually the change is marked by an abrupt difference in sculpture, a varix or occasionally digitations of the larval shell overlapping onto the first teloconch whorl.

Differing lengths of larval life may be marked by the number of protoconch whorls present. Larger numbers of whorls commonly indicate longer larval life. In some high-spired forms such as *Triphora* there may be as many as eight or nine whorls (Fretter and Graham, 1962, p. 471). In *Rissoa* different numbers of protoconch whorls between species mark greater or lesser larval stages (Fretter and Graham, 1962, p. 472).

Studies of the protoconch-teloconch relationships of fossil shells therefore should provide a reasonable measure for gauging the potential of individual species for wide dispersal. It is not sufficient, however, to depend upon knowledge of the length of larval life of living species to estimate the dispersal capabilities of related fossil forms. Thus, if we extrapolate the fact that living *Turritella communis* possesses a short larval life to all fossil *Turritella*, we would be faced with abundant anomalous situations. For example, the turritellid *Haustator trilira* (Conrad) ranged from Mexico to New Jersey during Late Cretaceous time (Sohl, 1960), and the *Turritella forgemoli* Coquand species group is found in rocks of Maastrichtian age in such widely scattered areas as Texas, Alabama, North Africa, Transcaucasia, Russia, Iran, and Baluchistan. Surely this is not indicative of an especially restrictive larval existence. As a second example, MacNeil (1965) indicated that the cool-water-inhabiting snail *Neptunea*, although being entirely an epifaunal crawler after hatching, still has managed to accomplish the same spatial distribution in the northern hemisphere as have such forms as the bivalve *Mya*, which has a planktonic larval stage. He does, however, provide evidence to indicate that the rate of migration of *Mya* was more rapid than *Neptunea*. Each species, both fossil and living, must be viewed on its own merits as there seems to be significant variance even between closely related species as to the length of their larval life.

In summary, it is obvious that, with the exception of the prosobranchs that live in the high-latitude Arctic seas, most gastropods undergo a pelagic larval stage of a few weeks to a few months duration. Such time intervals are sufficient under advantageous current, temperature, and bottom conditions to allow transoceanic distribution of species that in terms of biostratigraphic levels of resolution therefore qualify as organisms of high potential for stratigraphic purposes.

Rates of Speciation

Refinement in biostratigraphic zonation of any group is closely related to the rate of evolution within the lineages utilized. However, little attention has been paid to speciation rate and its relationship to biostratigraphic potential among the gastropods. As indicated in Text-figure 2, species of the gastropod genus *Drepanochilus*, found in the Upper Cretaceous (Campanian) Pierre Shale in the Western Interior, range through about four ammonite zones. Gill and Cobban (1966, p.

A2) have calculated these ammonite zones to average 500,000 years each; thus the gastropod rate of speciation in this example approaches 2 million years. Other species in the same rocks (Sohl, 1967, p. B6) vary greatly even within the same genus. For example, *Anisomyon borealis* (Morton) ranges through seven ammonite zones equating with a species range of 3.5 million years, whereas *Anisomyon centrale* Meek appears to have a more restricted range of 1 million years. As presently understood, many other common Cretaceous species, such as *Haustator trilira* (Conrad), of the Coastal Plains have a demonstrably longer species range, but one that might be restricted by detailed systematic investigation and subdivision such as that done by Fisher et al. (1964) for the Eocene gastropod *Athleta petrosa* (Conrad).

Limiting Factors

The biostratigraphic utilization of gastropods for the solution of problems of correlation does not approach the potential outlined above. They have been most commonly used in zonation and correlation of Cenozoic rocks (Teichert, 1958). Much less often they have been utilized for solution of Mesozoic and Paleozoic biostratigraphic problems. In a gross way, this pattern of use parallels the history of progressive increase in family diversity from the Late Cambrian to Recent time exhibited by the group (Text-fig. 1); The same pattern holds at higher and lower taxonomic levels. Although a broad relationship exists between taxonomic diversity with time and biostratigraphic utilization of the group, it is probably not the major reason for the infrequent use of gastropods in biostratigraphy.

A primary limiting factor is simply the state of knowledge of fossil gastropods. As Yochelson (1968, p. 1363) stated,

> Paleozoic gastropods do have inherent stratigraphic utility. The problem of applying them to dating is not so much poor preservation, now that the Pandora's Box of silicified faunas is open, as it is manpower. Many paleontologists have described Paleozoic Gastropoda, but the first half century produced only one dedicated student. . . .

In other words, the raw materials are available; they just have not been used. In addition, most studies have concentrated on the total assemblage of a small time interval and have not emphasized the study of lineages over a longer time span. Thus their biostratigraphic potential has not been realized.

The second factor has been the overall concentration of workers on the study of other fossil groups of proven biostratigraphic utility. For example, in discussing the Upper Cretaceous gastropods of the Western Interior of the United States, Sohl stated (1967, p. 1363),

> Paleontologists have long concentrated upon the study of the ammonites and inoceramids because of their proven biostratigraphic utility. . . . Although in the Interior the remainder of the mollusks far outstrip the ammonites and inoceramids in diversity,

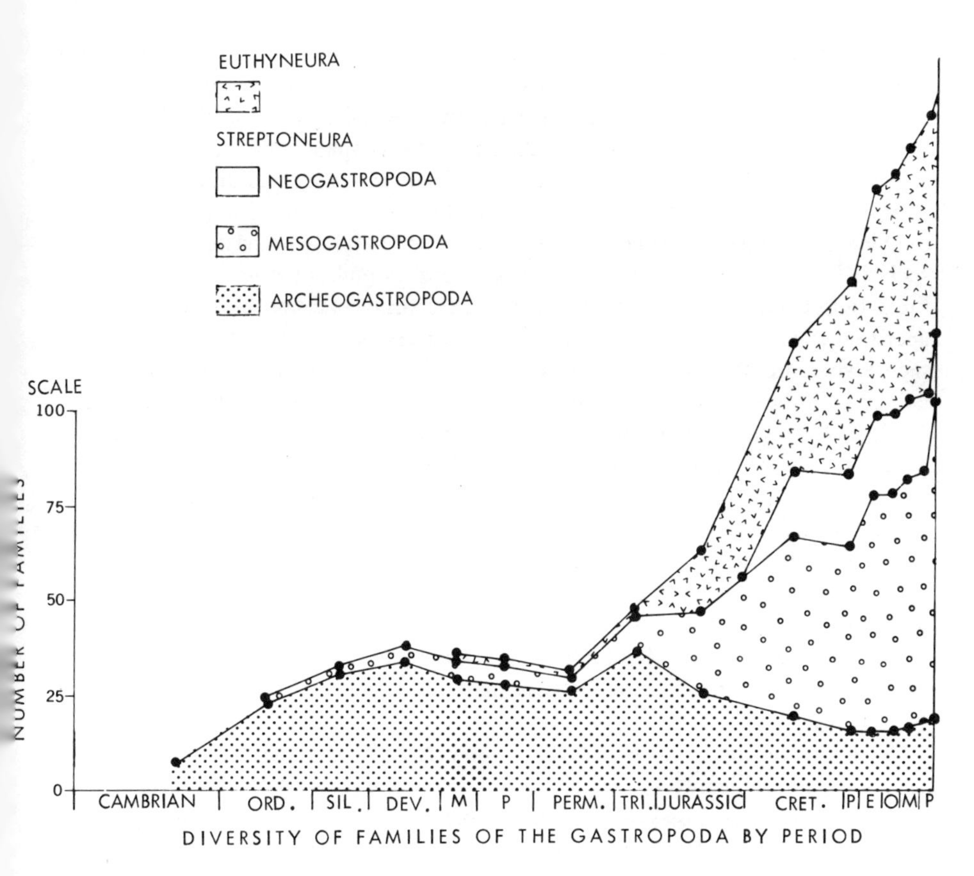

Text-Figure 1
Diversity of families of the Gastropoda by period.

the number of described species is proportionately small. This situation is especially true of the Gastropoda.

The third factor is that of preservation. Most fossil groups suffer from vagaries of the preservational factor, but gastropods suffer to a greater extent than most because almost all possess an aragonitic shell, which is easily dissolved below compensation depth or during diagenesis. In the Cretaceous chalk facies, aragonite commonly is not preserved and thus gastropods are represented only as internal molds. In the same facies, however, calcitic shells of such bivalve groups as the ostreids, pectens, and anomiids may be well preserved. Internal molds of most snails are valueless for precise identification. In many instances they cannot even be assigned to a superfamily with absolute confidence. In contrast, much of ammonite taxonomy is based upon internal molds in which such features as the suture are readily discernable, and the main features of external ornamentation, such as nodes, shape of the venter, etc., are reflected.

Similarly, musculature and hinge features are commonly represented on internal molds of bivalves. Thus gastropods are of little use in such facies as the Cretaceous chalks, and workers have had to rely upon the partially or wholly calcitic shelled bivalves or forms such as the ammonites to formulate a biostratigraphic zonation.

The shape of the shell has also often proved a disadvantage and discouragement to work on the snails. For example, many of the better preserved older fossil assemblages are embedded in concretions of very well indurated sandstones. When such rocks are split, regular rounded bivalves, brachiopods, or ammonites may break out easily, but the irregular assymetrical shells of snails normally exfoliate, apertural expansion or high spires break off, and one is normally left only with a recoverable internal mold. Thus the recovery of snails from such matrices is difficult and requires much time-consuming preparation (Sohl, 1967, p. B3; Popenoe, 1957, p. 430). Such situations have caused some workers to concentrate and depend on the more easily recoverable taxa for biostratigraphic information.

Convergence in form and homeomorphy are features characteristic of many groups of snails that cause confusion in systematic assignment and, potentially, dating and correlation. For example, the patelliform shape is repeated in such systematically widely separated groups as the Fissurellacea, Patellacea, Cocculinacea, and Trochacea of the Archeogastropoda; Neritidae, Phenoclepidae, Hipponicacea, and Lamellariacea of the Mesogastropoda; Siphonariacea, Acroloxidae, Planorbidae, Lymnaeidae, and Ancylidae of the Bassomatophora; and Umbraculacea of the Notaspidea.

Such phenomena of convergence and homeomorphy at higher taxonomic levels are a decided problem to the systematist in determining the proper taxonomic affinities of his species, and may also present problems to the biogeographer and paleoecologist, but they are of less concern to the biostratigrapher. For him, the fact that species and perhaps lineages can be distinguished is a more critical point than that their higher systematic position be absolutely known. However, convergence at lower taxonomic levels is a greater problem to the biostratigrapher. For example, Merriam (1941, p. 10) in his study of fossil turritellids from the Pacific Coast cited a number of convergent species. Although many of these species, some of which are zonal indexes, had been confused for one another, he maintained that "knowledge of the growth line trace and nepionic development will make possible the immediate separation of almost all of them."

When gastropods have been utilized, as in the Cenozoic, often the erected zonation rests upon a faulty base. In areas such as the Caribbean, in spite of its rich molluscan faunas (about 20,000 species), molluscan biostratigraphy is hampered by large gaps in stratigraphic knowledge coupled with the propensity for mollusks to occur in separated shell beds. Jung (1974, pp. 33-34) has pointed out that in this region when species range zones are plotted out they terminate at levels where there are such stratigraphic gaps. The erected chart may look impressive and delimit what appears to be good potential assemblage zones, but in effect the zones so defined may only reflect lack of information. The fault lies, however, not with the fossils but with the paleontologist.

Intercontinental correlation utilizing Cenozoic mollusks has often been attempted with varying degrees of success. In such endeavors a main stumbling block is the element of endemism. For example, Addicott (1970, p. 32) in discussing correlation of the California Miocene faunas stated,

It seems unlikely that correlation of benthic molluscan assemblages will ever permit accurate positioning of the European Oligocene-Miocene boundary in the Pacific coast middle Tertiary sequence. Part of the uncertainty is based upon disagreement among workers as to the original designation of the type Miocene in Europe. Secondly, and more importantly, is the problem of endemism in the later Tertiary molluscan faunas. Because of this characteristic, it is difficult to correlate even between adjacent mid-Tertiary molluscan faunal provinces along the Pacific Coast.

As with other groups of benthic organisms, the effect endemism plays in the biostratigraphic utilization of gastropods varies greatly through the Phanerozoic.

In spite of all the above mentioned factors, gastropods have been successfully utilized at almost all Phanerozoic levels. That they could be used to a greater extent is obvious, but such use rests primarily with the development of a greater number of students of the group.

PALEOZOIC MARINE GASTROPODS

Diverse and abundant gastropod assemblages have been described from many areas at almost all post-Cambrian levels, yet knowledge of Paleozoic gastropod faunas must be considered primarily in the descriptive phase. The framework of basic systematic studies is still insufficient to provide a biostratigraphic structure of much more than local utility. The fact that Paleozoic gastropod assemblages are so thoroughly dominated by Archeogastropoda (Text-fig. 1) might lead one to assume that their biostratigraphic potential might be limited. Most living archeogastropods are restricted in feeding habit to browsing and grazing on algae or rasping algae from rock surfaces. This limitation of feeding habit coupled with a respiratory system generally considered less capable of handling heavier loads of sediment than the later Mesogastropoda and Neogastropoda leads to an assumption that Paleozoic archeogastropods may have been facies contricted—mainly to firmer substrates. Such a factor would limit their biostratigraphic potential, especially in regional correlation. Much additional investigation is needed, but the fossil record does yield some information that indicates at least certain Paleozoic archeogastropods probably tolerated environmental conditions and developed feeding habits of a wider variety than are represented in the group today. For example, *Platyceras* (Ordovician-Triassic) was a cophrophagous symbiont on crinoids (Bowsher, 1955). Batten (1958) has noted the association of pleurotomariaceans with sponges on muddy bottoms of the West Texas Permian seas, and Chronic (1957, p. 106) has suggested similar soft bottom conditions for molluscan associations in the Permian of Arizona that include numerous archeogastropods. In general, as Sloan (1955) and others have shown, Paleozoic gastropod assemblages are most diverse in nearshore environments, especially estuarine and lagoonal situations. Occasionally, as noted by Batten (1973, p. 596), they even dominate some preserved assemblages in Upper Carboniferous black shales. These few examples serve only to illustrate that Paleozoic archeogastropods were probably more ecologically diversified than their modern counterparts.

There has been little direct attention paid to how closely individual groups of Paleozoic gastropods were tied to facies. Obviously, restriction to only certain facies will limit utilization of any form for regional correlation. Some information relative to the Paleozoic gastropods as a group has, however, been presented by Sloss (1958). The data, presented as percent representation of total fauna, are difficult to interpret. In general, they indicate that gastropods are more common in the Lower Paleozoic in black phosphatic shales and dolomites. In the Upper Paleozoic, gastropods are associated most closely with sandstones, although they continue as proportionally well represented in dolomites and shales. Such generalizations can only suggest that the biostratigraphic potential for snails may be low in certain types of sequences. The increase in representation of gastropods in sands during the Late Paleozoic cited by Sloss may be related to the rise of the Mesogastropoda.

The occasional utilization of gastropods as zonal indexes in the Paleozoic gives some indication of their potential. The Upper Carboniferous *Omphalotrochus* zone of the Moscow basin in Russia (Gorsky, 1939) serves as such an example. Yochelson (1954) has discussed the Late Pennsylvanian and Early Permian distribution of the genus, suggesting that some species are of very wide, perhaps intercontinental, distribution. Similarly, Cooper (1956, p. 126) proposed the *Palliseria* (*Mitrospira*) zone for rocks of part of the Lower Ordovician Whiterock Stage in Nevada. Yochelson (1957) has demonstrated that this zone can be carried north from Nevada to Alberta. Recent investigations of opercula of gastropods from the Lilydale Limestone (Early Devonian) of Australia by Yochelson and Linsley (1972, p. 9) even suggest conspecific relationships of taxa geographically separated as widely today as Australia and Austria.

Although Paleozoic gastropods have not often been utilized, these examples serve to indicate that the potential for use is significant. At species level, Paleozoic gastropods have served for zonation of parts of stages and are of at least basinal utility. At the generic range-zone level, they may be of use for broader intercontinental correlations, as in the case of *Omphalotrochus.*

MESOZOIC GASTROPODS

The Mesozoic Era marks a period of great radiation and diversification among the Gastropoda (Text-fig. 1). According to Batten (1973, pp. 604-606), the Triassic faunas are marked by the presence of few genera during the earliest Scythian and Anisian stages, but during the Ladinian there is an abrupt diversification. Overall, however, these Triassic faunas retain their Paleozoic aspect. Major changes occur in the Jurassic, by the end of which most of the major living superfamilies not previously present had been introduced. The Cretaceous is marked by great diversification at the familial and generic level (Text-fig. 1), especially among the Mesogastropoda, Neogastropoda, and the opisthobranchs. The appearance of the cypraeids, cones, turrids, strombs, and others all lend a modern aspect to the Upper Cretaceous gastropod assemblages that strongly contrasts with the earlier assemblages of the Jurassic and Triassic.

Among the larger Mesozoic invertebrate fossils, the greatest biostratigraphic reliance has been placed on the ammonites, and the study of snail assemblages has been primarily only an adjunct to larger faunal monographs. Few workers have tested their biostratigraphic potential. For example, in North America reports of Triassic gastropod assemblages are few and described species very rare. The works of Kittl (1891-1894), Leonardi and Fiscona (1959), and Haas (1953), among others, have demonstrated great local abundance and diversity; however, no significant attempt has been made at a biostratigraphic synthesis of Triassic gastropods.

Jurassic gastropods are of at least great local abundance and considerable diversity, as examplified by the monographs by Huddleston (1887-1896) on the Inferior Oolite of England. Monographs such as this normally give some indication of stratigraphic position of the species, but do not synthesize their occurrence into a usable zonal scheme. The very detailed studies of the Jurassic stratigraphy in Great Britain summarized by Arkell (1933) contain abundant reference to such informal units as the "*Nerinea* Bed" near Condicote, the *Viviparus* Marl and Pebble Bed, the *Natica* band, or the *Exillissa* limestone. These notations indicate the occurrence of gastropods in such an abundance that they physically characterize parts of stratigraphic sequences. Their biostratigraphic use, however, is restricted to very localized areas, and their use in correlation reflects an ecologic rather than an evolutionary correlation of similar assemblages. A few workers have proceeded beyond this level. For example, Wendt (1968), in a study of the planispiral gastropod *Discohelix*, demonstrated restricted ranges of successive species in the Lower and Middle Jurassic of Sicily and maintained that some retained the same range in the northern Calcareous Alps.

Knowledge of North American Jurassic gastropod assemblages does not approach that of European assemblages, but as indicated by Sohl (1965) they are moderately diverse and geographically widespread from Mexico to Alaska. He indicated several species, such as *Lyosoma powelli* White, a neritid, and *Cossmannea imlayi* Sohl, a nerineid, to be of sufficiently restricted stratigraphic range to be of at least basinal value for correlation. Frebold (1957, p. 58) described *Turbo ferniensis* from Upper Oxfordian "Green Beds" of the Fernie Shale of Alberta occurring over several hundred miles at the same stratigraphic level.

Considerably more is known about the Cretaceous gastropod assemblages than for any older period. As a generality, most work on Cretaceous gastropoda has been descriptive and has appeared as parts of larger monographs on total molluscan or invertebrate faunas from formational or chronostratigraphic units of a specific area or locality. Aside from faunal charts indicating species occurring in common between areas, little synthesis exists that attempts to place snails within a formal biostratigraphic framework. For the most part, stratigraphic ranges of described species have been determined by their relationship to sequences zoned by other fossils (ammonites, planktonic foraminifera, etc.); they have seldom been used as the primary tool of zonation. As associated taxa of distinctive assemblages, they have, however, been used as secondary elements for biostratigraphic purposes. Even in such facies as the warm-water Tethyan shelf carbonates where ammonites are rare indeed, and where nerineid and actaeonellid gastropods may be dominant elements in many limestones, their biostratigraphic use has remained secondary to such groups as the rudist bivalves.

During the Cretaceous, gastropod species in general have their greatest geographic distribution during the Lower Cretaceous, any many of the nerineid gastropod species of the Carribbean region, for example, are virtually impossible to distinguish from those of the Mediterranean Tethys. Thus in Albian rocks they offer the potential for use in intercontinental correlation, enhanced, perhaps, by short dispersal distance across the Atlantic before the Cenomanian acceleration and isolation. Provinciality seems to increase markedly with time, especially during the Upper Cretaceous. Many species, however, are widespread throughout individual provinces during the Cretaceous (Sohl, 1969). Similar distributional patterns have been described for the bivalves by Kauffman (1973).

Rather detailed phylogenies have been worked out for some genera of the Tethyan realm, such as *Itruvia, Trochactaeon, Plesioptyxis,* and others, by Pchelentsev (1953, 1954, figs. 18, 26, 30, etc.). At least locally within Transcaucasia and Central Asia, the species of *Itruvia* appear to offer the possibility for subdivision of the Turonian into a series of phylozones. More commonly, however, critical evaluation of gastropod taxa and the changes they manifest in sections collected in detail is lacking. This has meant that gastropods have been used only for the broadest stage or greater level of correlation. That such detailed collection of samples can provide information amenable to the erection of a refined zonation of either basinal or provincial value may be seen in several examples from the Cretaceous of North America.

The only really formalized zones in the typical sense are those Popenoe (1942) proposed for parts of the Holz Shale of California in which *Turritella chicoensis perrini* and *Turritella chicoensis chicoensis* characterize the Santonian and Lower Campanian rocks. As proposed, these represent occurrence range zones and are used in the sense of an index or guide species.

As most Cretaceous paleontological monographs have dealt primarily only with the faunas of certain stratigraphic intervals, studies of the evolution of species groups are rare; thus utilization of evolving lineages for the erection of *phylozone* schemes has received little attention. Popenoe's (1957) study of the West Coast ringiculid genus *Biplica* is a noteworthy exception. He distinguished five successive species extending from approximately middle Albian to some part of the Maastrichtian. His studies provided a degree of zonal refinement of about stage level, assuming an 8 million year longevity for each of the *Biplica* species. As known, *Biplica* is restricted to Pacific Coast Cretaceous rocks.

Other genera of gastropods have been shown to have a more rapid rate of evolution and hence a potential for biostratigraphic resolution at less than stage magnitude. Text-figures 2 and 3 show examples of such segments of evolutionary sequences. In the first instance, as exemplified by the genus *Drepanochilus* in the Pierre Shale and Fox Hills formations of the Western Interior, four successive

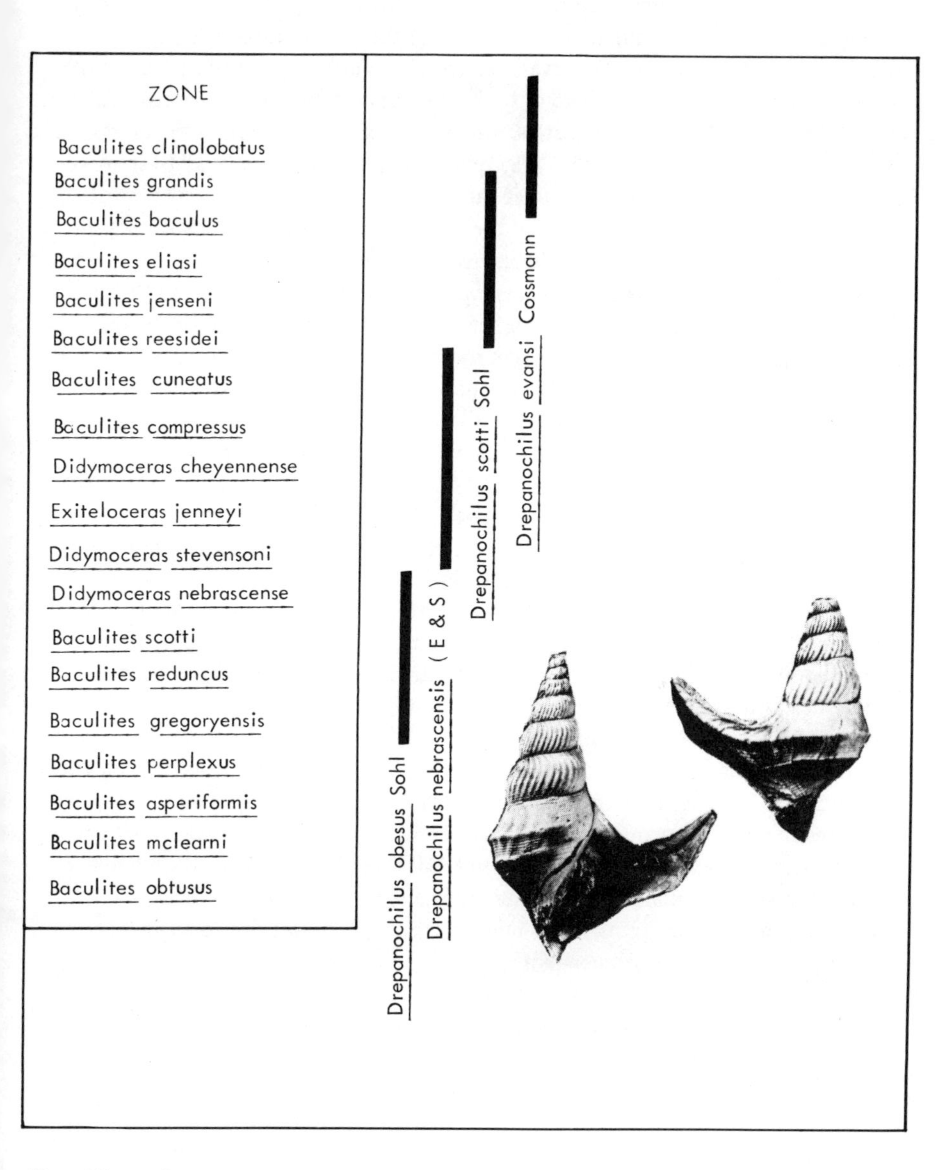

Text-Figure 2
Stratigraphic distribution of *Drepanochilus* in the Pierre Shale of the Western Interior.

species, one evidently derived from the other, can be delineated. They can be morphologically distinguished from each other readily in the earlier three species, but in the younger two, *D. scotti* and *D. evansi*, there is an overlap of form, which is present through the zone of *Baculites grandis* (Sohl, 1967).

The time interval represented for each zone in this example ranges from approximately 2 million to 2.5 million years/species. In the Gulf Coastal Plain during approximately the same time interval (Navarroan), another group of species of this same genus occurs. However, here only three successive species can be distinguished (Sohl, 1960, fig. 10). In either instance the species are to be found over several hundred miles of outcrop and thus may qualify as phylozone indexes. Even more complicated lineages may be delineated, as shown by Sohl (1964, p. 258) in the Campanian and Maastrichtian rocks of the Gulf Coastal Plain. Text-figure 3 illustrates the suggested phylogeny of the genus *Liopeplum*. As known, the genus ranges from below the occurrence of *Scaphites hippocrepis* type III of Cobban (1969), and therefore must originate in rocks of earliest Campanian or latest Santonian age, and ranges into the uppermost Maastrichtian rocks of the area. Two main species groups are present in the earliest Campanian (Upper Austinian). In one, as typified by the end species *Liopeplum canalis* (Conrad), the whorls of the spire possess a basal callus ridge that tends to develop a canaliculate or overhung suture. The other main line is characterized by species such as *L. rugosum*, possessing transverse ribs at the shoulder angulations. Representatives of this genus are known throughout the coastal plain Cretaceous outcrop belt from New Jersey to Texas and northward into the southern part of the Western Interior. Thus, in a true sense, these species ranges represent phylozones extending throughout a biotic province. Many other genera in the Cretaceous of this area would surely provide similar information if studied in detail. Phylozones such as these are biostratigraphically important and powerful tools if one accepts the basic premise that the rate of speciation and dispersal within the lineage was the same in all cases. Within a single biotic province, although it be several thousand miles long, this premise seems relatively safe within the time frame or speciation rate of about 2 million years.

In dealing with most organisms, stage-of-evolution correlations have more practical than theoretical drawbacks; this seems, for several reasons, especially true of the gastropods. The main problems are those already discussed for the group in the section on limiting factors. It is inherent in the index species concept that recognition of the zone is dependent on only one organism. When this species is absent because of facies control or other variables, no determination is possible. For example, the geographic utility of the *Liopeplum* lineage illustrated in Text-figure 3 is given in Text-figure 4. In this figure the horizontal bars indicate the geographic extent of the distribution of the successive species. The patterns

---

Text-Figure 3
Suggested phylogeny of *Liopeplum* in the Upper Cretaceous of the Gulf and Atlantic coastal plains.

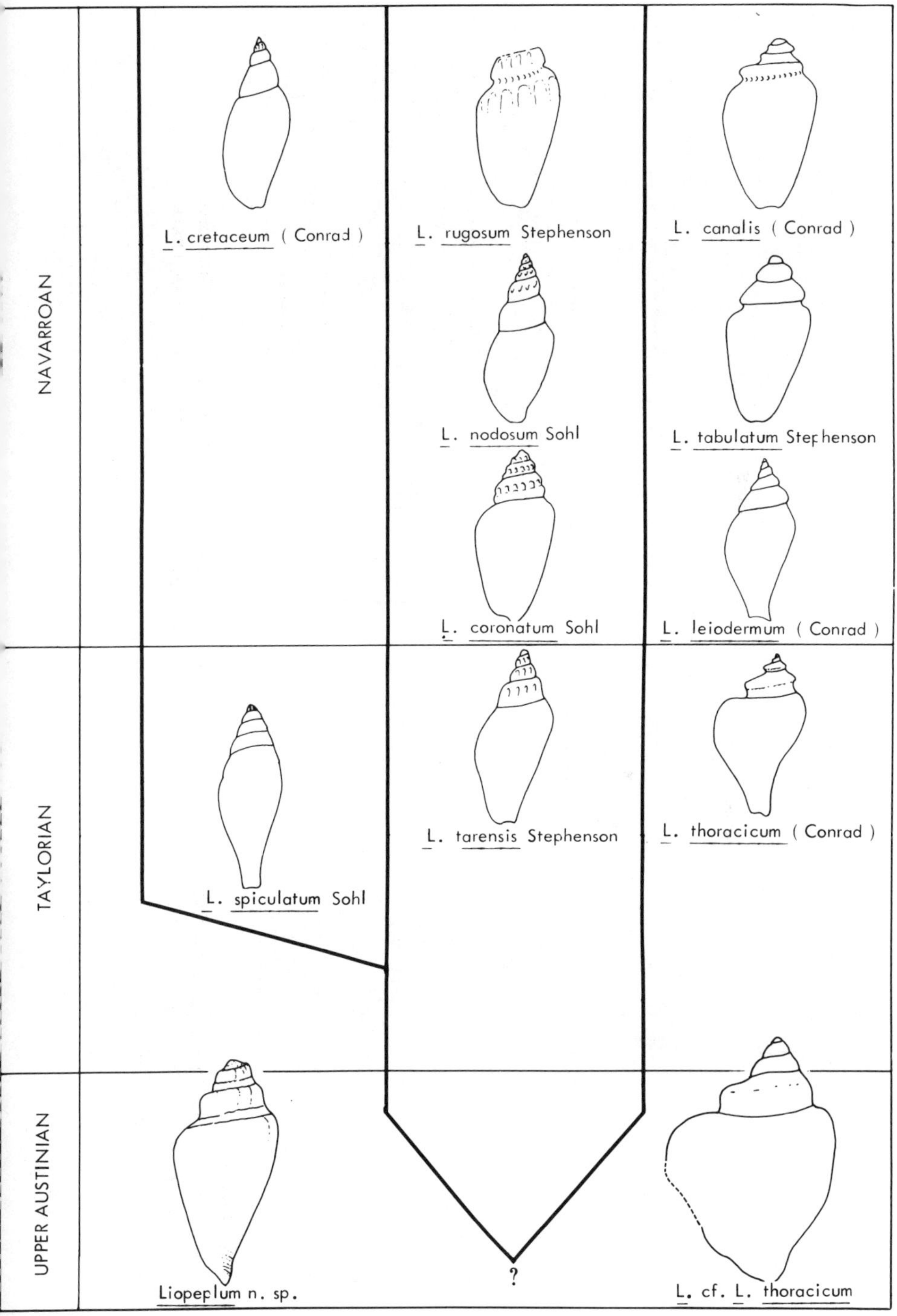

NAVARROAN

L. cretaceum (Conrad)

L. rugosum Stephenson

L. canalis (Conrad)

L. nodosum Sohl

L. tabulatum Stephenson

L. coronatum Sohl

L. leiodermum (Conrad)

TAYLORIAN

L. tarensis Stephenson

L. thoracicum (Conrad)

L. spiculatum Sohl

UPPER AUSTINIAN

Liopeplum n. sp.

?

L. cf. L. thoracicum

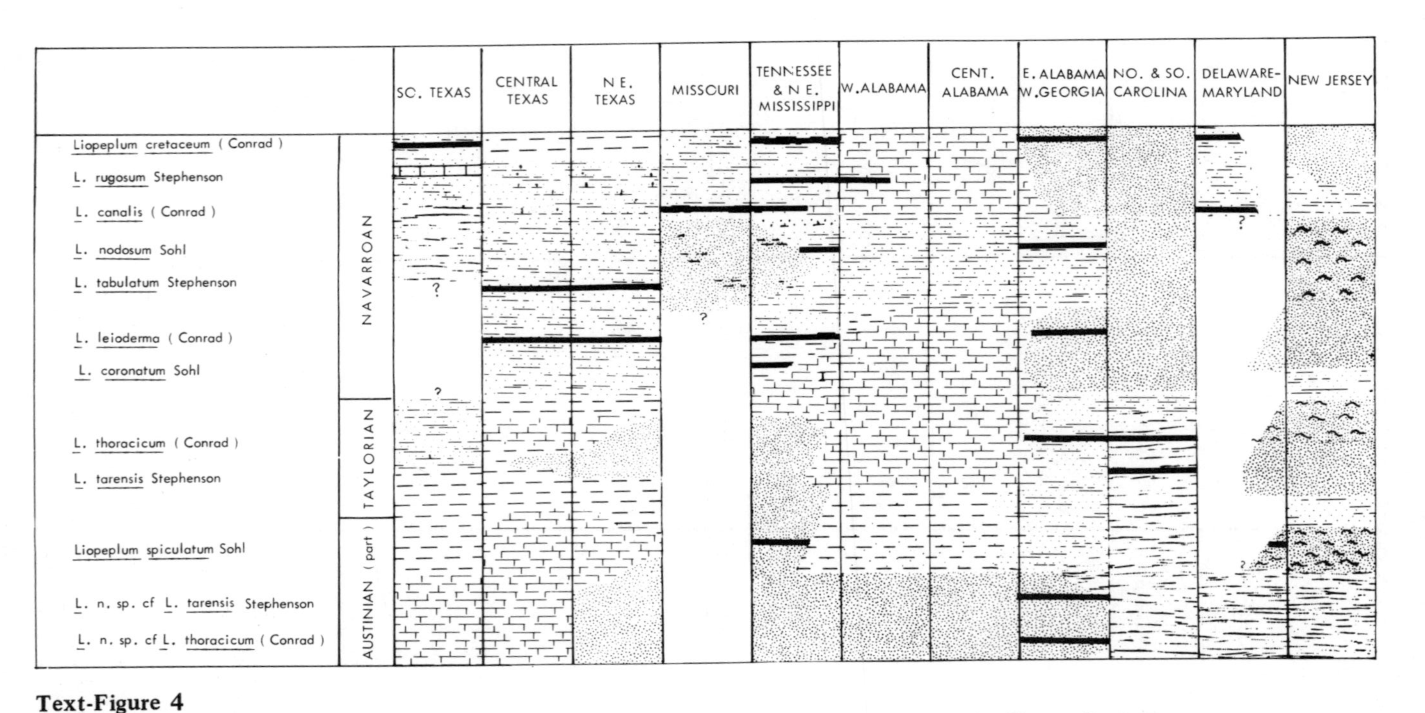

Text-Figure 4
Relationship of the geographic distribution (black bars) of the species of *Liopeplum* in the Upper Cretaceous (Campanian-Maastrichtian) rocks of the Atlantic and Gulf coastal plains to the generalized lithologic character of the stratigraphic units.

within the vertical columns indicate the general lithic character of the units in the cited geographic areas. It is apparent from this diagram that the distribution of *Liopeplum* is linked to lithic type; the genus is virtually absent from the calcareous facies of Texas and Alabama. In other areas of clastic sedimentation that are shown to lack *Liopeplum* its absence may in part be correlated with general lack of information about the faunas. Thus the use of *Liopeplum* for erection of a widely applicable system of phylozones has distinct limitations.

An alternative approach that utilizes gastropods in conjunction with other mollusks in erection of a framework of assemblage zones appears to circumvent most of the problems encountered in dealing with phylozones. In studies of the Cretaceous gastropod faunas of the Mississippi Embayment Region, Sohl (1960) felt that the approximately 300 ft of Campanian and Maastrichtian (Navarroan) age deposits present could be divided into five zones based solely upon the ranges of certain gastropod taxa. Continuation of these studies to the east in the Chattahoochee River region (Georgia-Alabama) showed, however, that such a zonation was not applicable in this area. Only three of the five embayment zones could be recognized in eastern Alabama and western Georgia. These three zones with varying degrees of confidence, however, can be carried to most other parts of the coastal plain. The level of confidence can be increased significantly if other characteristic molluscan species are included as part of the assemblage. At present, based upon assemblages of mutually occurring gastropods, bivalves, and ammonites whose range zones are demonstrably restricted stratigraphically, it is possible to divide the Campanian and Maastrichtian rocks into a successive series of seven assemblage zones that are identifiable in all marine facies on the coastal plains from New Jersey into northern Mexico. Thus the zones are recognizable throughout a whole biotic province, a criteria used by some as a prerequisite for formal designation.

An example of one of these zones is that which encompasses the range of the turritellid *Haustator bilira* (Text-fig. 5). Characteristic of this stratigraphic interval are, among others, the ammonites *Baculites columna* Morton, *B. tippahensis* Conrad, *B. carinatus* Morton, and *Discoscaphites conradi* (Morton); the bivalves *Trigonia cerulia* Whitfield (includes *T. haynesensis* Stephenson), *Trigonia angulicostata* Gabb, *Arctostrea aguilerae* Böse, *Anomia ornata* Gabb, *Diploschiza melleni* Stephenson, *Camptonectes bubonis* Stephenson, and *Titanosarcolites oddsensis* Stephenson; and the gastropods *Haustator bilira* (Stephenson), *Stantonella interrupta* (Conrad), *Piestochilus curviliratus* Conrad, *Eoancilla acutula* Stephenson, *Liopeplum cretaceum* (Conrad), *Liopeplum canalis* (Conrad), and *Ringicula clarki* Gardner.

As is obvious from Text-figure 5, few species occurs in all the facies and none in all representative formations on the Atlantic and Gulf Coastal plains. Ammonites are proportionally more common in the chalks; bivalves and gastropods have their greatest representation in the clastics. The characteristic species common to this zone appear to be less in number at either end of the geographic extent of the biotic province (i.e., south Texas and New Jersey). In part, this is an artifact of both marginal marine facies and lack of study of the Escondido Formation of southern

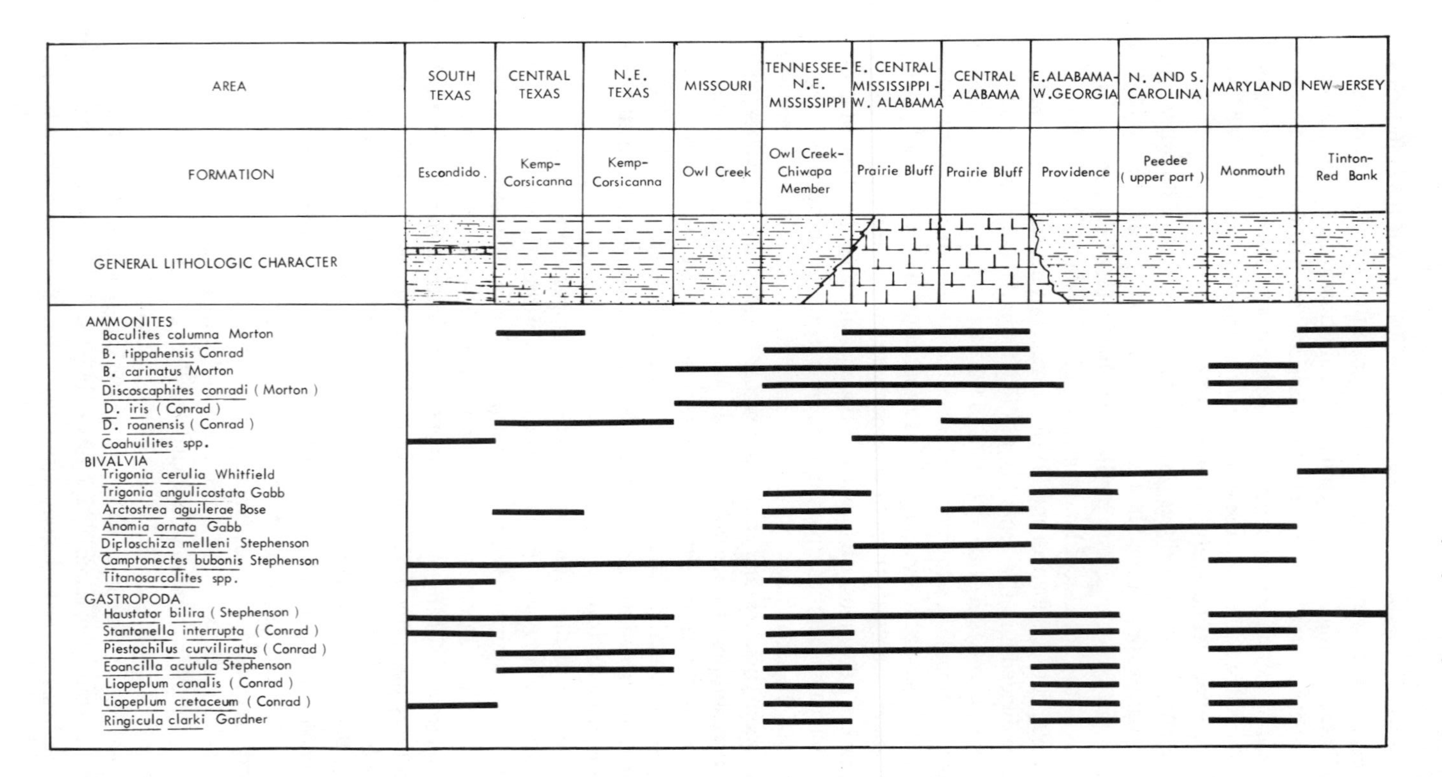

Text-Figure 5
Geographic distribution (black bars) of some characteristic taxa of the *Haustator bilira* assemblage zone (Maastrichtian) on the Atlantic and Gulf coastal plains.

Texas and poor state of preservation of the Tinton and Red Bank Sand faunas of New Jersey, and not necessarily a lack of distinctive species.

The obvious and positive feature shown in Text-figure 5 is that the zone can be traced with integrity for some 2,000 miles from Texas to New Jersey throughout a spectrum of marine deposits ranging from fine- to coarse-grained clastics to carbonates. Subdivision of this zone is possible at least locally. For example, within the Mississippi Embayment region and the east Alabama and western Georgia areas three distinguishable forms of *Trigonia angulicostata* are present. They occur in stratigraphic succession and may be viewed as chronologic subspecies. Thus in these areas the broader assemblage zone may be subdivided into phylozones of local or basinal utility, but not of geographic extent suitable for more long-range correlation. Similarly, several taxa within the assemblage are to be found in other biotic provinces and provide a mechanism for at least tentative intraprovince correlation. *Titanosarcolites*, a rudist pelecypod, is widespread in Maastrichtian-age rocks throughout the Greater Antilles and Mexico. *Arctostrea aguilerae* is known from Mexico (Cardenas area) and Cuba (Habana Formation). *Baculites columna* is found in the Fox Hills Formation of South Dakota. Thus this assemblage zone provides a broad time frame for comparison of depositional patterns and tectonic events over a large portion of the North American continent.

Because of their exceptional abundance and distribution, several groups of gastropods possess significant but as yet untested capabilities for the solution to Mesozoic stratigraphic problems. Both the Nerinellacea and Actaeonellacea are individually abundant and taxononically diverse in lagoonal and nearshore deposits throughout the warm-water Tethyan realm. However, correlation between such deposits has mainly rested upon utilization of the rudist bivalves.

Another major group of gastropods, the Opisthobranchia, which include the acteonids, ringiculids, bullids, and others, could also prove to be especially useful. Perhaps because of their small size they have often been overlooked by collectors, and thus our knowledge of their distribution is limited at present. They are widely distributed and often common elements in the Upper Cretaceous deposits of the North American Coastal plains and no doubt elsewhere. Such groups of small-sized gastropods possess the additional advantage of potential recovery from well cores and feasibly could play a role in correlation of subsurface units.

In summary, Mesozoic gastropods have been utilized for biostratigraphic purposes in a variety of ways. Aside from local or basinal correlation, however, I am convinced that like most benthic organisms they are best utilized in conjunction with other mollusks, or even other groups, as parts of broader groupings into assemblage zones.

CENOZOIC GASTROPODS

The Cenozoic Era has classically been a time interval in which both gastropods and the Bivalvia have played an exceptional part in the growth of both marine and nonmarine biostratigraphy. Several factors may account for their increased use in

Cenozoic rocks. The first is related to their great increase in diversity during this period of time (Text-fig. 1), and the second to the loss by extinction of forms (ammonites and inoceramids) utilized in zoning the Mesozoic rocks.

The literature abounds with examples of the multifarious ways in which Cenozoic gastropods have been applied to the solution of biostratigraphic problems. In point of fact, the classification of the Tertiary strata as Eocene, Miocene, and Pliocene proposed by Lyell in his *Principles of Geology* (1833) was based on a plan related to the percentage of living species present in the various dominantly molluskan faunas. This system has been widely used in past studies of Tertiary molluskan faunas. For example, nearly a century after Lyell's definitions, Woodring (1928) in his classic studies of the mollusks of the Bowden "shell bed" of Jamaica maintained that "percentage of living species, used with due caution, still is the fundamental basis for determining the age of a late Tertiary fauna." On the basis of percentage of living species, he considered the Bowden fauna Middle Miocene. Such methodology has been strongly criticized because it requires a full knowledge of both the living and fossil fauna. In spite of its drawbacks, the method has proved valuable as a "rule of thumb" at least in unexplored areas.

No system of zones based upon commonly occurring specific level molluskan taxa exists for the purposes of Tertiary intercontinental correlation. Yet over wide areas, and especially for the early Tertiary, molluskan faunas possess sufficiently distinctive taxonomic composition or characters (common species groups and the like) that they provide a key for at least gross international correlation at the series of somewhat finer level. For example, Gardner (1931) in searching for equivalents of her Midway molluskan assemblages from Texas found a significant relationship to described Paleocene faunas from Copenhagen; she states,

> Importance should be attached, however, to the similarity of *Fusus morchi* and *Levifusus trabeatus*, of *Pleurotoma johnstrupi* and members of the *mediavia* group . . . , and of *Voluta nodifera* to *Volutocorbis texana.* These species seem to represent groups widespread but of short stratigraphic duration (p. 151).
>
> The fauna of the Soldado formation of Trinidad, which includes *Ostrea pulaskiensis, Ostrea crenulimarginata, Cucullaea hartii, Calyptraphorus compressus,* and *Turritella nerinexa,* is definitely not only of Midway but of lower Midway age. *Mesalia pumila nettoana* is a link not only with the Midway of Alabama and Texas, but also with the Pernambuco beds of Brazil, from which it was first described, and through *Mesalia fasciata*, a similar species ubiquitous in the Tethyan province with the basal Eocene of northern Africa and India (p. 159).

In this instance the Midway fauna not only provided information pertinent for intercontinental correlation at least at series level, but correlations that transcended the biotic province boundaries from the Boreal province through the warm temperate to the tropical Tethyan province.

In spite of the general lack of precision for substage levels of refinement in global correlation, the Tertiary benthic mollusks remain a significant group for basinal and in some areas for interprovincial correlation. Of the gastropods, certain groups such as the Turritellidae have been more often utilized than others. For example, the correlation charts of Weaver et al. (1944) for the marine Cenozoic of western North America subdivide the column into 27 zones based on metazoans. Thirteen, or almost half, of these zones are based upon species of *Turritella*. These zones basically are founded on Merriam's (1941) definitive study of Pacific Coast Turritellidae in which the stratigraphic and geographic distribution of 13 stocks (species-group lineages) are delineated. As shown in Merriam's (1941, fig. 6) diagram (Text-fig. 6) of the stratigraphic distribution and suggested evolution of one of these "stocks," that of *Turritella uvasana*, lineage zones of both stage and substage level may be distinguished within the California, Oregon, and Washington area.

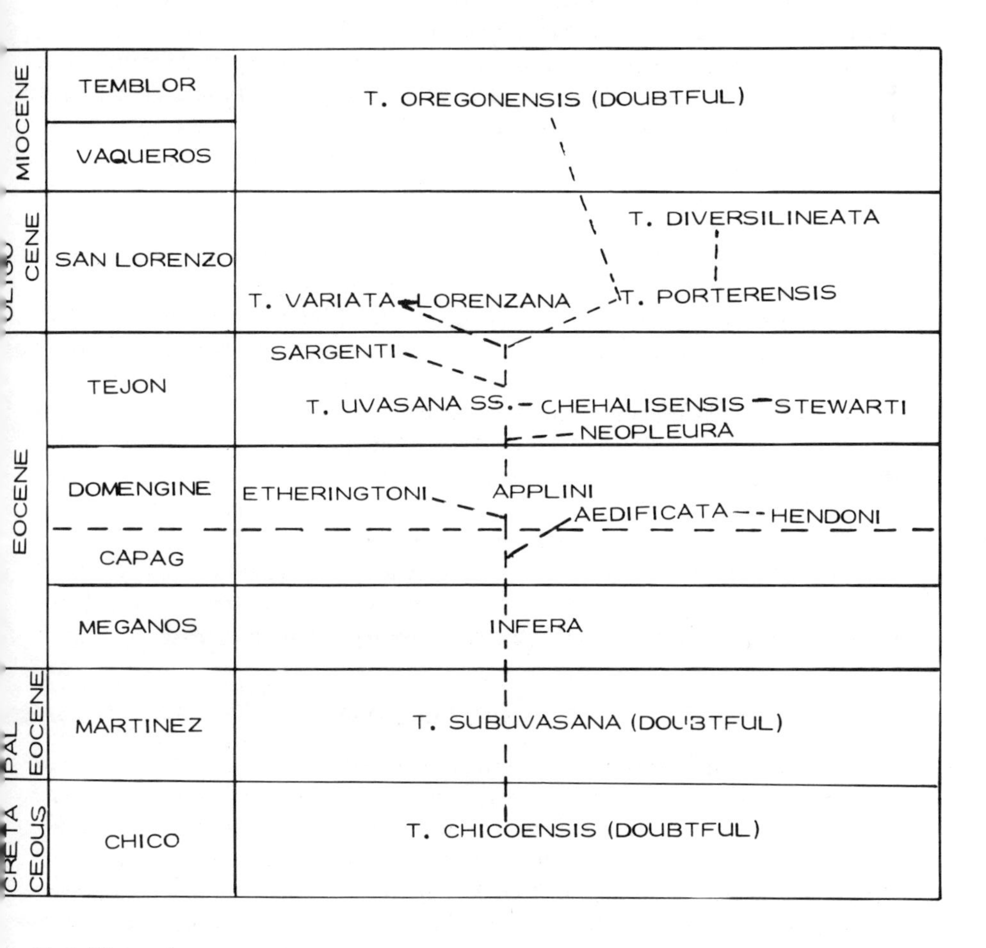

Text-Figure 6
Stratigraphic distribution and suggested lines of evolution of *Turritella uvasana* stock. (After Merriam, 1941.)

Similar studies on *Turritella* in other areas have shown that they provide a similar basis for biostratigraphic range zone or phylozone subdivisions (Bowles 1939; Stenzel 1940; Macsotay 1971). In some instances these *Turritella* species groups obtain a distribution rather exceptional for Tertiary benthic Mollusca, as demonstrated by Allison and Adegoke's (1969) study of the *Turritella rina* species group that is found in Middle to Late Eocene rocks of North Carolina south across the coastal plains to Mexico, and thence to Panama, Columbia, and Peru.

The demonstrable biostratigraphic utility of *Turritella* has, however, not precluded the recognition of other valuable index species or genera. Some are widespread but facies restricted. For example, the strombid *Orthaulax* is found in reef-associated deposits in Alabama, Georgia, Florida, the Greater Antilles, Anguilla, Antigua, Mexico, Guatemala, Panama, Venezuela, and Brazil. Woodring (1959, p. 190) has stated that three *Orthaulax* zones may be distinguished over its geographic range, one late Oligocene and two of early Miocene age. Some members of the volute genus *Athleta* have been treated in exceptional detail by Fisher et al. (1964). Their detailed biometric analyses delineate interrelationships among such features as height, width, columellar folds, denticulation, sculpture, and callus, which in turn describe the trends of an evolving lineage. This division of the Eocene *Athleta petrosa* stock into successional and morphologically transitional subspecies provides an especially fine example of the potential for distinction of stratigraphically useful gastropods.

The foregoing examples deal with instances of forms having biotic province utility. Very commonly Tertiary gastropods have in addition formed the basis of refined correlations of local or basinal extent. For example, Woodring et al. (1940), in dealing with the geology of the Kettleman Hills area of California, described a series of "zones" based upon certain types of mollusks (*Neverita* zone, Pecten zone, *Acila* zone, and *Siphonalia* zone). These "zones" are recognizable on criteria of either assemblage composition or dominance of single species, and may encompass a single bed a few feet or hundreds of feet of section. Some zones of this nature most likely equate with the ecozone concept and probably are of only very local biostratigraphic utility. For the most part, these local or basinal zonation schemes are tied together by species range zones or, where available, by phylozones. All too seldom, in most areas, true assemblage-zone concepts have been utilized.

CONCLUSIONS

Sufficient examples exist to indicate that gastropods possess the potential for use as either independent standards or integral parts of biostratigraphic zonation schemes at almost all levels. That this potential has normally not been fulfilled lies more in the realm of lack of knowledge of the fossil faunas and their distribution than in inherent weaknesses of the group. The biostratigraphic weaknesses that they possess are primarily those such as facies restriction shared by other benthic organisms. However, those snails, such as the cited nerineids and *Orthaulax* that are linked to reefal facies find wide distribution in such facies throughout the extent of biotic provinces. Others, such as the Cretaceous turritellid species *Haustator bilira* Stephenson, transcend facies boundaries and are found in sands, intermixed sands and shales, and in marly chalks over several thousand miles of outcrop.

The few comparative studies that have been done suggest that generally gastropods are somewhat more conservative in rate of evolution than such groups as the ammonites. Thus, for example, the degree of refinement of zonation with snails may never be that for Jurassic or for Western Interior Cretaceous ammonites. Yet, in nearshore sands or reefal or lagoonal facies wherein ammonites are rare, gastropods may be abundant and provide a prime tool for correlation. Factors such as patterns of abundance (dominance of taxa) may provide a method of local zonations at less than species range levels of biostratigraphic resolution.

Gastropods as instruments of intercontinental correlation have never been fully tested. They are used with varying success in broad or gross intercontinental correlations of Tertiary rocks. This situation is somewhat a paradox when viewed from the framework of existing theory of migrations of continents. If such correlations possess any validity for the Tertiary, when continents are widely separated, then gastropods should be significantly more useful during the Mesozoic because, for example, of the proximity of Europe and North America. Comparative studies of European and North American early Cretaceous gastropods have not been made. A literature exists, but relationships are clouded by a provincial nomenclature that may be an unrealized taxonomic artifact of the theory of continental stability.

To fully realize the biostratigraphic potential of gastropods, much work needs yet to be done. Obviously, additional information from carefully collected, stratigraphically documented, and taxonomically described faunas is much needed. Such studies form the backbone of development of viable assemblage zones. Beyond this, detailed studies of individual selected lineages through time may provide, where feasible, phylozone schemes of a refinement greater than that of the assemblage zone, although perhaps not so widely applicable in a geographic context. Such zonations based upon gastropods, when integrated with existing schemes based upon ammonites, planktonic Foraminiferida, and other groups, will indeed provide a powerful tool for the decipherment of earth history.

Concepts and Methods of Echinoid Biostratigraphy

Gundolf Ernst **Technische Universität Braunschweig**

Ekbert Seibertz

INTRODUCTION

Echinoids have been commonly employed in biostratigraphy. They are especially valuable in Cretaceous and Tertiary deposits, when the irregular echinoids underwent a major period of radiation. As early as the nineteenth century, species of the genus *Micraster* were used as biostratigraphic indexes for zonation of the Chalk facies of Western Europe. The zones of *Micraster cortestudinarium* and *M. coranguinum* are still widely used today in England, in place of the stage names Coniacian and Santonian.

No cosmopolitan guide species are found among fossil echinoids, and a biostratigraphic zonation such as that constructed from ammonites is not possible. However, in refined biostratigraphic systems employing modern assemblage zone techniques, and utilizing diverse groups of organisms, echinoids are valuable aids. This is especially true in the Upper Cretaceous, where they may occur in strata lacking Inoceramidae and cephalopods normally used in biostratigraphic zonation. In some cases, echinoids constitute the only preserved, biostratigraphically applicable macrofossils (e.g., in the North Spain *Micraster* marls of Coniacian and Santonian age).

The utilization of diverse fossil groups in modern assemblage zonation has remarkably enhanced the development of echinoid biostratigraphy: e.g., the works of N. Peake and C. J. Wood in England (in progress), M. Meijer (1965) in Belgium and the Netherlands, H. Raabe (1965) and F. Radig (1973) in Spain, G. Ernst (1967-1973) and M.-G. Schulz (Ernst and Schulz, 1971) in Germany, S. S. Maczynska (1958, 1968) and E. Popiel-Barczyk (1958) in Poland, G. N. Dshabarow (1964) and N. A. Poslavskaja et al. (1959) in the Soviet Union, and C. W. Cooke (1953, 1959), P. M. Kier (1962, 1972), and J. W. Durham et al. (1966) in the United States have all successfully utilized echinoids in biostratigraphy during the last two decades.

Whereas some workers persist in using classical biostratigraphic methods involving simple description and tracing of range zones based on echinoid guide fossils, more recently numerous specialists have attempted to interpret in detail the evolution, ecology, dispersal, and biogeographic spread of echinoids as they relate to biostratigraphic concepts and methods.

Rowe (1899) was the first to apply modern biostratigraphic methods to the development of an echinoid zonation. Inspired by evolutionary theory, Rowe described the gradual species transformations within the *Micraster* lineage, utilizing large collections from the English Chalk. In his study, Rowe abandoned traditional systematic and biostratigraphic methods, and utilized stages of development in evolutionary series as biochronologic indicators. Without being restricted by predetermined species concepts, Rowe was thus able to zone and correlate stratigraphic sequences accurately on the basis of the stage of evolution in coronal structures, like the ambulacra or plastra (Text-fig. 4).

Rowe's evolutionary and biostratigraphic scheme was fairly simplistic however, assuming orthogenetic evolution, and resulted in the definition of long-ranging and widespread *Micraster* zones. Subsequent work (Ernst, 1970c, 1972; Stokes, 1975; Wood, in progress) has shown more clearly the complex mosaic of *Micraster* evolution. Rowe studied only the orthophyletic phase of the main lineage, and disregarded side branches as well as its partly retrograde final phase of evolution. Only a comprehensive understanding of the complex evolution of *Micraster*, or any echinoid lineage, yields greater refinement in biostratigraphic zonation.

BIOSTRATIGRAPHIC VALUE OF ECHINOIDS

Biostratigraphers have defined the most important attributes for zonal fossils as (1) great abundance, (2) wide, rapidly attained biogeographical distribution, (3) broad environmental tolerance, and (4) short-lived species reflecting high rates of evolution. All these criteria are seldom realized in any one guide fossil; they are perhaps most completely attained in nektonic or planktonic organisms, which are not so tightly restricted by ecological and geographic barriers. Benthic organisms, including the echinoids, commonly possess only certain of these attributes, in many cases showing strong facies dependence and thus limited biogeographical range.

Abundance

The population density of echinoids may be extremely high in modern seas. This applies to both shelf-dwelling epibenthic forms and to endobenthic burrowers. For example, population densities of 80 individuals/m^2 are recorded for sand dollars

(*Dendraster excentricus*) and for *Echinocardium cordatum* (Fechter, 1970). The abundance of Recent and fossil echinoids has led to their use as identifier organisms in the classification of benthic faunal associations or communities (e.g., the *Echinocardium cordatum-Amphiura filiformis* association of the German Sea (Mortensen and Lieberkind, 1928).

Although *regular echinoids* are highly abundant in Recent seas, they are often rare in fossil deposits. This may be attributed to the fact that they prefer nearshore, hard, rocky to coarse clastic substrates, where their fragile tests are largely destroyed by wave action and currents. The potential for preservation of the massive spines of regular echinoids is better, so that they can be used for biostratigraphic purposes (e.g., *Tylocidaris* spines).

Irregular echinoids are more widespread and more abundant in fossil deposits than regular forms. They are better preserved in the quiet-water environments of the more seaward and deeper parts of seas, the preferred habitat, than in coastal shallow-water. Formerly, good preservation of irregular echinoids was attributed to fossilization of those mostly within the endobenthic habitat. But recent studies indicate that burrowing echinoids normally come to the substrate surface shortly before dying (Ernst et al., 1973). This is also proved by the rich epizoan fauna found on the tests of Recent and fossil burrowing echinoids. Fossil forms lacking epizoans and Recent dead echinoids still retained in living position generally indicate a sudden death (e.g., through rapid burial under quantities of sediment sufficient to smother them; Schäfer, 1962; Ernst, 1967). Some fossil deposits are extraordinarily rich in irregular echinoids (e.g., the "*Micraster* marls" of the North Spanish Upper Cretaceous and the "*Conulus* facies" of the Central European Turonian; Ernst, 1967). These occurrences parallel high population densities noted in Recent seas.

Regional Distribution and Environmental Tolerance

Practically no species of echinoids achieve worldwide distribution in modern oceans (Fechter, 1970). Only the biogeographic range of *Echinocardium cordatum* is relatively extended, but even this species is limited in its distribution by lithofacies (sand and sandy mud) and water depth (to around 230 m). Normally, wide biogeographic distribution is only attained where the preferred sediments and water temperatures are continuous over broad areas. A good ancient example exists in the Chalk facies of the North Temperate Upper Cretaceous in Europe, which contains numerous widely distributed echinoid species. The distribution of echinoid species is restricted by several geographic and ecologic factors. The most important are deep marine basins and bathymetric, climatic, and lithofacies barriers.

Bathymetric tolerance. Echinoids are mainly distributed in coastal and shallow shelf areas. The diversity of species gradually decreases below the shallow sublittoral zone. Below 4,000 m water depth, only 10 to 12 species are known; below 6,000 m, the number is reduced to 2 or 3 species. The maximal number of species (365) occurs within the range of 0 to 100 m water depth (Durham et al., 1966). These statements strongly suggest that most of the echinoid species are restricted in their distribution by bathymetric factors. The bathymetric tolerance of single species varies within more or less wide limits. The range of *Echinocyamus pusillus* is between 10 and 50 m water depth; that of *Echinus elegans* is from 50 to 2,000 m (Mortensen and Lieberkind, 1928). Geographic barriers created by deep-sea basins can be traversed by most species only in the larval stage. The spread of echinoids generally takes place along the shore or continental shelves, rather than offshore. The duration of the planktonic larval stage is too short (several days or even weeks) to surmount great distances while drifting in oceanic currents, and in succeeding developmental stages the larva drops down to the bottom seeking a suitable substrate for habitation. Thus most echinoids have not dispersed across the Pacific Basin, and New Zealand, which became isolated in the Cretaceous, has a highly endemic echinoid assemblage. These facts should be considered when interpreting the paleobiogeography of fossil species or assemblages.

Temperature tolerance. Many living echinoids are distinctly temperature controlled in their distribution. Numerous species prefer warm water and are only abundant near the equator (e.g., *Echinometra mathaei*). Others (e.g., *Pourtalesia jeffreysi*) are restricted to colder seas. On the other hand, there are a few echinoids occurring in the Atlantic from Northern Europe to Middle or South Africa where preferred substrates are available (e.g., *Brissopsis lyrifera* and *Spatangus purpureus*). Probably the low temperature tolerance of most echinoids has established completely distinct Arctic and Antarctic echinoid assemblages (Fechter, 1970).

According to Mortensen (1928), the distribution of cidarid echinoids is an example of climatic control. The cidarids prefer warm water and are widely spread in the Indo-Pacific today. The present Atlantic Ocean is remarkably poor in cidarid species, but in the climatically warmer Jurassic and Cretaceous, seas of the Atlantic margin of Europe were richly populated by cidarids.

In comparing different fossil echinoid faunas from equivalent substrates and stratigraphic levels, the biostratigrapher must therefore consider the role of temperature control. The differences in the echinoid faunas of the North Temperate and Mediterranean Upper Cretaceous, for instance, are best explained by habitation of different climatic zones (see p. 544).

Salinity tolerance. Echinoids are almost completely restricted to normal marine water. Very few species also penetrate into brackish water; for instance, in the Öresund (at the entrance to the Baltic Sea) *Echinocyamus pusillus* may live in water of about 20 parts per thousand salinity (Brattström, 1941). The use of echinoids as guide fossils, therefore, is generally confined to normal marine sediments.

Faunal provinces. In present seas several faunal provinces can be defined by means of echinoids. These provinces are sharply separated by the various geographic and environmental-ecological barriers described above. Biogeographic zonation was less marked during much of the geological past (e.g., the Cretaceous), when water temperatures were warmer and more stable, and province boundaries broad and transitional. Even so, noticeable biogeographic differences can still be found between the Cretaceous faunas of the North Temperate and the Tethyan Realms. The shelf sea of the Tethys allowed a high level of faunal interchange between

Eurasia and Africa and, prior to accelerated plate movements and isolation during the Upper Cretaceous, between Europe and America. But between Asia and western America the broad Pacific Basin inhibited species dispersal. The paleogeographic distribution of echinoids and other organisms reflect the plate spreading history of the Cretaceous. Africa and Europe show a number of identical species beginning in the upper part of Lower Cretaceous (e.g., *Holaster laevis* and *Hemiaster phrynus* of the Aptian from southeastern Africa and central Europe), reflecting the opening of the Strait of Madagascar and the possibility of migration of European stocks into South African areas. Faunal affinities also exist between Europe and America during Cretaceous time when they were still situated near enough to each other to allow a high level of faunal interchange within the Tethyan Realm. This is shown by numerous comparable Upper Cretaceous echinoid species from Europe, the American Atlantic and Gulf Coastal Plain, and the Antilles (Cooke, 1953). Some groups of European affinity even expanded their range from the Caribbean Tethys north into the Warm Temperate Western Interior Basin during the Cretaceous. With Late Cretaceous sea floor spreading and increased opening of the Atlantic, echinoid faunas of Europe and America became more distinct, and the ability to establish long-range correlations on echinoids diminished.

Despite broad regional affinities between Euramerican echinoid assemblages, it is possible to distinguish some faunal provinces in the European Upper Cretaceous by means of a few distinct echinoid species-groups. Stokes (1975), utilizing *Micraster* assemblages, proposes recognition of five echinoid "provinces" from the north to the south in Europe; these, of course, are not comparable in magnitude to modern multitaxa-based provinces. Most certainly these "provinces" were not sharply separated from each other, and generally they contain a few species in common. A steady faunal intermingling took place between them through migration.

Dependence of Lithofacies and Ecomorphology

One main disadvantage in using echinoids in biostratigraphy is the strong lithofacies control on distribution. As in many benthic animals, they show an extremely narrow range of adaptation to specific sediment types. Most regular echinoids are restricted to rock, stone, or gravel bottoms where they graze the algal crust. Even sands, which lack stable algal growths, may not be occupied by these sea urchins.

In sand, marl, and stabilized calcareous oozes, however, irregular echinoids are typical, commonly obtaining their food supply by burrowing (Text-fig. 1). Clays or marly clay substrates are rarely colonized or completely avoided by echinoids because of high turbidity, caused by fine suspended clay, which clogs their water vascular system.

Apart from some stratigraphically unimportant cidarids, virtually no echinoids are able to tolerate the whole spectrum of lithotopes. Application of echinoid-based

---

Text-Figure 1
Life positions and modes of adaptation among the holasteroids of the northwest European Upper Cretaceous. Enlargement scales variable.

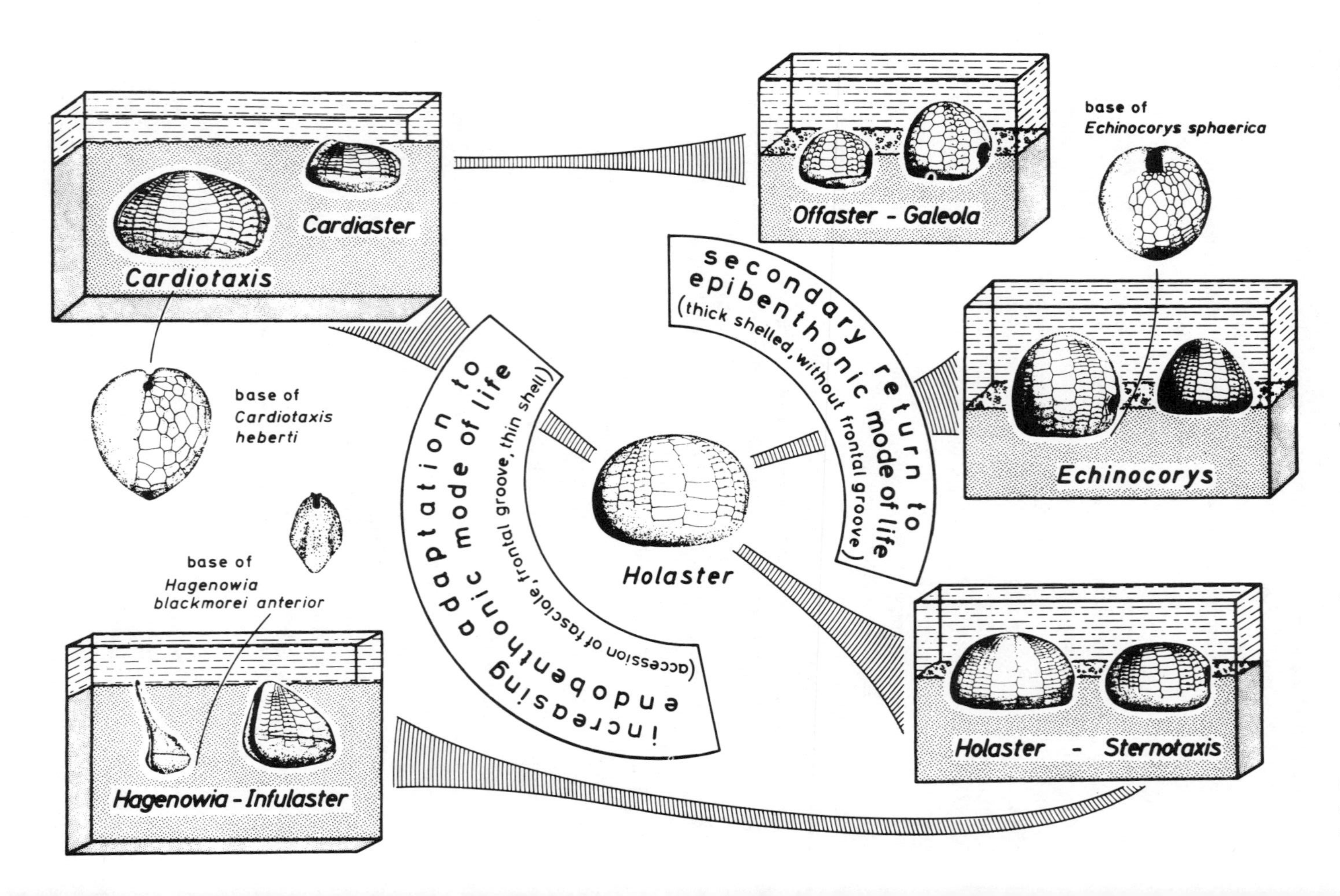
Cardiaster
Cardiotaxis
Offaster - Galeola
base of
Echinocorys sphaerica
secondary return to
epibenthonic mode of life
(thick shelled, without frontal groove)
Echinocorys
base of
Cardiotaxis
heberti
increasing
endobenthonic mode of life
adaptation to
(accession of fasciole, frontal groove, thin shell)
Holaster
base of
Hagenowia
blackmorei anterior
Hagenowia - Infulaster
Holaster - Sternotaxis

biostratigraphic zones is often strictly limited to certain facies, therefore. Commonly, forms that might link facies-restricted systems are lacking. Abrupt change of lithofacies is characterized by a sharp change in echinoid assemblages as well. In changing facies situations, therefore, changing echinoid distribution in space and time may more closely reflect an *ecological zonation* than a time-evolutionary zonation, and this places important constraints on stratigraphic correlations by echinoids between different facies.

An excellent Recent example of the strong lithofacies dependence of echinoids is shown by the arrangement of the echinoid assemblages in the Mediterranean Sea. Depending upon the substrate type, the upper sublittoral zone can be separated into four laterally successive echinoid ecozones, each typified by a different species association: (1) *Arbacia lixula* on rock bottoms, (2) *Paracentrotus lividus* on boulder bottoms, (3) *Sphaerechinus granularis* on gravel bottoms, and (4) *Spatangus purpureus* and *Echinocardium cordatum* on organic limy sand bottoms (Ernst et al., 1973). Firmness and particle size of the bottom deposits, water movement, and food resources primarily influence the distribution of these echinoid assemblages. Kier (1972, Text-fig. 1) provides another significant example of facies dependency and ecological zonation among echinoids, and applies Recent examples to interpretation of the diversity and assemblages of echinoids in the Late Miocene-Pliocene Yorktown Formation of the American Middle Atlantic Coast.

It is therefore practical to employ echinoids as guide fossils only within comparable facies areas, for example, within the widely distributed chalk and marly limestone lithotopes of the North Temperate Upper Cretaceous, or in the calcareous sandstone lithotopes of the Mediterranean Neogene. But no biostratigraphic comparison is possible, for example, between assemblages of the arenaceous, tuffaceous chalk facies of the Maastrichtian from Limbourg and the more offshore Maastrichtian pelitic chalk facies of northwestern Europe. These echinoid assemblages have only a few genera and no species in common (Ernst, 1970a).

Ecophenotypic variation. In addition to strong facies dependency, ecophenotypic variation *within* echinoid species further complicates their use in biostratigraphy by making accurate species determinations more difficult, especially for the nonspecialist. Small differences in ecological factors, especially in lithofacies, may influence the structure and morphologic variability of populations from closely related species or subspecies (e.g., recent studies on Cretaceous echinoids show that, in particular, size and shape of the tests reflect differences in lithofacies; Ernst 1970b). In pure calcareous lithotopes the tests of populations of the same, or closely related, species are larger and relatively higher on the average than tests found in marly limestones, or silty to arenitic limy marls. The "*rule of flatness*" (Ernst, 1970b) states that the relative height of the test decreases with increasing clay or sand components in the substrate. The large populations available from the evolutionary lineages of *Echinocorys, Galeola*, or *Galerites* provide significant examples of these ecophenotypic effects (Text-fig. 1). Such variation is especially common among epibenthonic species; populations of burrowers are ecomorphologically less variable.

Biometric investigations on large populations of Recent echinoids have demonstrated that, in addition to lithofacies, differences in water movement, exposure, or water depth may induce ecophenotypic variation (Ernst et al., 1973). Echinoid tests that are highly vaulted in quiet-water environments are more flattened in zones of turbulent surf and wave action. These factors pose a danger to the biostratigrapher; stratigraphic changes in population structure, even when determined through quantitative methods and correlation techniques, may reflect ecophenotypic rather than evolutionary changes; without thorough knowledge of regional population variation within species, they may lead to construction of artificial biostratigraphic divisions.

Stratigraphically Important Characters

Echinoids, especially irregular groups, commonly possess a great number of characters usable in evolutionary, and thus biostratigraphic, studies (Text-fig. 2). The shape of the test is generally considered to be of primary importance; it may range from flattened, to globular, conical, or pyramidal in a single lineage, allowing recognition of steps in evolution and providing valuable biostratigraphic data. In conjunction with evolution in the shape of the test, the shape, number, and convexity of the coronal plates may change. Other phylogenetically and biostratigraphically important features include the position of the peristome and periproct, the development and loss of fascioles, and the development of the labrum, rostrum, frontal groove and petals, or other characters (Text-fig. 2). Important microscopic features include intensity of granulation, and the shape, arrangement, and number of ambulacral pores. In some cases, the normally isolated spines also have stratigraphic importance. The best example is the glandiform spine of the genus *Tylocidaris* used by Ravn (1928) to identify zones in the Maastrichtian and Danian of Denmark.

Many characters useful in evolutionary and biostratigraphic studies closely reflect the mode of life of the echinoid and thus have paleoecologic value. Burrowing species may be characterized by, for example, their flattened form, fascioles, frontal groove, elongated pores, and thin-shelled test. On the other hand epibenthonic irregular echinoid species are commonly characterized by tall tests, rounded pores, and thicker shells (Text-fig. 1).

Generally, the phylogenetic history and biostratigraphic value of an echinoid is determined by the collective change in diverse characters, rather than by evaluation of single characters.

Tempo and Patterns in Evolution

The main lineages of the Irregularia presumably developed within the Triassic-Jurassic boundary zone. Biostratigraphically, however, the echinoids have no importance until the Cretaceous. Species diversity and the quantity of individuals are too low prior to that time. The major radiations of the echinoids took place in the Upper Cretaceous and Tertiary, mainly among the irregular groups, and during this time they attain biostratigraphic utility. Genus and species diversity rises

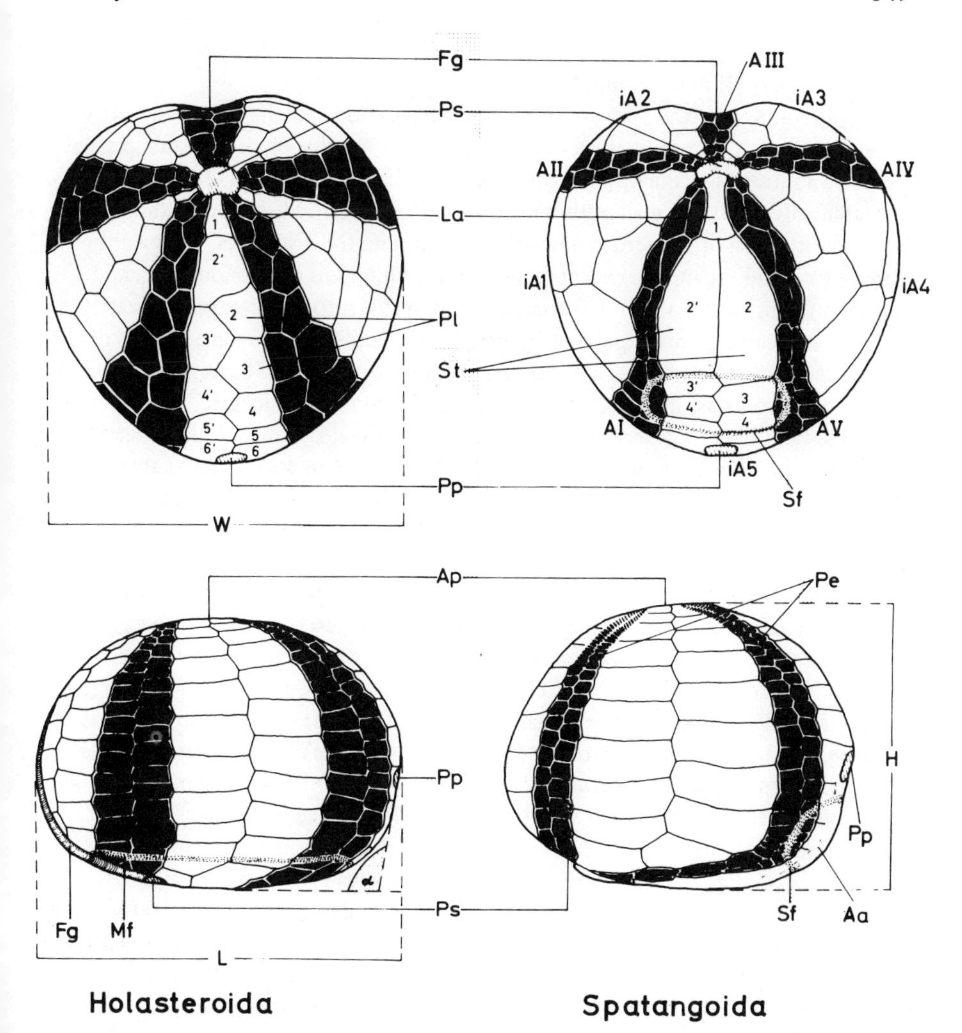

Text-Figure 2
Stratigraphically important test features in representatives of the holasteroids (*Holaster*) and spatangoids (*Micraster*). Ambulacral plates black, interambulacral plates white; most important characters in dotted screen (after Ernst, 1972). Aa, anal area; Ap, apex; AI - AV, ambulacrum I - V; Fg, frontal groove; H, height of the test; iA 1 - iA 5, interambulacrum 1 - 5; L, length of the test; La, labrum; Mf, marginal fasciole; Pe, petal; Pl, plastron with numbered plates; Pp, periproct; Ps, peristome; Sf, subanal fasciole; St, sternum with numbered sternal plates; W, width of the test; α, anal angle.

markedly in the Late Cretaceous, and the number of individuals increases. As Kauffman (1970) pointed out for bivalves (Inoceramidae, Ostreidae) and ammonites, the Turonian stage in particular is marked by rapid increase in rates of evolution and by high diversity of new echinoid groups.

Rapid radiation of the two orders of holasteroids and spatangoids in the early Upper Cretaceous primarily reflects creation of new benthic environments, possibly new food sources, and the opportunity for new adaptations (Ernst, 1972). The new, geographically extended coccolithophorid and oligostegenid lithotopes stimulated the differentiation of numerous new life types among the echinoids. Adaptation to these new and vacant ecological niches started slowly in the Cenomanian and attained maximum rates in the Turonian. During this initial phase, genus density gradually increased as principal new niches were occupied. In the later stages of the Upper Cretaceous, colonization of the principal new niches had been achieved, and evolutionary progression among holasteroids and spatangoids occurred primarily at the species and lower taxonomic levels. Through specialization and niche-partitioning, new small-scale lithotopes were colonized gradually by new species and subspecies. The addition and replacement of lower taxa during the late Upper Cretaceous provided a continuous succession of new guide species for zonation and correlation, and echinoid biostratigraphy attained its greatest potential.

As in other groups, evolution of related echinoid lineages was frequently characterized by convergent or parallel morphological changes, the "*evolutionary homeomorphies.*" Striking examples of such parallel development are found in the evolution of size, differentiation of coronal form, displacement of the periproct, peristome, and other features, especially in the groups of Holasteroidae and Spatangoidae (Ernst, 1972). The similarity of tests, for example, in the distantly related *Echinocorys* and *Offaster-Galeola* lineages (Text-fig. 1) reflects comparable changes in their mode of life from burrowing to a life on the surface of the substrate. Lack of knowledge concerning these phylogenetic relationships caused previous specialists to consider *Galeola papillosa* as a juvenile of *Echinocorys conica* (Text-fig. 1).

Heterochronous recapitulation of congeneric or similar types of tests in the same evolutionary lineage (*iterative evolution*) also may confuse the biostratigrapher. The genus *Echinocorys* shows particularly good examples of iteration. From one or two evolutionary lineages, short-lived gibbous, conical, or pyramidal offshoots were repeatedly differentiated in which the typical species attributes were progressively more strongly accentuated (Ernst, 1972). Without detailed knowledge of such phylogenetic relationships, the biostratigrapher cannot reliably use echinoids in biostratigraphy.

Another common evolutionary trend among Cretaceous echinoids is the splitting of two or more evolutionary *side branches* with different test shape from more generalized populations having a high variability in body form. A particularly characteristic pattern is the development of side branches with flattened or elevated tests, which subsequently continue to diverge from each other. For example, the evolution in lineages of *Echinocorys, Galeola* (Text-fig. 5), *Micraster* (Text-fig. 4), *Galerites*, and *Conulus* show this pattern. These side branches may be adapted to different facies, which they may gradually dominate, with increasing specialization during their evolution. This process could lead to genetic differentiation and fixation of new characters by regional isolation. An excellent example of such a course in evolution is provided by the genus *Micraster* (Text-fig. 4). Beside the flattened main lineage the group evolved two side branches, *Gibbaster* and *Isomicraster*, with conical differentiation of the test (Text-fig. 4). After their geographical separation in the German Late Turonian, the main lineage and the side branches overlap again in their geographic range during the Campanian, but retained their morphological distinctiveness.

Echinoids are useful in biostratigraphy only during times of rapid evolution. Mainly this was during episodes of "explosive" evolution in the Upper Cretaceous and Tertiary. During these intervals, species often are restricted to one substage, biozone, or less, and the evolutionary lineages can be used in refined zonation by successions of guide fossils. The *rate of evolution* of the *Offaster-Galeola* series (Text-fig. 5) was especially rapid, some species having a "life" span of about 1 to 2 million years. This is about equivalent to rates of evolution in other important groups of guide fossils. Normally, however, the duration of an echinoid species is longer. The evolutionary rates of species in the main *Micraster* lineage can be considered as average. In about 18 million years, from Turonian until Late Campanian time, at least six species arose in the lineage, and are usable for broad-scale biostratigraphic zonation (Text-fig. 4).

Summary

Generally, echinoids only partially fulfill the four main criteria of good guide fossils: abundance, wide rapidly attained geographic distribution, broad environmental tolerance, and rapid evolution. They are abundant mainly during the period of Cretaceous and Tertiary radiation, which was also characterized by increasing diversification and high rates of species evolution. Commonly, the course of evolution passes a certain multiphase evolutionary pattern that may be initiated by the creation of new substrate types. Typically, an initial phase of rapid and diverse radiation in new superspecific groups is followed by a phase of speciation. Many lineages demonstrate striking examples of parallel development. These homeomorphous trends, as well as the recapitulation of congenous types of tests in the same lineage (iteration), may complicate biostratigraphic zonation. Low levels of tolerance to environmental fluctuation constitute a major drawback to the development and use of echinoid-based biostratigraphic systems. Chiefly, echinoids are highly sensitive to changes in lithofacies, and thus the geographical distribution of guide species is commonly restricted.

METHODS OF ECHINOID BIOSTRATIGRAPHY

Modern biostratigraphic methods have been successfully used by several echinoid workers, especially in the Upper Cretaceous of northwestern Europe. These works, except where they are based on simple range zones and peak-occurrence zones, primarily rely on data from studied evolutionary lineages. Use of biometrical methods has enhanced the definition of species, zonal refinement, and correlation accuracy in lineage studies. Examples of these evolution-based systems and methods, given below, are based on studies of about 18,000 specimens of the northwestern European Cretaceous, mostly collected horizon by horizon within a detailed lithostratigraphic framework. Systematically, the material comprises approximately 100 species of the four orders of holasteroids, spatangoids, holectypoids, and cassiduloids (see Ernst, 1972).

Range Zones and Faunal Zones

The actual distribution of a species in a stratigraphic sequence is called its *range zone* (Eicher, 1968). They are frequently utilized units in zonal schemes employing echinoids. The disadvantages are obvious: at different localities, the zonal species may have different local range zones, appearing or disappearing at different stratigraphical levels due to local environmental factors, migration, or extinction differences, sampling error, preservational factors, etc., and this may seriously hinder precise correlation. Faunal zones constructed from two or more echinoid taxa are more useful in biostratigraphy than simple range zones.

Peak Zones

Peak zones (acme zones, epiboles) are defined as periods of maximum abundance of a guide fossil (Eicher, 1968). Peak zones are often considered better for regional correlation than range zones, as recognized early by micropaleontologists and utilized in their stratigraphical schemes with foraminifera. However, the geographical range of peak zones is generally limited, because the abundance of fossils may strongly depend on local ecological factors. Successful application of peak zones in biostratigraphy always requires that population densities contemporaneously increase over a wide geographic area, providing an excellent possibility of correlation of short-term events. A good example of a widespread peak zone is the Upper Turonian "*Micraster* marl" of the Münster Basin in Germany.

Still better "time-lines" in correlation are certain *echinoid layers* of limited thickness in which the density of settlement rapidly increased for a short time. Such concentrations are characteristic for certain echinoid taxa, particularly in the Chalk facies from England and northern Germany (e.g., *Offaster* and *Galeola*). Sometimes these forms may be completely absent in the intermediate strata. Echinoid beds often have an amazing horizontal range. The "*Micraster* key bed" in the Middle Turonian of Lower Saxony (Germany), for instance, can be traced over at least 45 km. The existence of two parallel tuff beds in the same stratigraphic level demonstrates that the echinoid bed represents really an isochronus event and not a time transgressive layer.

Not all echinoid beds or layers are of such a good biostratigraphic utility. The "*Conulus* layers" in the Turonian of central Europe (Ernst, 1967) only appear locally in exposures and are lacking in neighboring outcrops at the same stratigraphic level. Detailed mapping of the *Conulus* layers shows that they are ecologically restricted to the flanks of ancient salt structures, existing obviously as shallow ridges in the

Cretaceous seas. In the deeper channels between salt domes, *Conulus* only occurs sporadically. It is interesting that the few deeper-water specimens have a different, comparatively more elevated test. Presumably, this phenomenon reflects ecophenotypic variation in response to water depth.

Biometrical Methods Applied to Lineage Studies

Rowe (1899) first utilized biometrics in constructing his *Micraster* zonation by studying "populations" and population averages in place of single individuals from various levels. Subsequently, many attempts were made to consolidate the gradual phylogenetic and stratigraphic transformations in important echinoid lineages by means of quantitative techniques, or to examine the taxonomic justification and relationships of certain species on a statistical basis (e.g., Kongiel, 1938, 1949; Hayward, 1940; Kermack, 1954; Ernst, 1971). Most certainly, correlation schemes can be considerably refined through statistical analysis, particularly since population analyses allow a much more accurate zonation and correlation of sections than do techniques employing simple guide fossils.

Biometrical analyses of echinoids are made difficult by allometric growth tendencies and ecomorphologic influences (Text-fig. 3). Many biochronologically valuable features, such as granulation, density of the fasciole, etc., elude mathematical analysis and may be evaluated at the best by semiquantitative methods. The usual measurements, like length, width, height (Text-fig. 2), and their ratios, do not adequately represent the general shape of the test, which many biostratigraphers consider to be valuable in recognizing species. In part, this is because the test of an echinoid may have markedly different forms – globular, hemispherical, conical, roof-like, or pyramidal in extreme cases – and still give similar ratio data.

Many characters, however, may be measured with comparative ease. In the *Offaster-Galeola* lineage these are, for instance, the height of the periproct (Text-fig. 5), and in the *Infulaster-Hagenowia* lineage the elongation of the rostrum and changes of coronal angles (Text-fig. 6). The alteration of the plastronal structure and the coupled distance from the peristome to the frontal groove can be quantitatively evaluated in the *Micraster* lineage (Text-fig. 4). The gradual transformation of these characters within lineages will be subsequently described.

Because of *allometric growth tendencies* in ontogenetic development, most measured characters change their relationship to other characters with increasing test size, and this further complicates the biometric analysis of many echinoid lineages. As a consequence of *evolution in size*, the allometric growth tendencies in the first members of a lineage are usually continued as evolutionary trends in their descendants. In those cases, evolutionarily younger members of echinoid lineages have proportions in their juvenile growth stages resembling the adult stages of their phylogenetic ancestors. Only quantitative differences in the same size classes result, if the allometric changes of proportions are reprojected palingenetically on younger growth stages. Text-figure 3 gives an example of such relations in three stratigraphically successive "zonal populations" of the *Offaster-Galeola* lineage in which data on anal angles and distances from the periproct to the test base are plotted against the

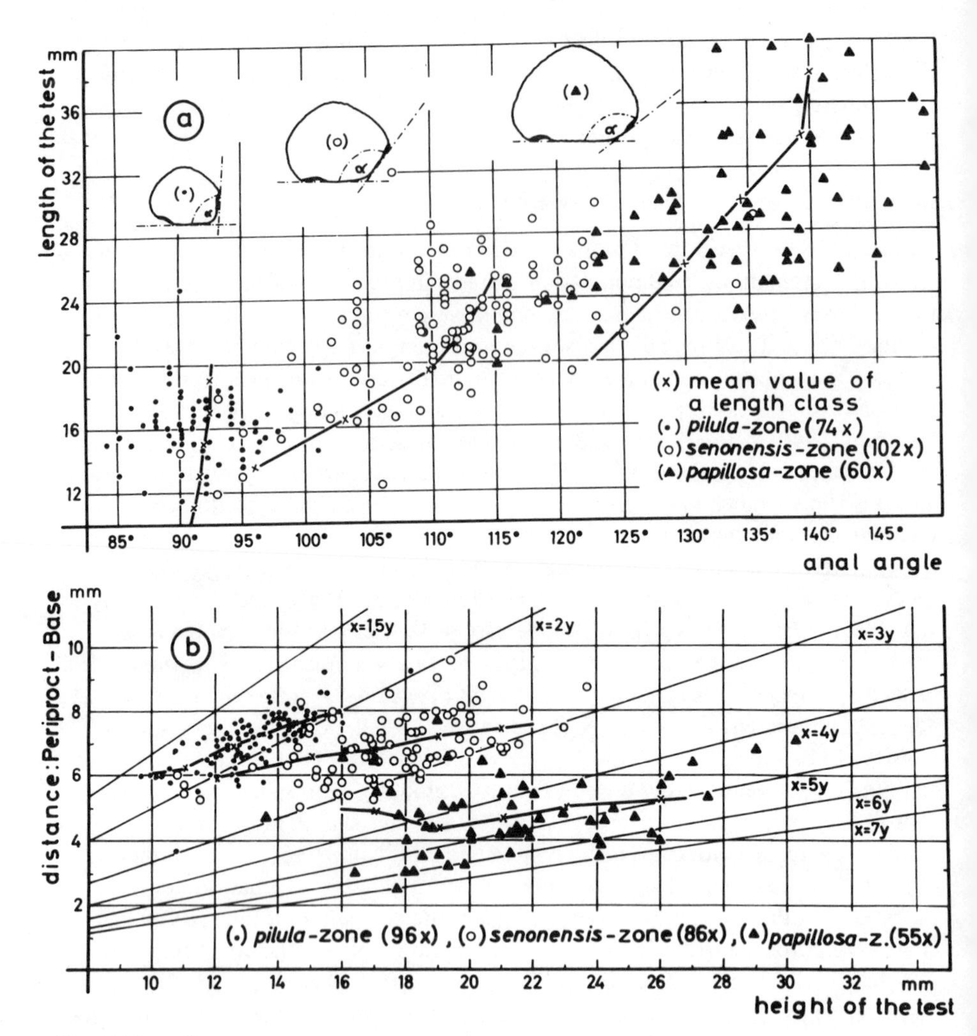

Text-Figure 3

Allometric and phylogenetic development of the anal angle (a) and the strongly related distance of periproct to base (b), demonstrated by three stratigraphically successive populations of the *Offaster-Galeola* lineage from the Lower Campanian of northwestern Germany (after Ernst, 1971). The mean values of each size or height class are connected to growth lines. In isometric growth, the curves in part a should be vertical and those in part b parallel to the straight lines of isometry ($x = ay$). This is only approximately the case in the *Offaster* population from the *O. pilula* zone. The growth lines have a strong allometric trend in the other two populations, and it is obvious that the juvenile tests of each younger population are similar to the adult tests of the stratigraphically older populations.

length or height of the test. In each case the juvenile tests of each successively younger population are biometrically similar to the adult tests of the stratigraphically older population.

The biometric analysis of lineages in biostratigraphy becomes further handicapped by *ecological influences* on morphological characters (see previous discussion). Such processes have been recently investigated in detail with large populations of living and fossil material (Ernst, 1970b; Ernst et al. 1973). Populations of the same age but from different habitats show different mean values for the same character, especially with regard to relative height of the test. Such data could easily cause the working biostratigrapher to conclude that different stratigraphic levels were represented owing to the different mean values of the various populations, which in reality reflect only the effect of ecological factors on test structures (lithofacies, water depth, or exposure).

Evolutionary Lineages and Biostratigraphy

The application of echinoids to biostratigraphy was considerably stimulated by wide use of evolutionary stages within studied lineages, and their application to zonation and correlation of stratigraphical units following the lead of Rowe (1899). In comparison with simple guide fossil range zones, evolutionary stages as zonal indexes have one great advantage: their stratigraphic position and range boundaries do not vary so much from locality to locality. The structural changes in evolution seem to take place more or less contemporaneously throughout the range of the phyletic lineage. Progressive genetic variants are apparently able to disperse new genetic material rapidly, or to rapidly replace ancestral populations. Larval drift is important here. The comparatively short time necessary for phyletic replacement over a broad area enhances the biostratigraphic potential of echinoid-based zones, utilizing evolutionary series, and suggests that zonal boundaries can be near-isochronous.

On the other hand, zonal boundaries drawn in evolutionary lineages are usually more difficult to define than simple range zones because of their transitional nature. Only population analysis of a large number of specimens from several levels enables an accurate definition of taxonomic, and thus zonal boundaries. From the major period of Cretaceous and Tertiary radiation, numerous biostratigraphically interesting evolutionary series have been studied and applied to biostratigraphy. In the following sections, four important and well-studied echinoid lineages from the European Upper Cretaceous are given as examples.

Evolution in the Micraster lineage (Text-fig. 4). With the exception of *Echinocorys,* the Micrasters include the most common and geographically widespread guide fossils among Cretaceous echinoids (Text-fig. 4). They are of primary importance in many European biostratigraphic systems. *Micraster* first appeared in the Cenomanian or Early Turonian (Text-fig. 4). By the Late Turonian the genus had evolved considerably, giving rise to many different forms and dividing into several morphologically parallel branches. Micrasters were the dominant group of spatangoids until the Late Campanian. Afterwards, the number of species decreased rapidly (Text-fig. 4), until their extinction in the Eocene. According to recent investigations by Ernst (1970c, 1972), Stokes (1975), C. J. Wood (in progress), and others, the radiation of the genus in time and space seems to be extremely complicated. High levels of ecological and geographical species and subspecies diversity, complex migrations, and the effects of competition all complicate the study of their phylogenetic history. More than 100 species of *Micraster* have been described (see Lambert and Thiery, 1909-1925), a great number of them being, at the most, subspecies or synonyma.

In the *Micraster* phylogenetic complex (Text-fig. 4) only material from the northwest European Upper Cretaceous has been considered. In particular, this is presented to demonstrate gradual evolutionary changes in the structures of the plastra and the petaloid ambulacra and their use in biostratigraphy. Both coronal structures are superimposed onto the phylogenetic history of *Micraster* as enlarged drawings in Text-figure 4. Due to variation in these characters, the structures of several specimens were combined to formulate a typical composite.

Evolution of the plastronal structure. (For morphologic terms, see Text-figs. 2 and 4.) In the early evolutionary stages of the lineage, the plastra show the suture of the sternal plates trending obliquely. The labral plate is broadest where it joins the sternum and tapers to a point towards the anterior end. The number of primary tubercles on the labral plate is low. The anterior lip does not yet overlap the peristomal orifice. By Late Turonian time, the labral plates become narrower, the number of primary tubercles increases, and the lip shifts onto the posterior part of the peristome. These trends are more or less homeomorphous among contemporaneous lineages, and become more accentuated until the Campanian, when the very pronounced lip covers the peristome completely and partly projects into the frontal groove.

Evolution of petaloid ambulacra. (For morphological terms see Text-fig. 4.) The English *Micraster* specialists (Rowe, 1899; Nichols, 1959; Stokes, 1975; and others) distinguish five different evolutionary stages of petals, identifying them with the terms "smooth," "sutured," "inflated," "subdivided," and "divided" (see Text-fig. 4). In the smooth basic type of the Early Turonian, the plate sutures in the interporiferous area are extremely faint, and granulation, interporiferous ridges, and ambulacral furrows are not yet evolved. Phylogenetic trends during the Upper Cretaceous lead to the development of granulation pads, small interporiferous ridges, and broad central ambulacral furrows. The culmination point in evolution is reached in the Santonian with the development of a divided area (Text-fig. 4). Later, petals lose their stratigraphical value and show striking tendencies towards retrograde evolution to more primitive structures.

Biostratigraphy. As is common throughout the *Micraster* lineage, phylogenetically important characters are at the same time stratigraphically important characters in the lineages studied. Single individuals do not suffice to accurately determine evolutionary or stratigraphic position within the *Micraster* lineage; the biostratigrapher must consider the evolutionary stage of whole populations. Furthermore, because there are specimens which are advanced in some characters and retarded in the development of others in each population, it is necessary in applying *Micraster* to biostratigraphy to consider the sum of evolving characters. These facts being considered, there can be defined at least six *Micraster* zones within the main lineage (Text-fig. 4). By combining these zones with the biostratigraphic units defined in each of the evolutionary side branches of *Micraster*, the zonation will be even further refined.

Evolution in the Echinocorys lineage. The variety of forms in the genus *Echinocorys* is even greater than in *Micraster.* The genus splits off from the primitive *Holaster* group in the Cenomanian (Text-fig. 1). Its comparatively slow early phase of evolution in the lower part of Upper Cretaceous is followed by relatively high levels of radiation (Coniacian-Maastrichtian), in which the number of species and individuals considerably increases (Smiser, 1935). The final phase of the evolutionary history is marked by reduction in test size and decrease in the diversity of test shapes.

N. Peake (Norwich) presumes a specific pattern in the evolution of *Echinocorys* (pers. comm., 1971). According to his opinion, numerous short-lived offshoots diverged from one or two conservative lineages, most of them retaining intermediate morphological links with the parent stocks. Because of their brief span of existence, these offshoots, whose taxonomical position is difficult to evaluate, have considerable biostratigraphic value. But the regional dispersal of many of these "form varieties" is only limited, and they are only useful in local correlation. Another typical pattern in *Echinocorys* phylogeny has already been discussed, involving heterochronous parallel evolution of similar form types.

Biostratigraphy. An excellent example of the use of *Echinocorys* in biostratigraphy is given by Peake and Hancock (1961), who divided the Upper Campanian and Maastrichtian of Norfolk, England, into 15 biostratigraphic units based on this genus. However, the geographic range of these units is limited and differs from the Upper Cretaceous zonation with *Echinocorys* of northwestern Germany. For biostratigraphic work utilizing *Echinocorys*, excellent material is necessary, because coronal form is the main character used in determining taxa.

Evolution in the Offaster-Galeola lineage (Text-fig. 5). The principal trends in

---

Text-Figure 4 *(following two pages)*
Evolution and stratigraphic distribution of the genus *Micraster* and its subgenera in the North Temperate Upper Cretaceous, with special reference to German (hatched periods of expansion) and English forms (modified after Ernst, 1970c, 1972). Stratigraphic codes like $Krca_{2\delta}$ in Text-figures 4-6 indicate faunal zones or composite faunal zones of the German Upper Cretaceous (e.g., Kr, Cretaceous; ca_2, upper part of Lower Campanian; δ, *conica-papillosa* subzone of *Echinocorys conica* and *Galeola papillosa*; see Text-fig. 5). Besides the main *Micraster* lineage, characterized by flattened tests, the group evolved two important parallel side branches, the subgenera *Gibbaster* and *Isomicraster*, with conical differentiation of the test. The most important transformations of the ambulacra and the plastra are shown in enlarged detail drawings (for detailed explanation, see text).

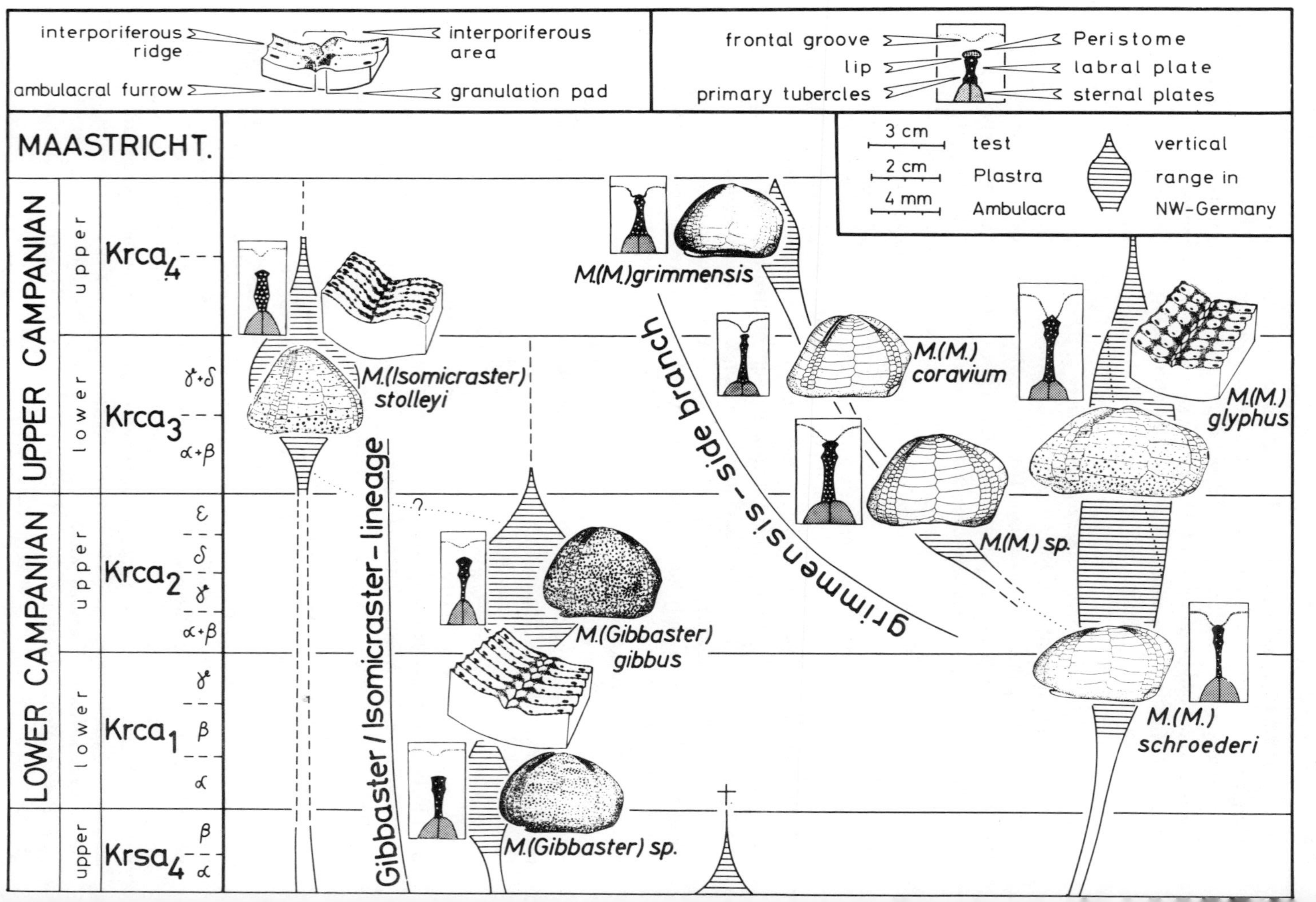
interporiferous ridge
ambulacral furrow
interporiferous area
granulation pad
frontal groove
lip
primary tubercles
Peristome
labral plate
sternal plates
3 cm
2 cm
4 mm
test
Plastra
Ambulacra
vertical range in NW-Germany
MAASTRICHT.
UPPER CAMPANIAN
LOWER CAMPANIAN
upper
lower
upper
lower
upper
Krca4
γ+δ
Krca3
α+β
ε
δ
Krca2
γ
α+β
γ
Krca1
β
α
β
Krsa4
α
M.(Isomicraster) stolleyi
?
Gibbaster / Isomicraster - lineage
M.(Gibbaster) gibbus
M.(Gibbaster) sp.
M.(M.)grimmensis
grimmensis - side branch
M.(M.) coravium
M.(M.) sp.
M.(M.) glyphus
M.(M.) schroederi

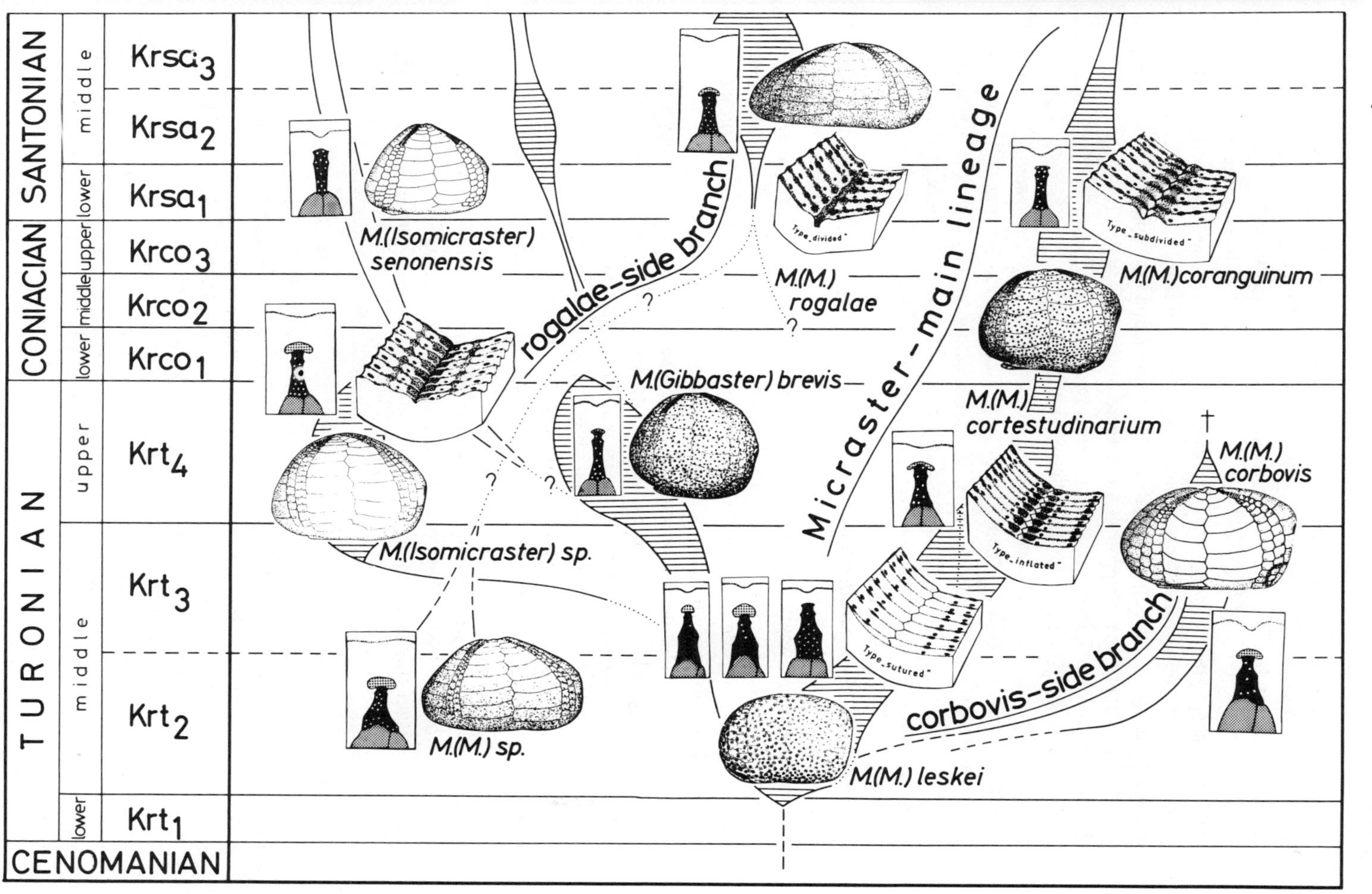
SANTONIAN
CONIACIAN
TURONIAN
CENOMANIAN
middle
lower
upper
middle
lower
upper
middle
lower
$Krsa_3$
$Krsa_2$
$Krsa_1$
$Krco_3$
$Krco_2$
$Krco_1$
Krt_4
Krt_3
Krt_2
Krt_1
M.(Isomicraster) senonensis
rogalae-side branch
M.(M.) rogalae
Type „divided"
Micraster-main lineage
Type „subdivided"
M.(M.)coranguinum
M.(Gibbaster) brevis
M.(M.) cortestudinarium
M.(M.) corbovis
M.(Isomicraster) sp.
Type „inflated"
Type „sutured"
corbovis-side branch
M.(M.) sp.
M.(M.) leskei

evolution of this echinoid lineage, which is important for Campanian zonation, are shown in Text-figure 5. The rapid morphological transformation through time reflects a change in life habit. *Offaster* is a burrower in mud, whereas *Galeola* acquires an epibenthonic mode of life (Text-fig. 1). In adaptating to this change in life habit, the fascioles are gradually reduced, the base of the test becomes flattened, and the periproct moves downward through time. Contemporaneously, a considerable evolutionary change in size, and a modification of test shape takes place. Closely coupled with the shifting of the periproct is an increase of the anal angle (Text-fig. 3). In the final phase, the *Galeola* lineage splits into two ecological subspecies, which gradually adapt to different types of lithofacies (Text-fig. 5).

Biostratigraphy. The greatest advantages of using the *Offaster-Galeola* lineage in zonation and correlation are their wide geographical distribution and their abundance in limy and marly sediments. In the Lower Campanian, at least six biostratigraphic zones can be defined within this evolutionary lineage. In conjunction with other echinoids and with belemnites, an even more refined assemblage zonation is possible (Text-fig. 5, $Krca_{1\gamma}$ to $Krca_{3\beta}$).

Evolution in the Infulaster-Hagenowia lineage (Text-fig. 6). Recently, the evolutionary stages of this interesting echinoid lineage, with their extraordinary shape, have become increasingly important in biostratigraphy. The original *Infulaster* lineage splits off in the Late Turonian into two *Hagenowia* lineages, whose final evolutionary members differ considerably (Text-fig. 6). From the elongated form, *Hagenowia rostrata,* a new lineage with more slender and more delicate rostra splits off in the Early Santonian. This lineage terminates in the Maastrichtian with the final species being highly specialized. In contrast to other echinoid lineages with a distinct evolutionary pattern involving increase in size, the *Infulaster-Hagenowia* lineage shows a striking reduction of size (note the variable enlargement scales in Text-fig. 6).

The development of an elongated rostrum is the most distinctive phylogenetic feature; in life position, the top of this structure presumably projected just above the substrate (Text-fig. 1). In the final phase of the evolution of the lineage, studied in chalk facies, the rostra become so elongated and thin that they occur nearly always separated from the body (*Hagenowia elongata*, Text-fig. 6, top; Schmid, 1972). Other important trends in the evolution of the *Infulaster-Hagenowia* lineage are the

---

Text-Figure 5

Evolution and stratigraphic distribution of the *Offaster-Galeola* lineage in the Upper Cretaceous of northwestern Europe (modified after Ernst, 1971, 1972). The lineage shows a remarkable evolution of size. The enlargement scales of the test drawings correspond to the mean values of populations of the northwestern German Campanian. These rapid morphological transformations, involving several features, reflect a change in mode of life. *Offaster* is adapted for burrowing in the mud, whereas *Galeola* acquires an epibenthic mode of life. In upper part of Lower Campanian, the *Galeola* lineage splits into two ecological subspecies, which gradually adapted to different types of lithofacies (arenaceous limy marl and chalk and marly limestone).

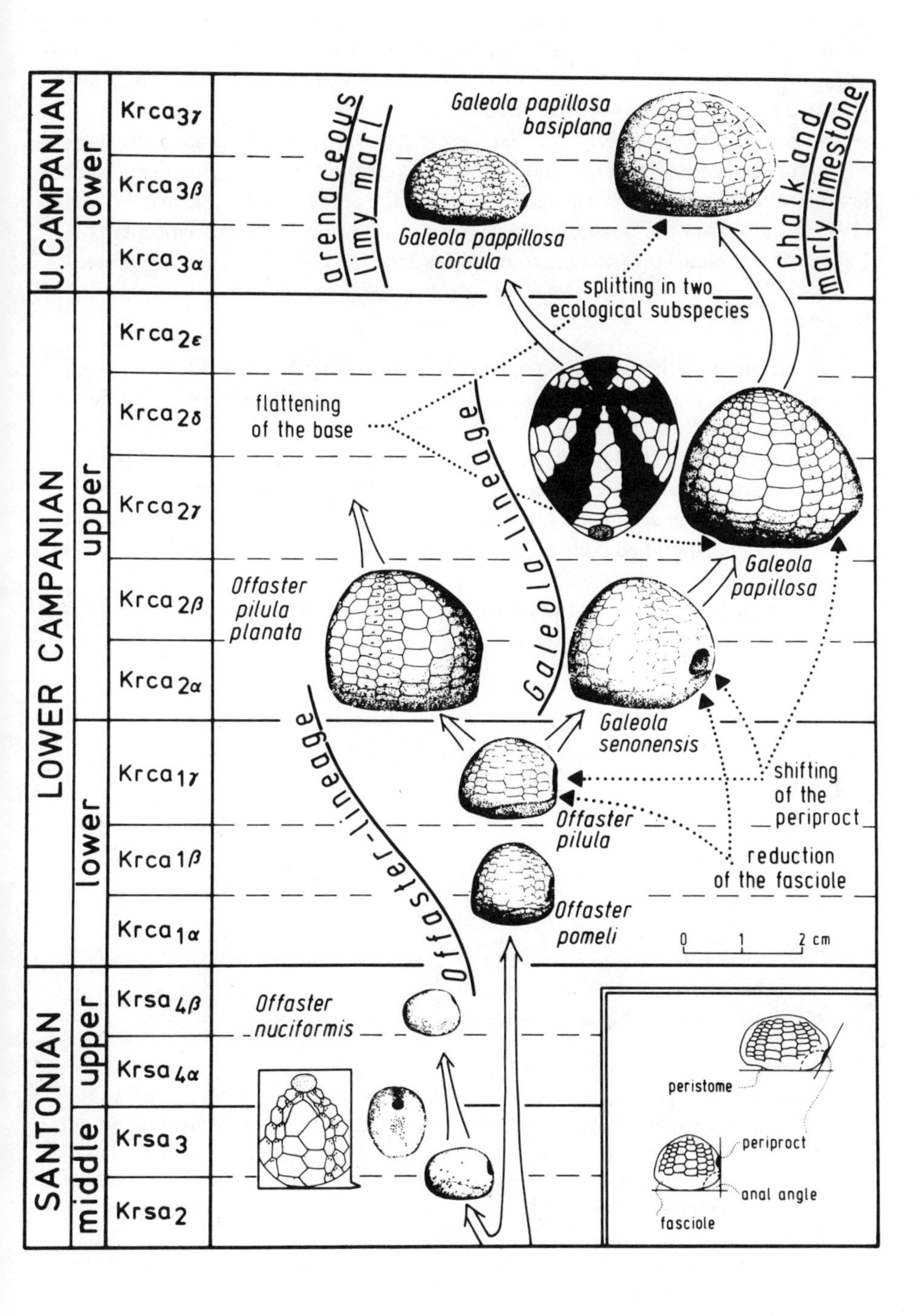
U. CAMPANIAN
lower
LOWER CAMPANIAN
upper
lower
SANTONIAN
upper
middle
Krca3γ
Krca3β
Krca3α
Krca2ε
Krca2δ
Krca2γ
Krca2β
Krca2α
Krca1γ
Krca1β
Krca1α
Krsa4β
Krsa4α
Krsa3
Krsa2
arenaceous limy marl
Galeola papillosa basiplana
Chalk and marly limestone
Galeola pappillosa corcula
splitting in two ecological subspecies
flattening of the base
Galeola-lineage
Offaster pilula planata
Galeola papillosa
Galeola senonensis
Offaster-lineage
shifting of the periproct
Offaster pilula
reduction of the fasciole
Offaster pomeli
0 1 2 cm
Offaster nuciformis
peristome
periproct
anal angle
fasciole

development of the fasciole, the shifting of the peristome to the base, and the development of an asymmetrical posterior area (Ernst and Schulz, 1971). The development of most of these morphologic structures is considered to be due to continuing adaption to a more efficient burrowing mode of life.

Biostratigraphy. Formerly, the *Hagenowia* group was biostratigraphically neglected because the tests were small and delicate. Recently, *Hagenowia* has been found in many Chalk occurrences, the facies to which they are restricted. Ernst and Schulz (1971) pointed out the possibility of subdividing the Coniacian and Santonian of Lägerdorf (Holstein) into at least four biostratigraphic zones utilizing *Hagenowia.*

Summary

The different methods of using echinoids in biostratigraphy are discussed using examples of irregular forms from the northwest European Cretaceous. It can be demonstrated that, for zonation, use of morphological stages in evolving lineages proves superior in biostratigraphy to the use of traditional range zones and peak zones based on broad species concepts. The phyletic series of *Micraster, Echinocorys,* and *Offaster-Galeola* attained extraordinary biostratigraphical importance. By means of large bed-by-bed collections it was possible to study these phyletic lineages through population systematic techniques. But, still, statistical studies of these groups and their use in biostratigraphy are hampered by strong ecomorphologic influences and allometric growth tendencies.

Text-Figure 6

Evolution and stratigraphic distribution of the *Infulaster-Hagenowia* lineage in the Boreal Upper Cretaceous of northwestern Europe (modified after Ernst and Schulz, 1971; Ernst, 1972; and with reference to new material collected by M. Kutscher from the Isle of Rügen, and M. -G. Schulz from Hamburg). The lineage is characterized by a striking phyletic reduction of size (note: in the evolutionary diagram this trend is diminished by different enlargement scales). The elongation of the rostrum is the most distinctive morphologic change; increased development of the fasciole, the shift in position of the peristome, and the evolution of an asymmetrical posterior part of the test are further important phylogenetic trends.

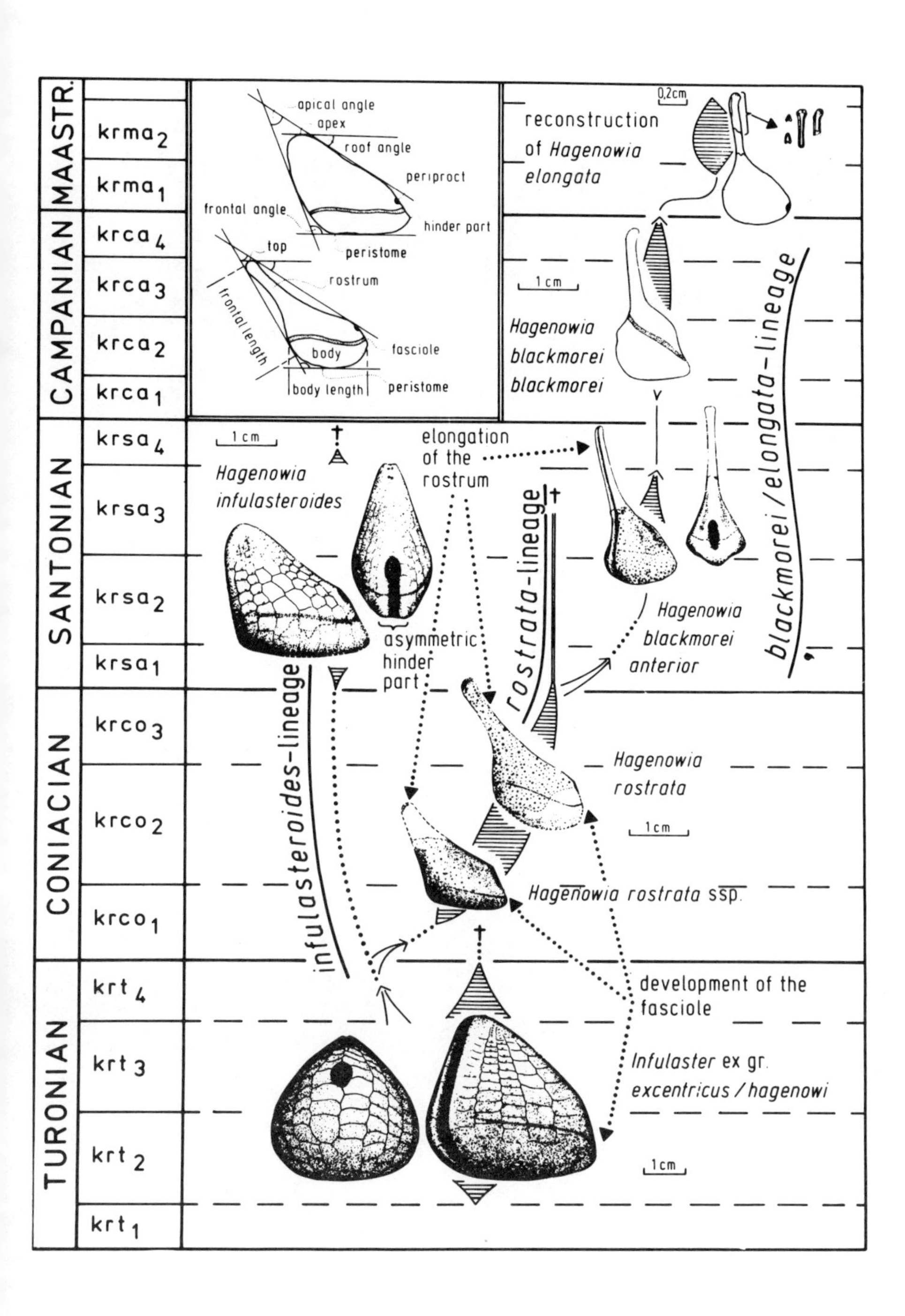
MAASTR.
krma2
krma1
CAMPANIAN
krca4
krca3
krca2
krca1
SANTONIAN
krsa4
krsa3
krsa2
krsa1
CONIACIAN
krco3
krco2
krco1
TURONIAN
krt4
krt3
krt2
krt1
apical angle
apex
roof angle
periproct
frontal angle
hinder part
top
peristome
rostrum
frontal length
body
fasciole
body length
peristome
0,2cm
reconstruction of Hagenowia elongata
1 cm
Hagenowia blackmorei blackmorei
blackmorei/elongata-lineage
1 cm
Hagenowia infulasteroides
elongation of the rostrum
rostrata-lineage
Hagenowia blackmorei anterior
asymmetric hinder part
infulasteroides-lineage
Hagenowia rostrata
1 cm
Hagenowia rostrata ssp.
development of the fasciole
Infulaster ex gr. excentricus / hagenowi
1 cm

References

Abich, H. 1858. Verleichende grundzuge der geologie des kaukasus wie der armenischen und nordpersischen gebirge. *Acad. Imp. Sci. St. Petersbourg, Mem., 6th ser. (Sci. Math. et Phys.) 7:* 439, 528, pl. 13.

Adams, T. D., M. Khalili, and A. K. Said. 1967. Stratigraphic significance of some oligosteginid assemblages from Lurestan Province, northwest Iran. *Micropaleont. 13* (1): 55-67.

Addicott, W. O. 1970. Miocene gastropods and biostratigraphy of the Kern River Area, California. *U.S. Geol. Surv. Prof. Paper 642:* 1-174, 21 pls.

——. 1973a. Giant Neogene pectinids of eastern North Pacific—chronostratigraphic and zoogeographic significance. *Amer. Assoc. Petol. Geol., Ann. Mtg. 57* (4): 766.

——. 1973b. Oligocene molluscan biostratigraphy and paleontology of the lower part of the Type Temblor Formation, California. *U.S. Geol. Surv. Prof. Paper 291:* 1-48, 9 pls.

——. 1974. Giant pectinids of the eastern North Pacific margin: significance in Neogene zoogeography and chronostratigraphy. *Jour. Paleont. 48* (1): 180-194.

Adkins, W. S. 1933. The Mesozoic systems in Texas. *Texas Univ. Bull. 3232:* 239-518.

Ager, D. V. 1963. Principles of Paleoecology. McGraw-Hill, New York, 371 p.

_____. 1964. The British Mesozoic Committee. *Nature 203:* 1059.

_____. 1971. Space and time in brachiopod history. *In* Faunal Provinces in Space and Time, F. A. Middlemiss, P. F. Rawson, and G. Newall, Eds. Seel House Press, Liverpool. p. 94-110.

Akers, W. H. 1965. Pliocene-Pleistocene boundary, northern Gulf of Mexico. *Science 149* (3685): 741-742.

_____, and J. H. Dorman. 1964. Pleistocene Foraminifera of the Gulf Coast. *Tulane Stud. Geol. 3* (1): 1-93.

_____, and C. W. Drooger. 1957. Miogypsinids, planktonic Foraminifera and Gulf Coast Oligocene-Miocene correlations. *Amer. Assoc. Petrol. Geol. Bull. 41*(4): 656-678.

_____, and A. J. J. Holck. 1957. Pleistocene beds near the edge of the continental shelf, southeastern Louisiana. *Geol. Soc. Amer. Bull. 68:* 983-992.

Alberdi, Maria Teresa. 1974. El género Hipparion en España. Neuvas formas de Castilla y Andalucía, revisión e historia evolutiva. Estud. Geol. (in press).

Aldridge, R. J., R. L. Austin, and S. Husri. 1968. Viséan conodonts from North Wales and Ireland. *Nature 219:* 255-258.

Alexandrowicz, S. W. 1969. *Thuramminoides sphaeroidalis* Plummer (Foraminifera) from the Cambrian beds of the vicinity of Sandomiere. *Rocnik. Polsk. Towarzystwa Geol. 39* (1-3): 27-34.

Allen, J. A., and R. S. Scheltema. 1972. The functional morphology and geographical distribution of *Planktomya henseni,* a supposed neotenous pelagic bivalve. *J. Mar. Biol. Assoc., U.K. 52:* 19-31.

Allison, R. C., and O. S. Adegoke. 1969. The *Turritella rina* group (Gastropa) and its relationship to *Torcula* Gray. *Jour. Paleont. 43* (5): 1248-1266, pls. 147, 148.

American Commission on Stratigraphic Nomenclature. 1961-1970. Code of Stratigraphic Nomenclature. Amer. Assoc. Petrol. Geol. 22 p.

_____. 1961. Code of Stratigraphic Nomenclature. *Amer. Assoc. Petrol. Geol. Bull. 45:* 645-660.

Anderson, J. M. 1973. The biostratigraphy of the Permian and Triassic. Part 2 (Charts 23-35) – A preliminary review of the distribution of Permian and Triassic strata in time and space. *Palaeont. Africana. 16:* 59-83, charts.

Anglada, R. 1971a. Sur la position du datum à Globigerinoides (Foraminiferida) la zone N4 (Blow 1967) et la limite oligo-miocène en Mediterranée. *C.R. Acad. Sci., Paris 272:* 1067-1070.

_____. 1971b. Sur la limite Aquitanien-Burdigalien, sa place dans l'échelle des Foraminifères planctoniques et sa signification dans le Sud-Est da la France. *C.R. Acad. Sci., Paris 272:* 1948-1950.

_____. 1972. Etude des petits Foraminifères. *Bull. B.R.-G.M. (2nd sér.), sect. 1, 4:* 29-35.

Antunes, M. T. 1969. Mamiferos nao marinhos do Miocénico de Lisboa: ecologia e estratigrafia. *Soc. Geol. Port., Bol. 17:* 75-85.

_____, L. Ginsburg, J. R. Torquato, and M. de L. Ubaldo. 1973. Age des couches à mammifères de la basse vallée du Tage (Portugal) et de la Loire moyenne (France). *C.R. Acad. Sci., Paris, Sér. D, 277:* 2313-2316.

Applin, Esther R., Alva E. Ellisor, and Hedwig T. Knicker. 1925. Subsurface stratigraphy of the coastal plain of Texas and Louisiana. *Amer. Assoc. Petrol. Geol. Bull. 9* (1): 79-122.

Arkell, W. J. 1933. The Jurassic System in Great Britain. Oxford University Press, Oxford. 681 + xii p., 41 pls.

——. 1946. Standard of the European Jurassic. *Geol. Soc. Amer. Bull. 57:* 1-34.

——. 1949. Jurassic ammonites in 1949. *Sci. Progress 147:* 401-417.

——. 1956a. Comments on stratigraphic procedure and terminology. *Amer. Jour. Sci. 254:* 457-467.

——. 1956b. Jurassic Geology of the World. Oliver & Boyd, Edinburgh. 806 p., 46 pls.

——, Bernhard Kummel, and C. W. Wright. 1957. Mesozoic Ammonoidea. *In* Treatise on Invertebrate Paleontology, R. C. Moore, Ed. Part L, Mollusca 4. Geol. Soc. Amer. and University of Kansas Press. L80-L490.

Armstrong, A. K. 1962. Stratigraphy and paleontology of the Mississippian System in southwestern New Mexico and adjacent southeastern Arizona. *N. Mex. Bur. Mines Min. Res. Mem. 8:* 1-99.

Armstrong, F. C. 1969. Geologic map of the Soda Springs quadrangle, southeastern Idaho. *U.S. Geol. Surv., Misc. Geol. Inv. Map I-557.*

Armstrong, W. G., and L. B. Tarlo. 1966. Amino-acid components in fossil calcified tissues. *Nature 210:* 481-482.

Atwater, Gordon I., and McLean J. Forman. 1959. Nature of growth of southern Louisiana salt domes and its effect on petroleum accumulation. *Amer. Assoc. Petrol. Geol. Bull. 43* (11): 2592-2622.

Austin, G. S. 1972. Paleozoic lithostratigraphy of southeastern Minnesota. *In* Geology of Minnesota: A Centennial Volume, P. K. Sims and G. B. Morey, Eds. Minnesota Geol. Surv., St. Paul. p. 459-473.

Austin, R. L., and C. R. Barnes. In Press. Carboniferous conodont paleoecology.

——, and F. H. T. Rhodes. 1969. A conodont assemblage from the Carboniferous of the Avon Gorge, Bristol. *Paleont. 12:* 400-405.

——, R. Conil, F. H. T. Rhodes, and M. Streel. 1970. Conodontes, spores et foraminifères du Tournaisien Inférieur dans la Vallée du Hoyoux. *Ann. Soc. Geol. Belgique 93:* 305-315.

Axelrod, D. I. 1952. A theory of angiosperm evolution. *Evolution 6:* 29-60.

——. 1959. Poleward migration of early angiosperm flora. *Science 130:* 203-207.

——. 1960. The evolution of flowering plants. *In* The Evolution of Life, S. Tax, Ed. University of Chicago Press, Chicago. p. 227-305.

——. 1970. Mesozoic paleogeography and early angiosperm history. *Botanical Rev. 36* (3): 277-319.

Ayala, F. J., D. Hedgecock, G. S. Zumwalt, and J. W. Valentine. 1973. Genetic variation in *Tridacna maxima,* an ecological analog of some unsuccessful evolutionary lineages. *Evolution 27:* 177-191.

——, J. W. Valentine, T. E. Delaca, and G. S. Zumwalt. 1975. Genetic variability of the Antarctic brachiopod *Liothyrella notorcadensis* and bearing on mass extinction hypotheses. *Jour. Paleont. 49(1):* 1-9.

Baldi, T. 1969. On the Oligocene-Miocene Stages of the Central Paratethys and on the Formations of the Egerian in Hungary. *Ann. Univ. Sc. Budapestinensis, Sec. Geo. 12:* 19-28.

——. 1973. Mollusc fauna of the Hungarian Upper Oligocene (Egerion).

——, and J. Senes. 1975. OM-Egerian. *Chronostratigraphy and Neostratigraphy 5:* 1-577.

Ballesio, R. 1971. Le Pliocène rhodanien. Ve Congress du Néogène Mediterranéen. Volume I. *Docum. Lab. Geol. Univ. Lyon. Hors series 1971*: 20-239.

Bamber, E. W. 1966. Type lithostrotionid corals from the Mississippian of western Canada. *Canada Geol. Surv. Bull. 135*: 1-28.

_____, and J. B. Waterhouse. 1971. Carboniferous and Permian Stratigraphy and Paleontology, northern Yukon Territory, Canada. *Bull. Canad. Petrol. Geol. 19* (1): 29-260, 15 figs., 27 pls.

Bandy, O. L. 1961. Distribution of foraminifera, radiolaria, and diatoms in the sediments of the Gulf of California. *Micropaleontology 7*: 1-26.

_____. 1967. Benthic Foraminifera as environmental indices. *In* Paleoecology: Short Course Lecture Notes, Orville L. Bandy, James C. Ingle, Robert R. Lankford, and Heinz A. Lowenstam, Eds. Amer. Geol. Inst. OB1-OB29B.

_____(Ed.). 1970. Radiometric Dating and Paleontologic Zonation. *Geol. Soc. Amer. Spec. Paper 124*: 247 p.

_____. 1972. Neogene planktonic foraminiferal zones, California, and some geologic implications. *Palaeogeogr. Palaeoclimatol. Palaeoecol. 12:* 131-150.

_____, and J. C. Ingle, Jr. 1970. Neogene planktonic events and radiometric scale, California. *In* Paleontologic Zonation and Radiometric Dating, O. L. Bandy, Ed. *Geol. Soc. Amer. Spec. Paper 124:* 133-174.

Barnes, C. R. 1967. A questionable natural conodont assemblage from Middle Ordovician limestone, Ottawa, Canada. *Jour. Paleont. 41:* 1557-1560.

_____, Daniel B. Sass, and Eugene A. Monroe. 1973a. Ultrastructure of some Ordovician Conodonts. *Geol. Soc. Amer. Spec. Paper 141:* 1-30.

_____, D. B. Sass, and M. L. S. Poplawski. 1973b. Conodont ultrastructure: the family Panderodontidae. *Life Sci. Contr., Royal Ont. Mus. 90:* 1-36.

_____, C. B. Rexroad, and James F. Miller. 1973c. Lower Paleozoic conodont provincialism. *Geol. Soc. Amer. Spec. Paper 141:* 157-190.

Barnes, H., M. Barnes, and W. Klepal. 1972. Some Cirripedes of the French Atlantic Coast. *J. Exp. Mar. Biol. Ecol. 8:* 107-194.

Barnett, Stockton G. 1971. Biometric determination of the evolution of *Spathognathodus remscheidensis:* A method for precise intrabasinal time correlations in the Northern Appalachians. *J. Paleont. 45:* 274-300.

_____. 1972. The evolution of *Spathognathodus remscheidensis* in New York, New Jersey, Nevada, and Czechoslovakia. *J. Paleont. 46:* 900-917.

Barrois, C. 1876. Recherches sur le terrain crétacé supérieur de l'Angleterre et de l'Irlande. *Soc. Géol. du Nord. Mém. 1:* 1-232, 3 pls.

Barron, J. A. 1976. Middle Miocene- Lower Pliocene Marine Diatom and Silicoflagellate Correlations in the California Area: The Neogene Symposium, Spring, 1976; Annual Meeting, Pacific Section SEPM, San Francisco, California, pp. 117-123.

Bassler, R. S. 1950. Faunal lists and descriptions of Paleozoic corals. *Geol. Soc. Amer. Mem. 44:* 1-315.

Batten, R. L. 1958. Permian Gastropoda of the southwestern United States, 2. Pleurotomariacea. *Amer. Mus. Nat. Hist. Bull. 114*(2): 159-246, pls. 32-42.

_____. 1973. The vicissitudes of the gastropods during the interval of Guadalupian-Ladinian time. *In* The Permian and Triassic Systems and Their Mutual Boundary, A. Logan and L. V. Hills, Eds. *Canadian Soc. Petrol. Geol. Mem. 2:* 596-607.

Bayne, B. L. 1965. Growth and delay of metamorphosis of the larvae of *Mytilus edulis* (L.). *Ophelia 2:* 1-47.

Bé, Alan W. H. 1965. The influence of depth on shell growth in *Globigerinoides sacculifer* (Brady). *Micropaleontology 11*(1): 81-97.

——, Stanley M. Harrison, and Leroy Lott. 1973. *Orbulina universa* d'Orbigny in the Indian Ocean. *Micropaleontology 19*(2): 81-97.

Beard, John H. 1969. Pleistocene paleotemperature record based on planktonic foraminifers, Gulf of Mexico. *Gulf Coast Assoc. Geol. Soc., Trans. 19:* 535-553.

Beede, J. W., and H. T. Kniker. 1924. Species of the genus *Schwagerina* and their stratigraphic significance. *Texas Univ. Bull. 2433:* 1-96, 9 pls.

Bell, W. A. 1929. Horton-Windsor district, Nova Scotia. *Canada Geol. Surv. Mem. 155:* 1-268.

Bell, W. C., O. W. Feniak, and V. E. Kurtz. 1952. Trilobites of the Franconia Formation, southeast Minnesota. *Jour. Paleont. 26*(2): 175-198.

——, R. R. Berg, and C. A. Nelson. 1956. Croixan type area—Upper Mississippi Valley. *20th Internat. Geol. Cong., Proc. el Sistema Cambrico, su Paleogeografia y el Problema de su Base* (J. Rodgers, Ed.) *2*(2): 415-446.

Benda, L., and J. E. Meulenkamp. 1972. Discussion on Biostratigraphic Correlations in the Eastern Mediterranean Neogene. *Z. Deutsch. Geol. Ges. 123:* 559-564, 1 table.

Benson, R. H. 1972. Ostracods as indicators of threshold depth in the Mediterranean during the Pliocene. *In* The Mediterranean Sea, D. J. Stanley, Ed. Dowden, Hutchinson & Ross, Stroudsburg, Pa. p. 63-73.

——. 1975. The origin of the psychrosphere as recorded in changes of deep-sea ostracode assemblages. *Lethaia 8*(1): 69-83.

Berger, E. M. 1973. Gene-enzyme variation in three sympatric species of *Littorina. Biol. Bull. 145:* 83-90.

Berggren, W. A. 1962. Some planktonic Foraminifera from the Maestrichtian and type Danian stages of southern Scandinavia. *Stockh. Contr. Geol. 9*(1): 1-106.

——. 1964. The Maestrichtian, Danian and Montian stages and the Cretaceous Tertiary boundary. *Stockh. Contr. Geol. 11*(5): 103-176.

——. 1969. Paleogene biostratigraphy and planktonic foraminifera of northern Europe. *In* Proc. 1st Internat. Conf. Planktonic Microfossils, Geneva, 1967, P. Bronnimann and H. H. Renz, Eds. E. J. Brill, Leiden. p. 121-160.

——. 1972a. A Cenozoic Time-Scale: Some implications for regional geology and paleobiogeography. *Lethaia 5:* 195-215.

——. 1972b. Cenozoic biostratigraphy and paleobiogeography of the North Atlantic. *In* Initial Reports of the Deep Sea Drilling Project, Volume 12, A. S. Laughton, W. A. Berggren et al., Eds. Government Printing Office, Washington, D. C. p. 965-1001, 13 pls., 6 text-figs.

——. 1973a. The Pliocene time-scale: Calibration of planktonic foraminiferal and calcareous nannoplankton zones. *Nature 243*(5407): 391-397.

——. 1973b. Biostratigraphy and Biochronology of the Late Miocene (Tortonian and Messinian) of the Mediterranean. *In* Messinian Events in the Mediterranean: C. W. Drooger, Ed. North-Holland Publishing Co., Amsterdam. p. 10-20

——, and C. D. Hollister. 1974. Paleogeography, paleobiogeography and the history of circulation in the Atlantic Ocean. *Soc. Econ. Paleont. Mineral., Spec. Publ. 20:* 126-186.

——, and J. D. Phillips. 1971. Influence of continental drift on the distribution of the Tertiary benthonic foraminifera in the Caribbean and Mediterranean regions. Symp. Geol. Libya, Fac. Sci. Univ. Libya, Beirut. p. 263-299.

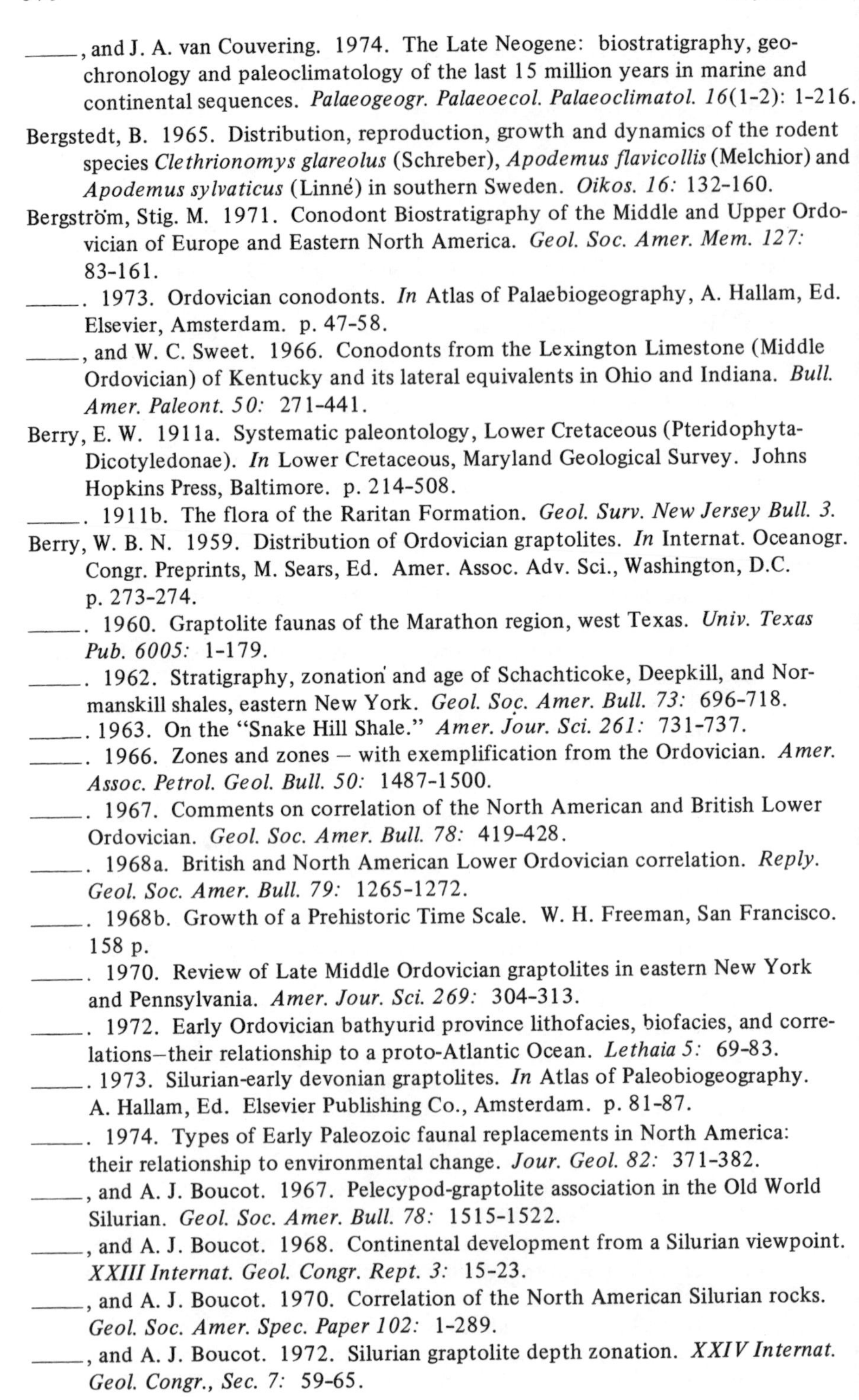

_____, and J. A. van Couvering. 1974. The Late Neogene: biostratigraphy, geochronology and paleoclimatology of the last 15 million years in marine and continental sequences. *Palaeogeogr. Palaeoecol. Palaeoclimatol. 16*(1-2): 1-216.

Bergstedt, B. 1965. Distribution, reproduction, growth and dynamics of the rodent species *Clethrionomys glareolus* (Schreber), *Apodemus flavicollis* (Melchior) and *Apodemus sylvaticus* (Linné) in southern Sweden. *Oikos. 16:* 132-160.

Bergström, Stig. M. 1971. Conodont Biostratigraphy of the Middle and Upper Ordovician of Europe and Eastern North America. *Geol. Soc. Amer. Mem. 127:* 83-161.

_____. 1973. Ordovician conodonts. *In* Atlas of Palaebiogeography, A. Hallam, Ed. Elsevier, Amsterdam. p. 47-58.

_____, and W. C. Sweet. 1966. Conodonts from the Lexington Limestone (Middle Ordovician) of Kentucky and its lateral equivalents in Ohio and Indiana. *Bull. Amer. Paleont. 50:* 271-441.

Berry, E. W. 1911a. Systematic paleontology, Lower Cretaceous (Pteridophyta-Dicotyledonae). *In* Lower Cretaceous, Maryland Geological Survey. Johns Hopkins Press, Baltimore. p. 214-508.

_____. 1911b. The flora of the Raritan Formation. *Geol. Surv. New Jersey Bull. 3.*

Berry, W. B. N. 1959. Distribution of Ordovician graptolites. *In* Internat. Oceanogr. Congr. Preprints, M. Sears, Ed. Amer. Assoc. Adv. Sci., Washington, D.C. p. 273-274.

_____. 1960. Graptolite faunas of the Marathon region, west Texas. *Univ. Texas Pub. 6005:* 1-179.

_____. 1962. Stratigraphy, zonation and age of Schachticoke, Deepkill, and Normanskill shales, eastern New York. *Geol. Soc. Amer. Bull. 73:* 696-718.

_____. 1963. On the "Snake Hill Shale." *Amer. Jour. Sci. 261:* 731-737.

_____. 1966. Zones and zones – with exemplification from the Ordovician. *Amer. Assoc. Petrol. Geol. Bull. 50:* 1487-1500.

_____. 1967. Comments on correlation of the North American and British Lower Ordovician. *Geol. Soc. Amer. Bull. 78:* 419-428.

_____. 1968a. British and North American Lower Ordovician correlation. *Reply. Geol. Soc. Amer. Bull. 79:* 1265-1272.

_____. 1968b. Growth of a Prehistoric Time Scale. W. H. Freeman, San Francisco. 158 p.

_____. 1970. Review of Late Middle Ordovician graptolites in eastern New York and Pennsylvania. *Amer. Jour. Sci. 269:* 304-313.

_____. 1972. Early Ordovician bathyurid province lithofacies, biofacies, and correlations–their relationship to a proto-Atlantic Ocean. *Lethaia 5:* 69-83.

_____. 1973. Silurian-early devonian graptolites. *In* Atlas of Paleobiogeography. A. Hallam, Ed. Elsevier Publishing Co., Amsterdam. p. 81-87.

_____. 1974. Types of Early Paleozoic faunal replacements in North America: their relationship to environmental change. *Jour. Geol. 82:* 371-382.

_____, and A. J. Boucot. 1967. Pelecypod-graptolite association in the Old World Silurian. *Geol. Soc. Amer. Bull. 78:* 1515-1522.

_____, and A. J. Boucot. 1968. Continental development from a Silurian viewpoint. *XXIII Internat. Geol. Congr. Rept. 3:* 15-23.

_____, and A. J. Boucot. 1970. Correlation of the North American Silurian rocks. *Geol. Soc. Amer. Spec. Paper 102:* 1-289.

_____, and A. J. Boucot. 1972. Silurian graptolite depth zonation. *XXIV Internat. Geol. Congr., Sec. 7:* 59-65.

———, and A. J. Boucot. 1973. Glacio-eustatic control of Late Ordovician-Early Silurian platform sedimentation and faunal changes. *Geol. Soc. Amer. Bull. 84:* 275-284.

Beu, A. G. 1970a. The mollusca of the subgenus *Monoplex* (Family Cymatiidae). *Trans. Roy. Soc. New Zealand (Biol. Sci.) 11:* 225-237.

———. 1970b. The mollusca of the genus *Choronia* (Family Cymatiidae). *Trans. Roy. Soc. New Zealand (Biol. Sci.) 11:* 205-332.

Beyrich, H. E. 1856. Über den Zusammenhang der norddeutschen Tertiärbildungen zur Erläuterung einer geologischen Übersichtskarte. *Abh. K. Akad. Wissensch., Physik. Abh. 1855:* 1-20.

Bidder, A. M. 1962. Use of the tentacles, swimming and buoyancy control in the pearly *Nautilus. Nature 196:* 451-454.

Bieri, Robert. 1959. The distribution of the planktonic Chaetognatha in the Pacific and their relationship to the water masses. *Limnol. and Oceanogr. 4:* 1-28.

Bird, J. M., and J. F. Dewey. 1970. Lithosphere plate–continental margin tectonics and the evolution of the Appalachian orogen. *Geol. Soc. Amer. Bull. 81:* 1031-1060.

Birkeland, C., Fu-Chiang Chia, and R. R. Strathmann. 1971. Development, substrate selection, delay of metamorphosis and growth in the sea-star *Mediaster aequalis* Stimpson. *Biol. Bull. Woods Hole. 141:* 99-108.

Birkelund, T. 1957. Upper Cretaceous belemnites from Denmark. *Biol. Skr. Dan. Vid. Selsk. 9*(1): 1-69.

———. 1966. Die Entwicklung der jüngsten Scaphiten und ihre stratigraphische Bedeutung im baltischen Gebiet. *Ber. deutsch. Ges. Geol. Wiss. A. Geol. Paläont. 11:* 737-744.

———. 1967. Submicroscopic shell structure and early growth-stage of Maestrichtian ammonites (*Saghalinites* and *Scaphites*). *Dansk Geol. Foren. Medd. 17:* 95-101.

———, and H. J. Hansen. 1968. Early growth stages and structure of septae and the siphunclar tube in some Maestrichtian ammonites. *Dansk Geol. Foren. Medd. 18:* 71-78.

Bishop, W. W., J. A. Miller, and F. J. Fitch. 1969. New potassium-argon age determinations relevant to the Miocene fossil mammal sequence in East Africa. *Amer. Jour. Sci. 267:* 669-699.

Bissell, H. J. 1939. The use of fusulinids in zoning the Utah Pennsylvanian strata [Abst.]. *Utah Acad. Sci., Proc. 16:* 85.

Black, M. 1953. The constitution of the chalk. *Proc. Geol. Soc. Lond. 1499:* 81-86.

Blackith, R. E., and R. A. Reyment. 1971. Multivariate Morphometrics. Academic Press, London. 412 p.

Blakely, R. J. 1974. Geomagnetic reversals and crustal spreading rates during the Miocene. *J. Geophys. Res. 79*(20): 2979-2985.

Blanton, J. 1971. Exchange of Gulf Stream water with North Carolina shelf water in Onslow Bay during stratified conditions. *Deep-Sea Res. 18:* 167-178.

Blow, W. H. 1957. Transatlantic correlation of Miocene sediments. *Micropaleontology 3:* 77-79.

———. 1969. Late Middle Eocene to Recent planktonic foraminiferal biostratigraphy. *In* Proceedings of the 1st International Conference on Planktonic Microfossils, Geneva, 1967, P. Bronnimann and H. H. Renz, Eds. E. J. Brill, Leiden. 1: p. 199-421, pl. 1-54.

_____, and A. H. Smout. 1968. The Bormidian stage and the base of the Miocene. C.M.N.S. Proc. IV Session, Bologna, 1967. *Giorn. Geol. 35(III):* 307-314.

Bogush, O. I., and O. V. Juferev. 1966. Carboniferous and Permian Foraminifera of the Carboniferous deposits in Kara-Tau and Talasskiy Ala-Tau [in Russian]. Akad. Nauk S.S.S.R., Sib. Otdel, Inst. Geol. Geophys. 234 p., 9 pls.

Bolli, Hans M. 1957. Planktonic Foraminifera from the Oligocene-Miocene Cipero and Lengua Formations of Trinidad. *B.W.I. U.S. Nat. Mus. Bull. 215:* 97-123.

_____. 1959. Planktonic foraminifera as index fossils in Trinidad, West Indies and their value for world wide stratigraphic correlation. *Eclog. Geol. Helv. 52:* 627-637.

_____. 1964. Observations on the stratigraphic distribution of some warm water planktonic foraminifera in the young Miocene to Recent. *Eclog. Geol. Helv. 57:* 541-552.

_____. 1966. Zonation of Cretaceous to Pliocene marine sediments based on planktonic Foraminifera. *Bol. Inform. Assoc. Venezolana Geol. Min. Petr. 9:* 3-32.

Bolshakov, V. N. 1968. On the relationship of clinal variability and species structure. *Zool. Zh. 47:* 807-815.

Bonham-Carter, G. F. 1967. FORTRAN IV program for Q-mode cluster analysis of nonquantitative data using IBM 7090/7094 computers. *Kansas Geol. Surv. Computer Contr. 17:* 1-28.

Bornhauser, Max. 1947. Marine sedimentary cycles of Tertiary in Mississippi Embayment and central Gulf Coast area. *Amer. Assoc. Petrol. Geol. Bull. 31* (4): 698-712.

Boss, K. J. 1971. Critical estimate of the number of Recent Mollusca. *Mus. Comp. Zool. Occasional Papers 3*(40): 81-135.

Boucot, A. J., J. G. Johnson, and J. A. Talent. 1969. Early Devonian brachiopod zoogeography. *Geol. Soc. Amer. Special Paper 119:* 1-113.

Bowen, R. 1966. Paleotemperature Analysis. Elsevier, Amsterdam. 265 p.

Bowles, Edgar. 1939. Eocene and Paleocene Turritellidae of the Atlantic and Gulf Coastal Plain of North America. *Jour. Paleont. 13(3):* 267-336, pls. 31-34.

Bowsher, A. L. 1955. Origin and adaptation of platyceratid gastropods. *Univ. of Kansas Paleont. Contr. Mollusca 5:* 1-11.

_____. 1961. The stratigraphic occurrence of some Lower Mississippian corals from New Mexico and Missouri. *Jour. Paleont. 35*(5): 955-962.

_____, and J. T. Dutro, Jr. 1957. The Paleozoic section in the Shainin Lake area, central Brooks Range, Alaska. *U.S. Geol. Surv. Prof. Paper 303-A:* 1-39.

Boyer, J. F. 1974. Clinal and size-dependent variation at the LAP locus in *Mytilus edulis. Biol. Bull. Woods Hole 147:* 535-549.

Branson, C. C. 1949. Bibliographic index of Permian invertebrates. *Geol. Soc. Amer. Mem. 26:* 1-1049.

Brattström, H. 1941. Studien über die Echinodermen zwischen Skagerrack und Ostsee, besonders des Öresundes, mit einer Übersicht uber die physische Geographic. *Undersökningar över Öresund. 27:* 1-329.

Braunstein, Jules (Ed.). 1970. Bibliography of Gulf Coast geology. *Gulf Coast Assoc. Geol. Soc., Spec. Pub. 1,* I Bibliography. 1-682, II Index 683-1045.

Brenner, G. J. 1963. The spores and pollen of the Potomac Group of Maryland. *Maryland Dept. Geol., Mines, Water Res. Bull. 27:* 1-215.

———. 1967. Early angiosperm pollen differentiation in the Albian to Cenomanian deposits of Delaware (U.S.A.). *Review of Palaeobotany and Palynology 1:* 219-227.

———. 1968. Middle Cretaceous spores and pollen from northeastern Peru. *Pollen et Spores 10*(2): 341-383.

———. 1976. Middle Cretaceous floral provinces and early migrations of angiosperms. *In* The Origin and Early Evolution of the Angiosperms. C. B. Beck, Ed. Columbia University Press, New York. p. 23-47.

Bretsky, P. W. 1973. Evolutionary patterns in the Paleozoic Bivalvia: Documentation and some theoretical considerations. *Geol. Soc. Amer. Bull. 84:* 2079-2096.

———, and D. M. Lorenz. 1969. Adaptive response to environmental stability: A unifying concept in paleoecology. *Proc. North Amer. Paleont. Conf.* 522-550.

Brinkmann, R. 1929a. Statistisch-biostratigraphische Untersuchungen an mittel jurassischen Ammoniten über Artbegriff und Stammesentwicklung. *Abh. Gesell. Wiss. Göttingen, Math. Phys. Kl. 13*(3): 1-249.

———. 1929b. Monographie der Gattung *Kosmoceras. Abh. Gesell. Wiss. Göttingen, Math. Phys. Kl. 13*(4): 1-124.

Bromley, R. G. In press. Trace fossils at omission surfaces. *In* Studies in Trace Fossils, R. W. Frey, Ed. Springer-Verlag, New York.

Brongniart, A. 1823. Mémoire sur les terrains de sédiment supérieurs calcaréo-trappéens du Vicentin, et sur quelques terrains d'Italia, de France, d'Allemagne, etc., qui peuvent se rapporter à la même époque. Paris. 86 + iv p., 6 pls.

———. 1829. Tableau des Terrains qui composent l'Écorce du Globe, ou Essai sur la structure de la partie connue de la Terre. Paris and Strasbourg. 435 + xxxii p.

Bronn, H. G. 1831. Italiens Tertiär-Gebilde und deren organische Einschlüsse. Heidelberg. 176 + xii p., 1 pl., 6 tables.

Bronnimann, P., E. Martini, J. Resig, W. R. Riedel, A. Sanfilippo, and T. Worsley. 1971. Biostratigraphic synthesis – Late Oligocene and Neogene of the western tropical Pacific. *In* Initial Reports of the Deep Sea Drilling Project, E. L. Winterer et al., Eds. Government Printing Office, Washington. IX(2): p. 1723-1745.

Brooks, J., P. R. Grant, M. Muir, P. van Gijzel, and G. Shaw (Eds.). 1971. Sporopollenin. Academic Press, London. 718 p.

Brotzen, F. 1945. De geologiske resultaten från borrningarna vid Höllviken. *Del I: Kritan. Sver. Geol. Unders. ser. C, 465, Årsbok. 36:* 1-64.

Bruijn, H. de. 1973. Analysis of the data bearing on the correlation of the Messinian with the succession of land mammals. *In* Messinian Events in the Mediterranean, C. W. Drooger, Ed. North-Holland, Amsterdam. p. 260-262.

———, and J. E. Meulenkamp. 1972. Late Miocene Rodents from the Pandanassa Formation (Prov. Rethymnon), Crete, Greece. *Koninkl. Nederl. Akad. Wetensch., Proc., Ser. B. 75:* 54-60, 1 pl.

———, P. Y. Sondaar, and W. J. Zachariasse. 1971. Mammalia and foraminifera from the Neogene of Kastellios Hill (Crete): A correlation of continental and marine biozones. *I. K. Ned. Akad. Wetensch. Proc. Ser. B. 74*(5): 1-22.

Bruun, A. F. 1957. Deep sea and abyssal depths. *In* Treatise on Marine Ecology and Paleoecology, Vol. 1, Ecology, J. W. Hedgpeth, Ed. *Geol. Soc. Amer. Mem. 67:* 641-672.

Buckland, W. 1839. Geology. *In* The Zoology of Captain Beechey's Voyage, J. Richardson et al., Eds. London. p. 157-180.

Buckman, S. S. 1893. The Bajocian of the Sherborne District. *Geol. Soc. London, Quart. Jour. 49:* 479-522.

_____. 1898. On the grouping of some divisions of so-called "Jurassic" time. *Geol. Soc. London, Quart. Jour. 54:* 422-462, 2 tables.

_____. 1902. The term "Hemera." *Geol. Mag. 9:* 554-557.

Buday, T., I. Cicha, and J. Senes. 1965. Miozän der Westkarpaten. *Geol. Ustav Dionyza Stura.* 295.

Bukry, E., E. E. Brabb, and J. G. Vedder. (in press). Correlation of tertiary nannoplankton assemblages from the coast and peninsular ranges of California. Proc. 2nd Latin Am. Geol. Cong.

Bulman, O. M. B. 1958. The sequence of graptolite faunas. *Palaeontology 1:* 159-173.

_____. 1970. Graptolithina. *In* Treatise on Invertebrate Paleontology, Part V (Revised), Curt Teichert, Ed. Geol. Soc. Amer. and Univ. Kansas Press, Boulder, Colorado and Lawrence, Kansas. 163 p.

Bumpus, D. F. 1965. Residual drift along the bottom on the contental shelf in the middle Atlantic Bight area. *Limnol. Oceanogr. 10* (Suppl): R50-53.

_____. 1973. A description of the circulation on the continental shelf of the east coast of the United States. *Prog. Oceanogr. 6:* 111-157.

_____, and L. M. Lauzier. 1965. Surface circulation on the continental shelf off eastern North America between Newfoundland and Florida. *In* Atlas of the Marine Environment, Wilfred Webster, Ed. *Amer. Geogr. Soc. 7:* 1-4, 8 pls., Appendix iii.

Burckle, L. H. 1972. Late Cenozoic planktonic diatom zones from the eastern equatorial Pacific. *Nova Hedwigia. 39:* 217-245, 3 pls.

Burger, D. 1970. Early Cretaceous angiospermous pollen grains from Queensland. *Bureau of Mineral Resources. Geol. and Geophys. Bull. 116:* 1-10.

Burke, K., and J. B. Waterhouse. 1973. Saharan Glaciation dated in North America. *Nature 241*(5387): 267-268.

Burnaby, T. P. 1965. Reversed coiling trend in *Gryphaea arcuata. Geol. Jour. 4:* 257-278.

Burton, C. J., and N. Eldredge. 1974. Two new subspecies of *Phacops rana* (Trilobita) from the Middle Devonian of North-West Africa. *Palaeontology 17*(2): 349-363.

Butts, C. 1917. Descriptions and correlation of the Mississippian formations of western Kentucky. Kentucky Geol. Surv. 1-119.

_____. 1922. The Mississippian series of eastern Kentucky; a regional interpretation of the stratigraphic relations of the Subcarboniferous group based on new and detailed field examinations. *Kentucky Geol. Surv. 7:* 1-188.

_____. 1926. Geology of Alabama: The Paleozoic rocks. *Alabama Geol. Surv. Spec. Report. 14:* 41-430.

_____. 1929. Some issues in Chester stratigraphy in Kentucky and Illinois. *Jour. Geol. 37*(1): 30-46.

———. 1940. Geology of the Appalachian Valley in Virginia. *Virginia Geol. Surv. Bull. 52*(1): 1-568.

Callomon, J. H. 1963. Sexual dimorphism in Jurassic ammonites. *Leicester Lit. Philos. Soc. Trans. 57:* 21-56.

———. 1964. Notes on the Callovian and Oxfordian Stages. *In* Colloque du Jurassique à Luxembourg, 1962, P. L. Maubeuge, Ed. Ministère des Arts et des Sciences, Luxembourg. p. 269-291.

———. 1968. The Kellaway Beds and the Oxford Clay. *In* The Geology of the East Midlands, P. C. Sylvester-Bradley and T. D. Ford, Eds. Leicester University Press, Leicester. p. 264-290.

———. 1969. Dimorphism in Jurassic ammonites. *In* Sexual dimorphism in fossil metazoa and taxonomic implications, G. E. G. Westermann, Ed. *Internat. Union Geol. Sci. 1:* 111-125.

———, and D. T. Donovan. 1966. Stratigraphic classification and terminology. *Geol. Mag. 103:* 95-97.

Cameron, A. W. 1965. Competitive exclusion between rodent genera *Microtus* and *Clethrionomys. Evolution 18:* 630-634.

Cameron, R. A., and R. T. Hinegardner. 1974. Initiation of metamorphosis in laboratory cultured sea urchins. *Biol. Bull. Woods Hole 146:* 335-342.

Carbonnel, Gilles. 1973. L'analyse de groupe en paleoecologie et en biostratigraphie. Application aux ostracodes (Crustacea) miocenes. *Arch. Sci. Geneve 26:* 23-68.

Cariou, E. 1973. Ammonites of the Callovian and Oxfordian. *In* Atlas of Palaeobiogeography, A. Hallam, Ed. Elsevier, Amsterdam. p. 287-295.

Carloni, G. C., P. Marks, R. F. Rutsch, and R. Selli, Eds. 1971. Stratotypes of Mediterranean Neogene Stages. *Giorn. Geol. 37*(II): 1-266.

Carlson, H. L. 1960. Genetic conditions which promote or retard the formation of species. *Cold Spring Harbor Symp. Quant. Biol. 24:* 87-105.

Casey, Raymond. 1963. The dawn of the Cretaceous Period in Britain. *Southeastern Union Sci. Socs. Bull. 117:* 1-15.

Černajsek, T. 1971. Die Entwicklung und Abgrenzung der Gattung Aurild Pokorny (1955) im Neogen Osterreichs (Vorbericht). *Verh. Geol. Bundesanst. Jg. 1971, H. 3:* 571-575.

Chalishev, V. I. 1966. Upper Permian beds of the northern-pre Urals. Paleozoic beds of the northern pre-Urals. Acad. Sci. USSR. Geol. Inst. Moscow. p. 73-95, 3 figs. [in Russian].

Chaloner, W. G. 1970a. The evolution of miospore polarity. *Geoscience and Man (American Association of Stratigraphic Palynologists Proceedings) 1:* 47-56.

———. 1970b. The rise of the first land plants. *Biol. Reviews 45*(3): 353-377.

———, and M. Muir. 1968. Spores and floras. *In* Coal and Coal-bearing Strata, D. G. Murchison and T. S. Westoll, Eds. Oliver & Boyd, Edinburgh. p. 127-146.

Cheetham, A.H. 1960. Time, migration, and continental drift. *Bull. Amer. Assoc. Petrol. Geol. 44:* 244-251.

———, and P. B. Deboo. 1963. A numerical index for biostratigraphic zonation in the mid-Tertiary of the eastern Gulf. *Gulf Coast Assoc. Geol. Socs. Trans. 13:* 139-147.

———, and J. E. Hazel. 1969. Binary (presence-absence) similarity coefficients. *Jour. Paleont. 43*(5): 1130-1136.

Chen, S. 1934. Fusulinidae of south China; Part 1. *China Geol. Survey, Palaeont. Sinica, ser. B. 4*(2): 1-185, 16 pls., 36 graphs.

_____. 1956. Fusulinidae of south China; Part 2. *China Geol. Survey, Palaeont. Sinica, new ser., ser. B. 140:* 13-71, pls. 1-14.

_____, and J. C. Sheng. 1964. The fusulinid zones in the Chinese Carboniferous. Cinquieme Cong. Internat. Stratigraphie et Geologie Carbonifere 5th, Paris, 1963. *Compte Rendu 1:* 321-324.

Chlupáč, I., H. Jaeger, and J. Zikmundova. 1972. The Silurian-Devonian Boundary in the Barrandian. *Canadian Petrol. Geol. Bull. 20:* 104-174.

Choi, D. R. 1973. Permian fusulinids from the Setami-Yahagi district, southern Kitakami Mountains, N. E. Japan. *Hokkaido Univ. Fac. Sci. Jour., ser. 4, Geol. Min. 16*(1): 1-90, pls. 1-20, 12 text-figs.

Choubert, G., R. Charlot, A. Faure-Muret, L. Hottininger, J. Marcais, D. Tisserant, and P. Vidal. 1968. Stratigraphie et Micropaléontologie du Néogène au Maroc septentrional. Inst. Lucas Malladas, Cursillos Conf. 9: 229-259.

Christopher, R. A., and G. F. Hart. 1971. A statistical model in palynology. *Geoscience and Man (Amer. Assoc. Strat. Palynol. Proc.). 3:* 49-56.

Chronic, H. P. 1957. Molluscan fauna from the Permian Kaibab Formation, Walnut Canyon, Arizona, *Geol. Soc. Amer. Bull. 63:* 95-166, 10 pls.

Chronic, J. 1958. Pennsylvanian paleontology in Colorado. *In* Rocky Mountain Association of Geologists, Symposium on Pennsylvanian Rocks of Colorado and Adjacent Areas, Denver. p. 13-16, 2 text-figs.

Chun, C. 1896. Atlantis, Biologische Studien über pelagische Organismen II. *Auricularia nudibranchiata. Bibl. Zool. XIX*(1): 53-76.

Churkin, Michael, Jr., and E. E. Brabb. 1965. Ordovician, Silurian and Devonian biostratigraphy of east-central Alaska. *Amer. Assoc. Petrol. Geol. Bull. 49:* 172-185.

Cicha, I. 1970. Stratigraphical problems of the Miocene in Europe. *Rozpravy Ustred. Ust. Geol. 35:* 1-134, 12 pls.

_____, and J. Senes. 1968. Sur la position du Miocène de la Paratethys Centrale dans le cadre du Tertiaire de l'Europe. *Geol. Zborn. Slov. Akad. Vied. 19:* 95-116, 4 figs., 2 tables.

_____, and J. Senes. 1973. Neogene of the West Carpathian Mountains. *10 Congr. Carpath.-Balkan Geol. Assoc., Exkursion-Guide F:* 1-46, 7 text-figs.

_____. J. Senes, and J. Tejkal. 1967. Chronostratigraphie und Neostratotypen, bd. 1; M3 (Karpatian) Slovokian Akad. Sci., Bratislava 312 p.

_____, H. Hagn, and E. Martini. 1971a. Das Oligozän and Miozän der Alpen und der Karpaten: Ein Vergleich mit Hilfe Planktonischen Organismen. *Mitt. Bayer, Staatssamml. Paläont. Hist. Geol., Munich. 11:* 279-293.

_____, J. Senes, A. Papp, and F. F. Steininger. 1971b. Badenian. F. Steininger and L. Neveskaja, Eds. *Stratotypes of Mediterranean Neogene Stages 2:* 43-49.

_____, V. Fahlbusch, and O. Fejfar. 1972. Biostratigraphic correlation of some late Tertiary vertebrate faunas in central Europe. *N. Jb. Paläont. Abh. 140:* 129-145.

_____, J. Senes, and F. F. Steininger. 1975. Karpatian. F. Steininger and L. Neveskaja Eds. *Stratotypes of the Mediterranean Neogene Stages 2:* 93-100.

Cifelli, Richard. 1969. Radiation of Cenozoic planktonic Foraminifera. *System. Zool. 18*(2): 154-168.

Cita, M. B. 1972. Pliocene biostratigraphy and chronostratigraphy. *In* Initial Reports of the Deep-Sea Drilling Project, Vol. XIII, W. B. F. Ryan, K. T. Hsü, et al., Eds. Government Printing Office, Washington, D.C. p. 1343-1364.

_____. 1974. The Miocene-Pliocene boundary. History and definition *In* Late Neogene Epoch Boundaries, T. Saito, Ed. Micropaleontology Press.

_____, and W. H. Blow. 1969. The biostratigraphy of the Langhian, Serravallian and Tortonian stages in the type-sections of Italy. *Riv. Ital. Paleont. 75*(3): 549-603.

Clapham, W. B., Jr. 1970. Nature and paleogeography of Middle Permian floras of Oklahoma as inferred from their pollen record. *Jour. Geol. 78*(2): 153-171.

Closs, Darcy. 1967. Goniatiten mit Radula und Kieferapparat in der Itararé Formation von Uraguay. *Paläont. Zeitschr. 41:* 19-37.

Coates, A. G. 1973. Cretaceous Tethyan coral-rudist biogeography related to the evolution of the Atlantic Ocean. *In* Organisms and Continents through Time, N. F. Hughes, Ed. *Spec. Papers Palaeontology 12:* 169-174.

Cobban, W. A. 1951. Scaphitoid cephalopods of the Colorado Group. *U.S. Geol. Survey Prof. Paper 239:* 1-39.

_____. 1958. Late Cretaceous fossil zones of the Powder River Basin, Wyoming and Montana. *Wyo. Geol. Assoc. Guidebook, 13th Ann. Field Conf., 1958, Powder River Basin.* p. 114-119.

_____. 1969. The Late Cretaceous ammonites *Scaphites leei* Reeside and *Scaphites hippocrepis* (DeKay) in the western interior of the United States. *U.S. Geol. Survey Prof. Paper 619:* 1-28.

_____, and Glenn R. Scott. 1972. Stratigraphy and Ammonite Fauna of the Graneros Shale and Greenhorn Limestone near Pueblo, Colorado. *U.S. Geol. Survey Prof. Paper 645:* 1-108.

Cocks, C. R. M., and W. C. McKerrow. 1973. Brachiopod distributions and faunal provinces in the Silurian and Lower Devonian. *In* Organisms and Continents through Time, N. F. Hughes, Ed. *Spec. Papers Palaeont. 12 and Syst. Assoc. Pub. 9:* 291-304.

Colani, M. 1924. Nouvelle contributions a l'etude des Fusulinides de l'Extreme-Orient. *Indochina Serv. Géol. Mém. 11*(1): 1-191, 29 pls., 28 graphs.

Collin, A. E., and M. J. Dunbar. 1964. Physical oceanography in Arctic Canada. *Oceanogr. Mar. Biol., Ann. Rev. 2:* 45-75.

Collinson, Charles, C. B. Rexroad, and T. L. Thompson. 1971. Conodont Zonation of the North American Mississippian. *Geol. Soc. Amer. Mem. 127:* 353-394.

_____, M. J. Avcin, R. D. Norby, and G. K. Merrill. 1972. Pennsylvanian Conodont assemblages from LaSalle County, northern Illinois. *Illinois. State Geol. Survey Guidebook Ser. 10:* 1-37.

Conil, R., and M. Lys. 1964. Matériaux pour l'étude micropaléontologique du Dinantien de la Belgique et de la France (Avesnois). Part 1, Algues et Foraminifères and Part 2, Foraminifères. *Mém. Inst. Géol. Univ. Louvain 23:* 1-335, 42 pls.

Connell, J. H. 1961. The influence of interspecific competition and other factors on the distribution of the barnacle *Chthamalus stellatus. Ecology 42:* 710-723.

_____. 1973. Population ecology of reef-building corals. *In* Biology and Geology of Coral Reefs, E. A. Jones and R. Endean, Eds. Biology 1. 2: 205-245. Academic Press, New York.

Conybeare, W. D., and W. Phillips. 1822. Outlines of the Geology of England and Wales, Part I. Williams Phillips, London. 470 + ixi p., 2 pls., 1 map.

Cook, H. E., and M. E. Taylor. 1975. Early Paleozoic continental margin sedimentation, trilobite biofacies, and the thermocline, western United States. *Geology 3*(10): 559-562.

_____, P. N. McDaniel, E. W. Mountjoy, and L. C. Pray. 1972. Allochthonous carbonate debris flows at Devonian bank ('reef') margins, Alberta, Canada. *Bull. Canadian Petrol. Geol. 20*(3): 439-497.

Cooke, C. W. 1953. American Upper Cretaceous Echinoidea. *U.S. Geol. Survey Prof. Paper 254-A:* 1-44, 16 pls.

_____. 1959. Cenozoic Echinoids of Eastern United States. *U.S. Geol. Survey Prof. Paper 321:* 1-106, 43 pls.

Coomans, H. F. 1962. The marine mollusk fauna of the Virginian area as a basis for defining zoogeographical provinces: Beaufortia, v. 9, no. 98, p. 83-104.

Cooper, B. N. 1948. Status of Mississippian stratigraphy in the central and northern Appalachian region. *Jour. Geol. 56*(4): 255-263.

Cooper, G. A. 1956. Chazyan and related brachiopods. *Smithsonian Misc. Coll. 127*(1): 1-1024.

_____, and R. E. Grant. 1973. Dating and Correlating the Permian of the Glass Mountains in Texas. *In* The Permian and Triassic Systems and Their Mutual Boundary, A. Logan and L. V. Hills, Eds. *Canadian Soc. Petrol. Geol. Mem. 2:* 363-377.

Corbet, G. B. 1961. Origin of the British insular races of small mammals and of the 'Lusitanian' fauna. *Nature 191:* 1037-1040.

_____. 1963. An isolated population of the Bank Vole *Clethrionomys glareolus* with aberrant dental pattern. *Proc. Zool. Soc. London 140:* 316-319.

_____. 1964. Regional variation in the bank-vole *Clethrionomys glareolus* in the British Isles. *Proc. Zool. Soc. London 143:* 191-219.

Couper, R. A. 1958. British Mesozoic microspores and pollen grains. *Palaeontographica 103B:* 75-179.

Cowen, Richard, Richard Gertman, and Gail Wiggett. 1973. Camouflage patterns in *Nautilus*, and their implications for cephalopod paleobiology. *Lethaia 6*(2): 201-213.

Cowie, J. W. 1964. The Cambrian Period. *In* The Phanerozoic Time-Scale, A Symposium. *Geol. Soc. London, Quart. Jour. 120*(s): 255-258.

_____, A. W. A. Rushton, and C. J. Stubblefield. 1972. A Correlation of Cambrian rocks in the British Isles. *Geol. Soc. (G.B.), Spec. Report 2:* 1-42.

Cox, A., and G. B. Dalrymple. 1967. Statistical analysis of geomagnetic reversal data and the precision of potassium-argon dating. *Jour. Geophys. Res. 72*(10): 2603-2621.

Cracraft, J. 1973. Continental drift, paleoclimatology, and the evolution and biogeography of birds. *Jour. Zool. London. 169:* 455-545.

Crick, G. C. 1898. On the muscular attachment of the animal to its shell in some fossil Cephalopoda (Ammonoidea). *Linnean Soc. Trans. 7:* 71-113.

Crisp, D. J. 1958. The spread of *Elminius modestus* Darwin in Northwest Europe. *Jour. Mar. Biol. Assoc. U.K. 37:* 483-520.

_____. 1974. Factors influencing the settlement of marine invertebrate larvae. *In* Chemoreption in Marine Organisms, P. T. Grant and A. N. Mackie, Eds. Academic Press, New York. p. 177-265.

Croneis, Cary. 1941. Micropaleontology, past and future. *Amer. Assoc. Petrol. Geol. 25*(7): 1308-1355.

Crouch, Robert W. 1955. A practical application of paleoecology in exploration. *Gulf Coast Assoc. Geol. Soc., Trans. 5:* 89-96.

Cserna, Zoltan de. 1972. Essay Review (of O.H. Schindewolf's Stratigraphie und Stratotypus). *Am. Jour. Sci. 272:* 189-194.

Culver, H. E. 1922. Note on the occurrence of Fusulinas in the Pennsylvanian rocks of Illinois. *Illinois State Acad. Sci., Trans. 15:* 421-425 [1923].

Curray, Joseph R. 1960. Sediments and history of Holocene transgression, continental shelf, northwest Gulf of Mexico. *In* Recent Sediments, Northwest Gulf of Mexico, F. P. Shepard, F. B. Phleger, and Tj. H. van Andel, Eds. *Amer. Assoc. Petrol. Geol.:* 221-266.

Curtis, Doris M. 1970. Miocene deltaic sedimentation, Louisiana Gulf Coast. *In* Deltaic Sedimentation, Modern and Ancient, J. P. Morgan, Ed. *Soc. Econ. Paleont., Mineral., Spec. Pub. 15:* 293-308.

Cuvier, G. L. C. F. C. 1817. Essay on the Theory of the Earth, with Mineralogical Notes, and an Account of Cuvier's Geological Discoveries by Professor Jameson. 3rd Ed. William Blackwood, Edinburgh. 348 + xxiv p., 5 pls. [translated from the French by R. Kerr].

_____. 1818. Essay on the Theory of the Earth, with Mineralogical Notes, and an Account of Cuvier's Geological Discourses, by Professor Jameson. To which are added observations on the geology of North America by S. L. Mitchill. New York. 431 p., 8 pls.

_____, and A. Brongniart. 1808. Essai sur la géographie minéralogique des environs de Paris. *Ann. Mus. Hist. Nat. Paris 11:* 293-326.

_____, and A. Brongniart. 1822. Description géologiques des couches des environs de Paris, parmi lesquelles se trouvent les gypses a ossemens. *In* Recherches sur les Ossemens Fossiles, G. Cuvier, Ed. G. Dufour and E. d'Ocagne, Paris 2. p. 239-648, 18 pls.

Dall, W. H. 1890. Contributions to the tertiary fauna of Florida. *Trans. Wagner Free Inst. Sci. 3* (pt. 1-6): 1-1654.

Dalrymple, G. B. 1972. Potassium-argon dating of geomagnetic reversals and North American glaciations. *In* Calibration of Hominoid Evolution, W. W. Bishop and J. A. Miller, Eds. Edinburgh. p. 107-134.

Darwin, C. 1859. The Origin of Species. Murray, London. 490 p.

Davis, John C. 1973. Statistics and Data Analysis in Geology. John Wiley, New York. 550 p.

Davis, R. A., W. M. Furnish, and B. F. Glenister. 1969. Mature modification and dimorphism in late Paleozoic ammonoids. *In* Sexual Dimorphism in Fossil Metazoa and Taxonomic Implications, G. E. G. Westermann, Ed. *Internat. Union Geol. Sci., ser. A. 1:* 101-110.

Dawson, G. M. 1878. Report on explorations in British Columbia, chiefly in the basins of the Blackwater, Salmon, and Nechacco rivers, and on François Lake. *Canada Geol. Survey, Rept. 1876-1877.* 17-94, pl. 3.

Dayton, P. K. 1971. Competition, disturbance and community organization: the provision and subsequent utilization of space in a rocky intertidal community. *Ecol. Monogr. 41:* 351-389.

Dean, W. T., D. T. Donovan, and M. K. Howarth. 1961. The Triassic ammonite zones and sub zones of the northwest European province. *British Mus. (Nat. Hist.) Bull. 4:* 438-505.

Dear, J. F. 1971. Strophomenoid Brachiopods from the higher Permian faunas of the Back Creek Group in the Bowen Basin. *Geol. Survey Qld., Pub. 347, Palaeont. Paper 21:* 1-39, 7 pls.

Deboo, P. B. 1965. Biostratigraphic correlation of the type Shubuta Member of the Yazoo Clay and Red Bluff Clay with their equivalents in southwestern Alabama. *Alabama Geol. Survey Bull. 80:* 1-84.

De La Beche, H. T. 1839. Report on the Geology of Cornwall, Devon and West Somerset. Longman, and the Geol. Survey England and Wales, London. 648 + xxviii p., 13 pls. [index by Reid, C. 1903, Mem. Geol. Survey England and Wales].

Denton, E. J., and J. B. Gilpin-Brown. 1966. On the buoyancy of the pearly *Nautilus. Mar. Biol. Assoc. U.K. Jour. 36:* 723-759.

Depéret, C. 1893. Sur la classification et le parallélisme du système Miocene. *Soc. Géol. France Bull. 3* (21): 170-266.

Deprat, J. 1912. Etude des Fusulinides de Chine et d'Indochine et classification des calcaires a Fusulines. *Indochina Serv. Géol. Mém. 1* (pt. 3): 1-76, pls. 1-9, text-figs. 1-30.

_____. 1913. Etude des Fusulinides de Chine et d'Indochine et classification des calcaires carboniferiens et permiens du Tonkin, du Laos et du Nord-Annam. *Indochina Serv. Géol. Mém. 2* (1): 1-74, pls. 1-10, text-figs. 1-25.

_____. 1914. Etude des Fusulinides du Japon, de Chine et d'Indochine et classification des calcaires à Fusulines (III Mémoire), Etude comparative des Fusulinides a Akasaka (Japon) et des Fusulinides de Chine et d'Indochine. *Indochina Serv. Géol. Mém. 3* (1): 1-45, pls. 1-8.

_____. 1915. Etude des Fusulinides de Chine et d'Indochine et classification des calcaires à Fusulines (IV Mémoire), Les Fusulinides des calcaires carboniferiens et permiens du Tonkin, du Laos et du Nord-Annam. *Indochina Serv. Géol. Mém. 4* (1): 1-30, pls. 1-3, text-figs. 1-11.

Deshayes, P. G. 1830. Tableau comparatif des espèces de coquilles vivantes avec les espèces de coquilles fossiles des terrains tertiaires de l'Europe, et des espèces de fossiles de ces terrains entr'eux. *Soc. Géol. Fr. Bull. 1:* 185-187.

Dewalque, G. 1882. Sur L'unification de la nomenclature géologique résumé et conclusions. *Inter. Geol. Cong. 2 (Bologne):* 549-559.

Dickson, C. N., and F. Evans. 1956. The *Petula*'s meteorological logbook. *Marine Observer Oct. 1956:* 215-218.

Diener, Carl. 1912. Lebensweise und Verbreitung der Ammoniten. *Neues Jahrb. Mineral., Geol., Paläont., Jahrg. 1912 2:* 67-89.

Dietl, G. 1973. Middle Jurassic (Dogger) heteromorph ammonites. *In* Atlas of Palaeobiogeography, A. Hallam, Ed. Elsevier, Amsterdam. p. 283-285.

Dixon, W. J. 1971. BMD biomedical computer programs. *California Univ. Pubs. Automatic Computation 2:* 1-600.

Dollo, L. 1910. La paléontologie éthologique. *Soc. Belge Géol. Paléont. Hydrol. 23:* 377-421.

Donovan, D. T. 1954. Upper Cretaceous fossils from Traill and Geographical Society ϕer, East Greenland. *Meddelelser om Grϕnland 72* (6): 1-33.

_____. 1958. The Lower Jurassic ammonite fauna from the fossil bed at Langeneckgrat, near Thun (median Pre-Alps). *Schweizerische Palaeont. 74:* 1-58.

Douglass, R. C. 1970. Morphologic studies of fusulinids from the Lower Permian of West Pakistan. *U. S. Geol. Survey Prof. Paper 643-G:* 1-11, 7 pls., 6 text-figs.

Douvillé, Henri. 1931. Les ammonites de Salinas. *In* Contribution à la Géologie de l'Angola. *Lisboa Univ. Mus. e Lab. Mineral. e Geol. Bol. 1* (1): 17-46.
Downey, M. E. 1973. Starfishes from the Caribbean and the Gulf of Mexico. *Smithsonian Contr. Zool. 126:* 1-158.
Doyle, J. 1945. Developmental lines in pollination mechanisms in the Coniferales. *Scientific Proceedings of the Royal Dublin Society 24:* 43-62.
Doyle, J. A. 1969a. Cretaceous angiosperm pollen of the Atlantic Coastal Plain and its evolutionary significance. *Jour. Arnold Arboretum 50:* 1-35.
——. 1969b. Angiosperm pollen evolution and biostratigraphy of the basal Cretaceous formations of Maryland, Delaware, and New Jersey. *Geol. Soc. Amer. Abstr. with Progr., 1969, part 7:* 51 [abstract].
——. 1970. Evolutionary and stratigraphic studies on Cretaceous angiosperm pollen. Unpublished Ph.D. thesis, Harvard University.
——. 1973. Fossil evidence on early evolution of the monocotyledons. *Quart. Rev. Biol. 48* (3): 399-413.
——. 1974. Angiosperm pollen and correlation of Potomac megafossil localities. *American Association of Stratigraphic Palynologists Proceedings, Geoscience and Man 9:* 73 [abstract].
——, and L. J. Hickey. 1972. Coordinated evolution in Potomac Group angiosperm pollen and leaves. *Amer. Jour. Botany 59* (6): 660 [abstract].
——, and L. J. Hickey. 1976. Pollen and leaves from the mid-Cretaceous Potomac Group and their bearing on early angiosperm evolution. *In* The Origin and Early Evolution of the Angiosperms, C. B. Beck, Ed. Columbia University Press, New York. 139-206 p.
Doyle, R. W. 1972. Genetic variation in *Ophiumusium lymani* (Echinodermata) populations in the deep sea. *Deep-sea Res. 19:* 661-664.
Druce, Edric C. 1969. Upper Paleozoic Conodonts from the Bonaparts Gulf Basin North Western Australia. *Bull. Bur. Miner. Resour. Geol. Geophys. Australia 98:* 1-242.
——. 1970. Upper Paleozoic Conodont Distribution. *Abstr. Proc. 4th Ann. Meet. N. Cent. Sect., Geol. Soc. Amer.:* 386.
——. 1973. Upper Paleozoic and Triassic conodont distribution and the recognition of biofacies. *Geol. Soc. Amer. Special Paper 141:* 191-237.
——, F. H. T. Rhodes, and R. L. Austin. 1971. Statistical analysis of British Carboniferous conodont faunas. *Geol. Soc. London, Quart. Jour. 128:* 53-70.
——, F. H. T. Rhodes, and R. L. Austin. 1974. Recognition, evolution and taxonomy of Lower Carboniferous conodont assemblages. *Jour. Paleont. 48:* 387-407.
Dshabarow, G. N. 1964. Werchnemelowyje Morskije éshi zentralnogo Kopet-Daga: 1-71, 7 figs., 20 pls.
Dubar, J. R., and J. R. Solliday. 1963. Stratigraphy of the Neogene deposits, lower Neuse Estuary, North Carolina. *Southeastern Geol. 4* (4): 213-233.
Dumont, A. 1850. Rapport sur la carte géologique de royaume. *Acad. Roy. Sci. Lettres, Beaux-Arts, Belgique Bull. 16:* 351-373.
Dunbar, C. O. 1945. The geologic and biologic significance of the evolution of the Fusulinidae. *New York Acad. Sci. Trans., ser. 2, 7:* 57-60.
——. 1963. Trends of evolution in American fusulines. *In* Evolutionary Trends in Foraminifera, G. H. R. von Koenigswald et al., Eds. Elsevier, Amsterdam. p. 25-44, 12 text-figs.

_____, and G. E. Condra. 1927. The Fusulinidae of the Pennsylvanian system in Nebraska. *2nd ser., Nebraska Geol. Survey Bull. 2:* 1-135, 15 pls., 13 text-figs.

_____, and L. G. Henbest. 1942. Pennsylvanian Fusulinidae of Illinois with section on stratigraphy by J. M. Weller, L. G. Henbest, and C. O. Dunbar. *Illinois State Geol. Survey Bull. 67:* 1-218, 23 pls., 11 text-figs.

_____, and J. W. Skinner. 1937. Permian Fusulinidae of Texas. *Texas Univ. Bull. 3701* (pt. 2): 517-825, pls. 42-81, text-figs. 89-97.

_____ et al. 1960. Correlation of the Permian formations of North America. *Geol. Soc. Amer. Bull. 71* (12): 1763-1805.

Dunbar, M. J. 1968. Ecological Development in Polar Regions. Prentice-Hall, Englewood Cliffs, N.J. 119 p.

Durham, C. O., Jr. 1957. The Austin Group in central Texas. Ph.D. dissertation, Columbia University.

_____. 1961. Austin Group in Central Texas. *In* Geology of the Atlantic and Gulf Coastal Province of North America, G. E. Murray, Ed. Harper & Row, New York. 351 p.

Durham, J. W. 1950. Cenozoic marine climates of the Pacific coast. *Geol. Soc. Amer. Bull. 61:* 1243-1264.

_____. 1967. Presidential address: The incompleteness of our knowledge of the fossil record. *J. Paleont. 41:* 559-566.

_____, H. B. Fell, A. G. Fischer, P. M. Kier, R. V. Melville, D. L. Pawson, and C. D. Wagner. 1966. Echinoids. *In* Treatise on Invertebrate Paleontology, R. C. Moore, Ed. Part U, Echinodermata. 3(1) and 3(2): U211-U640, figs. 152-517.

Dutro, J. T., Jr., and W. J. Sando. 1963. New Mississippian formations and faunal zones in Chesterfield Range, Portneuf Quadrangle, southeast Idaho. *Amer. Assoc. Petrol. Geol. Bull. 47* (11): 1963-1986.

Dymond, J. T. 1966. Potassium-argon geochronology of deep-sea sediments. *Science 152:* 1239-1241.

Eames, F. E. 1970. Some thoughts on the Neogene/Paleogene boundary. *Palaeogeography, Palaeoclimatol., Palaeoecol. 8:* 37-48.

Echols, Dorothy J., and Doris M. Curtis. 1973. Paleontologic evidence for mid-Miocene refrigeration from subsurface marine shales, Louisiana Gulf Coast. *Gulf Coast Assoc. Geol. Soc., Trans. 23:* 422-426.

Edinburgh Oceanographic Laboratory. 1973. Continuous plankton records: A plankton atlas of the North Sea. *Bull. Mar. Ecol. 7:* 1-174.

Ehrlich, P. R., and P. H. Raven. 1969. Differentiation of populations. *Science 165:* 1228-1232.

Eicher, D. L. 1968. Geologic Time. *In* Foundations of Earth Science Series, A. L. McAlester, Ed. J. Wiley and Sons. 150 p., 63 figs.

_____. 1969. Cenomanian and Turonian planktonic Foraminifera from the Western Interior of the United States. Proc. 1st Internat. Conf. Planktonic Microfossils. 163-174.

Einarsson, H. 1948. Echinoderms. *Zoology of Iceland 4* (pt. 70): 1-67.

Ekman, S. 1953. Zoogeography of the Sea. Sidgwick & Jackson, London. 417 + xiv p.

Eldredge, N. 1971. The allopatric model of phylogeny in Paleozoic invertebrates. *Evolution 25:* 156-167.

———. 1972. Systematics and evolution of *Phacops rana* (Green, 1832) and *Phacops iowensis* Delo, 1935 (Trilobita) in the Middle Devonian of North America. *Bull. Amer. Mus. Nat. Hist. 47:* 45-114.

———. 1974. Stability, diversity, and speciation in Paleozoic epeiric seas. *Jour. Paleont. 48:* 540-548.

———, and S. J. Gould. 1972. Punctuated equilibria: an alternative to phyletic gradualism. *In* Models in Paleobiology, T. J. M. Schopf, Ed. Freeman, Cooper, San Francisco. p. 82-115.

———, and S. J. Gould. 1974. Evolutionary Models in Biostratigraphic Strategies (this volume).

———, and S. J. Gould. 1975. Morphological transformation, the fossil record, and the mechanisms of evolution: a debate. Part II, the reply. *In* Evolutionary Biology, T. Dobzhansky et al., Eds. Plenum, New York. Vol. 7, p. 303-308.

———, and I. Tattersall. 1975. Evolutionary models, phylogenetic reconstruction, and another look at hominid phylogeny. *In* Approaches to Primate Paleobiology, F. S. Szalay, Ed. *Contrib. Primat. 5:* 218-242.

Elias, M. K. 1960. Marine Carboniferous of North America and Europe. Congres Av. Etudes Stratigraphie et Géologie Carbonifère, 4th, Heerlen, 1958, *Compte Rendu, Fusulinids 1:* 151-161, 4 text-figs.

Ellerman, J. R. 1951. Order Rodentia. *In* Checklist of Palaearctic and Indian Mammals, 1758 to 1946, by J. R. Ellerman and T. C. S. Morrison-Scott. Brit. Mus. (Nat. Hist.), London. p. 456-712.

Elles, G. L. 1939a. The stratigraphy and faunal succession in the Ordovician rocks of the Builth-Llandrindod inlier, Radnorshire. *Geol. Soc. London Quart. Jour. 95:* 383-445.

———. 1939b. Factors controlling graptolite successions and assemblages. *Geol. Mag. 76:* 181-188.

———, and E. M. R. Wood. 1901-1918. A monograph of British Graptolites. Published in 11 parts by the Palaeontographical Society: (1) 1901, 1-54, pls. 1-4; (2) 1902, i-xxviii, 55-102, pls. 5-13; (3) 1903, xxix-lii, 103-134, pls. 14-19; (4) 1904, liii-lxxii, 135-180, pls. 20-25; (5) 1906, lxxiii-xcvi, 181-216, pls. 26-27; (6) 1907, xcvii-cxx, 217-272, pls. 28-31; (7) 1908, cxxxi-cxlviii, 272-358, pls. 32-35; (8) 1911, 359-414, pls. 36-41; (9) 1913, 415-486, pls. 42-49; (10) 1914, 487-526, pls. 50-52; (11) 1918, cxlix-clxxi, 527-539.

Ellisor, Alva C. 1933. Jackson group of formations in Texas, with notes on Frio and Vicksburg. *Amer. Assoc. Petrol. Geol. Bull. 17*(10): 1293-1350.

Elton, C. S. 1942. Voles, Mice and Lemmings. Clarendon Press, Oxford. 496 p.

———. 1958. The ecology of invasions by animals and plants. Methuen, London. 181 p.

———. 1966. The Pattern of Animal Communities. Methuen, London. 432 p.

Eltringham, S. K. 1971. Life in Mud and Sand. English Universities Press, London. 218 p.

Enay, R. 1973. Upper Jurassic (Tithonian) ammonities. *In* Atlas of Palaeobiogeography, A. Hallam, Ed. Elsevier, Amsterdam. p. 297-307.

Endler, J. A. 1973. Gene flow and population differentiation. *Science 179:* 243-250.

Engesser, B. 1972. Die Obermiozäne Säugetierfauna von Anwil (Baselland). Tätigkeitsber. *Naturforsch. Ges. Baselland 28:* 35-363, 6 pls., 134 text-figs., 6 tables, 38 diagrams.

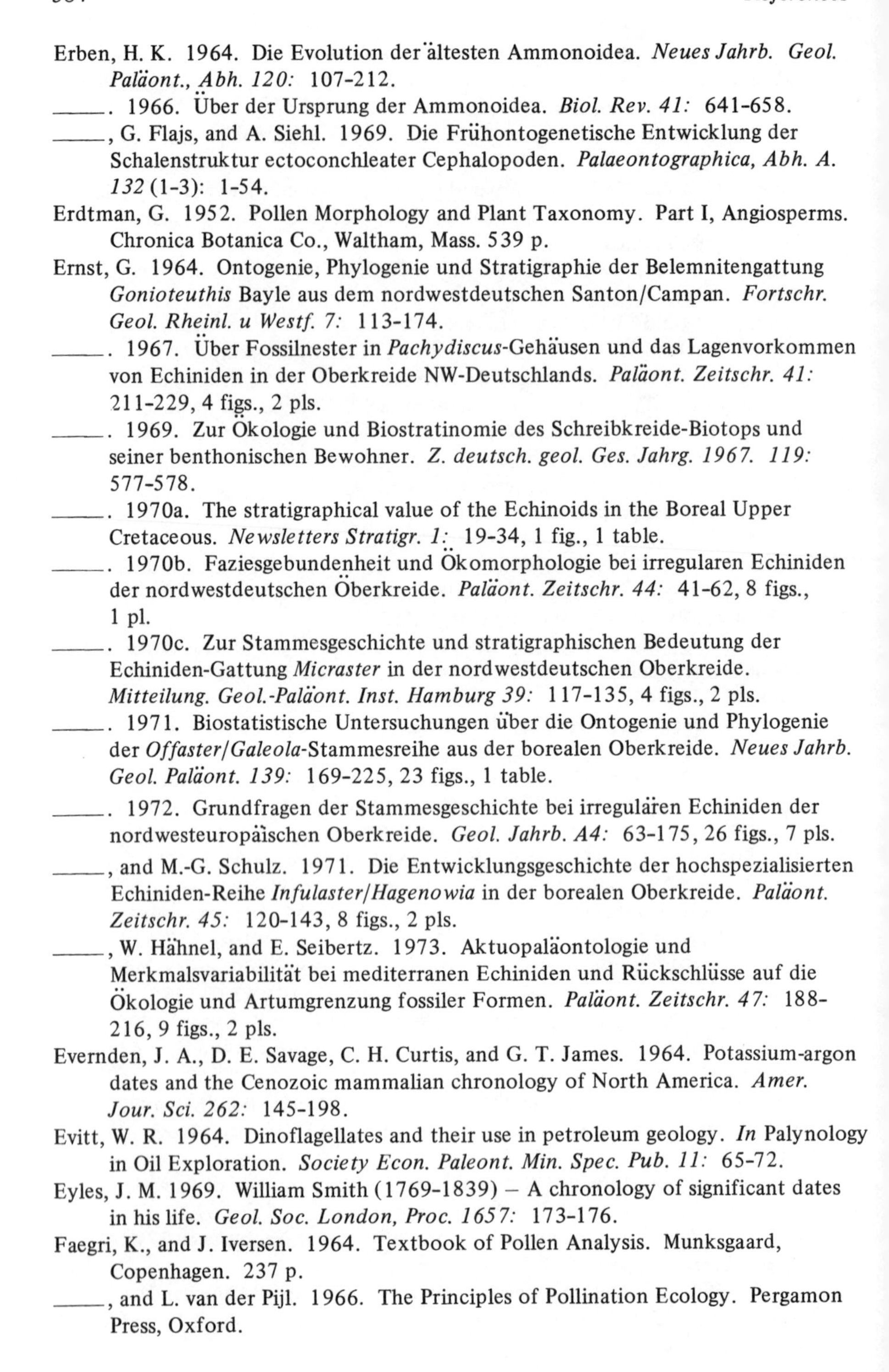

Erben, H. K. 1964. Die Evolution der ältesten Ammonoidea. *Neues Jahrb. Geol. Paläont., Abh. 120:* 107-212.

_____. 1966. Über der Ursprung der Ammonoidea. *Biol. Rev. 41:* 641-658.

_____, G. Flajs, and A. Siehl. 1969. Die Frühontogenetische Entwicklung der Schalenstruktur ectoconchleater Cephalopoden. *Palaeontographica, Abh. A. 132* (1-3): 1-54.

Erdtman, G. 1952. Pollen Morphology and Plant Taxonomy. Part I, Angiosperms. Chronica Botanica Co., Waltham, Mass. 539 p.

Ernst, G. 1964. Ontogenie, Phylogenie und Stratigraphie der Belemnitengattung *Gonioteuthis* Bayle aus dem nordwestdeutschen Santon/Campan. *Fortschr. Geol. Rheinl. u Westf. 7:* 113-174.

_____. 1967. Über Fossilnester in *Pachydiscus*-Gehäusen und das Lagenvorkommen von Echiniden in der Oberkreide NW-Deutschlands. *Paläont. Zeitschr. 41:* 211-229, 4 figs., 2 pls.

_____. 1969. Zur Ökologie und Biostratinomie des Schreibkreide-Biotops und seiner benthonischen Bewohner. *Z. deutsch. geol. Ges. Jahrg. 1967. 119:* 577-578.

_____. 1970a. The stratigraphical value of the Echinoids in the Boreal Upper Cretaceous. *Newsletters Stratigr. 1:* 19-34, 1 fig., 1 table.

_____. 1970b. Faziesgebundenheit und Ökomorphologie bei irregularen Echiniden der nordwestdeutschen Oberkreide. *Paläont. Zeitschr. 44:* 41-62, 8 figs., 1 pl.

_____. 1970c. Zur Stammesgeschichte und stratigraphischen Bedeutung der Echiniden-Gattung *Micraster* in der nordwestdeutschen Oberkreide. *Mitteilung. Geol.-Paläont. Inst. Hamburg 39:* 117-135, 4 figs., 2 pls.

_____. 1971. Biostatistische Untersuchungen über die Ontogenie und Phylogenie der *Offaster/Galeola*-Stammesreihe aus der borealen Oberkreide. *Neues Jahrb. Geol. Paläont. 139:* 169-225, 23 figs., 1 table.

_____. 1972. Grundfragen der Stammesgeschichte bei irregulären Echiniden der nordwesteuropäischen Oberkreide. *Geol. Jahrb. A4:* 63-175, 26 figs., 7 pls.

_____, and M.-G. Schulz. 1971. Die Entwicklungsgeschichte der hochspezialisierten Echiniden-Reihe *Infulaster/Hagenowia* in der borealen Oberkreide. *Paläont. Zeitschr. 45:* 120-143, 8 figs., 2 pls.

_____, W. Hähnel, and E. Seibertz. 1973. Aktuopaläontologie und Merkmalsvariabilität bei mediterranen Echiniden und Rückschlüsse auf die Ökologie und Artumgrenzung fossiler Formen. *Paläont. Zeitschr. 47:* 188-216, 9 figs., 2 pls.

Evernden, J. A., D. E. Savage, C. H. Curtis, and G. T. James. 1964. Potassium-argon dates and the Cenozoic mammalian chronology of North America. *Amer. Jour. Sci. 262:* 145-198.

Evitt, W. R. 1964. Dinoflagellates and their use in petroleum geology. *In* Palynology in Oil Exploration. *Society Econ. Paleont. Min. Spec. Pub. 11:* 65-72.

Eyles, J. M. 1969. William Smith (1769-1839) – A chronology of significant dates in his life. *Geol. Soc. London, Proc. 1657:* 173-176.

Faegri, K., and J. Iversen. 1964. Textbook of Pollen Analysis. Munksgaard, Copenhagen. 237 p.

_____, and L. van der Pijl. 1966. The Principles of Pollination Ecology. Pergamon Press, Oxford.

Fahlbusch, V., von. 1970. Populationsverschiebungen bei tertiären Nagetieren, eine Studie an oligozänen und miozänen Eomyiden Europas. *Abh. Bayer. Akad. Wiss., Math.-Naturwiss. Kl., N.F. 145:* 1-136, 42 text-figs., 26 tables, 11 pls.

———. 1973. Die stammesgeschichtlichen Beziehungen zwischen den Eomyiden (Mammalia, Rodentia) Nordamerikas und Europas. *Mitt. Bayer. Staatssamml. Paläont. Hist. Geol. 13:* 141-175.

Fechter, H. 1970. Die Seeigel. *In* Crzimeks Tierleben 3, Weichtiere und Stachelhäuter. Kindler-Verlag, Zürich. p. 326-356, 8 pls.

Fell, H. B. 1967a. Cretaceous and Tertiary surface currents of the oceans. *Oceanogr. Mar. Biol., Ann. Rev. 5:* 317-341.

———. 1967b. Echinoderm Ontogeny. *In* Treatise on Invertebrate Paleontology, Pt. S, Echinodermata 1, R. C. Moore, Ed. Geol. Soc. Amer. and University of Kansas Press. p. 60-85.

Fischer de Waldheim, G. 1829. Sur les Cephalopodes fossils de Moscou et de ses environs, en montrant des objets en nature. *Imp. Soc. Naturalistes Moscou Bull. 1:* 300-331.

———. 1837. Oryctographic du Gouvernment de Moscou. p. 126-127, pl. 13, figs. 1-11.

Fisher, R. V., and J. M. Rensberger. 1972. Physical stratigraphy of the John Day Formation, central Oregon. *Univ. Calif. Publ. Geol. Sci. 101:* 1-33.

Fisher, W. L., P. U. Rodda, and J. W. Dietrich. 1964. Evolution of *Athleta petrosa* stock (Eocene Gastropoda) of Texas. *Texas Univ. Pub. 6413:* 1-117, 7 pls.

Fitton, W. H. 1827. Observations on some of the strata between the Chalk and the Oxford Oolite, in the Southeast of England. *Geol. Soc. London, Trans. Ser. 2. 4:* 104-400, pls. 7-23 [re-issued as a separate publication 1836; + viii p. in 1836].

Florin, R. 1951. Evolution in cordaites and conifers. *Acta Horti Bergiani 15* (11): 285-388.

Flower, R. H. 1964. The nautiloid Order Ellesmeroceratida (Cephalopoda). *New Mexico Bur. Mines and Min. Res. Mem. 12:* 1-234.

———. 1968. The first great expansion of the Actinoceroids and some additional Whiterock cephalopods. *New Mexico Inst. Mining & Tech. Mem. 19, Parts I and I:* 1-55.

Fontaine, W. M. 1889. The Potomac or Younger Mesozoic flora. *U.S. Geol. Surv. Mono. 15:* 1-375.

Foster, J. H., and N. D. Opdyke. 1970. Upper Miocene to Recent Magnetic Stratigraphy in Deep-Sea Sediments. *Jour. Geophys. Res. 75* (23): 4465-4473.

Francis, E. H., and A. W. Woodland. 1964. The Carboniferous period. *In* The Phanerozoic Time-Scale, W. B. Harland, A. G. Smith, and B. Wilcock, Eds. *Geol. Soc. London Quart. Jour. 120s:* 221-232.

Frazer, P. 1886. The work of the International Congress of Geologists and of its committees. Amer. Comm. Intern. Geol. Congr. 1-109, 1 pl.

Frebold, H. 1924. Ammonitenzonen und Sedimentationszyklen in ihrer Beziehung zueinander. *Centrb. für Mineralogie, etc. (for 1924):* 313-320.

———. 1957. The Jurassic Fernie Group in the Canadian Rocky Mountains and Foothills. *Canada Geol. Surv. Mem. 287:* 1-197, 44 pls.

Frentzen, Kurt. 1936. Fossiler Mageninhalt aus dem Lias Delta (Amaltheen-Schichten) von Reichenbach. *Beitr. Naturk. Forschung. 1:* 293-303.

Fretter, Vera, and Alastair Graham. 1962. British Prosobranch Molluscs. Royal Society, London. 755 p.

Freund, Raphael, and Menahem Raab. 1969. Lower Turonian ammonites from Israel. *Palaeont. Assoc. London Spec. Paper 4:* 1-83.

Fritz, W. H. 1972. Lower Cambrian trilobites from the Sekwi Formation type section, Mackenzie Mountains, northwestern Canada. *Geol. Surv. Canada Bull. 212:* 1-90, 20 pls.

Fujimoto, H., and H. Igo. 1958. The fusulinid zones in the Japanese Carboniferous. *Tokyo Kyoiku Daigaku, Sci. Repts., Sec. C. 6* (53): 127-146.

Gabb, Wm. M. 1874-1881. Description of a collection of fossils made by Doctor Antonio Raimundi, in Peru. *Philadelphia Acad. Nat. Sci. Jour., 2nd Ser., 8:* 3, 263-336.

Gaines, M. S., and C. J. Krebs. 1971. Genetic changes in fluctuating Vole populations. *Evolution 25:* 702-723.

Ganelina, R. A., L. P. Grozdilova, N. S. Lebedeva, and M. I. Sosnina. 1972. Taxonomic significance of the test uncoiling in Paleozoic Foraminifera [in Russian]. *Akad. Nauk. S.S.S.R., Voprosy Mikropal. 15:* 30-39, 4 pls.

Gardiner, J. S. 1904. Notes and observations on the distribution of the larvae of marine animals. *Ann. Mag. Nat. Hist., Ser. 7, 14:* 403-410.

Gardner, Julia. 1931. Relation of certain foreign faunas to Midway fauna of Texas. *Bull. Amer. Assoc. Petrol. Geol. 15* (2): 149-160.

Garrett, J. B., Jr. 1938. The Hackberry assemblage, an interesting foraminiferal fauna of post-Vicksburg age from deep wells in the Gulf Coast. *Jour. Paleont. 12:* 309-317.

_____, and A. D. Ellis, Jr. 1937. Distinctive Foraminifera of the genus *Marginulina* from Middle Tertiary beds of the Gulf Coast. *Jour. Paleont. 11* (8): 629-633.

Gartner, S., Jr. 1969. Correlation of Neogene planktonic foraminifera and calcareous nannofossil zones. *Trans. Gulf Coast Assoc. Geol. Soc., Houston 19:* 585-599.

_____. 1973. Absolute chronology of the late Neogene nannofossil succession in the equatorial Pacific. *Geol. Soc. Amer. Bull. 84* (6): 2021-2034.

Geikie, A. 1903. Text-book of Geology, 4th edition. Macmillan, London. 1472 p.

George, T. N. et al. 1967. Report of the stratigraphical code sub-committee. *Geol. Soc. London, Proc. 1638:* 75-87.

_____, Chairman, and Members of the Stratigraphy Committee of the Geological Society of London. 1969. Recommendations on stratigraphical usage. *Proc. Geol. Soc. London 1656:* 139-166.

Germeraad, J. H., C. A. Hopping, and J. Muller. 1968. Palynology of Tertiary sediments from tropical areas. *Review of Palaeobotany and Palynology 6:* 189-348.

Gigot, Patrick, and Pierre Mein. 1973. Découvertes de mammifères aquitaniens dans la molasse burdigalienne du Golfe de Digne. *C. R. Acad. Sci. Paris 276:* 3293-3294.

Gill, J. R., and W. A. Cobban. 1966. The Red Bird section of the Upper Cretaceous Pierre Shale in Wyoming, with a section on A new echinoid from the Cretaceous Pierre Shale of eastern Wyoming, by P. M. Kier. *U.S. Geol. Surv. Prof. Paper 393-A:* 1-73.

Gilluly, J. 1949. Distribution of mountain building in geologic time. *Bull. Geol. Soc. Amer. 60:* 561-590.
Ginkel, A. C. van. 1965. Carboniferous fusulinids from the Cantabrian Mountains (Spain). *Leidse Geol. Meded. 34:* 1-225, 53 pls., 13 text-figs., 63 tables, 9 maps.
_____. 1973. Carboniferous fusulinids of the Sama Formation (Asturia, Spain) (1. Hemifusulina). *Leidse Geol. Meded. 49:* 85-123, 10 pls., 12 text-figs.
Ginsburg, Leonard. 1967. Une faune de Mammifères dans l'Helvétien marin de Sos (Lot-et-Garonne) et de Rimbez (Landes). *Soc. Géol. France Bull. 1* (7): 5-18.
Ginsburg, R. N., and H. S. Lowenstam. 1958. The influence of marine bottom communities on the depositional environment of sediments. *Jour. Geol. 66:* 310-318.
Girty, G. H. 1904. *Triticites*, a new genus of Carboniferous Foraminifera. *Amer. Jour. Sci., 4th Ser. 17:* 234-240.
_____. 1923. Observations on the faunas of the Greenbrier Limestone and adjacent rocks. *West Virginia Geol. Survey, Tucker County Rept.:* 450-488.
Glaser, Gerald C., and Andrew C. Jurasin. 1971. Paleoecology, stratigraphy, production – getting it all together in offshore Louisiana. *Gulf Coast Assoc. Geol. Soc., Trans. 17:* 428-480.
Glintzboeckel, C., and J. Rabaté. 1964. Microfaunes et microfaciès du Permo-Carbonifère du Sud Tunisien. E. J. Brill, Leiden. 45 p., 108 pls., 6 text-figs.
Gobbett, D. J. 1967. Paleozoogeography of the Verbeekinidae (Permian Foraminifera). *Systematics Assoc. Pub. 7:* 77-91, 3 text-figs.
Göes, A. 1883. Om *Fusulina cylindrica* Fischer fran Spitzbergen. *Svenska Vetenskapsakademien, Stockholm, Ofversigt af . . . Forhandlinger 40:* 29-35.
Gooch, J. L., and J. M. Schopf. 1973. Genetic variability in the deep sea: Relation to environmental variability. *Evolution 26* (4): 545-552.
Gordon, M., Jr. 1970. Carboniferous ammonoid zones of the south-central and western United States. 6th International Congress on Carboniferous Stratigraphy and Geology, Sheffield, 1967, *Compte Rendu 2:* 817-826.
Gorsky, I. I. 1939. The Atlas of the Leading Forms of the Fossil Faunas of the U.S.S.R., Vol. 5, The Middle and Upper Carboniferous. Central Geol. Prospecting Inst., Leningrad.
Gould, S. J. 1972. Allometric fallacies and the evolution of *Gryphaea:* a new interpretation based on White's criterion of geometric similarity. *In* Evolutionary Biology, Th. Dobzhansky, et al., Eds. Appleton-Century-Crofts, New York. 6: p. 91-118.
Gower, J. C. 1966. Some distance properties of latent root and vector methods used in multivariate analysis. *Biometrika 53:* 325-338.
_____. 1967. A comparison of some methods of cluster analysis. *Biometrics 23* (4): 623-637.
Grabau, A. W. A. 1913. Principles of Stratigraphy. New York. 1185 p.
Grant, P. R. 1969. Experimental studies of competitive interaction in a two-species system. I. *Microtus* and *Clethrionomys* species in enclosures. *Can. Jour. Zool. 47:* 1059-1082.
_____. 1970. Experimental studies of competitive interaction in a two-species system. II. The behaviour of *Microtus, Peromyscus* and *Clethrionomys* species. *Animal Behavior 18:* 411-426.

Grant, R. E. 1962. Trilobite distribution, Upper Franconia Formation (Upper Cambrian), southeastern Minnesota. *Jour. Paleont. 36* (5): 965-998, pl. 139, 10 text-figs.

_____. 1965. Faunas and stratigraphy of the Snowy Range Formation (Upper Cambrian) in southwestern Montana and northwestern Wyoming. *Geol. Soc. Amer. Mem. 96:* 1-171, 15 pls.

_____, and G. A. Cooper. 1973. Brachiopods and Permian correlations. *In* The Permian and Triassic Systems and Their Mutual Boundary, A. Logan and L. V. Hills, Eds. *Canad. Petrol. Geol. Soc. Mem. 2:* 572-595, 7 figs.

Grassle, J. F. 1972. Species diversity, genetic variability and environmental uncertainty. *In* Fifth European Marine Biological Symposium, Piccin, Padua. B. Battaglia, Ed. p. 19-26.

_____, and J. P. Grassle. 1974. Opportunistic life histories and genetic systems in marine benthic polychaetes. *J. Mar. Res. 32:* 253-284.

Gray, J., and A. J. Boucot. 1971. Early Silurian spore tetrads from New York: earliest New World evidence for vascular plants? *Science 173:* 918-921.

Gray, J. S. 1974. Animal-sediment relationships. *Oceanogr. Mar. Biol. Ann. Rev. 12:* 223-261.

Gressly, A. 1838. Observations geologiques sur le Jura Soleurois. *Nouv. Mem. Soc. Helvet. Sci. Nat. (Nauchatel) 2* (paper 6): 1-112, 5 pls.

Grill, R. 1941. Stratigraphische Untersuchungen mit Hilfe von Mikrofaunen im Wiener Becken und den benachbarten Molasse-Anteilen. *Oel und Kohle 37:* 12.

Groot, J. J., and C. R. Groot. 1961. Plant microfossils and age of the Raritan, Tuscaloosa, and Magothy formations of the eastern United States. *Palaeontographica 108B:* 121-140.

_____, and J. S. Penny. 1960. Plant microfossils and age of nonmarine Cretaceous sediments of Maryland and Delaware. *Micropaleont. 6* (2): 225-236.

Gruber, U., and H. Kahmann. 1968. Eine biometrische Untersuchung an alpinen Rotelmausen (*Clethrionomys glareolus* Schreber, 1780). *Säugetierk. Mitt. 16:* 310-338.

Gubler, J. 1935. Less fusulinides du Permien de l'Indochine. *Soc. Géol. France, Mém., N.S., T.N., 4* (Mém. 26): 1-69, 24 pls., 54 text-figs.

Guex, J. 1968. Note préliminaire sur le dimorphism sexuel des Hildoceratacea du Toarcian moyen et supérieur de l'Aveyron France. *U.S. Geol. Survey Prof. Paper 70* (part 2): 57-84.

_____. 1971. Sur la classification des Dactylioceratidae (Ammonoidea) du Toarcien. *Eclog. Geol. Helv. Basel 64:* 225-243.

Gümbel, W. 1874. Ausland. p. 479.

Guppy, H. B. 1917. Plants, Seeds and Currents in the East Indies and Azores. William Norgate.

Haas, Otto. 1942. Recurrence of morphologic types and evolutionary cycles in Mesozoic ammonites. *Tour. Paleont. 16:* 643-650.

_____. 1953. Late Triassic gastropods from central Peru. *In* Mesozoic Invertebrate Faunas of Peru. *Amer. Mus. Nat. Hist. Bull. 101* (1-2): 1-328, 18 pls.

Håkansson, E. 1974. Adaptive strategies among soft bottom cheilostomes from the Danish chalk (Maastrichtian). Unpublished thesis, University of Copenhagen. 92 p.

_____, R. Bromley, and K. Perch-Nielsen. (In Press). Maastrichtian chalk of North-West Europe – a pelagic shelf sediment. *In* Pelagic Sediments: On Land and Under the Sea, K. J. Hsü and H. C. Jenkyns, Eds. *Spec. Publ. Int. Assoc. Sediment 1:* 211-234.

Halbouty, Michel T. 1967. Salt domes – Gulf region, United States and Mexico. Gulf Publishing Co., Houston. 425 p.

Haldane, J. B. S. 1957. The cost of natural selection. *Jour. Genet. 55:* 511-524.

Hall, J. 1852. Geology and paleontology. *In* Exploration and Survey of the Valley of the Great Salt Lake of Utah, Including a Reconnaissance of a New Route through the Rocky Mountains, H. Stansbury, Ed. U.S. 32nd Cong. Spec. Sess., *Senate Executive Doc. 3:* 399-414.

_____. 1858. *In* Report on the Geological Survey of the State of Iowa, Embracing the Results of Investigations made during Portions of the Years 1855, 1856 and 1857, J. Hall and J. D. Whitney, Eds. *Palaeontology, Albany 1* (part 2): 473-724.

_____, and J. M. Clarke. 1888. Palaeontology: Vol. VII. Containing descriptions of the trilobites and other Crustacea of the Oriskany, Upper Helderberg, Hamilton, Portage, Chemung, and Catskill Groups. New York Geol. Survey, Albany. 236 p.

Hall, T. S. 1894. Note on the distribution of the graptolites in rocks of Castlemaine. *Australian Assoc. Adv. Sci. Rept. 11:* 374.

_____. 1896. On the occurrence of graptolites in North Eastern Victoria. *Royal Soc. Victoria, Proc. 9:* 183.

_____. 1897. Victorian graptolites, Part I. *Royal Soc. Victoria, Proc. 10:* 13-16.

_____. 1898. Victorian graptolites, Part II. The graptolites of the Lancefield beds. *Royal Soc. Victoria, Proc. 11:* 164-178.

_____. 1899. The graptolite-bearing rocks of Victoria, Australia. *Geol. Mag. 6:* 438-451.

Hallam, A. 1959. On the supposed evolution of *Gryphaea* in the Lias. *Geol. Mag. 96:* 99-108.

_____. 1965. Observations on marine Lower Jurassic stratigraphy of North America, with special reference to United States. *Bull. Amer. Assoc. Petrol. Geol. 49:* 1485-1501.

_____. 1967a. Sedimentology and palaeogeographic significance of certain red limestones and associated beds in the Lias of the Alpine region. *Scottish Jour. Geol. 3:* 195-220.

_____. 1967b. The bearing of certain palaeozoogeographic data on continental drift. *Palaeogeog., Palaeoclim., Palaeoecol. 3:* 201-241.

_____. 1969. Faunal realms and facies in the Jurassic. *Palaeontology 12:* 1-18.

_____. 1971. Provincially in Jurassic faunas in relation to facies and palaeogeography. *In* Faunal Provinces in Space and Time, F. A. Middlemiss and P. R. Rawson, Eds. *Geol. Jour., Spec. Issue 4:* 129-152.

_____, Ed. 1973. Atlas of Paleobiogeography. Elsevier, Amsterdam. 531 + xii p.

Hamilton, W. R. 1973a. The Lower Miocene ruminants of Gebel Zelten, Libya. *Brit. Mus. Nat. Hist., Bull. Geol. 21* (3): 73-150.

_____. 1973b. North African Lower Miocene Rhinoceroses. *Brit. Mus. Nat. Hist., Bull, Geol. 24* (6): 352-395.

Hancock, J. M. 1966. Theoretical and Real Stratigraphy. *Geol. Mag. 103:* 179.

Hanna, G. D. 1966. Introduced mollusks of western North America. *Calif. Acad. Sci., Occas. Pap. 48:* 1-108, 4 pls.

Hanzawa, Shoshiro. 1954. Stratigraphical distribution of the fusulinid Foraminifera in Japan. Internat. Geol. Congr. 19th, Algeria, 1952. *Compte Rendu, Sect. 13. 15:* 129-137, 1 table.

_____, and M. Murata. 1963. The paleontologic and stratigraphic considerations on the *Neoschwagerininae* and *Verbeekininae*, with the descriptions of some fusulinid Foraminifera from the Kitakami Massif, Japan. *Tohoku Univ. Sci. Repts., Ser. 2 (Geology). 35* (1): 1-31, pls. 1-20, 10 text-figs. (graphs), 5 tables.

Harbaugh, J. W., and D. F. Merriam. 1968. Computer applications in stratigraphic analysis. John Wiley, New York. 282 p.

Harker, P., and G. O. Raasch. 1958. Megafaunal zones in the Alberta Mississippian and Permian. *In* Jurassic and Carboniferous of Western Canada, A. J. Goodman, Ed. Amer. Assoc. Petrol. Geol., John Andrew Allan Memorial Volume. p. 216-231.

Harland, N. B., C. H. Holland, M. R. Rouse, A. B. Reynolds, M. J. Rudwick, G. E. Satterthwaite, L. G. Tarlo, and E. C. Willey. 1967. The fossil record. *Geol. Soc. London, Spec. Publ. 2:* 1-827.

Harland, W. B., and E. H. Francis. 1971. The Phanerozoic time scale – a supplement. *Geol. Soc. London, Spec. Publ. 5.*

Harman, H. H. 1967. Modern Factor Analysis, 2nd Ed. University of Chicago Press, Chicago. 474 p.

Harris, J. M. 1973. *Prodeinotherium* from Gebel Zelten, Libya. *Brit. Mus. Geol. Hist., Bull. Geol. 23* (5): 283-348.

Harrison, W., J. J. Norcross, N. A. Pore, and E. M. Stanley. 1967. Circulation of shelf waters off Chesapeake Bight – surface and bottom drift of continental shelf waters between Cape Henlopen, Delaware, and Cape Hatteras, North Carolina, June 1963-December 1964. *U.S. ESSA, Prof. Papers 3:* 1-82.

Haug, E. 1908-1911. Traité de Géologie II Les Périodes géologiques. Armand Colin, Paris. p. 539-2024.

Hay, W. W. 1972. Probablistic stratigraphy. *Ecologae Geol. Helv. 65* (2): 255-266.

_____, H. P. Mohler, P. H. Roth, R. R. Schmidt, and Joseph E. Boudreaux. 1967. Calcareous nannoplankton zonation of the Cenozoic of the Gulf Coast and Caribbean-Antillean area and transoceanic correlation. *Gulf Coast Assoc. Geol. Soc., Trans. 17:* 428-480.

Hayami, I., and T. Ozawa. 1975. Evolutionary models of lineage-zones. *Lethaia 8* (1): 1-14.

Haynie, A. V., Jr. 1957. The Worm Creek Quartzite Member of the St. Charles Formation, Utah-Idaho. Unpublished M.S. thesis, Utah State University, Logan. 39 p.

Hays, J. D., T. Saito, N. D. Opdyke, and L. H. Burckle. 1969. Pliocene-Pleistocene sediments of the equatorial Pacific – their paleomagnetic, biostratigraphic and climatic record. *Geol. Soc. Amer. Bull. 80:* 1481-1514.

Hayward, J. F. 1940. Some variations in *Echinocorys* in South-Eastern England. *Proc. Geol. Assoc. 51* (4): 291-310, 5 figs.

Hazel, J. E. 1963. Part 1, Systematics of some Gulfian trachyleberids from Texas and Arkansas; Part 2, Ostracode biostratigraphy in some Austinian-Tayloran rocks (abs.). *Dissert. Abs. 24* (6): 2423.

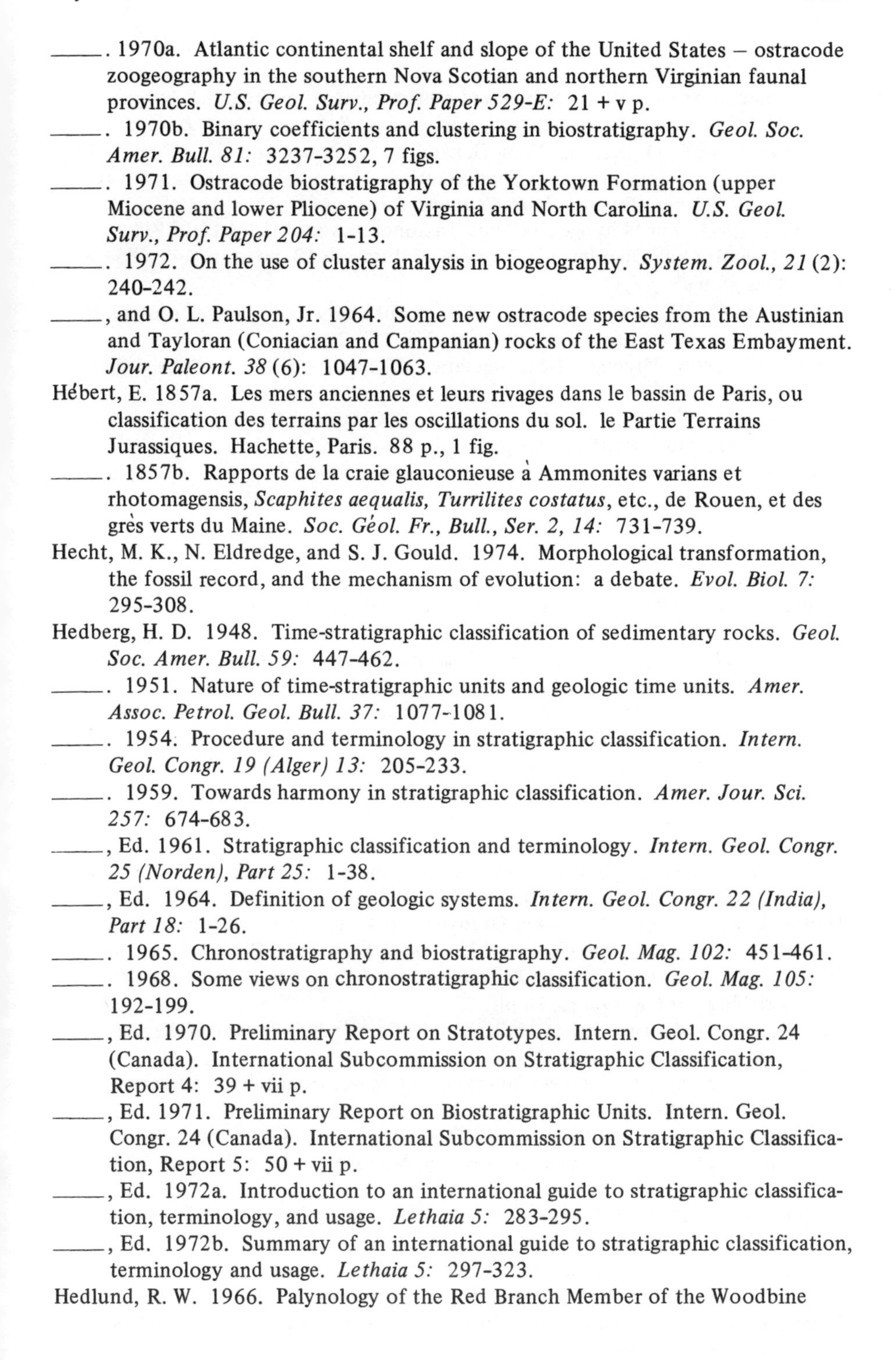

———. 1970a. Atlantic continental shelf and slope of the United States – ostracode zoogeography in the southern Nova Scotian and northern Virginian faunal provinces. *U.S. Geol. Surv., Prof. Paper 529-E:* 21 + v p.

———. 1970b. Binary coefficients and clustering in biostratigraphy. *Geol. Soc. Amer. Bull. 81:* 3237-3252, 7 figs.

———. 1971. Ostracode biostratigraphy of the Yorktown Formation (upper Miocene and lower Pliocene) of Virginia and North Carolina. *U.S. Geol. Surv., Prof. Paper 204:* 1-13.

———. 1972. On the use of cluster analysis in biogeography. *System. Zool., 21* (2): 240-242.

———, and O. L. Paulson, Jr. 1964. Some new ostracode species from the Austinian and Tayloran (Coniacian and Campanian) rocks of the East Texas Embayment. *Jour. Paleont. 38* (6): 1047-1063.

Hébert, E. 1857a. Les mers anciennes et leurs rivages dans le bassin de Paris, ou classification des terrains par les oscillations du sol. le Partie Terrains Jurassiques. Hachette, Paris. 88 p., 1 fig.

———. 1857b. Rapports de la craie glauconieuse à Ammonites varians et rhotomagensis, *Scaphites aequalis, Turrilites costatus*, etc., de Rouen, et des grès verts du Maine. *Soc. Géol. Fr., Bull., Ser. 2, 14:* 731-739.

Hecht, M. K., N. Eldredge, and S. J. Gould. 1974. Morphological transformation, the fossil record, and the mechanism of evolution: a debate. *Evol. Biol. 7:* 295-308.

Hedberg, H. D. 1948. Time-stratigraphic classification of sedimentary rocks. *Geol. Soc. Amer. Bull. 59:* 447-462.

———. 1951. Nature of time-stratigraphic units and geologic time units. *Amer. Assoc. Petrol. Geol. Bull. 37:* 1077-1081.

———. 1954. Procedure and terminology in stratigraphic classification. *Intern. Geol. Congr. 19 (Alger) 13:* 205-233.

———. 1959. Towards harmony in stratigraphic classification. *Amer. Jour. Sci. 257:* 674-683.

———, Ed. 1961. Stratigraphic classification and terminology. *Intern. Geol. Congr. 25 (Norden), Part 25:* 1-38.

———, Ed. 1964. Definition of geologic systems. *Intern. Geol. Congr. 22 (India), Part 18:* 1-26.

———. 1965. Chronostratigraphy and biostratigraphy. *Geol. Mag. 102:* 451-461.

———. 1968. Some views on chronostratigraphic classification. *Geol. Mag. 105:* 192-199.

———, Ed. 1970. Preliminary Report on Stratotypes. Intern. Geol. Congr. 24 (Canada). International Subcommission on Stratigraphic Classification, Report 4: 39 + vii p.

———, Ed. 1971. Preliminary Report on Biostratigraphic Units. Intern. Geol. Congr. 24 (Canada). International Subcommission on Stratigraphic Classification, Report 5: 50 + vii p.

———, Ed. 1972a. Introduction to an international guide to stratigraphic classification, terminology, and usage. *Lethaia 5:* 283-295.

———, Ed. 1972b. Summary of an international guide to stratigraphic classification, terminology and usage. *Lethaia 5:* 297-323.

Hedlund, R. W. 1966. Palynology of the Red Branch Member of the Woodbine

Formation (Cenomanian), Bryan County, Oklahoma. *Oklahoma Geol. Surv. Bull. 112:* 1-69.

———, and G. Norris. 1968. Spores and pollen grains from Fredericksburgian (Albian) strata, Marshall County, Oklahoma. *Pollen et Spores 10* (1): 129-159.

Heirtzler, J. R., G. O. Dickson, E. M. Herron, W. C. Pitman, and X. Le Pichon. 1968. Marine magnetic anomalies, geomagnetic field reversals and motions of the ocean floor continents. *Jour. Geophys. Res. 13* (6): 2119-2136.

Helms, J. 1963. Zur "Phyogenese" und Taxionomie von PALNATOLEPIS (Conodontida Oberdevon): Geologie, Jahrg. vol. 12, no. 4, p. 449-485. 3 Figs., 5 pls.

Henbest, L. G., Ed. 1952. Distribution of evolutionary explosions in geological time: A symposium. *Jour. Paleont. 26:* 298-394.

Hennig, W. 1966. Phylogenetic Systematics. University of Illinois Press, Urbana. 263 p.

Henningsmoen, G. 1957. The Trilobite Family Olenidae. Norske Videnskaps-Akademi Oslo Skrifter. 303 p., 31 pls.

Hertlein, L. G. 1937. A note on some species of marine mollusks occurring in both Polynesia and the western Americas. *Proc. Amer. Philos. Soc. 78:* 303-312.

Heywood, V. H. 1959. The taxonomic treatment of ecotypic variation. *Syst. Assoc. Publ. 3:* 87-112.

Hibbard, C. W. 1949. Techniques of collecting micro-vertebrate fossils. *Contrib. Mus. Paleont. Univ. Michigan 8:* 7-19.

Hill, D. 1957. The sequence and distribution of upper Palaeozoic coral faunas. *Australian Jour. Sci. 19* (3a): 42-61.

Hinde, G. J. 1879. On conodonts from the Chazy and Cincinnati group of the Cambro-Silurian and from the Hamilton Genesee Shale Divisions of the Devonian, in Canada and the United States. *London Quart. Jour. Geol. Soc. 35:* 351-369.

Hinton, M. A. C. 1910. A preliminary account of the British fossil voles and lemmings. *Proc. Geol. Assoc. London 21:* 489-507.

———. 1926. Monograph of the Voles and Lemmings (Microtinae) Living and Extinct, Vol. 1. Brit. Mus. (Nat. Hist.), London. 487 p.

Hintze, L. F. 1952 (1953). Lower Ordovician trilobites from western Utah and eastern Nevada. *Utah Geol. and Mineralog. Surv. Bull. 48:* 1-249, 26 pls.

Hoelder, H. 1960. Geologie und Palaeontologie in Texten und ihrer Geschichte. Frieburg. 566 & xviii p., 16 pls.

Hoernes, M. 1856. Die fossilen Mollusken der Tertiar-Beckens von Wien. *Abh. K.K. Geol. Reichsanstalt 3:* 1-733.

Hoffmeister, W. S. 1959. Lower Silurian plant spores from Libya. *Micropaleontology 5* (3): 331-334.

Hoppin, R. A. 1953. Oscillations in the Vicksburg Stage as shown by the Foraminifera from a well in George County, Mississippi. *Jour. Paleont. 27* (4): 577-584.

Hopping, C. A. 1967. Palynology and the oil industry. *Review of Paleobotany and Palynology 2:* 23-48.

Hornaday, G. R. 1972. Oligocene smaller foraminifera associated with an occurrence of *Miogypsina* in California. *Jour. Foram. Res. 2*(1): 35-46.

House, M. R. 1973. An analysis of Devonian goniatite distributions. *Palaeontology Spec. Paper 9:* 305-317.

Howarth, M. K. 1973a. Lower Jurassic (Pleinsbachian and Toarcian) ammonites. *In* Atlas of Palaeobiogeography, A. Hallam, Ed. Elsevier, Amsterdam. p. 275-282.

_____. 1973b. The stratigraphy and ammonite fauna of the Upper Liassic Grey Shales of the Yorkshire coast. *Bull. Brit. Mus. Nat. Hist. Geol. 24:* 237-277.

Howe, W. B., V. E. Kurtz, and K. H. Anderson. 1972. Correlation of Cambrian strata of the Ozark and upper Mississippi Valley regions. *Missouri Geol. Surv. and Water Res., Rept. 52:* 1-60.

Howell, B. F., Chairman. 1944. Correlation of the Cambrian formations of North America. *Bull. Geol. Soc. Amer. 55:* 993-1003.

Hsü, K. J., M. B. Cita, and W. B. F. Ryan. 1972. The origin of the Mediterranean evaporites. *In* Initial Reports of the Deep-Sea Drilling Project, Vol. XIII, W. B. F. Ryan, K. J. Hsü, et al., Eds. Government Printing Office, Washington, p. 1203-1231.

Hubbs, C. L. 1952. Antitropical distribution of fishes and other organisms. *Proc. Seventh Pacific Sci. Congr. 3:* 324-329.

Huddleston, W. H. 1887-1896. A monograph of the Inferior Oolite Gastropoda. Beint Part 1 of the British Jurassic Gastropoda. *Paleontographical Soc. 40* (1887), *41* (1888), *42* (1889), *43* (1890), *45* (1892), *46* (1893), *48* (1894), *49* (1895), *50* (1896): 519, 44 pls.

Hudson, J. D. 1967. Speculations on the depth relation of calcium carbonate in Recent and ancient seas. *Mar. Geol. 5:* 473-480.

Hughes, N. F. 1961. Fossil evidence and angiosperm ancestry. *Sci. Progress 49:* 84-102.

_____. 1970. The need for agreed standards of recording in palaeopalynology and palaeobotany. *Paläontologische Abhandlungen 3B* (3/4): 357-364.

_____. 1973a. Towards effective data-handling in palaeopalynology. *In* Morfologiya i sistematika iskopaemykh spor i pyl'tsy (Trudy III Mezhdunarodnoy Palinologicheskoy Konferentsii). Izdatel'stvo "Nauka", Moscow. p. 9-14.

_____. 1973b. Environment of angiosperm origins. *In* Palinologiya Mezofita (Trudy III Mezhdunarodnoy Palinologicheskoy Konferentsii). Izdatel'stvo "Nauka", Moscow. p. 135-137.

_____, and C. A. Croxton. 1973. Palynologic correlation of the Dorset "Wealden". *Palaeontology 16* (3): 567-601.

_____, and J. C. Moody-Stuart. 1967. Proposed method of recording pre-Quaternary palynological data. *Rev. Palaeobot. Palynol. 3:* 347-358.

_____, and J. C. Moody-Stuart. 1969. A method of stratigraphic correlation using Early Cretaceous spores. *Palaeontology 12* (1): 84-111.

_____, and B. Pacltová. 1972. Fresh-water to marine time-correlation potential of Cretaceous and Tertiary palynomorphs. *24th Intern. Geol. Congr., Sect. 7:* 397-401.

Hunt, R. M., Jr. 1972. Miocene Amphicyonidae (Mammalia, Carnivora) from the Agate Spring Quarries, Sioux County, Nebraska. *Amer. Mus. Nat. Hist., Novit. 2506:* 1-39.

Hupé, P. 1960. Les zones stratigraphiques. *Bull. Trimestr. Serv. Inform. Géol. B.R.G.M. 49:* 1-20.

Hurley, Patrick J., Ted E. Jacques, and N. Eugene Swick. 1973. Developments in Louisiana Gulf Coast in 1972. *Amer. Assoc. Petrol. Geol. Bull. 57*(8): 1532-1541.

Hutchinson, G. E. 1957. Concluding remarks. *Cold Spring Harbor Symp. Quart. Biol. 22:* 415-427.

Hutt, J. E., R. B. Rickards, and W. B. N. Berry. 1972. Some major elements in the evolution of Silurian and Devonian graptoloids. *XXIV Internat. Geol. Congr., Sect. 7:* 163-173.

Huxley, J. S. 1959. Clades and grades. *Syst. Assoc. Publ. 3:* 21-22.

Huxley, T. H. 1862. The Anniversary Address. *Geol. Soc. London, Quart. Jour. 18:* xl-liv.

Igo, Hisayoshi. 1957. Fusulinids of Fukuji, southeastern part of the Hida Massif, central Japan. *Tokyo Kyoiku Daigaku, Sci. Repts., Sec. C, 5*(47): 153-246, 15 pls., 2 text-figs.

_____. 1964. Fusulinids from the Nabeyama Formation (Permian) Kuzu, Tochigi Prefecture, Japan. *Mejiro Gakuen Woman's Junior College Mem. 1:* 1-28, pls. 1-10.

Ikebe, N. 1973. Neogene biostratigraphy and radiometric time scale. *J. Geo. Sc. Osaka City Univ. 16:* 51-67.

_____, Takayanagi Tokichi, Chiji Manzo, and Chinzei Kiyotaka. 1972. Neogene biostratigraphy and radiometric time scale of Japan – an attempt at intercontinental correlation. *Pac. Geol. 4:* 39-78.

Imbrie, John, and E. G. Purdy. 1962. Classification of modern Bahamian carbonate sediments. *Amer. Assoc. Petrol. Geol. Mem. 1:* 253-272.

Imlay, R. W. 1971. Jurassic ammonite succession in the United States. Extr., Colloq. Jur. Luxembourg, 1967; Mem. B.R.G.M., Fr. 75. Pub. Inst. Grand-Ducal Sect. Sci. Nat., Phys., Math. 709-724.

Ishii, Kenichi. 1958. On the phylogeny, morphology and distribution of *Fusulina*, *Beedeina* and allied fusulinid genera. *Osaka City Univ., Inst. Polytech. Jour., Ser. G. 4:* 29-70, pls. 1-4, 5 text-figs.

Israelsky, Merle C. 1949. Oscillation Chart. *Amer. Assoc. Petrol. Geol. Bull. 33* (1): 92-98.

Jaanusson, V. 1972. Biogeography of the Ordovician Period. *In* Treatise on Invertebrate Paleontology, Part A, R. C. Moore, Ed. University of Kansas Press, Lawrence.

_____. 1973. Ordovician articulate brachiopods. *In* Atlas of Palaeobiogeography, A. Hallam, Ed. Elsevier, Amsterdam. p. 19-25.

Jackson, D. E. 1964. Observations on the sequence and correlation of Lower and Middle Ordovician graptolite faunas of North America. *Geol. Soc. Amer. Bull. 75:* 523-534.

_____. 1966. Graptolite facies of the Canadian Cordillera and Arctic Archipelago: A Review. *Canadian Petrol. Geol. Bull. 14:* 469-485.

_____. 1973. *Amplexograptus* and *Glyptograptus* isolated from Ordovician limestones in Manitoba. *In* Contributions to Canadian Paleontology. *Geol. Surv. Can. Bull. 22:* 1-8.

_____, and A. C. Lenz. 1962. Zonation of Ordovician and Silurian graptolites of northern Yukon, Canada. *Amer. Assoc. Petrol. Geol. Bull. 46:* 30-45.

_____, G. Steen, and D. Sykes. 1965. Stratigraphy and graptolite zonation of the Kechika and Sandpile Groups in northwestern British Columbia. *Canadian Petrol. Geol. Bull. 13:* 139-154.

Jackson, J. B. C. 1972. The ecology of molluscs of *Thalassia* communities, Jamaica, West Indies. II. Molluscan population variability along an environmental stress gradient. *Mar. Biol. 14:* 304-337.

———. 1973. The ecology of molluscs of *Thalassia* communities, Jamaica, West Indies. I. Distribution, environmental physiology, and ecology of common shallow-water species. *Bull. Mar. Sci. 23:* 313-350.

———. 1974. Biogeographic consequences of eurytopy and stenotopy among marine bivalves and their evolutionary significance. *Amer. Natur. 108:* 541-560.

———. 1975. Comparative distributions of solitary and colonial marine Metazoa attached to hard substrates: the importance of competition and disturbance. Manuscript submitted to *Amer. Natur.*

———, T. F. Goreau, and W. D. Hartman. 1971. Recent brachiopod-coralline sponge communities and their paleoecological significance. *Science 173:* 623-625.

Jaeckel, S. 1958. Nachtrag. *In* Handbuch der Zoologie, W. Kukenthal and Th. Krumbach, Eds. 5: p. 259-275.

Jaeger, J. J., J. Michaux, and B. David. 1973. Biochronologie du Miocène moyen et supérieur continental du Maghreb. *C.R. Acad. Sci. Paris, Sér. D, 277:* 2477-2480.

Jardine, Nicholas, and Robin Sibson. 1971. Mathematical taxonomy. John Wiley, London. 286 p.

Jardiné, S., and L. Magloire. 1965. Palynologie et stratigraphie du Crétacé des bassins du Sénégal et de Côte d'Ivoire. *Mémoires du Bureau de Recherches Géologiques et Minières 32:* 187-245.

Jefferies, R. P. S. 1962. The palaeoecology of the *Actinocamax plenus* (lowest Turonian) in the Anglo-Paris Basin. *Palaeontology 4, Part 4:* 609-647.

———. 1963. The stratigraphy of the *Actinocamax plenus* subzone in the Anglo-Paris Basin. *Geol. Assoc. (London) Proc. 74:* 1-33.

Jeffrey, C. 1962. The origin and differentiation of the archegoniate land plants. *Botaniska Notiser 115:* 446-454.

Jekhowsky, B. de. 1958. Méthodes d'utilisation stratigraphique des microfossiles organiques dans les problèmes pétroliers. *Revue de l'Institut Français du Pétrole 8*(10): 1391-1418.

Jeletzky, J. A. 1951. Die Stratigraphie und Belemnitenfauna des Obercampan und Maastricht Westfalens, Nordwestdeutschlands und Dänemarks sowie einige allgemeine Gliederungs-Probleme der jüngren borealen Oberkreide Eurasiens. *Geol. Jahrb. Beiheft 1:* 1-142.

Jenkins, D. G. 1966. Planktonic foraminifera from the type Aquitanian-Burdigalian of France. *Cushman Found. Foram. Res., Contrib. 17:* 1-15.

Jeppsson, Lennart. 1971. Element arrangement in conodont apparatuses of *Hindeodella* type and in similar forms. *Lethaia 4:* 101-123.

Jepsen, G. L. 1963. Eocene vertebrates, coprolites, and plants in the Golden Valley Formation of western North Dakota. *Bull. Geol. Soc. Amer. 74:* 673-684.

———, and M. O. Woodburne. 1969. Paleocene hyracothere from Polecat Bench Formation, Wyoming. *Science 164:* 543-547.

Jiricek, J. 1973. Central Paratethys Correlation Tables. Unpublished.

Johnson, J. G. 1971 A quantitative approach to faunal province analysis. *Amer. Jour. Sci. 270:* 257-280.

Jones, D. L. 1961. Muscle attachment impressions in a Cretaceous ammonite. *Jour. Paleont. 35:* 502-503.
Jordan, R. R. 1962. Stratigraphy of the sedimentary rocks of Delaware. *Delaware Geol. Surv. Bull. 9:* 1-51.
——. 1968. Zur Anatomie mesozoischer Ammoniten nach der Strukturelementen der Gehäuse-Innenwand. *Geol. Jahrb. Beihefte 77:* 1-64.
——, and W. Stahl. 1970. Isotopische Palaotemperatur-Bestimmungen an jurassischen Ammoniten. *Geol. Jahrb. 9:* 33-62.
Jöreskog, K. G., J. E. Klovan, and R. A. Reyment, 1976. Geological Factor Analysis: Methods in Geomathematics, No. 1. Elsevier, Amsterdam. 178 p.
Jørgensen, N. O. 1970. The ostracods of the Danish White Chalk and their stratigraphical and ecological significance. Unpublished thesis, University of Copenhagen. 232 p. [in Danish].
Juferev, O. V. 1967. The most important problems of paleogeography and the role of foraminifers in the Carboniferous and the Permian. *In* New Data for the Devonian and Upper Paleozoic Biostratigraphy of Siberia. Akad. Nauk S.S.S.R., Sibirsk Otdeleniye, Inst. Geol., Geof. p. 61-76 [in Russian].
——. 1973. Carboniferous deposits of the Siberian biogeographical realm. *Akad. Nauk. S.S.S.R., Sibirsk Otdeleniye, Trudy Inst. Geol. Bull. 162:* 1-278 [in Russian].
Juignet, Pierre, and W. J. Kennedy. 1977. Faunes d'Ammonites et Biostratigraphie compareé du Cénomanien du nord-ouest de la France (Normandie) et du sud d'Angleterre. Bull. Soc. Géol. Normand Amic Muséum du Havre, vol. 63, 192 pp, 32 pls.
Jukes-Browne, A. J. 1903. The Term "Hemera." *Geol. Mag. 10:* 36-38.
——, and W. Hill. 1896. A delimitation of the Cenomanian: Being a comparison of the corresponding beds in South-western England and Western France. *Geol. Soc. London Quart. Jour. 52:* 99-178, pl. 5.
Jung, Peter. 1974. The problems of working with mollusks in the Caribbean Tertiary. Abstract in recusil des resumes de communications, Seventh Conference on the Geology of the Caribbean. 33-34.
Kahler, Franz. 1962. Stratigraphische Vergleiche im Karbon und Perm mit Hilfe der Fusuliniden. *Geol. Gesell. Wien, Mitt. 54:* 147-161, 1 text-fig., 2 tables.
——, and Gustava Kahler. 1937a. Beiträge zur Kenntnis der Fusuliniden der Ostalpen: Die Pseudoschwagerinen der Grenzlandbänke und des oberen Schwagerinenkalkes. *Palaeontographica 87*(1-2): 1-44, pls. 1-3, 2 text-figs.
——, and Gustava Kahler. 1937b. Stratigraphische und fazielle Untersuchungen im Oberkarbon und Perm der Karnischen Alpen. Cong. Avanc. Étude Strat. Carb. 2d Heerlen, 1935, *Comptes Rendus 1:* 445-487, pls. 59-62, 3 text-figs.
——, and Gustava Kahler. 1938. Beobachtungen an Fusuliniden der karnischen Alpen. Uber die Einbettung von Pseudoschwagerinen im roten Trogkofelkalk. B. Frasspuren in Fusulinidenschalen. C. Trimorphismus bei Paraschwagerinen. *Zentralbl. f. Mineralogie, Geologie u. Palaontologie 4:* 101-115, 1 pl., 2 text-figs.
——, and Gustava Kahler. 1941. Beiträge zur Kenntnis der Fusuliniden der Ostalpen; Die Gatung *Pseudoschwagerina* und ihre Vertreter im unteren Schwagerinenkalk und im Trogkofelkalk. *Palaeontographica 92:* 59-98, pls. 10, 11.

Kaiser, Peter, and Ulrich Lehmann. 1971. Vergleichende Studien zur Evolution des Kieferapparates rezenter und fossiler Cephalopoden. *Paläont. Zeitsch. 45:* 18-32.

Kapounek, J., A. Kröll, A. Papp, and K. Turnovsky. 1965. Die Verbreitung von Oligozän, unter-und Mittelmiozän in Niederösterreich. *Erdoel-Erdgas Z. 81:* 109-115.

Kato, M., K. Nakamura, E. Hirai, and Y. Kakinuma. 1961. The distribution pattern of Hydrozoa on seaweed with some notes on the so-called coaction among hydrozoan species. *Bull. Biol. Stat. Asamushi, Tôhoku Univ. 10:* 195-202.

Kauffman, E. G. 1964. A new subgenus of *Lima* from the Cretaceous of the Gulf and Atlantic Coast Province. *Tulane Studies Geol. 2:* 89-101.

———. 1965. Middle and Late Turonian oysters of the *Lopha lugubris* group. *Smithsonian Misc. Coll. 148*(6): Pub. 4602: 1-93.

———. 1967a. Coloradoan macroinvertebrate assemblages, cental Western Interior, United States. *In* A symposium on paleoenvironments of the Cretaceous seaway in the Western Interior. Geol. Soc. Amer., Rocky Mtn. Sec., 20th Ann. Mtg., 1967, Golden, Colo., Colorado School of Mines. 67-143.

———. 1967b. Cretaceous *Thyasira* from the Western Interior of North America. *Smithsonian Misc. Coll. 152*(1): Pub. 4695: 1-159.

———. 1969. Cretaceous Marine Cycles of the Western Interior. *Mountain Geologist 6*(4): 227-245.

———. 1970. Population systematics, radiometrics, and zonation – a new biostratigraphy. *Proc. N. Amer. Paleont. Conv., Pt. F:* 612-666, text-figs. 1-10.

———. 1972. Evolutionary rates and patterns of North American Cretaceous mollusca. *Internat. Geol. Congr. 24th session, Sect. 7:* 174-189.

———. 1973. Evolutionary rates and biostratigraphy. Ann. Meeting Geol. Soc. Amer. (abstract). 688.

———. 1975. The value of benthic Bivalvia in Cretaceous biostratigraphy of the Western Interior. *In* The Cretaceous System in the Western Interior of North America – Selected Aspects, W. G. E. Caldwell, Ed. *Geol. Assoc. Canada, Spec. Paper 13.*

———. 1977 (in press). Cretaceous geochronology of the Western Interior United States; new data and applications. Submitted to Geol. Soc. Amer., Geol. 31 ms. pages, 4 text-figs.

———, and R. V. Kesling. 1960. An Upper Cretaceous ammonite bitten by a mosasaur. *Michigan Univ. Mus. Paleont. Contr. 15*(9): 193-248.

———, and N. F. Sohl. 1974. Structure and evolution of Antillean Cretaceous rudist frameworks. *Verhandl. Naturf. Ges. Basel 84:* 399-497.

Kay, M., and N. Eldredge. 1968. Cambrian trilobites in central Newfoundland volcanic belt. *Geol. Mag. 105:* 372-377.

Keen, A. M. 1963. Marine molluscan genera of western North America. Stanford University Press, Stanford, Calif. 126 p.

Kemp, E. M. 1968. Probable angiosperm pollen from British Barremian to Albian strata. *Palaeontology 11*(3): 421-434.

———. 1970. Aptian and Albian miospores from southern England. *Palaeontographica 131B:* 73-143.

Kennedy, W. J. 1971. Trace fossils in the chalk environment. *In* Trace Fossils, T. P. Crimes and J. C. Harper, Eds. p. 263-282.

_____, and Pierre Juignet. 1973. First record of the ammonite family Binneyitidae in western Europe. *Jour. Paleont. 47:* 900-902.

Kepper, J. C. 1972. Paleoenvironmental patterns in middle to lower Upper Cambrian interval in eastern Great Basin. *Amer. Assoc. Petrol. Geol. Bull. 56:* 503-527.

Kermack, K. A. 1954. A biometrical study of *Micraster coranguinum* and *M. (Isomicraster) senonensis. Philos. Trans. Roy. Soc. London, Ser. B, Biol. Sci. No. 649 237:* 375-428, 11 figs., 3 pls., 20 tables.

Kessler, P. 1923. "Konchinbänder," "Haftlinie," "Hohlkiel" und "Streifenbüschel" bei Ammoniten. *Neues Jahrb. Min. Geol. Paläont,, Monatsh., Jahrg. 1923:* 499-511.

Khalfin, L. L. (Ed.). 1971. Classification in Stratigraphy. Series: Stratigraphy and Paleontology. Proceeding Siberian Sci. Research Inst. Geol., Geophy. Min. Res., Ministry of Geol. U.S.S.R. 178 p. (English translation from the original 1969, Russian edition available from the U.S. Department of Commerce, National Technical Information Service, Springfield, Va., 22151.

Kier, P. M. 1962. Revision of the Cassiduloid Echinoids. *Smithsonian Misc. Coll. 144*(3): 1-262, 184 figs., 44 pls., 6 tables.

_____. 1972. Upper Miocene Echinoids from the Yorktown Formation of Virginia and Their Environmental Significance. *Smithsonian Contr. Paleobiol. 13:* 1-41, 7 figs., 10 pls., 2 tables.

Kimura, M. 1960. Optimum mutation rate and degree of dominance as determined by the principle of minimum genetic load. *Jour. Genet. 57:* 21-34.

Kindle, C. H., and H. B. Whittington. 1958. Stratigraphy of the Cow Head region, western Newfoundland. *Geol. Soc. Amer. Bull. 69:* 315-342.

King, W. E. 1973. Fusulinids *Millerella* and *Eostaffella* from the Pennsylvanian of New Mexico and Texas. *New Mexico Bur. Mines and Min. Res., Mem. 26:* 1-34, pls. 1-4, 6 text-figs.

Kinzie, R. A., Jr. 1968. The ecology of the replacement of *Pseudosquilla ciliata* (Fabricus by *Gonodactylus falcatus* (Forskal) (Crustacea; Stomatopoda) recently introduced into the Hawaiian Islands. *Pacific Sci. 22:* 465-475.

Kirkegaard, J. B. 1969. A quantitative investigation of the central North Sea Polychaeta. *Spolia Zool. Mus. Haun. 29:* 1-285.

Kittl, Ernst. 1891-1894. Die Gastropoden der Schichten von St. Cassian der südalpinean Trias. *K. K. Naturhist. Hofmus. Annalen 6* (1) [1891]: 166-262; *7* (2) [1892]: 35-97; *9* (3) [1894]: 143-277.

Kjellström, G. 1973. Maastrichtian microplankton from the Höllviken borehold no. 1 in Scania, southern Sweden. *Sver. Geol. Unders. ser. C, 688:* 1-59.

Knopf, A. 1949. Time in earth history. *In* Genetics, Paleontology and Evolution, G. L. Jepsen et al., Eds. Princeton University Press, Princeton, N.J. 9 p.

Knudsen, J. 1967. The deep sea Bivalvia. *John Murray Exped. Rept. 11:* 239-343.

Kobayashi, T. 1967. The Cambrian of eastern Asia and other parts of the continent. The Cambro-Ordovician formations and faunas of South Korea, Part 10, sec. C. *Tokyo Univ., Faculty Sci., Jour., Sec. 2, 16* (pt. 3): 381-534.

Kochansky-Devidé, Vanda. 1959. Die fusuliniden Foraminiferen aus dem Karbon und Perm im Velebit und in der Lika (Kroatien). *Unteres Perm. Palaeont. Jugoslav. Akad. (Zagreb) 3:* 1-62, 8 pls.

_____. 1965. Die fusuliniden Foraminiferen aus dem Karbon und Perm im Velebit und in der Lika (Kroatiens) Mittlers und oberes Perm. *Jugoslav Akad. Znanosti, Umjetnosti Acta Geol. 5:* 101-150, pls. 1-14.

———. 1969. Parallel tendencies in the evolution of the fusulinids. *Geol. Soc. Pologne, Ann. 39:* 35–40, pl. 1.

Koehn, R. K., and J. B. Mitten. 1972. Population genetics of marine pelecypods. I. Ecological heterogeneity and evolutionary strategy at an enzyme locus. *Amer. Nat. 106:* 47-56.

Kohut, J. 1969. Determination, statistical analysis and interpretation of recurrent conodont groups in Middle and Upper Ordovician strata of the Cincinnati Region (Ohio, Kentucky, and Indiana). *J. Paleont. 43:* 392–412.

Kollmann, K. 1965. Jungtertiär im Steirischen Becken. *Mitt. Geol. Ges. Wien 57* (1964): 479-632, 6 pls.

Kongiel, R. 1938. Rozwazania nad zwiennościa jezowców (Considérations sur la variabilité des Échinides). *Ann. Soc. Géol. Pologne 13:* 194-250, 5 figs., 2 pls., 38 tables.

———. 1949. Les Echinocorys du Danien de Danemark de Suède de Pologne. *Trav. Serv. Geol. Pologne 5:* 1-89, 61 figs., 18 pls.

Kowalski, K. 1966. The stratigraphic importance of Rodents in the studies on the European Quaternary. *Folia Quatern. 22:* 1-16.

———. 1970. Variation and Speciation in Fossil Voles. *Symp. Zool. Soc. London 26:* 149-161.

Kraeuter, J. N. (1974,) [1976]. Offshore currents, larval transport, and establishment of southern populations of *Littorina littorea* along the U.S. Atlantic Coast. Thalassia Jugoslavica 10: 159-170.

Krebs, W. N. 1974. Convergence of the living isopod *Serolis* Leach with the trilobites. *Geol. Soc. Amer., Abs. with Programs* 6(3): 203-204.

Kreuzer, H., C. H. von Daniels, F. Gramman, W. Harre, and B. Mattiat. 1973. K/Ar dates of some glauconites of the northwest German Tertiary Basin. *Fortschr. Mineral.*

Kruskal, J. B. 1964. Multidimensional scaling by optimizing goodness of fit to a nonmetric hypothesis. *Psychometrika 29:* 1-27.

———. 1972. Multidimensional scaling in archaeology: Time is not the only dimension. *In* Mathematics in the Archaeological and Historical Sciences, F. R. Hodson, D. G. Kendall, and P. Tantu, Eds. Edinburgh University Press, Edinburgh. p. 119-132.

Kuhn, T. S. 1962. The Structure of Scientific Revolutions. University of Chicago Press, Chicago. 172 p.

Kühne, W. G. 1971. Collecting vertebrate fossils by the Henkel process. *Curator 14:* 175-179.

Kullman, Jürgen, and Jost Wiedmann. 1970. Significance of sutures in phylogeny of Ammonoidea. *Kansas Univ. Paleont. Contr., Paper 47:* 1-32.

Kummell B., and D. Raup. 1965. Handbook of Paleontological Techniques. W. H. Freeman, San Francisco. 852 p.

Kurtén, B. 1957. Mammal migrations, Cenozoic stratigraphy, and the age of Peking man and the australopithecines. *J. Paleont. 31:* 215-227.

———. 1958. A differentiation index, and a new measure of evolutionary rates. *Evolution 12:* 146-157.

———. 1959. Rates of evolution in fossil mammals. *Cold Spring Harbor Symp., Quant. Biol. 24:* 205-215.

———. 1968. Pleistocene Mammals of Europe. Wiedenfeld & Nicholson, London. 317 p.

Kussakin, O. G. 1973. Peculiarities of the geographical and vertical distribution of marine isopods and the problem of deep-sea fauna origin. *Mar. Biol. 23*(1): 19-34.

Kuyl, O. S., J. Muller, and H. T. Waterbolk. 1955. The application of palynology to oil geology, with special reference to western Venezuela. *Geologie en Mijnbouw, n.s. 17*(3): 49-76.

Labarbera, Michael. 1974. Larval and post-larval development of five species of Miocene bivalves (Mollusca). *J. Paleont. 48*(2): 256-277.

Lagaaij, R., and P. L. Cook. 1973. Some recent Tertiary to recent Bryozoa. *In* Atlas of Palaeobiogeography, A. Hallam, Ed. Elsevier, New York. p. 489-498.

Lamb, James L., and John H. Beard. 1972. Late Neogene planktonic foraminifers in the Caribbean, Gulf of Mexico, and Italian stratotypes. *Univ. Kansas Paleont. Contr., Art. 57* (Protozoa 8): 1-67.

_____, and R. L. Hickernell. 1972. The Late Eocene to Early Miocene passage in California. *In* Proc. Pacific Coast Miocene Biostratigraphy Sympos., Pacific Sect, E. H. Stinemeyer, Ed. Soc. Econ. Paleont., Mineral., Bakersfield, Calif. p. 62-88.

Lambert, J., and P. Thiery. 1909-1925. Essai de nomenclature raissonée des Echinides. Libraire L. Ferrière Chaumont. 607 p., 15 pls.

Lang, J. 1971. Interspecific aggression by scleractinian corals. 1. The rediscovery of *Scolymia cubensis* (Milne-Edwards and Haime). *Bull. Mar. Sci. 21:* 952-959.

_____. 1973. Interspecific aggression by scleractinian corals. 2. Why the race is not only to the swift. *Bull. Mar. Sci. 23:* 260-279.

Lange, F. 1968. Conodonten-Gruppenfunde aus Kalken des tieferen Oberdevon. *Geol. Paleont. 2:* 37-57.

Lapparent, A. de. 1900. Traité de Géologie, 4th edition. Masson, Paris. 1912 + vii p.

Lapworth, C. 1878. The Moffat Series. *Geol. Soc. London, Quart. Jour. 34:* 240-346.

_____. 1879. On the tripartite classification of the Lower Paleozoic rocks. *Geol. Mag. 6:* 1-15.

_____. 1879-1880. On the Geological Distribution of the Rhabdophora. *Ann. Mag. Nat. Hist. ser. 5, 1879, 3:* 245-257, 449-455; *4:* 333-341, 423-431; 1880, *5:* 45-62, 273-285, 358-369; *6:* 16-29, 185-207.

Larson, M. L., and D. E. Jackson. 1966. Biostratigraphy of the Glenogle Formation (Ordovician) near Glenogle, British Columbia. *Can. Petrol. Geol. Bull. 14:* 486-503.

Laskarev, V. 1924. Sur les equivalents du Sarmatien Supérieur en Serbie. Rec. trav. M.I. Cvijic, Belgrade. 5 p.

Laudon, L. R. 1931. The stratigraphy of the Kinderhook series of Iowa. *Iowa Geol. Surv. 35:* 333-451.

_____. 1933. The stratigraphy and paleontology of the Gilmore City formation of Iowa. *Iowa Univ. Studies in Nat. Hist. 15*(2): 1-74.

_____, and A. L. Bowsher. 1941. Mississippian formations of Sacramento Mountains, New Mexico. *Amer. Assoc. Petrol. Geol. Bull. 25*(12): 2107-2160.

Laugel. 1858. Un tableau résumé de la classification du terrain jurassique, établie par M. le docteur Albert Oppel. *Soc. Géol. Fr., Bull. ser. 2, 15:* 657-664.

Lee, A. 1963. The hydrography of the European Arctic and Subarctic seas. *Oceanogr. Mar. Biol., Ann. Rev. 1:* 47-76.

Lee, J. S. 1927. Fusulinidae of North China. *China Geol. Surv., Paleont. Sinica, ser. B, 4*(1): 11-123, pls. 1-24.

Lehmann, Ulrich. 1967a. Ammoniten mit Kieferapparat und Radula aus Lia-Geschieben. *Kansas Univ. Paleont. Contr., Paper 41:* 38-45.

——. 1967b. Ammoniten mit Tintenbeutel. *Kansas Univ. Paleont. Contr., Paper 41:* 132-136.

——. 1968. Stratigraphie und Ammonitenführung der Ahrensburger Glazial-Geschiebe aus dem Lias epsilon (= Unt. Toarcium). *Mitt. Geol. Stinst. Hamb. 37:* 41-68.

——. 1971. New aspects in ammonite biology. *North American Paleont. Conv. (Chicago), Proc., Part 1:* 1251-1269.

——, and Wolfgang Weitschat. 1973. Zur Anatomie und Ökologie von Ammoniten; Funde von Kropf und Kiemen. *Paläont. Zeitschr. 47*(1/2): 69-76.

Lehner, Peter. 1969. Salt tectonics and Pleistocene stratigraphy on continental slope of northern Gulf of Mexico. *Amer. Assoc. Petrol. Geol. Bull. 53*(12): 2431-2479.

Leonardi, Piero, and Flavia Fiscon. 1959. La Fauna Cassiana di Cortina d'Ampezzo, pt. 3, Gasteropodi. *Univ. Padovia Inst. Geol. Mineral. Mem. 21:* 1-103, 9 pls.

Leppäkoski, E. 1971. Benthic recolonization of the Bornholm Basin (Southern Baltic) in 1969-71. *Thalassia Jugosl. 7:* 171-179.

Leven, E. Ya. 1963. O filogenii vysshikh fuzulinid i raschlenenii verkhnepermskikh otlozhenij Tetisa. *Voprosy Mikropaleontologii 7:* 57-70, 2 text-figs.

——. 1967. Stratigraphy and fusulinids of the Pamirs Permian deposits. *Akad. Nauk. SSSR Geol. Inst., Trudy 167:* 1-215, 39 pls. [in Russian].

Levins, R. 1964. The theory of fitness in a heterogeneous environment. IV. The adaptive significance of gene flow. *Evolution 17:* 635-638.

Levinton, J. 1973. Genetic variation in a gradient of environmental variability: Marine Bivalvia (Mollusca). *Science 180: 75-76.*

——. 1974. Trophic group and evolution in bivalve molluscs. *Palaeontology. 7* Pt. 3: 579-585.

Lewis, H. P. 1935. The Lower Carboniferous corals of Nova Scotia. *Ann. Mag. Nat. Hist., ser. 10, no. 91. 16:* 118-142.

Likharev, B. K. 1966. The Permian System. Stratigraphy of the SSSR, Moscow. 536 p., 18 tables [in Russian].

Lillegraven, J. A. 1969. Latest Cretaceous mammals of upper part of Edmonton Formation of Alberta, Canada, and review of marsupial-placental dichotomy in mammalian evolution. *Paleont. Contr. Univ. Kansas 50:* 1-122.

Lindsay, E. H. 1972. Small mammal fossils from the Barstow Formation, California. *Univ. Calif. Publ. Geol. Sci. 93:* 1-104.

Lindström, Maurits. 1970. Faunal provinces in the Ordovician North Atlantic areas. *Nature 255:* 1158-1159.

——. 1973. On the affinities of conodonts. *Geol. Soc. Amer. Spec. Paper 141:* 85-102.

——, and W. Ziegler. 1965. Ein Conodontentaxon aus vier Morphologisch Verschiedenen Typen. *Fortschr. Geol. Rheinl. Westfal. 9:* 209-218.

Linnarsson, G. 1876. On the vertical range of the graptolite types in Sweden. *Geol. Mag. 13:* 241-245.

Lipina, O. A. 1973. Tournaisian stratigraphy and paleobiogeography after the Foraminifera. *Akad. Nauk. SSSR, Voprosy Mikropaleontologii 16:* 3-35 [in Russian].

Lippolt, H. J., W. Gentner, and W. Wimmenauer. 1963. Altersbestimmungen nach der Kalium-Argon-Methode an Tertiären Eruptivgesteinen Südwestdeutschlands. *Jh. Geol. Landes. Baden-Wurtt. 6:* 507-538.

Lipps, J., and M. Kalisky. 1972. California Oligo-Miocene calcareous nannoplankton biostratigraphy and paleoecology. *In* Proc. Pacific Coast Miocene Biostratigraphy Symposium, E. H. Stinemeyer, Ed. Pacific Sect. Soc. Econ. Paleont., Mineral., Bakersfield, Calif. p. 239-254.

Lochman, C. 1950. Upper Cambrian faunas of the Little Rocky Mountains, Montana. *Jour. Paleont. 24*(3): 322-349.

_____. 1964. Upper Cambrian faunas from the subsurface Deadwood Formation, Williston Basin, Montana. *Jour. Paleont. 38*(1): 33-60.

_____, and C. -H. Hu. 1959. A *Ptychaspis* faunule from the Bear River Range, southeastern Idaho. *Jour. Paleont. 33*(3): 404-427.

Lochman-Balk, C. 1970. Upper Cambrian faunal patterns on the craton. *Geol. Soc. Amer. Bull. 81*(11): 3187-3224.

_____. 1971. The Cambrian of the craton of the United States. *In* Lower Palaeozoic Rocks of the World, Vol. 1 Cambrian of the New World, C. H. Holland, Ed. Wiley-Interscience, New York. p. 79-167.

_____. 1972. Cambrian System. *In* Geologic Atlas of the Rocky Mountain Region, United States of America. Rocky Mountain Assoc. of Geologists, Denver. p. 60-75.

_____. 1974. Late Dresbachian (Late Cambrian) biostratigraphy of North America. *Geol. Soc. Amer. Bull. 85*(1): 135-140.

_____, and J. L. Wilson. 1958. Cambrian biostratigraphy in North America. *Jour. Paleont. 32*(2): 312-350.

_____, and J. L. Wilson. 1967. Stratigraphy of Upper Cambrian-Lower Ordovician subsurface sequence in Williston Basin. *Amer. Assoc. Petrol. Geol Bull. 51*(6): 883-917.

Longacre, S. A. 1970. Trilobites of the Upper Cambrian Ptychaspid Biomere, Wilberns Formation, central Texas. *Paleont. Soc., Mem. 4 (Jour. Paleont. 44*(1): supp.): 1-70.

Lorenz, C. 1972. Étude des Lépidocyclines et Miogypsines. *Bull. B.R.G.M. (2d ser.), Sect. 1, No. 4:* 37-44.

Lowman, S. W. 1949. Sedimentary facies in Gulf Coast. *Amer. Assoc. Petrol. Geol. Bull. 33*(12): 1939-1997.

Lu, Yen-Hao. 1954. Upper Cambrian trilobites from Santu, Southeastern Kueichou. *Acta Paleont. Sinica 2*(2): 109-152.

Luyendyk, B. P., D. Forsyth, and J. D. Phillips. 1972. Experimental approach to the paleocirculation of the oceanic surface waters. *Geol. Soc. Amer. Bull. 83:* 2649-2664.

Lyell, C. 1833. Principles of geology, being an attempt to explain the former changes of the earth's surface, by reference to causes now in operation. Vol. 3. John Murray, London, 398 + xviii p., 1 pl. (Appendix I: Tables of fossil shells by G. P. Deshayes, 109 p., 3 pls.).

———. 1865. Elements of Geology, 6th Edition. London. 794 p.

Lys, Maurice, and A. F. de Lapparent. 1971. Foraminiferes et microfacies du Permien de l'Afghanistan Central. *Paris Natl. Mus., Nat. Hist. Notes and Mem. 12* (pt. 1): 47-133, pls. 7-22, 13 text-figs.

MacNeil, F. S. 1965. Evolution and distribution of the genus *Mya*, and Tertiary migrations of mollusca. *U.S. Geol. Surv. Prof. Paper 483G:* G1-G51, pls. 1-11.

MacPherson, A. H. 1965. The origin of diversity in mammals of the Canadian Arctic Tundra. *Syst. Zool. 14:* 153-173.

Macqueen, R. W., and E. W. Bamber. 1967. Stratigraphy of Banff Formation and lower Rundle Group (Mississippian), southwestern Alberta. *Canada Geol. Surv. Paper 67-47:* 1-37.

———, and E. W. Bamber. 1968. Stratigraphy and facies relationships of the Upper Mississippian Mount Head Formation, Rocky Mountains and foothills, southwestern Alberta. *Canadian Petrol. Geol. Bull. 16*(3): 225-287.

———, and E. W. Bamber, and B. L. Mamet. 1972. Lower Carboniferous stratigraphy and sedimentology of the southern Canadian Rocky Mountains. *24th Inter. Geol. Congr. Guidebook for Excursion C17:* 1-62.

Macsotay, Oliver. 1971. Zonacion del Post-Eoceneo de la paleoprovincia Caribe-Antillana a base de taxa de *Turritella* (Molusco: Gasteropodo). *Bol. Informativo Assoc. Venezolana Geol., Min. y Pet. 12*(2): 8-60, 1 pl.

Maczynska, S. S. 1958. Jezowce rodzaju *Discoidea* z cenomanu i turonu okolic Krakowa, Miechowa i Wolbromia (Cenomanian and Turonian Echinoids of genus *Discoidea* from the vicinity of Krakow, Miechow and Wolbrom). *Prace Muzeum Ziemi 2:* 81-115, 36 figs., 9 pls.

———. 1968. Echinoids of the Genus *Micraster* L. Agassiz from the Upper Cretaceous of the Cracow-Miechow area. *Prace Muzeum Ziemi 12:* 87-168, 1 fig., 22 text-pls., 28 pls.

Maglio, Vincent J. 1973. Origin and evolution of the Elephantidae. *Amer. Phil. Soc. Trans. 63*(3): 1–149.

Makarov, R. R. 1969. Transport and distribution of Decapoda larvae in the plankton of the western Kamchatka shelf. *Okeanologyi 9:* 306-317. [Translated in: *Oceanology 9:* 251-261, Scripta Technica.]

Makowski, Henryk. 1962 [1963]. Problem of sexual dimorphism in ammonites. *Palaeontologica Polonica 12:* 1-92.

Malakhova, N. P. 1973. On the stratigraphic position of the Gusikin suite in the southern Urals. *Akad. Nauk. SSSR, Ural. Nautch. Centr., Trudy Inst. Geol. Geok. Bull. 82:* 127-170, 15 pls. [in Russian].

Mamet, B. 1962. Remarques sur la microfaune de Foraminifères du Dinantien. *Bull. Soc. Belge Géol. Paléont. Hydrol. 70*(2): 166-173.

———. 1972. Un essai de reconstitution paléoclimatique basé sur les microflores algaires du Viséen. 24th Inter. Geol. Congr., Sect. 7, Palaeont. 282-291.

———, and B. Skipp. 1970. Lower Carboniferous Foraminifera: Preliminary zonation and stratigraphic implication for the Mississippian of North America. 6th Inter. Congr. Carboniferous Stratigraphy, Sheffield. 3: 1129-1146.

Mangold-Wirz, K. 1963. Biologie des cephalopodes benthiques et nektoniques de la Mer Catalane. *In* Vie et Milieu. *Paris Univ. Lab. Anago Publ., Suppl. 13.*

Manning, T. H. 1956. The northern red-backed mouse *Clethrionomys rutilus* (Pallas), in Canada. *Nat. Mus. Can. Bull. 144:* 1-67.

Mantell, G. 1822. The fossils of the South Downs; or illustrations of the geology of Sussex. Jupton Relfe, London. 327 + xvi p., 42 pls.

Marche-Marchad, I. 1968. Remarques sur le développement chez les Cymba, Prosobranches Volutides et l'hypothèse de leur origine sud américaine. *Bull. Inst. Fond. Afr. Noire 30A:* 1028-1037.

Marcou, J. 1848. Recherche géologique sur le Jura salinois. *Soc. Géol. Fr., Mém. Ser. 2, 3:* 1-151, 2 pls.

Martini, E. 1971. Standard Tertiary and Quaternary calcareous nannoplankton zonation. *In* Proceedings of the II Planktonic Conference, Rome, 1970, A. Farinacci, Ed. Ediz. Tecnoscienza. p. 739-785.

_____. 1972. Der stratigraphischen Wert von Silicoflagellaten im Jungtertiär von Kalifornien und des östlichen Pacifischen Ozeäns. *Deutsch. Geol. Ges. Nachrichten 5:* 47-49.

Mashkova, Tamara V. 1972. *Ozarkodina steinhornensis* (Ziegler) apparatus, its conodonts and biozone. *Geol. Palaeont. SB 1:* 81-90.

Massé, H. 1971. Étude quantitative de la macrofaune de peuplements des sables fins infralittoraux. II. La Baie du Prado (Golfe de Marseille). *Tethys 3:* 113-158.

Matsumoto, Tatsuro. 1973. Late Cretaceous Ammonoidea. *In* Atlas of Palaeobiogeography, A. Hallam, Ed. Elsevier, Amsterdam. p. 421-429.

Mayer-Eymar, K. 1857-1858. Versuch eine synchronistischen Tabelle der Tertiär-Gebilde Europas. *Verh. Schweiz. Naturf. Ges. 6:* 7-23.

Mayr, E. 1951. Speciation in birds. Proc. Xth Inter. Ornithol. Congr., Uppsala, 1950. 91-131.

_____. 1963. Animal species and evolution. Harvard University Press, Cambridge, Mass. 797 p.

McFarlan, Edward J. 1961. Radiocarbon dating of late Quaternary deposits, South Louisiana. *Geol. Soc. Amer. Bull. 72*(1): 129-158.

McGrew, P. O., and R. Sullivan. 1970. The stratigraphy and paleontology of Bridger A. *Univ. Wyo. Contrib. Geol. 9:* 66-85.

McKenna, M. C. 1962. Collecting small mammals by washing and screening. *Curator 5:* 221-235.

Meek, F. B. 1864. Description of the Carboniferous fossils. *California Geol. Surv., Paleont. 1:* 1-16, pls. 1, 2.

_____, and F. V. Hayden. 1858. Remarks on the Lower Cretaceous beds of Kansas and Nebraska, together with descriptions of some new species of Carboniferous fossils from the valley of the Kansas River. *Acad. Nat. Sci. Philadelphia Proc. 1858 10:* 256-266.

_____, and A. H. Worthen. 1873. Description of invertebrates from Carboniferous System. *Illinois Geol. Surv. 5:* 321-619.

Meijer, M. 1965. The stratigraphical distribution of Echinoids in the Chalk and Tuffaceous Chalk in the neighbourhood of Maastricht (Netherlands). *Mededelingen Geol. Stichting, N.S. 17:* 21-25, 1 fig.

Mein, P. 1975. Resultats du Group de Travail des Vertébrés. *Rep. Activity R.C.M.N.S. Work. Gr. (1971-1975):* 78-81.

_____, and M. Freudenthal. 1971. Une nouvelle classification des Cricetidae (Mammalia, Rodentia) du Tertiaire de l'Europe. *Scripta Geol. 2:* 1-37.

———, G. Truc, and G. Demarq. 1971. Micromammifères et gastropodes continentaux des biozones de Paulhiac et de La Romien dans le Miocène de La Bastidonne et de Mirabeau (Vaucluse, Sud est de la France). *C.R. Acad. Sci. Paris 273:* 566-568.

Meischner, D. 1968. Perniciöse, Epökie von *Placunopsis* auf *Ceratites. Lethaia:* 156-174.

Melton, William, and Harold Scott. 1973. Conodont-bearing animals from the Bear Gulch Limestone, Montana. *Geol. Soc. Amer. Spec. Paper 141:* 31-65.

Menzies, R. J., and R. Y. George. 1969. Polar faunal trends exhibited by Antarctic isopod Crustacea. *Antarctic Jour. 4:* 190-191.

———, and R. Y. George, and G. T. Rowe. 1973. Abyssal environment and ecology of the world oceans. Wiley-Interscience, New York. 488 p.

Merkt, J. 1966. Über Austern und Serpeln als Epöken auf Ammonitengehäusen. *Neues Jahrb. Geol. Paläont. Abh. 125:* 467-479.

Merriam, C. W. 1941. Fossil turritellas from the Pacific Coast Region of North America. *Calif. Univ. Pub. Geol. 26*(1): 1-214, 41 pls.

Merrill, Glen K. 1971. North American Pennsylvanian conodont biostratigraphy. *Geol. Soc. Amer. Mem. 127:* 395-414.

———. 1973. Pennsylvanian Conodont paleoecology. *Geol. Soc. Amer. Spec. Paper 141:* 239-274.

Metcalf, W. G., A. D. Voorhis, and M. C. Stalcup. 1962. The Atlantic Equatorial Undercurrent. *J. Geophys. Res. 67:* 2499-2508.

Meyerhoff, A. A. (Ed.). 1968. Geology of natural gas in south Louisiana. *In* Natural gases of North America–Pt. 1, Natural gases in rocks of Cenozoic age. *Amer. Assoc. Petrol. Geol. Mem. 9:* 1: 376-581.

Middlemiss, F. A. 1973. The geographical distribution of Lower Cretaceous Terebratulacea in Western Europe. *In* The Boreal Lower Cretaceous, R. Casey and P. F. Rawson, Eds. *Geol. Jour. Spec. Issue 5:* 111-120.

———, and P. F. Rawson. 1971. Faunal provinces in space and time–some general considerations. *In* Faunal Provinces in Space and Time, F. A. Middlemiss, P. F. Rawson and G. Newall, Eds. *Geol. Jour. Spec. Issue 4:* 199-210.

Miklukho-Maklai, A. D. 1959a. The system and phylogeny of the fusulinids (the genus *Triticites* and other genera similar to it). *Leningrad Univ., Vestnik, 6, Geol. Geogr. 1:* 5-23, 1 text-fig. [in Russian].

———. 1959b. On the stratigraphic role, taxonomy, and phylogenesis of *Staffella*-like foraminifers. *Akad. Nauk. SSSR, Doklady 125*(3): 628-631, 2 text-figs. [in Russian].

———. 1963a. K. Obosnovaniyn osnovnogo deleniya Kamennougol'noy Permskoy sistem. *Leningrad Univ., Vestnik, 18, Geol. Geogr. 3:* 54-56, 2 figs., 1 chart.

———. 1963b. Verkhniy Paleozoy Sredney Azii (Upper Paleozoic of Central Asia). *Leningrad Univ. Izdatel'stvo:* 1-328, 8 pls., 18 figs., 13 tables.

Mileikovsky, S. A. 1968. Distribution of pelagic larvae of bottom invertebrates of the Norwegian and Barents Sea. *Mar. Biol. 1:* 161-167.

———. 1971. Types of larval development in marine bottom invertebrates, their distribution and ecological significance: a re-evaluation. *Mar. Biol. 10:* 193-213.

Milkman, R., R. Zeitler, and J. F. Boyer. 1972. Spatial and temporal genetic variation in *Mytilus edulis:* natural selection and larval dispersal. *Biol. Bull. Mar. Biol. Lab., Woods Hole 143:* 470 [Abstract].

Miller, A. H. 1956. Ecological factors that accelerate formation of races and species of terrestrial vertebrates. *Evolution 10:* 262-277.

Miller, T. G. 1965. Time in stratigraphy. *Palaeontology 8:* 113-131.

Mitey, D. 1968. Studies on the Systematization of *Clethrionomys glareolus* Schreber from the Rhodopes and Balkan Range. Ec. norm. sup. "Paissi Hilendarski" Plovdiv, *Trav. sci. Biol. 6:* 179-184.

Mojsisovics, E. V. 1873. Das Gebirge um Hallstatt, Part 1, Die Cephalopoden der Hallstatter Kalke. *Wien, Geol. Reichsanst.* 6(1): 1-82.

Möller, V. von. 1880. Uber Einige Foraminiferenfuhrende Gesteine Persiens. *Geol. Reichsanstalt (Wien), Jahrbuch 30* (pt. 4): 573-586, pls. 9, 10.

Montenat, C., and M. Crusafont. 1970. Découverte de Mammiféres dans le Néogène et le Pleistocène du Levant espagnol (Provinces d'Alicante et de Murcia). *Acad. Sci. Paris, C. R. Hebd. Sér. D, 270:* 2434-2437.

Moore, R. C. 1948. Paleontological features of Mississippian rocks in North America and Europe. *Jour. Geol. 56*(4): 373-402.

_____. 1954. Evolution of late Paleozoic invertebrates in response to major oscillations of shallow seas. *Bull. Mus. Comp. Zool. 112:* 259-286.

_____. 1955. Expansion and contraction of shallow seas as a causal factor in evolution. *Evolution 9:* 482-483.

_____, and P. C. Sylvester-Bradley. 1957. Suggested new article: Proposed recognition of the concept "parataxon" and the provision of rules for the nomenclature of units of this category. *Bull. Zool. Nomenclature 15:* 5-13.

_____ et al. 1965. Treatise on Invertebrate Paleontology, Part H. Brachiopoda, vols. 1, 2. Geol. Soc. Amer. and University of Kansas Press.

Moore, R. E. et al. 1944. Correlation of Pennsylvanian formation of North America. *Geol. Soc. Amer. Bull. 55:* 657-706, pl. 1.

Morris, H. T., and T. S. Lovering. 1961. Stratigraphy of the East Tintic Mountains Utah. *U.S. Geol. Survey Prof. Paper 361:* 1-145.

Morrison, D. F. 1967. Multivariate statistical methods. McGraw-Hill, New York. 338 p.

Mortensen, Th. 1898. Die Echinodermlarven der Plankton-Expedition nebst einer Systematischen Revision der bisher bekannten Echinodermenlarven. *Ergebn. Plankton Exped. ii, J:* 1-120.

_____. 1921. Studies of the Development and Larval Forms of Echinoderms. Copenhagen. 266 p.

_____. 1928. A monograph of the Echinoidea. I. Cidaroidea. C. A. Reitzel, Copenhagen. 551 p., 173 figs., 88 pls.

_____, and I. Lieberkind. 1928. Echinoderma. *In* Die Tierwelt der Nord- und Ostsee, Grimpe and Wagler, Eds. 8: 128 p., 126 figs.

Mouterde, R., R. Enay, E. Cariou, D. Contini, S. Elmi, J. Gabilly, C. Mangold, J. Mattei, M. Rioult, J. Thierry, and H. Tintant. 1971. Les zones de Jurassique en France. Extr., *C. R. Sommaire des Seances, Soc. Geol. France 6:* 1-27, text-figs. 1-11.

Mu, A. T. 1963. Research in graptolite faunas of Chilianshan. *Scient. Sinica 12:* 347-371.

Müller, A. H. 1969. Ammoniten mit "Eirbeutel" und die Frage nach dem Sexual-Dimorphismus der Ceratiten (Cephalopoda). *Berlin Deutsch. Akad. Wiss. 11:* 411-420.

Müller, C. 1974. Calcareous nannoplankton from mid-Tertiary stratotypes. V Congrès Néogène Mediterranéen, Lyon, (1971) Bur. Rech. Géol. Min., no. 78, tome 1, 427–432.

Müller, C. 1971. Calcareous nannoplankton from mid-Tertiary stratotypes. Ve Congrès Néogène Mediterranéen, Lyon, 1971 [Preprint].

Müller, H. 1966. Palynological investigations of Cretaceous sediments in northeastern Brazil. Proceedings of the Second West African Micropaleontological Colloquium, Ibadan. Leiden. p. 123-136.

Muller, J. 1959. Palynology of Recent Orinoco delta and shelf sediments. *Micropaleontology 5:* 1-32.

———. 1970. Palynological evidence on early differentiation of angiosperms. *Biological Reviews 45* (3): 417–450.

Murray, Grover E. 1961. Geology of the Atlantic and Gulf Coastal Province of North America. Harper & Row, New York. 692 p.

Murray, John W. 1973. Distribution and ecology of living benthic foraminiferids. Crane, Russak, New York. 274 p.

Mutvei, Harry, and R. A. Reyment. 1973. Buoyancy control and siphuncle function in ammonoids. *Palaeontology 16:* 623-636.

Muus, K. 1973. Settling, growth and mortality of young bivalves in the Øresund. *Ophelia 12:* 79–116.

Nalivkin, D. V. 1957. The zoogeographical provinces of the Devonian period of the USSR territory. Trudy I sessii vses. paleont. obsh. p. 77–80 [in Russian].

Natland, M. L. 1957. Paleoecology of the West Coast Tertiary sediments. *In* Treatise on Marine Paleoecology, H. S. Ladd, Ed. *Geol. Soc. Amer. Mem. 67*(2): 543-572.

Nelson, C. A. 1951. Cambrian trilobites from the St. Croix Valley. *Jour. Paleont. 25*(6): 765-784.

Nelson, S. J. 1959. Mississippian *Syringopora* of western Canada. *Alberta Soc. Petrol. Geol. Jour. 7*(4): 91-92.

———. 1960. Mississippian lithostrotionid zones of the southern Canadian Rocky Mountains. *Jour. Paleont. 34*(1): 107-126.

———. 1961. Mississippian faunas of western Canada. *Geol. Assoc. Canada Spec. Paper 2:* 1-39.

———. 1962. Analysis of Mississippian *Syringopora* from the southern Canadian Rocky Mountains. *Jour. Paleont. 36*(3): 442–460.

Nestler, H. 1965. Die Rekonstruktion des Lebensraumes der Rügener Schreibkreide-Fauna (Unter-Maastricht) mit Hilfe der Paläoökologie und Paläobiologie. *Geologie, Jahrg. 14, Beih. 49:* 1-147.

———. 1967. Die quantitative Verteilung der Fauna in einem Profil der weissen Schreibkreide (Unter-Maastricht) an der Ernst-Moritz-Arndt-Sicht auf Rügen. *Ber. deutsch. Ges. geol. Wiss. A. Geol. Paläont. 12:* 535-547.

Newman, K. R. 1974. Palynomorph Zones in Early Tertiary Formations of the Piceance Creek and Uinta Basins, Colorado and Utah: Rocky Mountain Assoc. Geologists–1974 Guidebook, p. 47-55.

Nichols, D. 1959. Changes in the Chalk Heart-Urchin *Micraster* interpreted in relation to living Forms. *Philos. Transact. Roy. Soc. London, Ser. B, Biol. Sci. 693, 242:* 347–437, 46 figs., 13 tables.

Nicollet, J. N. 1843. Report intended to illustrate a map of the hydrographical basin of the upper Mississippi River. U.S. 26th Cong., 2d sess., Senate Document 237 (House Document 52): 1-170.

Nielsen, C. 1971. Entoproct life-cycles and entoproct-ectoproct relationships. *Ophelia 9:* 209-341.

Noetling, F. 1893. Carboniferous fossils from Tennasserim. *India Geol. Survey Records 26:* 96-100, text-figs. 1-1b.

Norris, G. 1967. Spores and pollen from the Lower Colorado Group (Albian-?Cenomanian) of central Alberta. *Palaeontographica 120B:* 72-115.

Nuttall, T. 1821. Observations on the geological structure of the valley of the Mississippi. *Philadelphia Acad. Nat. Sci. Jour. 2:* 14-52.

Obradovich, J. D., and W. A. Cobban. 1973. A Time-Scale for the Late Cretaceous of the Western Interior of North America. *Prog., Abstr., Ann. Mtng. Geol. Assoc. Canada, Cretaceous Colloq.:* 42-43.

_____, and W. A. Cobban. 1975. A time-scale for the Late Cretaceous of the Western Interior of North America. *Geol. Assn. Canada, Spec. Pap. 13:* 31-54, 3 text-figs.

Ocamb, R. D. 1961. Growth faults of south Louisiana. *Gulf Coast Assoc. Geol. Soc., Trans. 11:* 139-175.

Ockelmann, K. W. 1958. The Zoology of East Greenland. Marine Lamellibranchiata. *Medd. Grønland 122:* 1-256.

_____. 1965. Developmental types in marine bivalves and their distribution along the Atlantic coast of Europe. *Proc. First European Malacological Congr. 1962:* 25-35.

Odin, G. S. 1973a. Sur les datations radiométriques du Miocène de la Paratéthys. *Bull. Soc. Geol. France (7), 15:* 2 p.

_____. 1973b. Resultats de datations radiométriques dans des séries sédimentaires du Tertiaire de l'Europe occidentale. *Rev. Géogr. Phys. et Géol. Dynam. (2), XV*(3): 317-330.

_____, Marcel Gulinck, Jacques Bodelle, and Claude Lay. 1969. Géochronologie de niveaux glauconieux Tertiaires du bassin de Belgique (méthode potassium-argon). *C.R. Séances Soc. Géol. France 5*(6): 1969: 198-199.

_____, Jacques Bodelle, Claude Lay, and Charles Pomerol. 1970. Géochronologie de niveaux glauconieux paléogènes d'Allemagne du Nord (méthode potassium-argon). Résultats préliminaires. *C.R. Somm. Séances Soc. Géol. France 1970:* 220-221.

Opdyke, N. D. 1972. Paleomagnetism of deep-sea cores. *Rev. Geophys. Space Phys. 10*(1): 213-249.

_____, L. H. Burckle, and A. Todd. 1974. The extension of the magnetic time-scale in sediments of the Central Pacific Ocean. *Earth and Planetary Sci. Newsletters* (in press).

Oppel, Albert. 1856-1858. Die Juraformation Englands, Frankreichs und des sudwestlichen Deutschlands, nach ihren einzelnen gliedern eingetheilt und verglichen. von Ebner and Seubert, Stuttgart. (Originally published in three parts in Abdruck der Württemb. naturw. Jahreshefte 12-14; 1856, 1-438; 1857, 439 bis-694 + map; 1858, 695-857 + table).

_____. 1863. Ueber jurassische Cephalopoden. *Paläont. Mitt. Mus. Konigl. Bayer Staats. 1:* 127-266.

Orbigny, A. d'. 1842-1843. Paléontologie Française. Description zoologique et géologique de tous les animaux mollusques et rayonnés fossiles de France. 2 (Gastropoda). Victor Masson, Paris. 456 p., pls. 149-236bis.

———. 1842-1851. Paléontologie Française; description des Mollusques et Rayonnés fossiles. Terrains jurassiques 1, Céphalopodes. Victor Masson, Paris. 642 p., 234 pls.

———. 1849-1852. Cours élémentaire de Paléontologie et de Géologie stratigraphique. Victor Masson, Paris. 1: 299 p.; 2(1): 382 p.; 2(2): p. 383-847.

Orlowski, S. 1967. The Stratigraphy of the Upper Cambrian of the Holy Cross Mountains. *Acad. Polonaise Sci. Bull., Sér. Sci. Géol. et Géogr. 15*(1): 47-50.

———. 1968. Upper Cambrian fauna of the Holy Cross Mountains. *Acta Geol. Polonica 18*(2): 257-292, pls. 1-8.

Osborne, F. F., and W. B. N. Berry. 1966. Tremadoc rocks at Levis and Lauzon. *Naturaliste Canadien 93:* 133-143.

Ostrom, M. E. 1970. Sedimentation cycles in the Lower Paleozoic rocks of western Wisconsin. *In* Field Trip Guidebook for Cambrian-Ordovician Geology of Western Wisconsin, M. E. Ostrom, R. A. Davis, Jr., and L. M. Cline, Eds. *Wisconsin Univ., Geol. Nat. History Survey, Inf. Circ. 11:* 10-34.

Ota, Shigeru. 1961. Identification of the larva of *Pinna atrina japonica* (Reeve). *Bull. Jap. Soc. Sci. Fish 27:* 107-111.

Owen, D. D. 1852. Report of a geological survey of Wisconsin, Iowa, and Minnesota; and incidentally of a portion of Nebraska territory made under instructions from the United States Treasury Department, Philadelphia. 638 p.

Owens, J. P. 1969. Coastal Plain rocks. *In* The Geology of Harford County, Maryland. Maryland Geol. Survey, Baltimore. p. 77-102.

———, and N. F. Sohl. 1969. Shelf and deltaic paleoenvironments in the Cretaceous-Tertiary formations of the New Jersey Coastal Plain. *In* Geology of Selected Areas in New Jersey and Eastern Pennsylvania and Guidebook of Excursions, S. Subitzky, Ed. Rutgers University Press, New Brunswick, N.J. p. 235-278.

Ozawa, Tomowo. 1967. *Pseudofusulinella*, a genus of Fusulinacea. *Palaeont. Soc. Japan, Trans. Proc., n.s., 68:* 149-173, pls. 14-15, 5 text-figs.

———. 1970. Notes on the phylogeny and classification of the superfamily Verbeekinoidea. *Kyushu Univ., Fac. Sci., Mem., ser. D, 20*(1): 17-58, 9 pls., 13 text-figs.

Ozawa, Yoshiaki. 1925a. Paleontological and stratigraphical studies on the Permo-Carboniferous limestone of Nagato. Pt. 2. Paleontology. *Tokyo, Imp. Univ. Jour. 45*(6): 1-90, pls. 1-14.

———. 1925b. On the classification of Fusulinidae. *Tokyo Imp. Univ., Coll. Sci., Jour. 45*(4): 1-26, pls. 1-4, text-figs. 1-3.

———. 1927. Stratigraphical studies of the Fusulina limestones of Akasaka, Province of Mino. *Tokyo Imp. Univ., Coll. Sci., Jour., ser. 2, 2:* 121-164.

Pacltová, B. 1961. Zur Frage der Gattung *Eucalyptus* in der bohmischen Kreideformation. *Preslia 33:* 113-129.

———. 1971. Palynological study of Angiospermae from the Peruc Formation (?Albian-Lower Cenomanian) of Bohemia. *Ústřední ústav geologický, Sborník geologických věd, paleontologie, řada P13:* 105-141.

Paden-Phillips, P., and C. J. Felix. 1971. A study of Lower and Middle Cretaceous spores and pollen from the southeastern United States. II. Pollen. *Pollen et Spores 13*(3): 447-473.

Page, R. W., and I. McDougall. 1970. Potassium-argon dating of the Tertiary F1-2 stage in New Guinea and its bearing on the geological time-scale. *Amer. Jour. Sci. 269:* 321-342.

Paine, R. T. 1969. The *Pisaster-Tegula* interaction: prey patches, predator food preference, and intertidal community structure. *Ecology 50:* 950-961.

_____. 1971. A short-term experimental investigation of resource partitioning in a New Zealand rocky intertidal habitat. *Ecology 52:* 1096-1106.

Palframan, D. F. B. 1967. Variation and ontogeny of some Oxford Clay ammonites; *Distichoceras bicostatum* (Stahl) and *Horioceras baugieri* (d'Orbigny), from England. *Palaeontology 10, Pt. 1:* 60-94.

Palmer, A. R. 1960. Some aspects of the early Upper Cambrian stratigraphy of White Pine County, Nevada and vicinity. Intermountain Assoc. Petrol. Geol., 11th Ann. Field Conf. 53-58.

_____. 1965a. Biomere – A new kind of biostratigraphic unit. *Jour. Paleont. 39*(1): 149-153.

_____. 1965b. Trilobites of the Late Cambrian Pterocephaliid Biomere in the Great Basin, United States. *U.S. Geol. Survey, Prof. Paper 493:* 1-106, 20 pls.

_____. 1968. Cambrian trilobites of east-central Alaska. *U.S. Geol. Survey, Prof. Paper 559-B:* 1-115, 15 pls.

_____. 1971. The Cambrian of the Great Basin and adjacent areas, western United States. *In* Lower Palaeozoic Rocks of the World, vol. 1, Cambrian of the New World, C. H. Holland, Ed. Wiley-Interscience, New York. 78 p.

_____. 1973. Cambrian Trilobites. *In* Atlas of Palaeobiogeography, A. Hallam, Ed. Elsevier, New York. p. 3-11.

Pander, C. H. 1856. Monographie der fossilen Fische des silurischen Systems der russisch-baltischen Gouvernements. St. Petersburg. 91 + x p.

Papp, A. 1951. Das Pannon des Wiener Beckens. *Mitt. Geol. Ges. Wien. 39-41 (1946-1948):* 99-193, 7 text-figs., 4 tables.

_____. 1958. Morphologisch-genetische Studien an Mollusken des Sarmats von Wiesen (Burgenland). *Wiss. Arb. Burgenld. 22:* 1-39, 11 text-figs.

_____. 1959. Tertiär. 1. Pt. Grundzüge regionaler Stratigraphie. *Handb. Strat. Geol. 3/1:* 1-411, 89 text-figs., 63 tables.

_____. 1963a. Die biostratigraphische Gliederung des Neogens im Wiener Becken. *Mitt. Geol. Ges. Wien. 56:* 255-317, 17 pls.

_____. 1963b. Das Verhalten neogener Molluskenfaunen bei verschiedenen Salzgehalten. *Fortschr. Geol. Rheinld. and Westf. 10:* 35-48, 3 pls.

_____, and K. Küpper. 1954. The Genus Heterostegina in the Upper Tertiary of Europe. *Contr. Cush. Found. Foram. Res. 5:* 107-127, 8 pls.

_____, and M. E. Schmid. 1971. Zur Entwicklung der Uvigerinen im Badenien des Wiener Beckens. Verh. Geol. Bundesanst. 47-58.

_____, and J. Senes. 1974. M 5 – Sarmatien. (Sensu E. Suess, 1866). *Chronostrat. and Neostrat. 4:* 1-707.

_____, and F. Steininger. 1973. Die stratigraphischen Grundlagen des Miozäns der Zentralen Paratethys und die Korrelationsmöglichkeiten mit dem Neogen Europas. Verh. Geol. Bundesanst. 59-65.

_____, and F. F. Steininger. 1975a. Pannonian (sensu Stevanovic, 1951). F. F. Steininger and L. Neveskaja, Eds. *Stratotypes of Mediterranean Neogene Stages 2:* 121-126.

_____, and F. F. Steininger. 1975b. Sarmatian sensu stricto (Suess, 1866) F. F. Steininger and L. Neveskaja, Eds. *Stratotypes of the Mediterranean Neogene Stages 2:* 139-148.

_____, and E. Thenius. 1954. Vösendorf – ein Lebensbild aus dem Pannon des Wiener Beckens. *Mitt. Geol. Ges. Wien. 46:* 1-109, 15 pls.

_____, and K. Turnovsky. 1953. Die Entwicklung der Uvigerinen im Vindobon (Helvet und Torton) des Wiener Beckens. *Jb. Geol. Bundesanst. 96:* 117-142, 1 pl.

_____ et al. 1968. Nomenclature of the Neogene of Austria. Verh. Geol. Bundesanst. 9-27, 1 table.

_____, F. Rögl, and F. Steininger. 1970. Führer zur Paratethys – Exkursion 1970 in die Neogen-Gebiete Osterreichs. 57 p.

_____, F. F. Steininger, and F. Rögl. 1971. Bericht über die Ergebnisse der 3. Sitzung der Arbeitsgruppe Paratethys des Committee Mediteranean Neogene Stratigraphy 1970 in Wien. Verh. Geol. Bundesanst. 59-62.

_____, F. Rögl, and J. Senes. 1973. M 2 – Ottnangien. *Chronostrat. & Neostrat. 3:* 1-841, 44 text-figs., 82 pls.

Parker, Robert L. 1974. Stacking marine magnetic anomalies: a critique. *Geophys. Res. Lett. 1*(6): 259-260.

Parks, J. M. 1951. Corals from the Brazer Formation (Mississippian) of northern Utah. *Jour. Paleont. 25*(2): 171-186.

Paulson, O. L., Jr. 1960. Ostracoda and stratigraphy of Austin and Taylor equivalents of northeast Texas (abstract). *Dissert. Abs. 20*(11): 4370.

Pchelentsev, V. F. 1953. Gastropod fauna of the Upper Cretaceous deposits of Transcaucasia and Central Asia. Izvestiya Akad. Nauk USSR. 388 p., 51 pls.

_____. 1954. Gastropods from the Upper Cretaceous beds of the Armyansku SSR and overlying parts of Azerbaidzhan SSR, Izvestiya Akad. Nauk USSR. 178 p., 23 pls.

Peake, N. B., and J. M. Hancock. 1961. The Upper Cretaceous of Norfolk. *Transact. Norfolk Norwich Nat. Soc. 19*(6): 293-339, 7 figs., 1 table.

Perch-Nielsen, K. 1968. Der Feinbau und die Klassifikation der Coccolithen aus dem Maastrichtien von Dänemark. *Biol. Skr. Dan. Vid. Selsk. 16*(1): 1-96.

Perner, Jaroslav. 1894-1899. Etudes sur les Graptolites de Boheme, Pts. 1-3: Raimund Gerhard, Prague, pt. 1 (1894), 1-14; pt. 2 (1895), 1-31; pt. 3a (1897), 1-25; pt. 3b (1899), 1-24.

Pessagno, E. A., Jr. 1969. Upper Cretaceous stratigraphy of the western Gulf Coast area of Mexico, Texas, and Arkansas. *Geol. Soc. Amer. Mem. 111:* 1-139.

Petocz, R. C. 1970. Biostratigraphy and Lower Permian Fusulinidae of the Upper Delta River area, east-central Alaska Range. *Geol. Soc. Amer., Spec. Paper 130:* 1-94, 10 pls., 7 text-figs.

Phillips, J. 1829. Illustrations of the Geology of Yorkshire; or, a description of the strata and organic remains of the Yorkshire Coast. The Author, York, 192 + xvi p., 23 pls., 1 map.

_____. 1844. Memoirs of William Smith, L.L.D., author of the 'Map of the Strata of England and Wales'. John Murray, London. 150 + ix p.

Phillips, J. D. and D. Forsyth. 1972. Plate tectonics, paleomagnetism, and the opening of the Atlantic. *Geol. Soc. Amer. Bull. 83:* 1579-1600.

Phleger, Fred B., and Frances L. Parker. 1951. Ecology of Foraminifera, northwest Gulf of Mexico. *Geol. Soc. Amer. Mem. 46*(Pt. 1): 1-88; (Pt. 2): 1-64.

Pierce, R. L. 1961. Lower Upper Cretaceous plant microfossils from Minnesota. *Minnesota Geol. Survey Bull. 42:* 1-86.

Pietzner, Horst, Johanna Vahl, Hans Werner, and Willi Ziegler. 1968. Zur Chemischen Zusammensetzung und Micromorphologie der conodonten. *Paleontographica 128(A):* 115-152.

Pilkington, M. C., and V. Fretter. 1970. Some factors affecting the growth of prosobranch veligers. *Helgoländer wiss. Meeresunters. 20:* 576-593.

Pjastolova, O. A. 1972. The role of rodents in the energetics of biogeocenosess of forest-tundra and southern tundra. *In* Tundra Biome, F. E. Wielgolaski and T. Rosswall, Eds. Swedish IBP Committee, Stockholm. p. 128-130.

Poag, C. Wylie. 1971. A reevaluation of the Gulf Coast Pliocene-Pleistocene Boundary. *Gulf Coast Assoc. Geol. Soc., Trans. 21:* 291-308.

——. 1972a. Shelf-edge submarine banks in the Gulf of Mexico: paleoecology and biostratigraphy. *Gulf Coast Assoc. Geol. Soc., Trans. 22:* 267-287.

——. 1972b. Correlation of early Quaternary events in the U.S. Gulf Coast. *Quat. Res. 2: 447-469.*

——. 1972c. Gulf Coast submarine banks as potential hydrocarbon traps. *Gulf Coast Assoc. Geol. Soc., Trans. 22:* 73-83.

——. 1973. Late Quaternary sea levels in the Gulf of Mexico. *Gulf Coast Assoc. Geol. Soc., Trans. 23:* 394-400.

——, and W. H. Akers. 1967. *Globigerina nepenthes* Todd of Pliocene age from the Gulf Coast. *Cushman Found. Foram Res., Contr. 18*(4): 168-175.

——. and B. R. Sidner. 1976. Foraminiferal biostratigraphy of the shelf edge: a key to late Quaternary paleoenvironments. *Paleogeogr., Palaeoclimatol., Palaeoecol. 19:* 17-37.

——, and William E. Sweet, Jr. 1972. Claypile Bank, Texas continental shelf. *In* Contributions on the Geological and Geophysical Oceanography of the Gulf of Mexico, Richard Rezak and Vernon J. Henry, Eds. *Texas A&M Univ. Oceanogr. Stud. 3:* 223-261.

——. and P. C. Valentine. 1976. Biostratigraphy and ecostratigraphy of the Pleistocene Basin: Texas-Louisiana Continental Shelf: Gulf Coast Assoc. *Geol. Soc. Trans. 26:* 185-256.

Pollock, C. A. 1968. Questionable Silurian natural conodont assemblages from Indiana. *Abstr. Prog. N. Cent. Sect. Geol. Soc. Amer.:* 49.

Popenoe, W. P. 1942. Upper Cretaceous Formations and Faunas of Southern California. *Bull. Amer. Assoc. Petrol. Geol. 26*(2): 162-187.

——. 1957. The Cretaceous genus *Biplica*, its evolution and biostratigraphic significance. *California Univ. Pubs. Geol. 30*(6): 425-454, pls. 50-51.

Popiel-Barczyk, E. 1958. Jezowce rodzaju *Conulus* z turonu okolic Krakowa, Miechowa i Wolbromia (The Echinoid genus *Conulus* in the vicinity of Krakow, Miechow and Wolbrom). *Prace Muzeum Ziemi 2:* 41-79, 36 figs., 17 tables, 5 pls.

Poslavskaja, N. A. et al. 1959. Echinoidea *In* Atlas Werchnemelowoj Fauny sewernowo Kawkasa i Kryma, M. M. Moskvin, Ed. Gostoptechisdat Pub., Moskow. p. 242-304, figs. 41-109, 1 table, 20 pls.

Post, L. von. 1967. Forest tree pollen in south Swedish peat bog deposits. *Pollen et Spores 9*(3): 375-401. [translation by M. B. Davis and K. Faegri, introduction by K. Faegri and J. Iversen]

Powell, J. R. 1971. Genetic polymorphism in varied environments. *Science 174:* 1035-1036.

Pujol, C. 1970. Contribution à l'étude des Foraminifères planctoniques néogènes dans le Bassin Aquitaine. *Bull. Inst. Géol. Bassin Aquitaine 9:* 201-219.
Quenstedt, F. A. 1856-1858. Der Jura. Tubingen, H. Lauppschen. 842 + vi p., 101 pls. (published in parts: April 1856, 1-208, pls. 1-24; September 1856, 209-368, pls. 25-49; December 1856, 369-576, pls. 50-72; May 1857, 577-823, pls. 73-100; 1858, title pages and indexes).
Raabe, H. 1965. Die irregulären Echiniden aus dem Cenoman und Turon der Baskischen Depression (Nordspanien) in ihrer stratigraphischen Stellung. *Neues Jahrb. Geol. Paläont., Abh. 121:* 95-110, 19 figs.
Raasch, G. O. 1951. Revision of Croixan dikelocephalids. *Illinois State Acad. Sci., Trans. 44:* 85-128 (reprinted 1952; *Illinois State Geol. Surv. Circ. 179:* 137-151).
Rabeder, G., and F. F. Steininger. 1975. Die direkten Biostratigraphischen Korrelationsmöglichkeiten von Säugetierfaunen aus dem Oligo/Miozän der Zentralen Paratethys. Proc. 6th Congr. R.C.M.N.S.Bratislava, Sept. 4-7. 1975, 1: 177-183.
Radig, F. 1973. Beiträge zur Kenntnis der höheren Oberkreide der Baskischen Depression (Nordspanien) und ihrer Echinozoen-Fauna. *Erlanger Geol. Abh. 94:* 1-68, 92 figs., 4 tables, 11 pls.
Radinsky, L. 1966. The adaptive radiation of phenacodontid condylarths and the origin of the Perissodactyla. *Evolution 20:* 408-417.
Radwin, G. E., and J. L. Chamberlin. 1973. Patterns of larval development in Stenoglossan gastropods. *Trans. San Diego Soc. Nat. Hist. 17:* 107-117.
Rafinesque, C. S., and J. D. Clifford. 1820. Prodrome d'une monographie des turbinolies fossiles du Kentucky (dans l'Ameriq. septentr.). *Ann. Génér. Sci. Phys., Bruxelles 5:* 231-235.
Rasetti, F. 1948. Middle Cambrian trilobites from the conglomerates of Quebec (exclusive of the Ptychopariidea). *Jour. Paleont. 22:* 315-339.
——. 1959. Trempealeauian trilobites from the Conococheague, Frederick, and Grove limestones of the Central Appalachians. *Jour. Paleont. 33*(3): 375-398.
Rasmussen, H. W. 1950. Cretaceous Asteroidea and Ophiuroidea, with special reference to the species found in Denmark. *Danm. Geol. Unders. II rk. 77:* 1-134.
——. 1961. A monograph on the Cretaceous Crinoidea. *Biol. Skr. Dan. Vid. Selsk. 12*(1): 1-428.
Raup, D. M. 1966. Geometric analysis of shell coiling; general problems. *Jour. Paleont. 40:* 1178-1190.
——. 1967. Geometric analysis of shell coiling; coiling in ammonoids. *Jour. Paleont. 41:* 43-65.
——. 1972. Taxonomic diversity during the Phanerozoic. *Science 177:* 1065-1071.
——, and J. A. Chamberlain. 1967. Equations for volume and center of gravity in ammonoid shells. *Jour. Paleont. 41:* 566-574.
—— , and S. M. Stanley. 1971. Principles of Paleontology. Freeman, San Francisco. 388 p.
Rauzer-Chernousova, D. M. 1940. Stratigraphy of the upper Carboniferous and Artinskian stage on the western slope of the Urals and materials concerning

the faunas of fusulinids. *Akad. Nauk SSSR, Inst. Geol. Namk, Trudy, 7, Geol. Ser. 2:* 37-104, pls. 1-6, 6 text-figs., 3 tables [in Russian; English summary].

——. 1961. Biostratigraficheskoe raschlenenie po foraminiferam srednekamennougol' nykh otlozhenij Samarskoj Luki i srednego Zavolzh'ja. *Akad. Nauk SSSR, Geol. Inst., Regionl'nya Stratigrafija SSSR 5:* 149-212, 2 pls., 2 text-figs.

——. 1965. Foraminifery stratotipicheskogo razreza Sakmarskogo Yarusa (Foraminifers in the stratotypical section of the Sakmarian Stage). *Akad. Nauk SSSR, Geol. Inst., Trudy 135:* 1-79, 6 pls., 5 figs., 5 tables.

——, and E. L. Kulik. 1949. Relationships between fusulinids and facies and on the periodicity in their evolution. *Akad. Nauk SSSR, Izv., Ser. Geol. 6:* 131-148, 6 text-figs.

Ravn, J. P. J. 1928. De Regulaere Echinider i Danmarks Kridtaflejringer. *Kgl. Danske Vidensk. Selsk. Skrifter, naturv. og mathem. Afd 9.R. 1*(1): 1-63, 12 figs., 6 pls.

Rees, C. B. 1952. Continuous plankton records: first report on the distribution of Lamellibranch larvae in the North Sea. *Hull Bull. Mar. Ecol. 3:* 105-133.

——. 1954. Continuous plankton records: the distribution of echinoderm and other larvae in the North Sea. *Hull Bull. Mar. Ecol. 4:* 47-67.

Reeside, J. B., Jr., and W. A. Cobban. 1960. Studies of the Mowry Shale (Cretaceous) and contemporary formations in the United States and Canada. *U.S. Geol. Surv. Prof. Paper 355:* 1-126.

——, and A. A. Weymouth. 1931. Mollusks from the Aspen Shale (Cretaceous) of southwestern Wyoming. *U.S. Natl. Mus., Proc. 78, art. 17:* 1-24.

Renevier, E. 1901. [Rapport de la] commission internationale de classification stratigraphique. *Int. Geol. Congr. 8 (Paris):* 192-203.

Rensberger, J. M. 1971. Entoptychine pocket gophers (Mammalia, Geomyoidea) of the early Miocene John Day Formation, Oregon. *Univ. Calif. Publ. Geol. Sci. 90:* 1-209.

Renz, Carl, and M. Reichel. 1900. Beitrage zur Stratigraphie und Palaontologic des ostmediterranean Jungpalaozoikums und dassen Einordnung im griechischen Gebirgssystem. 1 und 2 Teil Geologic und Stratigraphie von Carl Renz.

Renz, H. H. 1948. Stratigraphy and fauna of the Agua Salada group, State of Falcon, Venezuela. *Geol. Soc. Amer. Mem. 32:* 1-219.

Reyment, R. A. 1955. Some examples of homeomorphy in Nigerian Cretaceous ammonites. *Geol. Fören. Stockholm Forh. 77*(4): 567-594.

——. 1958. Some factors in the distribution of fossil cephalopods. *Stockholm Contr. Geol. 1:* 97-184.

——. 1973. Factors in the distribution of fossil cephalopods. 3, Experiments with exact models of certain shell types. *Uppsala Univ. Geol. Inst. Bull., n. ser., 4:* 7-41.

Rhoads, D. C., and D. K. Young. 1970. The influence of deposit-feeding organisms on sediment stability and community trophic structure. *Jour. Mar. Res. 28:* 150-178.

Rhodes, F. H. T. 1952. A classification of Pennsylvanian conodont assemblages. *Jour. Paleont. 26:* 886-901.

——. 1954. The zoological affinities of the conodonts. *Cambridge Philos. Soc., Biol. Reviews 29:* 419-452.

_____. 1957. Comment on the Moore-Sylvester-Bradley "Parataxa Plan". *Bull. Zool. Nomenclature 15:* 305-312.

_____. 1962. Recognition, interpretation, and taxonomic position of conodont assemblages. Treatise on Invert. Paleont. Geol. Soc. Amer. W70-W83.

_____. 1973. Conodont Paleozoology. *Geol. Soc. Amer. Spec. Paper 141:* 1-296.

_____, and R. L. Austin. 1971. Carboniferous conodont faunas of Europe. *Geol. Soc. Amer. Mem. 127:* 317-352.

_____, R. L. Austin, and E. C. Druce. 1969. British Avonian (Carboniferous) conodont faunas, and their value in local and intercontinental correlation. *Brit. Mus. Nat. Hist. Bull. Geol. Supp. 5:* 1-313.

Richard, M. 1946. Contribution a l'étude du bassin Aquitaine. Les gisements de Mammiferes Tertiaires. *Soc. Geol. France, Mem. N.S. 52:* 1-380.

Rickards, R. B. 1967. The Wenlock and Ludlow succession in the Howgill Fells, Northern England (north-west Yorkshire and Westmorland). *Quart. Jour. Geol. Soc. London 123:* 215-251.

_____. 1970. The Llandovery (Silurian) graptolites of the Howgill Fells, Northern England. Palaeontographical Soc. (Monograph): 1-108.

Ricketts, E. F., and J. Calvin. 1968. Between Pacific Tides, 4th ed., revised by J. W. Hedgpeth. Stanford University Press, Stanford, Calif. 614 p.

Riedel, W. R., and A. Sanfilippo. 1970. Radiolaria, Leg 4, Deep Sea Drilling Project. *In* Initial Reports of the Deep Sea Drilling Project, Volume IV, R. G. Bader et al., Eds. Government Printing Office, Washington, D.C. p. 503-575.

Riva, John. 1969. Middle and Upper Ordovician graptolite faunas of St. Lawrence lowlands of Quebec and of Anticosti Island. *In* North Atlantic – Geology and Continental Drift, Marshall Kay, Ed. *Amer. Assoc. Petrol. Geol. Mem. 12:* 513-556.

Robertson, Robert. 1964. Dispersal and wastage of larval *Philippia krebsii* (Gastropoda: Architectonicidae) in the North Atlantic. *Acad. Nat. Sci. Philadelphia, Proc. 116*(1): 1-27.

_____. 1971. Scanning electron microscopy of planktonic larval marine gastropod shells. *Veliger 14*(1): 1-12, 9 pls.

_____. 1973. On the fossil history and intrageneric relationships of *Philippia* (Gastropoda: Architectonicidae). *Proc. Phil. Acad. Sci. 125:* 37-46.

_____. 1974 [1976]. Marine prosobranch gastropods: larval studies and systematics. Thalassia Yugoslavica 10: 213-238.

Robinson, P. 1966. Fossil mammalia of the Huerfano Formation, Eocene, of Colorado. *Bull. Peabody Mus. Nat. Hist., Yale Univ. 21:* 1-95.

Robison, R. A. 1964a. Middle-Upper Cambrian boundary in North America. *Geol. Soc. Amer. Bull. 75*(10): 987-994.

_____. 1964b. Upper Middle Cambrian stratigraphy of western Utah. *Geol. Soc. Amer. Bull. 75*(10): 995-1010.

Rögl, F. 1975. Die planktonischen Foraminiferen der Zentralen Paratethys. Proc. 6th Congr. R.C.M.N.S. Bratislava, Sept. 4-7, 1975. 1: 113-120.

Rohlf, F. J. 1970. Adaptive hierarchical clustering schemes. *System. Zool. 19:* 58-82.

_____. 1972. An empirical comparison of three ordination techniques in numerical taxonomy. *System. Zool. 21:* 271-280.

Roll, Artur. 1935. Über Frasspuren an Ammonitenschalen. *Neues Jahrb. Min., Geol., Paläont., Monatsh., Jahrg. 1935, Pt. B:* 120-124.

Romer, A. S. 1966. Vertebrate Paleontology. 3rd Ed. University Chicago Press, Chicago.

Rosenkrantz, A., and H. W. Rasmussen. 1960. South-eastern Sjaelland and Mön, Denmark. Int. Geol. Congr., XXI Sess., Norden. Guide to Excursions Nos. A4 and C37, Pt. 1: 1-17.

Rosewater, J. 1961. The family Pinnidae in the Indo-Pacific. *Indo-Pacific Mollusca 1:* 175-186.

Rosovskaya, S. Ye. 1969. K revisii otryada Fusulinida (Revision of the Order Fusulinida). *Paleont. Zhur. 3:* 34-44.

Ross, C. A. 1961. Fusulinids as paleoecological indicators. *Jour. Paleont. 35*(2): 398-400, 1 text-fig.

———. 1962. The evolution and dispersal of the Permian fusulinid genera *Pseudoschwagerina* and *Paraschwagerina. Evolution 16*(3): 306-315, 4 text-figs.

———. 1963. Standard Wolfcampian Series (Permian), Glas Mountains, Texas. *Geol. Soc. Amer. Mem. 88:* 1-205, 29 pls., 11 text-figs., maps, correlation charts, columnar secs., geol. sec., 4 tables.

———. 1967. Development of fusulinid (Foraminiferida) faunal realms. *Jour. Paleont. 41*(6): 1341-1354, 9 text-figs.

———. 1969. Paleoecology of Triticites and Dunbarinella in Upper Pennsylvanian strata of Texas. *Jour. Paleont. 43:* 298-311, 8 text-figs.

———. 1970. Concepts in Late Paleozoic correlation. *In* Radiometric Dating and Paleontologic Zonation, O. L. Bandy, Ed. *Geol. Soc. Amer., Spec. Paper 124:* 7-36, 11 text-figs.

Ross, R. J., Jr. 1951. Stratigraphy of the Garden City Formation in northeastern Utah, and its trilobite faunas. *Yale Univ., Peabody Mus. Nat. Hist. Bull. 6:* 1-161, 36 pls.

———, and W. B. N. Berry. 1963. Ordovician graptolites of the Basin Ranges in California, Nevada, Utah and Idaho. *U.S. Geol. Surv. Bull. 1134:* 1-177.

———, and J. K. Ingham. 1970. Distribution of the Toquima-Table Head (Middle Ordovician Whiterock) faunal realm in the Northern Hemisphere. *Geol. Soc. Amer. Bull. 81:* 393-408.

Rossolimo, O. L. 1964. On the intraspecies variability of the bank vole. *Zool. Zh. 43:* 749-756.

Rowe, A. W. 1899. An analysis of the genus *Micraster,* as determined by rigid zonal collection from the zones of *Rhynchonella cuvieri* to that of *Micraster cor-anguinum. Quart. Jour. Geol. Soc., London 55:* 494-547, 1 fig., 1 table, 5 pls.

Rowe, G. T., and R. J. Menzies. 1969. Zonation of large benthic invertebrates in the deep-sea off the Carolinas. *Deep-Sea Research 16:* 531-537.

Rowell, A. J., D. J. McBride, and A. R. Palmer. 1973. Quantitative study of Trempealeauian (latest Cambrian) trilobite distribution in North America. *Geol. Soc. Amer. Bull. 84*(10): 3429-3442.

Rozovskaia, S. E. 1963. The earliest fusulines and their ancestors. *Akad. Nauk. SSSR, Inst. Paleont., Trudy 97:* 1-127, 22 pls.

Rudwick, M. J. S. 1970. Living and fossil brachiopods. Hutchison University Library, London. 200 p.

Ruedemann, Rudolf. 1904. Graptolites of New York. Part 1; Graptolites of the lower beds. *New York State Museum Memoir 7.*

_____. 1908. Graptolites of New York. Part 2; Graptolites of the higher beds. *New York State Museum Memoir 11.*

_____. 1947. Graptolites of North America. *Geol. Soc. Amer. Mem. 19:* 1-652.

Runnegar, Bruce, and N. D. Newell. 1971. Caspian-like relict molluscan fauna; in the South American Permian. *Bull. Amer. Mus. Nat. Hist. 146:* 1-66.

Russell, D. E. 1964. Les Mammifères paléocènes d'Europe. *Mem. Mus. Nat. d'Hist. Natur., Paris (C), 13:* 1-324.

Rützler, K. 1970. Spatial competition among Porifera; solution by epizoism. *Oecologia (Berl.) 5:* 85-95.

Ruzhentsev, V. E., and T. G. Sarycheva. 1965. The development and change of marine organisms at the Paleozoic-Mesozoic boundary. Akad. Nauk. SSSR, Trudy 108. [translated by D. A. Brown, 1968; Australian Natl. Univ. Publ. 117]

Ryan, W. B. F., M. B. Cita, M. Dreyfus Rawson, L. H. Burckle, and T. Saito. 1974. A Paleomagnetic Assignment of Neogene Stage Boundaries and the Development of Isochronous Datum Planes between the Mediterranean, the Pacific and the Indian Oceans. In Order to Investigate the response of the World Ocean to the Mediterranean "Salinity Crisis". *Riv. Ital. Paleont. 80:* 631-688.

Ryland, J. 1965. Polyzoa (Bryozoa) order Cheilostomata, Cyphonautes larvae. *Fiche Ident. Zooplancton, I.C.E.S. 107:* 1-6.

Sachs, Jules B., and Hubert C. Skinner. 1973. Calcareous nannofossils and late Pliocene-early Pleistocene biostratigraphy Louisiana Continental Shelf. *Tulane Stud. Geol. Paleont. 10*(3): 113-162.

Sada, Kimiyoshi. 1964. Carboniferous and lower Permian fusulines of the Atetsu Limestone in west Japan. *Hiroshima Univ., Jour. Sci., ser. C, 4*(3): 225-269, pls. 21-28, 23 tables.

Sage, N. M., Jr. 1954. The stratigraphy of the Windsor Group in the Antigonish quadrangles and the Mahone Bay-St. Margaret Bay area, Nova Scotia. *Nova Scotia Dept. Mines Mem. 3:* 1-168.

Salter, J. W. 1855. Account of the Arctic Carboniferous fossils. *In* The Last of the Arctic Voyages, E. Belcher, Ed. L. Reeve, London, *Palaeontology 2:* 377-389, pl. 36.

Sandberg, Charles A., and Willi Ziegler. 1973. Refinement of Standard Upper Devonian conodont zonation based on sections in Nevada and West Germany. *Geol. Palaeont. 7:* 97-122.

Sanders, H. L. 1968. Marine benthic diversity: a comparative study. *Amer. Natur. 102:* 243-282.

_____, and R. R. Hessler. 1969. Ecology of the deep-sea benthos. *Science 163:* 1419-1424.

Sanderson, G. A., and G. J. Verville. 1970. Morphologic variability of the genus *Schwagerina* in the Lower Permian Wreford Limestone of Kansas. *Palaeontology 13*(pt. 2): 175-183, pls. 35-36, 8 text-figs.

Sando, W. J. 1960. Distribution of corals in the Madison Group and correlative strata in Montana, western Wyoming, and northeastern Utah. *In* Geological Survey Research 1960. *U.S. Geol. Survey Prof. Paper 400-B:* B225-B227.

_____. 1964. Stratigraphic importance of corals in the Redwall Limestone, northern Arizona. *In* Geological Survey Research 1964. *U.S. Geol. Survey Prof. Paper 501-C:* C39-C42.

——. 1967a. Mississippian depositional provinces in the northern Cordilleran region. *In* Geological Survey Research 1967. *U.S. Geol. Survey Prof. Paper 575-D:* D29-D38.

——. 1967b. Madison Limestone (Mississippian), Wind River, Washakie, and Owl Creek Mountains, Wyoming. *Amer. Assoc. Petrol. Geol. Bull. 51*(4): 529-557.

——. 1969. Corals. *In* History of the Redwall Limestone of Northern Arizona, E. D. McKee and R. C. Gutschick, Eds. *Geol. Soc. Amer. Mem.114:* 257-342 [1970].

——, and J. T. Dutro, Jr. 1960. Stratigraphy and coral zonation of the Madison Group and Brazer Dolomite in northeastern Utah, western Wyoming, and southwestern Montana. Wyoming Geol. Assoc. 15th Ann. Field Conf. Guidebook: 117-126.

——, B. L. Mamet, and J. T. Dutro, Jr. 1969. Carboniferous megafaunal and microfaunal zonation in the northern Cordillera of the United States. *U.S. Geol. Survey Prof. Paper 613-E:* 1-25.

——, E. W. Bamber, and A. K. Armstrong. 1975. Endemism and similarity indices: clues to the zoogeography of North American Mississippian corals. *Geology, 3*(11): 661-664.

Sapper, K. T. 1894a. Grundzüge der physikalischen Geographie von Guatemala. *Petermanns Mitt. 24*(113): 1-59, pls. 1-4.

——. 1894b. Informe sobre la geografía física y la geologia de los Estados de Chiapas y Tabasco. *Bol. Agr. Min. e Ind. 3*(9): 187-211.

Sará, M. 1970. Competition and co-operation in sponge populations. *Symp. Zool. Soc. London 25:* 273-284.

Sarytcheva, T. G., and J. B. Waterhouse. 1972. New Brachiopod species of the family Retariidae on the Permian of North Canada. *Paleont. Zhurn. Moscow 4:* 62-74.

Savage, Donald E., and Lawrence G. Barnes. 1972. Miocene vertebrate chronology of the West Coast of North America. *In* West Coast Miocene. AAPG-SEPM Symp., Bakersfield, Calif., March 9-10, 1972 (cyclostyled): 124-145.

Say, Thomas. 1823. *In* Account of an expedition from Pittsburgh to the Rocky Mountains . . . , Edwin James, Ed. Carey & Lea, Philadelphia. p. 146-152.

Schaeffer, B., M. K. Hecht, and N. Eldredge. 1972. Phylogeny and paleontology. *In* Evolutionary Biology, Th. Dobzhansky et al., Eds. Appleton-Century-Crofts, New York 6. p. 31-46.

Schäfer, W. 1962. Aktuo-Paläontologie nach Studien in der Nordsee. Frankfurt a.M: 666 p., 277 figs., 36 pls.

Schellwien, E. 1898. Die Fauna des karnischen Fusulinenkalks. II: Foraminifera. *Palaeontographica 44:* 237-282, pls. 17-24.

——, and G. Dyhrenfurth. 1909. Monographie der Fusulinen. Teil II. Die asiatischen Fusulinen. A. Die Fusulinen von Darwas. *Palaeontographica 56:* 137-176, pls. 13-16.

——, and H. von Staff. 1908. Monographie der Fusulinen. Teil I. Die Fusulinen des russischartischen Meeresgebietes. *Palaeontographica 55:* 145-194, pls. 13-20.

——, and H. von Staff. 1912. Monographie der Fusulinen. Teil III. Die Fusulinen Nordamerikas. *Palaeontographica 59:* 157-192, pls. 15-20, text-figs. 1-17.

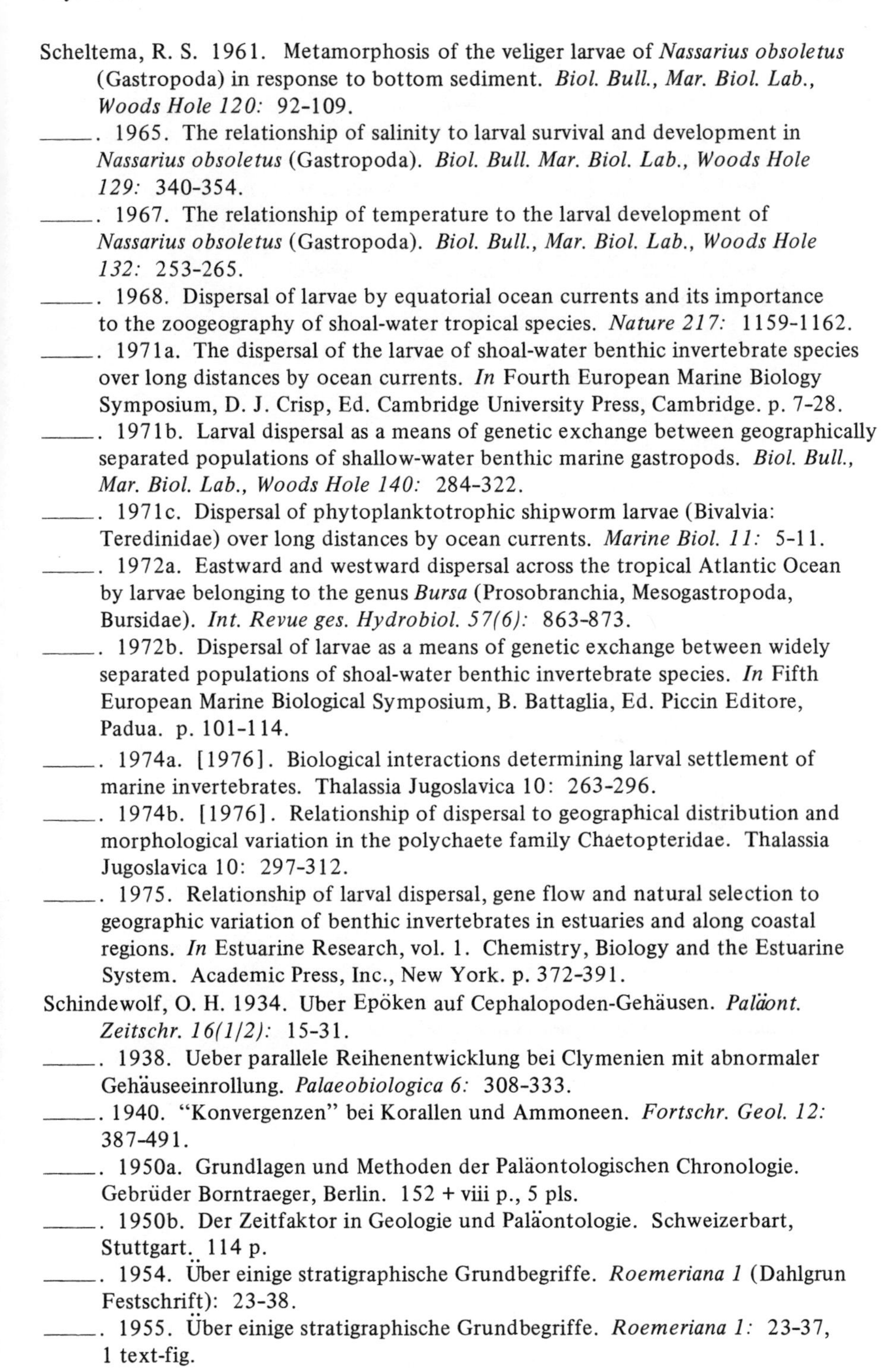

Scheltema, R. S. 1961. Metamorphosis of the veliger larvae of *Nassarius obsoletus* (Gastropoda) in response to bottom sediment. *Biol. Bull., Mar. Biol. Lab., Woods Hole 120:* 92-109.

______. 1965. The relationship of salinity to larval survival and development in *Nassarius obsoletus* (Gastropoda). *Biol. Bull. Mar. Biol. Lab., Woods Hole 129:* 340-354.

______. 1967. The relationship of temperature to the larval development of *Nassarius obsoletus* (Gastropoda). *Biol. Bull., Mar. Biol. Lab., Woods Hole 132:* 253-265.

______. 1968. Dispersal of larvae by equatorial ocean currents and its importance to the zoogeography of shoal-water tropical species. *Nature 217:* 1159-1162.

______. 1971a. The dispersal of the larvae of shoal-water benthic invertebrate species over long distances by ocean currents. *In* Fourth European Marine Biology Symposium, D. J. Crisp, Ed. Cambridge University Press, Cambridge. p. 7-28.

______. 1971b. Larval dispersal as a means of genetic exchange between geographically separated populations of shallow-water benthic marine gastropods. *Biol. Bull., Mar. Biol. Lab., Woods Hole 140:* 284-322.

______. 1971c. Dispersal of phytoplanktotrophic shipworm larvae (Bivalvia: Teredinidae) over long distances by ocean currents. *Marine Biol. 11:* 5-11.

______. 1972a. Eastward and westward dispersal across the tropical Atlantic Ocean by larvae belonging to the genus *Bursa* (Prosobranchia, Mesogastropoda, Bursidae). *Int. Revue ges. Hydrobiol. 57(6):* 863-873.

______. 1972b. Dispersal of larvae as a means of genetic exchange between widely separated populations of shoal-water benthic invertebrate species. *In* Fifth European Marine Biological Symposium, B. Battaglia, Ed. Piccin Editore, Padua. p. 101-114.

______. 1974a. [1976]. Biological interactions determining larval settlement of marine invertebrates. Thalassia Jugoslavica 10: 263-296.

______. 1974b. [1976]. Relationship of dispersal to geographical distribution and morphological variation in the polychaete family Chäetopteridae. Thalassia Jugoslavica 10: 297-312.

______. 1975. Relationship of larval dispersal, gene flow and natural selection to geographic variation of benthic invertebrates in estuaries and along coastal regions. *In* Estuarine Research, vol. 1. Chemistry, Biology and the Estuarine System. Academic Press, Inc., New York. p. 372-391.

Schindewolf, O. H. 1934. Uber Epöken auf Cephalopoden-Gehäusen. *Paläont. Zeitschr. 16(1/2):* 15-31.

______. 1938. Ueber parallele Reihenentwicklung bei Clymenien mit abnormaler Gehäuseeinrollung. *Palaeobiologica 6:* 308-333.

______. 1940. "Konvergenzen" bei Korallen und Ammoneen. *Fortschr. Geol. 12:* 387-491.

______. 1950a. Grundlagen und Methoden der Paläontologischen Chronologie. Gebrüder Borntraeger, Berlin. 152 + viii p., 5 pls.

______. 1950b. Der Zeitfaktor in Geologie und Paläontologie. Schweizerbart, Stuttgart. 114 p.

______. 1954. Über einige stratigraphische Grundbegriffe. *Roemeriana 1* (Dahlgrun Festschrift): 23-38.

______. 1955. Über einige stratigraphische Grundbegriffe. *Roemeriana 1:* 23-37, 1 text-fig.

_____. 1970. Stratigraphie und Stratotypus. *Akad. Wiss. & Lit. Mainz, Abh. Math.-Naturwiss. Kl:* 1-134.

Schmid, F. 1972. *Hagenowia elongata* (Nielson), ein hochspezialisierter Echinide aus dem höheren Untermaastricht NW-Deutschlands. *Geol. Jb. A4:* 177-195, 2 figs., 4 pls.

Schmidt, H. 1934. Conodonten-Funde in ursprünglichen Zusammenhang. *Paläontol. Z. 16:* 76-85.

_____, and K. J. Müller. 1964. Funde von Conodonten-Gruppen aus dem oberen Karbon des Sauerlandes. *Paläontol. Z. 38:* 105-135.

Schopf, T. J. M. 1966. Conodonts of the Trenton Group (Ordovician) in New York, Southern Ontario and Quebec. *Bull. New York St. Mus. Sci. Serv. 405:* 1-93.

_____, and J. L. Gooch. 1972. A natural experiment to test the hypothesis that loss of genetic variability was responsible for mass extinction of the fossil record. *Jour. Geol. 800:* 481-483.

_____, David M. Raup, Stephen Jay Gould, and Daniel S. Simberloff. 1975. Genomic Versus morphologic rates of evolution: Influence of morphologic complexity. *Paleobiology 1(1):* 63-70.

Schwager, C. 1883. Karbonische Foraminiferen aus China und Japan. *In* von Richthofens, F. F., China, vol. 4, Palaeont. Theil, Abhandl. 7: 106-159, pls. 15-18. Reimer, Berlin.

_____. 1887. Salt Range fossils, Protozoa. *Palaeontologia Indica, ser. 13, 1:* 983-990, pls. 126-128.

Schwarzbach, Martin. 1936. Zur Lebensweise der Ammoniten. *Natur. u. Volk 66* (1): 8-11.

Scott, Gayle. 1940. Paleoecological factors controlling the distribution and mode of life of Cretaceous ammonoids in the Texas area. *Jour. Paleont. 14* (4): 299-323.

Scott, G. H. 1972. Phyletic trends for trans-Atlantic lower Neogene *Globigerinoides. Rev. Esp. Micropal. 3* (3): 283-292.

Scott, H. W. 1934. The zoological relationships of the conodonts. *J. Paleont. 8:* 438-445.

_____. 1942. Conodont assemblages from the Heath Formation, Montana. *J. Paleont. 16:* 293-300.

_____. 1969. Discoveries bearing on the nature of the conodont animal: Micropaleontology. *Geol. Soc. Amer. Special Paper 141:* 420-426.

Scupin, Hans. 1912. Welche Ammoniten waren benthonisch, welche Swimmer? *Deutsch Geol. Gesell. Zeitschr. 21:* 350-367.

Seddon, G. 1970. Frasnian conodonts from the Sadler Ridge-Bugle Gap area, Canning Basin, Western Australia. *J. Geol. Soc. Australia 16:* 723-753.

_____, and W. C. Sweet. 1971. An ecologic model for conodonts. *J. Paleont. 45:* 869-880.

Seilacher, Adolf. 1960. Epizoans as a key to ammonite ecology. *Jour. Paleont. 34:* 189-193.

_____. 1963. Umlagerung und Rolltransport von Cephalopoden-Gehäusen. *Neues Jahrb. Geol. Paläont., Monatsh., Jahrg. 1963:* 593-615.

_____. 1966. Lobenlibellen und Fullstruktur bei Ceratiten. *Neues Jahrb. Geol. Paläont., Abh. 125:* 480-488.

_____. 1968. Sedimentations Prozesse in Ammoniten-gehäusen. *Akad. Wiss. u. Literatur Abh., Math.-Naturw. Kl. 9:* 191-203.

_____. 1971. Preservational history of Ceratite shells. *Palaeontology 14:* 593-615.

Selli, R. 1970. Report on the absolute age. CMNS Proc. IV Session, Bologna, 1967. *Biorn. Geol. 25* (1): 51-59.

Senes, J. 1959. Sucasne znalosti o stratigrafii centralnej Paratetydy. *Geol. Prace 55.*

_____. 1960. Entwicklungsphasen der Paratethys. *Mitt. Geol. Ges. Wien 52:* 181-187.

_____. 1961. Paläogeographie des Westkarpathischen Raumes in Beziehung zur übrigen Paratethys im Miözan. *Geol. Prace 60:* 159-195.

_____. 1967. M 3 - Karpatien. *Chronostrat. & Neostrat. 1:* 1-312, 57 pls.

_____et al. 1971. Korrelation des Miozäns der Zentralen Paratethys. *Geol. Sbornik 22:* 3-9, 5 tables.

Serafinski, W. 1969. Ecological structure of the species in mammals. 2. The intra-specific differentiation of the red bank vole (*Clethrionomys glareolus* (Schreb.)) in the light of environmental conditions. *Ekol. pol. 16A:* 193-211.

Seward, A. C. 1931. Plant Life through the Ages. Cambridge University Press, Cambridge.

Shaw, A. B. 1955. Paleontology of northwestern Vermont. V. The Lower Cambrian fauna. *Jour. Paleont. 29:* 775-805.

_____. 1964. Time in Stratigraphy. McGraw-Hill, New York: 365 + xiv p.

_____. 1969. Presidential address: Adam and Eve, paleontology and the non-objective arts. *Jour. Paleont. 43:* 1085-1098.

Shcherbovich, S. F. 1969. Fusulinides of the late Gzhelian and Asselian time of the Pre Caspian syneclise. *Akad. Nauk. SSSR Geol. Inst. Trudy 176:* 1-82, 18 pls. [in Russian].

Sheng, J. C. 1963a. Permian fusulinids of Kwangsi, Kueichow and Szechuan. *Palaeont. Sinica, n.s. B, 10:* 1-247, 36 pls.

_____. 1963b. The marine Permian formations and their fusulinid zones of southwest China. *Scientia Sinica 12* (6): 885-890.

Shumard, B. F. 1858. Notice of new fossils from the Permian strata of New Mexico and Texas, collected by Dr. Geo. G. Shumard, Geologist of the U.S. Govt. Exped. for obtaining water by means of artesian wells along the 32nd parallel, under the direction of Capt. John Pope, U.S. Corps Top. Eng. *St. Louis Acad. Sci., Trans. 1:* 290-297.

_____. 1859. Notice of fossils obtained from the Permian strata of Texas and New Mexico . . . *Acad. Sci., St. Louis, Trans. 1:* 387-403.

Shuto, T. 1974. Larval ecology of prosobranch gastropods and its bearing on biogeography and paleontology. *Lethaia 7:* 239-256.

Sickenberg, O. 1939. Die Insectenfresser. Fledenmäuse und Nagetiere der Höhlen von Goyet (Belgien). *Bull. Mus. Nat. Hist. Nat. Belg. 15:* 1-23.

Sidner, Bruce R., and C. Wylie Poag. 1972. Late Quaternary climates indicated by foraminifers from the southwestern Gulf of Mexico. *Gulf Coast Assoc. Geol. Soc., Trans. 22:* 305-313.

Simpson, G. G. 1944. Tempo and Mode in Evolution. Columbia Univ. Press, New York. 237 p.

_____. 1947. Holarctic Mammalian Faunas and Continental Relationships during the Cenozoic. *Bull. Geol. Soc. Amer. 68:* 613-688, 6 text-figs., 9 tables.

_____. 1953a. Evolution and geography, an essay on historical biogeography with special reference to mammals. Oregon State System of Higher Education, Eugene. 64 p.

_____. 1953b. The Major Features of Evolution. Columbia University Press, New York. 434 p.

Skevington, David. 1969. Graptolite faunal provinces in Ordovician of Northwest Europe. *In* North Atlantic - Geology and Continental Drift, Marshall Kay, Ed. *Amer. Assoc. Petrol. Geol. Mem. 12:* 557-562.

——. 1973. Ordovician graptolites. *In* Atlas of Palaeobiogeography, A. Hallam, Ed. Elsevier, Amsterdam. p. 27-35.

Skinner, Hubert C. (Ed.). 1972. Gulf Coast stratigraphic correlation methods with an atlas and catalogue of principal index Foraminiferida. Louisiana Heritage Press, New Orleans. 213 p.

Skinner, J. W., and G. L. Wilde. 1954a. Fusulinid wall structure. *Jour. Paleont. 28:* 445-451, pls. 46-52.

——, and G. L. Wilde. 1954 b. New early Pennsylvania fusulinids from Texas. *Jour. Paleont. 28:* 796-803, pls. 95-96.

——, and G. L. Wilde. 1965. Permian biostratigraphy and fusulinid faunas of the Shasta Lake area, northern California. *Kansas Univ., Paleont. Contr. [no. 39], Protozoa, Art. 6:* 1-98, 65 pls., 3 figs.

——, and G. L. Wilde. 1966. Permian fusulinids from Pacific Northwest and Alaska. *Kansas Univ., Paleont. Contr., Paper 4:* 1-64, 49 pls., 10 figs.

Sliter, W. V. 1972. Cretaceous foraminifers - depth habitats and their origin. *Nature 239* (5374): 514-515.

Sloan, R. F. 1955. Paleoecology of the Pennsylvanian marine shales of Palo Pinto County, Texas. *Jour. Geol. 63:* 412-428.

Sloane, B. J. 1971. Recent developments in the Miocene *Planulina* gas trend of South Louisiana. *Gulf Coast Assoc. Geol. Soc., Trans. 21:* 199-210.

Sloss, L. L. 1958. Paleontologic and lithologic associations. *Jour. Paleont. 32* (4): 715-729.

Smart, J., and N. F. Hughes. 1972. The insect and the plant: progressive palaeoecological integration. *In* Insect/Plant Relationships, H. F. van Emden, Ed. *Symposia of the Royal Entomological Society of London 6:* 143-155.

Smiser, J. S. 1935. A Revision of the Echinoid Genus *Echinocorys* in the Senonian of Belgium. *Mém. Mus. Roy. d'Hist. Nat. Belgique 67:* 1-52, 25 figs., 2 pls.

Smith, A. G., J. C. Briden, and G. E. Drewry. 1973. Phanerozoic world maps. *In* Organisms and Continents Through Time, N. F. Hughes, Ed. *Spec. Papers in Palaeontology 12 and Syst. Assoc. Pub. 9:* 1-42.

Smith, D. J. 1948. Miocene Foraminifera of the "Harang sediments" of southern Louisiana. *Louisiana Geol. Surv., Geol. Bull. 26:* 23-76.

Smith, J. M. 1962. Disruptive selection, polymorphism and sympatric speciation. *Nature 195:* 60-62.

——. 1970. The causes of polymorphism. *Symp. Zool. Soc. London 26:* 371-383.

Smith, Lee A. 1965. Paleoenvironmental variation curves and paleoeustatics. *Gulf Coast Assoc. Geol. Soc., Trans. 15:* 47-60.

——, and Jan Hardenbol. 1973. Proceedings of symposium on calcareous nanno-fossils. Gulf Coast Section, Soc. Econ. Paleont., Mineral., Houston. 151 p.

Smith, W. 1815. A Memoir to the Map and Delineation of the Strata of England and Wales, with part of Scotland. John Cary, London. 51 + xii p., 2 tables.

——. 1816-1819. Strata Identified by Organized Fossils, Containing Prints on Coloured Paper of the Most Characteristic Specimens in each Stratum. The author, London. 32 + ii p., 19 pls. [Parts 1 and 2 in 1816; Part 3 in 1817; Part 4 in 1819.]

——. 1817. Stratigraphical System of Organized Fossils, with reference to the specimens of the original geological collection in the British Museum: explaining their state of preservation and their use in identifying the British Strata. E. Williams, London. 118 + xii p., 2 tables.

Sneath, P. H. A., and R. R. Sokal. 1973. Numerical taxonomy. W. H. Freeman, San Francisco. 573 p.

Snyder, T. P., and J. L. Gooch. 1973. Genetic differentiation in *Littorina saxatilis*. *Mar. Biol. 22:* 177-182.

Sohl, N. F. 1960 [1961]. Archaeogastropoda, Mesogastropoda and stratigraphy of the Ripley, Owl Creek, and Prairie Bluff Formations. *U. S. Geol. Surv. Prof. Paper 331A:* 1-151, pls. 1-18.

——. 1964. Neogastropoda, Opisthobranchia and Bassomatophora from the Ripley, Owl Creek, and Prairie Bluff Formations. *U.S. Geol. Surv. Prof. Paper 331B:* 152-344, pls. 19-52.

——. 1965. Marine Jurassic gastropods central and southern Utah. *U. S. Geol. Surv. Prof. Paper 503D:* D1-D25, 5 pls.

——. 1967a. Post-Paleozoic Gastropoda. *In* Developments, Trends, and Outlooks in Paleontology, R. C. Moore, Ed. *Jour. Paleont. 42* (6): 1363-1364.

——. 1967b. Upper Cretaceous gastropods from Pierre Shale at Red Bird, Wyoming. *U. S. Geol. Surv. Prof. Paper 393B:* B1-B46.

——. 1971. North American biotic provinces delineated by gastropods. *North Amer. Paleont. Conv. Proc. Pt. L:* 1610-1637.

Sokal, R. R., and T. J. Crovello. 1970. The biological species concept: a critical evaluation. *Amer. Natur. 104:* 127-153.

——, and C. D. Michener. 1958. A statistical method for evaluating systematic relationships. *Kansas Univ. Sci. Bull. 38:* 1409-1438.

——, and F. J. Rohlf. 1962. The comparison of dendrograms by objective methods. *Taxon 11:* 33-40.

——, and P. H. A. Sneath. 1963. Principles of Numerical Taxonomy. W. H. Freeman, San Francisco. 359 p.

Sorgenfrei, T. 1966. Strukturgeologischer Bau von Dänemark. *Geologie 15:* 641-660.

Spangler, W. B., and J. J. Peterson. 1950. Geology of Atlantic Coastal Plain in New Jersey, Delaware, Maryland and Virginia. *Amer. Assoc. Petrol. Geol. Bull. 34:* 1-100.

Spath, L. F. 1923-1943. A monograph of the Ammonoidea of the Gault. *Palaeont. Soc. Monogr.:* 787 + xix p., 72 pls.

——. 1937. The Canadian ammonite genus *Gastroplites* in the English Gault. *Annals and Mag. Nat. Hist., ser. 10, 19:* 257-260.

Spencer, R. S. 1970. Evolution and geographic variation of *Neochonetes granulifer* (Owen) using multivariate analysis of variance. *Jour. Paleont. 44:* 1009-1028.

Spiller, J. 1973. Character variation and microevolutionary accelerations. *Geol. Soc. Amer. Abstrs. 5* (7): 817-818.

Spjeldnaes, N. 1961. Ordovician climatic zones. *Norsk Leol. Tidsskr. 41:* 45-77.

Stacy, M. C. 1953. Stratigraphy and paleontology of the Windsor Group (Upper Mississippian) in parts of Cape Breton Island, Nova Scotia. *Nova Scotia Dept. Mines Mem. 2:* 1-143.

Stahl, W., and R. Jordan. 1969. General considerations on isotopic paleotemperature determinations and analyses on Jurassic ammonites. *Earth and Planetary Sci. Letters 6:* 173-178.

Stalcup, M. C., and W. G. Metcalf. 1966. Direct measurements of the Atlantic Equatorial Undercurrent. *Jour. Mar. Res. 24:* 44-55.

Stanley, S. M. 1973. Effects of competition on rates of evolution, with special reference to bivalve mollusks and mammals. *Syst. Zool. 22:* 486-506.

Stebbing, A. R. D. 1973. Competition for space between the epiphytes of *Fucus serratus* L. *Jour. Mar. Biol. Assoc. U.K. 53:* 247-261.

Steeves, M. W. 1959. The pollen and spores of the Raritan and Magothy Formations (Cretaceous) of Long Island. Unpublished Ph.D. thesis, Radcliffe College.

Stehli, F. G. 1965. Paleontological technique for defining ancient ocean currents. *Science 148:* 943-946.

_____. 1968. Taxonomic diversity gradients in pole location: The Recent model. *In* Evolution and environment. E. T. Drake, Ed. Yale Univ. Press, New Haven. p. 163-227.

Steinich, G. 1965. Die artikulaten Brachiopoden der Rügener Schreibkreide (Unter-Maastricht). *Paläont. Abh. A. 2* (1): 1-220.

_____. 1972. Endogene Tektonik in den Unter-Maastricht-Vorkommen auf Jasmund (Rügen). Geologie, Jahrg. 20, Beih. 71/72: 1-205.

Steininger, F. F. 1963. Die Molluskenfauna aus dem Burdigal (Unter Miozän) von Fels am Wagram in Niederösterreich. *Denkschr. Österr. Akad. Wiss. math. - naturwiss. Kl. 110:* 87 p.

_____. 1969. Das Tertiär des Linzer Raumes. *In* Geologie und Paläontologie des Linzer Raumes. p. 35-53, 14 pls.

_____. 1973. Neogene stratigraphy of Paratethys of Central Europe and its correlation with other areas. *Amer. Assoc. Petrol. Geol. Ann. Mtg. 57* (4): 807 [abstract].

_____. 1975a. Egerian, F. F. Steininger and L. Neveskaja, Eds. *Stratotypes of Mediterranean Neogene Stages 2:* 71-81.

_____. 1975b. Eggenburgian, F. F. Steininger and L. Neveskaja Eds. *Stratotypes of Mediterranean Neogene Stages 2:* 83-91.

_____, and L. Neveskaja, 1975. *Stratotypes of the Mediterranean Neogene Stages 2:* 1-364.

_____. and A. Papp. 1973. Die stratigraphischen Grundlagen des Miozäns der zentralen Paratethys und die Korrelationsmöglichkeiten mit dem Neogen Europas.-Verh. Geol. Bundesanst. Wien, p. 39-65, 3 text-fig.

_____, and J. Senes. 1971. M 1 - Eggenburgien. *Chronostrat. and Neostrat. 2:* 1-827, 112 pls.

_____, and E. Thenius. 1965. Eine Wirbeltierfaunula aus dem Sarmat (Ober-Miozän) von Sauerbrunn (Burgenland). *Mitt. Geol. Ges. Wien 57:* 449-467.

_____, F. Rögl, and E. Martini. 1976. Current Oligocene/Miocene Biostratigraphic Concept of the Central Paratethys (Middle Europe). *Newsl. Stratigr. 4:* no. 3, p. 174-202, Gebrüden, Borntraegen, Berlin.

Stenestad, E. 1969. The genus *Heterohelix* Ehrenberg, 1843 (Foraminifera) from the Senonian of Denmark. Proc. 1st Internat. Conf. Planktonic Microfossils: 644-662.

_____. 1971. Upper Cretaceous in Rönde No. 1. *In* The Deep Test Well Rønde No. 1 in Djursland, Denmark, L. B. Rasmussen et al., Eds. *Danm. Geol. Unders. III rk. 39:* 56-60.

Steno, N. 1669. De solido intra solidum naturaliter contento dissertationis prodromus. Printing Shop under the Sign of the Star, Florentiae. 79 p., 1 pl.

Stenzel, H. B. 1940. New zone in Cook Mountain Formation, *Crassatella texalta* Harris – *Turritella cortezi* Bowles zone. *Amer. Assoc. Petrol. Geol. Bull.* *24* (9): 1663-1675.

_____. 1971. Oysters – Dispersal. *In* Treatise on Invertebrate Paleontology, R. C. Moore, Ed. Part N, vol. 3, Mollusca 6, Bivalvia: 1035-1036.

Stephanesco, Gr. 1880. Sur l'uniformité de la nomenclature gélogique dans tous les pays en ce qui regarde les terrains et les étages. Intern. Geol. Cong. 1 (Paris): 82-84.

Stephenson, L. W. 1937. Stratigraphic relations of the Austin, Taylor, and equivalent formations in Texas. *U. S. Geol. Surv. Prof. Paper 186-G:* 133-146, pl. 44, fig. 7.

_____, P. B. King, W. H. Monroe, and R. W. Imlay. 1942. Correlation of the outcropping formations of the Atlantic and Gulf Coastal Plain and Trans-Pecos, Texas. *Geol. Soc. Amer. Bull. 53:* 435-448, pl. 1, chart 9.

Stephenson, T. A., and A. Stephenson. 1972. Life between tidemarks on rocky shores. San Francisco. 425 p.

Steven, D. M. 1953. Recent evolution in the genus *Clethrionomys. Symp. Soc. Exp. Biol. 7:* 310-319.

_____. 1955. Untersuchungen über die britischen Formen von *Clethrionomys. Z. Saugetierk. 20:* 70.

Stewart, J. H., and F. G. Poole. 1974. Lower Paleozoic and uppermost Precambrian Cordilleran miogeocline, Great Basin, western United States. *Soc. Econ. Paleont. and Mineralog., Spec. Publ. 22:* 28-57.

Stewart, W. J. 1968. The stratigraphic and phylogenetic significance of the fusulinid genus *Eowaeringella*, with several new species. *Cushman Found. for Foram. Res., Spec. Publ. 10:* 1-29, 7 pls., 2 text-figs.

Stitt, J. H. 1971a. Repeating evolutionary pattern in Late Cambrian trilobite biomeres. *Jour. Paleont. 45* (2): 178-181.

_____. 1971b. Cambrian-Ordovician trilobites, western Arbuckle Mountains. *Oklahoma Geol. Surv. Bull. 110:* 1-83, 8 pls.

_____. 1973. Repeated mass extinctions and adaptive radiations in Cambrian cratonic trilobites of North America. *Geol. Soc. Amer., Abstracts with Programs 5* (7): 823.

_____. 1975. Adaptive radiation, trilobite paleoecology and extinction, Ptychaspid Biomere, Late Cambrian of Oklahoma. Fossils and Strata, n. 4, p. 381-390, Oslo.

Stokes, R. B. 1975. Royaumes et provinces fauniques du Crétacé, établis sur la base d'une étude systématique du genre *Micraster. Mém. Mus. Nat. d'Hist. Natur. C, 31:* 1-94, 30 figs., 12 pls.

Stover, L. E. 1964. Comparison of three Cretaceous spore-pollen assemblages from Maryland and England. *In* Palynology in Oil Exploration. *Society of Economic Paleontologists and Mineralogists, Spec. Publ. 11:* 143-152.

Strachan, Isles. 1971. A synoptic supplement to "A Monograph of British Graptolites by Miss G. L. Elles and Miss E. M. R. Wood." Palaeont. Soc. Monogr.: 1-130.

Strauch, F. 1970. Die Thule-Landbrucke als Wanderweg und Faunenscheid zwischen Atlantik und Skandik im Tertiär. *Geol. Rundschau 60:* 381-417.

Struhsaker, J. W. 1968. Selection mechanisms associated with intraspecific shell variation in *Littorina picta* (Prosobranchia; Mesogastropoda). *Evolution 22:* 459-480.

Stude, Gerald R. 1970. Application of paleobathymetry in exploration. *Gulf Coast Assoc. Geol. Soc., Trans. 20:* 194-200.

Stumm, E. C. 1953. Trilobites of the Devonian Traverse Group of Michigan. *Contrib. Mus. Paleont. Univ. Michigan 10* (6): 101-157.

Suchanek, T. H., and J. Devington. 1974. Articulate Brachiopod Food. *Jour. Paleont. 48:* 1-6.

Sudbury, Margaret. 1958. Triangulate monograptids from the *Monograptus gregarius* zone (Lower Llandovery) of the Rheidol Gorge (Cardinganshire). *Phil. Trans. Roy. Soc. London, Ser. B, 241:* 485-555.

Suess, E. 1870. Uber das Verkommen von Fusulinen in den Alpen. *K. K. Geol. Reichs. Wien Verh. 24:* 4-5.

Surlyk, F. 1969. A study on the articulate brachiopods of the Danish white chalk (U. Campanian and Maastrichtian) with a review of the sedimentology of the white chalk and the flora and fauna of the chalk sea. Unpublished prize dissertation, University of Copenhagen: 1-319 [in Danish].

_____. 1970. Die Stratigraphie des Maastricht von Dänemark und Norddeutschland aufgrund von Brachiopoden. *Newsl. Stratigr. 1:* 7-16.

_____. 1971. Skrivekridtklinterne på Møn. Geologi på Øerne. *Varv Ekskursionsfører 2:* 5-24.

_____. 1972. Morphological adaptations and population structures of the Danish chalk brachiopods (Maastrichtian, Upper Cretaceous). *Biol. Skr. Dan. Vid. Selsk. 19* (2): 1-57.

_____. 1973. Autecology and taxonomy of two Upper Cretaceous craniacean brachiopods. *Bull. Geol. Soc. Denmark 22:* 219-243.

_____. 1974. Life habit, feeding mechanism and population structure of the Cretaceous Brachiopod Genus Aemula. *Palaeogeog., Palaeoclim., Palaeoecol.* 15: 185-203.

_____, and W. K. Christensen. 1974. Epifaunal zonation on an Upper Cretaceous rocky coast. *Geology. 2* (11): 529-534.

Sweet, W. C., and S. M. Bergstrom. 1971a. The Upper American Upper Ordovician standard, XIII. A revised time-stratigraphic classification of North American Upper Middle and Upper Ordovician rocks. *Geol. Soc. Amer. Bull. 82:* 499.

_____, and S. M. Bergstrom. 1971b. Symposium on Conodont Biostratigraphy. *Geol. Soc. Amer. Mem. 127:* 499.

_____, C. A. Turco, E. Warner, and L. C. Wilkie. 1959. The American Upper Ordovician standard. I. Eden conodonts from the Cincinnati region of Ohio and Kentucky. *Jour. Paleont. 33:* 1029-1068.

_____, R. C. Ethington, and C. R. Barnes. 1971. North American Middle and Upper Ordovician conodont faunas. *Geol. Soc. Amer. Mem. 127:* 163-193.

Sylvester-Bradley, P. C. 1951. The Subspecies in Palaeontology. *Geol. Mag. 88:* 88-102.

_____. 1954. The Superspecies. *System. Zool. 3:* 145-146, 173.

_____(Ed.). 1956. The species concept in palaeontology. The Systematics Assoc. Pub. 2, Systematics Assoc., London: 1-145.

_____. 1958. The description of fossil populations. *Jour. Paleont. 32:* 214-235.

———. 1959. Iterative evolution in fossil oysters. *XV Internat. Congr. Zool. Proc.:* 193-197.
———. 1960. Geology and the History of Life. Leicester University Press, Leicester. 27 p.
———. 1962. The taxonomic treatment of phylogenetic patterns in time and space, with examples from the Ostracoda. *System. Assoc. Publ. 4:* 119-133.
———. 1964. Type sections of Bathonian, Portlandian and Purbeckian stages, and the problem of the Jurassic-Cretaceous boundary. *In* Colloque du Jurassique à Luxembourg, 1962, P. L. Maubeuge, Ed. Ministere des Arts et des Sciences, Luxembourg. p. 259-263.
———. 1967. Towards an International Code of Stratigraphic Nomenclature. *In* Essays in Paleontology and Stratigraphy, C. Teichert and E. L. Yochelson, Eds. Kansas University Press, Lawrence. p. 49-56.
———. 1971. Dynamic factors in animal palaeogeography. *In* Faunal Provinces in Space and Time, F. A. Middlemiss, P. F. Rawson, and G. Newell, Eds. *Geol. Jour. Special Issue 4:* 1-18.
Szöts, E. 1965. Le stratotype de l'Aquitanien (Mayer-Eymar 1857-1858). *Soc. Géol. France, Bull. VII* (7): 743-746.
Takhtajan, A. L. 1969. Flowering Plants: Origin and Dispersal. Smithsonian Institution, Washington, D.C. 310 p.
Talwani, M., C. C. Windisch, and M. G. Langseth, Jr. 1971. Reykjanes Ridge crest: a detailed geophysical study. *Jour. Geophys. Res. 76* (2): 473-517.
Tappan, H., and A. R. Loeblich, Jr. 1972. Fluctuating rates of protistan evolution, diversification and extinction. 24th Internat. Geol. Congr., Sect. 7, Paleontology: 205-213.
Taraz, Hushang. 1973. Correlation of uppermost Permian in Iran, Central Asia, and South China. *Amer. Assoc. Petrol. Geol. Bull. 57:* 1117-1133, 4 text-figs.
Tate, R. 1865. On the Correlation of the Cretaceous Formations of the Northeast of Ireland. *Geol. Soc. London, Quart. Jour. 21:* 15-44, pls. 3-5.
———. 1867. On the Fossiliferous Development of the Zone of *Ammonites angulatus*, Schloth., in Great Britain. *Geol. Soc. London, Quart. Jour. 23:* 305-314.
Taylor, J. B. 1971. Reef associated molluscan assemblages in the western Indian Ocean. *Symp. Zool. Soc. Lond. 28:* 501-534.
Taylor, J. D. 1968. Coral reef and associate invertebrate communities (mainly molluscan) around Mahé, Seychelles. *Phil. Trans. Roy. Soc. London (B) 254:* 129-206.
Taylor, M. E. 1971. Biostratigraphy of the Upper Cambrian (Upper Franconian-Trempealeauan stages) in the central Great Basin, Nevada and Utah. Unpublished Ph.D. dissertation, University of California, Berkeley: 1-428, 19 pls.
———. 1976. Indigenous and redeposited trilobites from Late Cambrian basinal environments of central Nevada. *Jour. Paleont. 50*(4): 668-700.
———, and Cook, H. E. 1976. Shelf to slope facies transition in the early Paleozoic of Nevada. *In* Cambrian of Western North America, R. A. Robison and A. J. Rowell, Eds. Brigham Young University, Geol. Studies, v. 2, pt. 2, pp 181-214.
———, and R. B. Halley. 1974. Systematics, environment, and biogeography of some Late Cambrian and Early Ordovician trilobites from eastern New York state. *U.S. Geol. Survey Prof. Paper 834:* 1-38, 4 pls.

Tedford, R. H. 1970. Principles and practices of mammalian geochronology in North America. Proc. North Amer. Paleont. Convention, 1969. Allen Press, Lawrence, Kansas 1. p. 666-703.

Teichert, Curt. 1958. Some biostratigraphic concepts. *Geol. Soc. Amer. Bull. 69* (1): 99-119.

Thaler, Louis. 1965. Une échelle de zones biochronologiques pour les mammifères du Tertiaire d'Europe. *C. R. Somm. S.G.P.F., 1965:* 118.

———. 1966. Les Rongeurs Fossiles du Bas Languedoc dans leur Rapport avec l'Histo des Faunes et la Stratigraphie du Tertiaire d'Europe. *Mém. Mus. Nat. Hist. Natur., N.S., Ser. C, 17:* 1-295, 25 text-figs., 15 tables, 27 pls.

Thenius, E. 1952. Die Säugetierfauna des Torton von Neudorf a.d. March (CSR). *N. Jb. Geol. Paläont., Abh., 96:* 27-136.

———. 1959. Tertiär. 2. Pt. Wirbeltierfaunen. *Handb. Strat. Geol. 3/2:* 1-328, 32 tables, 10 pls.

———. 1972. Grundzüge der Verbreitungsgeschichte der Säugetiere. G. Fischer, Jena. 355 p.

Theyer, F. 1974, In press. Paleomagnetic polarity sequence and radiolarian zones, Brunhes Epoch to Polarity Epoch 20. *Earth and Planetary Sci. Newsletters.*

———, and S. R. Hammond. 1974. The Cenozoic Magnetic Time Scale in Deep-Sea Cores: Completion of the Neogene. Geol. (in press).

Thiede, J. 1974. Marine bivalves: Distribution of meroplanktonic shell-bearing larvae in eastern North Atlantic surface waters. *Palaeogeogr. Palaeoclimat. Palaeoecol. 15:* 267-290.

Thompson, M. L. 1948. Studies of American fusulinids. *Kansas Univ., Paleont. Contr. 4, Protozoa, Art. 1:* 1-184, 38 pls., 7 text-figs.

———. 1954. American Wolfcampian fusulinids. *Kansas Univ., Paleont. Contr. 14, Protozoa, Art. 5:* 1-225, 52 pls., 14 text-figs.

———. 1957. Northern Midcontinent Missourian fusulinids. *Jour. Paleont. 31* (2): 289-328, pls. 21-30.

———. 1964. Fusulinacea. *In* Treatise on Invertebrate Paleontology, Pt. C, Vol. 1, R. C. Moore, Ed. Geol. Soc. Amer. and University of Kansas Press, Lawrence. p. 358-436, figs. 274-328.

———. 1967. American fusulinacean faunas containing elements from other continents. *Kansas Univ., Dept. Geol., Spec. Publ. 2:* 102-112, 1 text-fig.

———, G. J. Verville, and D. H. Lokke. 1956. Fusulinids of the Desmoinesian-Missourian contact. *Jour. Paleont. 30* (4): 793-810, pls. 89-93, 1 text-fig.

Thorson, Gunnar. 1936. The larval development, growth, and metabolism of Artic marine bottom invertebrates. *Medd. Grønland 100* (6): 1-155.

———. 1946. Reproduction and larval development of Danish marine bottom invertebrates. *Medd. Fra. Kom. Danmarks Fiskeri-ogttavunder. Serie Plankton 4* (1): 1-523.

———. 1950. Reproduction and larval ecology of marine bottom invertebrates. *Biol. Rev. 25:* 1-45.

———. 1957. Bottom communities. *In* Treatise on Marine Ecology and Paleoecology, J. W. Hedgpeth, Ed. *Geol. Soc. Amer. Mem. 67, 1:* 461-534.

———. 1961. Length of pelagic larval life in marine bottom invertebrates as related to larval transport by ocean currents. *In* Oceanography, Mary Sears, Ed. Amer. Assoc. Adv. Sci., Washington, D.C. p. 455-474.

———. 1962. The distribution of benthic marine mollusca along the N.E. Atlantic Shelf from Gibraltar to Murmansk. Proc. 1st Europ. Malacol. Congr. (London). Conch. Soc. Gt. Britain and Malac. Soc. London: 5-23.

———. 1964. Light as an ecological factor in the dispersal and settlement of larvae of marine bottom invertebrates. *Ophelia 1:* 167-208.

Thurman, J. 1836. [Letter to E. de Beaumont]. *Soc. Géol. France, Bull. 7:* 207-211.

Tikhvinskiy, I. N. 1965. K. ekologii shvagerin [Schwagerinid ecology]. *Paleont. Zhur 2:* 18-22 (translated in *Internat. Geol. Rev. 8* (1): 60-63).

Tintant, H. 1963. Les Kosmocératidés du Callovien Inférieur et Moyen d'Europe Occidentale. *Publ. Univ. Dijon 29:* 1-500, 58 pls.

Tipsword, Howard L. 1962. Tertiary Foraminifera in Gulf Coast petroleum exploration and development. *In* Geology of the Gulf Coast and central Texas. Houston Geol. Soc. Guidebk. Ann. Mtg.: 16-57.

———, F. M. Setzer, and Fred L. Smith, Jr. 1966. Interpretation of depositional environment in Gulf Coast petroleum exploration from paleoecology and related stratigraphy. *Gulf Coast Assoc. Geol. Soc., Trans. 16:* 119-130.

Tobien, Heinz. 1973. On the evolution of the Mastodonta (Proboscidea, Mammalia). Pt. 1: the bunodont trilophodont group. *Hess. L.-Amt. Bondenforsch. Hotizbl. 101:* 202-276.

Toghill, Peter. 1968. The graptolite assemblages and zones of the Birkhill Shales (Lower Silurian) at Dobb's Linn. *Palaeont. 11:* 654-668.

———. 1970. Highest Ordovician (Hartfell Shales) graptolite faunas from the Moffat area, South Scotland. *Brit. Mus. Nat. Hist. Geol. Bull. 19:* 1-26.

Tomczykowa, E. 1968a. The Cambrian-Ordovician boundary in Poland and its correlation with the Scandinavian and Great Britain areas. *23rd Internat. Geol. Congr., Proc. 9:* 43-51.

———. 1968b. Stratigraphy of the uppermost Cambrian deposits in the Swietokrzyskie Mountains. *Instytut Geologiczny, Prace. 54:* 71-91. [in Polish with English summary].

Toriyama, Ryuzo. 1958. Geology of Akiyoshi, Pt. 3, Fusulinids of Akiyoshi. Kyushu Univ., Fac. Sci., Mem., ser. D. Geol. 7: *1-264, 48 pls.*

———. 1967. The fusulinacean zones of Japan. *Kyushu Univ., Fac. Sci., Mem., ser. D, Geol. 18* (1): 35-260, 11 text-figs., 2 correlation charts.

———. 1973. Upper Permian Fusulininan Zones: *In* The Permian and Triassic Systems and their mutal Boundaries. A. Logan and L. V. Hills, Ed. Canadian Society of Petroleum Geologists. Memoir 2, p. 498-512, 2 text-figs.

———, T. Sato, and T. Hamada. 1966. *Nautilus pompilius* drift on the west coast of Thailand. *Japanese Jour. Geol. Geogr. 36:* 149-161.

Tornquist, S. L. 1897. On the Diplograptidae and Heteroprionidae of the Scanian *Rastrites* beds. *Kgl. Fysiogr. Sallsk. Lund, Handl. n. ser. 8:* 1-24.

———. 1899. Monograptidae of the Scanian *Rastrites* beds. *Lunds Univ. Arsskr. 35, pt. 2, 1: 1-25.*

———. 1901-1904. Graptolites of the lower zones of the Scanian and Vestrogothian *Phyllo-Tetragraptus* beds, pts. 1 and 2. *Lunds Univ. Arsskr. 37* [pt. 2] (5): 1-26; *40* [pt. 1] (2): 1-29.

Townsend, J. 1813. The Character of Moses established for veracity as an historian, recording events from the Creation to the Deluge. Bath: M. Gye; London: Longman, Hurst, Rees, Orne and Brown. 436 + vi p.

Trautschold, H. 1877. Der Französische Kimmeridge und Portland verglichen mit den Gleichaltrigen Moskauer Schichten. *Soc. Imp. Nat. Moscou, Bull. 50 [for 1876]:* 381-392.

Traverse, A., K. H. Clisby, and F. Foreman. 1961. Pollen in drilling-mud "thinners", a source of palynological contamination. *Micropaleontology* 7(3): 375-377.

Troedsson, G. T. 1937. On the Cambro-Ordovician faunas of western Quruq tagh, eastern T'ien-shan. *Palaeontologia Sinica, n.s. B, 2, Whole Series 106:* 1-74, 10 pls.

Troelsen, J. 1937. Om den stratigrafiske inddeling af skrivekridtet i Danmark. *Meddr. dansk, geol. Foren. 9:* 260-263.

Trueman, A. E. 1922 The use of *Gryphaea* in the correlation of the Lower Lias. *Geol. Mag. 59:* 256-268.

_____. 1923. Some theoretical aspects of correlation. *Geol. Assoc. London, Proc. 34:* 193-206.

Tschudy, R. H., and R. A. Scott (Eds.). 1969. Aspects of Palynology. John Wiley, New York. 510 p.

Tullberg, S. A. 1883. Skanes graptoliter, I and II. *Sverige Geol. Undersok., ser. C, 50:* 1-44; 55: 1-43.

Turesson, G. 1922. The species and variety as ecological units. *Hereditas 3:* 100-113.

Turner, Donald L. 1970. Potassium-argon dating of Pacific Coast Miocene foraminiferal stages. *In* Radiometric Dating and Paleontologic Zonation, O. L. Bandy, Ed. *Geol. Soc. Amer., Spec. Paper 124:* 91-129.

Turner, R. D. 1946. The family Tonnidae in the western Atlantic. *Johnsonia 2*(26): 165-192.

_____. 1966. A Survey and Illustrated Catalogue of the Teredinidae (Mollusca; Bivalvia). Mus. Comp. Zool., Harvard University, Cambridge, Mass.

_____. and J. Rosewater. 1958. The family Pinnidae in the western Atlantic Johnsonia. 3(38): 285-326.

Tyndale-Biscoe, H. 1973. Life of Marsupials. Edward Arnold, London.

Ulrich, E. O., and R. S. Bassler. 1926. A classification of the tooth-like fossils, conodonts, with descriptions of American Devonian and Mississippian species. *Proc. U.S. Nat. Mus. Washington 68:* 1-63.

_____, and C. E. Resser. 1930. The Cambrian of the upper Mississippi Valley. Pt. 1, Trilobita; Dikelocephalinae and Osceolinae. *Milwaukee Public Mus., Bull. 12* (1): 1-122.

_____, and C. E. Resser. 1933. The Cambrian of the Upper Mississippi Valley. Pt. 2, Trilobita; Saukiinae. *Milwaukee Public Mus., Bull. 12*(2): 123-306.

Upshaw, C. F., W. E. Armstrong, W. B. Creath, E. J. Kidston, and G. A. Sanderson. 1974. Biostratigraphic Framework of Grand Banks. *Amer. Assoc. Petrol. Geol. Bull. 58*(6): 1124-1132.

Ustritz, V. I. 1967. Subdivision of the arctic basins in the Late Paleozoic. *Trudy Vses Paleont. Obsh. 96*(9): 56-69 [Voprosy paleontografiteskoio raionirovaniia v svette dannykh paleontologii; in Russian].

Valentine, J. W. 1963. Biogeographic units as biostratigraphic units. *Bull. Amer. Assoc. Petrol. Geol. 47*(3): 457-466.

_____. 1966. Numerical analysis of marine molluscan ranges on the extra tropical northeastern Pacific Shelf. *Limnology and Oceanography 11:* 198-211.

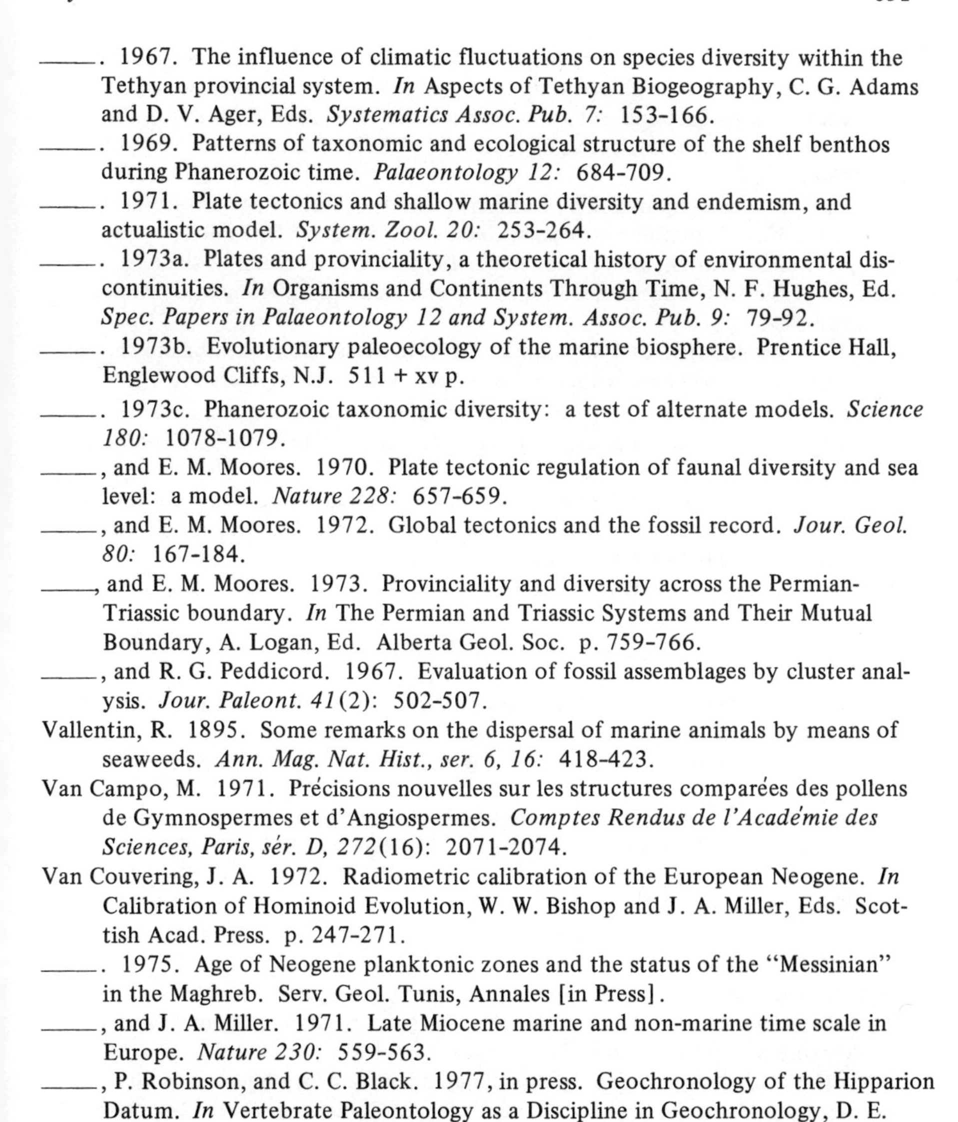

———. 1967. The influence of climatic fluctuations on species diversity within the Tethyan provincial system. *In* Aspects of Tethyan Biogeography, C. G. Adams and D. V. Ager, Eds. *Systematics Assoc. Pub. 7:* 153-166.

———. 1969. Patterns of taxonomic and ecological structure of the shelf benthos during Phanerozoic time. *Palaeontology 12:* 684-709.

———. 1971. Plate tectonics and shallow marine diversity and endemism, and actualistic model. *System. Zool. 20:* 253-264.

———. 1973a. Plates and provinciality, a theoretical history of environmental discontinuities. *In* Organisms and Continents Through Time, N. F. Hughes, Ed. *Spec. Papers in Palaeontology 12 and System. Assoc. Pub. 9:* 79-92.

———. 1973b. Evolutionary paleoecology of the marine biosphere. Prentice Hall, Englewood Cliffs, N.J. 511 + xv p.

———. 1973c. Phanerozoic taxonomic diversity: a test of alternate models. *Science 180:* 1078-1079.

———, and E. M. Moores. 1970. Plate tectonic regulation of faunal diversity and sea level: a model. *Nature 228:* 657-659.

———, and E. M. Moores. 1972. Global tectonics and the fossil record. *Jour. Geol. 80:* 167-184.

———, and E. M. Moores. 1973. Provinciality and diversity across the Permian-Triassic boundary. *In* The Permian and Triassic Systems and Their Mutual Boundary, A. Logan, Ed. Alberta Geol. Soc. p. 759-766.

———, and R. G. Peddicord. 1967. Evaluation of fossil assemblages by cluster analysis. *Jour. Paleont. 41*(2): 502-507.

Vallentin, R. 1895. Some remarks on the dispersal of marine animals by means of seaweeds. *Ann. Mag. Nat. Hist., ser. 6, 16:* 418-423.

Van Campo, M. 1971. Précisions nouvelles sur les structures comparées des pollens de Gymnospermes et d'Angiospermes. *Comptes Rendus de l'Académie des Sciences, Paris, sér. D, 272*(16): 2071-2074.

Van Couvering, J. A. 1972. Radiometric calibration of the European Neogene. *In* Calibration of Hominoid Evolution, W. W. Bishop and J. A. Miller, Eds. Scottish Acad. Press. p. 247-271.

———. 1975. Age of Neogene planktonic zones and the status of the "Messinian" in the Maghreb. Serv. Geol. Tunis, Annales [in Press].

———, and J. A. Miller. 1971. Late Miocene marine and non-marine time scale in Europe. *Nature 230:* 559-563.

———, P. Robinson, and C. C. Black. 1977, in press. Geochronology of the Hipparion Datum. *In* Vertebrate Paleontology as a Discipline in Geochronology, D. E. Savage, Ed. *Geol. Soc. Amer. Spec. Paper.*

Van Couvering, Judith H., and J. A. Van Couvering. 1974, in press. Geological setting and faunal analysis of the African Early Miocene. *In* Perspectives on Human Evolution, Vol. III, G. Ll. Issac, Ed.

Van Hoepen, E. C. N. 1965 [1966]. New and little known Zululand and Pondoland ammonites. *South African Geol. Surv. Annals 4:* 157-181.

Van Houten, F. B. 1945. Review of latest Paleocene and early Eocene mammalian faunas. *Jour. Paleont. 19:* 421-461.

Van Tuyl, F. M. 1925. The stratigraphy of the Mississippian formations of Iowa. *Iowa Geol. Surv. 30:* 33-349.

Vasilyuk, N. P., Ye. I. Kachanov, and I. V. Pyzh'yanov. 1970. Paleobiogeograficheskiy ocherk kamennougol'nykh i permskikh tselenterat (Paleobiogeographical review of Carboniferous and Permian coelenterates). Distribution and sequence of Paleozoic corals of the USSR, USSR Acad. Sci., Moscow, tr. 3: 45-60.

Vass, D., G. P. Bagdasarjan, and V. Konecný. 1970. Absolute age of several stages of the West-Carpathian Miocene. *Geol. Prace Soravy 51:* 71-97.

_____, G. P. Bagdasarjan, and V. Konecný. 1971. Determination of the absolute age of the West Carpathian Miocene. *Föld. Közl., Bull. Hungar. Geol. Soc. 101:* 321-327.

_____, G. P. Bagdasarjan, and S. Bajanik. 1974, in press. Contributions to the Neogene radiometric time-scale from Northern Africa. *Geol. Mag.*

Vaughan, T. Wayland. 1923. On the relative value of species of smaller Foraminifera for the recognition of stratigraphic zones. *Amer. Assoc. Petrol. Geol. Bull. 7* (5): 517-530.

Vdovenko, M. V. 1961. Zoogeographic study of the Carboniferous period in the USSR, according to the Foraminifera. *Visn. Kiev Univ., ser. geol. geograf., Bull. 2*(3): 21-29 [in Russian].

Verbeek, R. D. M. 1875. On the geology of central Sumatra. *Geol. Mag., n.s., 2:* 477-486.

Vermeij, G. J. 1972. Endemism and environment: some shore molluscs of the tropical Atlantic. *Amer. Natur. 106:* 89-101.

Verneuil, E. de. 1846a. Extract of letter on *Fusulina* in the Carboniferous of Ohio. *Amer. Quart. Jour. Agricul. Sci. 4:* 166.

_____. 1846b. On the *Fusulina* in the coal formation of Ohio. *Amer. Jour Sci., 2d ser.:* 293.

Verwoerd, W. J. 1964. [1966]. Stratigraphic classification: a critical review; [with discussion]. *Geol. Soc. S. Afr. Trans. 67:* 263-282, 304-316, illus.

_____. 1967. Stratigraphic Classification: A Critical Review. *Geol. Soc. S. Africa, Trans. 67:* 263-282.

Vohra, F. C. 1971. Zonation on a tropical sandy shore. *Jour. Anim. Ecol. 40:* 679-708.

Voigt, E. 1929. Die Lithogenese der Flach- und Tiefwasser-sedimente des jüngeren Oberkreidemeeres. *Jahrb. Halleschen Verband zur Erforschung d. mitteldeutschen Bodenschätze u ihre Verwertung 8:* 1-136.

_____. 1963. Über Randtröge von Schollenrändern und ihre Bedeutung im Gebiet der mitteldeutsche Senke und an grenzender Gebiete. *J. Deut. Geol. Ges. 114:* 378-418.

Von Bitter, Peter H. 1972. Environmental control of conodont distribution in the Shawnee Group (Upper Pennsylvanian) of eastern Kansas. *Univ. Kansas Paleont. Contr. Art. 59:* 1-105.

Voronov, N. N. 1961. Ekologicseskie i nekotorye morfologicseskie osobennosti ryzhikh polevok (*Clethrionomys* Tilesius) europeiskogo severo-vostoka. *Trudy Zool. Inst., Leningrad 29:* 101-136.

Waagen, W. 1870. Ueber die Ansatzstelle der Haftmuskeln beim *Nautilus* und der Ammoniten. *Palaeontographica 17:* 185-210.

Walker, E. P. 1968. Mammals of the World. 2nd Ed., Johns Hopkins Press, Baltimore.

Walker, K. R., and L. F. Laporte. 1970. Congruent fossil communities from Ordovician and Devonian carbonates of New York. *Jour. Paleont. 44:* 928-944.

Wallace, A. R. 1876. The Geographical Distribution of Animals, Vol. 1. Harper, New York. 503 p.

Walliser, O. H. 1964. Conodonten des Silurs. *Hess. Landesamt. Bodenforsch, Abh. 41:* 1-106.

Walther, Johannes. 1897. Ueber die Lebensweise fossiler Meeresthiere. *Deutsch. Geol. Gesell. Zeitschr. 49:* 209-273.

Walton, W. R. 1964. Recent foraminiferal ecology and paleoecology. *In* Approaches to Paleoecology, John Imbrie and N. D. Newell, Eds. John Wiley, New York. p. 151-237.

Wang, C.-S. 1964. In Defense of Traditional Stratigraphy. *Geol. Soc. China, Proc. 7:* 40-47.

Warner, R. 1801. The History of Bath. R. Cruttwell, Bath; and J. Robinson, London. 402 + 123 + vii p.

Warren, B. A. 1971. Antarctic deep water contribution to the world ocean. *In* Research in the Antarctic, L. O. Quam, Ed. *Amer. Assoc. Adv. Sci., Publ. 93:* 631-643.

Warren, P. T. 1971. The sequence and correlation of graptolite faunas from the Wenlock-Ludlow rocks of North Wales. *Bur. Res. Geol. Min. 73:* 451-460.

Waterhouse, J. B. 1964. The Permian of New Zealand. Internat. Geol. Congr. 22 India, Pt. 9, Gondwanas: 203-224.

———. 1967. Cool-water faunas from the Permian of the Canadian Arctic. *Nature 216:* 47-49.

———. 1971. The Permian Brachiopod genus *Terrakea* Booker, 1930. *In* Paleozoic Perspectives: A Paleontological Tribute to G. Arthur Cooper, T. Dutro, Ed. *Smithsonian Contr. Paleobiol. 3:* 347-361.

———. 1972. The evolution, correlation and paleogeographic significance of the Permian ammonoid family Cyclolobidae Zittel. *Lethaia 5:* 230-250.

———. 1973a. Permian Brachiopod Correlations for South-East Asia. *Bull. Geol. Soc. Malaysia 6:* 187-210.

———. 1973b. Communal Hierarchy and Significance of Environmental Parameters for Brachiopods: The New Zealand Permian Model. *Life Sci. Contr. Roy. Ont. Mus. 92:* 1-49, 16 figs.

———. 1974. Paleozoic Era, Upper. *Encyclopedia Britannica 13:* 921-930.

———, and G. Bonham-Carter. 1972. Permian paleolatitudes judged from brachiopod diversities. 24 Inter. Geol. Congr. 24(7): 350-361, 6 figs.

———, and G. Bonham-Carter. 1975. Global distribution and character of Permian biomes based on brachiopod assemblages. *Can. Jour. Earth Sci. 12*(7): 1085-1146.

———, and J. Waddington. (in press). Systematic descriptions, paleoecology and correlations of the late Paleozoic brachiopoda genera *Spiriferella, Eridmatus, Timaniella* and *Elivina* from Yukon Territory and Canadian Arctic Archipelago. *Canadian Geological Survey Bulletin.*

Waters, B. T., and J. H. Hutchison. 1975. Modified bulk sampling of fossils in continental stratal successions. *PaleoBios.*

Watkins, R. M., W. B. N. Berry, and A. J. Boucot. 1973. Why "communities"? *Geology 1:* 55-58.

Watson, R. B. 1886. Report on the Scaphopoda and Gastropoda collected by H.M.S. Challenger during the years 1873-1876. *Rept. Sci. Res. Expl. Voy. Challenger, Zoology 15:* 756 + v p.

Weaver, C. E. et al. 1944. Correlation of the marine Cenozoic formations of western North America. *Geol. Soc. Amer. Bull. 55:* 569-598.

Weaver, K. N., E. T. Cleaves, J. Edwards, and J. D. Glaser. 1968. Geologic Map of Maryland. Maryland Geol. Survey, Baltimore.

Webers, G. F. 1966. The Middle and Upper Ordovician conodont faunas of Minnesota. *Minnesota Geol. Surv. Spec. Pub. SP-4:* 1-123.

Wegmann, E. 1963. L'exposé original de la notion de faciès par A. Gressley (1814-1865). *Science de la Terre 9:* 83-119, 2 pls.

Weisbord, N. E. 1962. Late Cenozoic gastropods from northern Venezuela. *Bull. Amer. Paleont. 42:* 672.

Weller, J. M. 1931. The Mississippian fauna. *In* The Paleontology of Kentucky, W. R. Jillson, Ed. *Kentucky Geol. Surv. ser. 6, 36:* 251-267.

_____. 1960. Stratigraphic Principles and Practice. Harper, New York. 725 p.

_____(Chairman) et al. 1948. Correlation of the Mississippian formations of North America [chart no. 5]. *Geol. Soc. Amer. Bull. 59*(2): 91-196.

Weller, S. 1926. Faunal zones in the standard Mississippian section. *Jour. Geol. 34* (4): 320-335.

Wells, H. W., and I. E. Gray. 1960. The seasonal occurrence of *Mytilus edulis* on the Carolina coast as a result of transport around Cape Hatteras. *Biol. Bull. mar. biol. Lab., Woods Hole 118:* 550-559.

Wells, J. W. 1957. Corals. *In* Treatise on Marine Ecology and Paleontology, J. W. Hedgepeth, Ed. *Ecology 1:* 1087-1104.

_____. 1963. Early investigations of the Devonian System in New York, 1656-1836. *Geol. Soc. Amer., Spec. Pub. 74:* 1-74, 11 pls.

Wendt, J. 1968. *Discohelix* (Archaeogastropoda, Euomphalocea) as an index fossil in the Tethyan Jurassic. *Paleontology 11*(4): 554-575.

Westermann, G. E. G. 1966. Covariation and taxonomy of the Jurassic ammonite *Sonninia adicra* (Waagen). *Neues. Jahrb. Geol. Paläont., Abh. 124:* 289-312.

_____. 1971. Form, structure and function of shell and siphuncle in coiled Mesozoic ammonoids. *Royal Ontario Mus. Life Sci. Contr. 78:* 1-39.

_____. 1973. The late Triassic bivalve *Monotis. In* Atlas of Palaeogeography, A. Hallam, Ed. Elsevier, New York. 251-258.

Wetzel, W. 1969. Seltene Wohnkammer Inhalte von Neoammoniten. *Neues Jahrb. Geol. Paläont., Monatsh., Jahrg. 1969:* 46-53.

White, C. A. 1883. Contributions to invertebrate paleontology, no. 8: Fossils from the Carboniferous rocks of the interior states. *U.S. Geol. Geog. Surv. Terr. (Hayden), Ann. Rept. 12*(pt. 1): 155-171.

Whittington, H. B. 1963. Middle Ordovician trilobites from Lower Head, western Newfoundland. *Harvard Univ. Mus. Comp. Zool. Bull. 129*(1): 1-118.

_____. 1966. Phylogeny and distribution of Ordovician trilobites. *Jour. Paleont. 40:* 696-737.

_____. 1968. Zonation and correlation of Canadian and Early Mohawkian Series. *In* Studies of Appalachian Geology: Northern and Maritime, E-an Zen, W. S. White, J. B. Hadley, and J. B. Thompson, Jr., Eds. John Wiley, New York. p. 49-60.

———. 1973. Ordovician trilobites. *In* Atlas of Palaeobiogeography, A. Hallam, Ed. Elsevier, Amsterdam. 13-18.

———, and C. P. Hughes. 1972. Ordovician geography and faunal provinces deduced from trilobite distribution. *Phil. Trans. Roy. Soc. London 263:* 235-278.

———, and C. P. Hughes. 1973. Ordovician trilobite distribution and geography. *In* Organisms and Continents Through Time, N. F. Hughes, Ed. *Spec. Papers Palaeont. 12 and Syst. Assoc. Pub. 9:* 235-240.

———, and C. H. Kindle. 1963. Middle Ordovician Table Head Formation, western Newfoundland. *Geol. Soc. Amer. Bull. 74:* 745-758.

Wiedmann, Jost. 1963a. Entwicklungsprinzipien der Kreideammoniten (Notizen zur Systematik der Kreideammoniten IV). *Paläont. Zeitschr. 37:* 103-121.

———. 1963b. Unterkreide-Ammoniten von Mollarca. 2, Lieferung, Phylloceratina. *Akad. Wiss. u. Literatur Abh., Math.-Naturw. Kl., Jahrg. 1963, 4:* 151-264.

———. 1966. Stammesgeschichte und System der post-triadischen Ammonoideen; ein Überblick. *Neues Jahrb. Geol. Paläont., Abh. 125:* 49-79; 127: 13-81.

———. 1968. Das Problem stratigraphischer Grenzziehung und die Jura/Kreide-Grenze. *Eclog. Geol. Helvet. 61:* 321-386, 4 tables.

———. 1969. The heteromorphs and ammonoid extinction. *Biol. Rev. 44:* 563-602.

———. 1970a. Problems of stratigraphic classification and the definition of stratigraphic boundaries. *Newsl. Stratigr. 1:* 35-48.

———. 1970b. Über den Ursprung den Neoammonoideen - Das Problem einer Typogenese. *Eclogae Geol. Helvetiae 63*(3): 923-1020.

———. 1973. Ancyloceratina (Ammonoidea) at the Jurassic/Cretaceous boundary. *In* Atlas of Palaeobiogeography, A. Hallam, Ed. Elsevier, Amsterdam. p. 309-316.

Wilde, G. L. 1965. Abnormal growth conditions in fusulinids. *Cushman Found. Foram. Res., Contr. 16*(pt. 3): 119, 121-124, 130-131, pls. 18-20.

———. 1971. Phylogeny of *Pseudofusulinella* and its bearing on Early Permian stratigraphy. *In* Paleozoic Perspectives: A paleontological tribute to G. Arthur Cooper, J. T. Dutro, Jr., Ed. *Smithsonian Contr. Paleobiol. 3:* 363-379, 1 pl., 8 text-figs.

Williams, A. 1969. Ordovician faunal provinces with reference to brachiopod distribution. *In* The Precambrian and Lower Paleozoic Rocks of Wales, A. Wood, Ed. University of Wales Press, Cardiff. p. 117-154.

———. 1973. Distribution of brachiopod assemblages in relation to Ordovician palaeogeography. *In* Organisms and Continents Through Time, N. G. Hughes, Ed. *Spec. Paper Palaeont. 12:* 241-269, 12 figs.

———, Isles Strachan, D. A. Bassett, W. T. Dean, J. K. Ingham, A. D. Wright, and H. B. Whittington. 1972. A correlation of Ordovician rocks in the British Isles. *Geol. Soc. London Spec. Rept. 3:* 1-74.

Williams, H. S. 1901. The discrimination of time values in geology. *Jour. Geol. 9:* 570-585.

Willis, B. 1912. Index to the stratigraphy of North America. *U.S. Geol. Surv. Prof. Paper 71:* 1-894.

Wilson, D. P. 1952. The influence of the nature of the substratum on the metamorphosis of the larvae of marine animals, especially the larvae of *Ophelia bicornis* Savigny. *Ann. Inst. Oceanogr., Monaco 27:* 49-156.

Wilson, G. J. 1971. Observations on European late Cretaceous dinoflagellate cysts. Proc. 2nd Planktonic Conf. Roma, 1970; 1259-1275.

Wilson, J. L. 1957. Geography of olenid trilobite distribution and its influence on Cambro-Ordovician correlation. *Amer. Jour. Sci. 255:* 321-340.

——. 1969. Microfacies and sedimentary structures in "deeper water" lime mudstones. *In* Depositional Environments in Carbonate Rocks. *Soc. Econ. Paleont. Mineralog., Spec. Paper 14:* 4-19.

Wilson, L. G. 1972. Charles Lyell – the years to 1841: the revolution in geology. Yale University Press, New Haven, Conn. 553 + xiii p.

Wilson, R. W. 1968. Insectivores, Rodents and Intercontinental Correlation of the Miocene. 23rd Intern. Geol. Congr. 10: 19-25.

Winston, D., and H. Nicholls. 1967. Late Cambrian and Early Ordovician faunas from the Wilberns Formation of central Texas. *Jour. Paleont. 41*(1): 66-96.

Wittenkindt, H. 1965. Zur conodontenchronologie des Mitteldevons. *Fortschr. Geologie Rhinland u. Westfalen 9:* 621-646.

Wodehouse, R. P. 1935. Pollen Grains. McGraw-Hill, New York.

Wolfe, J. A. 1972. Phyletic significance of Lower Cretaceous dicotyledonous leaves from the Patuxent Formation, Virginia. *Amer. Jour. Botany 59* (6, pt. 2): 664 [abstract].

——, and H. M. Pakiser. 1971. Stratigraphic interpretations of some Cretaceous microfossil floras of the Middle Atlantic states. *U.S. Geol. Surv. Prof. Paper 750-B:* B35-B47.

——, J. A. Doyle, and V. Page. 1975. Paleobotany. *In* The Bases of Angiosperm Phylogeny, J. W. Walker, Ed. *Ann. Missouri Botanical Garden 62*(3): 801-824.

Wood, H. E., et al. 1941. Nomenclature and correlation of the North American Continental Tertiary. *Bull. Geol. Soc. Amer. 52:* 1-48, 1 table.

Woodburne, M. O., R. H. Tedford, M. S. Stevens, and B. E. Taylor. 1974. Early Miocene mammalian faunas, Mojave Desert, California. *J. Paleont. 48:* 6-26.

Woodbury, H. O., I. B. Murray, Jr., P. J. Pickford, and W. H. Akers. 1973. Pliocene and Pleistocene depocenters, outer continental shelf, Louisiana and Texas. *Amer. Assoc. Petrol. Geol. Bull. 57*(12): 2428-2439.

Woodford, A. O. 1965. Historical Geology. W. H. Freeman & Co., San Francisco.

Woodring, W. P. 1928. Miocene Mollusca from Bowden, Jamaica, pt. II, Gastropoda. *Carnegie Inst. Washington Pub. 385:* 1-594, 40 pls.

——. 1959. Geology and paleontology of Canal Zone and adjoining parts of Panama. *U.S. Geol. Survey Prof. Paper 306B:* 147-239, pls. 24-38.

——, Ralph Steward, and R. W. Richards. 1940. Geology of the Kettleman Hills oil field, California. *U.S. Geol. Survey Prof. Paper 195:* 1-170, 50 pls.

Woodward, H. B. 1876. The Geology of England and Wales. Longman, Green, London. 476 + xx p., map, 28 figs.

——. 1887. The Geology of England and Wales, 2nd Edition. George Phillip, London. 670 + xv p., 1 pl., 1 map.

——. 1892. On geological zones. *Geol. Assoc. London, Proc. 12:* 295-315.

Worthen, A. H. 1866. Geology of Illinois. *Illinois Geol. Survey 1:* 1-152.

Worzel, J. L., W. R. Bryant et al. 1973. Initial Reports of the Deep Sea Drilling Project, X. Government Printing Office, Washington, D.C. 748 p.

Wray, John L., and C. Howard Ellis. 1965. Discoaster extinction in neritic sediments, northern Gulf of Mexico. *Amer. Assoc. Petrol. Geol. Bull. 49*(1): 98-99.

Wüst, G. 1950. Blockdiagramme der atlantischen Zirkulation auf Grund der "Meteor"-Ergebniss. *Kieler Meeresforschungen. 7:* 24-34.

Yabe, Hisakatsu. 1949. Fusulinid zones in the Carboniferous of Japan. *Japanese Acad. Proc. 25:* 168-174.

Yeatman, H. C. 1962. The problem of dispersal of marine littoral copepods in the Atlantic Ocean including some redescriptions of species. *Crustaceana 4:* 253-272.

Yochelson, E. L. 1954. Some problems concerning the distribution of the Paleozoic gastropod *Omphalotrochus. Science 120:* 233-234.

———. 1957. Notes on the gastropod *Palliseria robusta* Wilson. *Jour. Paleont. 31* (3): 648-650.

———. 1968. Paleozoic Gastropoda. *In* Developments, Trends, and Outlooks in Paleontology, R. C. Moore et al., Eds. *Jour. Paleont. 46*(6): 1363.

———, and J. T. Dutro, Jr. 1960. Late Paleozoic Gastropoda from northern Alaska. *U.S. Geol. Surv. Prof. Paper 334-D:* 111-147.

———, and R. M. Linsley. 1972. Opercula of two gastropods from the Lilgdale limestone (Early Devonian) of Victoria, Australia. *National Mus. Victoria Mem. 33:* 1-14.

Young, G., and J. Bird. 1822. A Geological Survey of the Yorkshire Coast: describing the strata and fossils occurring between the Humber and the Tees, from the German Ocean to the Plain of York. George Clark, Whitby. 332 + iv p., 20 pls.

Young, K. 1963. Upper Cretaceous ammonites from the Gulf Coast of the United States. *Texas Univ. Pub. 6304:* vii-373, 34 figs., 82 pls.

———. 1965. A revision of Taylor nomenclature, Upper Cretaceous, central Texas. *Texas Univ. Bur. Econ. Geol., Geol. Circ. 65-3:* 1-11.

———. 1966. Mexas Mojsisovicziinae (Ammonoidea) and the zonation of the Fredericksburg. *Geol. Soc. of Amer. Mem. 100:* 1-225.

Zapfe, H. 1948. Die Säugetierfauna aus dem Unterpliozän von Gaiselberg bei Zistersdorf in Niederosterreich. *Jb. Geol. Bundesanst 94:* 83-97.

———. 1949. Eine mittelmiozäne Säugetierfauna aus einer Spaltenfüllung bei Neudorf a.d. March (CSR). *Anz. Österr. Akad. Wiss., Math.-naturw. Kl. 7:* 173-181.

———. 1951. Die Fauna der miozänen Spaltenfüllung von Neudorf a.d. March (CSR); Insectivora. *Sitz. Ber. Österr. Akad. Wiss., Math.-naturwiss. Kl. 160:* 449-480.

———. 1953. Das geologische Alter der Spaltenfüllung von Neudorf a.d. March (CSR). *Verh. Geol. Bundesanst.:* 195-202.

———. 1960. Die Primatenfunde aus der miozänen Spaltenfüllung von Neudorf a.d. March (Devinska Nova Ves), Tschechoslowakei. *Schweiz. Paläont. Abh. 78:* 1-293.

Zejda, J. 1960. The influence of age on the formation of third upper molar in the bank-vole *Clethrionomys glareolus* (Schreber, 1780) (Mammalia: Rodentia). *Zool. Listy 9:* 159-166.

Zeuner, F. E. 1958. Dating the Past – An Introduction to Geochronology. 4th Ed. Methuen, London.

Ziegler, A. M., L. R. Cocks, and W. S. McKerrow. 1968. The Llandovery Transgression of the Welsh Borderland. *Palaeontology 11:* 736-782.

Ziegler, Bernhard. 1967. Ammoniten-Ökolokie am Beispiel des Oberjura. *Geol. Rundschau 56*(2): 439-464.

Ziegler, Willi. 1962. Taxionomie und Phylogenie Oberdevonischer Conodonten und ihre stratigraphische Bedeutung. *Hess. Landesamt. Bodenforsch. Abh. 38:* 1-166.

_____. 1971. Conodont stratigraphy of the European Devonian. *Geol. Soc. Amer. Mem. 127:* 227-284.

Zimmerman, E. C. 1960. Possible evidence for rapid evolution in Hawaiian moths. *Evolution 14:* 137-138.

Zingula, R. P. 1968. A new breakthrough in sample washing. *Jour. Paleont. 42* (4): 1092.

Zittel, K. von. 1899. Geschichte der Geologie und Paläontologie bis Ende des 19. Jahnhunderts. R. Oldenbourg, Munich and Leipzig. 868 + xi p. [an abridged translation was published in 1901, History of Geology and Palaeontology to the End of the Nineteenth Century. Walter Scott, London. 562 + xiv p., 13 pls.].

Zolotova, V. P., B. I. Graifer, L. A. Fedorova, and A. P. Shiryeva. 1966. Faunal characteristics of the Kungurian Stage of the Permian of the Kama River region. Stratigraphy, Lithology, Facies and Fauna of the Upper Proterozoic and Paleozoic of the Volga-Urals Oil and Gas beds. *Geol. Oil Gas Inst. Vnigri 51:* 247-253 [in Russian].

Index